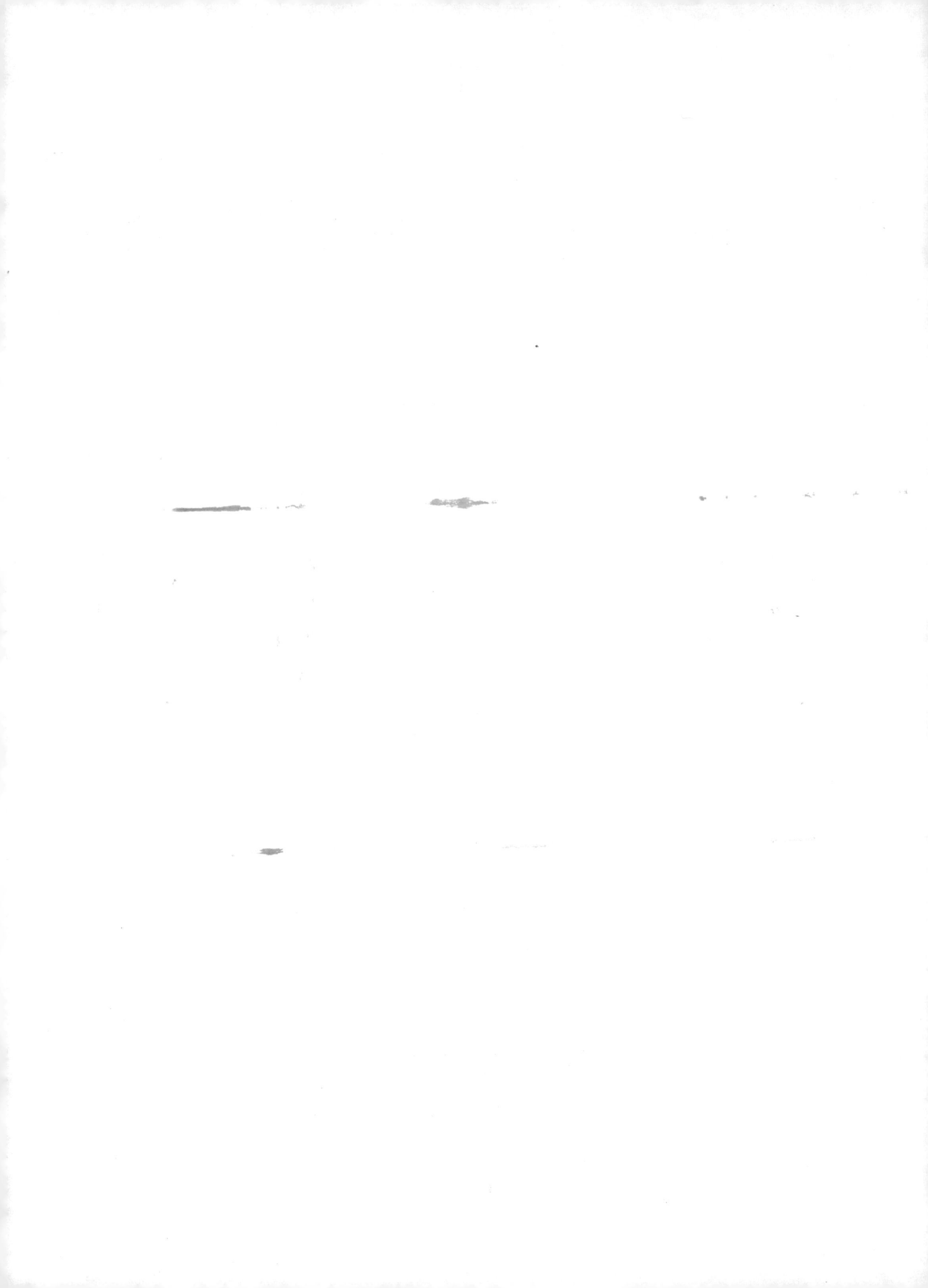

Networking with Windows 2000

Patrick Regan

Intel Corporation

Prentice
Hall

Upper Saddle River, New Jersey
Columbus, Ohio

Library of Congress Cataloging in Publication Data

Regan, Patrick E.
   Networking with Windows 2000 / by Patrick Regan.
      p. cm.
   ISBN 0-13-014558-0
    1. Computer networks. 2. Microsoft Windows (Computer file) I. Title.
   TK5105.5 .R445 2001
   005.7′13769—dc21   00-045216

Editor in Chief: Stephen Helba
Assistant Vice President and Publisher: Charles E. Stewart, Jr.
Assistant Editor: Delia K. Uherec
Production Editor: Alexandrina Benedicto Wolf
Production Coordination: Carlisle Publishers Services
Design Coordinator: Robin G. Chukes
Cover Designer: Tanya Burgess
Production Manager: Matthew Ottenweller

This book was set in 10/12 Times Roman by Carlisle Communications Ltd. It was printed and bound by Courier-Kendallville. The cover was printed by Phoenix Color Corp.

Prentice
Hall

10 9 8 7 6 5 4 3 2 1
ISBN 0-13-014558-0

This book is dedicated to my best friend, Jessica. She has always been there for me and has always been supportive of what I do.

Preface

Since I first got involved in local area networks and earned my first certification as a Novell Certified Network Engineer and a Novell Certified Network Instructor, I have learned how complex networking can be. As the Internet was introduced to the general public, new technology has also been introduced at a staggering pace. As I maintain the Novell certifications, gained the Microsoft certifications, and work as a network administrator, it is evident that essential topics in many books are skipped, and often topics that are included are incomplete and lack the ability to challenge me. Because I often encounter gaps in this information, it takes a great deal of effort to piece it all together.

This book is designed to help the reader learn the basic concepts of networking, starting with the basic hardware components that affect server performance and reliability. The first part of *Networking with Windows 2000* introduces the process of how to build a simple LAN by connecting the PCs with the various wiring and topology. The second part of the book introduces Windows 2000 as a PC operating system and a network operating system. The third part of the book focuses on today's basic features of a network server including DNS and DHCP services, directory services (Active Directory), disk and file management, and print management. The book then finishes with network topics including remote access, WAN technology, routing between networks, and web servers.

Although this book does not hold all the answers, it will give you the basic foundations that you need to succeed as a network administrator and in completing some beginning certifications including Comptia's Network+ Exam and the three Microsoft MSCE exams.

To reflect my method of teaching any of the PC or networking classes, each chapter in this book often repeats material that is related to basic, yet essential, concepts. In addition, each chapter includes review questions and hands-on exercises to reinforce those basic skills and to immediately apply what the student is reading. Lastly, I always try to use real-world examples so that you can see how basic skills are applied and show how to take those basic skills to the next level of learning.

ACKNOWLEDGMENTS

I would like to thank the following reviewers for their invaluable feedback: Mike Awwad, DeVry Institute–North Brunswick, NJ; Larry Knupp, DeVry Institute–Kansas City, MO; Darrell G. Kohr, Renton Technical College, Renton, WA; Rick Miller, Ferris State University, Big Rapids, MI; and David Oliver, Johnston Community College, Smithfield, NC.

Brief Contents

Contents

Unit 1

Network Basics

1   Introduction to Networks

INTRODUCTION

A network is a group of computers connected by cables or some method of wireless technology. In today's business world, a network is an important tool for many businesses. It allows users to access large databases, centralized files, and expensive equipment.

OBJECTIVES

1. Define *network*.
2. Define common networking terms for LANs and WANs.
3. List the components that comprise a network.
4. Compare a file-and-print server with an application server.
5. Compare a client-server network with a peer-to-peer network.
6. List and define the various network services.
7. Define *protocols*.
8. List the benefits of networks.
9. Differentiate between client-server and peer-to-peer networks.
10. List and explain the different server roles.
11. Differentiate between LANs, MANs, and WANs.
12. Define a NOS and explain how it relates to the LAN.
13. List the responsibilities of the network administrator.

3

1.1 NETWORK BEGINNINGS

Early computers were called **mainframes,** which were large centralized computers used to store and organize data. To access a mainframe, you used a **dumb terminal,** which consisted of a monitor to display the data and a keyboard to input the data. The dumb terminal did not process the data; instead, all processing was done by the mainframe computer (centralized computing).

Eventually, as computers became smaller and less expensive, the **personal computer (PC)** was introduced. Unlike the mainframe computer, a PC is meant to be used by one person and contains its own processing capabilities. As PCs became more popular, the need for people to move data from one computer to another became a reality. This process was initially performed with "sneaker net," in which a person would hand-carry the data from computer to computer via a floppy disk. Unfortunately, this method was relatively expensive and time consuming when the disk had to be transported over a long distance.

As the need to share data between different computers grew, distributed computing was developed, which consists of taking several computers and connecting them by cable. Different from centralized computing, the processing in a distributed computing system is done by the individual PCs. In fact, the distributed computing system is typically more powerful than the centralized computing system, because the sum of the processing power of the individual PCs is more than a single mainframe.

1.2 NETWORK DEFINED

A **network** is two or more computers connected to share resources such as files or a printer. See Figure 1.1. To function, a network requires a service to share or access a common medium or pathway and thus connect the computers. To bring it all together, protocols give the entire system common communication rules.

1.2.1 Network Services

The two most common services provided by a network are file sharing and print sharing. **File sharing** allows a user to access files on another computer without using a floppy disk or other forms of removable media. To ensure that files are secure, most networks can limit the access to a directory or file and to what type of access (permissions or rights) a user or users may have. For example, if you have full access to your home directory (personal directory on the network to store files), you can list, read, execute, create, change, and delete files in your home directory. Depending on the contents of the directory or file, you could specify who has access to it and what permissions or rights those users have over it. In this way, you may specify one group of people who will not be able to see or execute the file, and give a second group of people the ability to see or execute the file but not change or delete the file. Lastly, you may give rights to a third group of users to see, execute, and change the file.

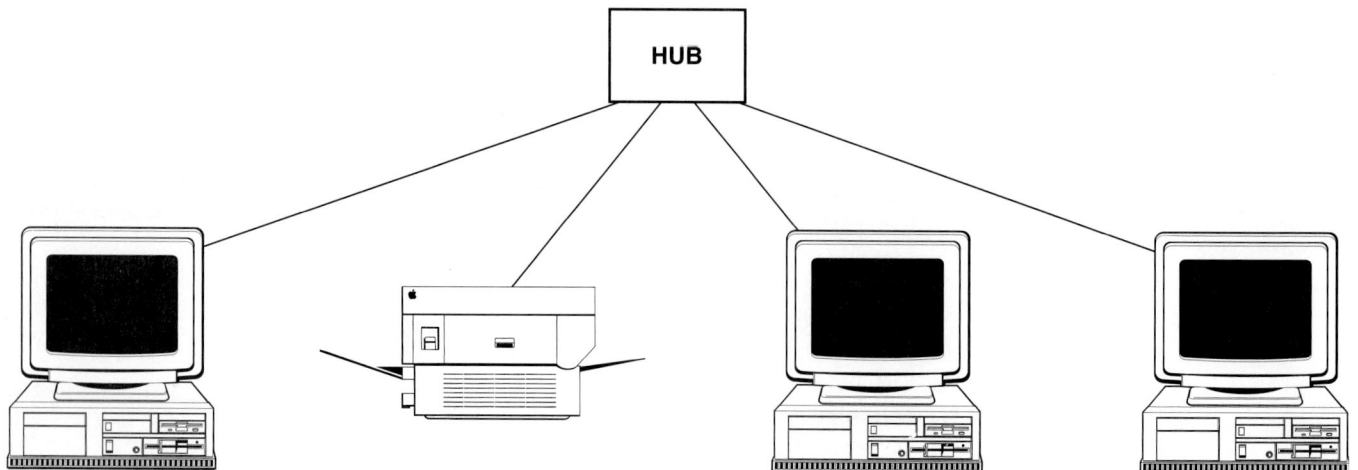

FIGURE 1.1   Computer network consisting of computers connected together with a HUB

4

Print sharing allows several people to send documents to a centrally located printer in the office, so not everyone requires a personal laser printer. Much like files, networks can limit who has access to the printer. For example, if you have two laser printers (a standard laser printer and an expensive high-resolution color laser printer), you can assign everyone access to the standard laser printer but only a handful of people access to the expensive printer.

Internet services provide important tools to business. E-mail and the World Wide Web (WWW) are two popular services. **Electronic mail, or e-mail,** is a powerful, sophisticated tool that allows users to send text messages and file attachments (documents, pictures, sound, and video) to another e-mail address. See Figure 1.2. Approximately 24% of all Internet users use e-mail. In early 1999, there were over 263 million e-mail bases. In the U.S., there were 3.4 trillion e-mail messages a year.

Like mail from the post office, e-mail is delivered to a mailbox (delivery location or holding area for electronic messages). An Internet mail address will include the user name, followed by the @ symbol and the name of the mail server. When users connect to the network, they can access their e-mail messages. Other features may include return receipt (to verify that the e-mail was read or delivered) and reply. To reply to an e-mail message, click on the reply button or option, sending the e-mail message to several people simultaneously or forwarding the message to someone else.

Because the Internet is essentially a huge network, it is possible to make a single network part of the Internet or to provide a common connection to the Internet for many users. For example, you can create your own Web page on the Internet to provide products and services to the public or you can perform research on the Internet.

1.2.2 Network Media

When users share a file or printer, they must have a pathway to access the network resource. Computers connect to the network by using a special expansion card called a **network interface card (NIC).** The network card will then communicate by sending signals through a cable (twisted pair, coaxial, and fiber optic) or by using wireless technology (infrared wavelengths, microwaves, or radio waves). The role of the network card is to prepare and send data to another computer, receive data from another computer, and control the data flow between the computer and the cabling system. See Figure 1.3.

FIGURE 1.2   E-mail with Microsoft Outlook

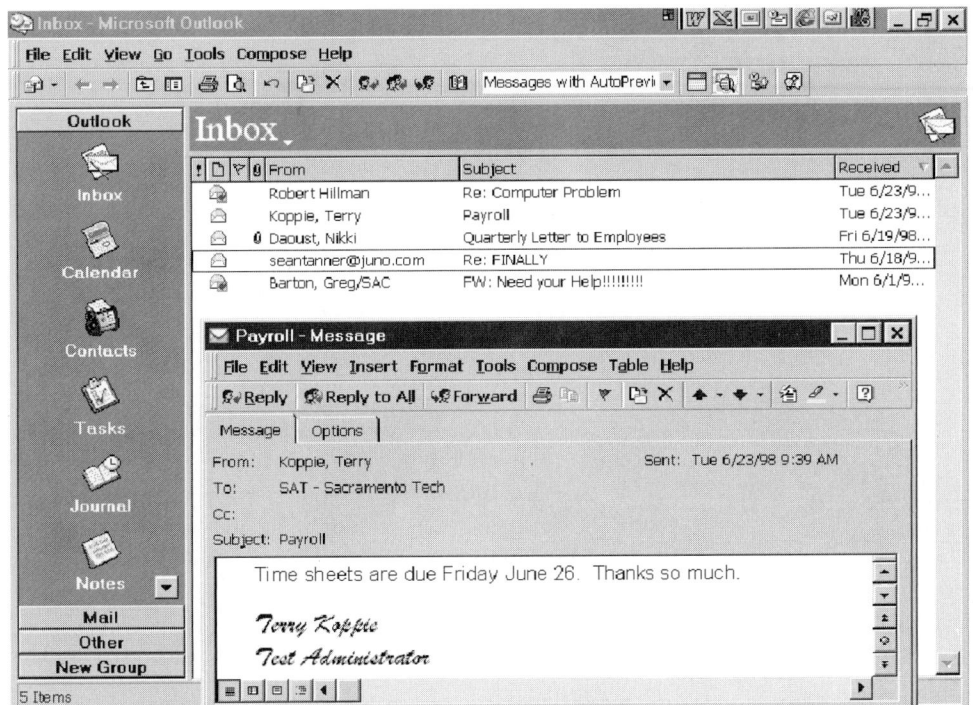

FIGURE 1.3 A network card with an unshielded twisted-pair (UTP) cable attached to a RJ-45 connector and an unused BNC connector

1.2.3 Protocols

Protocols are the rules or standards that allow the computer to connect to one another and enable computers and peripheral devices to exchange information with as little error as possible. Common protocol suites are TCP/IP, IPX, and NetBEUI. A **suite** is a set of protocols that works together.

1.3 NETWORK CHARACTERISTICS

All networks can be characterized by the following:

1. Client-server network or peer-to-peer network
2. Area network (LAN, MAN, or WAN)

1.3.1 Servers, Clients, and Peers

A network computer can either provide or request services. A **server** is a service provider that provides access to network resources, and a **client** is a computer that requests services. A network of servers and clients is called a **client-server network,** which is typically used on medium or large networks. A server-based network is the best network for sharing resources and data, as it provides network security for such information. Windows NT, Windows 2000, and Novell NetWare networks are primarily client-server networks.

A **peer-to-peer network,** sometimes referred to as a **workgroup,** has no dedicated servers. Instead, all computers are equal, and thus they both provide and request services. By keeping resources on their own machines, users manage their own shared resources. Windows 95, Windows 98, and Windows for Workgroups can be used to form a peer-to-peer network.

1.3.2 Server Roles

Because companies have different needs for their networks, servers can assume several different roles. A single server also could have several roles. These roles are as follows:

- **File server**—Manages user access to files stored on a server. When a file is accessed on a file server, the file is downloaded to the client's RAM. For example, if you are creating a report on a word processor, the word-processing files will be executed from your client computer and the report will be stored on the server. As you access the report from the server, it will be downloaded or copied onto the RAM of the client's computer. Note that all processing on the report is done by the client's microprocessors. Other advantages of a file server are it allows easy access to the data files from any computer and it allows for easy backup of data files on the server.
- **Print server**—Manages user access to printer resources connected to the network, allowing one printer to be used by many people
- **Application server**—Is similar to a file-and-print server, but also does some of the processing
- **Mail server**—Manages electronic messages (e-mail) between users
- **Fax server**—Manages fax messages sent into and out of the network through a fax modem
- **Communication server**—Handles data flow and e-mail messages from one network to another using modems or other dial-up technology
- **Web server**—Runs WWW and FTP (File Transfer Protocol) services for access via the intranet or Internet
- **Proxy server**—Provides local intranet clients with access to the Internet while keeping the local intranet free from intruders
- **Directory services server**—Locates information about the network such as domains (logical divisions of the network) and other servers

1.3.3 Area Networks

Modern networks are broken into three main categories: a **local area network (LAN),** a **metropolitan area network (MAN),** or a **wide area network (WAN).** A LAN consists of computers that are connected within a geographically close network such as a room, building, or group of adjacent buildings. A MAN is designed for a town or city, using such high-speed connections as fiber optics. A WAN uses long-range telecommunication links to connect the network computers over long distances and often consists of two or more smaller LANs. See Figure 1.4. Typically, LANs are connected through public networks (e.g., the public telephone system).

A WAN can be broken down into either an enterprise WAN or a global WAN. An **enterprise WAN** is owned by one company or organization. A **global WAN** is not owned by

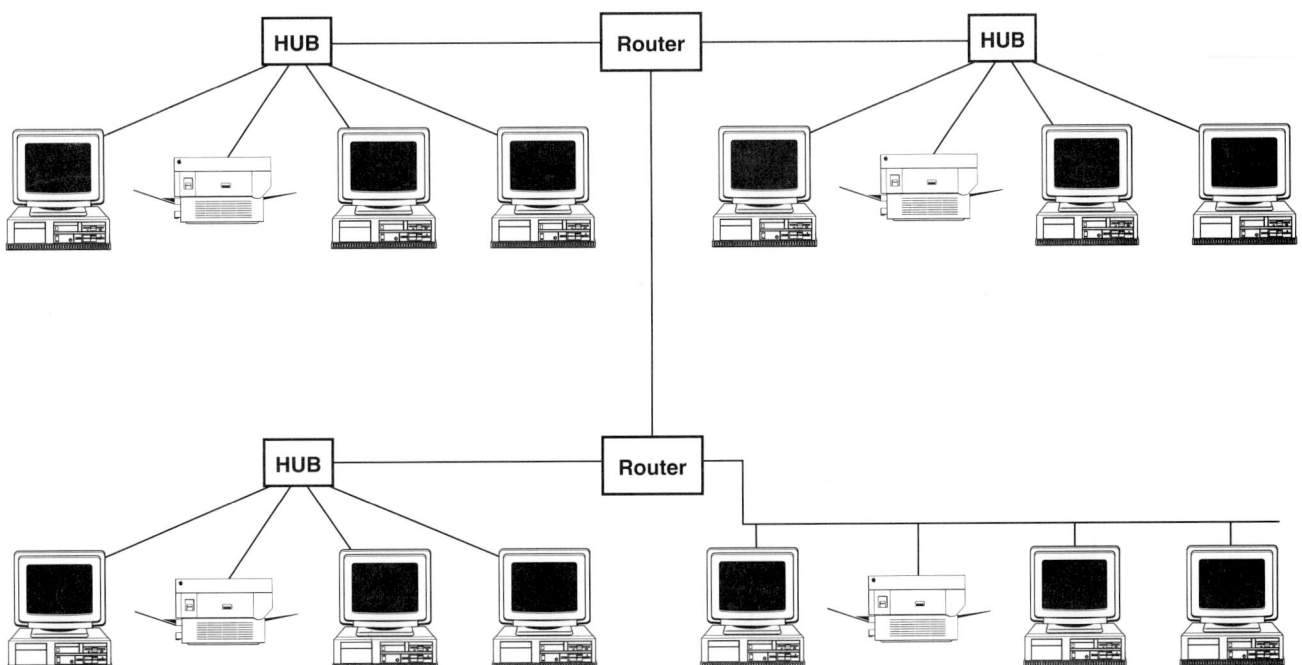

FIGURE 1.4   A wide area network (WAN)

FIGURE 1.5   The Internet

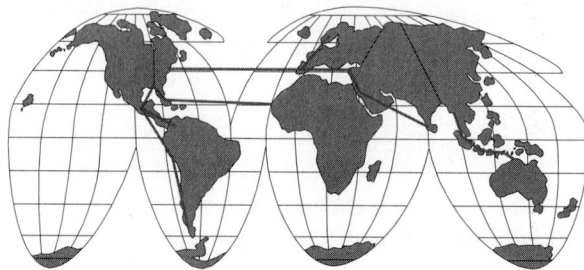

any one company and may cross national boundaries. The best known example of a global WAN is the Internet, which connects millions of computers. It is estimated that by the year 2002, the Internet has more than 490 million users in over 100 countries and the number is growing rapidly. See Figure 1.5.

1.4   NETWORK SOFTWARE

A **network operating system (NOS)** includes special functions for connecting computers and devices into a LAN, to manage the resources and services of the network and provide network security for multiple users. The most common client-server network operating systems are Windows NT Server, Novell NetWare, and UNIX. Some operating systems such as Windows for Workgroups, Windows 95/98, and Windows NT Workstation also can provide network resources as file and printer access although they are not servers.

The workstations that are attached to the network communicate through the use of **client software** called **shells, redirectors,** or **requesters.** While the network protocol enables data transmission across the LAN, the client software resides on top of the network protocol so that it can communicate with other network computers. In Windows 95/98 and Windows NT Workstation, the client software is installed and configured using the network dialog box accessed by clicking on the network applet in the control panel or by accessing the shortcut menu of the Network Neighborhood and selecting the Properties option.

1.4.1   UNIX

UNIX, a popular multiuser, multitasking operation, is the grandfather of network operating systems. It was developed at Bell Laboratories in the early 1970s. Although UNIX is a mature, powerful, reliable operating system, it has been traditionally known for its cryptic commands and its general lack of user friendliness. It is designed to handle high usage loads while having support for common Internet services such as web server, FTP server, terminal emulation (Telnet), and database access. In addition, it can use the network file system

---

OTHER NETWORK TERMS

Enterprise—Any large organization that utilizes computers, usually multiple LANs.

Internetwork—A network consisting of several LANs linked together. The smaller LANs are known as **subnetworks** or **subnets.**

Intranet—A network based on the TCP/IP protocol—the same protocol used by the Internet. Unlike the Internet, the intranet belongs to a single organization and is accessible only by that organization's members. An intranet's websites look and act like any other websites, but they are isolated by a firewall to stop illegal access. Note that an intranet could have access to the Internet, but does not require it.

Small office/home office (SOHO)—Small networks used primarily in home offices that might be part of a larger corporation but yet remain apart from it. SOHO networks are usually peer-to-peer networks.

Value-added network (VAN)—A network with special services such as electronic data interchange (EDI) or financial services such as credit card authorization or ATM transactions.

(NFS), which allows various network clients running different operating systems to access shared files stored on a UNIX machine.

Different from other network operating systems, UNIX is produced by many manufacturers. The two main dialects of UNIX are AT&T's System V and Berkeley University's BSD4.x. Popular manufactured versions include Digital Equipment Corporation UNIX, Hewlett-Packard HP-UX, SCO OpenServer, and Sun Microsystems Solaris. Another popular NOS, Banyan Vines, is based on a highly modified UNIX System V core. The official trademarked UNIX is now owned by The Open Group, an industry standards organization that certifies and brands UNIX implementations.

LINUX is a freely distributable implementation of UNIX that runs on several hardware platforms including Intel and Motorola microprocessors. Its features have made it extremely popular over the last couple years.

POSIX is a set of IEEE and ISO standards that defines an interface between programs and UNIX. By following the POSIX standard, developers have some assurance that their software can be easily ported or translated to a POSIX-compliant operating system. The POSIX standard is now maintained by an arm of the IEEE called the Portable Applications Standards Committee (PASC).

1.4.2 Novell NetWare

Several years ago, **Novell NetWare** was the standard LAN-based NOS. It was one of the primary players to bring networking to the PC arena, and so helped replace the dumb terminals with a PC and thus allowed a much easier way to share data files, applications, and printers. Novell NetWare is known as a strong file and print server. Unlike the other primary network operating systems, Novell NetWare runs as a dedicated stand-alone server.

Novell NetWare dominance was lost because of the popularity of the Windows GUI interface versus NetWare's command/menu interface. In addition, users cannot perform common administration tasks such as creating users and giving a user access to a network resource at the server, rather only at a client computer. Lastly, Novell initially missed key opportunities during the development of the Internet.

Early versions of NetWare (NetWare 3.11, 3.12, and 3.2) use the NetWare Bindery for security. The Bindery is a flat-based database that resides on a single server and contains profiles of the network users for that server. See Table 1.1.

Newer versions of NetWare use **Novell Directory Services (NDS),** a global, distributed, replicated database that tracks users and resources and provides controlled access to network resources. NDS is global because it spans an enterprise or multiple server network. It is distributed because the database is kept close to the users rather than in a single, central location and it is replicated to several servers for fault tolerance. Since the database exists for multiple servers, a user has to be created only once and has to log in to access all servers. Within the NDS design, the network objects such as users, printers, and storage units are

TABLE 1.1 The various NetWare versions

	Default Protocol	Type of Server-based Access	Interface	Comments
NetWare 3.11	IPX (Ethernet uses 802.3 frame)	Bindery (single server database)	Command interface	
NetWare 3.12/3.2	IPX (Ethernet uses 802.2 frame)	Bindery (single server database)	Command interface	
NetWare 4.11/4.2	IPX (Ethernet uses 802.2 frame)	NDS (distributed replicated database)	Command interface	Introduced directory services
NetWare 5.0	TCP/IP	NDS (distributed replicated database)	Command/GUI interface	Added many Internet and NT-like utilities and allowed management of network resources at the server

grouped into containers (used to organize the network objects) much like files are divided into subdirectories or folders on a user's hard drive.

The latest version of NetWare—NetWare 5—comes with support for both Novell's own IPX network protocol and the Internet's TCP/IP protocol. NetWare has integrated its own Novell Directory Services (NDS) with the TCP/IP's domain name system (DNS), dynamic host configuration protocol (DHCP), and application-level support for a web server. In addition, NetWare 5.0 supports Java applications and includes an enhanced file system, enhanced printing services, and advanced security.

1.4.3 Windows NT

Microsoft's early attempt at a network operating system began as a combined effort between Microsoft and IBM as OS/2. After Microsoft realized that OS/2 had some serious long-term flaws, Microsoft abandoned it to develop Windows NT. **Windows NT** (NT stands for new technology) is an advanced, high-performance NOS. It is robust in features and services, security, performance, and upgradability. Windows NT provides a good file and print server, and it makes an excellent application server. Today it is one of the fastest growing network operating systems.

The early versions of Windows NT (such as Windows NT 3.51) used the Windows 3.XX Program Manager interface. Newer versions of Windows NT use the popular Windows 95 or the Internet Explorer Active Desktop interface.

The newest version of Windows NT is **Windows 2000,** formerly known as NT 5.0. The four versions of Windows 2000 are Windows 2000 Professional, Windows 2000 Server, Windows 2000 Advanced Server, and Windows 2000 DataCenter Server. Different from older versions of Windows NT, Windows 2000 provides the following:

- **Microsoft Management Console (MMC)**—Provides a fully customizable administrative console used as a common interface for most administrative tasks
- **Distributed file system (DFS)**—Lets users see a distributed set of files in a single file structure across departments, divisions, or an entire enterprise. DFS is a significant improvement to the drive mapping process and allows multiple network locations to be mapped to a single drive.
- **Active directory (AD)**—Similar to NDS in Netware, uses the "tree" concept for managing resources on a network. Its benefits will be realized in enterprise applications in which network administration and management will be almost painless. Everything is treated as an object that can be moved or edited across servers and domains.
- **Common drivers**—Offers driver standardization, which means Windows 2000 and 98 can use the same device drivers
- **Device manager**—Makes managing hardware similar to that of Windows 9x (This feature was another that NT 4.0 was sorely lacking.)
- **Disk quotas**—Allows administration to designate a certain amount of disk space to a user or group of users for storage
- **Dynamic DNS (DDNS)**—Replaces WINS as Microsoft's name resolution system. Like WINS, DDNS is dynamic; however, it will be fully compatible with UNIX and other DNS-based systems.
- **Intellimirror**—A beautiful new management tool, replaces the NT 4.0 administration of user profiles. It provides roaming access capabilities and allows users to access their files, even when the network or server is down.
- **Internet connection sharing (ICS)**—Like Windows 98SE, allows a single dial-up connection to be shared across the network. This feature has great implications for SOHOs and home users with multiple machines as they will no longer need to purchase such an application.
- **Kerberos security**—Provides a security protocol for distributed security within a domain tree/forest. This allows for transitive trusts and a single logon to provide access to all domain resources.
- **NTFS 5.0**—A new file system, supports encryption on individual files and folders. NTFS 5.0 is now FAT32 compatible.
- **Multiple-monitor support**—Borrowed from Windows 98, allows use of multiple video cards in a single computer to expand viewing acreage

- **Plug-and-play**—Handles peripheral installation such as the Windows 9x series. A definite improvement over NT 4.0
- **Remote installation service (RIS)**—Allows for remote, automated installation of Windows 2000 Professional workstations and the management of Windows 2000 networks. Simply log on to the network either with a boot disk or a remote boot-enabled client computer.
- **Clustering**—Enables two or more servers to work together to keep server-based applications available. Windows 2000 Advanced Server and Windows 2000 DataCenter Server support clustering.
- **Total cost of ownership (TOC)**—A major issue in evaluating whether to upgrade to or install Windows 2000. Initial costs play a big part in the total costs. Windows 2000 includes many tools to enable more with less costs.

1.5 CAREERS IN NETWORKING

Networks are typically found only within companies and organizations. An information technology (IT) department typically maintains these networks, and the person who is responsible for maintaining them is the **administrator.** Job responsibilities include the following:

- Install and configure the server.
- Install and configure the network clients.
- Establish and maintain a network security system to protect the network resources.
- Document the network including the server configuration, the cabling architecture, and the network problems and solutions. Network documentation is a valuable tool in troubleshooting network problems.
- Establish backup systems and back up important data files on a regular basis.
- Minimize the number and severity of network failures.
- Monitor network performance to determine trends, foresee network failure, and provide troubleshooting.

Although the previous list summarizes the network administrator's duties, this person's primary responsibility is to fulfill the needs of the network clients. This means the administrator must make the network resources available to the user and make access easy, while keeping the network secure. Lastly, the network administrator must utilize proper time and priority management.

1.6 WHERE TO GET HELP

Network problems can be caused by hardware problems, electromagnetic interference, compatibility problems, protocol and software configurations, rights and permission problems, and so forth. As a system administrator, when faced with a new error message or when researching a network problem, you have several places to look for help. These are as follows:

- Documentation
- Resource kit
- Microsoft Web site
- Microsoft's TechNet and Microsoft's Knowledge Base
- Internet

Anything that deals with computers involves lots of documentation (hardware and software), which generally provides a wealth of information and often little tips or warnings that solve many problems. Microsoft and a few other companies offer resource kits for their operating systems and other software packages. For example, Microsoft has resources for DOS, Windows 3.xx (3.0, 3.1.0, 3.1.1), Windows 95, Windows 98, Windows NT, Windows 2000, Windows CE, and Office. These packages offer more in-depth help than manuals and address many issues that deal with networking.

When documentation and resource kits do not help, then users may go to the Internet. Excellent Web sites that have a wealth of troubleshooting information are Microsoft's TechNet and Microsoft's Knowledge Base. If you order Microsoft's TechNet on CD, you will receive monthly updates. After you become a MSCE (Microsoft Certified Engineer), you will receive TechNet free for a year. Of course, when all else fails, administrators must go to any search engine on the Internet and perform a basic search to find worthwhile information that someone else may have gathered.

SUMMARY

1. Large, centralized computers used to store and organize data are called mainframes, which were accessed by dumb terminals (a monitor to display data and a keyboard to input data).
2. A personal computer (PC) is meant to be used by one person and contains its own processing capabilities.
3. Distributed computing consists of taking several computers (computers that do their own processing) and connecting them by cable.
4. A distributed computing system is typically more powerful than a centralized computing system.
5. A network is two or more computers that are connected to share resources.
6. For a network to function, it requires network service (something to share), common media (pathways for contacting each other), and protocol (common communication rules).
7. File sharing allows users to access files, which are on another computer, without using a floppy disk or other forms of removable media.
8. Print sharing allows several people to send documents to a centrally located printer in the office.
9. Electronic mail (e-mail) is a powerful, sophisticated tool that allows users to send text messages and file attachments.
10. Computers connect to the network by using a special expansion card called a network interface card (NIC).
11. Protocols are the rules or standards that allow the computer to connect to one another.
12. A server is a service provider that provides access to network resources.
13. A client is a computer that requests services.
14. A network that consists of servers and clients is a client-server network.

15. A peer-to-peer network, or workgroup, has no dedicated servers.
16. A file server manages user access to files stored on a server.
17. A print server manages user access to printer resources connected to the network, thus allowing one printer to be used by many people.
18. An application server is similar to a file-and-print server but also does some of the processing.
19. A mail server manages electronic messages (e-mail) between users.
20. A directory services server is used to locate information about the network such as domains (logical divisions of the network) and other servers.
21. A local area network (LAN) has computers that are connected within a geographically close network.
22. A wide area network (WAN) uses long-range telecommunication links to connect the network computers over long distances and often consists of two or more smaller LANs.
23. An enterprise WAN is owned by one company or organization.
24. A global WAN is not owned by any one company and may cross national boundaries.
25. An internetwork consists of several LANs, which are linked together. The smaller LANs are known as subnetworks or subnets.
26. An intranet is a network based on the TCP/IP, the same protocol used by the Internet, usually belonging to a single organization and accessible only by that organization's members.
27. A network operating system (NOS) includes special functions for connecting computers and devices into a LAN, to manage the resources and services of the network and provide network security for multiple users.
28. The most common client-server network operating systems are Windows NT Server, Novell NetWare, and UNIX.
29. Workstations that are attached to the network communicate through client software called shells or requesters.
30. The information technology (IT) department maintains networks and computers for a company.
31. An administrator is an individual who is responsible for maintaining the network.
32. The network administrator should fulfill the needs of the network clients.

QUESTIONS

1. A network needs all of the following components except _____ .
   a. a protocol
   b. media
   c. services
   d. a dedicated server

2. The _____ is used to connect to a network's media.
   a. hard drive controller
   b. motherboard
   c. SCSI card
   d. network card

3. What are the rules or standards that allow computers to communicate with each other?
   a. media
   b. services
   c. client software
   d. protocols

4. Which type of network is made for a single building or a campus?
   a. LAN
   b. MAN
   c. global WAN
   d. enterprise WAN

5. A _____ is a network that uses long-range telecommunication links to connect the network computers over long distances.
   a. LAN
   b. MAN
   c. WAN
   d. NOS

6. An operating system that includes special functions for connecting computers and devices into a local area network (LAN), to manage the resources and services of the network and provide network security for multiple users, is a _____ .
   a. client
   b. server
   c. NOS
   d. WAN

13

7. You are the IT administrator for a small company. You currently have 15 users with plans to hire another 10 users. You have files that need to be located by all 15 users and you must maintain different security requirements for the users. What type of network would you implement?
   a. peer-to-peer network
   b. server-based network
   c. workgroup network
   d. client-server network with each computer acting as both a client and a server

8. A peer-to-peer network can sometimes be a better solution than a server-based network. Which of the following statements best describes a peer-to-peer network?
   a. It requires a centralized and dedicated server.
   b. It provides extensive user and resource security.
   c. It provides each user the ability to manage his own shared resources.
   d. It requires at least one administrator to provide centralized resource administration.

9. In a client-server environment, tasks are divided between the client computer and the server. Which role does the server play in a client-server environment?
   a. The server satisfies requests from the client computer for data and processing resources.
   b. The server stores data for the client, but all data processing occurs on the client computer.
   c. The server stores data and performs all data processing so that the client computer functions primarily as an intelligent display device.
   d. The server satisfies requests from the client for remote processing resources, but data are stored on the client computer.

10. Which one of the following statements is true of peer-to-peer networks?
    a. They are also called workgroups.
    b. They are best suited for sharing many resources and data.
    c. They can support thousands of users in different geographical areas.
    d. They provide greater security and more control than server-based networks.

11. Most networks operate in the client-server environment. Which of the following is an example of client-server computing?
    a. A terminal accesses information in a mainframe database.
    b. A workstation application accesses information in a remote database.
    c. A workstation application accesses information in a database on the local hard drive.
    d. A workstation application processes information obtained from a local database.

12. Which of the following is not a NOS?
    a. Windows NT   d. Novell NetWare
    b. UNIX   e. Windows 98
    c. Windows 2000   f. LINUX

13. For an operating system to connect to a network resource, the PC must _____ .
    a. load the appropriate server driver
    b. share a drive or directory
    c. be an administrator or server operator
    d. load the appropriate client software

2

PC Fundamentals

INTRODUCTION

Because networks consist of two or more computers, having a good understanding of the personal computer is a necessity. A computer is a machine composed of electronic devices that are used to process data. The performance of the server is partly based on its hardware, how the hardware is configured, the function of the server, and the load of the server.

OBJECTIVES

1. List and describe each of the major components of the microprocessor.
2. Describe the 386 protection mechanism.
3. Install and uninstall RAM.
4. Describe how virtual memory affects the server.
5. List the functions of the system ROM BIOS.
6. List and describe the types of expansion slots.
7. List and describe the resources of an expansion card.
8. Given an expansion card, install it into a system.
9. Compare and contrast the IDE and SCSI hard drives.
10. Given a hard drive, install it into a system and prepare it for use.
11. Install a CD-ROM drive and its drivers.
12. List and describe factors in the server room that affect the server and network.

2.1 UNDERSTANDING THE PC

When choosing the hardware components for Windows 2000, ensure that they are listed in the Microsoft Hardware Compatibility List (HCL) to avoid compatibility problems and much grief. For example, certain network interface cards (NICs) work with DOS, Windows 3.xx, Windows 95, and Novell NetWare, yet do not work for Windows NT or Windows 2000. The hardware items that are not listed on the HCL may still be compatible, as many devices emulate more common devices. For example, many sound cards emulate the Sound Blaster card and many network cards emulate the NE2000 network card.

2.2 MICROPROCESSORS

The **microprocessor,** also known as the central processing unit (CPU), is an integrated circuit that acts as the brain of the computer. It is made from a silicon wafer consisting of many transistors that act like tiny on–off switches. Unlike most integrated chips, the microprocessor is programmable, so it can be made to perform many tasks. The older 8088 and 8086 microprocessors had 29,000 transistors; the newer Pentium III microprocessor has 9.5 million transistors (not including the L2 cache).

Typically, the speed of the processor is determined by its clock speed, which is measured in megahertz (MHz)—1 MHz equals 1 million cycles per second and 1 gigahertz (GHz) equals 1 billion cycles per second. The faster the clock speed, the more instructions the microprocessor can execute per second. The original IBM PC had a clock speed of 4.77 MHz. Today's machines exceed 1 GHz.

A **register** is a high-speed internal storage area that acts as the microprocessor's short-term memory and work area. For a microprocessor to manipulate data, the data must be put in a register before they are processed. For example, if two numbers are to be added, both numbers are put in two different registers, added, and the result stored back into one of the two registers. The result can then be used for other processing or stored in the RAM.

The number of registers and the size of each register (word size or number of bits) help determine the power and speed of the microprocessor. For example, a 32-bit microproces-

---

The HCL is located on the Windows 2000 installation CD or can be located at the following Microsoft website:

http://www.microsoft.com/hcl/default.asp

To find a readiness analyzer program, product compatibility search, and Windows 2000 system requirements, go to the following website:

http://www.microsoft.com/windows2000/upgrade/compat/default.asp

sor, such as the 386DX or 486, has 32-bit registers, which can manipulate 32 bits of data. If the software is not written to use the entire register or the extra registers, the microprocessor will not perform at its peak.

The **data bus** carries the instructions and data into and out of the CPU. The size of the data bus is measured in bits. Each bit on the bus is a wire that allows one piece of information to flow into or out of the microprocessor. A microprocessor with a data bus size of 32 bits has 32 wires connected to it for data transfer. If a microprocessor has more wires, it can transfer more data at the same time, making the microprocessor faster. The Intel 8088 microprocessor used in the IBM PC has an 8-bit data bus while the Intel Pentium, Pentium Pro, Pentium II, and Pentium III have 64-bit data buses.

Today, the processor is the fastest component on the system. Since the other components could not keep up, the data bus that connects to the RAM runs at one speed and the processor runs at a faster speed (multiple of the data bus speed).

The signals of the **address bus,** controlled by the address control unit, define where the data bus signals are going to or coming from. Similar to the data bus, the address bus is measured in bits. For every wire or bit added, the amount of RAM that is addressable by the microprocessor is doubled. Therefore, the amount of RAM the address bus can access is determined by the following equation:

$$\text{Amount of RAM} = 2^{\text{Size of Address Bus}}$$

EXAMPLE 1 If you have a 20-bit address bus, the largest binary number that you can access is 1111111111 1111111111. If this figure is converted to decimals, then it is equivalent to 1048575 bytes (1 megabyte, MB) of RAM. For a 20-bit address, the highest address that can be accessed is 1,048,575—not 1,048,576—because there are 1,048,576 bytes altogether, but the numbering of bytes begins at 0, not 1.

Note that usually the motherboard (not the microprocessor) limits the amount of RAM that a system can recognize.

While all processors are integer (whole number) processors, all modern processors also include a floating-point processor. This type of processor specializes in performing calculations on float-point numbers (numbers that have a decimal number or an exponent).

The microprocessor is the fastest component within the computer. For example, the microprocessor is several times faster than the motherboard's system bus and RAM. To help bridge the speed gap, the computer uses a level 1 (L1) and level 2 (L2) RAM cache (a form of high-speed RAM). Since most programs access the same data and instructions over and over, RAM cache improves performance by keeping as much of this information as possible. When the processor needs to access the slower RAM, it will first look in the faster L1 cache in the hope of finding the needed instruction or data. If it fails to find it in the L1 cache, it will then look in the L2 cache. If it finds the needed information in the cache, the information is retrieved faster than if it went to the slower RAM. L1 cache is typically kept very small [8 kilobytes (KB) to 32 KB] and is always kept inside all modern processors. The larger L2 cache (128 KB to 2 MB) is typically kept inside the processor, but some older systems may have it located on the motherboard.

2.2.1 Superscalar Architecture

In 1993, Intel introduced the Pentium microprocessor (see Figure 2.1). The Pentium processor is the minimum processor needed to load Windows 2000. It is based on an advanced superscalar architecture that allows the processor to execute two instructions simultaneously. To keep the processor working at full capacity, the processor uses an assembly-line approach where it will start processing another instruction before finishing the current instruction. In addition, when it reaches a point in the program where it needs to decide on which direction to take, the processor uses dynamic branch prediction in an attempt to keep the processor working instead of waiting for the rest of the processor to catch up.

2.2.2 P6 Architecture

The P6 architecture uses a longer pipeline than the Pentium processor. The pipeline is divided into three units: the in-order front unit, the out-of-order execution unit, and the

FIGURE 2.1 A motherboard
with the Pentium processor, two
PCI slots (center), and three ISA
slots (bottom left). (Courtesy of
Intel Corporation)

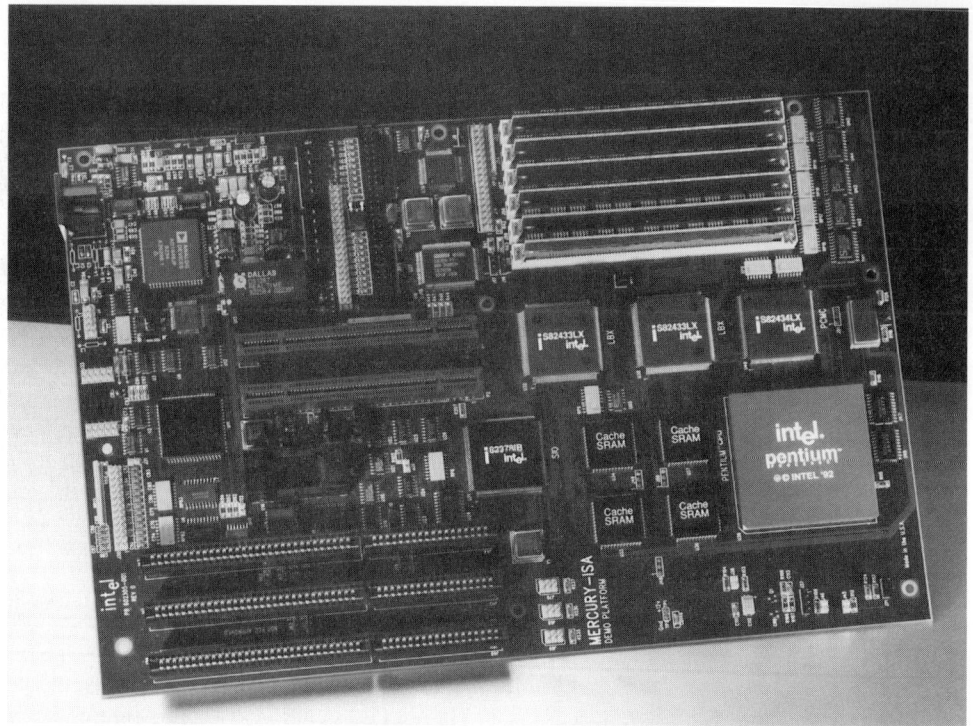

in-order retirement unit. The in-order front unit brings and decodes the instructions in the correct program sequence. The execution unit determines the optimal order and executes the instructions. The retirement unit then saves the results of the instruction in the correct order.

To make the processor work more efficiently than previous processors, P6 architecture uses advanced algorithms (multiple-branch prediction, data flow analysis, and speculative execution) and a dual independent bus (DIB) architecture. The advanced algorithms are designed to keep the assembly lines working while they wait for certain parts of the processor to catch up. The DIB uses two buses (one from the processor to the RAM and one from the processor to the L2 cache) that can be accessed simultaneously. With these features, P6 architecture has approximately 50% higher performance than a Pentium of the same clock speed. The Pentium Pro, Pentium II, Pentium III, and Intel Celeron processors are based on P6 architecture.

The Pentium Pro processor is constructed with a dual-cavity pin grid array (PGA) package, which includes an internal 256 KB or 512 KB, L2 RAM cache running at the full speed of the processor core.

Pentium processors with MMX technology and the Pentium II/III processors use 57 new instructions specifically designed to manipulate and process video, audio, and graphical data. They are oriented to the highly parallel, repetitive sequences often found in multimedia operations. In addition, they can use single instruction, multiple data (SIMD) that reduce the number or repetitive actions by loading several pieces of data into the microprocessor and performing the same command on them simultaneously. For example, if the microprocessor has several 16-bit numbers that need to have the same action performed on them, a Pentium can load four 16-bit numbers ($4 \times 16 = 64$ bits) and perform the same action on all four numbers at the same time.

The Pentium II processor uses the same P6 architecture as the Pentium Pro and includes MMX technology. It includes 512 KB of cache that runs at half the speed of the processor core. Instead of using a PGA package, it uses a single-edge contact (SEC) package, which has a single-edge connector that is inserted into slot 1 on the motherboard.

Intel's mainstream family of microprocessors is the Pentium III, which starts at 450 MHz and currently goes up to 1 gigahertz (GHz), which equals 1 billion cycles per second. The biggest performance increase comes from the new Internet streaming SIMD extensions, which are 70 multimedia instructions used in the chip to improve performance of three-dimensional graphics, video performance, audio performance, and speech recogni-

FIGURE 2.2   Pentium III processor (SECC2 and FCPGA packages)

tion. It can be found in the SECC2 package and the flip-chip PGA (FCPGA) package. See Figure 2.2.

The flagship of the Pentium II and III processor families includes the Pentium II Xeon and Pentium III Xeon processors. See Figure 2.3. It incorporates an L2 cache ranging from 256 KB to 2 MB, which operates at the same speed as the processor core. In addition, it has a 36-bit address that can allow the processor to address up to 64 GB of RAM. Different from the Pentium II and III processors, the Xeon processors connect to the motherboard via slot 2. Today's Xeon processors are typically found in high-power workstations or servers and so contain additional management features that make it easier to monitor performance and reliability of the processor.

FIGURE 2.3   Pentium III Xeon
processor

2.2.3   386 Protection Mechanism

Processors since the time of the 386 processor have the **386 protection mechanism,** whereby it assigns data and instructions to a privilege ring or level. See Figure 2.4. The four privilege levels are 0, 1, 2, and 3. The privilege level 0 is the highest and level 3 is the lowest. Programs in ring 0 can access programs in all rings, but programs in ring 3 cannot access the programs in rings 0, 1, and 2. Today, some operating systems such as Windows NT and Windows 2000 assign the operating system to level 0 and other programs to level 3. If a program becomes corrupt in ring 3, then the program will not affect the operating system.

2.2.4   Symmetric Multiprocessing

Symmetric Multiprocessing (SMP) involves a computer that uses two or more microprocessors that share the same memory. If software is written to use the multiple microprocessors, then several programs can be executed simultaneously or multithreaded applications can be executed faster. A multithreaded application is broken into several smaller

FIGURE 2.4   386 Protection
mechanism

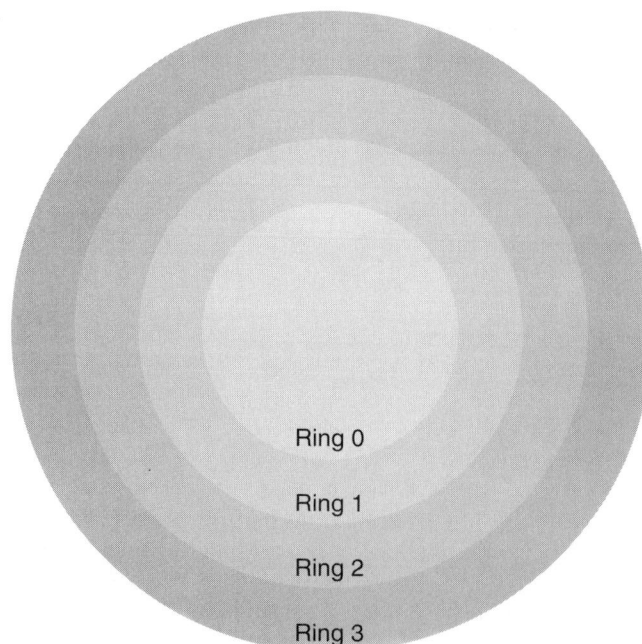

Ring 0

Ring 1

Ring 2

Ring 3

parts and executed simultaneously. For example, Microsoft Word uses a thread to respond to keys typed on the keyboard by the user to place characters in a document. Other threads are used to check spelling, paginate the document as it is typed, and spool a document to the printer in the background. The ability of an OS to use additional microprocessors is known as multiprocessing scalability. To use multiple processors, the processors need to use the same speed, cache, and steppings (internal version of the processor).

2.3 RANDOM ACCESS MEMORY

Random access memory, more commonly referred to as **RAM,** consists of electronic memory chips, which store information inside the computer. It is called random access memory because the information is accessed nonsequentially. Unlike a reel of storage tape, which is accessed starting at one end until it gets to the data, information in the RAM can be accessed directly without looking at the preceding information.

RAM is **volatile memory,** which means it loses its contents when the power is turned off. Much like a blackboard, information can be stored and overwritten. In addition, RAM is temporary memory, because when power is no longer supplied to the RAM, the contents of the RAM is lost. This explains why the user must save the contents of RAM to a disk.

The RAM is considered the main memory or the primary memory, because it is the memory that the microprocessor accesses directly. This means the program instruction that the CPU is executing must be first copied to the RAM from a device such as a disk. In addition, the data such as letters, reports, and charts, which are generated by programs, must be in the RAM to be manipulated. Therefore, it is the amount of RAM that determines how many programs can be executed at one time and how much data can be available to a program. In addition, the amount of memory is a major factor in determining how fast the computer will operate.

2.3.1 Types of RAM

The most common RAM chips are dynamic RAM (DRAM) chips, which use a storage cell consisting of a tiny solid-state capacitor and a MOS transistor. A capacitor is a simple electrical device, similar to a battery, which is capable of storing a charge of electrons. The charge or lack of charge represents a single bit of data. If it is charged, it has a logic state of 1. If it is discharged, it has a logic state of 0. The storage cells are organized into a large, two-dimensional array or table of rows and columns.

At the time of the early PCs, the typical access speed for a DRAM chip was 120 nanoseconds (ns) or slower. Newer machines (up through 1995), used **fast-page mode RAM (FPM RAM),** which had typical speeds of 70 ns or 80 ns. A lower number indicates a faster speed.

The FPM RAM works in a similar way to the normal DRAM chip except it tries to reduce the number of wait states. A wait state is a period of time during which the microprocessor has to wait for the RAM chips to catch up. If data are accessed in the same row (sometimes referred to as a page) as the preceding data, time is saved by not deactivating the row after the first piece of data and reactivating it for the second piece of data. Ideally, a read from a 70-ns FPM memory can achieve a burst cycle timing of 5-3-3-3, down from 5-5-5-5. The 5-3-3-3 indicates that it takes the RAM five clock cycles to access the first piece of data in a row. After that row is activated, it takes an additional three clock cycles each for the next three pieces of data contained within the same row.

Extended data output RAM, or **EDO RAM,** takes the fast-page mode RAM one step further. To make it even faster, EDO DRAM uses a two-stage pipeline, which lets the memory controller read data off the chip while it is being reset for the next operation. Instead of deactivating the column and turning off the output buffer, EDO RAM keeps the output data buffer on. If the data to be transferred are sequential to the data just sent, the EDO RAM will do it more efficiently. As a result, the burst cycle timing is reduced to 5-2-2-2, giving EDO RAM approximately 10% to 15% better performance over FPM RAM. Therefore, the speed of an EDO RAM chip is either 50 ns or 60 ns.

A faster form of EDO RAM is **burst extended data output (BEDO) RAM.** It is faster because it allows the page-access cycle to be divided into two components. To achieve faster access time, data are read from the memory array at the same time that data are being transferred through the data bus. In addition, a counter on the chip keeps track of the next address so that sequential data can be accessed faster. As a result, EDO RAM offers an average of 15% to 25% increase in efficiency over FPM RAM and reduces the transfer of sequential data transfer to one clock (5-1-1-1).

Although EDO RAM and BEDO RAM are faster than FPM RAM, their popularity has already decreased, because they do not work well with bus speeds higher than 66 MHz, which some motherboards are already using. Instead, you will find synchronous dynamic RAM (SDRAM), double data rate–synchronous DRAM (DDR–SDRAM), synchronous link DRAM (SLDRAM), and rambus DRAM (RDRAM). A brief overview of each follows.

The more popular form of RAM is **synchronous dynamic RAM (SDRAM).** Unlike the previous forms of RAM, SDRAM is synchronized to the external clock used on the motherboard. While working with bus speeds up to 100 MHz, its fastest access speed in CPU cycles is 5-1-1-1. All of this is made possible by using an older technique in speeding RAM access called interleaving. Interleaving is when two banks work together to service a bus. When one bank is getting ready for access, the other bank is being accessed. In addition, SDRAM uses internal pipelining (bringing in additional data while the previous data are being processed) to improve throughput. Both the interleaving and pipelining are designed to keep the SDRAM working at full capacity at all times.

During the next couple of years, newer forms of RAM will replace SDRAM, including DDR SDRAM, SLDRAM, and RDRAM. **Double data rate–synchronous DRAM (DDR–SDRAM),** also referred to as SDRAM II, is a type of SDRAM that supports data transfers on both the rise and fall of each clock cycle, effectively doubling the memory chip's data throughput.

Synchronous link DRAM (SLDRAM) uses a 64-bit bus running at a 200-MHz clock speed. Like the DDR–SDRAM, the transfers are made on the rise and fall of each clock cycle for an effective speed of 400 MHz. This yields a net theoretical bandwidth of about 3.2 gigabytes per second (GB/S), which is double of DDR–SDRAM.

Rambus DRAM (RDRAM) is based on a rambus channel, a high-speed 16-bit bus running at a clock rate of 400 MHz. Like the SLDRAM and DDR SDRAM, transfers are accomplished on the rising and falling edges of the clock, yielding an effecting theoretical bandwidth of approximately 1.6 GB/s for a single rambus channel and 3.2 GB/s for dual rambus channels.

2.3.2 RAM Chip Packaging

Many modern PCs use a **single inline memory module,** or **SIMM.** A SIMM uses a small circuit board consisting of several soldered Dual Inline Pin (DIP) chips. To connect to the motherboard, the SIMM has a row of tin or gold metal plates (contacts) along the bottom of the module. As they are inserted into the motherboard, metal pins make contact with the metal plates.

SIMMs are available in a wide range of capacities (from 256 KB to 64 MB and beyond) and come in two pin configurations: 30 pin (8/9 bits) and 72 pin (32/36 bits). If a 30-pin SIMM uses parity for error control, it will have 9 bits; if it does not use parity, it will have 8 bits. If a 72-pin SIMM uses parity, it will have 36 bits; if not, it will have 32 bits.

The 72-pin SIMM is approximately 25% larger than the 30-pin SIMM and it includes a notch in the center to prevent connecting a 30-pin SIMM into a 72-pin socket. Both SIMMs have notches to prevent a SIMM from being connected backwards. In addition, 4 of the 72 pins are used to indicate the speed rating of the chips.

Dual inline memory modules, or **DIMMs,** closely resemble SIMMs. See Figure 2.5. Similar to the SIMM, DIMMs are installed vertically into memory expansion sockets. One difference between SIMMs and DIMMs is the SIMMs opposing pins on both sides of the circuit board are connected as one, whereas the pins of the DIMMs are electrically isolated to form two separate contacts. Another difference is the DIMMs support 64-bit pathways (72 bit with parity). As with 72-pin SIMMs, 168-pin DIMMs include electrical provisions for telling a PC the speed rating of the module. The module connector provides 8 pins for signaling this information.

The **rambus inline memory module (RIMM)** conforms to the standard DIMM form factor, but is not pin compatible. See Figure 2.6. Its architecture is based on the electrical

FIGURE 2.5   A DIMM

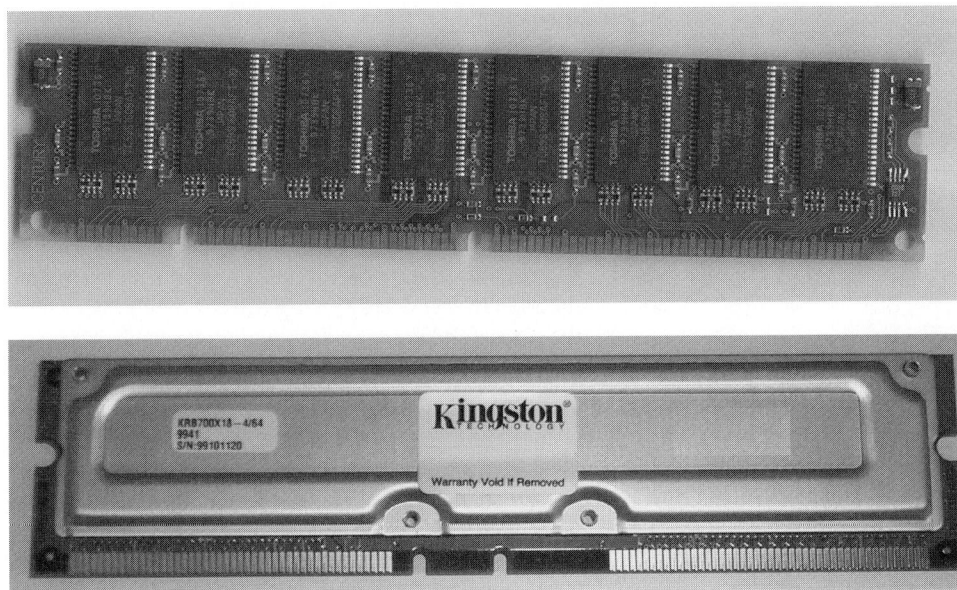

requirements of the direct rambus channel. Since the RAM runs at a faster speed than the DIMMs, RIMMs typically include a heat spreader to help dissipate heat faster. The motherboards will support two or more RIMMs. Since the RIMMs fit into a channel, low-cost continuity modules must be used to maintain channel integrity in systems having less than three RIMM modules.

2.3.3   Virtual Memory

Virtual memory is disk space pretending to be RAM, which allows the operating system to load more programs and data. For all programs and data to be accessed, parts of each are constantly swapped back and forth between RAM and disk. As far as the programs are concerned, the virtual memory looks and acts like regular RAM, which is beneficial to the user because disk memory is far cheaper than RAM. Unfortunately, a disk is a thousand times slower than RAM, and since the disk consists of mechanical parts and pieces, the disk has a higher failure rate than RAM.

The Intel microprocessors use **demand paging,** which swaps data between the RAM and disk only on the demand of the microprocessor. It does not try to anticipate the needs of the microprocessor.

The RAM and virtual memory are divided into chunks of information called **pages,** which are monitored by the operating system. When the RAM becomes full, the virtual memory system copies the least recently used programs and data to the virtual memory. Since this frees part of the RAM, it then has room to copy something from virtual memory, load another program, or load more data. Windows 3.xx and Windows 95/98 virtual memory take the form of an ordinary file called a *swap* file, whereas Windows NT and Windows 2000 call virtual memory a *paging* file.

2.3.4   Memory Error Control

RAM chips are found in a set of either 8 or 9 bits. Eight of the 9 bits were used for the data while bit 9 was used for parity to provide an error check for the other 8 bits. The DIPs used in the IBM PC, IBM XT, and IBM AT were arranged in neat rows of nine chips while some SIMMs and DIMMs use bit 9 for parity.

Parity chips ensure accurate reading and writing of data, particularly on systems released prior to the introduction of Pentium motherboards. When a PC uses parity checking, it is using odd parity checking. As the 8 individual bits in a byte are stored in memory, a parity generator/checker, which is either part of the microprocessor or is located in a special chip on the motherboard, evaluates the data bits by counting the number of 1s in a byte.

If an even number of 1s is in the byte, the parity generator/checker creates a 1 and stores it in the parity bit. This makes the total sum for all 9 bits an odd number. If the original sum of the 8 data bits is an odd number, the parity bit created is 0, again keeping the 9-bit sum an odd number. The value of the parity bit is always chosen so that the sum of all 9 bits is an odd number.

When the system reads memory, it checks the parity information. If the 9 bits equal an even number of bits, that byte must have an error. When a parity-check error is detected, the motherboard parity-checking circuits generate a **nonmaskable interrupt (NMI),** which halts processing and diverts the system's attention to the error. The NMI causes a routine in the system ROM BIOS chip (discussed in Section 2.4) to be executed. The routine clears the screen and then displays a message such as *Parity Error* or *Parity Check 1* in the upper-left corner of the screen.

There are two disadvantages of using parity for error control. First, RAM will only discover errors when bits 1, 3, 5, or 7 have changed. It will not discover a problem if 2 bits have been changed, since the 2 bits cancel each other out. Second, it detects errors but does not correct them.

Newer systems do not use parity checking, but rather **error checking and correcting (ECC) memory.** When data are stored in memory, this code is responsible for generating check bits, which are stored with the data. When the contents of a memory location are referenced, the ECC memory logic uses the check bit information and the data to generate a series of "syndrome bits." If these syndrome bits are all zeros, then the data are valid and the operation continues. If any bits are 1s, then the data have an error and the ECC memory logic isolates any errors and reports them to the operating system. In the case of a correctable error, the ECC memory scheme can detect single and double bit errors and correct single bit errors. For example, you need 1 bit for every 8 bits for parity, whereas you need 7 bits to check 32 bits using ECC.

2.4 MOTHERBOARDS

The **motherboard,** also referred to as the **main board** or the **system board,** is the primary printed circuit board located within the PC. It includes connectors for attaching additional boards (expansion slots) and additional devices (ports). It also contains the microprocessor, the RAM chips, the RAM cache, several ROM BIOS chips, CMOS RAM, the real-time clock, and several support chips. The CMOS RAM is a low-power RAM that holds basic PC configuration. It is powered by a battery so its contents are not lost when the PC is off.

2.4.1 Chipsets

The **chipset** is a number of integrated circuits designed to perform one or more related functions. Chipsets control the flow of bits between the processor, system memory, and the motherboard bus. Efficient data transfers, fast expansion bus support, and advanced power management features are just a few of the system chipset's responsibilities. It is the chipset that determines the functionality and capabilities of the motherboard. The common Intel chipset includes the 440BX and the 810, 820, 840, and 850 chipsets.

The advanced configuration and power interface (ACPI) is a relatively new power management specification that was developed by Intel, Microsoft, and Toshiba. Before ACPI, there was advanced power management (APM). It controls the amount of power given to each device attached to the computer. With ACPI, the operating system can turn off peripheral devices such as a controller card or network card when they are not in use. As another example, ACPI will enable manufacturers to produce computers that automatically power up at the touch of the keyboard.

2.4.2 ROM BIOS Chips

Computer ROM chips provide instructions and data to the microprocessor. The instructions and data kept in the ROM chip that control the boot process and hardware are known as the **basic input/output system (BIOS),** or sometimes known as **firmware.**

Every PC has several BIOS chips including the system ROM BIOS, the keyboard ROM, and the video ROM BIOS. The **system ROM BIOS,** which is located on the motherboard, directs the boot-up and allows basic control of the majority of the hardware. The **keyboard ROM BIOS,** which is also found on the motherboard, controls the keyboard. The **video ROM BIOS,** located on the video card or on the motherboard if the video card is built in to the motherboard, controls the video systems.

Other ROM chips, which are often found on expansion cards, supplement the instructions of the system ROM BIOS. They include instructions used to control new or nonstandard hardware that is not included in the system ROM BIOS. They include small computer systems interface (SCSI) controller cards, enhanced integrated drive electronics (IDE) controller cards, sound cards, proprietary CD-ROM drive controller cards, and network cards.

2.4.3 System ROM BIOS

The primary ROM BIOS chip is the system ROM BIOS, which has the following functions:

1. Controls the boot-up procedure.
2. Performs a series of diagnostic tests known as the power-on self-test (POST) during boot-up.
3. Generates hardware error (audio and video) codes if a problem is found during boot-up.
4. Finds other ROM BIOS chips from the expansion cards.
5. Finds boot sector/boot files to boot an operating system.
6. Provides the most basic commands to control hardware.
7. Provides the compatibility between the operating system and the hardware.
8. Contains the CMOS setup program.

During this time, the system ROM BIOS has instructions to search for additional ROM BIOS chips. Because the system ROM BIOS cannot possibly hold instructions for every piece of hardware or include instructions for new pieces of hardware, which are introduced every day, there was a need for additional instructions to control hardware. The additional hardware instructions are located in the other ROM BIOS chips found on the expansion cards.

The last function that the system ROM BIOS chip performs during boot-up is to find a boot device to load the operating system. If all goes well, the system will finish with a prompt waiting to input a command (DOS or UNIX) or will display a GUI interface (Windows 3.xx, Windows 95/98, Windows NT, and Windows 2000).

Today, several companies specialize in developing compatible IBM system ROM BIOS chips. See Table 2.1. The most popular are American Megatrends Incorporated (AMI), Phoenix, and Award. After a system ROM BIOS is developed, companies will license their own ROM BIOS chips to a motherboard manufacturer. The motherboard manufacturer then matches the hardware to a chosen system ROM BIOS or has one developed specifically for its motherboard.

TABLE 2.1 Popular BIOS manufacturers

BIOS Manufacturer	Description
IBM	IBM represents today's personal computer standard. All other ROM BIOSs need to be compatible with the IBM BIOS if they wish to be truly IBM compatible.
AMI	AMI is the most popular BIOS. Its success is due to its many features and enhancements.
Award	Award is unique among BIOS because it sells its BIOS code to other vendors and allows that vendor to customize the BIOS for its own particular system.
Phoenix	Phoenix was one of the first BIOS to design its own IBM-compatible BIOS and has been the standard for IBM-compatible BIOS. Phoenix ROM BIOS is efficient and reliable.

Note: IBM, AMI, Award, and Phoenix are by far the most popular, but many other companies sell their own BIOS. These BIOS are proprietary or compatible with common ROM BIOS. Some of the more popular are DTK, Epson, Hewlett-Packard, NCR, Compaq, Wang, and Zenith.

2.4.4 ROM BIOS Shortcomings

Although all PCs are compatible to the ROM BIOS level, ROM BIOS chips have two short-comings. First, the BIOS cannot hold every instruction for every hardware device or predict devices based on new technology. To overcome this shortcoming, PC designers use the following techniques to supplement the standard ROM BIOS:

1. Use additional ROM BIOS chips located on expansion cards to supplement the instructions of the system ROM BIOS.
2. Obtain software that is written to access and control the hardware directly.
3. Use device drivers to enhance the BIOS instructions and terminate-stay resident programs (TSRs) to modify the interrupt vector table.

EXAMPLE 2

> **Question:** Is every system ROM BIOS the same?
>
> **Answer:** No. First, when a company makes an IBM-compatible ROM BIOS chip, it must do so from scratch, because the IBM ROM BIOS program code is copyrighted. Second, the ROM BIOS codes have been curtailed for the hardware. Third, some BIOS are better optimized for hardware than other BIOS, which allows the PC to operate faster. Lastly, only some ROM BIOS chips include instructions to handle newer hardware.

The second shortcoming of ROM BIOS chips is occasionally you may need to upgrade the system ROM BIOS. Reasons to upgrade the system are as follows:

1. New hardware is introduced and the system ROM BIOS does not know how to work with the new hardware.
2. There are programming glitches or compatibility issues.
3. The newer system ROM BIOS runs more efficiently than the older system BIOS.

To upgrade a BIOS, access the Web site of the motherboard manufacturer and download the BIOS upgrade program. Then follow the instructions from the motherboard manufacturer or the downloaded files. **Be sure to have the most current version of the BIOS before installing Windows 2000.**

2.5 EXPANSION SLOTS AND CARDS

The **expansion slot,** also known as the **input/output (I/O)-bus,** extends the reach of the microprocessor so that it can communicate with peripheral devices. Expansion slots are so named because they allow the user to expand the system by inserting circuit boards called expansion cards. They are essential for the computer because a basic system cannot satisfy everyone's needs. In addition, the expansion cards allow the system to use new technology as it becomes available.

The expansion slots consist of connectors and metal traces that carry signals from the expansion card to the rest of the computer, specifically the RAM and CPU. These connections are used for power, data, addressing, and control lines.

The primary purpose of these slots is to carry data between the expansion cards and the CPU and RAM. The **data bus** carries the actual data between the expansion card, the RAM, and the CPU. The size (width) of a bus is the amount of data that can be transmitted at one time. For example, a 32-bit bus can transmit 32 bits of data over 32 traces or wires.

2.5.1 Hardware Interrupts

When a device (such as a modem, mouse, keyboard, hard drive controller, floppy drive controller, sound card, or network card) needs the attention of the microprocessor, it sends a signal through an interrupt line within the expansion bus to the interrupt controller. The interrupt controller will then signal the microprocessor so that the microprocessor can stop what it is doing to service the device. When the microprocessor is done servicing the device, it will return to its original task.

A single expansion card is set to use a single interrupt. When the device needs to interrupt the CPU, it sends a signal through the selected interrupt. If two devices are set to use the same interrupt, the cards will function improperly because the microprocessor will not know which device needs its attention. Computers since the IBM AT (286 microprocessor) have 16 interrupts (15 which are usable because IRQ 2 and IRQ 9 are cascaded). See Table 2.2.

With the wide assortment of devices available, it is possible to run out of interrupts; therefore, some bus designs are allowed for interrupt sharing. Interrupt sharing allows two different devices to use the same interrupt. The interrupt-handling software/firmware distinguishes for the microprocessor which device is making the request. Note that unless you have no other choice in sharing interrupts and the two devices are made to be shared with each other, it is best not to have two devices use the same interrupt.

2.5.2 Direct Memory Address

Direct memory address (DMA) channels are used by high-speed communication devices that must send and receive large amounts of information at high speed (e.g., sound cards, some network cards, and some SCSI cards). The DMA controller takes over the data bus and address lines to bring data from an I/O device to the RAM without any assistance or direction from the CPU. Since the CPU can perform other tasks while the data transfer is taking place, the PC performance will be greater. See Table 2.3.

A device using a DMA address is similar to a device using an IRQ (Interrupt). Each DMA line goes from the DMA controller to each of the expansion slots. When a device uses DMA to transfer data using a DMA channel, it will first send a signal along a single DMA channel to the DMA controller. The DMA controller will then send a request to the CPU. When the CPU acknowledges the request, the DMA controller takes control of the data bus and the address bus. The DMA controller then sends another signal to the device, which tells the device to start sending information (up to 64 KB of data). After the transfer is done, the DMA controller will release the data bus and address bus back to the CPU.

2.5.3 Bus Mastering

Bus mastering is when an expansion card with its own processor takes temporary control of the data and address bus to move information from one point to another. Consequently, the

TABLE 2.2 AT default interrupt assignments

IRQ	Default Use	Bus Slot
0	System Timer	No
1	Keyboard Controller	No
2	2nd IRQ Controller (Cascade)	No
3	Serial Port 2 (COM2)	Yes (8 bit)
4	Serial Port 1 (COM1)	Yes (8 bit)
5	Parallel Port 2 (LPT2)	Yes (8 bit)
6	Floppy Disk Controller	Yes (8 bit)
7	Parallel Port 1 (LPT1)	Yes (8 bit)
8	Real-Time Clock	No
9	Available	Yes (8/16 bit)
10	Available	Yes (8/16 bit)
11	Available	Yes (8/16 bit)
12	Bus Mouse (or Available)	Yes (8/16 bit)
13	Math Coprocessor	No
14	Primary IDE	Yes (8/16 bit)
15	Secondary IDE (or Available)	Yes (8/16 bit)

TABLE 2.3   AT default DMA assignments

DMA	Default Use	Bus Slot
0	Unused	Yes (8 bit)
1	Unused	Yes (8 bit)
2	Floppy Disk Controller	Yes (8 bit)
3	Unused	Yes (8 bit)
4	First DMA Controller	No
5	Unused	Yes (16 bit)
6	Unused	Yes (16 bit)
7	Unused	Yes (16 bit)

TABLE 2.4   Common IBM PC AT I/O addresses

Hex 000-0FF for System I/O Board		Common Expansion Cards and Ports	
000–01F	DMA Controller 1	1F0–1F8	Fixed Disk
020–03F	Interrupt Controller 1	200–207	Game Port
040–05F	Timer	278–27F	Parallel Port 2 (LPT2)
060–06F	Keyboard	2F8–2FF	Serial Port 2 (COM2)
070–07F	Real-Time Clock, NMI Mask	378–37F	Parallel Printer Port 1 (LPT1)
0A0–0BF	Interrupt Controller 2	3B0–3BF	Monochrome Display and Printer Adapter
0C0–0DF	DMA Controller 2		
0F8–0FF	Math Coprocessor	3C0–3CF	Enhanced Graphics Adapter
		3D0–3DF	CGA Monitor Adapter
		3F0–3F7	Diskette Controller
		3F8–3FF	Serial Port 1 (COM1)

PC is faster. This differs from DMAs in which the DMA controller takes control of the buses. If you have several devices trying to use the data and address bus at the same time, the data being transferred would become corrupted. To prevent two devices from taking over the buses at the same time, the bus is managed by integrated system peripheral (ISP) chips.

Much like IRQs and DMAs, the devices that use the bus are prioritized as follows:

1. RAM refresh
2. DMA transfers
3. CPU
4. Bus master expansion cards

EISA, MCA, and PCI expansion slots support bus mastering.

2.5.4   I/O Addresses

As the microprocessor communicates with a device, it will use either an OUT command or an IN command. The OUT command is an assembly language command used to send data or a command to an I/O device. The IN command is used to read data from the device or to check the status of a device. In either case, the IN and OUT commands must include the **I/O address,** sometimes called **base I/O port address,** which will identify the device. To ensure that the PC does not confuse an I/O address with a RAM address, a special signal accompanies the I/O address.

Each device is configured to respond to a range of addresses known as **ports.** Generally, a device will respond to a range of addresses. For example, the COM1 port responds to addresses 03F8 to 03FF. The address 03F8 handles all data. Four addresses are used to configure the line (speed and parity), control the telephone (hang-up and begin), check modem

TABLE 2.5   Common expansion slots

Name	Date	Bus Width	Operating Speed	Maximum Bandwidth	Comments
PC Bus	1981	8	4.77 MHz	2.385 MB/s	Used in IBM PC and XT
ISA	1984	16	8.33 MHz	8.33 MB/s	Used in IBM AT
MCA	1987	32	10 MHz	20 MB/s	Used in PS/2
EISA	1988	32	8.33 MHz	16.6 MB/s	Created to compete against MCA
VL Bus	1992	32	33 MHz	105.6 MB/s	Used mostly in 486 computers
PCI*	1992	32	33 MHz	66.6 MB/s or higher (nonburst mode) 133 MB/s (burst mode)	Used in 486 and some Pentium machines
PCI*	1992	64	66 MHz (or higher)	264 MB/s or higher (nonburst mode) 528 MB/s (burst mode)	Used in Pentium machines
PC Card	1990	16	8 MHz	16 MB/s	Used in notebook computers
CardBus	1994	32	33 MHz	133 MB/s	Used in notebook computers

*As the PCI bus matured, the bus speed has increased and is still being increased today

status, and perform other housekeeping tasks. (Note: The addresses are expressed in hexadecimal form.)

Therefore, the I/O address is a memory address used to identify the input/output (I/O) device much like a street address identifies a house or building. Similarly, no two devices can be set to use the same I/O address or range of addresses.

Intel designed its microprocessors with 65,536 I/O ports, each address being 16 bits long. The engineers did not think they would use all available ports, so they limited the addresses to 1024. See Table 2.4.

2.5.5   Memory Addresses

Many expansion cards use a range of **memory addresses,** or **base memory addresses,** in the reserve memory between 640 KB and 1 MB of RAM. The reserve memory will either be used as a working area for the expansion card or by the ROM BIOS chips.

2.6   TYPES OF EXPANSION SLOTS

Through the years there have been several types of expansion slots. The original IBM PC used the 8-bit PC slot and the original IBM AT used the 16-bit **Industry Standard Architecture (ISA)** slot. Later, the Micro Channel Architecture (MCA) and the Extended Industry Standard Architecture (EISA) were introduced for use within 386 machines. Today's desktop computers use a combination of ISA slots and local bus slots (VESA and PCI, to be discussed later in the chapter) while notebooks use a PC slot/CardBus slot. This chapter will only focus on the ISA, VESA, PCI, and PC slots. See Table 2.5.

2.6.1   ISA Slot

In the early 1980's, most technology used for personal computers were based on 8-bit technology. Therefore, IBM chose 8088 microprocessor, which had an external 8-bit data bus. This became known as the **PC slot** or the **8-bit ISA bus.**

In 1984, Intel released the 286 microprocessor. The 286 used an external 16-bit data bus and a 24-bit address bus, which allowed the microprocessor to see up to 16 MB of RAM.

This gave IBM two choices. First, IBM could choose to develop a new motherboard from scratch to take advantage of the new technology. Unfortunately, this means that the large base of 8-bit cards could not be used. The second choice was to somehow modify the older design to use both the 8-bit cards and the new 16-bit cards.

IBM's solution was introduced as the original IBM AT. The AT motherboard still used the older 8-bit connector to accommodate the 8-bit cards. To accommodate a 16-bit card, the 8-bit slots were extended by adding a second 36-pin connector. The second connector contained the additional data and address lines. In addition, there were five more interrupt lines and four additional DMA channels added.

By 1987, the Institute of Electrical and Electronic Engineers (IEEE) committee approved the AT bus as the Industry Standard Architecture (ISA). Although the ISA slot is considered old technology, even the most advanced motherboards contain some ISA slots. Other names include the classic bus, the AT bus, or the Legacy bus.

The ISA bus has not had a significant change until 1993 when Intel and Microsoft introduced the **plug-and-play (PNP) ISA.** The PNP ISA bus allows a PNP ISA card to be inserted and its resources automatically assigned without any need to use jumpers or DIP switches to configure the card. ISA cards that are not plug-and-play are called **Legacy cards.**

Instead of physically changing the ISA slot, a plug-and-play manager (software) identifies the card and assigns its needed resources. This is made possible because the PNP cards contain a ROM chip, which contains a number. Since the number identifies the card, not its model number, several of the same cards can be used within the same system, such as two network cards or two hard drive controllers. After the card is identified, the plug-and-play manager assigns the needed resources (I/O and memory addresses, IRQs, and DMAs) to the card.

2.6.2 MCA and EISA Slots

By 1985, the 386DX was introduced. It ran at 16 MHz and used a 32-bit data bus. Again, IBM was faced with designing a computer that would utilize all features of the new microprocessor. In addition, as programs incorporated graphics, it was becoming evident that video performance was a major concern in PC performance. Lastly, IBM tried to regain its dominance in the PC market back from companies such as Compaq.

As a result, IBM introduced the PS/2. On the outside, it was a sleek, modern-looking PC. On the inside, it was redesigned to accommodate the new features of the 386DX and overcome some of the limitations established by the original IBM PC. It integrated the display adapter, diskette controller, serial port, parallel port, and mouse port onto the motherboard. In addition, it used the 3.5-inch diskette drive and it introduced the new VGA graphics adapter standard. Lastly, it introduced the Micro Channel Architecture (MCA). To avoid the use of jumpers and DIP switches to configure the MCA expansion cards, cards used a PS/2 reference disk to configure the card.

The MCA was a 32-bit expansion bus. Its standard operating speed was 10 MHz, but had the capability of increasing speed to match the top speed of the card up to the speed of the microprocessor. To improve performance, MCA increased the speed of the DMA channel up to 2.5 times and it used a form of bus mastering called central arbitration point. This allowed the microprocessor to complete additional jobs while other devices used the data and address lines. IBM also used a burst mode that allowed a device to send a steady flow of data up to 12 milliseconds (ms).

Unfortunately, the PS/2 and MCA had three disadvantages. First, the MCA did not accommodate the older 8-bit and 16-bit ISA cards. Second, IBM tried to charge heavy licensing fees to companies that wanted to use the new technology, which in turn drove them away. Third, although the expansion slots were faster, system performance was not increased significantly. Therefore, customers did not want to pay the extra money for such a small increase in performance.

Since companies were not in favor of IBM's hefty licensing fees, several of them formed the "gang of nine" to design a bus to compete against the MCA. The gang was led by Compaq and consisted of AST Research, Epson, Hewlett-Packard, NEC, Olivetti, Tandy, Wyse, and Zenith Data Systems. The newly developed bus was the **Extended Industry Standard Architecture (EISA),** as mentioned earlier.

The EISA consists of two slots in one. The 16-bit slot, used by ISA cards, exists on top while the 32-bit slot exists on the bottom. Because the ISA card connectors are shorter and thicker, they only go part way into the slot. The EISA cards are longer and thinner, so they can be inserted to make contact with the 32-bit bus. Although the EISA slot initially ran at 8 MHz, newer technology increased the performance of the slots. First, instead of only reacting when the clock signal rose, EISA cards would lock their address when the clock signal rose and transfer data when the clock signal fell. In addition, much like the MCA, the EISA offered bus mastering.

EISA systems were an early attempt to produce a plug-and-play system. To avoid IRQ and I/O address conflicts, the EISA setup software would automatically configure the expansion cards. Unfortunately, not many EISA cards were developed, because many companies believed they did not have to develop such cards for a machine that already supported their ISA cards. In addition, instead of designing a complete line of EISA cards, companies waited to see if the EISA standard would become successful. As a result, only a handful of EISA cards was created, mostly for network servers (disk array controllers and network cards).

2.6.3   Local Bus Slot (VESA and PCI)

The **local bus** is a modern bus similar to one used in the original PC. To maximize performance, it has the same data path size of the microprocessor, which connects directly to the microprocessor, running at the same speed. After some early attempts, two standards emerged: the VESA local bus and the PCI local bus.

The first widely accepted local bus was the VESA local bus (VL bus). It was created by the Video Electronics Standards Association (VESA), which originally formed to standardize super VGA monitors. Its interest in developing a local bus was to improve video performance.

The VESA local bus was built around the 486 microprocessor, aimed at supplementing the existing slots (ISA, EISA, or MCA). When coupled with the ISA slot, the VESA slot was mounted directly in line with the ISA slot on the system board. Because it was built around the 486, it used a 32-bit data bus to match the microprocessor and ran at the same speed as the microprocessor.

The local bus can run in a faster **burst mode,** in which sequential data are transferred faster by specifying one memory address followed by transferring up to four pieces of data. Thus, when the bus is in burst mode running at 33 MHz, it could transfer up to 105.6 MB/s while nonburst mode can only transfer up to 66 MB/s. In either case, both speeds are far greater than the ISA's 8.33 MB/s. In addition, VESA supports bus mastering.

VESA provided improvements over ISA of bus mastering and speeds of up to 66 MHz. However, its direct connection to the 486 processor led to losses in data integrity and introduced timing problems. As a result, the maximum speed was usually 33 MHz for up to three devices (video, disk controller, and network card).

Today, when you hear the term *local bus,* it probably refers to the **peripheral component interconnect (PCI)** local bus. The PCI slot was developed by Intel and was designed to eventually replace the older bus designs. Unlike the VESA, the PCI local bus is not an extension of any other slot; it is mounted on the system board offset from the normal ISA, MCA, or EISA connector. The PCI design provides between three and five slots. To increase performance, the PCI bus has a built-in burst mode. While in burst mode, the transfer rate is 132 MB/s for the 32-bit PCI design and 264 MB/s for the 64-bit PCI design.

PCI supports plug-and-play technology. Different from other expansion bus designs, the expansion cards have their own storage space to store the cards' configuration information. Therefore, most PCI cards are not configured using jumpers or DIP switches, but are configured using software or a true plug-and-play system.

A newly developed high-performance PCI standard is the Compact PCI, which uses the rugged Eurocard packaging with a high-quality 2-mm metric pin and socket connector. The new Compact PCI supports hot swap capabilities, which allow inserting and extracting the card while the system is turned on.

2.7 CONFIGURING EXPANSION CARDS

When an expansion card is inserted into a system, it must be configured to use the proper resources. The resources include I/O addresses (including COMx/LPT), IRQs, DMAs, and memory addresses. When configuring a card, one general rule should apply: No two devices can use the same resource. Therefore, two expansion cards should not be set to use the same DMA channel or the same I/O address. If two devices are using the same setting, the devices will either not work properly or not work at all. To determine the available resources, use diagnostic software such as QA Plus or Checkit Pro or use Utilities, which comes with the operating system such as Windows Device Manager located within the control panel. It is possible to share an IRQ among more than one device, but only under limited conditions. If you have two devices that are seldom used, and that you never use simultaneously, you may be able to have them share an IRQ. Unfortunately, this can lead to more problems if you are not careful.

After determining which resources the card can use, the card itself can be configured one of three ways:

1. DIP switches and/or jumpers
2. Software setup program
3. Plug and play (PNP)

DIP switches are found on many older computers, either on the motherboard or on the expansion card. They consist of a bank of tiny on–off switches. See Figure 2.6.

A **jumper** is a small, plastic-covered metal clip that is used to connect two pins protruding from an expansion card. The jumper (same as an on switch) connects the pins, which closes the circuit and allows current to flow. The two types of pins that need to be jumped are the 2-pin configuration and the 3-pin configuration. See Figure 2.7.

The 2-pin configuration will have a jumper that either connects both pins or not. If the jumper is placed over the two pins, the documentation will say "On," "Closed," or show a small line diagram of where the jumper should be placed. If the jumper is not to be placed over the two pins, the documentation will say "Off," "Open," or show a small line diagram of where the jumpers should not be placed. Jumpers that are not in use can be placed over one pin to avoid losing them.

The 3-pin configuration has a high/low capability. If the documentation says "High," you will jumper the two pins on the side indicated with an *H*. If the documentation says "Low," you will jumper the two pins on the side indicated with an *L*.

To determine which resources the card can use and how to select the resources, you most likely need the expansion board documentation. Most documentation is written in a manual, but some is silk-screened on the card itself or placed within a file on a disk.

After installing the expansion card, it may require a software driver to function. DOS drivers are typically loaded by running the install program from disks that come with the expansion card or by copying the appropriate files to the hard drive and modifying the CONFIG.SYS and AUTOEXEC.BAT files. Windows 3.xx drivers are typically loaded by running the install program from disks that come with the expansion card. Windows 95/98 drivers are typically loaded by running the Add New Hardware Driver icon or by using the installation disks that come with the expansion card.

With no documentation, installation will usually require much trial and error before figuring out the correct settings. If the documentation or drivers become lost, look in the following places for help:

1. Internet
2. Manufacturer of the device
3. Microhouse Technical Library/On-line Service

FIGURE 2.6   DIP switches

32

FIGURE 2.7   NE2000
Compatible card with jumpers

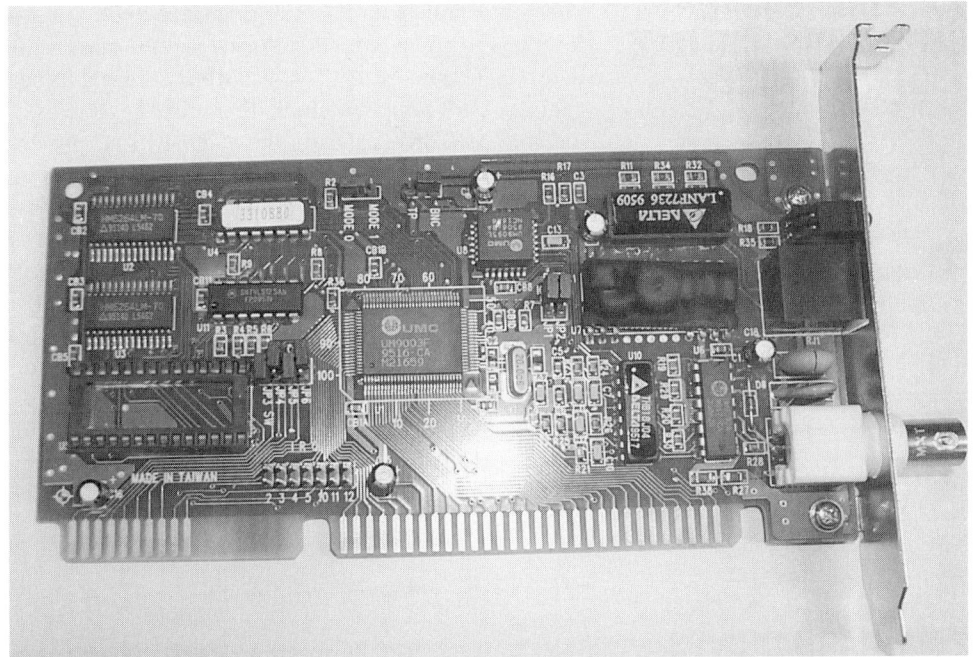

The Internet is a great resource on which companies post their documentation for quick and easy access, or the manufacturer of the device. Another option is to contact Microhouse Incorporated, which publishes a technical library (book/CD-ROM and online service) of motherboard and hard drive documentation and expansion cards.

2.8 UNIVERSAL SERIAL BUS AND IEEE 1394 (FIREWIRE)

A new technology that is emerging is the **universal serial bus (USB).** This external port allows you to connect up to 127 external PC peripherals in series (daisy chain) and offers a data transfer rate up to 12 Mbps. Mbps = megabits per second. The USB connector will accept any USB peripheral including mice, keyboards, printers, modems, network cards, and external disk drives. Using the standard four-pin connector, seven devices may be connected directly. The seven devices can be increased up to 127 devices by connecting external hubs (each hub accommodates another seven devices) in a chain. The universal serial bus allow devices to be attached while the computer is running and require no rebooting or reconfiguring every time a peripheral is added or removed.

Another external bus standard is IEEE 1394, sometimes called FireWire, which supports data transfer rates up to 400 MB/s. A single 1394 port can be used to connect up to 63 external devices. It can deliver data at a guaranteed rate, which makes it ideal for devices that need to transfer large amounts of data in real time, such as video cameras and other video devices. Like USB, 1394 supports plug-and-play and hot plugging and provides power to the IEEE 1394 devices. IEEE 1394 uses a tree topology (physical layout) in which any device can be connected to any other device so long as there are no loops. Unfortunately, IEEE 1394 costs considerably more than USB.

2.9 HARD DRIVES

Hard drives are half electronic and half mechanical devices, which use magnetic patterns to store information onto rotating platters. They are considered long-term storage devices because they do not forget their information when power is disconnected. They are also considered the primary mass storage system, because they hold all programs and data files, which are fed into RAM. The first hard drive used in the PC had a capacity of only 5 MB. Today, hard drives can be 50 GB or larger.

Hard drives are sometimes referred to as fixed disks because they usually are not removed from the PC as easily as a floppy disk. Today, this term does not describe all hard drives because there are external (disks that rest outside of the case) hard drives and removable hard drives. Hard drives communicate with the rest of the computer via a cable connected to a controller card. The controller card could be an expansion card or could be built on the motherboard.

2.9.1 Hard Drive Performance

Because programs and data files are larger than ever, hard drives are required to store greater amounts of information. Therefore, the overall performance of the PC is partially dependent on how long it takes to get the information from the hard drive into the RAM so that it can be processed by the microprocessor. Hard drive performance can be measured by access time (consisting of seek time and latency period), data transfer rate, drive RPMs, and PC data handling. See Figure 2.8.

Access time is the average amount of time it takes to move the read/write head to the requested sector. It is the sum of seek time and latency period. **Seek time** is the average time it takes the read/write heads to move to the requested track—usually to move one-third of

FIGURE 2.8 Hard drive performance factors

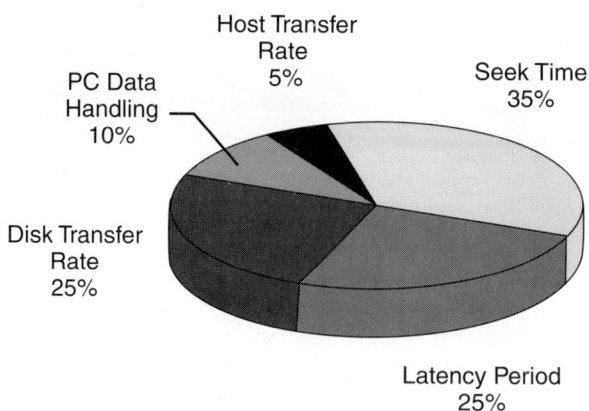

Host Transfer Rate 5%

PC Data Handling 10%

Seek Time 35%

Disk Transfer Rate 25%

Latency Period 25%

the way across the platter. The **latency period** is the time, after the read/write heads move to the requested track, for the requested sector to spin underneath the read/write head. The latency period is usually half the time required for a single revolution of the disk platter. All times are measured in milliseconds (ms). The **disk transfer rate** is the speed at which data are transferred to and from the platters, which is usually measured in bits per second (bps) or bytes per second.

The host transfer rate is defined by the method of data transfer through the hard drive interface (IDE, EIDE, SCSI, and so on). IDE hard drives use either a processor input/output mode or a direct memory address (DMA) mode while SCSI drives use either SCSI I, SCSI II, or SCSI III. The different versions of SCSI are explained later in this chapter.

The PC data handling occurs after the data have been transferred to the controller card. It is the time it takes for the data to be transferred to the RAM. It is not necessarily a factor of the hard drive, but rather of the PC itself. It is dependent on the type and speed of the microprocessor, the speed and amount of RAM, the speed and amount of RAM cache, and the speed of the hard drive controller card (expansion card).

Another way to increase disk performance is to use some form of software disk cache such as DOS's SMARTDRV.EXE or Windows 95/98 VCACHE. The software disk cache sets a cache area in RAM to store both data and instructions. One method is to keep a copy of information that has been recently accessed. If it has already been accessed, then it probably will be accessed again. Another method will read additional sectors after a sector has been accessed before those sectors are requested. Regardless of the method, when the disk needs to be accessed, it will look in the cache area first. If it cannot find it, the disk will access the slower hard drive. The time to search the cache area (RAM) is almost negligible compared with the time required to access the hard drive. Therefore, if it finds the needed information in the RAM, PC performance will be greatly increased. Since servers are providing files for use by the server's clients, any RAM not being used by an application will normally be used to cache the disk.

Another form of cache is the hardware cache located on the hard drive controller card or the hard drive logic board. This cache is not to be confused with the L1 or L2 cache used to buffer the microprocessor and the RAM. Much like its software counterparts, it fulfills the same functions as the software disk cache. Instead of using a hard drive software interrupt, the controller card physically completes the task. Hard drive controllers with cache are quite common for servers.

2.9.2   IDE Hard Drives

Today, there are two main types of hard drives, IDE and SCSI. The **integrated drive electronics (IDE)** interface was developed in 1988 as a fast, low-cost hard drive interface. The IDE hard drive connects to a controller card (via an expansion card or built into the motherboard) using a 40-pin cable. For example, the Ultra DMA-66 mode requires a special 40-pin cable that has 80 wires, 40 pins for the signal, and 40 ground pins to reduce noise. The newest IDE interface supports up to four IDE devices by using two IDE cables. In addition to hard drives, it supports nondisk peripherals, which follow the ATA packet interface (ATAPI). The ATAPI is the protocol used on the enhanced IDE devices (IDE CD-ROM drives, IDE tape backup drives, and other IDE storage devices).

When connecting the ribbon cable, be sure the cable has the correct orientation. Pin 1 of the cable (marked with a red or blue stripe along one edge) must be connected to pin 1 of the drive and to pin 1 of the cable connector on the motherboard or expansion card. Pin 1 of the drive is designated by a small number 1 or 2 printed on the drive's circuit board. Pin 1 of the expansion card/motherboard connector can be identified with a small number 1 or 2 printed on the circuit board or will use a square solder (other pins use a round solder). Some drives and cables will have a notch, which will prevent the cable from being inserted incorrectly. If the cable is connected incorrectly, the hard drive may cause a short which will then cause the computer not to boot.

Up to two hard drives can be connected to one IDE cable. See Figures 2.9 and 2.10. If one drive is installed, then the drive is known as a stand-alone drive. If two hard drives are installed, then the first physical drive is the master drive (drive C) and the second physical drive is the slave drive (drive D). The master drive got its name because the controlling

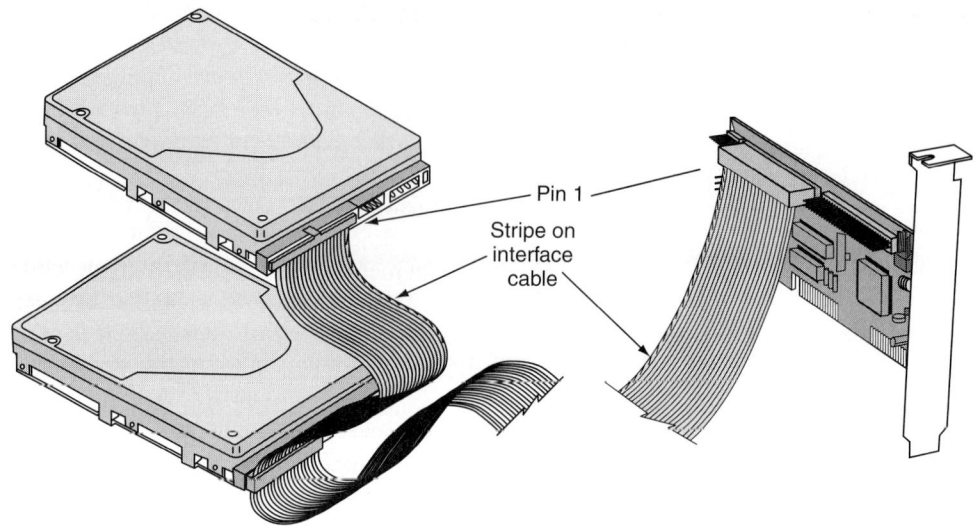

FIGURE 2.9   IDE hard drive connected to a controller card

Pin 1

Stripe on interface cable

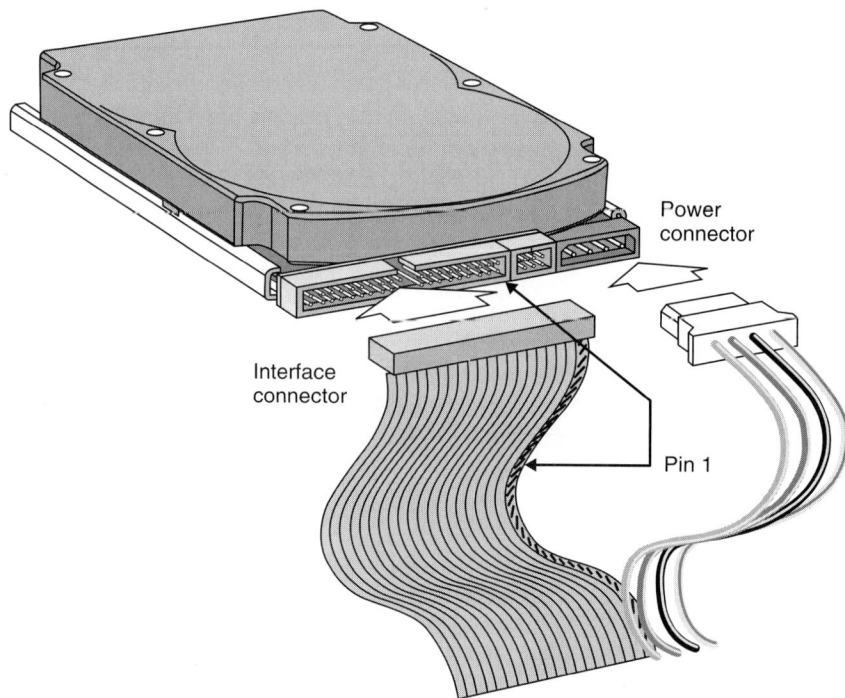

FIGURE 2.10   Connector ribbon cable and power connector (Courtesy of Seagate Corporation)

Power connector

Interface connector

Pin 1

electronics on the logic board of the master drive control both the master drive and the slave drive. The stand-alone, master, and slave drives are determined with jumpers on the hard drive. It does not matter where the drive and controller card is connected to the cable. The most common jumpers that are used to determine a drive as a stand-alone, master, and slave device are the Master (M/S) and the Slave Present (SP) jumpers.

The two methods in transferring data are **processor input/output (PIO)** and **direct memory access (DMA).** Depending on the transfer mode, a hard drive transfer rate can reach 66.6 MB/s. PIO mode has the transfers directly controlled by the processor whereas the DMA drives use a form of bus mastering. See Tables 2.6 and 2.7.

After the hard drive is configured and physically installed, the next step is to enter the CMOS setup program and input the proper hard drive parameters for the installed hard drive (such as the number of cylinders, the number of read/write heads, and the number of sectors/track or between various modes such as Large or LBA). Some CMOS setup programs have an Auto Detect feature which will automatically determine the hard drive parameters; others have an Auto choice which will automatically determine the hard drive parameters during boot-up.

TABLE 2.6   PIO modes used with EIDE devices

PIO Mode	Average Transfer Rate	Flow Controlled	Specification
0	3.3 MB/s	No	ATA
1	5.2 MB/s	No	ATA
2	8.3 MB/s	No	ATA
3	11.1 MB/s	Yes	ATA-2
4	16.6 MB/s	Yes	ATA-2

TABLE 2.7   DMA modes used with EIDE devices

DMA Mode	Average Transfer Rate	Requirements	Specifications
0—Single Word	2.08 MB/s		ATA
1—Single Word	4.16 MB/s		ATA
2—Single Word	8.33 MB/s		ATA
0—Multi-Word	4.16 MB/s	Local bus controller	ATA
1—Multi-Word	13.33 MB/s	Local bus controller	ATA-2
2—Multi-Word	16.6 MB/s	Local bus controller	ATA-3
3—Multi-Word	33.3 MB/s	Local bus controller	Ultra DMA-33
4—Multi-Word	66.6 MB/s	Local bus controller and 40-pin/80 conductor cable	Ultra DMA-66

2.9.3   SCSI Hard Drives

The other popular hard drive interface/system level interface is the **small computer systems interface** (**SCSI,** pronounced "skuzzy"). SCSI is a high-speed bus that is capable of supporting multiple devices in and out of the computer. It is a much more advanced interface than the ATA/ATA-2 IDE drives and is ideal for high-end computers including network servers. The standard SCSI interface allows up to 7 devices (hard drives, tape drives, CD-ROM drives, removable drives/disks, and scanners) connected to one SCSI adapter/controller, or 8 devices if including the controller card. Newer SCSI interfaces allow up to 15 devices (16 devices including the controller card). See Table 2.8.

The SCSI standard (SCSI-1) was approved in 1986 by the American National Standards Institute (ANSI). It specified an 8-bit bus with a 5 MB/s transfer rate. Unfortunately, the standard included only the hardware connection and did not specify the driver specification/common command set (CCS) required to communicate with the SCSI hard drive. Therefore, manufacturers used their own communication standard, which led to many compatibility problems between different devices, drivers, and adapters.

To overcome some of the shortcomings of the SCSI-1 interface, ANSI approved the SCSI-2 standard in 1992. As mentioned, it included a set of 18 basic SCSI commands called the common command set (CCS) used to support the different peripherals including CD-ROM drives, tape drives, removable drives/disks, and scanners. In addition, SCSI-2 used **command queuing,** which allows a device to accept multiple commands and execute them in an order that is more efficient than the order in which they were received. Consequently, this increases the performance of computers running multitasking operating systems and makes it ideal for servers. In addition, SCSI-2 established faster SCSI variations such as fast SCSI and fast-wide SCSI, which increased the data path and improved data transfer up to 20 MB/s.

SCSI-3 specifies a high-speed synchronous transfer called ultra SCSI and ultrawide SCSI. These allow up to a type A cable (8-bit data bus). The ultrafast-20 SCSI uses the P cable (16-bit data bus). SCSI-3 allows a data transfer rate up to 160 MB/s and up to 16 devices using the same cable.

SCSI Type	SCSI Specification	Bus Width (Bits)[1] and Cable Type	Maximum Throughput	Maximum Number of Devices	Maximum Length of Single-Ended Cable (meters)	Maximum Length of Differential Cable (meters)
Standard	SCSI-1/ SCSI-2	8 (Type A)	5 MB/s	8	6	25
Fast SCSI (also know as fast narrow)	SCSI-2	8 (Type A)	10 MB/s	8	3	25
Fast-Wide SCSI (also known as wide SCSI)	SCSI-2	16 (Type P)	20 MB/s	16	3	25
Ultra SCSI (also known as Ultra Narrow or Fast-20 SCSI)	SCSI-3	8 (Type A)	20 MB/s	8	1.5	25
Ultra-Wide SCSI (also known as Fast-Wide 20)	SCSI-3	16 (Type P)	40 MB/s	16	1.5	25
Ultra2 SCSI	SCSI-3	8 (Type A)	40 MB/s	8	NA	12[2]
Wide Ultra2 SCSI	SCSI-3	16 (Type P)	80 MB/s	16	NA	12[2]
Wide Ultra3 SCSI (also known as Ultra 160 SCSI)	SCSI-3	16 (Type P)	160 MB/s	16	NA	12[2]
Serial SCSI (also known as fiber channel SCSI)	SCSI-3	Serial (Coaxial, fiber, or special 6-wire cable)	100 MB/s	126	NA	10,000

[1]Narrow=8 bits and Wide=16 bits
[2]Requires a low-voltage differential cable that allows for lengths up to 12 meters.

TABLE 2.8   Different types of SCSI devices/interfaces

A miscellaneous SCSI interface is the serial SCSI and fiber channel SCSI. The fiber channel SCSI is a serial connection made of fiber or coaxial cable using lengths up to 10,000 meters (m). In addition, it has a 100 MB/s data transfer rate and support up to 126 devices.

The SCSI devices are connected in series forming a chain (daisy chain). The cables are characterized by the number of pins (50, 68, or 80) and if they are made for inside (internal) or outside (external) the computer. The 50-pin connector (type A) has an 8-bit data path, whereas the 68-pin (type P) and 80-pin cables have a 16-bit (wide) data paths. The 8-bit SCSI bus is known as narrow SCSI and the 16-bit SCSI bus is known as wide SCSI.

SCSI's high speeds and external cabling cause concern about signal integrity on the bus. In addition, a longer cable creates more potential for signal degradation or interference. Therefore, SCSI has defined two different electrical signal systems, single-ended SCSI and differential SCSI. **Single-ended SCSI** uses a conventional signaling in which a positive voltage indicates a 1, and ground or 0 volt indicates a zero. Each signal is carried on one wire. A **differential SCSI** uses differential signaling, in which each signal is actually carried by two different wires, each the mirror image of the other. In other words, when one of the two wires is a positive voltage, the other wire is at 0 volt. As a result, the signal is much more resilient to interference and allows for a much longer cabling. Unfortunately for this long cable, the cost is much higher. Note also that since single-ended and differential SCSI devices are incompatible at the electrical level, you should not mix single-ended and differential SCSI devices on the same bus.

Each device, including the adapter/controller card, is identified with a **SCSI ID number.** Depending on which SCSI standard is being used, the numbers will range from 0 to 7 or 0 to 16. The SCSI ID numbers are selected with jumpers, DIP switches, or a thumb wheel. To find out how to configure a SCSI device, refer to the device documentation, although when using jumpers or DIP switches, documentation will usually consist of binary counting. The SCSI adapter is usually set to ID 7, and the primary SCSI hard drive (or any other boot device) is set to ID 0. The SCSI ID numbers do not have to be in order, nor do they have to be sequential, but no two devices within the same chain can use the same SCSI ID number.

When using SCSI devices, the two devices at both ends of the chain must be terminated, but not the other devices. For example, if you have only an internal cable, the adapter card and the device at the end of the cable need to be terminated, but not all other devices in between. If you have an internal cable and an external cable, the two devices located at the end of each cable will be terminated, but not the other devices including the adapter/controller card. See Figures 2.11, 2.12, and 2.13.

To terminate or not terminate a device, either insert or remove the terminators on the end device or enable/disable them with jumpers or DIP switches. Some SCSI devices have automatic termination. When installing a terminating resistor, ensure that pin 1 is oriented

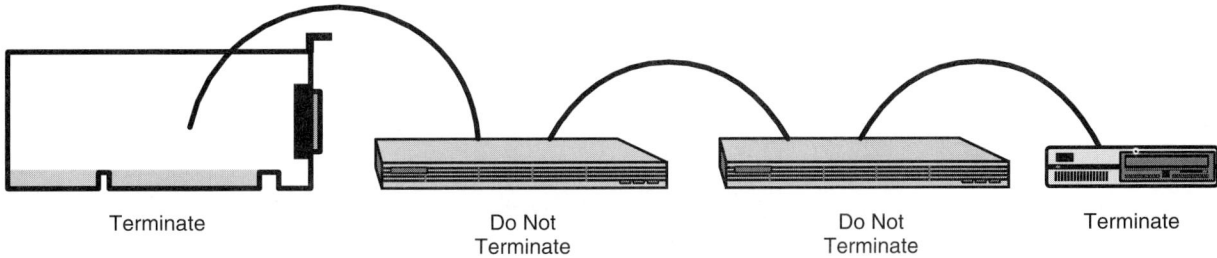

FIGURE 2.11   Internal daisy chain

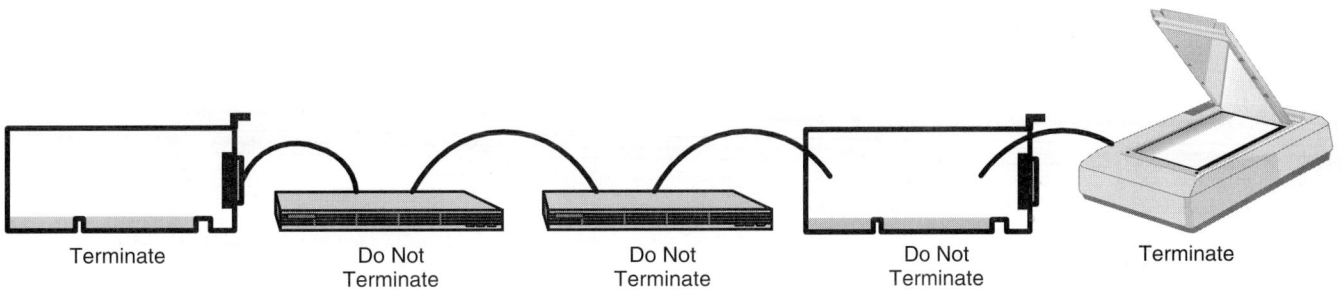

FIGURE 2.12   External daisy chain

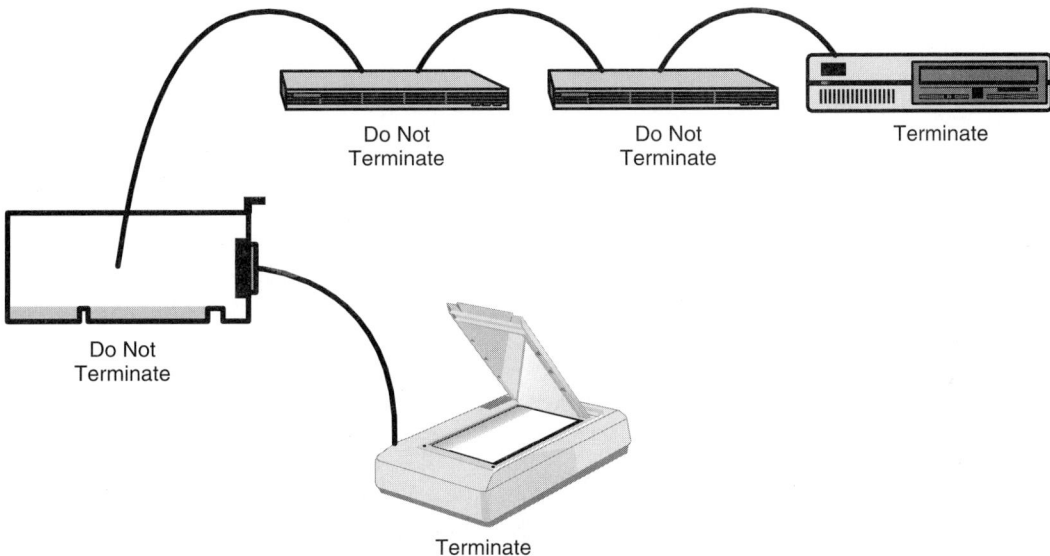

FIGURE 2.13   Internal and external daisy chain

properly. On the terminating resistor, pin 1 will be marked with a small square or rectangle, and the device will have a small number 1 or a small arrow. See Figures 2.14 and 2.15.

EXAMPLE 3

Question: What is a terminating resistor?

Answer: Terminating resistors are used to control the signal in the SCSI pathway. They are either passive or active.

The classic or standard terminator is the passive terminator. The **passive terminator** uses special electrical resistors to act as voltage dividers. They help ensure that the chain has the correct impedance load and thus prevent signals from reflecting or echoing when they reach the end of the chain. Passive terminating resistors work well for chains of short distances (2 to 3 feet) and slower speeds (SCSI-1 specification). The chain should never exceed 6 m.

A newer type of termination is active termination. **Active termination** acts as a voltage regulator to maintain a stable voltage through the chain by utilizing the termination power lines to compensate for voltage drops. Since the active termination helps reduce noise, it allows for longer cable lengths and faster speeds. The chain should never exceed 18 m.

FIGURE 2.14 Typical SCSI adapter

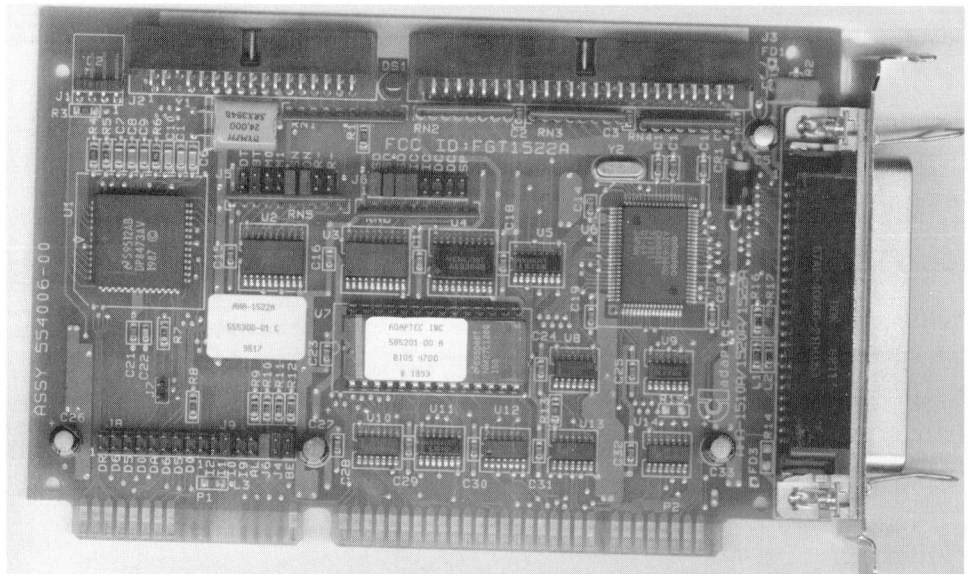

FIGURE 2.15 Terminating resistors and jumpers in a typical SCSI hard drive

The newest form of terminating resistor is the forced perfect terminator (FPT). The FPT attempts to remove reflections by automatically matching the line impedance, thus allowing "perfect" termination.

2.9.4 Hot Swappable Hard Drives

A new type of hard drive is the **hot swappable drive** which can be removed or installed while the computer is running. The ability to hot swap a disk drive is beneficial for those computers that cannot be shut off for even a few minutes, such as network servers. Currently, SCSI drives are the only drives available that are hot swappable.

2.9.5 Partitioning and Formatting a Hard Drive

The last two steps in installing a hard drive are to partition and format the hard drive. **Partitioning** is to define and divide the physical drive into logical volumes called *partitions*. Each partition functions as if it were a separate hard disk. See Figure 2.16. The physical drive is actually installed and configured, whereas the logical drive is only a drive that the system thinks it has. Every drive must include at least one partition. The reasons to partition hard drives are to overcome operating system size limits, to use a drive more efficiently, to have multiple operating systems, or to isolate data areas.

EXAMPLE 4   You are installing and configuring a 4-GB hard drive. The first steps would be to install the controller card, connect the cable, configure the hard drive, and set the CMOS setup program. If the drive requires a low-level format, it must be done with the proper software. The drive is then partitioned into two hard drives, C and D. Although one hard drive was installed, the operating system, including File Manager/Explorer, recognizes drives C and D.

The information about how a hard drive is partitioned is stored in the master boot record (MBR), which is located in the first sector of the disk. The MBR contains a small program that reads the partition table, checks which partition is active (the partition to boot from), and reads the first sector of the bootable partition.

The most common utility to partition a hard drive is the FDISK utility, which is found in DOS and Windows 95/98. When using FDISK, it can create both a primary and an extended partition. The primary partitions are the drives, which can be made active. The extended partitions can be used to designate logical drives, which are actually assigned the drive letter. Note that if you desire to change the partition size, then you will have to delete the partitions and re-create them. Unfortunately, all data are lost during this process. Also, Windows NT and Windows 2000 include a GUI program called Disk Manager.

The primary partition on the first hard drive is assigned as drive C. Any other primary partitions are assigned drive letters followed by any logical partitions. The assigned drive letters are not permanent, so, for example, if you install another hard drive and define a primary partition on that drive, it will grab the next drive letter after the first primary partition.

FIGURE 2.16   Partitioning (before and after)

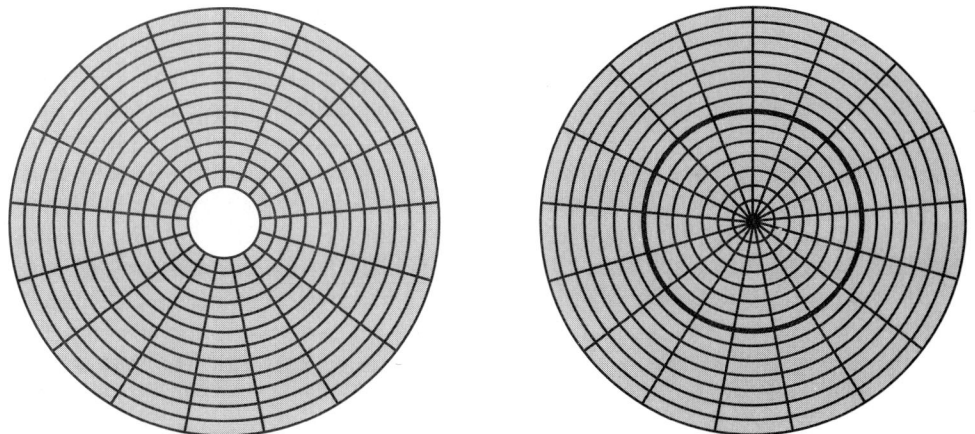

The logical drives also are reassigned new drive letters. Unfortunately, programs configured to look under a certain drive letter to find a particular directory and/or file will not be found and thus cause numerous software execution errors.

After the drive has been partitioned, it must then be formatted. **High-level formatting** is the process of writing the file system structure (e.g., FAT, FAT32, or NTFS) on the disk so that it can be used to store programs and data. This includes creating a file allocation table (an index listing all directories and files and where they are located on the disk) and a root directory. In addition, it creates a volume boot sector, which is used to store the boot files of an operating system. If you format a disk (high-level format), which already has files and a directory, you will usually re-create the file allocation tables; and without the index used to find the files, the previous information is inaccessible.

During the high-level formatting, the partition can be made bootable by copying the operating system's boot files such as DOS's IO.SYS, MSDOS.SYS, and COMMAND.COM. High-level formats are usually performed with the FORMAT.COM command. If the disk was not made bootable during formatting, it will have to be reformatted, which of course will lose all data on the partition, or will have to be done with a special operating system utility/command such as the SYS command.

Much of the installation can be simplified with the operating system's installation disks (or CD-ROM). If no partitions are created, then the installation program will usually partition the drive. If the partitions are not formatted, then it will format them individually. In addition, it will transfer the boot files to the disk, making it bootable, and copy the operating system files to the hard drive.

2.10 COMPACT DISKS

The **compact disk (CD)** is a 4.72-inch encoded platter, which is read by laser. A CD drive is required to read a CD. Because they can hold large amounts of information (680 MB) and are inexpensive, CDs are used to distribute programs, including:

1. Operating systems (such as Windows 95/98 and Windows NT)
2. Applications software (Office 97, Adobe Photoshop, various encyclopedias, and most games)
3. Device drivers
4. Multimedia libraries (pictures, sound clips, and video clips)

Through the years, there have been different kinds of compact disks available in the market. The more common ones include the audio CD, CD-ROM, CD-R and CD-W, and DVD.

Internal CD drives are half-height devices designed to fit into a standard 5.25-inch open drive bay. Like other drives, this one is typically held in place with four screws. External drive bays sit outside of the computer. Today, most CD drives use an IDE or SCSI interface. For example, if your drive uses an IDE interface, configure the drive as either a stand-alone, master, or slave. If it uses the SCSI interface, configure the drive's SCSI ID number and its terminating resistor.

After the drives are physically installed and configured, the final steps to activating the CD drive are to load the device drivers and to configure the operating system so it recognizes the CD drive. These steps are usually done with an installation disk that comes with the CD drive and/or controller card. Some operating systems will install the proper drivers automatically during installation.

Today, some system ROM BIOS chips support a bootable CD drive (El Torito standard) without loading any CD drivers. For example, you can insert a bootable compact disk in the drive and if the CMOS setup program is set to search the CD drive for a bootable disk, it will boot like the A drive or the C drive. For example, you can insert a bootable compact disk in the drive and if the CMOS setup program is set to search the CD drive for a bootable disk, it will boot like a bootable A drive or C drive.

For DOS and Windows 95/98 machines, EIDE CD drives require a device driver to be loaded in the CONFIG.SYS while SCSI drives usually require a device driver for the controller card and a device driver for the CD-ROM drive. The device drivers allow the operating system to control the CD-ROM drive. There are many drives available on the market, so be sure that the driver matches the CD-ROM drive.

Another driver needed to activate the CD drive is a file system extension. The file system extension enables the operating system to identify and use data from CDs that are attached to the system. For DOS, the file system extension is the Microsoft Compact Disk Extension (MSCDEX.EXE), which is loaded in the AUTOEXEC.BAT. For Windows 95/98, the file system extension is built into the operating system. If a user boots a Windows 95 or 98 machine to Windows DOS mode, the CONFIG.SYS will have necessary device drivers but will not have the file system extension (MSCDEX.EXE) loaded. Therefore, it needs to be executed at the command prompt.

2.11 DISKLESS WORKSTATIONS

Diskless computers that have no floppy drives or disk drives are known as **diskless workstations.** Because the computer does not have a hard drive, it boots by loading the instructions within a special ROM chip on the network card which connects to the network and loads all necessary files from it. The most obvious advantage of using a diskless workstation is its low cost compared with a fully equipped machine. Another reason to use a diskless workstation is for security. A person cannot copy sensitive files to a floppy disk or insert a floppy disk with a virus or a utility that could be used to bypass the network security.

2.12 SERVER OPTIMIZATION

When selecting components, a good starting point is to look at the minimum requirements needed to install the Windows 2000 Server on an Intel-based computer. They are as follows:

- Pentium microprocessor running at 133 MHz or higher
- 128 MB of RAM (256 MB or higher recommended for most network environments)
- 1 GB of free disk space on the partition that contains the system files
- VGA video card (PCI super VGA recommended) and monitor
- 3.5-inch floppy disk drive
- A CD-ROM drive (12X or faster recommended) Note: Windows 2000 could be installed over the network.
- Network card and cabling (if connected to the network)
- Microsoft mouse or other pointing device

To increase the performance of the server, you must look at the following components:

1. Speed and number of microprocessors
2. Speed and amount of RAM
3. Speed of the disk subsystem
4. Type and speed of the network card (discussed in Chapter 3)

Any of the previous components can cause a bottleneck. Sometimes relieving one bottleneck will cause another to occur. For example, when there is a relatively small number of clients, a 256-MB RAM configuration holds its own; but as the number of clients increases, the available cache shrinks and the file server starts to rely heavily on the disk subsystem. Adding RAM will increase the available file cache and let the server get the most out of its single microprocessor. If the microprocessor is saturated, however, there will be no significant improvement in performance by adding RAM. For example, if the hard drive is causing the bottleneck and you add a second microprocessor, you only make matters worse by inundating the already overworked disk with more requests. No matter how many microprocessors and RAM you add, you will not be able to increase server performance until you replace the disk subsystem with a faster one. When a server becomes overworked, an additional option would be to install another server and move some of the services from one server to the other.

Therefore, the question is, How many microprocessors and how much RAM is enough? Unfortunately, the answer depends on too many factors to provide a simple answer. Some factors include the number of clients, the type of server and type of services offered by the server, the type of files being accessed, and how the files are being accessed.

Therefore, when first installing a server, perform a best guess on hardware requirements based on prior experience. After the server is installed, keep monitoring the performance of the server to determine if any part needs to be upgraded. To monitor the server, most have intensive utilities that can help determine the server's performance and bottlenecks. For Windows 2000, it is Performance Monitor and Network Monitor and for Novell NetWare, it is the Monitor NLM and Novell's Lanalyzer. The Windows 2000 utilities will be discussed in detail later in the book.

For beginners and those with little experience, the author recommends the following hardware requirements for the Windows 2000 Server as a good starting point to service a moderate LAN of under 100 users:

- A minimum of a Pentium II processor running at 300 MHz or better if the main function of the server is a file-and-print server, and two Pentium II processors running at 366 MHz or better if it is a busy application server
- 256 MB of RAM (EDO, SDRAM, or better). Additional RAM can increase performance since it will be used to cache local request, network request, and the file system and it will reduce the use of virtual memory.
- 2 GB of free disk space plus a generous amount for the actual files that need to be stored on disk. The disks are connected to a fast PCI SCSI controller card.
- PCI or AGP Super VGA video card and monitor
- A sufficient power supply (300 watts or larger depending on the number of drives and processors)
- 3.5-inch floppy disk drive
- CD-ROM drive (Windows 2000 could be installed over the network.)
- PCI network card and cable
- Microsoft mouse or other pointing device

To learn Windows 2000, beginners do not need such a powerful system. When choosing equipment for a server, also avoid the newest or the fastest equipment. Buyers want to choose items that offer good performance, but it is also important to choose items that are reliable and have a good track record. Remember that a primary goal of a network is to minimize downtime.

2.13   SERVER ROOM

The **server room** is the work area of the IT department where the servers and most of the communication devices reside. The room should be secure with only a handful of people allowed access to it. In addition, the server also should be secure, so the department manager should take the following precautions:

- Lock the computer when not in use.
- Require user names and passwords.
- Log out when the server is not in use.
- Restrict accounts so that users cannot log on directly (interactive) to the server.
- Enable security monitoring, such as auditing.

Besides securing the servers in the server room, IT managers should also consider the following criteria:

- Separate environment controls for the server room, because the computers and telecommunication systems can generate much heat, and a dry room is more susceptible to electrostatic discharge which can damage electronic components. Electrostatic discharge is electricity generated by friction such as when your arm slides on a tabletop or when you walk across a carpet.
- An electrostatic discharge prevention program which would include using ESD wrist straps when opening computers and handling computer components, installing antistatic flooring, and/or placing the servers on an antistatic mat.
- An **uninterruptible power supply (UPS)** to protect the system against power fluctuations including outages.
- A tape backup system or some other method to back up all important data files.

If managers have several servers, then they also should consider a server rack to hold the servers. In addition, they also can purchase a switch box to connect a single keyboard, mouse, and monitor to several servers, to allow for less equipment and a more organized work environment.

SUMMARY

1. A computer is a machine made of electronic devices that are used to process data.
2. The performance of the server is partly based on hardware choices, hardware configuration, server function, and server load.
3. When choosing hardware components for Windows NT, ensure that they are listed in the Microsoft Hardware Compatibility List (HCL).
4. A microprocessor, also known as the central processing unit (CPU), is an integrated circuit that acts as the brain of the computer. The clock speed is the speed at which the microprocessor executes instructions.
5. Clock speeds (frequency) are expressed in megahertz (MHz); 1 MHz is equal to 1 million cycles per second.
6. The faster the clock speed, the more instructions the microprocessor can execute per second.
7. Symmetric multiprocessing (SMP) is a computer consisting of two or more microprocessors that share the same memory.
8. Random access memory, or RAM, consists of electronic memory chips which store information inside the computer. It is so named because the information is accessed non-sequentially.
9. ROM chips in the computer provide instructions and data to the microprocessor.
10. The system ROM BIOS that controls the boot process and hardware is known as the basic input/output system (BIOS), or firmware.
11. It is important to update the system ROM BIOS before installing Windows 2000.
12. Virtual memory is disk space pretending to be RAM, which allows the operating system to load more programs and data.
13. Virtual memory in Windows NT is called a paging file.
14. The expansion slot, also known as the I/O bus, extends the reach of the microprocessor so that it can communicate with peripheral devices. They are called expansion slots because they allow the user to expand the system by inserting circuit boards called expansion cards.
15. When a device needs the attention of the microprocessor, it sends a signal through an interrupt line within the expansion bus to the interrupt controller.
16. The DMA controller takes over the data bus and address lines to bring data from an I/O device to the RAM without any assistance or direction from the CPU.
17. Bus mastering is when an expansion card with its own processor takes temporary control of the data and address bus to move information from one point to another.
18. Each device is configured to respond to a range of addresses known as ports or I/O addresses.
19. Many expansion cards use a range of memory addresses, also known as the base memory address, in the reserve memory between 640 KB and 1 MB of RAM.
20. When an expansion card is inserted into a system, it must be configured to use the proper resources. The resources include I/O addresses (including COMx/LTPx), IRQs, DMAs, and memory addresses.
21. After determining which resources the card can use, the card itself can be configured by DIP switches and/or jumpers; electronically, using a software setup program; or by using plug-and-play (PNP).
22. Hard drives are half electronic and half mechanical devices, which use magnetic patterns to store information onto rotating platters. They are considered long-term storage devices because they do not forget their information when power is disconnected.
23. Hard drive performance can be measured by the access time (consisting of seek time and latency period), data transfer rate, drive RPMs, and PC data handling.
24. To increase the performance of the server, you must look at the speed and number of microprocessors, the speed and amount of RAM, the speed of the disk subsystem, and the type and speed of the network card.
25. The server room is the work area of the IT department where the servers and most of the communication devices reside.

1. Another term for *microprocessor* is _____ .
   a. motherboard   c. CPU
   b. system unit   d. RAM

2. The main integrated chip around which the PC is built is the _____ .
   a. UART   d. real-time clock
   b. processor   e. system ROM BIOS
   c. RAM

3. Computer clock speeds are expressed in _____ . (Select two answers)
   a. gigahertz   c. hertz
   b. megahertz   d. kilohertz

4. Which of the following uses a single-edge connector?
   a. Pentium   c. Pentium Pro
   b. Pentium with MMX   d. Pentium II

5. Microprocessors since the 386 microprocessor use the Intel Protection model. Which ring allows the greatest access to the CPU?
   a. 0   d. 3
   b. 1   e. 7
   c. 2

6. RAM is considered volatile memory. Which of the following best describes volatile memory?
   a. UV light is used to erase the contents of the chip.
   b. Data in RAM are permanent and cannot be erased.
   c. Data in RAM are lost when system is shut off.
   d. Data are maintained by a small battery when the power is shut off.

7. Virtual memory is _____ .
   a. extended memory pretending to be expanded memory
   b. ROM chips pretending to be memory
   c. extended memory being used for video RAM
   d. a disk pretending to be RAM

8. A small circuit board consisting of several DRAM chips and small metal contact plates, which are inserted into the motherboard, is known as a _____ . (Select two answers)
   a. DIPP   d. SRAM
   b. SIPP   e. DIMM
   c. SIMM   f. FIMM

9. Cache memory is _____ .
   a. spare storage
   b. virtual memory
   c. ultrafast memory, which buffers the microprocessor with the slower RAM
   d. shadow RAM

10. RAM cache inside the microprocessor is known as _____ .
    a. L1 cache   c. local cache
    b. fast cache   d. extended cache

11. COM1 is usually assigned which interrupt?
    a. IRQ 1   d. IRQ 4
    b. IRQ 2   e. IRQ 9
    c. IRQ 3

12. The DMA channel used for the floppy controller is _____ .
    a. DMA-1   c. DMA-3
    b. DMA-2   d. DMA-5

13. After installing a new sound card, the system fails to boot. Which of the following could be causing the problem?
    a. The operating system does not support the sound card.
    b. The cables to the sound card are incorrectly connected.

c. There is an IRQ conflict.

d. None of the above

14. How many hardware interrupts does a Pentium PC have?

a. 4   c. 12

b. 8   d. 16

15. Bus mastering is similar to (a) _____ .

a. I/O address   d. memory address

b. IRQ   e. none of the above

c. DMA

16. The most common expansion slot found on new and old machines is _____ .

a. MCA   d. PC

b. EISA   e. PCI

c. ISA   f. VESA

17. Which type of slot is the fastest?

a. ISA   d. PCI local bus

b. EISA   e. PCMCIA

c. MCA

18. Which of the following are *not* possible conflicts when installing an expansion card?

a. port conflict   d. power conflict

b. I/O address conflict   e. memory conflict

c. IRQ conflict

19. You are suspecting an IRQ conflict when installing a sound card. How would you check to see which IRQs are free?

a. Reboot the computer.

b. Install the operating system.

c. Run software diagnostic such as QA Plus or MSD.

d. Enter the CMOS setup program.

20. Which of the following describes IRQs?

a. an address that points to an input or output device

b. a line that allows access to memory without using the CPU

c. a line that allows the user to temporarily stop the CPU from doing something so that it can give its attention somewhere else

d. a line that supplies power to an interface card

21. Which of the following describes DMAs?

a. an address that points to an input or output device

b. a line that allows access to memory without using the CPU

c. a line that allows users to temporarily stop the CPU from doing something so that it can give its attention somewhere else

d. a line that supplies power to an interface card

22. The device I/O address _____ .

a. must be unique and correspond to a port or device

b. must be the same for each device

c. is not needed for some devices

d. is only needed for devices that do not require or use a DMA

23. Expansion boards are configured in which of the following ways? (Choose all that apply)

a. The board may have jumpers that need to be set.

b. The board may have DIP switches that need to be set.

c. The board may require a special preprogrammed ROM chip.

d. The board may be configured with software.

24. When installing an expansion board, what is the best method of determining the current settings?

a. Run MSD to view the IRQs after the board is installed.

b. Consult the documentation for the board.

c. Systematically change the DIP switch settings on the board.

d. Remove all jumpers on the expansion card and reinstall them one by one until the board functions.

25. What is the standard I/O address for parallel port 1?
    a. 3F8–3FF
    b. 2F8–2FF
    c. 3F0–3F7
    d. 278–27F
    e. 378–37F

26. What is the standard I/O address for serial port 1?
    a. 3F8–3FF
    b. 2F8–2FF
    c. 3F0–3F7
    d. 278–27F
    e. 378–37F

27. What is meant by the term *memory address* in relation to expansion boards?
    a. an area above 640 KB that a device can use exclusively for its operations
    b. an address in RAM that flags the CPU to an expansion board's presence
    c. a unique number assigned to a device to identify the expansion card to the computer
    d. ROM memory that specifies what drivers are needed

28. An NE2000 compatible card is configured to use IRQ 3 and I/O address: 0x300. With which device would the card most likely conflict?
    a. COM1
    b. COM2
    c. LPT1
    d. LPT2

29. Pat has just installed a network adapter card in his computer, but the operating system is unable to detect the network adapter card. Which of the following is the most likely cause of the problem?
    a. wrong protocol
    b. resource conflict such as an IRQ or DMA conflict
    c. faulty cable
    d. faulty terminating resistor

30. You want to install an NE2000 network card. Which of the following IRQs is most likely free?
    a. IRQ 3
    b. IRQ 4
    c. IRQ 7
    d. IRQ 11
    e. IRQ 14

31. A ribbon cable has a red stripe along one edge. What does this mean?
    a. This is the positive lead.
    b. This is the negative lead.
    c. The cable carries hazardous voltage.
    d. The conductor with the stripe connects to pin 1 of the connector.

32. The time it takes for the hard drive to find data is called the _____ .
    a. rotational latency period
    b. access time
    c. data transfer rate
    d. seek time

33. The time required to position the head over the proper track is the _____ .
    a. seek time
    b. access time
    c. latency period
    d. data transfer rate

34. Which of the following should be done *before* replacing a hard drive?
    a. Back up the hard drive.
    b. Run the CMOS setup program.
    c. Run CHKDSK.
    d. Erase the hard drive.

35. Which program partitions a hard drive?
    a. SCANDISK
    b. DISKCOPY
    c. FORMAT
    d. FDISK
    e. SETUP

36. When configuring a system with preloaded software, formatting the primary hard disk will _____ .
    a. improve system performance
    b. erase the partitions
    c. compress preloaded programs
    d. provide additional partition and format options

37. When replacing an IDE hard drive, you must _____ .
    a. set the ID jumper on the drive
    b. set the master/slave jumper on the drive
    c. set the speed jumper on the drive
    d. set the speed jumper on the motherboard

38. How many devices does EIDE support?
    a. 2   c. 7
    b. 4   d. 8

39. Which of the following IRQs would probably be free so that it can be used by a SCSI card if the system has a standard IDE controller card? (Choose two answers)
    a. IRQ 7   d. IRQ 14
    b. IRQ 6   e. IRQ 15
    c. IRQ 10

40. When installing a SCSI CD-ROM drive, you must configure the drives'. (Choose all that apply)
    a. terminating resistors   c. stand-alone, master, or slave jumpers
    b. SCSI ID number   d. frequency modulation

HANDS-ON EXERCISES

EXERCISE 1: IDENTIFY THE CPU

Note: When performing this exercise, follow an ESD program and use an ESD-grounded strap.

1. What kind of CPU do you have?
2. Look to see how pin A1 is designated on the motherboard and on the CPU.
3. If the microprocessor has a heat sink, carefully remove it.
4. If the microprocessor is not soldered onto the motherboard, carefully remove it from the motherboard. Be careful to not bend or mangle the pins of the microprocessors.
5. Reinsert the microprocessor, reinstall the heat sink, and test the system.

EXERCISE 2: IDENTIFY THE RAM CHIPS AND RAM CACHE CHIPS

Note: When performing this exercise, follow an ESD program and use an ESD-grounded strap.

1. Open your computer.
2. Remove the RAM chips.
3. Identify the size and speed of each RAM chip.
4. Reinsert the RAM chips.
5. Identify the RAM cache on the motherboard (if any).
6. Identify the microprocessor on the motherboard. Does the microprocessor have any internal L1 or L2 RAM cache? If yes, how much cache memory does the microprocessor have?

EXERCISE 3: USING DIAGNOSTIC SOFTWARE

Using diagnostic software, such as QA Plus or Checkit Pro, answer the following questions.

1. What device is actually using IRQ 4?
2. What device is actually using IRQ 5?

3. What device is actually using IRQ 7?
4. What device is actually using IRQ 11?
5. What device is actually using DMA-2?
6. What device is actually using DMA-3?

EXERCISE 4: USING THE OPERATING SYSTEM INTERNAL SOFTWARE

Using DOS's Microsoft Diagnostic (MSD), Windows 95/98's system icon within the control panel, or Windows NT's Microsoft Diagnostic software, answer the following questions.

1. What device is actually using IRQ 4?
2. What device is actually using IRQ 5?
3. What device is actually using IRQ 7?
4. What device is actually using IRQ 11?
5. What device is actually using DMA-2?
6. What device is actually using DMA-3?

EXERCISE 5: CONFIGURING A MODEM OR I/O CARD

Note: When performing this exercise, follow an ESD program and use an ESD-grounded strap.

1. Make sure the computer is off. If your system has a modem, remove it.
2. By looking at the back of the computer, how many serial ports does your system have?
3. Turn on the computer. If you have a COM1 and a COM2, by running some form of diagnostic software that shows hardware resources, what are the IRQs for COM1 and COM2?
4. Install the mouse to COM1 and load the mouse driver in the appropriate configuration file.
5. Test the mouse to make sure that it is working properly.
6. Shut off the computer. By looking at the modem's documentation, set the modem to COM1 and insert the modem card into an available expansion slot.
7. Turn on the computer. Test the mouse. Does the mouse work? If not, what is the problem?
8. Shut off the computer. Using the I/O controller card documentation or the motherboard documentation if the serial ports are part of the motherboard, disable COM2. Note: If the serial ports are part of the motherboard, it is probably done with the CMOS setup program.
9. If possible, use the modem documentation to change the modem to COM2 and use the same IRQ as the mouse. Remember that the mouse is plugged into COM1.
10. Turn on the computer. Test the mouse. Does the mouse work? If not, what is the problem?
11. Shut off the computer. Using the modem documentation, change the modem's IRQ to an available one. Hint: Use the IRQ that is standard for COM2.
12. Turn on the computer. Test the mouse again. Does the mouse work?
13. Shut off the computer and remove the modem.
14. Enable COM2 on the I/O controller card and reinsert the card into the expansion slot.
15. Turn on the computer and test the mouse to make sure it is working properly.

EXERCISE 6: INSTALLING A NETWORK CARD

Note: When performing this exercise, follow an ESD program and use an ESD-grounded strap.

1. By using some form of software diagnostic, determine which I/O address, IRQs, DMAs, and memory address are available to install the network card.

2. By reading the network card documentation, configure the card to an available I/O address, IRQ, DMA (if needed), and memory address (if needed).
3. Insert the card in the expansion slot.
4. Turn on the computer. Did it boot?
5. If the card is plug-and-play, determine what resources the card is using.

EXERCISE 7: INSTALL AN IDE HARD DRIVE

Note: When performing this exercise, follow an ESD program and use an ESD-grounded strap.

1. Ensure that the computer can boot to drive C.
2. With the CMOS setup program, record the drive type, number of tracks, number of read/write heads, number of sectors per track, and hard drive mode (if any).
3. Using the FDISK utility, view the partitions (option 4). Look for the number and the type of partitions.

```
                    Microsoft Windows 98
                  Fixed Disk Setup Program
            (C)Copyright Microsoft Corp. 1983-1998

                       FDISK Options

Current fixed disk drive: 1

Choose one of the following:

1. Create DOS partition or Logical DOS Drive
2. Set active partition
3. Delete partition or Logical DOS Drive
4. Display partition information

                    Enter choice: [4]

Press Esc to exit FDISK
```

```
               Display Partition Information

Current fixed disk drive: 1

Partition  Status  Type      Volume Label  MB    System  Usage
C: 1       A       PRI DOS                  1039  FAT16   50%
   2               EXT DOS                  1024  FAT32   50%

Total disk space is 2067 MB (1 MB = 1048576 bytes)

The Extended DOS Partition contains Logical DOS Drives.
Do you want to display the logical drive information
(Y/N)......?[Y]

Press Esc to return to FDISK Options
```

4. Using the FDISK utility, delete all partitions (option 3) by deleting any logical partitions, then deleting the extended partition, and then deleting the primary partition.

```
                Delete DOS Partition or Logical DOS Drive

Current fixed disk drive: 1

Choose one of the following:

1. Delete Primary DOS Partition
2. Delete Extended DOS Partition
3. Delete Logical DOS Drive(s) in the Extended DOS Partition
4. Delete Non-DOS Partition

Enter choice: [3]


Press Esc to return to FDISK Options
```

5. After the partitions are deleted, press the ESC key several times until it asks to insert a system disk.
6. If the drive is set to a type 47, type 48, or user-definable type, set the number of tracks, number of read/write heads, and number of sectors per track to zero.
7. Go into the CMOS setup program and set the drive C to Type 1.
8. Save the changes to the CMOS setup program and shut down the computer.
9. Shut off the computer. Disconnect the power cable and remove the hard drive, cable, and controller card from the system. Look to see how the cable, particularly pin 1, is connected to the card and the drive.
10. Reinstall the hard drive, cable, and controller into the system and connect the power to the hard drive.
11. Set the proper CMOS settings for the hard drive.
12. Try to boot the computer with the C drive. Record the error message.
13. Boot the computer with the A drive.
14. Run FDISK to create (option 1) a primary DOS partition. Do not maximize the partition and do not make it active. Specify the size of the primary DOS partition to be 75%. Note: You must include the % sign or it will create the partition of 75 MB instead of 75%.

```
                Create DOS Partition or Logical DOS Drive

Current fixed disk drive: 1

Choose one of the following:

1. Create Primary DOS Partition
2. Create Extended DOS Partition
3. Create Logical DOS Drive(s) in the Extended DOS Partition

Enter choice: [1]


Press Esc to return to FDISK Options
```

15. Go back to the main menu. Notice the warning at the bottom of the screen.
16. Use the rest of the free disk space to create an extended DOS partition. Assign one logical drive (drive D).
17. Press the ESC key until it asks to insert a system disk.
18. Boot the computer using drive C. Record the error message.
19. Using FDISK, make the primary DOS partition active (option 2).
20. Boot the computer using drive C. Record the error message.

21. Boot the computer using drive A. Switch over to the C drive. Perform a DIR command. Record the error message.
22. Use the Abort/Fail option to move back to the A drive.
23. Boot the computer using drive A. Format the hard drive, but do not make it bootable.
24. Boot the computer using drive C and record the error message.
25. Boot the computer using drive A. Make the hard drive bootable with either a FORMAT command or a SYS command.
26. Format the D drive.
27. Boot the computer using drive C.
28. Using the CMOS setup program, reduce the number of cylinders by 50% and double the number of read/write heads. Note: You may need to use a user-definable type if you are not already doing so. Save the changes.
29. Boot the computer using drive C. Record any error messages or if it fails to boot.
30. Using the CMOS setup program, put back the original settings.
31. Boot the computer to drive C.
32. Shut off the computer. Remove the cable connector from the hard drive and place it on backwards.
33. Boot the computer using drive C. Record any error messages or if it fails to boot.
34. Shut off the computer. Reconnect the cable properly.
35. Disconnect the gray ribbon cable.
36. Boot the computer to drive C. Record any error messages or if it fails to boot.
37. Shut off the computer. Reconnect the gray ribbon cable and disconnect the power connector.
38. Boot the computer to drive C. Record any error messages or if it fails to boot.
39. Shut off the computer. Reconnect the power cable.
40. Go into the CMOS setup program and set the drive D to a Type 1 drive.
41. Boot the computer to drive C. Record any error messages or if it fails to boot.
42. Go into the CMOS setup program and disable the drive D.
43. Ensure that the computer boots properly.

EXERCISE 8: INSTALL A SECOND IDE HARD DRIVE

Note: When performing this exercise, follow an ESD program and use an ESD-grounded strap.

1. Ensure sure that the computer can boot to drive C.
2. With the CMOS setup program, record the drive type, number of tracks, number of read/write heads, and number of sectors per track.
3. Shut off the computer. Remove the hard drive, which is already in the computer. Examine the jumpers on the hard drive. It should be set to a stand-alone hard drive.
4. Set the jumpers on the original hard drive to master and on the second hard drive as slave.
5. Connect both hard drives to the cable and connect the power cables.
6. Set the CMOS parameters for drives C and D.
7. Boot the computer using drive C.
8. Using FDISK, switch to the second hard drive (option 5). Create a primary DOS partition (option 1).
9. Press the ESC key until it asks to insert a system disk.
10. Boot the computer using drive C. Change over to the D drive. Remember that all primary partitions are given drive letters first.
11. Use the Abort/Fail option to move back to the A drive. Format the D drive.
12. Change to the E drive, which should be the extended DOS partition on the first hard drive.
13. Remove the second hard drive.
14. Change the first drive back to a stand-alone drive.
15. Disable the drive D in the CMOS setup program.
16. Ensure that the computer boots properly.

EXERCISE 9: INSTALLING A SCSI HARD DRIVE

Note: When performing this exercise, follow an ESD program and use an ESD-grounded strap.

1. Remove any IDE drives from your system.
2. Disable the drive C in the CMOS setup program.
3. Set the resources of the SCSI expansion card. Use the card's documentation and software diagnostic to determine I/O addresses, IRQs, and DMA.
4. Locate the terminating resistors on the SCSI expansion card. Make sure that they are installed and enabled.
5. If the SCSI controller card has a floppy disk drive controller and one already exists in the computer, disable the one on the SCSI controller card.
6. Install the SCSI controller.
7. Using the drive's documentation, set the boot drive to SCSI ID 0. Make sure that the terminating resistor is installed and enabled.
8. Install the boot drive to the end of the chain (cable) and connect the power.
9. Using the drive's documentation, set a second SCSI hard drive or a SCSI CD-ROM to SCSI ID 5.
10. Disable or remove the terminating resistor on the second drive.
11. Connect the second drive to the middle of the chain and connect the power.
12. Boot the computer using drive A.
13. Create a primary DOS partition. Use the entire drive for the partition and make it active.
14. Format the hard drive and make it bootable.
15. Using the correct software, low-level format the first SCSI hard drive.
16. Remove all SCSI drives including the controller card.
17. Reinstall the IDE hard drive.
18. Enter the correct parameters in the CMOS setup program.
19. Ensure that the computer boots properly.

EXERCISE 10: INSTALLING AN IDE CD-ROM DRIVE

Note: When performing this exercise, follow an ESD program and use an ESD-grounded strap.

1. Configure the IDE CD-ROM drive as either a master, slave, or stand-alone, as appropriate.
2. If the system already has an IDE device, configure the device as either a master, slave, or stand-alone, as appropriate.
3. Connect the CD to the appropriate IDE channel.
4. Boot the system and install the appropriate drivers for the CD drive.
5. Insert a CD into the drive and test it.
6. Remove the CD-ROM drive and reconfigure the system.

EXERCISE 11: INSTALLING A SCSI CD-ROM DRIVE

Note: When performing this exercise, follow an ESD program and use an ESD-grounded strap.

1. Configure the SCSI drive's SCSI ID number and terminating resistors.
2. Connect the CD to the appropriate SCSI connector.
3. Boot the system and install the appropriate drivers for the CD drive.
4. Insert a CD into the drive and test it.
5. Remove the CD-ROM drive and reconfigure the system.

EXERCISE 12: MOTHERBOARD BIOS UPGRADE

1. Go to the motherboard manufacturer's Web site.
2. Find and download the newest BIOS update program or file to update the motherboard system ROM BIOS.
3. Find the instructions on how to update/flash the BIOS and update the BIOS. Note: when updating the BIOS, it is very important that you do not shut off the computer or interrupt the computer in any way until the update is complete. If not, the motherboard will not boot and you would have to replace the BIOS chip.

EXERCISE 13: INTERNET RESEARCH

1. Find the CMOS parameters for the Western Digital Caviar AC33200 (3.2-GB EIDE) hard drive.
2. Find the jumper settings to make the Western Digital Caviar AC33200 (3.2-GB EIDE) a master drive.
3. Find the jumper settings to make the Western Digital Caviar AC33200 (3.2-GB EIDE) a slave drive.
4. Find the jumper settings to make a Seagate ST-11900N (1.7-MB SCSI hard drive) SCSI ID 0.
5. Find the jumper settings to make a Seagate ST-11900N (1.7-MB SCSI hard drive) SCSI ID 5.
6. Find how to terminate the resistance on a Seagate ST-11900N (1.7-MB SCSI hard drive).
7. Find and download the software to make a Seagate Medallist Pro hard drive to overcome certain system BIOS and operating system limitations.
8. Find and download the Windows NT driver for the AHA-2940 SCSI card.
9. Find and download the installation guide for the Adaptec 2940 Ultra SCSI card.

3 Introduction to Networking Technology

INTRODUCTION

In the early days of networking, networking software was created in a haphazard fashion. When networks grew in popularity, the need to standardize the by-products of network software and hardware became evident. Standardization allows vendors to create hardware and software systems that can communicate with one another, even if the underlying architecture is dissimilar. For example, you can run the TCP/IP protocol to access the Internet on either an IBM-compatible PC that is running Windows or on an Apple Macintosh that is running System 8.

OBJECTIVES

1. Explain the need for standardization in networks.
2. List the differences between the de jure and the de facto standard.
3. Describe the OSI model and how it relates to networks.
4. List and define the layers of the OSI model.
5. Define the communication devices that communicate at each level of the OSI model.
6. Describe the various addresses and ports and how they relate to networks.
7. Define a virtual circuit.
8. Define session and explain how it relates to networks.
9. List and define the types of dialogs.
10. List and define the 802 Project standards.

11. Define signaling, modulation, and encoding.
12. Define bandwidth and compare baseband and broadband.
13. List and describe the different connection services.
14. Compare the implications of using connection-oriented communications with connectionless communications.

3.1 THE NEED FOR STANDARDIZATION

To overcome compatibility issues, hardware and software often follow **standards** (dictated specifications or the most popular specifications). Standards exist for operating systems, data formats, communication protocols, and electrical interfaces. If a product does not follow a popular standard, it usually will not be widely accepted in the computer market and will often cause problems within a PC. As far as the user is concerned, standards help determine what hardware and software to purchase and they allow the user to customize a network made of components from different manufacturers.

As new technology is introduced, manufacturers are usually rushing to get their products out so that their products have a better chance of becoming the standard. Often, competing computer manufacturers introduce similar technology at the same time. Until one is designated as the standard, other companies and customers are sometimes forced to take sides. Since it is sometimes difficult to determine what will emerge as the true standard and that the technology sometimes needs time to mature, it is best to wait and see what happens.

There are two main types of standards. The first is called the de jure standard. The **de jure standard,** or the bylaw standard, is dictated by an appointed committee such as the International Organization for Standardization for (ISO). Some common standard committees are shown in Table 3.1. The second type is the de facto standard. The **de facto standard** or "from the fact" standard is accepted by the industry because of its commonality. Such standards are not recognized by a standard committee. For example, the de facto standard for microprocessors is produced by Intel and the de facto standard for sound cards is produced by Creative Labs.

When a system or standard has an **open architecture,** it indicates that the specification of the system or standard is public. This includes both approved standards and privately designed architecture whose specifications are made public by the designers. The advantage of an open architecture is that anyone can design products based on the architecture and can design add-on products for it. This type also allows other manufacturers to duplicate the product, however.

The opposite of an open architecture is a proprietary system. A **proprietary system** is privately owned and controlled by a company and has not divulged specifications that would allow other companies to duplicate the product. Proprietary architectures often do not allow mixing and matching products from different manufacturers and may cause hardware and software compatibility problems.

TABLE 3.1   Common standard committees

American National Standards Institute (ANSI) http://www.ansi.org	ANSI is primarily concerned with software. ANSI has defined standards for several programming languages including C Language and the SCSI interface.
International Telecommunications Union (ITU) http://www.itu.int/	ITU defines international standards, particularly communications protocols, and was formerly called the Comité Consultatif Internationale Télégraphique et Téléphonique (CCITT).
Institute of Electrical and Electronic Engineers (IEEE) http://www.ieee.org	IEEE sets standards for most types of electrical interfaces including RS-232C (serial communication interface) and network communications.
International Organization for Standardization (ISO) http://www.iso.ch/	ISO is an international standard for communications and information exchange.

3.2 OSI REFERENCE MODEL

With the goal of standardizing the network world, the **International Organization for Standardization (ISO)** began development of the **Open Systems Interconnection (OSI) reference model.** The OSI reference model was completed and released in 1984. Today, the OSI reference model is the world's prominent networking architecture model. It is a popular tool for learning about networks. The OSI protocols, on the other hand, have had a long "growing-up" period. While OSI implementations are not unheard of, they have yet to attain the popularity of many proprietary and de facto standards.

The OSI reference model adopts a layered approach in which a communication subsystem is broken down into seven layers, each performing a well-defined function. The OSI reference model defines the functionality at each layer but does not specify actual services and protocols to be used at each layer. From this reference model, actual protocol architecture can be developed. Figure 3.1 shows the seven layers of the OSI model.

To facilitate this, ISO has defined internationally standardized protocols for each of the seven layers. The seven layers are divided into three separate groups: application-oriented (upper) layers, an intermediate layer, and the network-oriented (lower) layers.

3.2.1 Network-Oriented Layers

The network-oriented (lower) layers are concerned with the protocols associated with the actual physical part of the network which allows two or more computers to communicate. The **physical layer** is responsible for the actual transmission of the bits sent across a physical media. It allows signals (electrical signals, optical signals, or radio signals) to be exchanged among communicating machines. Therefore, it defines the electrical, physical, and procedural characteristics required to establish, maintain, and deactivate physical links.

FIGURE 3.1   OSI reference model

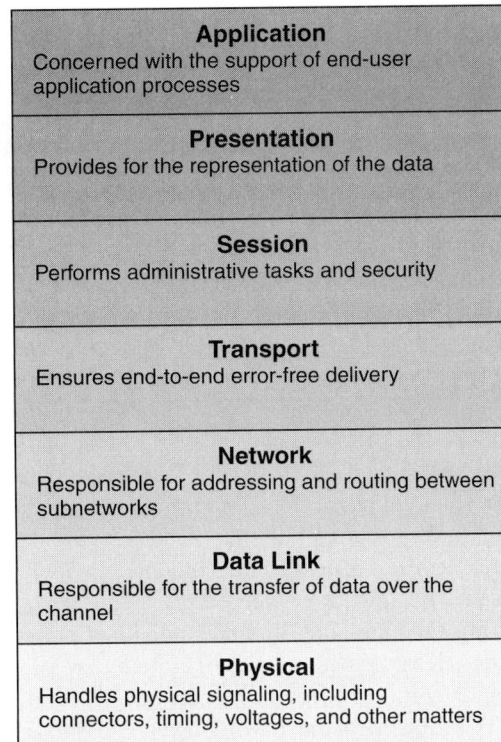

Application Concerned with the support of end-user application processes
Presentation Provides for the representation of the data
Session Performs administrative tasks and security
Transport Ensures end-to-end error-free delivery
Network Responsible for addressing and routing between subnetworks
Data Link Responsible for the transfer of data over the channel
Physical Handles physical signaling, including connectors, timing, voltages, and other matters

Please	Do	Not	Take	Sales	People's	Advice
Physical	Data Link	Network	Transport	Session	Presentation	Application

Note: An easy way to remember the order of the OSI reference model is to use mnemonics.

This includes how the bits (0s and 1s) of data are represented and transmitted. It is not concerned with how many bits comprise each unit of data, nor is it concerned with the meaning of the data being transmitted. In the physical layer, the sender simply transmits a signal and the receiver detects it. Lastly, the physical layer is also responsible for the physical topology or actual network layout. The physical layer includes the network cabling, hubs, repeaters, and network interface card (NIC). More information about the physical layer topics can be found in Chapter 4.

The **data link layer** is responsible for providing error-free data transmission and establishes local connections between two computers. This is achieved by packaging raw bits from the physical layer into blocks of data called frames and sending these frames with the necessary synchronization, error control, and flow control. Each frame includes a checksum (CRC) or some other form of error control information, a source address, a destination address, and data. A packet is a piece of a message that travels the network independently from other packets that contain the data and destination address. In IP networks, packets are often called datagrams.

The data link layer is divided into two sublayers, the logical link control (LLC) sublayer and the media access control (MAC) sublayer. The **media access control (MAC)** sublayer is the lower layer that communicates directly with the network adapter card. It defines the network logical topology, which is the actual pathway (ring or bus) of the data signals being sent. In addition, it allows multiple devices to use the same media and it determines how the network card gets access or control of the network media so that two devices do not trample over each other. Lastly, it maintains the physical device address, known as the MAC address, which is used to identify each network connection. Some examples of MAC sublayer protocols include CSMA/CA, CSMA/CD, token passing, and demand priority. The MAC sublayer topics are discussed further in Chapter 4.

The **logical link control (LLC)** sublayer manages the data link between two computers within the same subnet (a simple network or smaller network used to form a larger network) and defines **service access points (SAPs), or ports.** The ports are used to identify the upper-layer protocols and act as a switchboard to make sure the frames find their way to the right network layer process. For example, you are trying to read a Web page. The packet route from the Internet to your computer uses an Ethernet protocol/TCP/IP network. Because Windows is a multitasking environment that may be running many network services, it is the job of the LLC to route to the software that reads the Web page. This is done because the data packets have a port number assigned to them that identifies the packets as Web page data.

Ethernet, token ring, and ARCnet are three local area protocols that work at the physical and data link layers, those layers that are used to connect PCs to a LAN. The SLIP protocol works at the physical layer and the PPP protocol works at the physical and data link layers. Both of these protocols connect the PC to the Internet using the PC's modem. Frame relay is a WAN protocol that works at the physical and data link layers and is used to connect smaller networks to form a larger network. Most of these layers are discussed in Chapter 4.

The **network layer** addresses and routes data from one network (or subnet) to another. Responsibilities include establishing, maintaining, and terminating connections between networks; making routing decisions; and relaying data from one network to another. The OSI model classifies a **host** as the computer or device that connects to the network as the source or final destination of data, whereas **routers** (intermediate systems) perform routing and relaying functions that link the individual networks. Therefore, routers are network layer devices. For all networks to operate on an internetwork and identify themselves to each other, they must be assigned a network ID. Logical addresses such as those defined in IPX and TCP/IP work at the network layer, and are assigned by the administrator. The MAC address (physical address) is typically predefined for a network connection in the hardware.

When a packet is sent to a remote host several networks away, to begin it is sent to the first router. The router then determines which way to send the packet, strips off and rebuilds the data link layer source and destination address information, and sends the packet to the next router. When the packet gets to the next router, the router again determines which way to send the packet, strips off and rebuilds the data link layer source and the destination address information, and sends it to the next router. It will keep doing this until it gets to the destination network and finally to the destination computer.

Remember, the data link layer provides for the transmission of frames within the same LAN and the network layer provides for the much more complex task of transmitting packets between two network computers or devices in the network, regardless of how many data links/routers exist between the two. Examples of network layer protocols that connect the networks include IP and IPX; examples of protocols that determine routes include RIP and OSPF.

3.2.2 Intermediate Layer

The **transport layer** can be described as the middle layer that connects the lower and upper layers. It is also responsible for reliable, transparent transfer of data between two end points. Since it provides end-to-end recovery of lost and corrupted packets and flow control, it deals with packet handling, repackaging of messages, dividing messages into smaller packets, and error handling. In addition, if packets arrive out of order at the destination, the transport layer is responsible for reordering the packets back to the original order. Examples of transport layer protocols include SPX, TCP, UDP, and NetBEUI.

Virtual circuits establish formal communication between two computers on an internetwork using a well-defined path. This enables two computers to act as though they have a dedicated circuit between them, even though there is not. The path that the data take while being exchanged between the two computers may vary, but the computers do not need to know this. Virtual networks are sometimes depicted as a cloud, because the user does not worry about the path taken through the cloud, but rather about entering and exiting the cloud.

Computers and network devices often use long strings of numbers to identify themselves. The transport layer is responsible for name resolution, where it can take a more meaningful name and translate it to a computer or network address. Examples of name resolution protocols include domain name system (DNS) and Windows Internet Naming Service (WINS).

3.2.3 Application-Oriented Layers

The application-oriented layers are concerned with the protocols that provide user network applications and services. The **session layer** allows remote users to establish, manage, and terminate a connection (session). A **session** is a reliable dialog between two computers. It enables two users to organize and manage their data exchange and implement dialog control between the source and destination network devices, including the type of dialog (simplex, half duplex, full duplex) and the length of computer transmits. Simplex, half duplex, and full duplex are explained later in this chapter. When establishing these sessions, it creates checkpoints in the data streams. If a transmission fails, only the data after the last checkpoint have to be retransmitted. Lastly, it implements security, such as the validation of user names and passwords. Examples include NCP, Telnet, and FTP.

The **presentation layer** ensures that information sent by an application layer protocol of one system will be readable by the application layer protocol of a remote system. This layer, unlike its lower layers, is simply concerned with the syntax and semantics of the information transmitted. Therefore, it acts as the translator between different data formats, protocols, and systems. It also provides encryption/decryption and compression/decompression of data. Examples include NCP and SMB.

The **application layer** represents the highest layer of the OSI reference model. It is responsible for interaction between the operating system and provides an interface to the system. It provides the user interface to a range of network-wide distributed services including file transfer, printer access, and electronic mail. It does not include software applications such as word processors and spreadsheets. Some examples of upper-layer protocols include file transfer protocol (FTP), simple mail transfer protocol (SMTP), AppleTalk, and NFS.

3.3 LAYER INTERACTION

When one computer must communicate with another computer, it will start with a network service, which is running in the application layer. The actual data to be sent are generated by the software and sent to the presentation layer. The presentation layer then adds its own control information, called a header, which contains the presentation layer's requests and/or information. The packet then is sent to the session layer where another header is added. It keeps going down the OSI model until it reaches the physical layer, where the data are sent on the network media by the network interface card (NIC). See Figure 3.2 Layer Interaction of the OSI Reference Model.

When the data packet gets to the destination computer, the NIC sends the data packet to the data link layer. The data link layer then strips off the first header. As it goes up the model, each header is stripped away until it reaches the application layer. At this time only the original data are left which are processed by the network service.

What is so great about this system is that it allows you to communicate with different computer systems. For example, a Windows NT server can send information to a UNIX server or an Apple Macintosh client.

Let's follow the steps of communication between two computers from initial contact to data delivery. A computer wants to request a file. The process would start with a network application on the client computer. Before the request can be made, the client had to first determine the address of the computer. Therefore, it sent a request out to resolve a computer name to a network address.

If the client already knows the server's network address, it will make its request in the form of a data packet and send it to a presentation layer protocol. The presentation protocol would encrypt and compress the packet and send it to a session layer protocol, which would establish a connection with the server. During this time the packet would include the type of dialog such as half-duplex connection (discussed later in this chapter) and determine how long the computer can transmit to the server. The session protocol will then send the packet to the transport layer, which will divide the packet to smaller packets so that it can be sent over the physical network. The transport layer protocol will then send the packets to the network layer where the source network and destination network addresses are added. It will then send the packets to a data link layer so that it can add the source and destination addresses and the port number that identifies the service requested and prepare the packets to be sent over the media. As you can see in Figure 3.2, the packet is much bigger than when it is started. The packet is then converted to electrical or electromagnetic signals which are sent on the network media.

When the packets get to the destination server, the NIC sees the signals and interprets them as 0s and 1s. It then takes the bits and groups them into frames. It will then determine the destination address of the packet to see if the packet was meant for the server. After the server determines that it was, it will then remove the data link header and send the packet up to the network layer. The network layer will remove the network layer header and send the packet to the transport layer. If the data packets have reached the server out of order,

FIGURE 3.2 Layer Interaction of the OSI Reference Model

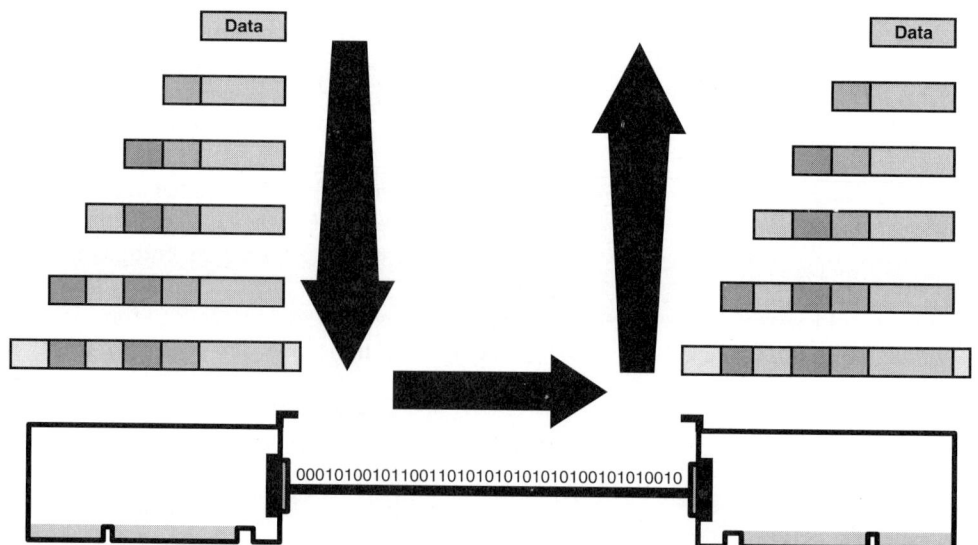

then the transport layer protocol will put them back into the proper order, merge them into one larger packet, and send the packet to the appropriate session layer protocol. The session layer protocol will authenticate the user and send the packet to the presentation layer protocol. The presentation layer protocol will decompress, decrypt, and reformat the packet so that it can be read by the application layer protocol. The application layer protocol will then read the request and take the appropriate steps to fulfill the request.

Before the packet is actually sent, the network layer must determine if the packet is to go to another computer within the same network or to a computer in another network. If it is local, the packet is sent to the computer. If it has to go to another network, the packet is sent to a router. The router will then determine the best way to get to its destination network and will send the packet from one router to another until it gets there. If one network route is congested, it can reroute the packet another way; and if it detects errors, it can slow the packet's transmission in the hopes that the link will become more reliable.

Many protocols will have the destination computer return an acknowledgment stating that the packet arrived intact. If the source computer does not receive an acknowledgment after a certain amount of time, then it will resend the packet.

3.4 THE 802 PROJECT MODEL

In the late 1970s, when LANs first began to emerge as a valuable potential business tool, the IEEE realized the need for certain LAN standards, specifically for the physical and data link layer of the OSI reference model. To accomplish this, the institute launched Project 802, which was named for the year and month it began (1980, February). These standards have several areas of responsibility including the network card, the wide area network components, and the media components, as shown in Table 3.2. Of these, the Project 802 standards discussed the most are the 802.2 (LLC), 802.3 (Ethernet), and 802.5 (token ring).

3.5 SIGNALING METHODS

Although this chapter will not go into the details of each layer, it will introduce some basic topics to help the reader understand how a network functions. Topics related to the OSI model will be discussed in subsequent chapters as needed.

Signaling is the method for using electrical or light energy to communicate. The process of changing a signal to represent data is often called **modulation** or **encoding.** The two forms of signaling are digital signaling and analog signaling.

TABLE 3.2   802 Project standards

Standard	Category
802.1	Overview and architecture of internetworking including bridging and virtual LAN (VLAN)
802.2	Logical link control (LLC)
802.3	Carrier-sense multiple access with collision detection (CSMA/CD) LAN (Ethernet)
802.4	Token bus LAN
802.5	Token ring LAN
802.6	Metropolitan area network (MAN)
802.7	Broadband technical advisory group
802.8	Fiber-optic technical advisory group
802.9	Integrated voice/data networks
802.10	Network security
802.11	Wireless networks
802.12	Demand priority access LAN, 100BaseVG-AnyLAN
802.14	Coaxial and fiber cable such as those found on cable television to support two-way communications

3.5.1 Digital Signals

A **digital signal** (the language of computers) is a system that is based on a binary signal system produced by pulses of light or electric voltages. The site of the pulse is either on/high or off/low to represent 1s and 0s. **Binary digits (bits)** can be combined to represent different values.

A digital signal can be measured in either of two ways. The first method, the **current state** method, periodically measures the digital signal for the specific state. The second method, the **transition state** method represents data by how the signal transitions from high to low or low to high. A transition indicates a binary 1 whereas the absence of a transition represents a binary 0. See Figure 3.3.

3.5.2 Analog Signals

An **analog signal** is the opposite of a digital signal. Instead of having a finite number of states, it has infinite number of values, which are constantly changing. Analog signals are typically sinusoidal waveforms, characterized by their amplitude and frequency. **Amplitude** represents the peak voltage of the sine wave. **Frequency** indicates the number of times that a single wave will repeat over any period. It is measured in hertz (Hz) or cycles per second. Another term used when talking about modulation is its time reference or phase. A **phase** is measured in degrees. Phase can be measured in 0°, 90°, 180°, and 270° or 0°, 45°, 90°, 135°, 180°, 225°, 270°, 315°, and 360°. See Figure 3.4.

FIGURE 3.3   Measuring data bits by using current state and state transition

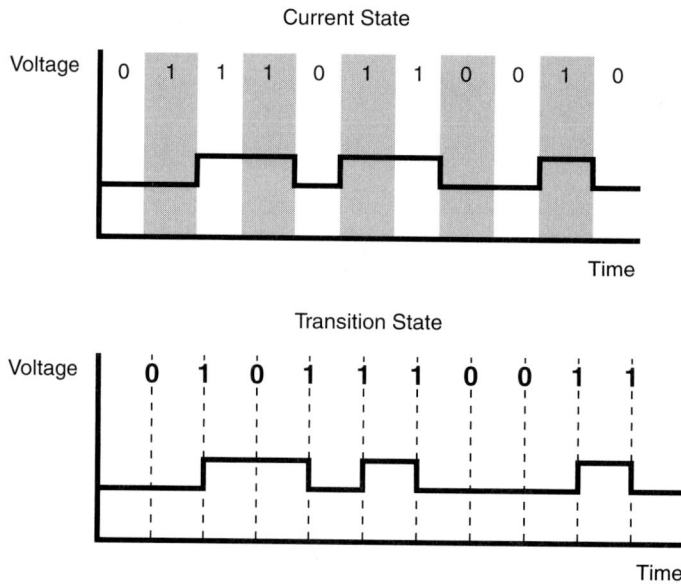

FIGURE 3.4   A sine wave showing the amplitude, frequency (cycles per second), and phase

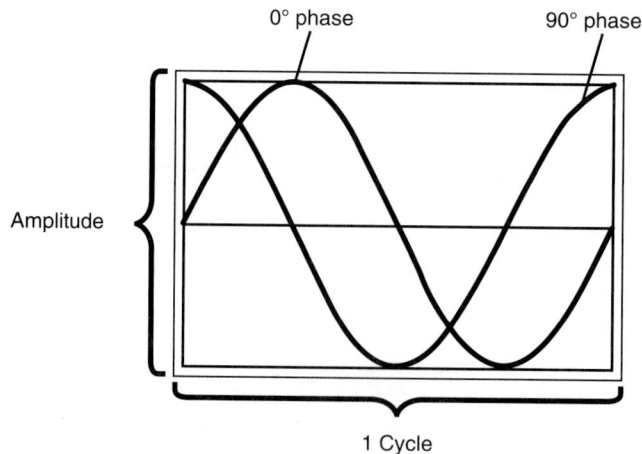

TABLE 3.3 Data bits interpreted from phase signals

Phase	Data Bits
0°	0 0
90°	0 1
180°	1 0
270°	1 1

By varying the amplitude, frequency, and phase, data are sent via telephone lines. The earliest form of encoding data over telephone lines is **frequency shift keying (FSK),** which is similar to frequency modulation used with FM radios. FSK sends a logical 1 at one particular frequency (usually 1750 Hz) and a logical 0 at another frequency (often 1080 Hz). FSK was primarily used with 300-baud modems.

Phase shift keying (PSK) is the varying of the phase angle (0°, 90°, 180°, and 270°) to represent data. Since the phase has four different values, it represents two bits of data. Therefore, a 1200-baud modem using PSK can transmit data at 2400 bps. See Table 3.3.

Much like digital signals, the analog signals can be measured as current state or state transition. For example, they can measure the amplitude as set intervals or the transition from one amplitude to another.

3.5.3 Synchronous and Asynchronous Connection

As mentioned earlier in the chapter, data bits are encoded on a network media and the receiving network card interprets the signal by measuring it. Therefore, the receiver network card must use a clock or a timing method to determine when to measure and decode the signal and when to decode the data bits. The two methods are either synchronous or asynchronous.

Synchronous devices use a timing or clock signal to coordinate communications between the two devices. If the send and receive devices were both supplied by exactly the same clock signal, then transmission could take place forever with the assurance that the signal sampling at the receiver was always in perfect synchronization with the transmitter.

In synchronous communications, pieces of data are not sent in individual bytes, but as frames of large data blocks. For example, frame sizes vary from a few bytes to many bytes. Ethernet uses 1500-byte packets. The clock is embedded in the data stream encoding, or is provided on a separate clock line so that the sender and receiver are always in synchronization during a frame transmission.

Asynchronous signals are intermittent signals, meaning they can occur at any time and at irregular intervals. They do not use a clock or timing signal. As the data frame is sent, it consists of a start signal, a number of data bits, and a stop signal. The start signal is sent to notify the other end that data are coming while the stop signal is sent to indicate the end of the data frame. Although the asynchronous signals are simpler electronic devices, they are not as efficient as synchronous signals because of the extra overhead of the start and stop signals. An example of an asynchronous device is a modem.

3.6 BANDWIDTH USAGE

Bandwidth refers to the amount of data that can be carried on a given transmission media. Larger bandwidth means greater data transmission capabilities. Bandwidth uses schemes that are based on the availability and utilization of the channels. A **channel** is part of the media's total bandwidth. It can be created by using the entire bandwidth for one channel or by splitting multiple frequencies to accommodate multiple channels. For example, if a medium can support 10 megabits per second (Mbps), two channels can be created at 5 Mbps each.

Baseband systems use the transmission medium's entire capacity for a single channel. Baseband networks can use either an analog or a digital signal, but digital is much more common.

A **broadband** system uses the transmission medium's capacity to provide multiple channels by using frequency-division multiplexing (FDM). Each channel uses a carrier signal, which runs at a different frequency than the other carrier signals used by the other channels. The data are embedded within the carrier channel. As data are sent onto the transmission channel, one multiplexer (mux) sends the several data signals at different frequencies. When the various signals reach the end of the transmission media, another multiplexer separates the frequencies so that the data can be read.

Although baseband can only support one signal at a time, it can be made into a broadband system by using time-division multiplexing (TDM). Time-division multiplexing divides the single channel into short time slots, thus allowing each multiple device to be assigned a time slot.

3.7 DIALOG CONTROL

Data can flow in one of three ways: simplex, half duplex, and full duplex. Simplex dialog allows communications on the transmission channel to occur in only one direction. Essentially, one device is allowed to transmit and all other devices receive. This process is often compared with a PA system where the speaker talks to an audience, yet the audience does not talk back to the speaker.

Half-duplex dialog allows each device to both transmit and receive, but not simultaneously, so only one device can transmit at a time. This process is often compared with a CB radio or walkie-talkies. A full-duplex dialog allows every device to both transmit and receive simultaneously. For networks, the network media channels would consist of two physical channels, one for receiving and one for transmitting. See Figure 3.5.

3.8 ADDRESSES

Because many network devices share the same transmission channel, the data link layer must have some way to identify itself from the other devices. The physical device address or MAC address is a unique hardware address (unique on the LAN/subnet) that is burned onto a ROM chip assigned by the hardware vendors or selected with jumpers or DIP switches.

At the network layer, networks are identified with a unique **network address.** This means that every computer on an individual LAN must use the same network address, but, if the LAN is connected to other LANs, its address must be different from all other LANs. A physical device (MAC address) and logical network addresses are used jointly to move data between devices on an internetwork. See Figure 3.6.

FIGURE 3.5 Simplex, half-duplex, and full-duplex communications

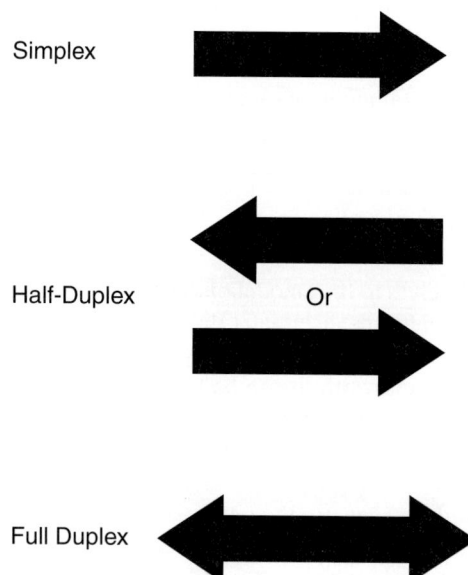

Simplex

Half-Duplex   Or

Full Duplex

NETWORK A

22-33-A3-45-23-32   22-33-A3-47-23-42   22-33-A3-34-43-43

Router

C3-5F-29-87-A4-8C

C3-5F-29-11-8D-CC

C3-5F-29-43-43-B2

22-33-A3-45-23-32

22-33-A3-64-87-34   **NETWORK B**

NETWORK C

22-33-A3-5B-AC-76   22-33-A3-34-63-12

22-33-A3-AA-BD-54   22-33-A3-B4-63-23

22-33-A3-34-43-43

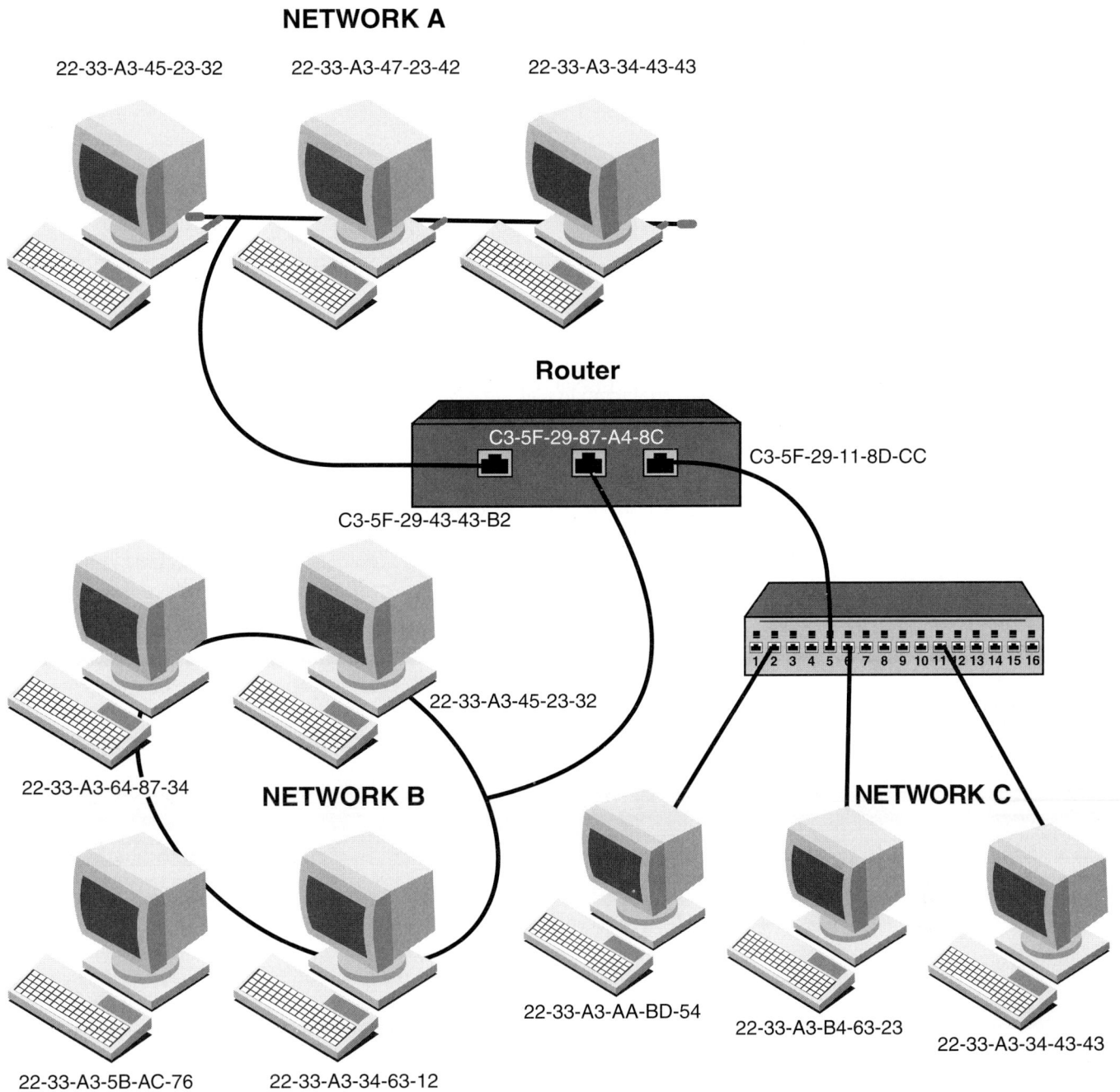

FIGURE 3.6   Three networks connected with a router. Each computer is identified by its eight-hexadecimal digit MAC address. Notice that the MAC address 22-33-A3-34-43-43 is used repeatedly but on different networks.

A network computer or other networked device can perform several roles simultaneously. The term **entity** identifies the hardware and software, which fulfills individual roles. Every entity must have a **service address** so that it can send and receive data. This address is usually referred to as a port, or **socket,** which is used to identify a specific upper-layer software process or protocol. Multiple service addresses can be assigned to any computer on which several network applications are running.

Since a network computer can handle multiple conversations with other computers simultaneously, the system must be able to keep track of the different conversations. The two methods of distinguishing conversations are by using either connection identifier or transaction identifier.

A common term frequently heard when dealing with Windows and networking is the Windows Socket or WinSock. A WinSock is a Windows implementation of the UC Berkeley

Sockets application programming interface (API), which is used in TCP/IP, IPX, and AppleTalk protocols to connect to the appropriate TCP/IP services such as HTTP, FTP, or Telnet. The WINSOCK.DLL file is the Windows interface for the TCP/IP protocol and the Internet.

3.9 CONNECTION SERVICES

Networks are either connection oriented or connectionless oriented. In a connection-oriented network, users must establish a connection or pathway between a source point and a destination point before they can transmit packets. Establishing a connection before transmitting packets is similar to making a telephone call. You must dial a number, the destination telephone must ring, and someone must pick up the telephone receiver before you can begin speaking.

The disadvantage of using a connection-oriented network is it takes time to establish a connection before transmitting packets. The advantage of using such a connection is it can reserve bandwidth for specific connections. As a result, connection-oriented networks can guarantee a certain **quality of service (QoS).** By using QoS to guarantee bandwidth, connection-oriented networks can provide sufficient bandwidth for audio and video without the jitters or pauses and the transfer of important data within a timely manner. Lastly, the connection-oriented network can better manage network traffic and prevent congestion by refusing traffic that it cannot handle.

Connection-oriented networks have devices that transmit packets on the wire. As the destination address is included in the packets, the network devices (such as hubs, switches and routers) find a pathway and deliver the packets to their destination.

The data link layer, network layer, and transport layer offer some form of connection services, which defines how two devices form a connection. The connection services determine the level of error detection and recovery and flow control that is used in communicating data between two network devices. The data link layer is concerned with connection services from one device to another within the LAN while the network layer is concerned with connection services from one device to another on different LANs. The transport layer is concerned with an end-to-end connection service (a type of overall quality assurance).

Connection services can be subdivided into the following three categories:

- Unacknowledged connection services
- Connection-oriented services
- Acknowledged-oriented services

Unacknowledged connection services (also known as connectionless services) send and receive data frames with no flow, error, or packet sequence control. For example, if you send data on a network that uses unacknowledged connection services, you have no guarantee that the data actually gets to the destination. Compare this process with a post office: When you send a letter, you assume that the letter gets to its destination. Yet, you have no guarantee that it actually gets there. This process is therefore considered unreliable, even though most data packets reach their destination.

Connection-oriented services provide flow, error, and packet sequence control through acknowledgments. An acknowledgment is a special message that is sent back when a data packet makes it to its destination. Compare this process with a letter sent from the post office with a return receipt that is sent back to the sender to indicate that the letter reached its destination.

Acknowledged connectionless services use acknowledgments to furnish flow and error control between point-to-point transmissions. Compare this process with a package sent by UPS or Federal Express that allows the senders to track the packet from point to point.

The error control refers to the notification of lost or damaged data frames. This includes the following:

- The destination does not receive the data packet.
- The checksum does not match. A checksum value is mathematically generated before the data packet is sent, attached to the data packet, and sent. When the data packet reaches its destination, the same mathematical calculation is performed. If the same value matches the value that was sent, the data are assumed intact.
- The packet size does not configure to a minimum or maximum size requirement for the frame type used.

- Noise, interference, and distortion scramble the data.
- The capacity of a channel or network device is exceedingly causing a buffer overflow.

SUMMARY

1. To overcome compatibility issues, hardware and software often follow standards (either dictated or the most popular specifications).
2. The de jure standard or the bylaw standard is dictated by an appointed committee.
3. The de facto standard or from the fact standard is accepted by the industry because of its commonality.
4. The OSI reference model is the world's prominent networking architecture model and is also the most popular tool for learning about networks.
5. The OSI reference model adopts a layered approach in which a communication subsystem is broken down into seven layers, each performing a well-defined function.
6. The layers of the OSI reference model are physical, data link, network, transport, session, presentation, and application.
7. The physical layer is responsible for the actual transmission of the bits sent across a physical circuit.
8. The data link layer is responsible for providing data transmission over a single connection from one system to another.
9. The network layer is concerned with making routing decisions and relaying data from one device to another through the network.
10. The transport layer can be described as the middle layer that connects the lower and upper layers together. It provides the session layer with a data transport service that hides the implementation details of the underlying network.
11. The session layer allows remote users to establish sessions.
12. The presentation layer ensures that information sent by an application layer protocol of one system will be readable by the application layer protocol of the remote system.
13. The application layer represents the highest layer of the OSI reference model.
14. IEEE created the 802 Project model to establish standards for the physical and data link layer.
15. Bandwidth refers to the amount of data that can be carried on a given transmission media.
16. Baseband systems use the transmission medium's entire capacity for a single channel.
17. A broadband system uses the transmission medium's capacity to provide multiple channels.
18. Simplex dialog allows communications on the transmission channel to occur in only one direction.
19. Half-duplex dialog allows each device to both transmit and receive, but not simultaneously.
20. A full-duplex dialog allows every device to both transmit and receive simultaneously.
21. The connection services determine the level of error detection and recovery and flow control used in communicating data between two network devices.
22. Unacknowledged connection services (also known as connectionless services) send and receive data frames with no flow, error, or packet sequence control.
23. Connection-oriented services provide flow, error, and packet sequence control through acknowledgments.
24. Acknowledged connectionless services use acknowledgments to furnish flow and error control between point-to-point transmissions.

QUESTIONS

1. Which standard is utilized because it is the most common?
   a. de jure standard
   b. de facto standard
   c. a company standard
   d. the standard most preferred

2. Which model is the most well-known model for networking architecture and technologies?
   a. OSI model
   b. client-server model
   c. peer-to-peer model
   d. IEEE model

3. Which describes the correct order of the OSI model layers from bottom to top?
   a. physical, data link, network, transport, session, presentation, and application
   b. data link, physical, network, transport, session, presentation, and application
   c. physical, data link, network, transport, presentation, session, and application
   d. application, presentation, session, transport, network, data link, and physical

4. Which one of the following connectivity devices typically works at the physical layer of the OSI model?
   a. routers   c. repeaters
   b. bridges   d. gateways

5. What happens to the data link layer source and destination addresses when packets are passed from router to router?
   a. They are stripped off and then re-created.
   b. They are stripped off and replaced with MAC (hardware) addresses.
   c. They are stripped off and replaced with NetBIOS names.
   d. They are reformatted according to the information stored in the routing table.

6. Which layer of the OSI model determines the route from the source to the destination computer?
   a. transport layer   c. network layer
   b. session layer   d. physical layer

7. Bridges are often called media access control bridges because they work at the media access control sublayer. In which OSI layer does the media access control sublayer reside?
   a. transport layer   c. network layer
   b. physical layer   d. data link layer

8. Which layer of the OSI model provides synchronization between user tasks by placing checkpoints in the data stream?
   a. transport layer   c. network layer
   b. session layer   d. physical layer

9. Which layer of the OSI model adds header information that identifies the upper-layer protocols sending the frame?
   a. presentation layer   c. MAC sublayer
   b. network layer   d. LLC sublayer

10. Which layer of the OSI model defines how cable is attached to a network adapter card?
    a. cable layer   c. hardware layer
    b. connection layer   d. physical layer

11. Many applications use compression to reduce the number of bits to be transferred on the network. Which layer of the OSI model is responsible for data compression?
    a. application layer   c. network layer
    b. session layer   d. presentation layer

12. Which of the following connectivity devices typically work at the data link layer of the OSI model?
    a. routers   c. repeaters
    b. bridges   d. gateways

13. Which layer of the OSI model packages raw data bits into data frames?
    a. physical layer   c. presentation layer
    b. network layer   d. data link layer

14. Which layer of the OSI model is responsible for translating the data format?
    a. application layer   c. presentation layer
    b. network layer   d. data link layer

15. Which layer of the OSI model provides flow control and ensures messages are delivered error free?
    a. transport layer   c. network layer
    b. session layer   d. physical layer

16. The data link layer of the OSI model is responsible for what tasks?
    a. creating, maintaining, and ending sessions, and encryption
    b. reliable delivery of data and error control
    c. transferring and routing of packets on the network
    d. addressing and reassembling frames

17. The IEEE Project 802 divides the data link layer into two sublayers. Which sublayer communicates directly with the network adapter card?
    a. logical link control sublayer
    b. logical access control sublayer
    c. media access control sublayer
    d. data access control sublayer

18. What is the term for two computers acting as if there is a dedicated circuit between them even though there is not?
    a. virtual circuit   c. gateway
    b. physical circuit   d. dialog path

19. The Project 802 model defines standards for which layers of the OSI model?
    a. physical layer and data link layer
    b. network layer and data link layer
    c. transport layer and network layer
    d. application layer and presentation layer

20. Who developed the various pieces to the 802 models?
    a. Institute of Electrical Engineers
    b. Institute of Electronic Engineers
    c. Institute of Engineering Electricians
    d. Institute of Electrical and Electronic Engineers

21. Which Project 802 model specification describes Ethernet?
    a. 802.2   c. 802.5
    b. 802.3   d. 802.10

22. Which Project 802 model specification adds header information that identifies the upper-layer protocols sending the frame and specifies destination processes for data?
    a. 802.2   c. 802.5
    b. 802.3   d. 802.10

23. What does the 802.5 specification specify?
    a. Ethernet   c. fiber optics
    b. token ring   d. ARCnet

24. Which system is based on a binary signal system produced by pulses of light or electric voltages?
    a. analog   c. signal propagation
    b. digital   d. synchronous

25. Which of the following provides multiple channels on a network medium?
    a. baseband   c. broadbase
    b. broadband   d. multiband

26. Which of the following allows for two devices to communicate at the same time?
    a. simplex   c. full duplex
    b. half duplex   d. complex

27. Which of the following allows for two devices to communicate to each other, but not at the same time?
    a. simplex   c. full duplex
    b. half duplex   d. complex

28. The address that identifies a network card to the network is known as the _____ .
    a. MAC address   c. port address
    b. MAC identity   d. I/O address

29. When different networks are not physically separated, you must have a unique configuration for each network. What has to be unique besides the node address on the network to tell the networks apart?
    a. network ID
    b. computer name
    c. IP address
    d. workgroup or domain name

30. Connection-oriented and connectionless communication are the two ways that communication can be implemented on a network. Which of the following is often associated with connectionless communication?
    a. fast but unreliable delivery
    b. error-free delivery
    c. fiber-optic cable
    d. infrared technology

31. The two ways to implement communication on networks are connection-oriented and connectionless communication. Which of the following is associated with connection-oriented communication?
    a. fast but unreliable delivery
    b. assured delivery
    c. fiber-optic cable
    d. infrared technology

32. What type of communication ensures reliable delivery from a sender to receiver without any user intervention?
    a. communication oriented
    b. connection oriented
    c. connectionless
    d. physical

33. Which of the following are connection-oriented protocols? (Choose all that apply.)
    a. TCP
    b. IP
    c. IPX
    d. SPX

4 Building a Network

INTRODUCTION

As mentioned in a previous chapter, a primary network component is the common media used to connect the individual computers into a network by using a network card. To allow the computers to communicate with each other, users have many options in cable types, cable layout, and protocols.

OBJECTIVES

1. Explain how a network card connects to the network and sends data through the network.
2. Define topology.
3. Differentiate physical and logical topology.
4. List and compare the various topologies.
5. List and compare the various cable types.
6. Select the appropriate media for various situations: twisted-pair cable, coaxial cable, and fiber-optic cable.
7. Build and test a UTP cable.
8. Build and test a coaxial (thinnet) cable.
9. Explain the purpose of the plenum cable.
10. List and explain the various cable access methods.
11. List the characteristics of a network card.
12. Define the purpose of a MAC address.
13. Explain the purpose of NDIS and Novell ODI network standards.
14. List and compare the different connectivity devices.
15. Given a network situation, choose the proper connectivity device.
16. Describe the characteristics and purpose of the media used in IEEE 802.3 and IEEE 802.5 standards.
17. Build a coaxial (thinnet) Ethernet network.
18. Build a UTP Ethernet network.
19. Explain how to build a token ring network.
20. Select the appropriate topology for various token ring and Ethernet networks.
21. List the different devices used for cable management.
22. Compare the Ethernet and token ring networks.
23. Select the appropriate connectivity devices for various token ring and Ethernet networks.
24. Given the manufacturer's documentation for the network adapter, install, configure, and resolve hardware conflicts for multiple network adapters in a token ring or Ethernet network.
25. Given a network situation, find the cable and topology problem.

26. List the different serial line protocols.
27. Given a network problem, troubleshoot a network cable problem.

4.1 DATA AND THE NETWORK

To communicate on LANs, a **network card,** or **network interface card (NIC),** found as an expansion card or integrated into the motherboard, uses transceivers (devices that both transmit and receive analog or digital signals). To get a better idea of the function of the network card, consider how data are sent from one computer to another. Data are generated by the microprocessor and sent to the network card using the computer's parallel data bus. At this point, the data consist of alphanumeric characters represented as a series of on (1) and off (0) signals.

Within the network card, data are processed and converted in order to be transmitted on the network media. If data are being sent via cable, they are turned into a serial bit stream, amplified, and sent on to the cable. The electrical signals that travel over the network cable must follow specific rules on how they gain sole access to the cable and what specific signals are sent to represent the data on the cable.

The network card on the receiving computer sees the packet and reads part of it to determine if the packet was meant for this computer. The network card then converts the data into digital signals and sends the data to the microprocessor to be processed.

4.2 TOPOLOGIES

Topology describes the appearance or layout of the network. Depending on how you look at the network, there is the physical topology and the logical topology. The **physical topology** describes how the network actually appears. The **logical topology** describes how the data flow through the physical topology or the actual pathway of the data.

4.2.1 Point-to-Point and Multipoint Connection

All physical topologies are variations of two fundamental methods of connecting devices, point-to-point and multipoint. See Figure 4.1. **Point-to-point topology** connects two nodes

FIGURE 4.1 (a) Point-to-point

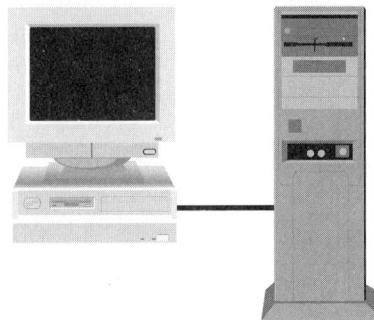

(b) multipoint connections

directly. Examples include two computers connected using modems, a PC communicating with a printer using a parallel cable, or WAN links connected by two routers using a dedicated T1 line. In a point-to-point link, the two devices monopolize the communication medium between the two nodes. Because the medium is not shared, nothing is needed to identify the two nodes. Whatever data are sent from one device are also sent to the other device.

Multipoint connection links three or more devices through a single communication medium. Because multipoint connections share a common channel, each device needs a way to identify both itself and the receiving device. The method used to identify senders and receivers is called **addressing.** SCSI devices connected on a single ribbon cable are identified with their SCSI ID numbers. Network cards connected on a network are identified with their media access control (MAC) address.

4.2.2 Physical Topology

As stated, the physical topology describes how devices on a network are wired or connected together. The four types of physical topologies used in local area networks are as follows:

- Bus topology
- Ring topology
- Star topology
- Hybrid topology

A **bus topology** resembles a line, where data are sent along the single cable. The two ends of the cable do not meet or form a ring or loop. See Figure 4.2. All **nodes** (devices connected to the computer including networked computers, routers, and network printers) listen to all traffic on the network, but only accept the packets addressed to them. The single cable is sometimes referred to as a **segment, backbone cable,** or **trunk.** Since all computers use the same backbone cable, the bus topology is easy to set up and install, and the cabling costs are minimized. Unfortunately, traffic easily builds up on this topology so it is not recommended for large networks. Examples of bus topologies include Ethernet 10Base2 and 10Base5.

Typically with a bus topology network, the two ends of the cable must be terminated, because when signals get to the end of a cable segment, they have a tendency to bounce back and collide with new data packets. If a break occurs anywhere or if one system does not pass the data along correctly, then the entire network will go down, because a break divides the trunk into two pieces, each with an end that is not terminated. In addition, these problems are difficult to troubleshoot since a break causes the entire network to go down with no indication of its location.

A **ring topology** has all devices connected to one another in a closed loop. See Figure 4.3. Each device is connected directly to two other devices. Typically in a ring, each node checks to see if the packet was addressed to it and acts as a repeater (duplicates the data signal which helps keep the signal from degrading) for the other packets. This process allows the network to span large distances. Though it might look inefficient, this topology sends data quickly, because each computer has equal access to communicate on the network.

Traditionally, a break in the ring will cause the entire network to go down and can be difficult to isolate. Today, some networks have overcome these pitfalls by allowing a computer to still communicate with its connected partners by using dual rings for fault tolerance and to act as a beacon if it notices a break in the ring. Lastly, since each node is a repeater, the networking device tends to be more expensive than the other topologies. IBM token ring and fiber distributed data interface (FDDI) are examples of ring topologies.

A **star topology** is the most popular topology in use. It connects each network device to a central point such as a hub, which acts as a multipoint connector. Other names for hub are concentrator, multipoint repeater, or media access unit (MAU). See Figure 4.4.

Star networks are relatively easy to install and manage, but may take some time to install, because each computer requires a cable that runs back to the central point. If a link fails (hub port or cable), then the remaining workstations are not affected like the bus and ring topologies. Unfortunately, bottlenecks can occur because all data must pass through the hub. Examples include Ethernet 10Base-T and 100Base-TX.

The **hybrid topology** scheme combines two of the traditional topologies, usually to create a larger topology. In addition, the hybrid topology allows users the strengths of the var-

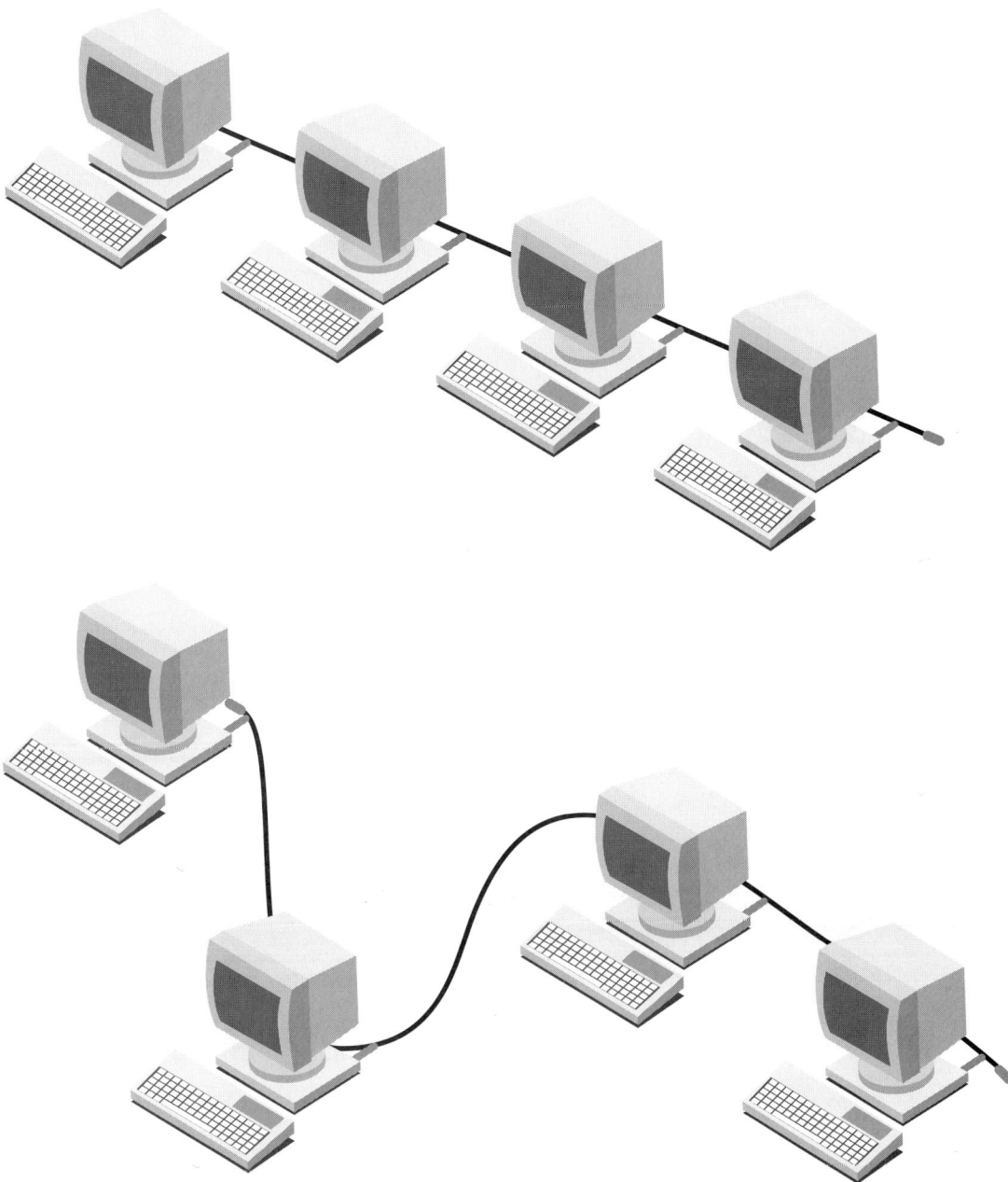

FIGURE 4.2   Bus topologies

ious topologies to maximize the effectiveness of the network. Examples of hybrid topologies are the star bus topology and the star ring topology. See Figure 4.5.

Another topology worth mentioning is the **mesh topology** in which every computer is linked to every other computer. Although this topology is uncommon in LANs, it is common in WANs where it connects remote sites over telecommunication links. This topology is the hardest to install and reconfigure, because the number of cables increases geometrically with each additional computer. Some networks will use a modified mesh topology, which has multiple links from one computer to another but does not necessarily have each computer linked to every other computer.

4.2.3   Logical Topology

The logical topology is not as easy to recognize as the physical topology. The logical topology describes how the data flow through the physical topology, but neither are always the same. The two logical topologies are bus and ring.

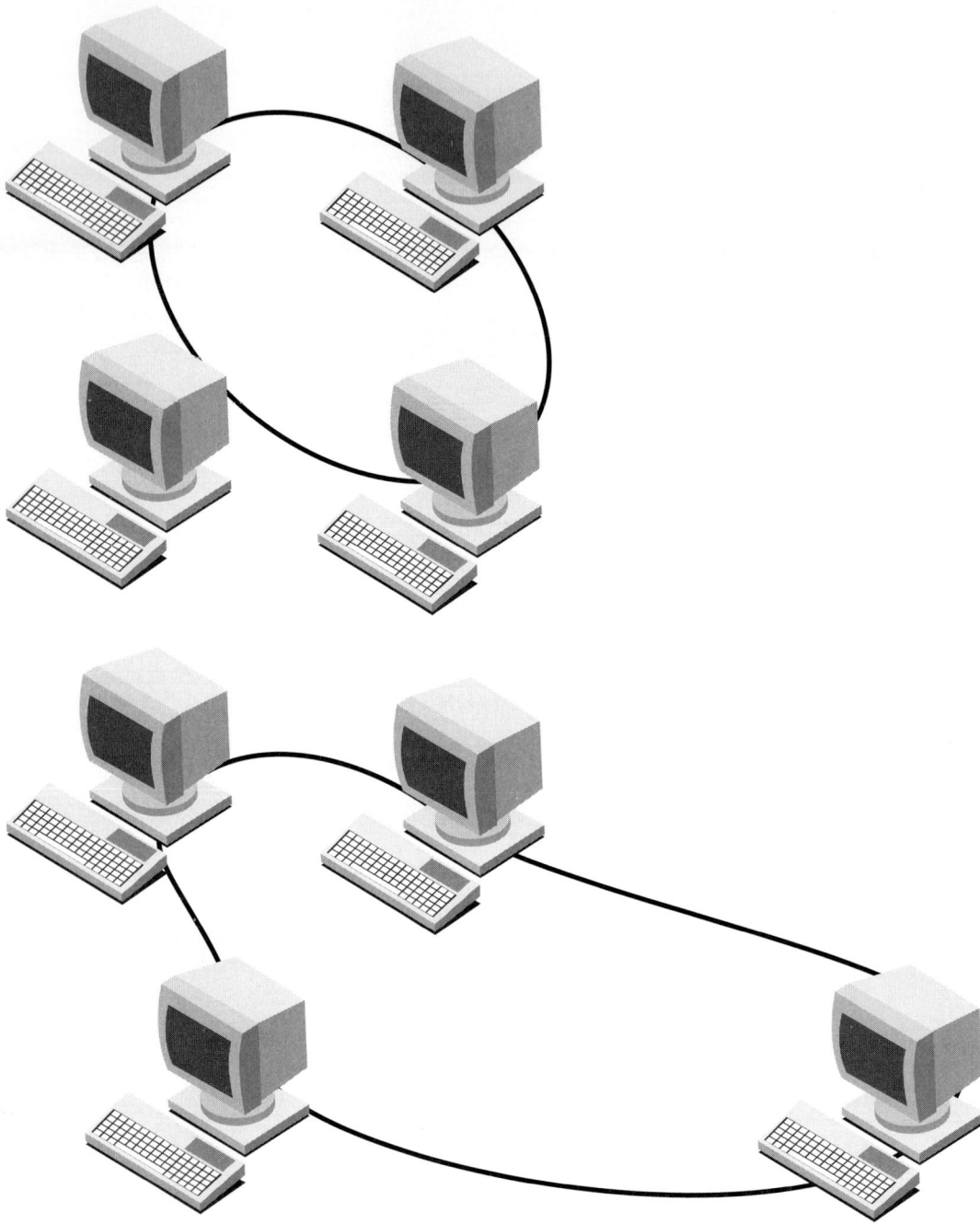

FIGURE 4.3   Ring topologies

For example, consider Ethernet using a backbone cable. Here you have a network that physically is a bus topology and the pathway of the signals or the logical topology is also a bus topology. If you compare that with an Ethernet network using a star topology, the physical network also is a star topology, yet the pathway or logical topology is a bus, because each cable attached to the hub has two wires. One wire carries data from the hub to the network device and the other wire carries data from the network device to the hub. The hub then connects the pairs of wires with each other, creating a large bus pathway. The bus has two ends which are both contained within the hub. See Figure 4.6.

Another example is IBM's token ring network. One form of token ring is connecting the network devices using a star topology. While this network has a physical star topology, the pathway or logical topology is a ring. The difference between the Ethernet hub and the token ring MAU is the hub contains two cable ends, whereas the token ring MAU connects the two ends to form a ring. See Figure 4.7.

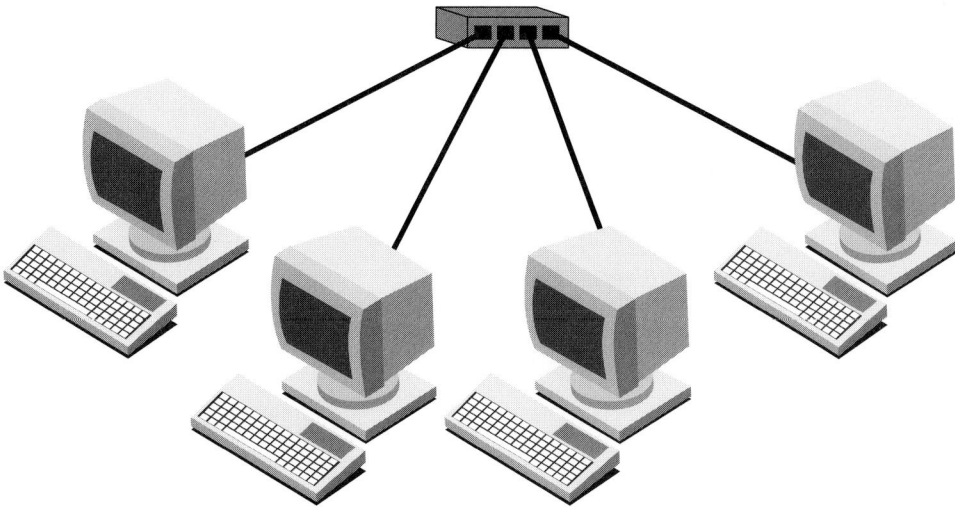

FIGURE 4.4   Star topology

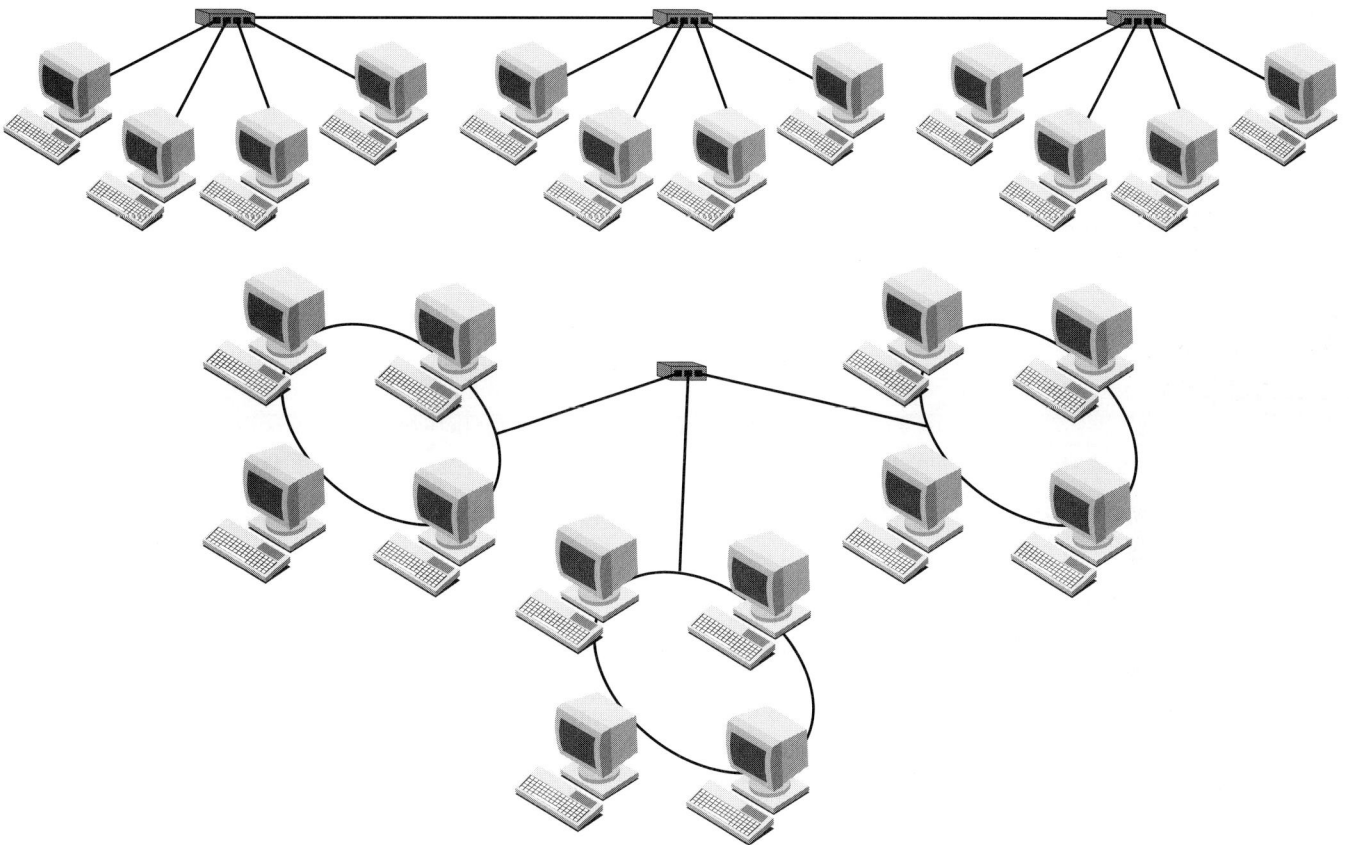

FIGURE 4.5   Hybrid topologies: (a) star bus and (b) star ring

FIGURE 4.6   Logical bus topologies: (a) bus and (b) star

FIGURE 4.7 Logical ring topologies using a token ring MAU

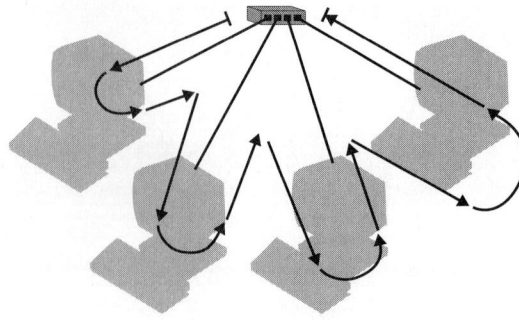

TABLE 4.1: Cable types

Cable Type	Cable Cost	Installation Cost	EMI Sensitivity	Data Bandwidth	Comments
UTP	Lowest	Lowest	Highest	Lowest to high	Used in 80% of LANs
STP	Medium	Moderate	Low	Moderate	Usually found in older networks
Coaxial	Medium	Moderate	Low	High	Often used as a backbone cable
Fiber Optic	Highest	Highest	None	Very High	Uses light instead of electrical signal

4.3 CABLING

The network cabling system is the circulatory system of the network in that it connects all computers and allows them to communicate with each other. The common types of cabling include unshielded twisted pair (UTP), shielded twisted pair (STP), coaxial, and fiber optics. See Table 4.1.

When choosing the type of cable, always consider the following factors. First, consider cost. For example, if you are installing a large network within a building, you will find that the cabling system can cost thousands of dollars. The cost of installing the cable (planning and hourly wages) actually is many times higher than the cost of the cable itself. Therefore, you must ensure that financial resources are available to install such a system. Next, look at the bandwidth (megabits per second—Mbps) to make sure that it will handle the predicted network traffic and to allow for some growth. Next, look at the ease of installation of the cabling system, as this affects labor costs and can indirectly affect the reliability of the network. In addition, look at the ease of troubleshooting, including how it is affected by media faults and if the cable system offers any fault tolerance. Lastly, look at two characteristics of the cabling system, attenuation and resistance to interference.

Attenuation occurs when the strength of a signal falls off with distance over a transmission medium. This loss of signal strength is caused by several factors, such as (1) the signal converted to heat due to the resistance of the cable, and (2) the energy is reflected as the signal encounters impedance changes throughout the cable.

Interference occurs when undesirable electromagnetic waves affect the desired signal. Interference can be caused by **electromagnetic interference (EMI),** which is caused by large electromagnets used in industrial machinery, and by **radio frequency interference (RFI),** which is caused by transmission sources such as a radio station. Another term used to describe instability in a signal wave is **jitter,** which is caused by signal interference.

4.3.1 Unshielded Twisted Pair

A **twisted pair** consists of two insulated copper wires twisted around each other. While each pair acts as a single communication link, a twisted pair is usually bundled together into a cable and wrapped in a protective sheath. See Figure 4.8.

EXAMPLE 1 **Question:** Why are the wires twisted around each other?

FIGURE 4.8 Twisted-pair cable

FIGURE 4.9 UTP cable

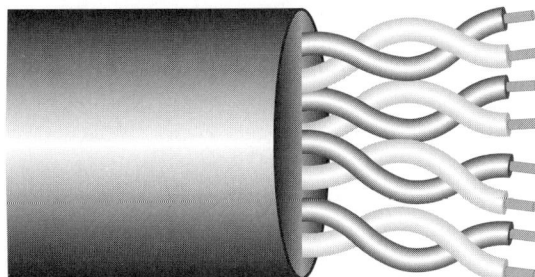

Answer: Copper wire does not constrain electromagnetic signals well. This means that if you have two copper wires next to each other, the signal will induct (law of induction) or transfer from one wire to the other. This phenomenon is called **crosstalk.** By twisting the wires around each other, the electromagnetic fields cancel each other and greatly reduce the crosstalk.

Today, crosstalk can be further divided into near-end crosstalk (NEXT) and far-end crosstalk (FEXT). Near-end crosstalk occurs at the end of the cable where the signal is inserted. Far-end crosstalk occurs at the other end of the cable where the signal is inserted. Because the signal is strongest when sent, near-end crosstalk is stronger than far-end crosstalk.

A term related to both near-end and far-end crosstalk is the power sum standard. The **power sum** measurements ensure that a cable reaches crosstalk performance requirements when all pairs are operating simultaneously.

Twisted pair comes in **unshielded twisted pair (UTP)** and **shielded twisted pair (STP).** Of these, unshielded twisted pair is the same type of cable as that used with telephones and is the most common cable used in networks. See Figure 4.9. UTP cable consists of four pairs of wires in each cable. Each pair of wires is twisted about each other and used together to make a connection. Compared with other cable types (unshielded twisted pair, shielded twisted pair, coaxial, and fiber optics), UTP is inexpensive and is the easiest to install. The biggest disadvantages of UTP are its limited bandwidth of 100 meters and its susceptibility to interference and noise. Traditional UTP has had a limited network speed, but more recently UTP can be used in networks running between 4 Mbps and 1 Gbps and there are some companies such as Hewlett-Packard that are working on a 10-Gbps network standard.

In 1995, UTP cable was categorized by the Electronic Industries Association (EIA) based on the quality and number of twists per unit. The UTP categories are published in EIA-568-A. See Table 4.2.

Early UTP networks typically used category 3, but today's high-speed networks typically use category 5 or enhanced category 5 cabling. Category 3 has three to four twists per foot and could operate up to 16 MHz, whereas category 5 uses three to four twists per inch, contains Teflon insulation, and can operate at 100 MHz. Enhanced category 5 is a higher-quality cable that is designed to reduce near-end crosstalk (NEXT) even further.

As network applications increased network traffic, eventually there came a need for faster networks. One way to increase the performance of the network is to use more expensive fast electronic devices. A cheaper way is to increase the amount of usable bandwidth by using all four pairs of the UTP cable instead of only two pairs.

Enhanced category 5 cabling has enhanced electrical performance attributes which support applications that require additional bandwidth. Although standard category 5 will support 1 GB (1000Base-T), enhanced category 5 will allow more usable bandwidth and higher data rates.

Currently, there are two new standards being developed, category 6 and category 7. Unfortunately, since these two standards have not been approved yet, no vendor can truthfully promise complete compatibility for future networks.

TABLE 4.2   Unshielded twisted-pair cable categories for networks

Cable Type	Bandwidth (Mhz)	Function	Attenuation	Impedance	Network Usage
Category 3	16	Data	11.5	100Ω	10Base-T (10 Mbps), token ring (4 Mbps), ARCnet, 100 VG-ANYLAN (100 Mbps)
Category 4	20	Data	7.5	100Ω	10Base-T (10 Mbps), token ring, ARCnet, 100VG-ANYLAN (100 Mbps)
Category 5	100	High-speed data	24.0	100Ω	10Base-T (10 Mbps), token ring, fast Ethernet (100 Mbps), gigabit Ethernet (1000 Mbps), and ATM (155 Mbps)
Category 5E (enhanced)	100	High-speed data	24.0	100Ω	10Base-T (10 Mbps), token ring, fast Ethernet (100 Mbps), gigabit Ethernet (1000 Mbps), and ATM (155 Mbps)
Category 6 (Not yet approved)	200	High-speed data			10Base-T (10 Mbps), token ring, fast Ethernet (100 Mbps), gigabit Ethernet (1000 Mbps), and ATM (155 Mbps)
Category 7 (Not yet approved)	600	High-speed data			

EXAMPLE 2   Question: When planning a network, what type of cable should you use?

Answer: Standard category 5 will handle standard network technologies for the next five years including fast Ethernet/100Base-TX and gigabit Ethernet 1000Base-TX. If you plan to support gigabit technology now or in the future, select enhanced category 5.

Telephones use cable with two pairs (four wires) and an RJ-11 connector; computer networks use cable with four pairs (eight wires) and an RJ-45 connector. See Figure 4.10. In a simple network, one cable end attaches to the network card on the computer and the other end attaches to a hub (multiported connection).

In a larger network, one cable end will connect the network card of the computer to a wall jack. The wall jack is connected to the back of a patch panel that is kept in a server room or wiring closet. A cable then is attached to the patch panel and connected to a hub. The cables that connect the computer to the wall jack and the cable that connects the patch panel to the hub are called patch cables.

UTP cable is commonly available in 22, 24, and 26 AWG (American Wire Gauge) using either solid or stranded cable. **Solid cable** is typically used for the cabling that exists throughout the building. This should include cables that lead from the wall jacks to the server room or wiring closet. **Stranded cable** is typically used as patch cables between patch panels and hubs and between the computers and wall jacks. Since stranded wire is not as firm as solid wire, it is a little easier to handle. Unfortunately, stranded cable has 20% more attenuation than solid wire.

4.3.2   Building UTP Cable

There are two types of Ethernet cables, a straight-through cable and a crossover cable. The **straight-through cable,** which can be used to connect a network card to a hub, has the same sequence of colored wires at both ends. A **crossover cable,** which can be used to connect one network card to another or one hub to another, reverses the transmit and receive wires.

FIGURE 4.10   UTP cable with RJ-45 connector

To create a category 5 UTP patch cable, obtain the following:

- Category 5 stranded cable
- Two RJ-45 round wire connectors
- Small wire cutter
- RJ-45 crimping tool

Step 1	Measure and cut the needed length of cable.	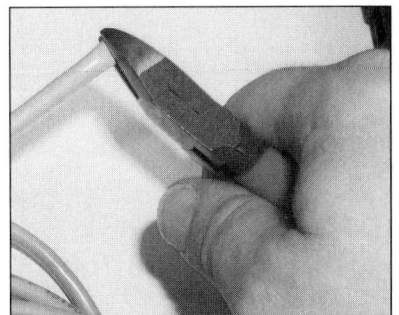
Step 2	Take one end of the wire and remove the outer sheathing to reveal about 3/4 inch of twisted wires.	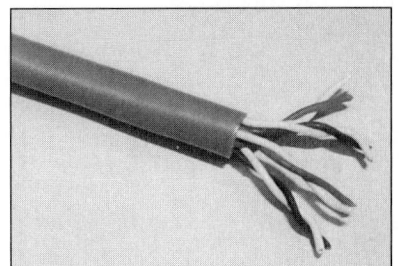

Step 3	Untwist all exposed wires and straighten them as much as possible. Rearrange the wires in a flat manner, as close together as possible in the proper color order. See Table 4.3.	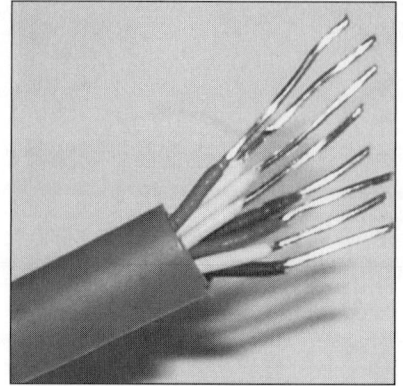
Step 4	Take all wires and cut them to the same length, approximately 1/2 inch exposed.	
Step 5	Hold an RJ-45 connector and insert the wires from the bottom, maintaining the correct color order. Each wire should slide into separate groves toward the tip of the connector. Push the wires in as far as they will go. In addition, ensure that the outer sheathing is up into the opening and under the crimping tab.	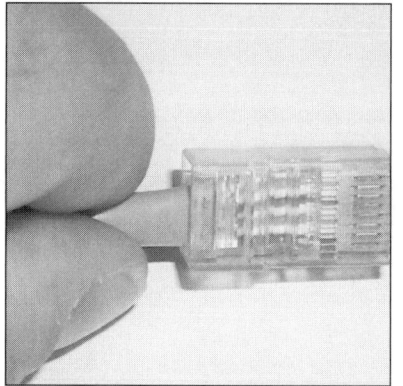
Step 6	Insert the RJ-45 connector into the crimping tool and crimp the connector.	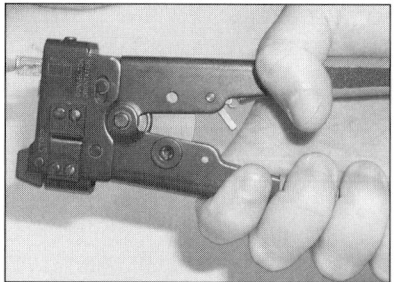
Step 7	Repeat steps 2 through 6 for the other end of the cable.	
Step 8	Using a cable tester or digital multimeter, test the cable.	

TABLE 4.3 Standard color scheme for UTP/RJ-45 cables

Wire (Pin) No.	Pair No.	IA/TIA 568A Color Standard	AT&T 258A or EIA/TIA 568B Color Standard
1	2	White with green strip	White with orange strip
2	2	Green	Orange
3	3	White with orange strip	White with green strip
4	1	Blue	Blue
5	1	White with blue strip	White with blue strip
6	3	Orange	Green
7	4	White with brown strip	White with brown strip
8	4	Brown	Brown

Note: To create a 100-Mbps crossover cable, reverse the pairs on the other end.

FIGURE 4.11 Shielded twisted-pair cable

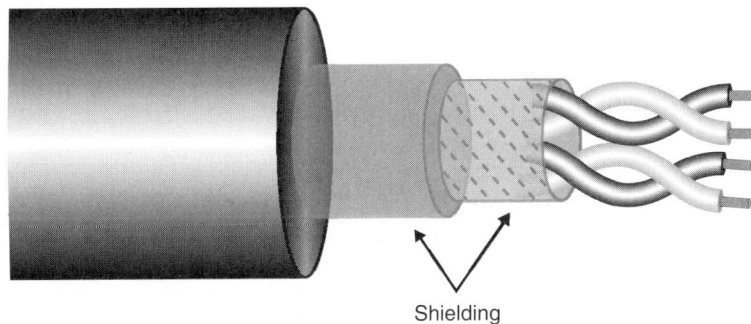

Shielding

FIGURE 4.12 IBM shielded twisted-pair cable

4.3.3 Shielded Twisted Pair

Shielded twisted pair (STP) is similar to unshielded twisted pair except that STP is usually surrounded by a braided shield that serves to reduce both EMI sensitivity and radio emissions. See Figure 4.11. Shielded twisted-pair cable was required for all high-performance networks such as IBM's token ring until a few years ago and is commonly used in IBM's token ring networks and Apple's LocalTalk network. See Figure 4.12. STP is relatively expensive compared with UTP and is more difficult to handle.

TABLE 4.4 IBM cabling system

IBM Type No.	Standard Label	Description
1	Shielded twisted paid	Two pairs of 22 AWG shielded wires surrounded by an outer braided shield. Used for computers and MAU. The impedance of type 1 cable is 150 ohms and is rated at 20 Mbps. Typically used in token ring applications.
1A	Shielded twisted pair	Same as type 1, except the cable is rated at 100 Mbps.
2	Voice and data cable	A voice and data shielded cable with two twisted pairs of 22 AWG wires for data, an outer braided shield, and then four twisted pairs of 26 AWG for voice (unshielded). The impedance of type 2 cable is 150 ohms. The shielded pairs are rated at 20 Mbps and the unshielded pairs at 4 Mbps.
2A	Shielded twisted pair	Same as type 2, except the cable is rated at 100 Mpbs.
3	Voice grade cable	Consists of four solid, unshielded twisted pair 22 or 24 AWG cables. The type 3 cable is rated up to speeds of 4 Mbps and an impedance of 100 ohms.
4	Not yet defined	
5	Fiber-optic cable	Consists of two 62.5/125-micron multimode optical fibers—the industry standard.
6	Data patch cable	Considered a thinner and more flexible version of a type 1 cable. Two 26 AWG twisted-pair stranded cables with a dual foil and braided shield. Type 6 is rated for speeds up to 20 Mbps and has an impedance of 150 ohms. Unfortunately, it has 1.33 times the loss of a type 1 cable. Typically used in token ring applications.
7	Not yet defined	A single-pair version of type 6. It is a 150-ohm cable and is rated for speeds up to 20 Mbps.
8	Carpet cable	Housed in a flat jacket for use under carpets. Two shielded twisted pair 26 AWG cables. Limited to one-half the distance of type 1 cable. Unfortunately, the type 8 cable has twice the loss of a type 1 cable.
9	Plenum	Fire safe. Two shielded twisted-pair cables with an impedance of 150 ohms and is rated up to 20 Mbps.

The STP cables used by IBM networks are shown in Table 4.4. Different from the other connectors, IBM connectors are hermaphroditic, which means they are neither male nor female. For example, if the cable does not connect properly, simply turn one around for it to connect.

4.3.4 Coaxial Cable

Coaxial cable, sometimes referred to as **coax** (pronounced 'kō-aks), is a cable that has a center wire surrounded by insulation and then a grounded shield of braided wire. See Figure 4.13. The shield minimizes electrical and radio-frequency interference. Coaxial cable is the primary type used by the cable television industry and is widely used for computer networks.

For computer networks, coaxial cables are primarily used as the backbone cable for Ethernet networks. The network devices are attached by cutting the cable and using a T-connector or by applying a tap (a mechanical device that uses conducting teeth to penetrate

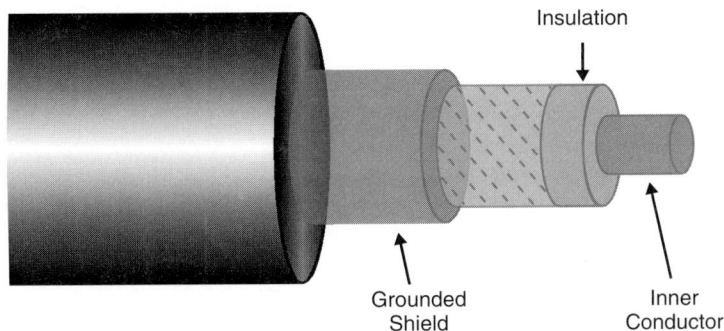

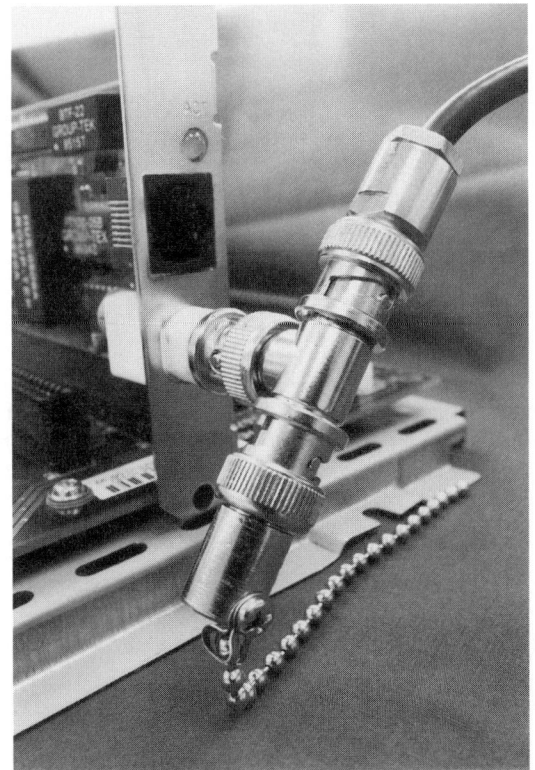

FIGURE 4.13   Coaxial cable

the insulation and attach directly to the wire conductor). To maintain the correct electrical properties of the wire, terminate the ends of the cable and ground it. The termination dampens signals that bounce back or reflect at the end of the cable and the ground completes the electrical circuit.

EXAMPLE 3

Question: You have two Ethernet hubs that you need to link together as one network. Each hub contains a single BNC connector, which allows you to connect the hubs. Therefore, you take a coaxial cable and connect it directly from the BNC connector to the other BNC connector. Unfortunately, the hubs do not function together. What is the problem?

Answer: Anytime that you use a coaxial cable with Ethernet, connect network devices using a T-connector. In addition, the two ends must be terminated with the proper terminating resistor.

There are several coaxial cable standards used in computer networking. The most common are as follows:

> 50-ohm RG-8 and RG-11 (used in 10Base5/thick Ethernet)
> 50-ohm RG-58 (used in 10Base2/thin Ethernet)
> 75-ohm RG-59 (used in cable television)
> 93-ohm RG-62 (used in ARCnet specifications)

4.3.5   Building a Coaxial Cable

To create a coaxial 10Base2 Ethernet cable, obtain the following:

- RG-58 A/U coaxial cable
- Two RG-58/U BNC connectors
- Racheting coax crimping tool
- Two-step rotating stripper
- Cable cutter (straight cut)

Step 1 Cut the coaxial cable (straight cut) to the coaxial cable using the cable cutter. Adjust the stripper to meet the desired cable diameter and stripping requirements. For best results, the stripper should be adjusted to expose 1/4 inch of the conductor and 1/4 inch of the insulation. Insert the coaxial cable into the coaxial wire stripper. Strip the ends of the coax cable using a rotary motion (three to five full turns) with the coax stripper. Always turn the stripper in the same direction. Do not cut completely through the jacket, to avoid nicking the shield. Flex the jacket to complete the separation. Pull the coax cable out of the stripper and inspect the cable for stripping quality. Ensure that the center conductor is not nicked or scored and the braid is pushed away from the conductor.

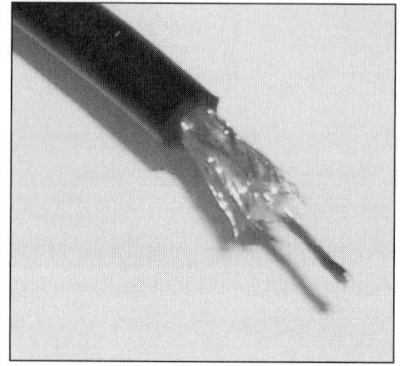

Step 2 Seat the center pin of the BNC connector on the exposed conductor. Crimp the center pin to the end of the center conductor using the small-diameter pin crimp die on the ratchet crimping tool. Ensure that all strands of the center conductor are in the hole in the center pin before crimping.

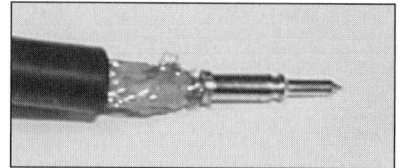

Step 3 Slide the sleeve ferrule over the pin and exposed insulation. Place the BNC connector body on the cable end. Align the connector body on the cable end so that its shaft fits over the pin and between the braid and the insulation.

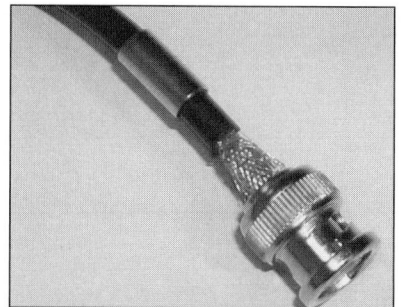

Step 4 Slip the connector body under the braided shield as far as it will go. Check for stray strands and push them out of the way. Ensure that the pin flange rests on the exposed insulation and the top of the pin is flush with the top of the BNC body. Slide up the crimp ferrule sleeve to cover the exposed braided shield up to the BNC body shoulder.

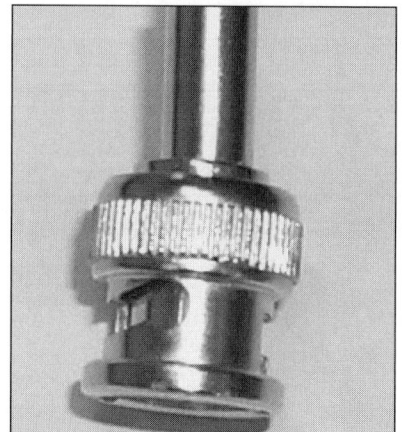

Step 5	Ensure that the ratcheting crimping tool is fitted with the proper hex die. Place the crimping tool over the ferrule sleeve and squeeze it to evenly and completely crimp the ferrule to the BNC body.	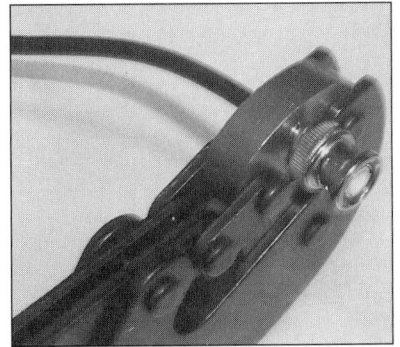
Step 6	Inspect for neatness and tightness of the connector. Pull and flex firmly on the BNC connector to make sure it is crimped tightly to the cable.	
Step 7	Repeat steps 1 through 6 for the other end of the cable.	
Step 8	Test the cable with a cable tester or digital multimeter.	

4.3.6 Plenum

Most cables contain halogens (chemicals that give off toxic fumes when they burn). These fumes release acid gases that sear a victim's eyes, nose, mouth, and throat. In a fire, a victim can easily become disoriented, which can prevent escape. Breathing halogen fumes can cause severe respiratory damage and ultimately lead to death.

In buildings, the **plenum** is the space above the ceiling and below the floors that is used to circulate air throughout the workplace. In the case of a fire, the plenum will also circulate toxic fumes generated by the burning of cabling. Some international governments have already standardized zero-halogen cabling. The U.S. National Electrical Code serves as the basis for local standards in most states to use halogen-sheathed cabling in the plenum spaces above ceilings and below floors in office buildings.

A **plenum cable** is a special type of cable that gives off little or no toxic fumes when burned. Depending on a building's fire code, wiring will include plenum cabling unless it exists in a special type of sealed conduit that would prevent the toxic fumes from escaping into the open air. The major disadvantage of using plenum cable is its high cost.

Riser cable is intended for use in vertical shafts that run between floors. Many buildings have a series of equipment rooms that are placed vertically in a reinforced shaft for the purpose of enclosing power distribution equipment, HVAC units, telephone distribution, and other utility services throughout the building. Cable placed in these shafts must not contribute to the spreading of fire from floor to floor. Since the shafts are not plenums, the riser rating is less stringent than the plenum rating.

4.3.7 Fiber Optics

A **fiber-optic** cable consists of a bundle of glass or plastic threads, each thread capable of carrying data signals in the form of modulated pulses of light. Although glass can carry the light pulses farther (several kilometers) than plastic, plastic is more workable. Because each thread can carry a signal in only one direction, a cable consists of two threads in separate jackets,

FIGURE 4.14 Fiber-optic cables

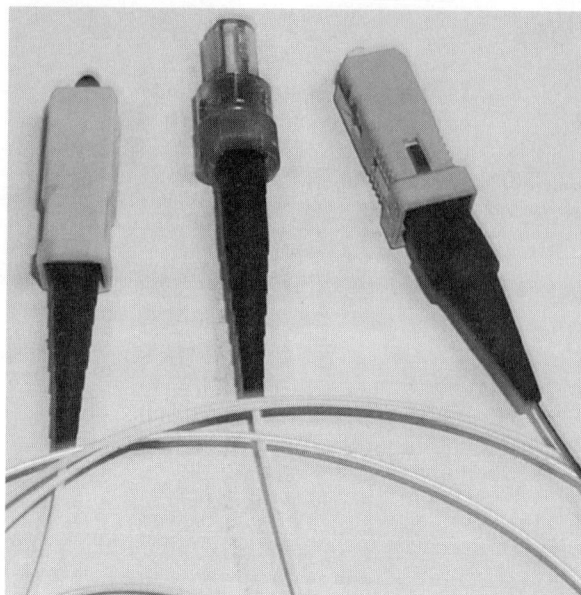

one to transmit and one to receive. The fiber-optic cable uses cladding that surrounds the optical fiber core, which helps reflect light back to the core and ensures that little of the light signal is lost. Lastly, the cable contains kevlar strands to provide strength. See Figure 4.14.

The light signals used in fiber-optic cables are generated by light-emitting diodes (LEDs) or injection laser diodes (ILDs). ILDs are similar to LEDs but produce laser light. Since laser light is purer than normal light, it can increase both the data rates and transmission distances. Signals are received by photodiodes, solid-state devices that detect variations in light intensity.

Fiber-optic cables come in multimode and single-mode fiber. **Multimode fiber** is capable of transmitting multiple modes (independent light paths) at various wavelengths or phases. Unfortunately, the core's greater diameter makes light more likely to bounce off the sides of the core, resulting in dispersion and limiting both the bandwidth to 2.5 Gbps and the transmission distance between repeaters. **Single-mode fiber** can transmit light in only one mode, but the narrower diameter yields less dispersion, resulting in longer transmission distances. Single-mode fiber costs are higher than multimode fiber costs, both for the fiber itself and for the electronics that transmit and receive the light beam.

The types of fiber-optic cables are differentiated by mode, composition (glass or plastic), and core/cladding size. The size and purity of the core determine the amount of light that can be transmitted. Common types of fiber-optic cables include:

- 8.3-micron core/125-micron cladding single mode
- 62.5-micron core/125-micron cladding multimode
- 50-micron core/125-micron cladding multimode
- 100-micron core/140-micron cladding multimode

Over the past five years, optical fibers have found their way into cable television networks, thus increasing reliability, providing a greater bandwidth, and reducing costs. In the local area network, fiber cabling has been deployed as the primary media for campus and building backbones, offering high-speed connections between diverse LAN segments.

Fiber has the largest bandwidth (up to 5 GHz) of any media available. It can transmit signals over the longest distance (20 times farther than copper segments) at the lowest cost, with the fewest repeaters and the least amount of maintenance. In addition, since it has such a large bandwidth, it can support up to 1000 stations and it will support faster speeds as introduced during the next 15 to 20 years.

Fiber optics is extremely difficult to tap, making it secure and highly reliable. Since fiber does not use electrical signals running on copper wire, interference does not affect fiber traffic. As a result, the number of retransmissions is reduced and the network efficiency is increased.

The main disadvantages of fiber optics are the cables are expensive to install and workers require special skills and equipment to split or splice cables. In addition, cables are more fragile than wire. Fortunately, in recent years, while fiber-optic products are being more mass produced, the cost gap between the high grades of UTP have closed significantly and many premade products are now available.

4.4 CABLE ACCESS METHODS

The set of rules defining how a computer puts data onto the network cable and takes data from the cable is called the **access method.** While multiple computers share the same cable system, only one device can access a cable at a time. If two devices use the same cable simultaneously, both data packets sent by the card become corrupted. The typical cable access methods are as follows:

- Contention
- Token passing
- Polling
- Demand priority

4.4.1 Contention

Contention occurs when two or more devices contend for network access. Any device can transmit whenever it needs to send information. To avoid data collisions (two devices sending data at the same time), specific contention protocols were developed that require the device to listen to the cable before transmitting data.

The most common form of contention is **carrier sense multiple access (CSMA)** network. Even though each station listens for network traffic before it attempts to transmit, it remains possible for two transmissions to overlap on the network or to cause a collision. As a result of collisions, access to a CSMA network is somewhat unpredictable, and CSMA networks can be referred to as random or statistical access networks. To avoid collisions, CSMA will use one of two specialized methods of collision management: collision detection (CD) and collision avoidance (CA).

The collision detection approach listens to the network traffic as the card is transmitting. By analyzing network traffic, it detects collisions and initiates retransmissions. **Carrier sense multiple access with collision detection (CSMA/CD)** is the access method utilized in Ethernet and IEEE 802.3. Collision avoidance uses time slices to make network access smarter and avoid collisions. **Carrier sense multiple access with collision avoidance (CSMA/CA)** is the access mechanism used in Apple's LocalTalk network.

Contention is a simple access method with low administrative overhead requirements. High traffic levels cause more collisions, however, which cause numerous retransmissions and therefore even slower network performance.

4.4.2 Token Passing

Token passing uses a special authorizing packet of information to inform devices that they can transmit data. These packets, called **tokens,** are passed around the network in an orderly fashion from one device to the next. Devices can transmit only if they have control of the token, which distributes the access control among all the devices. Token ring uses a ring topology, where each station passes the token to the next station in the ring. ARCnet uses a token passing bus as it passes the token to the next higher hardware address (MAC address), regardless of its physical location on the network.

Token passing is deterministic because users can calculate the maximum time before a workstation can grab the token and begin to transmit. In addition, users can assign priorities to certain network devices that will use the network more frequently. If a workstation has an equal or higher priority than the priority value in the token, it can take possession of the token.

4.4.3 Polling

Polling has a single device such as a mainframe front-end processor that is designated as the primary device. The primary device polls or asks the secondary devices, known as slaves, if they have information to be transmitted. Only when polled does the secondary computer have access to the communication channel. To ensure that a slave does not consume all of the bandwidth, each system has rules pertaining to how long each secondary computer can transmit data.

4.4.4 Demand Priority

Using the newest access method, **demand priority,** a device makes a request to the hub and the hub grants permission. High-priority packets are serviced before any normal-priority packets. To effectively guarantee bandwidth to time-sensitive applications such as voice, video, and multimedia applications, the normal-priority packets are promoted to a high priority after 200 to 300 ms.

4.5 NETWORK CARDS

To summarize briefly, network cards are media dependent, media access method dependent, and protocol independent. Thus, all network cards on the same network must use the same type of cable and must gain access to the cable in the same way. In addition, the protocol to be used is not important; in fact, several can be used at the same time.

For example, when installing a network you decide to start with a Novell NetWare 3.12 server. In this server you choose a NE2000 network card (a popular ISA choice) connected to a twisted-pair cable running (at 10 megabits) a protocol called IPX. If you install a second server such as Windows NT 4.0, you must choose a network card that also connects to a twisted-pair cable that supports 10 megabits speed. Therefore, you could choose an Intel E100B card, (one of today's popular PCI network cards), which supports both 10 megabits and 100 megabits speed. The Windows NT server uses the TCP/IP protocol. While the Novell server can communicate with any computer on the network using the IPX protocol and the Windows NT server can communicate with any computer on the network using the TCP/IP protocol, the two servers cannot communicate with each other even though they are using the same cabling system. See Figure 4.15.

4.5.1 Characteristics of a Network Card

When selecting a network card, first look at the type of network. For example, is the network Ethernet, token ring, or ARCnet or are you connecting your computer using a WAN connection? Next, look at the speed of the network. Are you going to be transmitting at 4 Mbps, 10 Mbps, 16 Mbps, 100 Mbps, or 1 Gbps? In addition, ensure that the network card has the proper network connector. If you are using unshielded twisted pair (UTP), you would most likely use the RJ-45 connector. If you are connecting to a 10Base2 coaxial cable, you will probably use

FIGURE 4.15 A typical network card with a 10Base2/BNC and 10Base-T/RJ-45 connectors

the BNC connector; or, if connecting to a 10Base5 coaxial cable, you would connect to an external transceiver using an adapter unit interface (AUI) connector, a female 15-pin D-connector. The AUI connector is also known as a Digital-Intel-Xerox (DIX) connector.

Next, select a card based on the type of expansion bus (ISA, MCA, EISA, VESA, PCI, PC-Card, or CardBus) available in the computer. Since the ISA slot can transfer up to 8.33 MB/s and the PCI card can transfer 264 MB/s, the PCI slot is needed for today's high-speed networks. If using a network card for a Windows NT or 2000 machine, ensure that the card appears on the Windows Hardware Compatibility List (HCL).

After selecting the basic characteristics of a network card, then choose a network card depending on its advanced features, such as how much processing the network card will do, to alleviate some of the microprocessor's work.

In addition, some cards offer various forms of fault tolerance such as using two cards on the same network link. Therefore, if one network card or connection fails, then the other card will still transmit and receive packets. Lastly, some cards offer port aggregation such as asymmetric port aggregation and symmetric port aggregation. Asymmetric port aggregation is the capability to evenly distribute outbound LAN traffic among multiple NICs while inbound traffic passes through a single NIC. Symmetric port aggregation is similar to the asymmetric type except that it combines the bandwidth of multiple NICs for both incoming and outgoing traffic to provide a single link of up to 800 Mbps.

4.5.2 MAC Address

A media access control (MAC) address is a hardware address that identifies a node on the network, much like a street address identifies a house or building within a city. Like a street address, two network cards or nodes cannot have the same MAC address on the same network (street). The same MAC address can be on two separate networks. For Ethernet and token ring cards, the MAC address is embedded onto the card's ROM chip. For ARCnet cards, the MAC address is selected by using jumpers or DIP switches.

4.5.3 Installing a Network Card

Installing a network card is similar to installing other expansion cards. First, configure the card's I/O address, IRQ, DMA, and memory addresses and any options that are specific to the type of network card selected. This would be done with jumpers, DIP switches, configuration software, or by plug-and-play. After the card is physically installed and the network cable is attached, then load the appropriate driver (included with the operating system, or a disk that comes with the network card, or it will have to be downloaded from the Internet) and client software. In addition, **bind** (logically attach) the protocol stack to the network card and install any additional network software needed. Depending on the type of card, the driver may need to have the hardware resources specified (I/O address, IRQ, DMA, and memory address). Specifications must match the settings of the cards. Lastly, ensure that the LED for the expansion card is active.

Network device interface specification (NDIS), developed by Microsoft, and **open data link interface (ODI),** developed by Apple and Novell, are sets of standards for protocols and network card drivers to communicate. Both allow multiple protocols to use a single network adapter card.

4.5.4 Diskless Workstations

A **diskless workstation** is a computer that does not have its own disk drive. Instead, the computer stores files on a network file server. Diskless workstations can reduce the overall cost of a LAN, because one large-capacity disk drive at the server is usually less expensive than several low-capacity drives at the workstations. In addition, workstations can provide an extra level of security because a user cannot (1) infect the network by inserting a virus-infected disk, (2) use a hackers/network snooper disk to bypass the network security, and (3) save confidential/secure information to a floppy disk. The disadvantage of using a diskless workstation is that if the network is down, the diskless workstation does not function.

To enable a diskless workstation to connect to the network and to boot an operating system, users require a network card with a boot PROM chip that is inserted into a socket located on the network card. Programmable ROM (PROM) is a memory chip on which data can be written only once. Once a program has been written onto a PROM, it remains there forever. Unlike RAM, PROMs retain their contents when the computer is turned off.

The PROM chip includes enough instructions to attach the network and to read the necessary boot files from the network. For a TCP/IP network, the bootstrap protocol (BOOTP) enables a diskless workstation to discover its own IP address, the IP address of a BOOTP server on the network, and a file to be loaded into memory to boot the machine.

A special type of diskless computer is a net PC, which was designed cooperatively by Microsoft and Intel. Different from most other diskless workstations, a net PC has a small hard disk to be used as a temporary cache to improve performance rather than to permanently store data. The net PC is designed to be an inexpensive computer and to discourage users from configuring the machines themselves. Configuration and management of a net PC is performed through a network server and a Microsoft Zero Administration Windows (ZAW) system.

4.6 CONNECTIVITY DEVICES

To connect networks, several devices can be used. These include repeaters, hubs, bridges, routers, brouters, and gateways.

4.6.1 Repeaters

A **repeater,** which works at the physical OSI layer, is a network device used to regenerate or replicate a signal or to move packets from one physical media to another. A repeater cannot connect different network topologies or access methods. It can be used to regenerate analog or digital signals that are distorted by transmission loss, and thus extend the length of a cable connection. Analog repeaters usually can only amplify the signal (including distortion) whereas digital repeaters can reconstruct a signal to near its original quality. See Figure 4.16.

4.6.2 Hubs

A **hub,** which works at the physical OSI layer, is a multiported connection point used to connect network devices via a cable segment. When a PC needs to communicate with another computer, it sends a data packet to the port to which the device is connected. When the packet arrives at the port, it is forwarded or copied to the other ports so that all network devices can "see" the packet. In this way, all stations see every packet just as they do on a bus network. A standard hub is inefficient on networks with heavy traffic, however, because it causes many collisions and leads to retransmission of packets. See Figure 4.17.

Hubs can be categorized as passive or active. A passive hub serves as a simple multiple connection point, which does not act as a repeater for the signal. An active hub, which always requires a power source, acts as a multiported repeater for the signal. [Note: When installing an active hub, make sure the fan (which provides cooling) is operating.]

The most advanced hub is called the intelligent hub, or manageable hub. An intelligent hub includes additional features that enable an administrator to monitor the traffic passing through the hub and to configure each port in the hub. For example, you can prevent certain computers from communicating with other computers or you can stop certain types of packets from being forwarded. In addition, you can gather information on a variety of network

FIGURE 4.16   A repeater

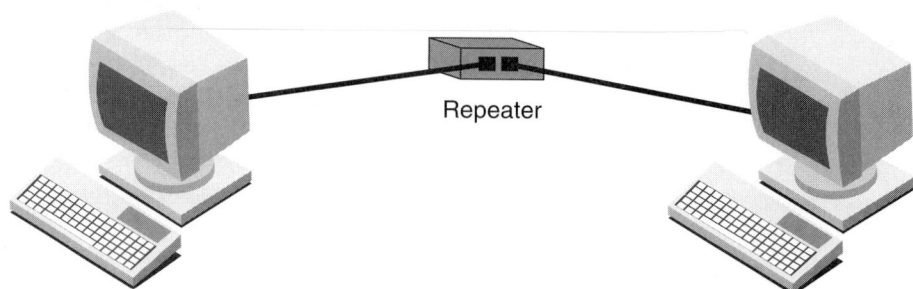

Repeater

FIGURE 4.17   A hub

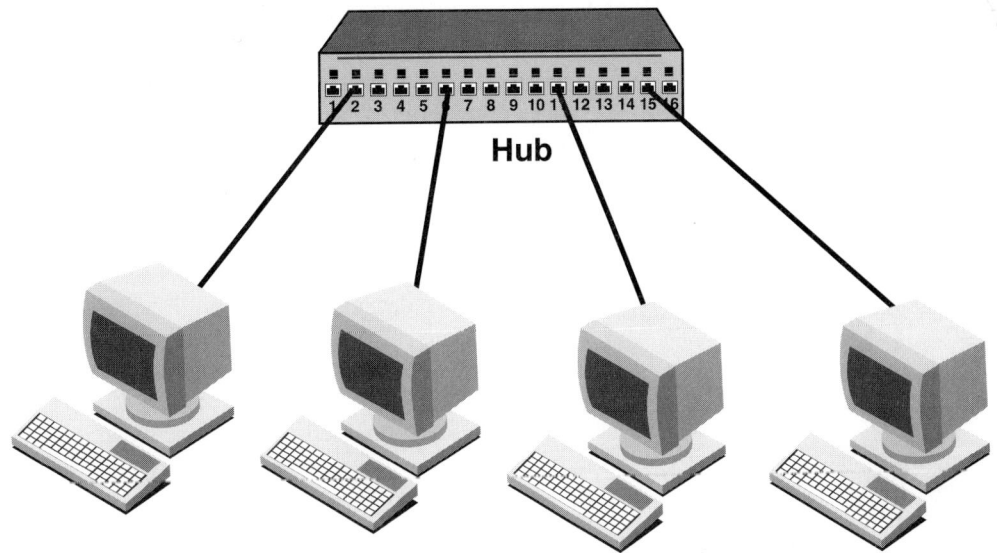

Hub

parameters, such as the numbers of packets that pass through the hub and each of its ports, the types of packets, whether the packets contain errors, and the number of collisions that have occurred.

4.6.3   Bridges and Switching Hubs

A **bridge,** which works at the data link OSI layer, is a device that connects two LANs and makes them appear as one or that connects two segments of the same LAN. The two LANs being connected can be alike or dissimilar such as an Ethernet LAN connected to a token ring LAN. Different from a repeater or hub, a bridge analyzes the incoming data packet and will forward the packet if its destination is on the other side of the bridge. Many bridges today filter and forward packets with very little delay, making them good for networks with high traffic. See Figure 4.18.

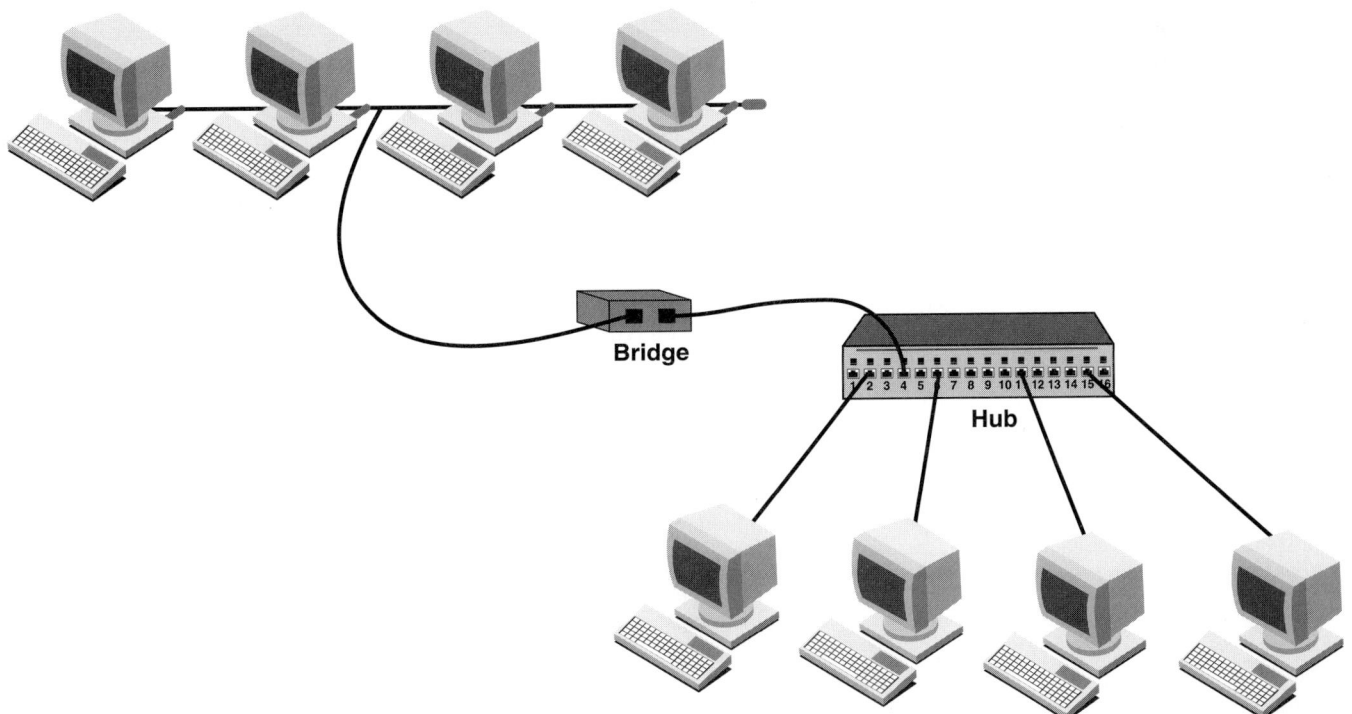

Bridge

Hub

FIGURE 4.18   A bridge

There are two kinds of bridges, basic and learning. A basic bridge is used to interconnect LANs using one (or more) of the IEEE 802 standards. Packets received on one port may be retransmitted on another port. Unlike a repeater, a bridge will not start retransmission until it has received the complete packet. As a consequence, stations on both sides of a bridge can transmit simultaneously without causing collisions. Bridges, like repeaters, do not modify the contents of a packet in any way.

A learning bridge examines the source field of every data packet it sees on each port and builds a table of (MAC) addresses of each port. It will then use this table to determine whether a data packet needs to be retransmitted to a destination address through the bridge. If the packet is addressed to a destination that is not listed in the address table, it retransmits the packet to every port except the one on which it was received. In addition, if a given address has not been heard from in a specified period of time, then the address is deleted from the address table.

If two separate LANs are located great distances apart, they may be joined into a single network. In this situation, users can implement two remote bridges that are connected with a modem via a dedicated telephone line. In case of multiple pathways, remote bridges use a spanning tree algorithm (STA) in which it can sense the existence of more than one route, determine which route would be the most efficient, and then configure the bridge to use that route.

Bridges that connect Ethernet networks are known as **source routing bridges.** Bridges that connect token ring networks are known as **spanning tree bridges.** Token ring bridges usually do no filtering, unlike Ethernet bridges.

A **switching hub** (sometimes referred to as a switch) is a fast multiported bridge, which actually reads the destination address of each packet and then forwards the packet to the correct port. A major advantage of using a switching hub is it allows one computer to open a connection to another computer (or LAN segment). While those two computers communicate, other connections between the other computers (or LAN segments) can be opened at the same time. Therefore, several computers can communicate simultaneously through the switching hub. As a result, the switches are used to increase performance of a network by segmenting large networks into several smaller, less congested LANs, while providing necessary interconnectivity between them. Switches increase network performance by providing each port with dedicated bandwidth, without requiring users to change any existing equipment such as NICs, hubs, wiring, or any routers or bridges that are currently in place.

Many switching hubs also support load balancing, so that ports are dynamically reassigned to different LAN segments based on traffic patterns. In addition, some include fault tolerance, which can reroute traffic through other ports when a segment goes down.

4.6.4 Routers and Brouters

A **router,** which works at the network OSI layer, is a device that connects two or more LANs. In addition, it can break a large network into smaller, more manageable subnets. As multiple LANs are connected, multiple routes are created to get from one LAN to another. Routers then share status and routing information to other routers so that they can provide better traffic management and bypass slow connections. Routers also provide additional functionality, such as the ability to filter messages and forward them to different places based on various criteria. Most routers are multiprotocol because they can route data packets using many different protocols. See Figure 4.19. Routers cannot, however, pass nonroutable protocols such as NetBEUI.

A **multihomed server** contains multiple network cards and acts as the router. This server can communicate with several networks at the same time and route data packets between them.

A **brouter** (short for "bridge router") is a device that functions as both a router and a bridge. A brouter understands how to route specific types of packets (routable protocols) such as TCP/IP packets. For other specified packets (nonroutable protocols), it acts as a bridge that simply forwards the packets to the other networks. For more information on routers and routing protocols, see Chapter 19.

4.6.5 Gateways

A **gateway** is hardware and/or software that links two different types of networks by repackaging and converting data from one network to another network or from one network

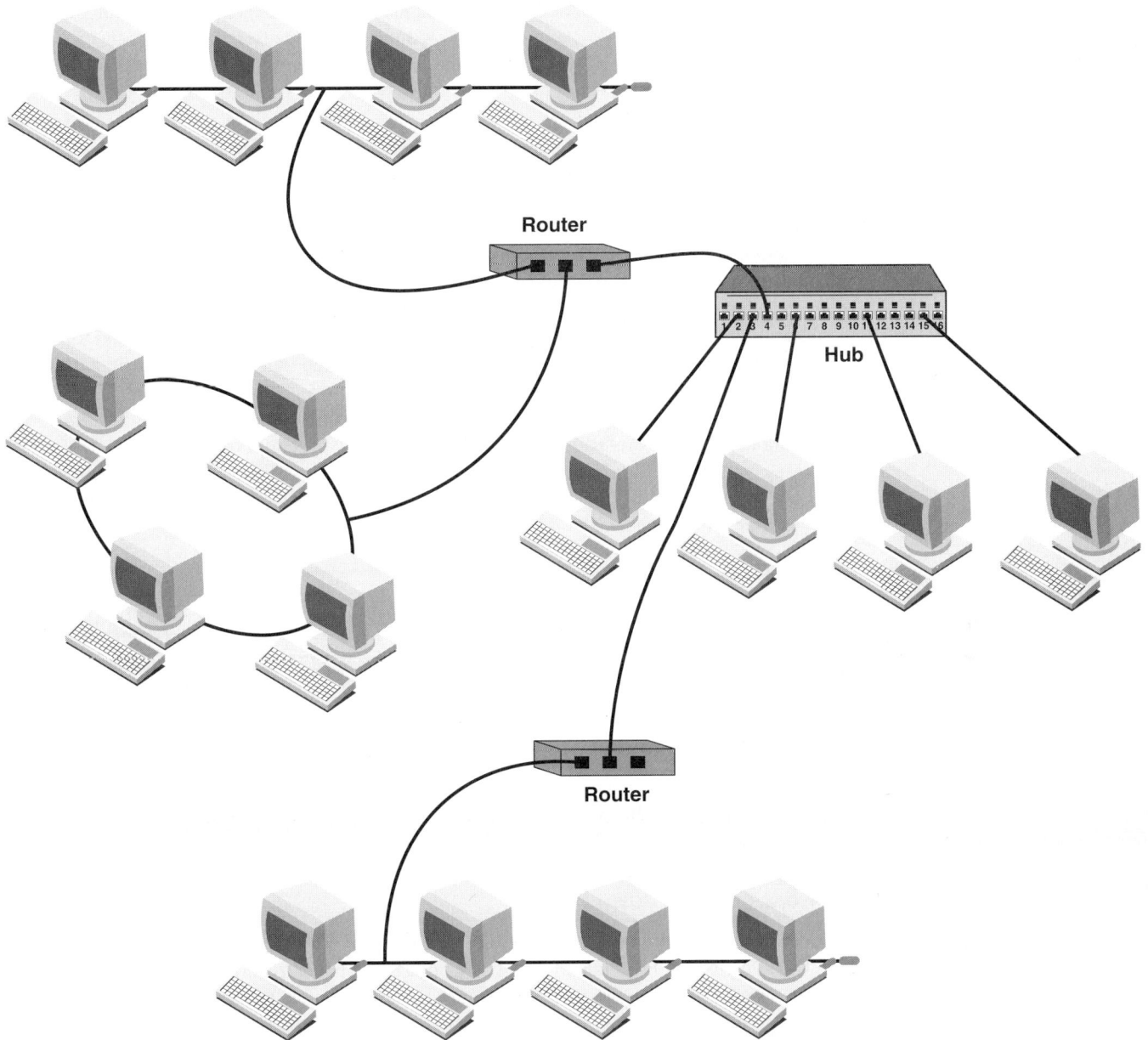

FIGURE 4.19   A router

operating system to another. An example of a gateway is a computer or device that connects a PC to a main frame or mini computer such as the AS400 midrange computer. This can be done by adding an Ethernet card or token ring card to the AS400 computer and adding software to the client computers, by adding a special expansion card to the PC so that it can communicate with the AS400 directly, or by using a gateway computer to act as a translator between the AS400 network and the client network. A gateway can be used at any layer of the OSI reference model, but is usually identified with the application layer.

4.7   VIRTUAL LAN

Virtual LANs (VLANs) provide LAN at WAN locations with costs as low as a typical Internet connection. A virtual (or logical) LAN is a local area network with a definition that maps workstations on some other basis than geographic location (e.g., by department, type of user, or primary application). In other words, it is a network of computers that behaves as if the computers are connected to the same wire even though they may actually be physically located on different segments of a LAN. Network management software keeps track of relating the virtual picture of the local area network with the actual physical picture. One big advantage of VLANs is that when a computer is physically moved to another location, it can stay on the same VLAN without any hardware reconfiguration.

4.8 CABLE MANAGEMENT

With a physical star topology used in Ethernet networks, all cables lead and connect to a hub or set of hubs. Whereas a small network will use only one or two hubs, a network consisting of a few hundred computers will use a few hundred cables. In these situations, it is important to use some form of cable management.

In a large network, users establish a horizontal wiring and/or backbone wiring system. The horizontal wiring system has cables that extend from wall outlets throughout the building to the wiring closet or server room. Backbone wiring, which often is designed to handle a higher bandwidth, uses cables to interconnect wiring closets, server rooms, and entrance facilities (telephone systems and WAN links from the outside world). The backbone wiring is sometimes referred to as the vertical cabling system. For example, the horizontal wiring system could be category 5 UTP (or better) whereas the backbone wiring system could be category 5 UTP (or better), coaxial, or fiber optics. Note that horizontal wiring and backbone wiring are usually done by a wiring contractor.

4.8.1 Wall Jacks

As the IT administrator, you may have to decide where to place the wall jacks that lead to the patch panel in the server room or wiring closet. If you plan out the location of the wall jacks when the building is being built, include jacks for any office computers (remember that some offices will be using two or more computers) that connect to the wall jacks directly or for any hubs (used to connect multiple computers using a single wall jack) that connect to the wall jacks directly. In addition, you should add extra wall jacks for any network printer that may be connected to the network. Since the network cabling system is being built along with the building, it is best to plan for growth by adding extra jacks throughout the building.

4.8.2 Punch-Down Block

To help manage the hundreds of cables that may enter a wiring closet or server room, the IT planner uses either a punch-down block or a patch panel. A **punch-down block** connects several cable runs to each other without going through a hub. Rather than using RJ-45 ports, the block attaches or is punched directly to the exposed wires by using a punch-down tool. These tools are typically used for permanent wire connections. For example, if you have category 5 cabling, you have four pairs or eight wires. Each wire would then connect to a row of connectors on the punch-down block. You can then connect additional wire from other cables by using the same connectors in a row. See Figure 4.20.

The most common punch-down blocks are the 110 block (commonly referred to as the insulation displacement connection, or IDC) and the 66 block. The 110 block is the termination of choice in category 5 high-bandwidth environments, and the 66 block is mostly used with telephone systems.

4.8.3 Patch Panel

Patch panels fill the same needs as a punch-down block. A patch panel has numerous RJ-45 ports. The wall jacks are connected to the back of the patch panel to the individual RJ-45 ports. (The ports are labeled with the corresponding wall jack.) Technicians can then use patch cables to connect the port in the front of the patch panel to a computer or a hub. As a result, they can connect multiple computers with a hub that is located in the wiring closet or server room. Compared with the punch-down block, the patch panel is easier to work with, allows for easier troubleshooting, makes temporary connection much easier to establish, and offers a much cleaner look. Therefore, patch panels are more common in LANs than punch-down blocks. See Figures 4.21 and 4.22.

FIGURE 4.20   Punch-down block

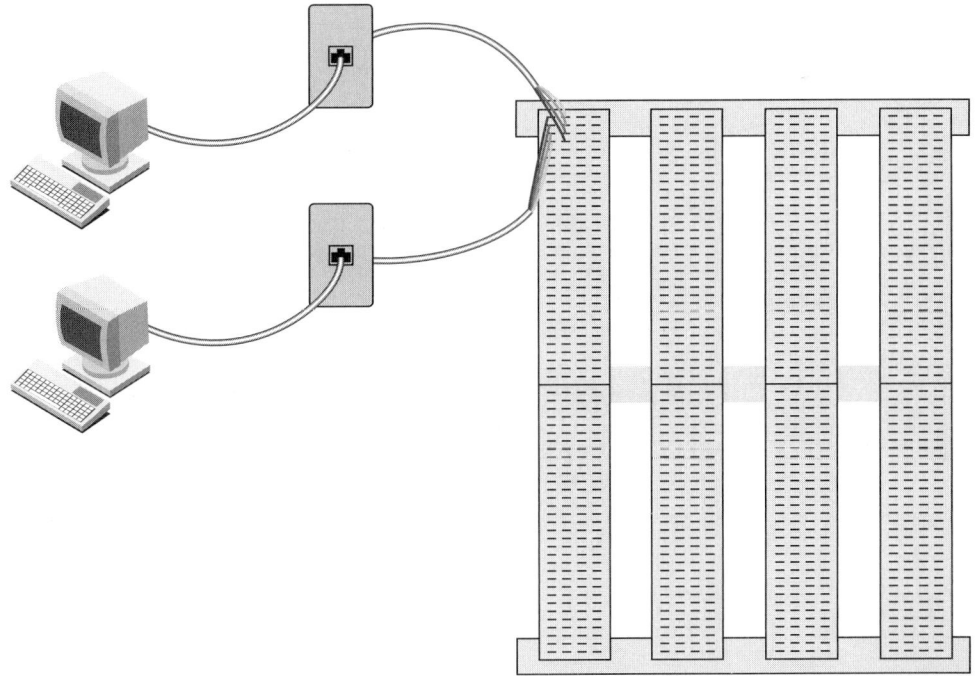

FIGURE 4.21   Patch panel (a) front and (b) back view

FIGURE 4.22 Patch panel
allows for easy connection of
computers

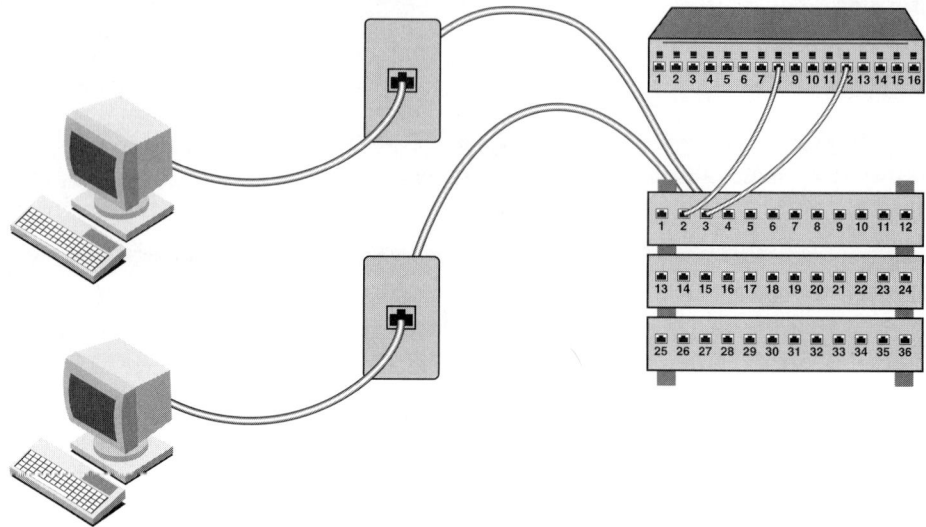

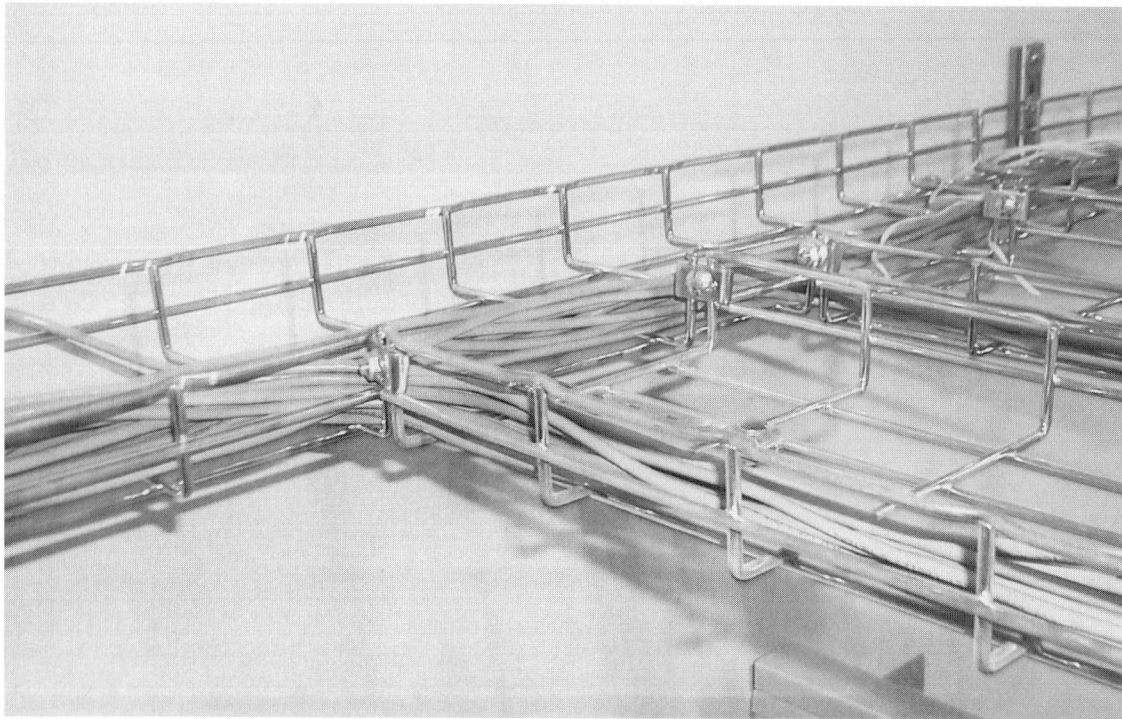

FIGURE 4.23 Basket system

4.8.4 Cable Ties, Basket Systems, and Management Racks

Cable ties and basket systems are tools used to help manage cables anywhere throughout the building. **Cable ties** are used to bundle cables traveling together and to pull the cables off the floor so that they will not get trampled or run over by office furniture. When installing or laying cable, take extreme care. Avoid stepping on cables, pinching them tightly with cable ties, or making sharp bends or kinks in the cable.

Basket systems utilize baskets or trays located on the back of office furniture. They are used to hold and hide network cables without using cable ties. See Figure 4.23.

Many cables come into the server room and wiring closets. Although a single cable weighs very little, the weight of many cables adds up quickly and causes much stress to the cables. For these situations, using some type of tray, conduit, trough, or ladder is recommended to help alleviate weight on the cables.

> **Note:** When dealing with any number of cables, it is important to always label both ends of the cables, especially those that are part of the backbone or horizontal wiring system. This practice will aid in quick troubleshooting of cable faults and make the job easier in the future. In addition, it is important to document the cable system, such as making a blueprint that shows where and how the cables are connected.

4.9   ETHERNET

A LAN technology defines topologics, packet structures, and access methods that can be used together on a segment. There are currently three LAN technologies being used: Ethernet, token ring, and ARCnet. Of these, **Ethernet** is the most widely used. It offers good balance between speed, price, reliability, and ease of installation. Approximately 83% of all LAN connections installed are Ethernet. All popular operating systems and applications are Ethernet compatible, as are upper-layer protocol stacks such as TCP/IP, IPX, and NetBEUI.

The original Ethernet network was developed by Digital, Intel, and Xerox in the early 1970s. It is currently referred to as Ethernet II. Today, the Ethernet standard is defined by the Institute of Electrical and Electronics Engineers (IEEE) in a specification commonly known as **IEEE 802.3.** The 802.3 specification covers rules for configuring Ethernet LANs, the types of media that can be used, and how the elements of the network should interact.

Ethernet was traditionally used on light to medium traffic networks, and performs best when a network's data traffic is sent in short bursts. Recently, newer and faster Ethernet standards, combined with switches, have significantly increased the performance of the network. Ethernet networks can be configured in either a star topology using unshielded twisted pair connected to a hub or a bus topology using a coaxial cable acting as a backbone. Of these two, UTP cabling is by far the most common. Ethernet cards can have one, two, or possibly all three of the following connectors:

- DIX (Digital-Intel-Xerox)/AUI connectors support 10Base5 external transceivers
- BNC (British Naval Connector) connectors support 10Base2 coax cabling
- RJ-45 connectors support 10Base-T/100Base-TX (UTP) cabling

When a network device wants to access the network, the method used is carrier sense multiple access with collision detection (CSMA/CD). When a computer wants to send data over the network, it will listen for any traffic on the network. If the network is clear, it will then broadcast. Unfortunately, it is possible for two network devices to listen and try to send data at the same time. As a result, a collision occurs and both data packets are corrupted. This situation is normal for an Ethernet network, and both network devices will wait a different random amount of time and try again. If the network is heavily congested, then more collisions will occur, resulting in more traffic and thus a slower network.

Within an Ethernet network, users are limited to 1024 nodes identified by their MAC/physical addresses, which are burned into the ROM chip on the network card. IEEE assigns the first three bytes of the six-byte address to the network card vendors. The vendor is then responsible for assigning the rest of the address to make sure that the MAC address is unique.

4.9.1   Ethernet Encoding Method

Although many methods are used for encoding the signal, the Manchester signal encoding method, used on an Ethernet medium, is the most common. When a device driver receives a data packet from the higher-layer protocols such as an IP or IPX packet, the device driver constructs a frame (much like an envelope) with the appropriate Ethernet header information and a frame check sequence at the end for error control. The circuitry on the adapter card then takes the frame and converts it into an electrical signal.

Data are measured as a transition state that incurs in the middle of each bit-time. To represent a binary 1, the first half of the bit-time is a low voltage and the second half is always the opposite of the first half. To represent a binary 0, the first half of the bit-time is a high voltage and the second half is a low voltage. See Figure 4.24.

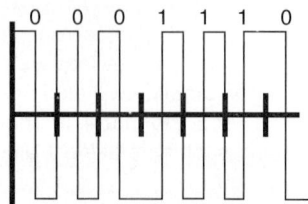

FIGURE 4.24   Ethernet using transition state to encode data

4.9.2   Ethernet Frame Types

When a packet is transmitted, information about the sender, receiver, and upper-layer protocols is attached to the data. The format for the completed packet is called the frame type. There are four Ethernet frame types (see Table 4.5). Consider Ethernet as being a language and the four different frame types as the dialects that exist for the language. They are all similar, yet different. The four frame types began with Novell, when Novell decided it could not wait for the emerging Ethernet standard to be ratified by IEEE.

The Ethernet packet starts with the preamble, which consists of 8 bytes of alternating ones and zeros, ending in 11. A station on an Ethernet network detects the change in voltage that occurs when another station begins to transmit, and uses the preamble to "lock on" to the sending station's clock signal. When it reads the 11 at the end, it then knows the preamble has ended.

Next, the MAC address of the target (destination) computer followed by the MAC address of the source of the sending computer is sent. Note that the address could be represented in a multicast (address sent to multiple computers) message. Most Ethernet adapters can be set into **promiscuous mode** in which they receive all frames that appear on the LAN regardless if the data are addressed to them. If this poses a security problem, a new generation of smart hub devices can filter out all frames, with private destination addresses belonging to another station.

Depending on the Ethernet frame type, either a type field or a length field is sent. For Ethernet 802.2, the length field describes the length of the data field including the LLC headers. The data field can be between 64 bytes and 1500 bytes. The LLC header is used to identify the process/protocol that generated the packet and the process/protocol for which the packet is intended. The data field is between 43 bytes and 1497 bytes, which includes the TCP/IP or IPX packet. The last 4 bytes that the adapter reads is the frame check sequence or CRC for error control. See Figure 4.25.

TABLE 4.5   Ethernet frame types

Novell Name	Cisco Name	Ethernet Type
Novell Ethernet_802.3	LLC	Also known as Ethernet Raw. Default Ethernet frame type of Novell NetWare 3.11 or earlier. Supports IPX.
Novell Ethernet_II	ARPA	Used for IPX and TCP/IP.
IEEE 802.2	Novell	Default Ethernet frame type of Novell NetWare 3.12 or later and Windows. Used for IPX.
IEEE 802.2 SNAP	SNAP	Used for IPX, TCP/IP, and AppleTalk.

FIGURE 4.25   Ethernet packet

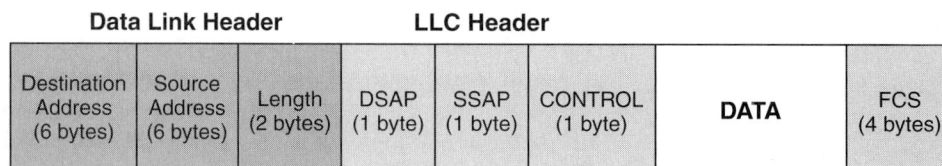

Data Link Header			LLC Header				
Destination Address (6 bytes)	Source Address (6 bytes)	Length (2 bytes)	DSAP (1 byte)	SSAP (1 byte)	CONTROL (1 byte)	**DATA**	FCS (4 bytes)

Ethernet_802.3 is an incomplete implementation of the IEEE 802.3 specification, sometimes called Ethernet Raw. Novell implemented the 802.3 header but not the fields as defined in the 802.2 specification. Instead, the IPX packet begins immediately after the 802.3 fields. The Novell frame type Ethernet_802.2 is a complete implementation; that is, it includes the 802.3 and 802.2 fields. Ethernet_II is the standard frame type for TCP/IP networks. Ethernet_SNAP includes the 802.3, 802.2, and SNAP (subnetwork access protocol) fields.

4.9.3 10Base5 (Thicknet)

The original Ethernet was called **10Base5** or **thicknet.** The name is derived because 10Base5 is a 10-Mbps baseband network that can have cable segments up to 500 meters long. It used a 50-ohm RG-8 and RG-11 as a backbone cable. 10Base5 uses physical and logical bus topology. Most network devices that are connected to the network use an external transceiver using a four-pair AUI cable (sometimes referred to as a drop cable) via a DIX connector (two-row, 15-pin female connector). Other network devices can connect to the backbone cable by using a vampire tap, which uses teeth that make contact with the inner conductor, or by using a BNC barrel connector. See Figure 4.26.

When building a 10Base5 network, the following rules apply:

- The minimum cable distance between transceivers is 8 feet or 2.5 meters.
- The AUI cable may be up to 164 feet or 50 meters.
- The maximum network segment length may not exceed 1640 feet or 500 meters.
- The entire network cabling scheme cannot exceed 8200 feet or 2500 meters.
- All unused ends must be terminated with 50-ohm terminating resistors.
- One end of the terminated network segment must be grounded.
- The maximum number of nodes per network segment, including repeaters, is 100.
- A signal quality error (SQE) test, which is present on some boards, must be turned off when repeaters are used.

4.9.4 10Base2 (Thinnet)

The **10Base2 (thinnet)** is a simplified version of the 10Base5 network. The name describes a 10-Mbps baseband network with a maximum cable segment length of approximately 200 meters (actually 185 meters). Instead of having external transceivers, the transceivers are on the network card, which attaches to the network using a BNC T-connector. The cable used is 50-ohm RG-58 A/U coaxial type cable. Different from the 10Base5 network, the 10Base2 does not use a drop cable. See Figure 4.27.

When building a 10Base2 network, the following rules apply:

- The minimum cable distance between workstations must be 1.5 feet or 0.5 meter.
- The maximum network segment limitation may not exceed 607 feet or 185 meters (not the 200 meters commonly stated).
- The entire network cabling scheme cannot exceed 3035 feet or 925 meters.
- The maximum number of nodes per network segment is 30 (including workstations and repeaters).
- All unused ends must be terminated with 50-ohm terminating resistors.
- A grounded terminator must be used on only one end of the network segment.
- The 10Base5 networks can be connected to 10Base2 networks using special boards.

4.9.5 10Base-T (Twisted-Pair Ethernet)

Different from 10Base2 and 10Base5, **10Base-T** uses UTP, which costs less, is smaller, and is easier to work with than coax cable. Whereas Ethernet is a logical bus topology, 10Base-T is a physical star topology, which has the network devices connected to a hub. It uses a category 3 (or greater) cable with two pairs of wires (pins 1, 2, 3, and 6) connected with RJ-45 connectors. See Figure 4.28.

When building a 10Base-T network, the following rules apply:

- The maximum number of network segments or physical LAN is 1024.
- The minimum unshielded cable segment length is 2 feet or 0.6 meter.

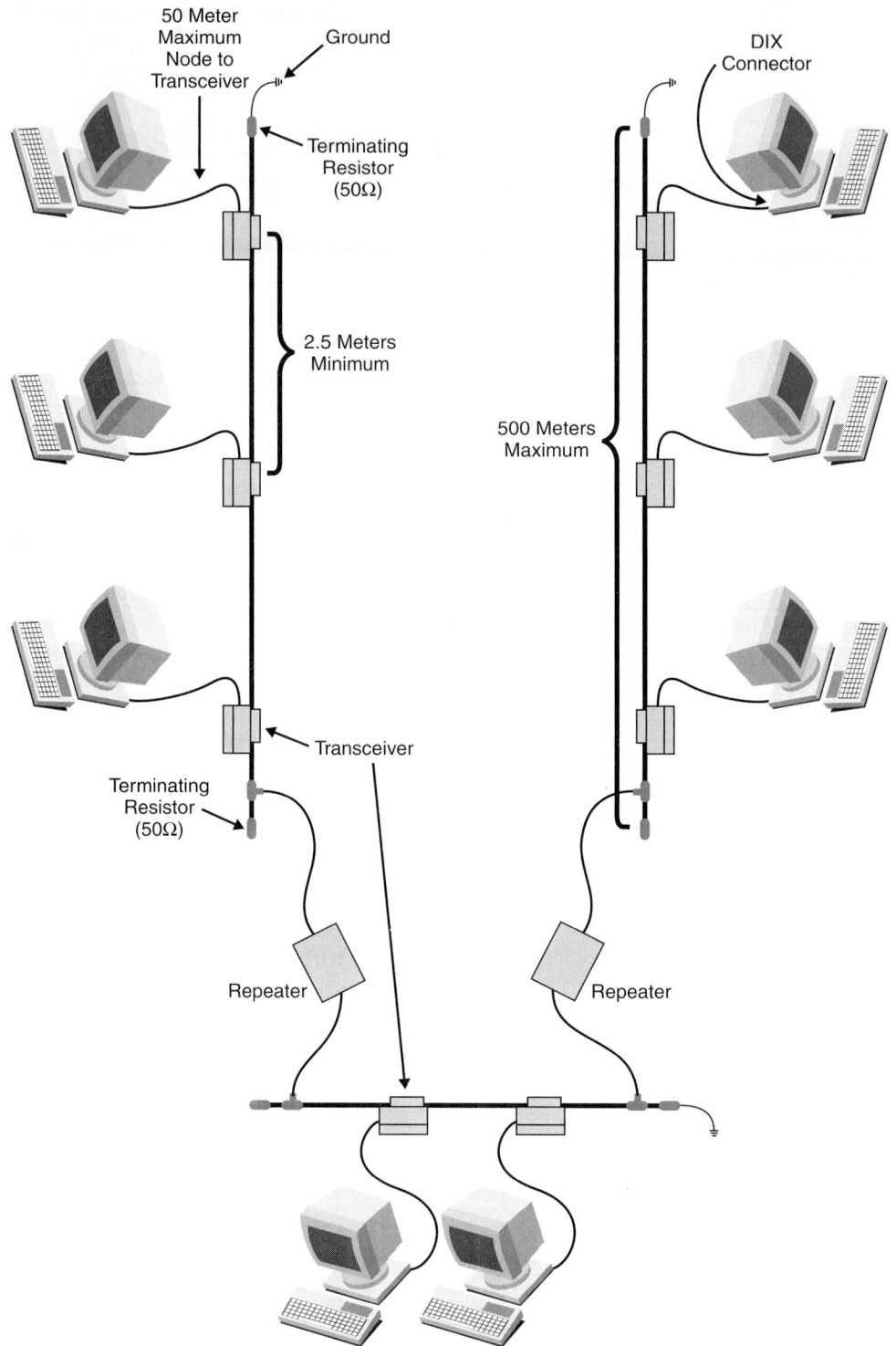

FIGURE 4.26 **10Base5 network**

- 50 Meter Maximum Node to Transceiver
- Ground
- Terminating Resistor (50Ω)
- 2.5 Meters Minimum
- Transceiver
- Terminating Resistor (50Ω)
- Repeater
- DIX Connector
- 500 Meters Maximum
- Repeater

- The maximum unshielded cable segment length is 328 feet or 100 meters.
- To avoid EMI, route UTP cable no closer than 5 feet to any high-voltage (power) wiring or fluorescent lighting.

One advantage of using 10Base-T is that if you need to add another workstation, you simply run an additional cable to the hub and plug it in. If there is a break in the cable, the hub has its own intelligence that will route traffic around the defective cable segment. As a result, the entire network is ineffective.

FIGURE 4.27   10Base2 network

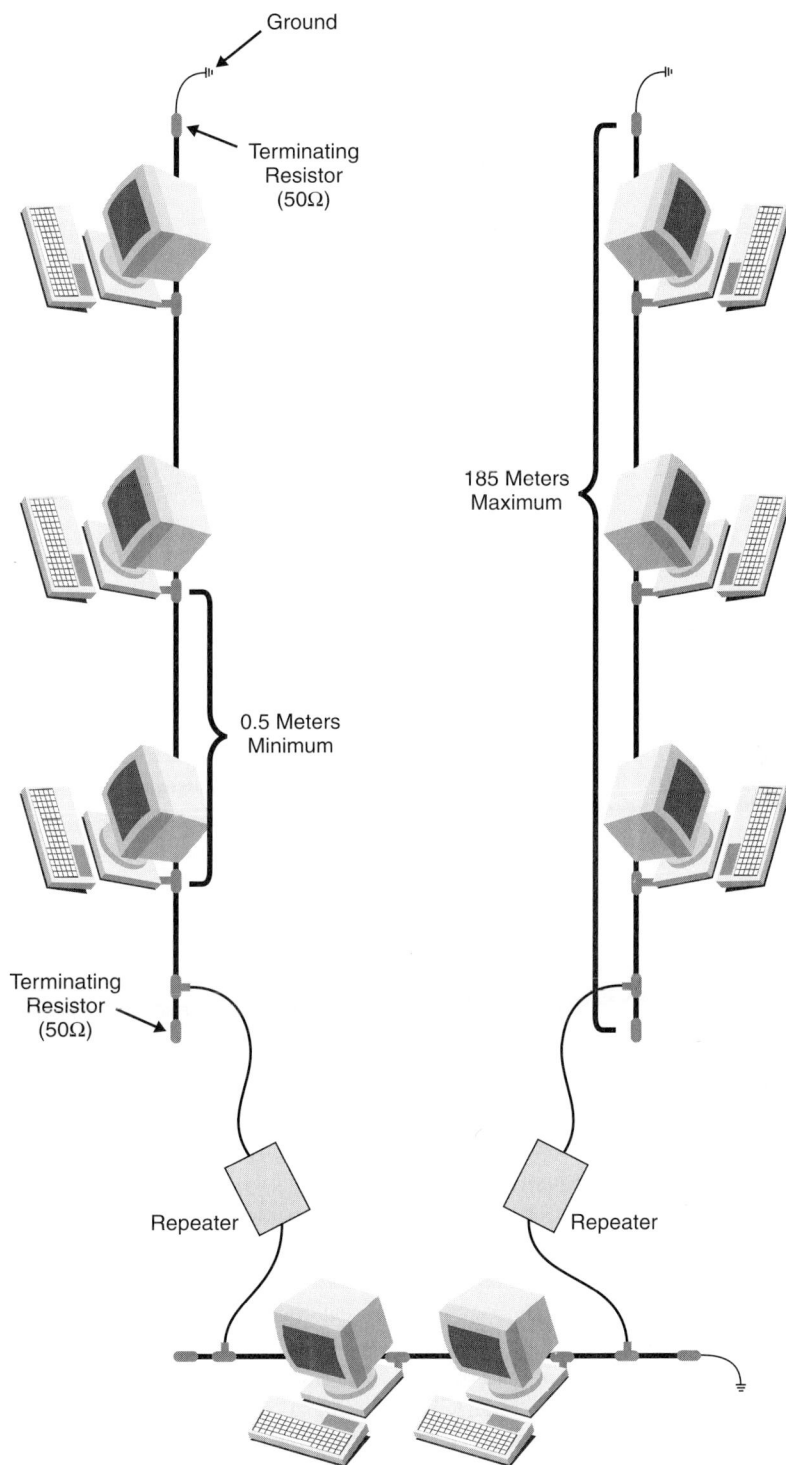

Ground

Terminating
Resistor
(50Ω)

185 Meters
Maximum

0.5 Meters
Minimum

Terminating
Resistor
(50Ω)

Repeater

Repeater

Note: Most networks use category 5 twisted-pair cabling (or enhanced category 5 twisted pair) in a star topology, so choose the same unless you have an overwhelming reason to choose differently.

4.9.6   100-Mbps Ethernet

Fast Ethernet is an extension of the 10Base-T Ethernet standard that transports data at 100 Mbps yet still keeps using the CSMA/CD protocol used by 10-Mbps Ethernet. The first type is **100Base-TX** which uses two pairs of the standard category 5 UTP. The second type is

FIGURE 4.28   10Base-T network

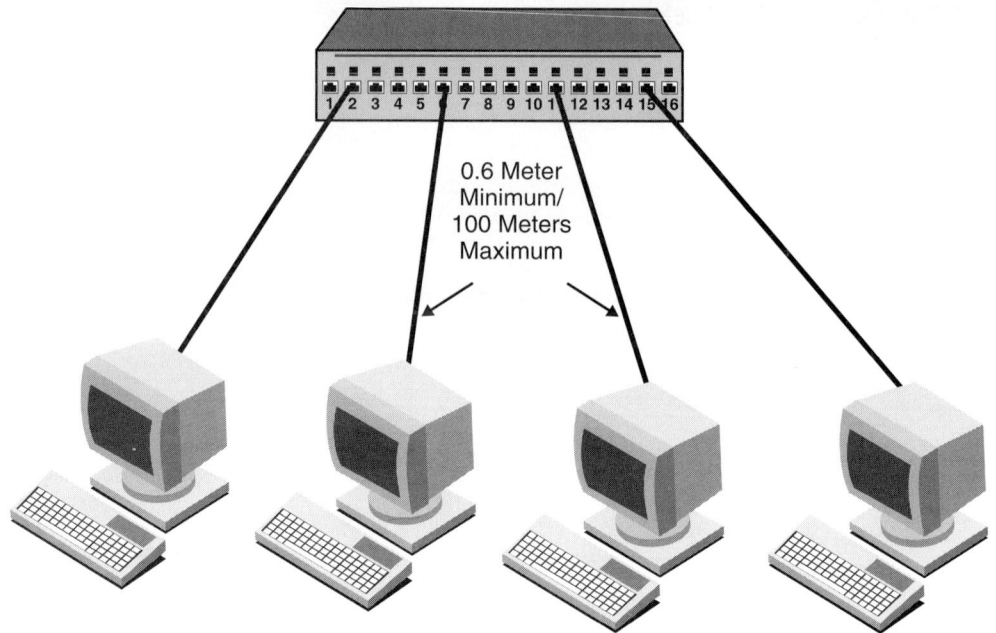

0.6 Meter
Minimum/
100 Meters
Maximum

100Base-T4, which runs over existing category 3 UTP by using all four pairs— three pairs are to transmit simultaneously and the fourth pair is for collision detection. If category 5 cable is used for backbone for fast Ethernet, it should not exceed 5 meters. The last standard is **100Base-FX,** which operates over multimode fiber-optic cabling.

100VG-ANYLAN was developed by Hewlett Packard as a 100-Mbps half-duplex transmission which allows 100 Mbps on a four-pair category 3 cabling system. Different from the previous Ethernet standards, it uses the demand priority protocol instead of the Ethernet CSMA/CD protocol. The demand priority protocol allows different nodes and data types to be assigned a priority. Unfortunately for 100VG-ANYLAN network, fast Ethernet has gained a greater market acceptance because it has been standardized by the original IEEE committee.

4.9.7   Gigabit Ethernet

Gigabit Ethernet has been demonstrated to be a viable solution for increased bandwidth requirements for growing networks as it is used for high-speed backbones and specialized needs. The gigabit Ethernet standards include 1000Base-SX (short wavelength fiber), 1000Base-LX (long wavelength fiber), 1000Base-CX (short-run copper), and 1000Base-T (100-meter, four-pair category 5 UTP).

The gigabit Ethernet CSMA/CD method has been enhanced to maintain a 100-meter collision at gigabit speeds. Without this enhancement, minimum-size Ethernet packets could complete transmission before the transmitting station senses a collision, thereby violating the CSMA/CD method. To resolve this issue, both the minimum CSMA/CD carrier time and the Ethernet slot time have been extended from their present value of 64 bytes to a new value of 512 bytes. Note that the minimum packet length of 64 bytes has not been affected. Packets smaller than 512 bytes have an extra carrier extension. Packets longer than 512 bytes are not extended. These changes, which can impact small-packet performance, have been offset by incorporating a new feature, packet bursting, into the CSMA/CD algorithm. Packet bursting will allow servers, switches, and other devices to send bursts of small packets in order to fully utilize available bandwidth.

4.9.8   5-4-3 Rule

When creating and expanding an Ethernet network, always follow the **5-4-3 Rule.** An Ethernet network must not exceed five segments connected by four repeaters. Of these segments, only three can be populated by computers, which means that the distance of a computer cannot exceed five segments and four repeaters when communicating with other computers on the network. See Figures 4.29, 4.30, and 4.31.

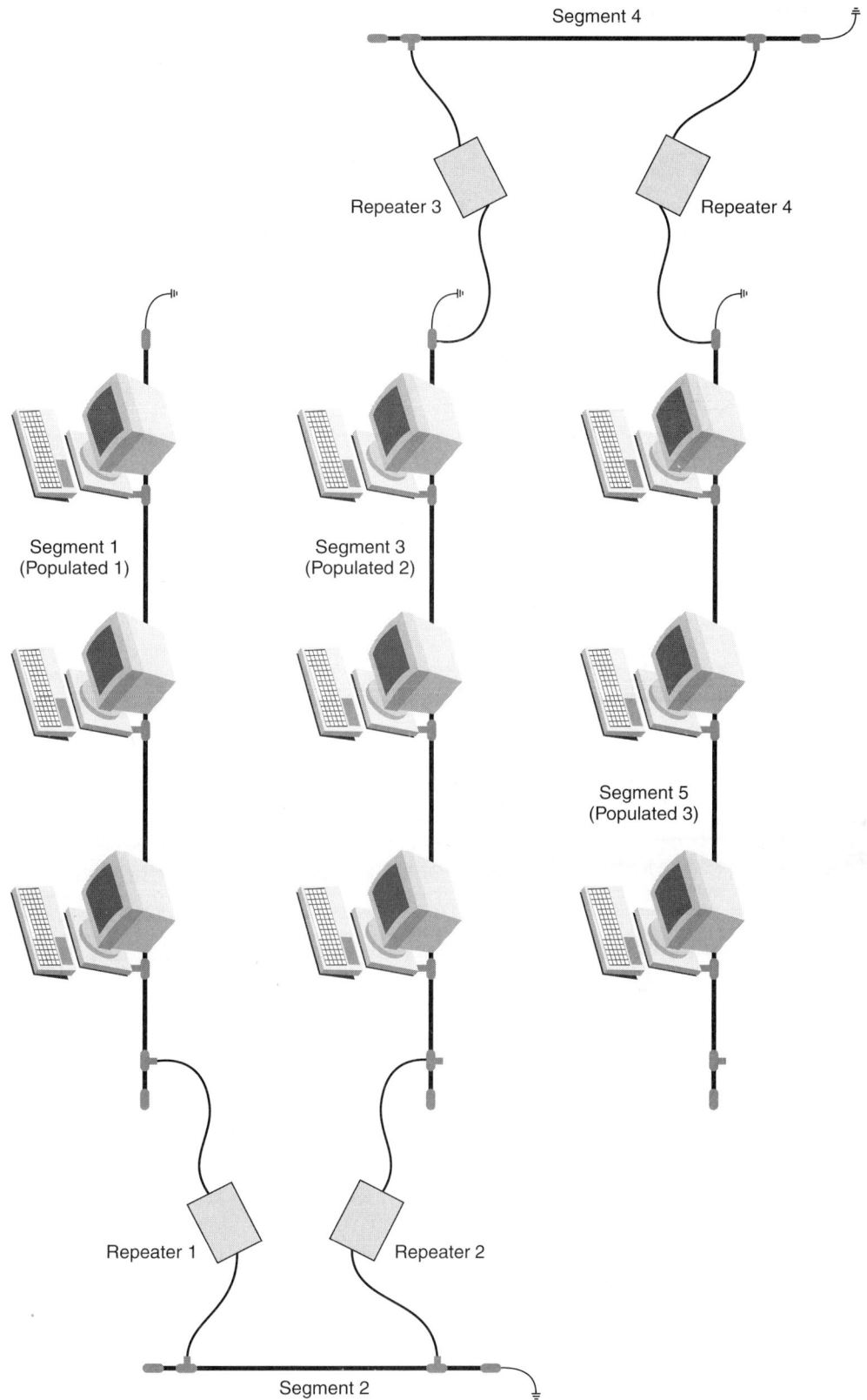

FIGURE 4.29   5-4-3 Rule for an Ethernet coaxial network

Segment 4

Repeater 3

Repeater 4

Segment 1
(Populated 1)

Segment 3
(Populated 2)

Segment 5
(Populated 3)

Repeater 1

Repeater 2

Segment 2

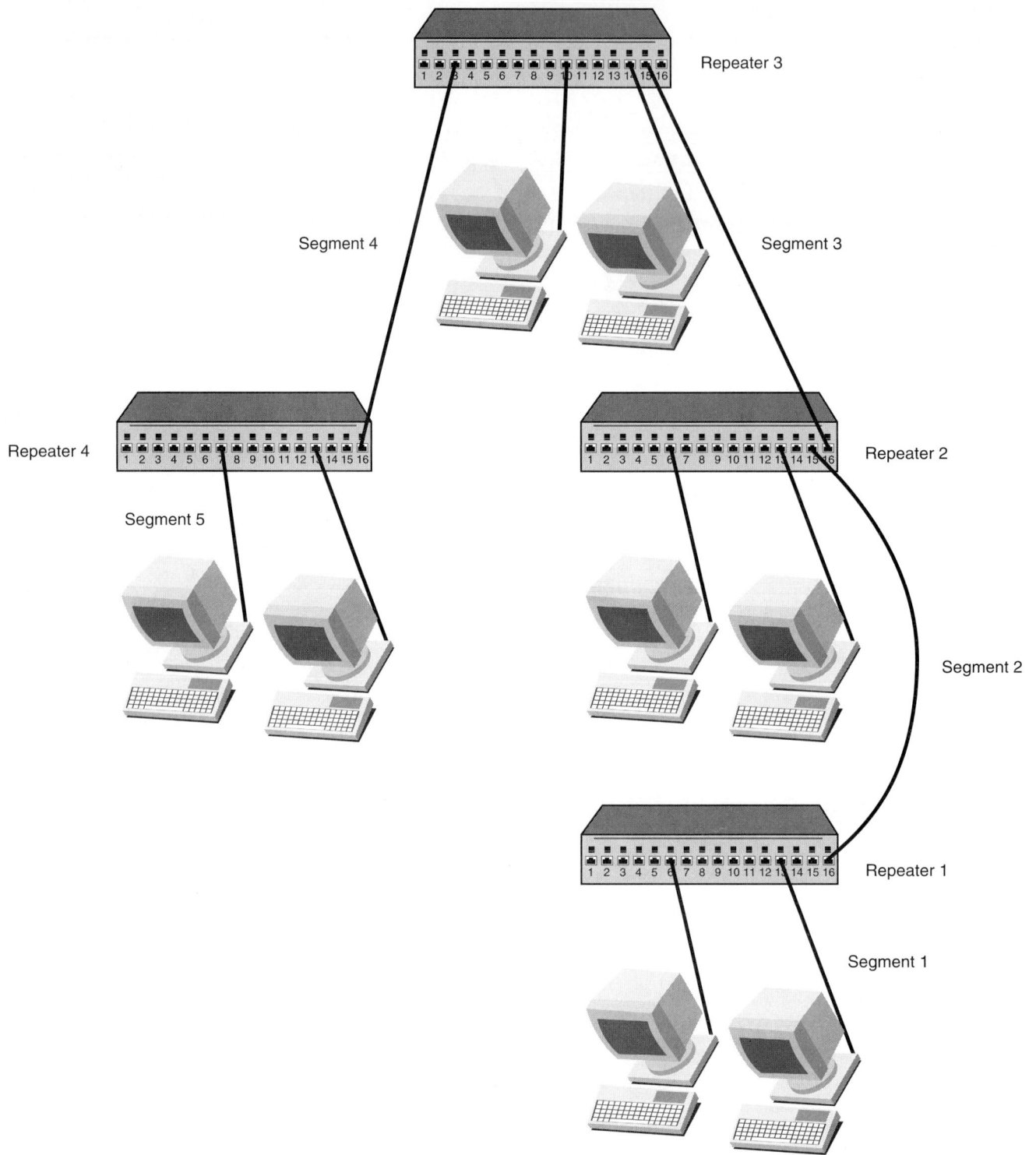

FIGURE 4.30   5-4-3 Rule for Ethernet UTP network

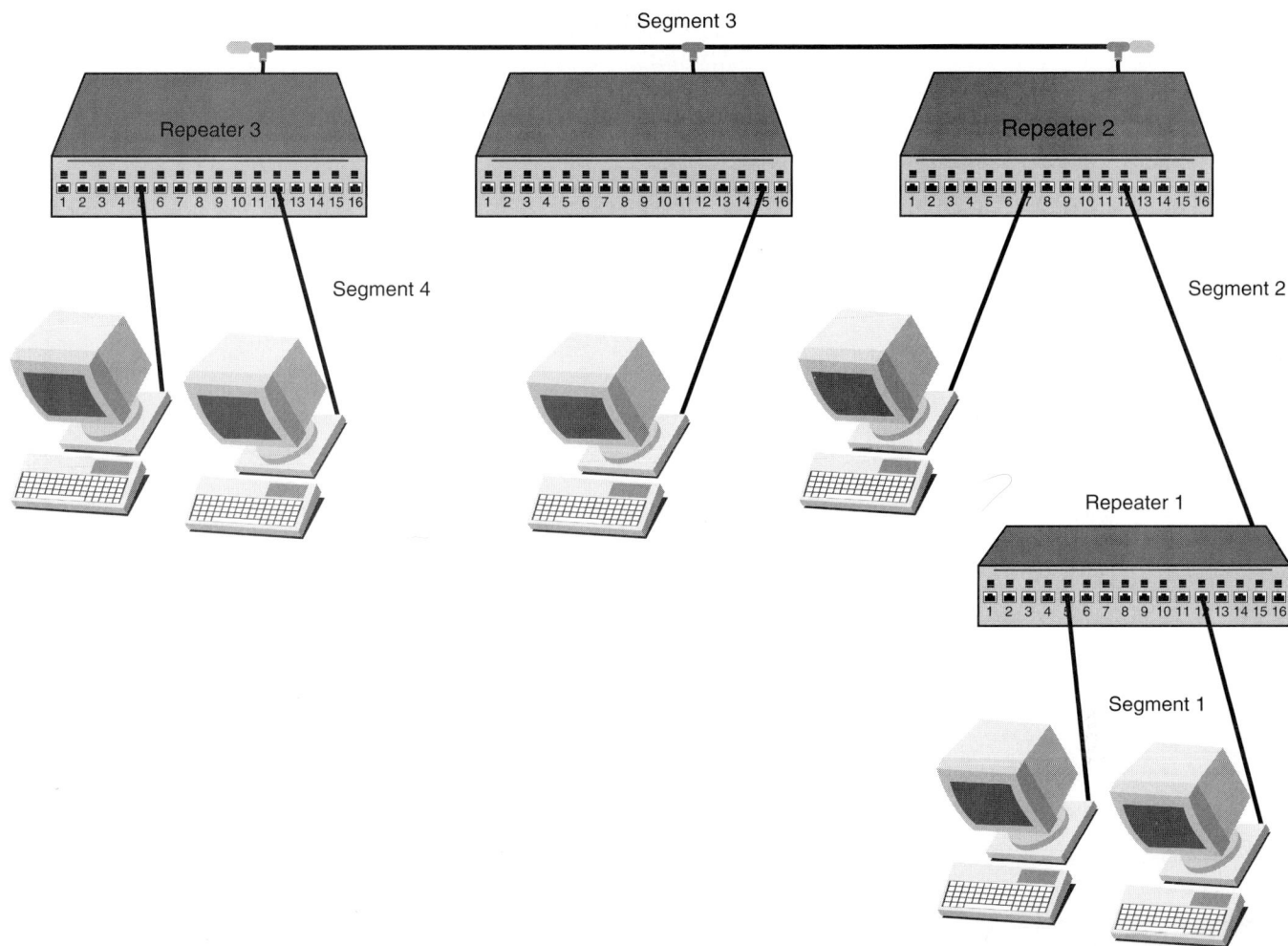

FIGURE 4.31   5-4-3 Rule for Ethernet UTP/coaxial network

SEGMENT AND BACKBONE

The terms *segment* and *backbone* have been used in different ways in this chapter.

A segment could be a single cable such as a backbone cable, or a cable that connects a hub to a computer. A logical segment contains all the computers on the same network and contains the same network address, such as networks with a backbone cable or a hub used in a logical bus topology.

A backbone can be used as a main cable segment such as that found in a bus topology network. This would include either a long, single cable with patch cables that are used to attach the computers or smaller cables connected with barrel and T-connectors. In addition, a backbone can refer to the main network connection through a building, campus, WAN, or on the Internet.

4.10 OTHER LOCAL AREA NETWORK TECHNOLOGIES

Ethernet networks comprise the majority of the LANs, but they are not the only network technologies available. Other networks are token ring, ARCnet, FDDI, and AppleTalk.

4.10.1 Token Ring

Another major LAN technology in use today is token ring. **Token ring** rules are defined in the IEEE 802.5 specification. As mentioned earlier in the chapter, the physical topology of token ring is a star, but the logical topology is a ring. Therefore, it is actually implemented in what can best be described as a collapsed ring. In token ring LANs, each station is connected to a token ring wiring concentrator called a **multistation access unit (MSAU)** using a shielded twisted pair or unshielded twisted pair. Like Ethernet hubs, MSAUs are usually located in a wiring closet.

The access method used on token ring networks is called **token passing.** In token passing, a network device only communicates over the network when it has the token (a special data packet that is generated by the first computer that comes online in a token ring network). The token is passed from one station to another around a ring. When a station gets a free token and transmits a packet, it travels in one direction around the ring, passing all other stations along the way.

Each node acts as a repeater that receives token and data frames from its nearest active upstream neighbor (NAUN). After a frame is processed by the node, the frame is passed or rebroadcast downstream to the next attached node. Each token makes at least one trip around the entire ring, and then returns to the originating node. Workstations that indicate problems send a beacon to identify an address of the potential failure. If a station has not heard from its upstream neighbor in 7 seconds, it sends a packet down the ring that contains its address and the address of its NAUN. Token ring can then reconfigure itself to avoid the problem area.

Token ring NICs can run at either 4 or 16 Mbps. The 4-Mbps cards can run only at that data rate; however, 16-Mbps cards can be configured to run at 4 or 16 Mbps. All cards on a given network ring must be running at the same rate. To run at 4 Mbps, use category 3 UTP or better, whereas 16 Mbps requires category 4 cable or higher, or Type 4 STP cable or higher.

Like Ethernet cards, the node address on each NIC is burned in at the manufacturer on a ROM chip and is unique to each card. For fault tolerance, a maximum of two token ring cards can be installed in any node, with each card being defined as the primary or alternate token ring card in the machine. The token ring card will use two types of connectors—one is equipped with a female (DB-9) nine-pin connector (which only four pins are used) while the second is an RJ-45 connector. MSAU repeaters, and most other equipment, use a special IBM Type 1 unisex data connector. To connect multiple MSAUs, the ring-out (RO) port of each MSAU must be connected to the ring-in (RI) port of the next MSAU so that it can complete a larger ring. See Figure 4.32.

When building a token ring network, the following rules apply:

- The maximum cable length between a node and a MSAU is 330 feet (100 meters) for Type 1 and 2 cables, 220 feet (66 meters) for Types 6 and 9 cable, and 150 feet (45 meters) for a category 3 UTP cable.
- Nodes must be separated by a minimum of at least 8 feet (2.5 meters).
- The maximum ring length is 660 feet (200 meters) for Types 1 and 2 cable, 400 feet (120 meters) for a Type 3 cable, 140 feet (45 meters) for a Type 6 cable, and 0.6 mile (1 km) for a fiber-optic segment.
- While IEEE 802.5 specifies a maximum of 250 nodes, if using a small movable token ring cable system, IBM STP specifies a maximum of 96 nodes and 12 MSAUs (IBM Model 8228) using IBM Type 6 cable.

FIGURE 4.32 MSAUs connected together. Notice the cables attached to the ring-in (RI) and ring-out (RO) are connected to form a larger ring.

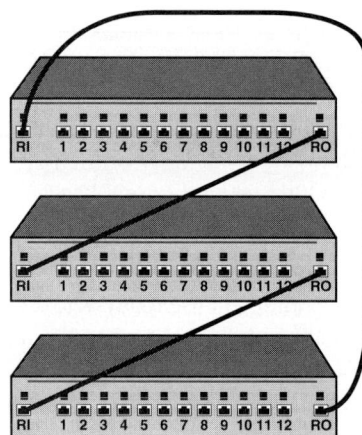

- While IEEE 802.5 specifies a maximum of 250 nodes, if using a large, nonmovable token ring cable system, users are limited to a maximum 260 nodes and 33 MSAUs (using IBM Types 1 and 2 cables).
- While IEEE 802.5 specifies a maximum of 250 nodes, IBM UTP specifies 72 nodes.
- To run at 4 Mbps, use category 3 UTP or better, while 16 Mbps requires category 4 cable or higher or Type 4 STP cable or higher.

4.10.2 ARCnet

ARCnet (Attached Resource Computer Network) was founded by the Datapoint Corporation. It was one of the early topologies in networking and is rarely used on current LAN environments. Standard ARCnet operates at 2.5 Mbps throughput using RG-62 A/U coax cable with 93-ohm terminating resistance or unshielded twisted-pair wiring. Although ARCnet can support up to 255 nodes on a single network, the 2.5 Mbps throughput is too slow to support 255 computers. Turbo ARCnet, however, is an ARCnet version that runs at 100 Mbps.

Typically, ARCnet uses a logical bus topology with token passing, but is wired as a star. The token is passed from one computer to another depending on its MAC address. Lower-number addresses get the token before the higher-number addresses.

Different from the Ethernet and token ring cards, each NIC on an ARCnet network is manually assigned a node address, typically using eight DIP switches, which allows binary counting between 0 and 255. Like the Ethernet and token ring cards, each number must be unique on each network.

When connecting the cabling to the hub, unused ports of a passive hub require terminating resistors. In addition, a passive hub can only connect to an active device, never to another passive hub.

4.10.3 Fiber Distributed Data Interface

Fiber distributed data interface (FDDI) is a MAN protocol that provides data transport at 100 Mbps (a much higher data rate than standard Ethernet or token ring) and can support up to 500 stations on a single network. Originally, FDDI networks required fiber-optic cable, but today they can also run on UTP. Of course, fiber offers much greater distances than UTP cable. Traditionally, FDDI was used primarily as a backbone to connect several department LANs within a single building or to link several building LANs in a campus environment.

FDDI communicates all of its information using symbols. Symbols are 5-bit sequences that, when taken with another symbol, make 1 byte. A bit can have one of two values: either a zero or a one. In FDDI, this is expressed by the change of state in the light on the other side. Approximately every 8 nanoseconds, the station will take a sample of the light coming from the other machine. The light will either be on or off. If it has changed since the last sample, that translates into a bit of one. If the light has not changed since the last sample, the bit is a zero.

Like token ring, FDDI is a token passing ring that uses a physical star topology. FDDI can be implemented in two basic ways: as a dual-attached ring and as a concentrator-based ring. In the dual-attached scenario, stations are connected directly to one another. FDDI's dual counter–rotating ring design provides a fail-safe alternative in case a node goes down. If any node fails, the ring wraps around the failed node. One limitation of the dual counter–rotating ring design, however, is that if two nodes fail, the ring is broken in two places, effectively creating two separate rings. Nodes on one ring are then isolated from those on the other ring. External optical bypass devices can solve this problem, but their use is limited because of FDDI optical power requirements.

Another way around this problem is to use concentrators to build networks. Concentrators are devices with multiple ports into which FDDI nodes connect. FDDI concentrators function like Ethernet hubs or token ring multiple access units (MAUs). Nodes are single-attached to the concentrator, which isolates failures occurring at those endstations. With a concentrator, nodes can be powered on and off without disrupting ring integrity. Since concentrators make FDDI networks more reliable, most FDDI networks are now built with them.

An extension to FDDI, called **FDDI-2,** supports the transmission of voice and video information as well as data. Another variation of FDDI, called **FDDI full-duplex technology (FFDT),** uses the same network infrastructure but can potentially support data rates up to 200 Mbps.

4.10.4 AppleTalk/LocalTalk

The Apple network architecture is called AppleTalk, and has been included in Apple Macintosh computers for several years. By using the computer's built-in network interface, users frequently connect the computers and Apple printers using shielded twisted pair, or unshielded twisted pair and fiber optics. LocalTalk is Apple's own network interface. Since AppleTalk can only operate at 1 Mbps, it is not commonly used. Instead, Macintosh computers primarily use EtherTalk or TokenTalk to communicate with larger and faster networks. Note that Appletalk also is a routable protocol.

4.11 SERIAL LINE PROTOCOLS

Serial line protocols enable a computer to connect to a server—such as those used by an Internet service provider (ISP)—via a serial line such as a modem to become an actual node on the Internet. Serial line protocols include the **serial line internet protocol (SLIP);** the **compressed serial line internet protocol (CSLIP),** which is simply a compressed version of SLIP; and the **point-to-point protocol (PPP).** SLIP is an older and simpler protocol. Typically with SLIP, an automated script is used to make the process of logging on more automatic. For more information, SLIP is described under RFC 1055. Requests for Comment (RFC) are documents that describe the TCP/IP standard and the Internet.

The protocol PPP is an advanced version of SLIP. It is a full-duplex protocol that can be used on various physical media, including phone lines, twisted-pair cables, fiber optic lines, or satellite transmission. It is usually preferred over SLIP because it is more stable and it can handle synchronous and asynchronous communication. In addition, it can share a line with other users and protocols such as TCP/IP, IPX, and AppleTalk, and it has error detection that SLIP lacks.

PPP can negotiate configuration parameters at the start of the connection, including automatic assigning of IP addresses, similar to the dynamic host configuration protocol (DHCP). PPP provides two methods with which logins can be automated—**password authentication protocol (PAP)** and **challenge-handshake authentication protocol (CHAP).** Both provide the means for a system to automatically send a login user ID and password information to the remote system. For more information on PPP, use RFC 1332 and 1661.

4.12 PLANNING AN ETHERNET NETWORK

Consider two network examples. The first example consists of a room with approximately 10 computers. This simple network would include purchasing network cards for each computer, one or more hubs, and enough UTP cabling to connect all of the network cards. The hub is located somewhere within the room and cables are stretched from each computer to the hub. Depending on the function of the network, standard 10-Mbps Ethernet using category 3 cabling (or better) will probably suffice.

The second example is a larger, more complicated network. To expand on our first example, place your server or servers in a server room where the server can be secured and have proper ventilation. Cables are then stretched from wall jacks throughout the building and lead to a patch panel located in the server room or in a wiring closet along with the hub or hubs. The cables in the walls should be eight-wire/four-pair category 5 or better solid wire cabling. Depending on the fire code for your area, some or all of the cabling will need to be plenum cabling. Patch cables are then made using eight-wire/four-pair category 5 stranded cabling. They connect the individual computers to the wall jacks. The port of the patch panel that connects to the wall jack is connected with another patch cable to the hub. The patch panel provides a convenient way to connect the network. In addition, it allows for easy reconfiguration. Again depending on the function and load of the network, you can take one of two options:

- Make the entire network either 1 Gbps or 100 Mbps.
- Install a backbone of 1 Gbps connected to 100-Mbps switches that are used to connect the computers.

4.13 TROUBLESHOOTING NETWORK CABLE PROBLEMS

Cables cause the most problems in networks. An understanding of how the cables are connected and how the signals travel through the cable system will provide IT managers with a solid foundation in isolating and correcting cable problems.

If the conductor such as a wire has a break (known as an **open**) and there is no other pathway for the electrons, the current will not flow. A **short** is when a circuit has zero or an abnormally low resistance path between two points, resulting in excessive current. A short on a network cable usually occurs when one wire mistakenly connects to another wire. To help troubleshoot cables, there are several tools that managers can use. These include the digital multimeter, time-domain reflectometers, oscilloscope, and cable tester.

Digital multimeters contain at least a voltmeter and an ohmmeter. An ohmmeter can be used to measure the resistance of a wire. If the wire is good, the ohmmeter should measure 0 ohm. If it measures infinite ohms, the wire usually has an open. Some multimeters will also contain a continuity checker. The checker will beep when it measures 0 ohm (indicating a good connection) but will not beep when it measures more than 0 ohm.

A **time-domain reflectometer (TDR)** sends a sonarlike pulse along a cable. Depending on the signal that is bounced back, the TDR can detect a break or short and its distance from the meter.

The **oscilloscope** is an electronic device that can measure the amount of signal voltage per unit of time and display the results on a monitor. When used with a TDR, it can display shorts, sharp bends or crimps in the cable, opens (breaks) in the cable, and loss of signal power due to attenuation.

A **cable tester** is a device specifically made to test a cable. An inexpensive cable tester will typically test for shorts and opens. More expensive cable testers (sometimes referred to as advanced cable testers) can display additional information about the condition of the physical cable as well as message frame counts, error frame counts, excess collisions, late collisions, congestion errors, and beaconing.

When experiencing a network failure, managers must first determine what is affected by that failure. If only one computer is affected, the problem is most likely either the cable that connects to the individual computer, a bad network card, or the driver and client software installed or configured incorrectly. If multiple computers have failed, however, look for a network item that is common to all of the computers. If using a star topology, look for a hub or MAU or a single cable that connects the hub or MAU to the rest of the network. If using a backbone cable such as a coax, look for an open. Remember that a break will cause ends that are not terminated.

SUMMARY

1. To communicate on LANS, network cards, or network interface card (NIC), found as an expansion card or are integrated into the motherboard, uses transceivers (devices that both transmit and receive analog or digital signals).
2. Topology describes the appearance or layout of the network.
3. The physical topology describes how the network actually appears.
4. The logical topology describes how the data flows through the physical topology or the actual pathway of the data.
5. Point-to-point topology connects two nodes directly together.
6. Multipoint connection links three or more devices together through a single communication medium. Because multipoint connections share a common channel, each device needs a way to identify itself and the device to which it wants to send information. The method used to identify senders and receivers is called addressing.
7. Network cards connected on a network are identified with their Media Access Control (MAC) address.
8. A bus topology looks like a line, where data is sent along the single cable.
9. A ring topology has all devices connected to one another in a closed loop.
10. A star topology is the most popular topology in use.
11. The mesh topology whereas every computer is linked to every other computer.
12. Attenuation is when the strength of a signal falls off with distance over a transmission medium.

13. Interference occurs when undesirable electromagnetic waves affect the desired signal.

14. A twisted pair consists of two insulated copper wires twisted around each other. Twisted pair comes in unshielded twisted pair (UTP) and shielded twisted pair (STP).

15. Solid cable is typically used for the cabling that exists throughout the building. Stranded cable is typically used as patch cables between patch panels and hubs and between the computers and wall jacks.

16. Coaxial cable, sometimes referred to as coax, is a cable that has a center wire surrounded by insulation and then a grounded shield of braided wire.

17. A plenum cable is a special cable that gives off little or no toxic fumes when burned.

18. Riser cable is intended for use in vertical shafts that run between floors.

19. A fiber optic cable consists of a bundle of glass or plastic threads, each of which is capable of carrying data signals in the forms of modulated pulses of light.

20. Network cards are media-dependent, media-access-method dependent, and protocol independent.

21. A repeater, which works at the physical OSI layer, is a network device used to regenerate or replicate a signal or to move packets from one physical media to another.

22. A hub, which works at the physical OSI layer, is a multiported connection point used to connect network devices via a cable segment.

23. A bridge, which works at the data link OSI layer, is a device that connects two LANs and makes them appear as one or is used to connect two segments of the same LAN.

24. A switching hub (sometimes referred to as switch) is a fast multi-ported bridge, which actually reads the destination address of each packet and then forwards the packet to the correct port.

25. A router, which works at the network OSI layer, is a device that connects two or more LANs. In addition, it can break a large network into smaller, more manageable subnets.

26. A gateway is hardware and/or software that links two different types of networks by repackaging and converting data from one network to another network or from one network operating system to another.

27. A virtual, or logical, LAN (VLAN) is a local area network with a definition that maps workstations on some other basis than geographic location (for example, by department, type of user, or primary application).

28. A patch panel is a panel with numerous RJ-45 ports. The wall jacks are connected to the back of the patch panel to the individual RJ-45 ports. You can then use patch cables to connect the port in the front of the patch panel to a computer or a hub.

29. Ethernet is the most widely used LAN technology in use today. It offers good balance between speed, price, reliability and ease of installation.

30. The original Ethernet was called 10Base5 or ThickNet. This name developed because 10Base5 is 10 Mbps baseband network that can have a cable segment up to 500 meters long. It used a 50-ohm RG-8 and RG-11 as a backbone cable.

31. The 10Base2 (Thinnet) is a simplified version of the 10Base5 network. The name describes a 10 Mbps baseband network with a maximum cable segment length of approximately 200 meters (actually 185 meters).

32. While Ethernet is a logical bus topology, 10Base-T is a physical star topology, which has the network devices connected to a hub. It uses a category 3 (or greater) cable with two pairs of wires connected with RJ-45 connectors.

33. Fast Ethernet is an extension of the 10Base-T Ethernet standard that transports data at 100 Mbps yet still keeps using the CSMA/CD protocol used by 10 Mbps Ethernet.

34. An Ethernet network must not exceed 5 segments connected by four repeaters. Of these segments, only three of them can be populated by computers.

35. Token ring, as defined in the IEEE 802.5 specification, has a physical topology of a star but the logical topology is a ring.

36. The access method used on token ring networks is called token passing. In token passing, a network device only communicates over the network when it has the token (a special data packet that is generated by the first computer that comes online in a token ring network.

37. ARCnet (Attached Resource Computer NETwork) was founded by the Datapoint Corporation. It was one of the topologies used early on in networking and is rarely used on current LAN environments.

38. Fiber Distributed Data Interface (FDDI) is a MAN protocol that provides data transport at 100 Mbps that is based on light carrying fiber.
39. The Apple network architecture is called AppleTalk, and has been included in Apple Macintosh computers for several years.
40. Serial Line Protocols are used for a computer to connect to a server (such as those used by an Internet Service Provider (ISP)) via a serial line, such as a modem, to become an actual node on the Internet.

QUESTIONS

1. Which physical topology describes a network that has computers connected to a long cable segment (the cable segment has two ends that do not meet)?
   a. bus
   b. ring
   c. star
   d. mesh

2. Which physical topology describes a network that has all of the computers connect to a central point?
   a. bus
   b. ring
   c. star
   d. mesh

3. Which of the following is used to connect two pieces of cable on a linear bus topology?
   a. BNC terminator
   b. BNC barrel connector
   c. network adapter card
   d. medium attachment unit

4. Crosstalk and other types of interference can affect network performance and security. Which type of cable is most susceptible to crosstalk?
   a. RG-58 A/U
   b. RG-58/U
   c. STP
   d. category 5 UTP

5. Which type of problem is most likely to be caused by increasing cable lengths?
   a. attenuation
   b. crosstalk
   c. beaconing
   d. jitter

6. Twisted-pair cables use different connection hardware than coaxial cables. Which type of connector is commonly used by twisted-pair cables?
   a. BNC
   b. AUI
   c. DIX
   d. RJ-45

7. When a signal jumps from one wire to an adjacent wire it is known as _____ .
   a. attenuation
   b. crosstalk
   c. jitter
   d. beaconing

8. Which one of the following types of cable is needed to connect a 100-Mbps backbone for five networks?
   a. category 3 UTP
   b. category 4 UTP
   c. fiber optic
   d. RG-58 A/U

9. Your company has offices in two separate buildings. Each office is networked. You are responsible for connecting the two separate networks. The distance between the two buildings is approximately 450 meters. The current capacity for each network is 10 Mbps. You could use 10Base5 cable to connect the two networks. What is one reason to consider using fiber-optic cable instead?
   a. Fiber-optic cable is light and flexible, and is easier to install than the heavy and inflexible coaxial cable used by 10Base5.
   b. Fiber-optic cable supports higher digital transmission rates and can provide additional capacity for future network expansion without the need to run new cable.
   c. Fiber-optic cable is the most widely used network cabling and is usually cheaper than plenum-grade coaxial cable.
   d. Fiber-optic cable is better suited for broadband transmission, providing a relatively inexpensive way to increase network capacity in the future.

10. Which cable is the most common in networks?
    a. UTP
    b. STP
    c. coaxial
    d. fiber optics

11. When planning a network, fire is always a concern. What type of cable emits little or no toxic fumes when it burns?
    a. polling
    b. plastic
    c. plenum
    d. glass

12. In a star topology, all network traffic goes through a single connection point called a hub. Which of the following statements best describes a star topology?
    a. It needs significantly more cabling than a bus topology.
    b. It offers equal access to all computers by implementing the CSMA/CD access method.
    c. It is more difficult to configure and troubleshoot than a ring topology.
    d. It is less reliable than a ring topology, because a break in a single cable segment can bring down an entire network.

13. Which of the following access methods checks for network traffic before sending data? (Choose all that apply.)
    a. CSMA/CD
    b. CSMA/CA
    c. token passing
    d. polling

14. Ethernet uses which of the following access methods?
    a. CSMA/CD
    b. CSMA/CA
    c. token passing
    d. polling

15. A network's topology directly affects its capabilities and the way the network is managed. Although real-world network designs can be quite complex, they all stem from three basic topologies: bus, star, and ring. Which of the following best describes a linear bus topology?
    a. It usually relies on CSMA/CD to regulate network traffic.
    b. It needs significantly more cabling than a star topology network.
    c. It is an active topology because each computer acts like a repeater to boost the signal and send it on to the next computer.
    d. It offers centralized resources and management because all computers are connected through the hub.

16. Token passing prevents data collisions on a token ring network by _____ .
    a. having multiple tokens take alternate paths
    b. broadcasting the intent to transmit before actually sending data
    c. assigning a priority to each token and transmitting the highest priority token first
    d. allowing only one computer at a time to use the token

17. There are three major media access methods: carrier sense multiple access, token passing, and demand priority. Which type of media access method is commonly used by Ethernet networks?
    a. token passing
    b. demand priority
    c. CSMA/CD
    d. CDMS/CA

18. If the network computers are connected to a cable that forms a loop, this topology is called a ring. Which of the following statements best describes a ring topology network?
    a. It offers equal access for all computers.
    b. It needs less cabling than a bus network.
    c. It requires terminators to function properly.
    d. It requires category 5 UTP to function properly.

19. Which one of the following connectivity devices can be used as a protocol translator between different networking environments?
    a. router
    b. bridge
    c. repeater
    d. gateway

20. Which of the following statements best describes the difference between bridges and routers?
    a. Bridges can segment network traffic but routers cannot.
    b. Routers can choose between multiple paths but bridges cannot.
    c. Bridges can only be installed on an Ethernet network whereas routers can only be installed on a token ring network.
    d. Routers can link unlike physical media but bridges cannot.

21. Your company has two local area networks. One is Windows NT and the other is Novell NetWare. Users need access to resources and the ability to transmit data between the LANs. Which one of the following devices would you use to enable communications between dissimilar LANs that use different protocols?
    a. bridge
    b. router
    c. gateway
    d. repeater

22. A repeater is a device that regenerates signals so that they can travel on additional cable segments. Which of the following statements is true of repeaters?
    a. Repeaters can be used to solve the problem of crosstalk.
    b. Repeaters can be used to solve the problem of attenuation.
    c. Segments joined by a repeater must use the same physical media.
    d. Segments joined by a repeater must use the same media access method.

23. Which connectivity device should be used in a complex Ethernet 10Base-T network that uses both TCP/IP and NetBEUI protocols?
    a. amplifier
    b. router
    c. brouter
    d. terminator

24. You want to have a 10Base-T segment that is greater than 300 meters. Which device will be required?
    a. repeater
    b. amplifier
    c. multiplexer
    d. RJ-45 connector

25. Which device can be used with all media types, provide flow control, broadcast management, and utilize multiple communication paths?
    a. bridge
    b. router
    c. gateway
    d. repeater

26. There is excessive traffic on your Ethernet 10Base2 network. Which of the following devices can be used to split the network and reduce the amount of traffic on each segment?
    a. bridge
    b. repeater
    c. gateway
    d. terminator

27. Which connectivity device operates at the data link layer and joins two different networks to make them look like one network?
    a. NIC card
    b. bridge
    c. router
    d. gateway

28. You have a network that has very heavy traffic so therefore you need to segment it. You want to keep certain workstations communicating with another group of workstations. What type of hardware would you implement to arrive at this configuration?
    a. router
    b. brouter
    c. bridge
    d. hub

29. How does bridging between networks differ from routing between networks? (Choose all that apply.)
    a. Routing protocols are designed to separate a physical network into multiple virtual networks.
    b. Bridging protocols are designed to separate a physical network into multiple virtual networks.
    c. Bridging protocols are designed to combine packets from multiple physical networks and consolidate them into one virtual network.
    d. Routing protocols are designed to combine packets from multiple physical networks and consolidate them into one virtual network.

30. You are setting up a network that uses both routable and nonroutable protocols. You need to minimize the amount of investment in equipment as well as the amount of equipment that will go into your rack. What type of connectivity device will you choose to implement?
    a. router
    b. brouter
    c. bridge
    d. switching hub

31. Routers, bridges, and gateways can be used to join various types of network segments. Which one of the following functions can be provided by a router?
    a. broadcast management
    b. flow control
    c. multiple paths
    d. multiplexing

32. At what layer of the OSI model does a router function?
    a. data link
    b. transport
    c. network
    d. physical

33. Which of the following network standards is known as fast Ethernet?
    a. 10Base2
    b. 10Base5
    c. 100Base-TX
    d. 100BaseVG-ANYLAN

34. What is the maximum cable segment length for thinnet Ethernet?
    a. 100 meters
    b. 185 meters
    c. 215 meters
    d. 285 meters

35. What is the distance limitation of a 10Base-T network?
    a. 100 meters
    b. 185 meters
    c. 500 meters
    d. 1 km

36. You have a thinnet bus network, which has been in use for about a year. You have just added three new client computers to the network. When you test the network after installation, none of the computers on the network can access the server. Which of the following can be preventing the network from functioning?
    a. The bus network is not properly terminated.
    b. A client computer on the network has failed.
    c. The new cables added to service the new computers are not compatible with the existing cable type.
    d. The network adapter cards in the new computers are not compatible with adapter cards in other computers.

37. A client machine cannot connect to the network, but all other computers can access network resources. Which of the following is the most likely cause of the problem?
    a. faulty cable terminator
    b. faulty network adapter card on the domain controller
    c. faulty network adapter card on the client computer
    d. excessive media collisions

38. You have been given the task of installing cables for an Ethernet network in your office building. The network cable will share the existing conduit with telephone cables. Cable segments will be up to 95 meters in length. Which cable is best suited for this installation?
    a. fiber optic
    b. category 1 UTP
    c. category 3 UTP
    d. thicknet coaxial

39. Which of the following networks is also known as standard Ethernet?
    a. 10Base2
    b. 10Base5
    c. 10Base-T
    d. 100Base-X

40. Mark's computer uses dial-up networking with TCP/IP. The dial-up protocol cannot provide automatic IP addressing. Which dial-up protocol is being used?
    a. DLC
    b. SLIP
    c. SMB
    d. PPP

41. Which of the following is a characteristic of the 10Base-T topology?
    a. It uses RJ-11 connectors.
    b. It uses either UTP or STP cabling.
    c. It uses BNC T-connectors.
    d. It uses 50-ohm BNC terminators.

42. What type of connector is normally used by a thicknet cable for connection to the network adapter card?

a. BNC T-connector   c. RJ-45 connector
b. AUI connector   d. BNC barrel connector

43. There are two types of coaxial cable, thinnet and thicknet. Both types are good choices for data transmission over long distances. Which of the following is true of thinnet and thicknet?
    a. Thinnet and thicknet are specified in IEEE Standard 802.5.
    b. The maximum cable segment length is 100 meters for thinnet and 500 meters for thicknet.
    c. Thicknet is generally associated with the 10Base5 topology, whereas thinnet is associated with the 10Base2 topology.
    d. Because of their different sizes and construction, thicknet and thinnet should not be used in the same network.

44. You need to terminate a 10Base2 network. What should be the value for the terminating resistors?
    a. 25 ohms   d. 93 ohms
    b. 50 ohms   e. 100 ohms
    c. 75 ohms   f. none of the above

45. When using 10Base2 and 10Base 5 Ethernet, which of the following statements are true?
    a. You should terminate one end and ground the other end.
    b. You should terminate both ends and ground one end.
    c. You should terminate one end and ground both ends.
    d. You should terminate and ground both ends of the cable.

46. You are installing a 100-Mbps Ethernet LAN where there is existing category 2 UTP cabling. Which one of the following types of cable will satisfy the installation requirement?
    a. existing cable   c. category 3 UTP
    b. thicknet   d. category 5 UTP

47. What type of connector is used by thinnet for connection to the network adapter card?
    a. AUI connector   c. BNC T-connector
    b. RJ-45 connector   d. BNC barrel connector

48. Ethernet uses a special type of data format called frames. Which of the following statements is true of Ethernet frames?
    a. All Ethernet frames are 1518 bytes long.
    b. All Ethernet frames contain preambles that mark the start of the frame.
    c. All Ethernet frames contain starting and ending delimiters.
    d. All Ethernet frames contain CRC fields that store the source and destination addresses.

49. 10Base5 is one of the topologies defined in the IEEE 802.3 specifications. Which of the following is another name for 10Base5 cabling?
    a. category 5 UTP   c. thinnet
    b. shielded twisted pair   d. thicknet

50. Which type of media access method is used by IBM-based LANs with MAUs?
    a. token passing   c. CSMA/CD
    b. demand priority   d. CDMS/CA

51. Each network architecture has standard cable types associated with it. What is the standard cable type associated with ARCnet networks?
    a. twisted pair   c. RG-58 A/U coaxial
    b. RG-58/U coaxial   d. RG-62 A/U coaxial

52. Which of the following are dial-up communication protocols? (Select all that apply.)
    a. FTP   c. TCP
    b. PPP   d. SLIP

119

53. Which one of the following tools can be used to check for cable breaks?
   a. digital volt meter
   b. time-domain reflectometer
   c. advanced cable tester
   d. multiplexer

54. NDIS and ODI are specifications that define interfaces for communication between the data link layer and protocol drivers. Which need do NDIS and ODI fulfill?
   a. the need to dynamically bind a single protocol to multiple MAC drivers
   b. the need to bind multiple protocols to a single network adapter card
   c. the need for monolithic protocols to conform to the OSI model
   d. the need for monolithic protocols to be loaded into the upper memory area

55. Open data link interface (ODI) is a specification defined by Novell and Apple to simplify driver development. Which one of the following statements is true of ODI?
   a. ODI specifies cable and network architecture standards.
   b. ODI specifies network topology standards.
   c. ODI provides support for multiple protocols on a single network adapter card.
   d. ODI provides support for multiple network topologies in a single networking environment.

56. You connect the leads from a digital multimeter (DMM) to each of the conductors in a twisted pair. Which of the following settings should you use on the DMM to check for a short?
   a. resistance
   b. capacitance
   c. DC voltage
   d. AC voltage
   e. current

57. Pat is not able to access network resources from her computer. When she plugs her Ethernet cable into a coworker's machine, she is able to access the network without any problems. Which of the following network components is causing the problem?
   a. server
   b. router or gateway
   c. cable
   d. network adapter

58. You have been contracted to build a network for a multimedia development firm that currently uses a 10-Mbps Ethernet network. The company requires a high bandwidth network for the multimedia team, which constantly views and manipulates large files across the network. The company is expecting moderate growth. You are to develop a solution to support the high bandwidth applications and growth potential of this company.

   Required result:
   • Increase network bandwidth

   Optional results:
   • Support future growth
   • Improve server response time

   Proposed solutions:
   • Install category 5 UTP cable in a star topology with the existing hub.
   • Upgrade the workstations' adapter cards to support 100 Mbps.
   • Increase the amount of RAM in the server.

   a. Achieves the required result and both optional results.
   b. Does not achieve the required result but achieves both optional results.
   c. Does not achieve the required result but does achieve one of the optional results.
   d. Achieves the required result but cannot achieve either of the optional results.
   e. Achieves neither the required result nor either of the optional results.

59. Your company is a medium-size firm that leases two buildings. The two buildings are 800 meters apart. Each building has a 100Base-T Ethernet network.

   Required result:
   • Connect the two separate networks.

Optional results:
- The network needs to support a transmission rate of 1000 Mbps.
- The network must be immune to interference.

Proposed solution:
- Implement fiber-optic Ethernet.

a. Achieves the required result and both optional results.
b. Does not achieve the required result but achieves both optional results.
c. Does not achieve the required result but does achieve one of the optional results.
d. Achieves the required result but cannot achieve either of the optional results.
e. Achieves neither the required result nor either of the optional results.

60. Look at the following figure and determine the problem with this Ethernet network.

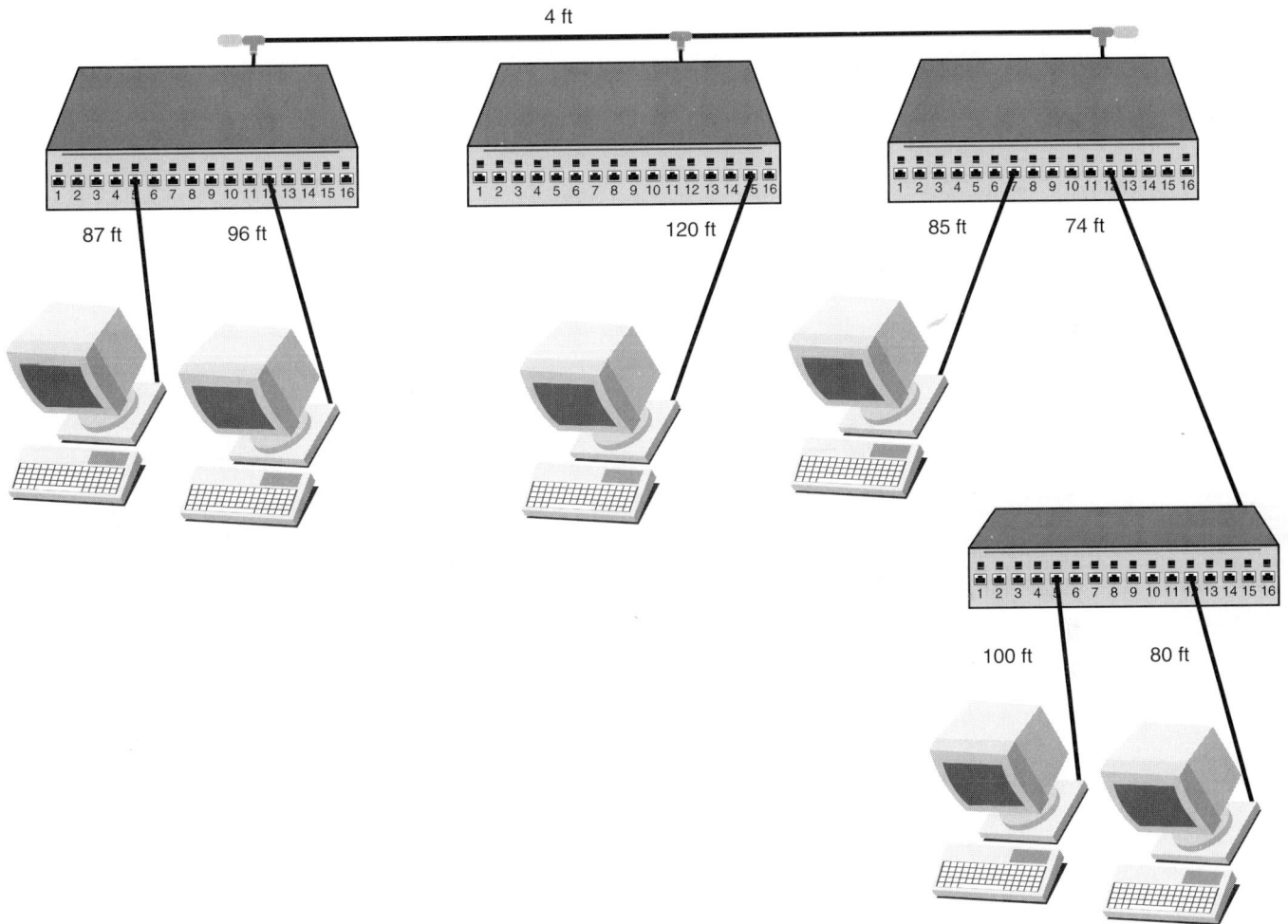

61. Look at the following figure and determine the problem with the network.

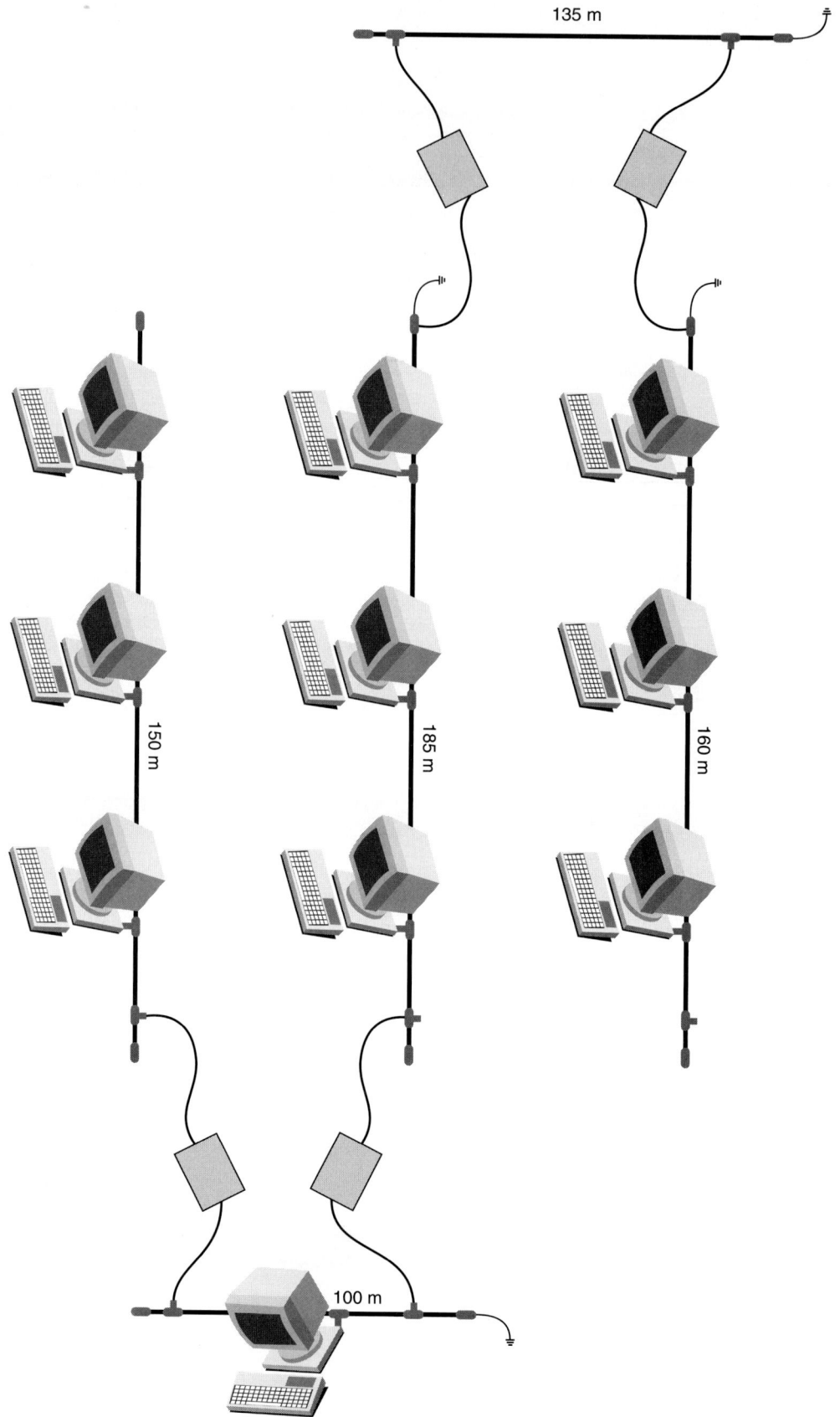

EXERCISE 1: CREATING A COAXIAL CABLE

1. Create a 6-foot 10Base2 coaxial cable with two BNC connectors.
2. With a voltmeter, test the BNC cable.
3. With a cable tester, test the BNC cable.

EXERCISE 2: CREATE AN UNSHIELDED TWISTED-PAIR CABLE

1. Create a standard 15-foot category 5 100Base-T with two RJ-45 connectors.
2. With a voltmeter, test the UTP cable.
3. With a cable tester, test the BNC cable.
4. Create a crossover 15-foot category 5 100Base-T with two RJ-45 connectors.
5. With a voltmeter, test the UTP cable.
6. With a cable tester, test the BNC cable.

> **Note:** To finish this lab you need an operating system that works on a network. Therefore, exercises dealing with building a network will continue with Chapter 6 Exercises.

5

Common Protocol Suites

INTRODUCTION

Up to now, we have discussed the physical part of the network. We have not talked about the software, in particular the protocols that make the network run. It is essential that any IT administrator understand these protocols to be able to configure and troubleshoot the network.

OBJECTIVES

1. Given a RFC number, find the RFC document on the Web.
2. List the protocols that can connect to a TCP/IP network.
3. List and describe each of the TCP/IP protocols.
4. Given an IP address, determine its class and default subnet mask.
5. Given an IP address and subnet mask, determine the network address and host address.
6. Given a network situation, determine a subnet mask that can be used for the network.
7. List and describe the ways to translate from the NetBIOS and host names to IP addresses.
8. Differentiate between ports and sockets.
9. Explain how classless interdomain routing differs from classful IP.
10. List the differences between IP next generation and classful IP.

11. Diagnose and troubleshoot a given IP problem.
12. Configure and troubleshoot the TCP/IP protocol.
13. Explain the difference between IPX and SPX.
14. Diagnose and troubleshoot a given IPX problem.
15. Explain how NetBEUI and NetBIOS relate to each other and how they can relate to TCP/IP and IPX protocols.
16. Describe DLC and AppleTalk protocols.

5.1 PROTOCOL AND PROTOCOL SUITE

A **protocol** is a set of rules or standards designed to enable computers to connect with one another, and peripheral devices to exchange information with as little error as possible. A **protocol stack,** sometimes referred to as a **protocol suite,** is a set of protocols that works together to provide a full range of network services.

5.2 TCP/IP AND THE INTERNET

The Internet that we know today began as a U.S. Department of Defense (DoD)–funded experiment to interconnect DoD-funded research sites in the United States. In December 1968, the Advanced Research Projects Agency (ARPA) awarded a grant to design and deploy a packet switching network (messages divided into packets, transmitted individually, and recompiled into their original forms). In September 1969, the first node of the ARPANET was installed at UCLA. By 1971, the ARPANET spanned the continental United States and had connections to Europe by 1973.

Over time, the initial protocols used to connect the hosts proved incapable of keeping up with the growing network traffic load. Therefore, a new TCP/IP protocol suite was proposed and implemented. By 1983, the popularity of the TCP/IP protocol grew as it was included in the communications kernel for the University of California's UNIX implementation, 4.2BSD (Berkeley Software Distribution) UNIX. Today, the TCP/IP protocol is the primary protocol used on the Internet and is supported by Novell NetWare, Microsoft Windows NT, Microsoft Windows 2000, UNIX, and other network operating systems.

The standards for TCP/IP are published in a series of documents called **Requests for Comments (RFC).** An RFC can be submitted by anyone. Eventually, if it gains enough interest, the submittal may evolve into an Internet standard. Each RFC is designated by an RFC number. Once published, an RFC never changes. Modifications to an original RFC are assigned a new RFC number.

RFCs are classified as one of the following: approved Internet standards, proposed Internet standards (circulated in draft form for review), Internet best practices, or For Your Information (FYI) documents. Administrators should always follow approved Internet standards.

5.2.1 IP, TCP, and UDP Protocols

With the TCP/IP protocol suite, TCP/IP does not worry about how the hosts (computers or any other network connection) connect to the network. Instead, TCP/IP was designed to operate over nearly any underlying local or wide area network. See Figure 5.1. This would include the following:

- LAN protocols: Ethernet, token ring, and ARCnet networks
- WAN protocols: ATM, frame relay, and X.25
- Serial line protocols: serial line internet protocol (SLIP) and point-to-point protocol (PPP)

When users send or receive data, the data are divided into little chunks called **packets.** Each of these packets contains both the sender's TCP/IP address and the receiver's TCP/IP

> **For more information on RFCs, visit the following websites:**
> http:/www.rfc-editor.org/
> http://www.cis.ohio-state.edu/hypertext/information/rfc.html

FIGURE 5.1 TCP/IP protocol suite

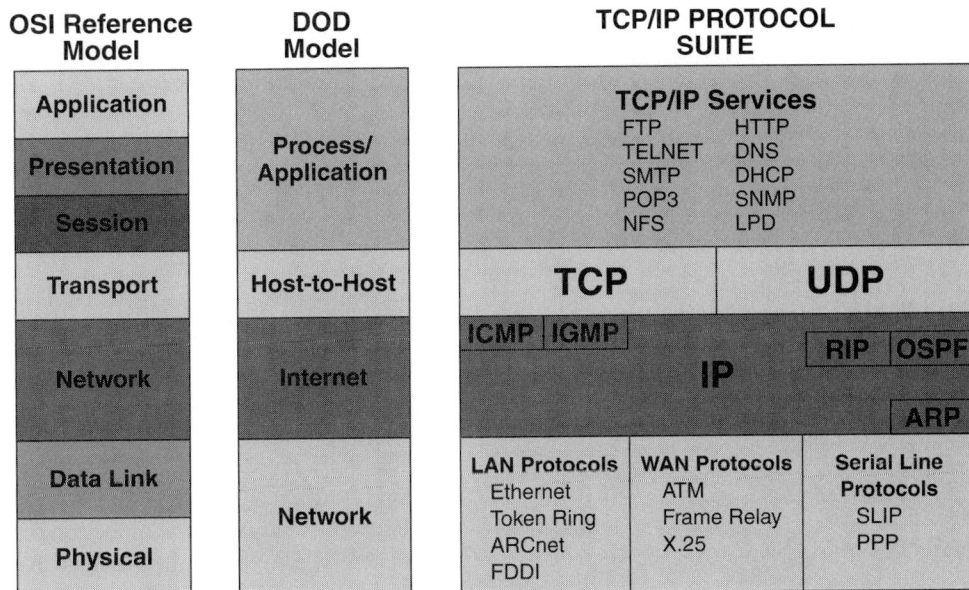

INTERNET PROTOCOLS

Routes data packets between different hosts or networks.

- **Internet Protocol (IP)**—Connectionless protocol primarily responsible for addressing and routing packets between hosts. (RFC 791)
- **Address Resolution Protocol (ARP)**—Used to obtain hardware addresses (MAC addresses) of hosts located on the same physical network. (RFC 826)
- **Internet Control Message Protocol (ICMP)**—Sends messages and reports errors regarding the delivery of a packet. (RFC 792)
- **Internet Group Management Protocol (IGMP)**—Used by IP hosts to report host group membership to local multicast routers. (RFC 1112)
- **Router Information Protocol (RIP)**—Distance vector route discovery protocol where the entire routing table is periodically sent to the other routers. (RFC 1723)
- **Open Shortest Path First (OSPF)**—Link State Route Discovery protocol where each router periodically advertises itself to other routers. (RFC 1245, 1246, 1247, and 1253)

HOST-TO-HOST PROTOCOLS

Maintains data integrity and sets up reliable, end-to-end communication between hosts.

- **Transmission Control Protocol (TCP)**—Provides connection/oriented, reliable communications for applications that typically transfer large amounts of data at one time or that require an acknowledgment for data received. (RFC 793)
- **User Datagram Protocol (UDP)**—Provides connectionless communications and does not guarantee that packets will be delivered. Applications that use UDP typically transfer small amounts of data at one. Reliable delivery is the responsibility of the application. (RFC 768)

PROCESS/APPLICATION PROTOCOLS

Acts as the interface for the user. Provides applications that transfer data between hosts.

- **File Transfer Protocol (FTP)**—Allows a user to transfer files between local and remote host computers. (RFC 959)
- **Telecommunication Network (TELNET)**—A virtual terminal protocol (terminal emulation) allowing a user to log on to another TCP/IP host to access network resources. (RFC 854)
- **Simple Mail Transfer Protocol (SMTP)**—The standard protocol for the exchange of electronic mail over the Internet. It is used between e-mail servers on the Internet or to allow an e-mail client to send mail to a server. (RFC 821 and 822)
- **Post Office Protocol (POP)**—Defines a simple interface between a user's mail client software and e-mail server. It is used to download mail from the server to the client and allows users to manage their mailboxes. (RFC 1460)
- **Network File System (NFS)**—Provides transparent remote access to shared files across networks. (RFC 1094)
- **Hypertext Transfer Protocol (HTTP)**—The basis for exchange over the World Wide Web (WWW). Web pages are written in the hypertext markup language (HTML), an ASCII-based, platform-independent formatting language. (RFC 1945 and 1866)
- **Domain Name System (DNS)**—Defines the structure of Internet names and their association with IP addresses. (RFC 1034 and 1035)
- **Dynamic Host Configuration Protocol (DHCP)**—Used to automatically assign TCP/IP addresses and other related information to clients. (RFC 2131)
- **Simple Network Management Protocol (SNMP)**—Defines procedures and management information databases for managing TCP/IP-based network devices. (RFC 1157 and 1441)
- **Line Printer Daemon (LPD)**—Provides printing on a TCP/IP network.

address. When the packet must go to another computer on another network, it is then sent to a gateway computer, usually a router. The gateway understands the networks to which it is connected directly. The gateway computer reads the destination address to determine in which direction the packet needs to be forwarded. It then forwards the packet to an adjacent gateway. The packet is then forwarded from gateway to gateway until it gets to the network to which the destination host belongs. The last gateway then forwards the packet directly to the computer whose address is specified. See Figure 5.2.

The lowest protocol within the TCP/IP suite is the **Internet protocol (IP).** It specifies the format of the packets (also called datagrams) and worries about their delivery. IP is a connectionless protocol, which means there is no established connection between the end points that are communicating. Each packet that travels through the Internet is treated as an independent unit of data. Therefore, they are not affected by other data packets. In addition, IP does not guarantee any deliveries, so packets can get lost, delivered out of sequence, or delayed.

The protocols that work on top of the IP are TCP and UDP. The **transport control protocol (TCP)** is a reliable, connection-oriented delivery service that breaks data into manageable packets. It establishes a virtual connection between the two hosts or computers so they can send messages back and forth for a period of time. A virtual connection appears to be always connected, but in reality it is made of many packets being sent back and forth independently.

The TCP has two other important functions. First, it uses acknowledgments to verify that the data were received by the other host. If an acknowledgment is not sent, the data are resent. Second, since the data packets can be delivered out of order, the TCP must put the packets back in the correct order.

FIGURE 5.2   TCP/IP packet

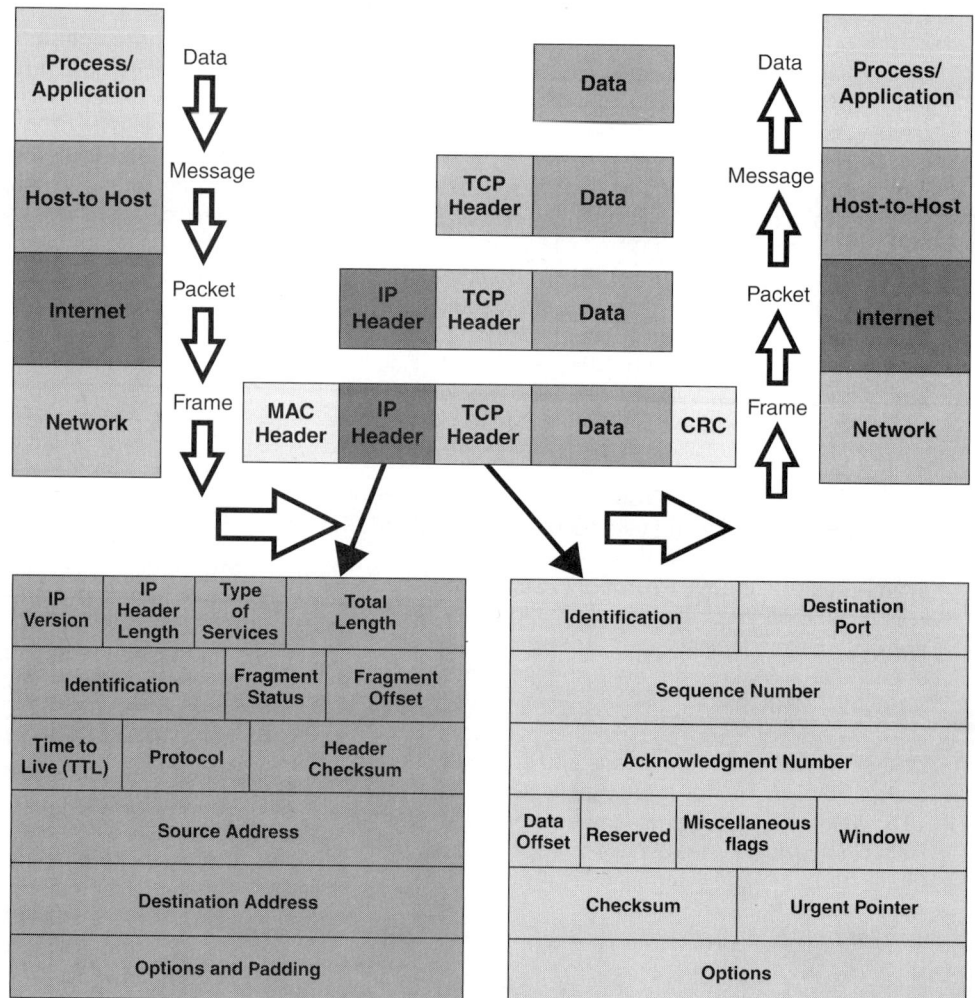

128

Another transport layer protocol is the **user datagram protocol (UDP).** Unlike the TCP, which uses acknowledgments to ensure data delivery, UDP does not. Therefore, the UDP is considered unreliable, "best effort" delivery. Since it is considered unreliable, UDP is used for protocols that transmit small amounts of data at one time or for broadcasts (packets sent to everyone). Being unreliable, however, does not mean that the packets will not be delivered, but rather there is no guarantee or check to ensure that they get to their destination.

5.2.2 TCP/IP Addressing

Each connection on a TCP/IP address is called a **host** (a computer or other network device that is connected to a TCP/IP network) and is assigned a unique **IP address.** A host is any network interface, including each network interface card or a network printer that connects directly onto the network. The format of the IP address is four 8-bit numbers divided by a period (.). Each number can be 0 to 255. For example, a TCP/IP address could be 131.107.3.1 or 2.0.0.1. Since the address is used to identify the computer, no two connections can use the same IP address or else one or both of the computers will not be able to communicate.

When connecting to the Internet, network numbers are assigned to a corporation or business. If the first number is between 1 and 126 (first bit is a 0), the network is a class A. If the first number is between 128 and 191 (first two bits are 1 0), the network is a class B. If the first number is between 192 and 223 (first three bits are 1 1 1), the network is a class C. See Table 5.1. Every computer on a TCP/IP network must have the same network number.

There are two additional classes that should be mentioned. These are used for special functions only and are not commonly assigned to individual hosts. Class D addresses may begin with a value between 224 and 239, and are used for IP multicasting. Multicasting is sending a single data packet to multiple hosts. Class E addresses begin with a value between 240 and 255, and are reserved for experimental use.

Since TCP/IP addresses are scarce for the Internet, a series of addresses have been reserved to be used by **private networks** (networks not connected to the Internet). They are:

Class A—10.x.x.x (1 class A addresses).
Class B—Between 172.16.x.x and 172.31.x.x (16 class B addresses).
Class C—192.168.0.x and 192.168.255.x (256 class C addresses).

If you are not connected to the Internet or are using a proxy server, it is recommended that you use private addresses to prevent a renumbering of your internet work when you eventually connect to the Internet.

To register for a TCP/IP network address, use the InterNIC Registration Service, which can be found at the following website:

http://www.internic.net/

TABLE 5.1: Standard IP classes for the IP address of w.x.y.z

Class Type	First Octet	Network Number	Host Number	Default Subnet Mask	Comments
A	1–126	w	x.y.z	255.0.0.0	Supports 16 million hosts on each of 126 networks.
B	128–191	w.x	y.z	255.255.0.0	Supports 65,000 hosts on each of 16,000 networks.
C	192–223	w.x.y	z	255.255.255.0	Supports 254 hosts on each of 2 million networks.

The TCP/IP address is broken down into a network number and a host number. The network number identifies the entire network while the host number identifies the computer or connection on the specified network. If it is a class A network, the first octet describes the network number while the last three octets describe the host address. If it is a class B network, the first two octets describe the network number while the last two octets describe the host address. If it is a class C, the first three octets describe the network number while the last octet describes the host number.

EXAMPLE 1   You have the following network address:

131.107.20.4

The 131 is between 128 and 191, identifying the address as a class B network. Therefore, the 131.107 identifies the network and the 20.4 identifies the host or computer on the 131.107 network.

EXAMPLE 2   You have the following network address:

208.234.23.4

The 208 is between 192 and 223, identifying the address as a class C network. Therefore, the 208.234.23 identifies the network and the 4 identifies the host or computer on the 208.234.23 network.

Several address values are reserved and/or have special meaning. The network number 127 is used for loopback testing, and the specific host address 127.0.0.1 refers to the **localhost,** or the actual host or computer currently in use.

Usually when defining the TCP/IP for a network connection, IT managers also would specify a subnet mask. The **subnet mask** is used to define which bits describe the network number and which bits describe the host address. Like the network address, every computer on a network must have the same subnet mask. The default subnet mask for a class A network is 255.0.0.0. Converting this to a binary equivalent gives 11111111.00000000.00000000.00000000, showing that the first 8 bits (first octet), marked with ones, is used to define the network address, while the last 24 bits, marked with zeros, are used to define the host address. The default subnet mask for a class B network is 255.255.0.0 (11111111.11111111.00000000.00000000) while the default subnet mask for a class C network is 255.255.255.0 (11111111.11111111.11111111.00000000). To convert decimal to binary numbers, see Appendix A.

If an individual network is connected to another network and users must communicate with any computers on the other network, they must also define the default gateway, which specifies the local address of a router. If the default gateway is not specified, users will not be able to communicate with computers on other networks. If the LAN is connected to more than two networks, users must specify only one gateway, because when a data packet is sent, the gateway will first determine if the data packet needs to go to a local computer or onto another network. If the data packet is meant to be sent to a computer on another network, the gateway will forward the data packet to the router. The router will then determine the best direction that the data packet must go to reach its destination. Occasionally, it will have to go back through the network to get to another gateway. See Figure 5.3.

5.2.3   Subnetting the Network

The subnet mask can be changed to take a large network and break it into small networks called **subnets.** This explains why the subnet mask must be defined, so that the network can calculate which bits represent the network address and which bits represent the host address. All subnets on a network must use the same network address. For a subnet, the subnet network number and the subnet mask must be the same for all computers.

EXAMPLE 3   Figure 5.4 shows two networks connected by a router. Every node (computer and router link) on LAN A has a network address of 141.171.50 with the subnet mask of 255.255.255.0. In addition, every node on LAN B has a network address of 132.11 with the subnet mask of 255.255.0.0. The router is actually a member of both networks. While every computer is not required to have the same gateway (router), since there is only one router

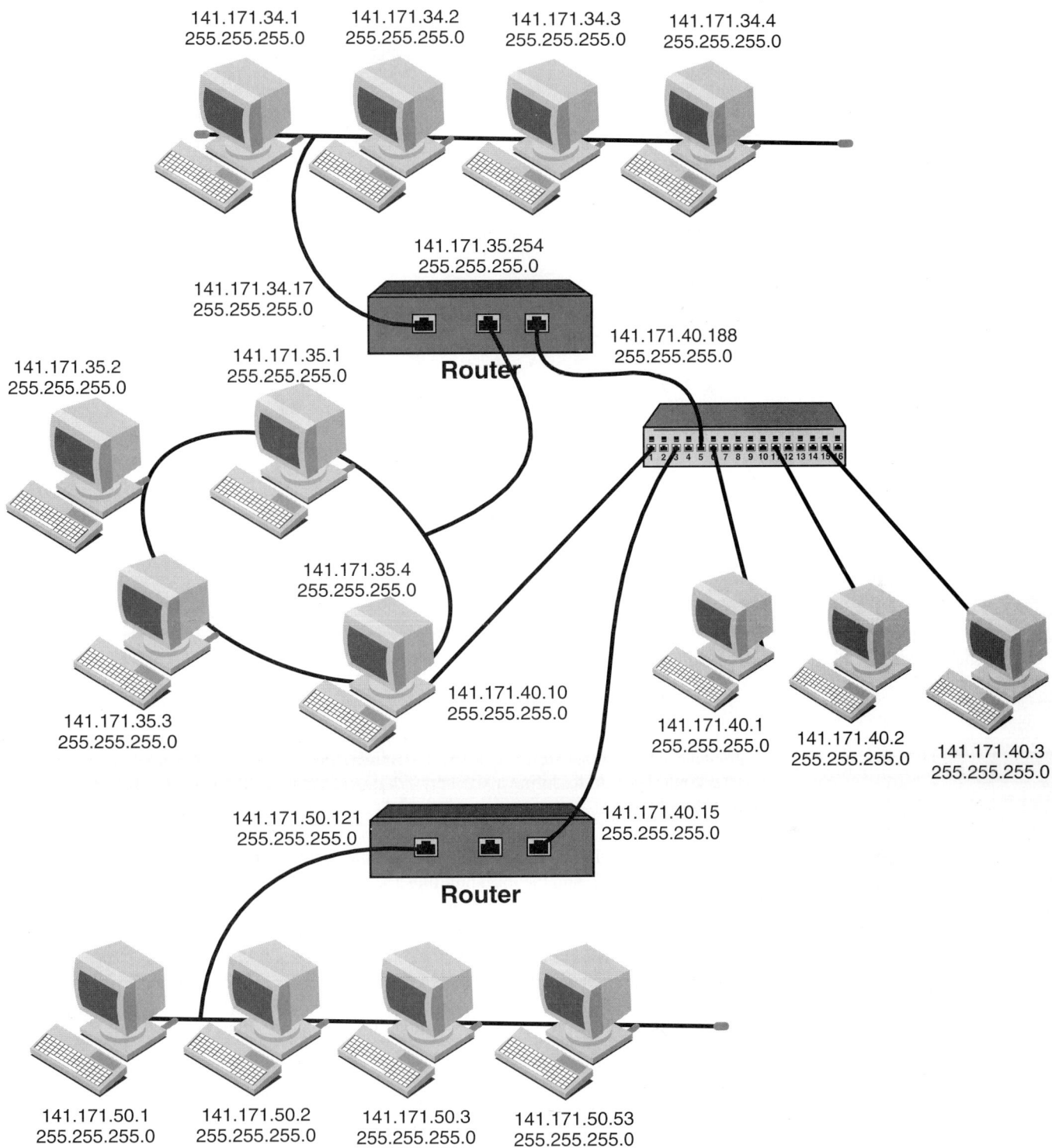

FIGURE 5.3   IP network with addresses and subnet masks. Notice the multihomed computer (computer with two network cards connected to two subnets).

connection on each network, the router link that is connected to the network is the gateway. If there were two routers connected to the network, then specify either one. Lastly, notice that the router does not use a default gateway because it *is* the gateway.

EXAMPLE 4   Your network is assigned a network number of 161.13. Since it is a class B network, it already has a default subnet mask of 255.255.0.0. Therefore, any network interface card that belongs to this network has a TCP/IP address beginning with 161.13. Since it is a class B network, it uses 16 bits to define the host address, which allows the network up to 65,534 computers. It's hard to imagine a single LAN with that many computers.

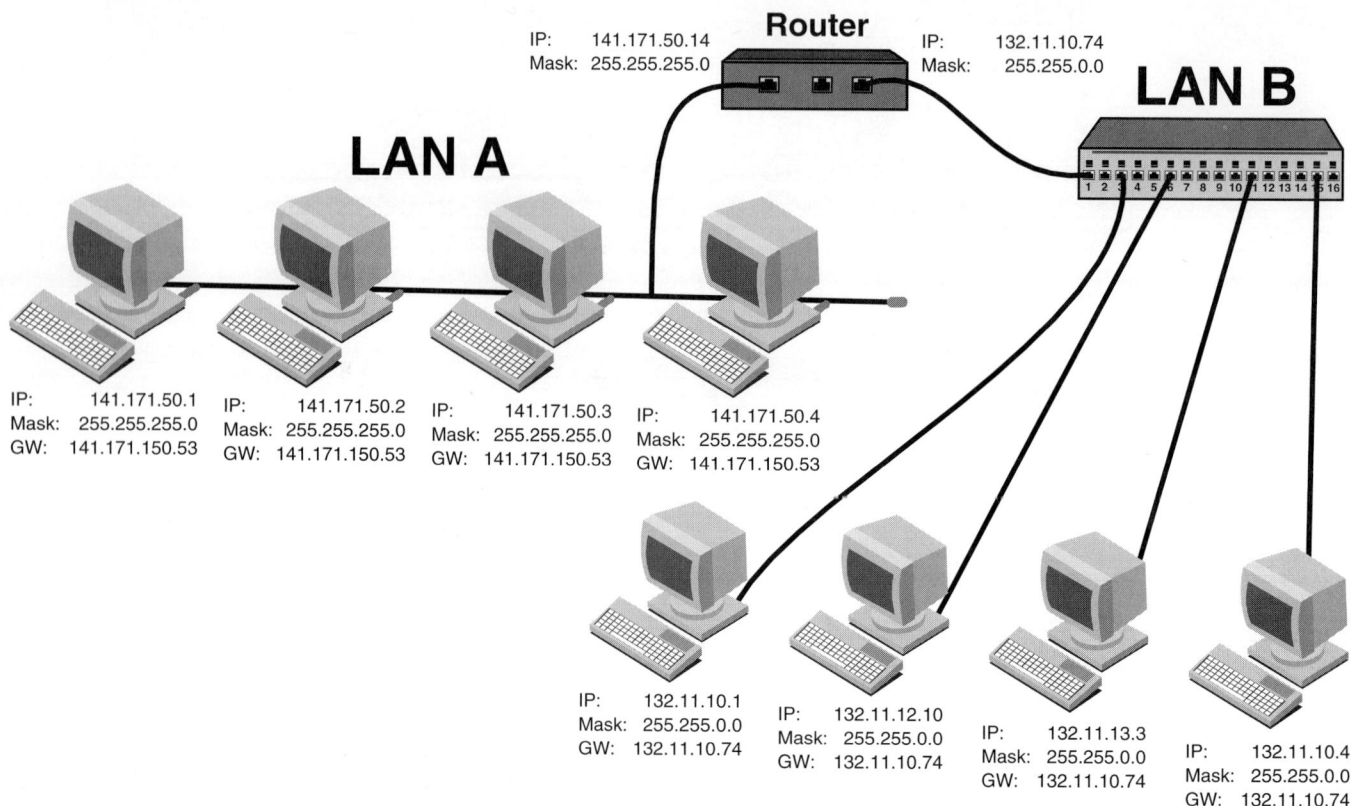

FIGURE 5.4   Two TCP/IP networks connected to a router

The network administrator could take this large network and divide it into several smaller subnets. For example, if 161.13 defines the entire corporation network, the third octet could be used to define a site network or individual LAN while the last octet defines the host address.

Therefore, if you have three individual LANs, use the third octet to define them. For example, the first building would have a network address of 161.13.1, the second building an address of 161.13.2, and the third building an address of 161.13.3. Since the last octet is used to define the host number and there are 8 bits to do so, there can be 254 hosts for each LAN. To let the network know that the 161.13 network is subnetted, the mask would have to be changed from 255.255.0.0 to 255.255.255.0 (11111111.11111111.11111111.00000000) to indicate that the first 24 bits are the network address.

To calculate the maximum number of subnets, use the following equation:

$$\text{Number of subnets} = 2^{\text{number of masked bits}} - 2$$

To calculate the maximum number of hosts, use the following equation:

$$\text{Number of hosts} = 2^{\text{number of unmasked bits}} - 2$$

The −2 is necessary because you cannot use a network number, subnet number, or host number of all zeros and all ones. For example, if you had an address of 131.107.3.4, then 131.107 is the network address and 3.4 is the host address. If you send a packet to the 131.107.0.0 (host address is all zeros), then you are sending the packet to the network itself, not to the individual computer on the network. If you send it to 131.107.255.255 (host address is all ones), then you are doing a broadcast to all computers on network 131.107. If you use the address of 0.0.3.4 (network number is all zeros), then it assumes 3.4 is on the current or local network.

EXAMPLE 5   In Example 3, you have 8 bits that are used to define the subnet/site number and 8 bits to define the host number. Therefore, the following equations apply:

$$\text{Number of subnets} = 2^{\text{number of masked bits}} - 2 = 2^8 - 2 = 254 \text{ subnets or sites}$$
$$\text{Number of hosts} = 2^{\text{number of unmasked bits}} - 2 = 2^8 - 2 = 254 \text{ hosts for each subnet}$$

EXAMPLE 6   Your network is assigned an address of 207.182.15. You choose to subnet your network into several smaller networks. Your largest network will have 25 computers. Therefore, how many bits can you mask to provide the largest number of subnets or sites?

First, you have only the 8 bits that were used for the host address. The easiest method is to use the formula that calculates number of hosts. To determine how many unmasked bits allow 25 or more computers, use the following calculations:

$$\begin{aligned}
\text{Number of hosts} &= 2^{\text{number of unmasked bits}} - 2 = 2^1 - 2 = 0 \\
&= 2^{\text{number of unmasked bits}} - 2 = 2^2 - 2 = 2 \\
&= 2^{\text{number of unmasked bits}} - 2 = 2^3 - 2 = 6 \\
&= 2^{\text{number of unmasked bits}} - 2 = 2^4 - 2 = 14 \\
&= 2^{\text{number of unmasked bits}} - 2 = 2^5 - 2 = 30
\end{aligned}$$

Since we used 5 of the 8 bits for the host number, we have 3 bits left that are masked, which gives us the following:

$$\text{Number of subnets} = 2^{\text{number of masked bits}} - 2 = 2^3 - 2 = 6$$

Lastly, the subnet mask would be 11111111.11111111.11111111. 11100000, which is equivalent to 255.255.255.224.

EXAMPLE 7   Using the network discussed in Example 5, what is the range of TCP/IP addresses for the second subnet?

The network number is 207.182.15 with a subnet mask of 255.255.255.224, which describes the first 3 bits of the forth octet used for the subnet number. The possible subnet numbers (in binary) are as follows:

```
0 0 0
0 0 1
0 1 0
0 1 1
1 0 0
1 0 1
1 1 0
1 1 1
```

You cannot use 0 0 0 or 1 1 1, which leaves six possible subnets.

Using binary counting, the second subnet is defined by 0 1 0. Since there are 5 bits left, they can range from 0 0 0 0 1 (remember, not be all zeros) to 1 1 1 1 0 (remember, not all ones). Therefore, the last octet will be 0 1 0 0 0 0 0 1 to 0 1 0 1 1 1 1 0. If you translate these to decimals, the last octet will be 65 to 94. Therefore, the address range is 207.182.15.65 to 207.182.15.94 with a subnet mask of 255.255.255.224.

5.2.4   ARP and Address Resolution

Early IP implementations ran on hosts commonly interconnected by Ethernet LANs. Every transmission on these LANs contained the MAC address of the source and destination nodes. Since there is no structure to identify different networks, routing could not be performed.

When a host needs to send a data packet to another host on the same network, the sender application must know both the IP and MAC addresses of the intended receiver, because the destination IP address is placed in the IP packet and the destination MAC address in the LAN's protocol frame (such as Ethernet or token ring). If the destination host is on another network, the sender will look instead for the MAC address of the default gateway or router. See Figure 5.5.

Unfortunately, the sender's IP process may not know the MAC address of the intended receiver on the same network. Therefore, the **address resolution protocol (ARP)** (RFC 826) provides a mechanism so that a host can learn a receiver's MAC address when knowing only the IP address.

NETWORK A

142.164.232.23
22-33-A3-45-23-32

142.164.232.56
22-33-A3-47-23-42

142.164.232.46
22-33-A3-34-43-43

Router

142.164.200.1
C3-5F-29-87-A4-8C

142.164.232.01
C3-5F-29-43-43-B2

142.164.205.1
C3-5F-29-11-8D-CC

142.164.200.34
22-33-A3-64-87-34

NETWORK B

NETWORK C

142.164.205.2
22-33-A3-AA-BD-54

142.164.200.37
22-33-A3-5B-AC-76

142.164.200.38
22-33-A3-34-63-12

142.164.205.5
22-33-A3-34-43-43

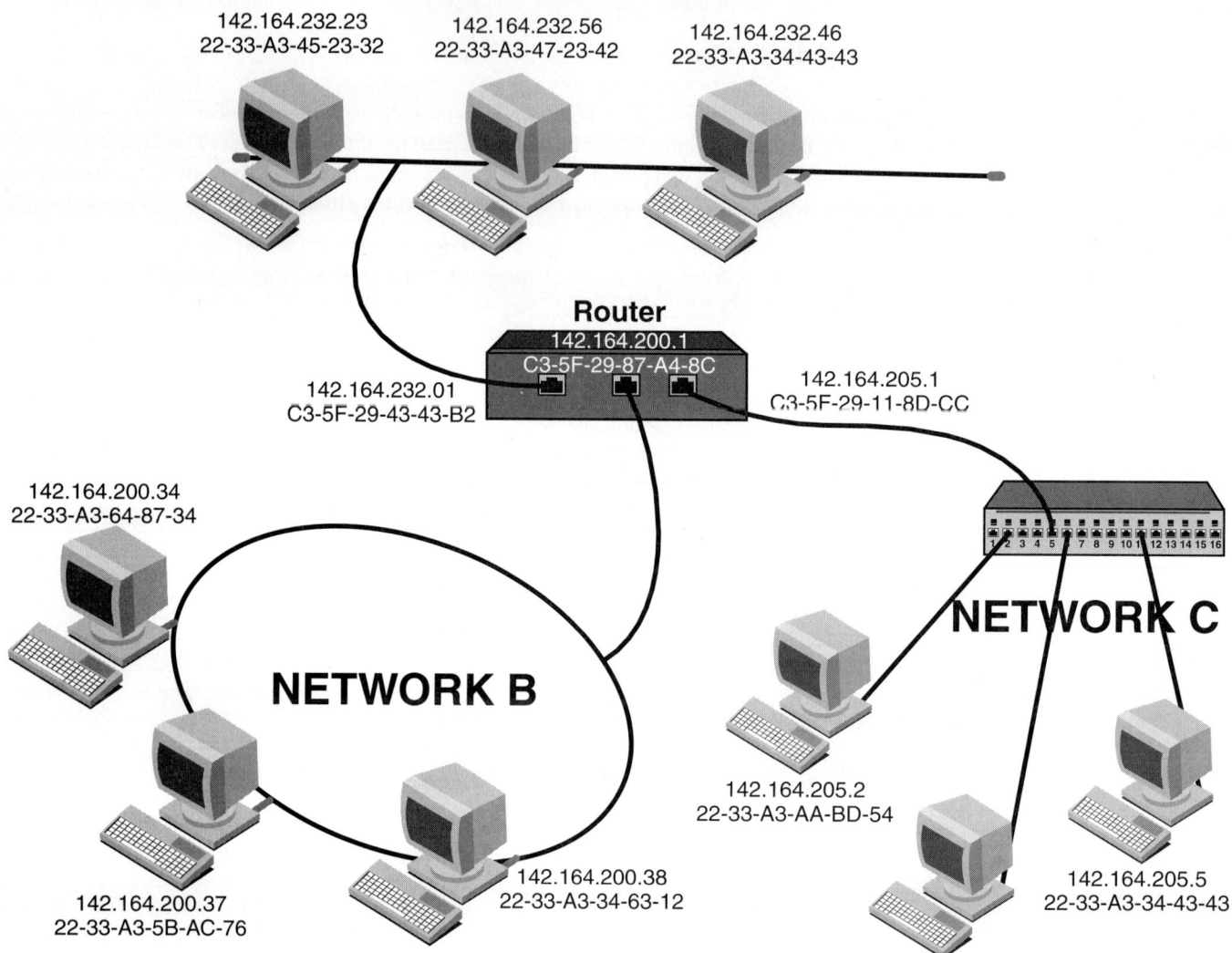

FIGURE 5.5   TCP/IP address and MAC addresses

Any time a computer needs to communicate with a local computer, it will first look in the ARP cache in memory to see if it already knows the MAC address of a computer with the specified IP address. If the address is not in the ARP cache, it will try to discover the MAC address by broadcasting an ARP request packet. The station on the LAN recognizes its own IP address, which then sends an ARP response with its own MAC address. Then both the sender of the ARP reply and the original ARP requester record each other's IP address and MAC address as an entry in ARP cache for future reference.

If a computer needs to communicate with another computer that is located on another network, it will do the same except it will send the packet to the local router. Therefore, it will search for the MAC address of the local port of the router or it will send a broadcast looking for it.

5.2.5   Navigating a TCP/IP Network

Fully qualified domain names (FQDN), sometimes referred to as simply *domain names,* are used to identify computers on a TCP/IP network. Examples include Microsoft.com and Education.Novell.com.

While IP addresses are 32 bits in length, most users do not memorize the numeric addresses of the hosts to which they attach. Instead, people are more comfortable with host

FIGURE 5.6 Sample HOSTS file

```
102.54.94.97          rhino.acme.com          # source server
38.25.63.10           x.acme.com              # x client host
127.0.0.1             localhost
```

names. Most IP hosts, then, have both a numeric IP address and a host name. Although this is convenient for people, the name must be translated back to a numeric address for routing purposes, by using either a HOSTS file or a DNS server.

The **HOSTS file** is a text file that lists the IP address followed by the host name. Each entry should be kept on an individual line. In addition, the IP address should be placed in the first column followed by the corresponding host name. A # symbol is used as a comment or REM statement, which means that anything after the # symbol is ignored. See Figure 5.6.

When a computer is using the host table shown in Figure 5.6, if rhino.acme.com is entered into a browser such as Internet Explorer or Netscape Navigator, it will find the equivalent address of 102.54.94.97 to connect to it. The HOSTS file is kept in the ETC directory on most UNIX machines, in the WINDOWS directory in Windows 95/98 machines, and in *%systemroot%*\SYSTEM32\DRIVERS\ETC directory in Windows 2000 machines.

Another way to translate the FQDN to the IP address is to use a domain name system (DNS) server. DNS is a distributed database (contained in multiple servers) containing host name and IP address information for all domains on the Internet. For every domain there is a single authoritative name server that contains all DNS-related information about the domain.

For example, if you type in the web address "Microsoft.com" in your browser, your computer will then communicate with your local area network's DNS server. If the DNS server does not know the address of Microsoft.com, another DNS server will be asked. This search will continue until it finds the address of Microsoft.com or it determines that the host name is not listed and will then reply back with "No DNS Entry."

5.2.6   TCP/IP Ports and Sockets

Every time a TCP/IP host communicates with another TCP/IP host, it will use the IP address and port number to identify the host and service/program running on the host. A **TCP/IP port number** is a logical connection place by client programs to specify a particular server program running on a network computer defined at the transport layer. Port numbers are from 0 to 65536. Ports 0 to 1024 are reserved for use by certain privileged services and popular higher-level applications. These are known as "well-known ports," which have been assigned by the Internet Assigned Numbers Authority. Some of the well-known protocols are shown in Table 5.2. For example, when using your browser to view a Web page, the default port to indicate the HTTP service is identified as port 80 and the FTP service is identified as port 21. Other application processes are given port numbers dynamically for

TABLE 5.2   Popular TCP/IP services and their default assigned port numbers

Network Program/Service	Default Assigned Port Number
FTP	21
Telnet	23
Simple Mail Transfer Protocol	25
Domain Name Server (DNS)	53
HTTP	80
Network News Transport Protocol (NNTP)	119
NetBIOS Session Service	139

each connection so that a single computer can run several services. When a packet is delivered and processed, the TCP (connection-based services) or UDP (connectionless-based services) will read the port number and forward the request to the appropriate program.

An application can create a socket and use it to send connectionless traffic to remote applications. An application creates a socket by specifying three items: the IP address of the host, the type of service, and the port the application is using. For Windows, the **Windows socket (WinSock)** provides the interface between the network program or service and the Windows environment.

5.2.7 Supernet Addressing and Classless Interdomain Routing

Because the class B network IDs would soon be depleted and, for most organizations, a class C network ID does not contain enough host IDs to provide a flexible subnetting scheme within the organization, supernetting was developed to gradually replace the original system of IP classes by using the available addresses more efficiently. In supernetting, when the address is assigned, neither the network number nor host number are made along the standard octet boundaries, but instead are made from a specific number of bits from the beginning of the address. A **supernet** is a group of networks that are identified by contiguous network addresses. IP service providers can assign to customers blocks of contiguous addresses in order to define supernets as needed. Each supernet has a unique address that consists of the upper bits shared by all addresses in the contiguous block. For example, a class C network address is 24 bits. Using classless interdomain routing (CIDR), as discussed later in this section, the network address of 26 bits can be assigned. This of course provides more networks but allows fewer host addresses. The number of bits in the network number is denoted by a slash. Therefore, supernetting using 26 bits could be 198.23.27.32 /26.

Consider the following block of contiguous 32-bit addresses (192.32.0.0 through 192.32.7.0 in decimal notation). The supernet address (network address) for this block is 192.32.0.0 (11000000 00100000 00000000 00000000), the 21 upper bits shared by the 32-bit addresses. The mask for the supernet address in this example is 255.255.248.0 (11111111 11111111 11111000 00000000).

```
192.32.0.0
192.32.1.0
192.32.2.0
192.32.3.0    Mask 255.255.255.0    ⟹    192.32.0.0    255.255.248.0
192.32.4.0
192.32.5.0
192.32.6.0
192.32.7.0
```

To prevent the Internet routers from becoming overwhelmed with routes from the individual class C network IDs that are allocated to an organization, **classless interdomain routing (CIDR)** was developed. CIDR collapses multiple network ID entries into a single entry. CIDR is now used by virtually all gateway hosts on the Internet's backbone network and is tied to the new, upcoming version of IPv6. For more information, see RFC 1517, 1518, 1519, and 1520.

5.2.8 IP Next Generation

Since the TCP/IP and the Internet became popular, the growth of the Internet has and still continues to grow at an exponential rate. At this growth rate, it is easy to see how the Internet will eventually run out of network numbers. To assist with this problem, a new IP, currently called **IP next generation (IPng),** formerly called IPv6, has been created. The current version is IPv4.

The changes to the IPv6 include the following:

- An increase of IP address size from 32 bits to 128 bits. To shorten the time it takes to enter the address, it will be reconfigured in hexadecimal instead of decimal format.
- The introduction of anycast address. An anycast address is assigned to more than one interface. When a packet is sent to an anycast address, it is routed to the nearest interface having that address.
- An IPng multicast address, used to identify a group of interfaces.
- A simplified header format that reduces the processing requirements.
- The ability to indicate real-time traffic, used in multimedia presentations so that routers can handle it differently (quality of service).
- The addition of extensions to provide support for authentication, data integrity, and confidentiality. This should eliminate a significant class of network attacks, including host masquerading attacks.

The process of upgrading from IPv4 to IPng will be done by completing a software upgrade. Because the Internet is so large, a controlled rollout will be difficult to achieve, so the new IPng will be made backward compatible with IPv4.

5.2.9   Troubleshooting a TCP/IP Network

Several utilities can be used to test and troubleshoot the TCP/IP network. To verify the TCP/IP configuration, use either the IP configuration program (WINIPCFG.EXE) command (available in Windows 95 and 98) or the IPCONFIG.EXE command (available in Windows 98, Windows NT, and Windows 2000). In addition, the ARP and TRACERT commands may be used. The utilities are as follows:

1. WINIPCFG or IPCONFIG
2. PING 127.0.0.1 (loopback address)
3. PING IP Address of the Computer
4. PING IP Address of Default Gateway (Router)
5. PING IP Address of Remote Host

Both IPCONFIG.EXE and WINIPCFG.EXE will display addressing and settings of the TCP/IP. The PING utility can be used to verify a TCP/IP connection, while the TRACERT will show all router addresses as it connects to another computer. The ARP protocol can be used to display the MAC address. See Table 5.3 for TCP/IP commands that are available in Windows 2000 for troubleshooting TCP/IP networks.

To execute the Windows IP configuration utility, type WINIPCFG.EXE from the Windows Command prompt or from the Run option located on the Start button. See Figure 5.7. When there is more than one network interface, select the appropriate adapter from the pulldown menu and view the IP address, the subnet mask, and the default gateway/router for that interface. By clicking on the More Info>> button, you can view other information such as addresses and configurations of the WINS, DNS, and DHCP servers. See Figure 5.8.

The IPCONFIG.EXE command can only be executed at the Command prompt. See Figure 5.9. Much like the IP configuration utility, it also shows the IP address, the subnet mask, and the default gateway/router address. To see additional information, type in IPCONFIG/ALL.

To make the address assignments easier, users can automatically assign the IP addresses to the computers using a DHCP server. To release the address assigned by the DHCP server, click on the IP Configurations Renew or Renew All button or type in IPCONFIG/RELEASE. To request new addresses from the DHCP server, click on the Renew or Renew All button. The Release All and Renew All buttons have all settings, including the WINS and DNS server addresses.

When using the IP configuration utility or the IPCONFIG.EXE command and are having trouble connecting to the network, you should verify the IP address (to make sure it is

TABLE 5.3 Windows 2000 troubleshooting TCP/IP commands

Commands	Description
ARP	Displays and modifies the IP-to-MAC address translation tables used by the Address Resolution Protocol (ARP). See Chapter 9.
IPCONFIG	Displays all current TCP/IP configuration, including the IP address. This command is of particular use on systems that run DHCP, allowing users to determine which TCP/IP configuration values have been configured by DHCP. See this chapter.
PATHPING	The PATHPING command is a route-tracing tool that combines features of the PING and TRACERT commands with additional information that neither of those tools provides. See Chapter 9.
PING	Verifies connections to a remote computer or computers by verifying configurations and testing IP connectivity. See this chapter.
TRACERT	Traces the route that a packet takes to a destination and displays the series of IP routers that are used in delivering packets to the destination. If the packets are unable to be delivered to the destination, the TRACERT command displays the last router that successfully forwarded the packet. See this chapter.
NBTSTAT	Displays the local NetBIOS name table, which is a table of user-friendly computer names mapped to IP addresses. See Chapter 10.
NETDIAG	Command line, diagnostic tool that helps isolate networking and connectivity problems by performing a series of tests to determine the state of the network client used in TCP/IP and IPX networks. See Chapter 9.
NETSTAT	Displays protocol statistics and current TCP/IP network connections. See this chapter.

FIGURE 5.7 WINIPCFG utility. If there is more than one network adapter, the various cards can be selected from the pulldown box.

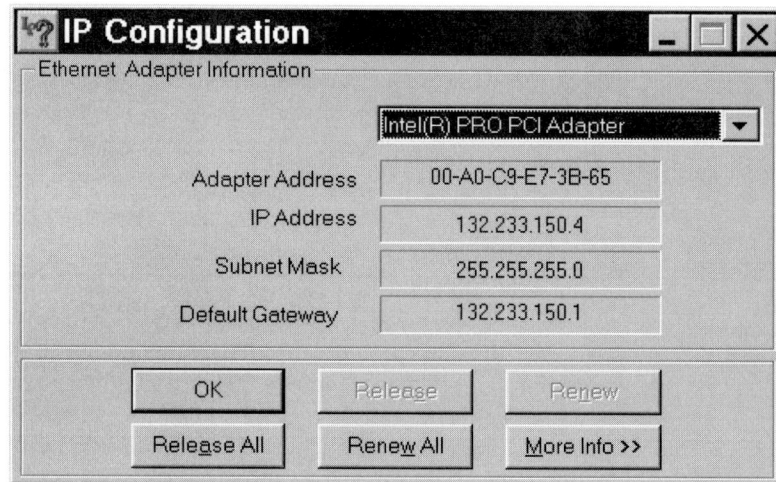

on the correct subnet), the subnet mask, and the router. If the address is not in the correct subnet, you would not be able to communicate with any other computer on the network. If the subnet mask is wrong, you may not be able to communicate with other computers. If the gateway address is wrong, you can communicate with computers on your subnet, but not on other subnets.

Another useful command to verify a network connection is PING.EXE. It sends three packets to a host computer and receives a report on the round trip. For example, if you ping the loopback address of 127.0.0.1 and the ping fails, verify that the TCP/IP software is installed correctly. If you ping your own IP address and it is unsuccessful, check the IP address. If you cannot ping the IP address of the default gateway/router, verify the IP address (make sure the address is in the correct subnet) and subnet mask. If you cannot ping a

**FIGURE 5.8 WINIPCFG utility
in detailed mode**

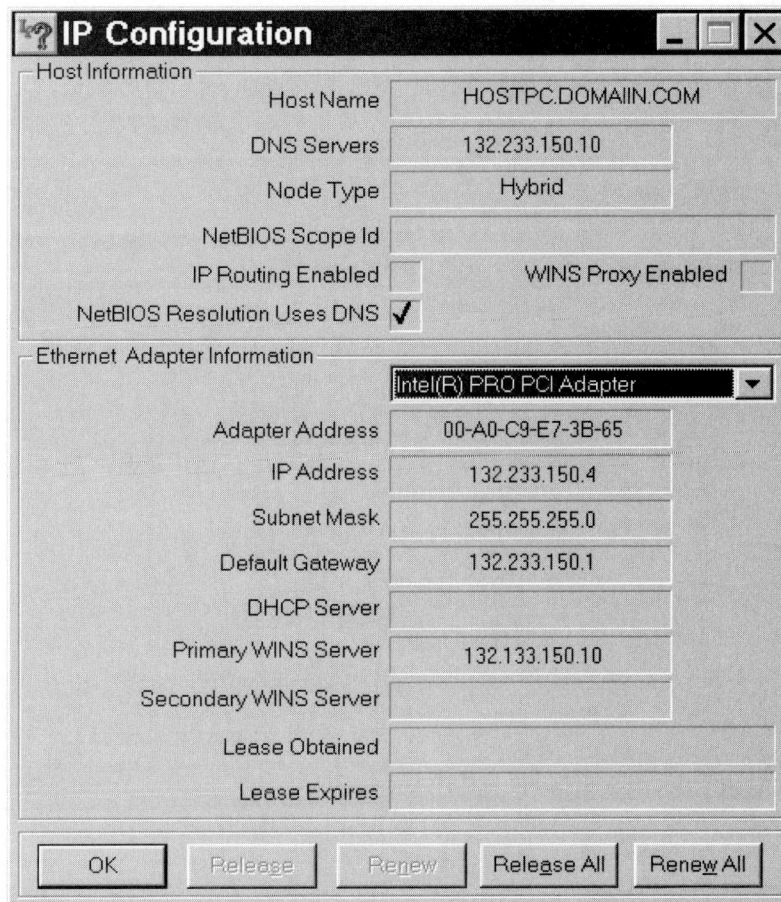

```
IP Configuration                                    _ □ X
┌─ Host Information ──────────────────────────────────┐
│           Host Name    HOSTPC.DOMAIIN.COM           │
│          DNS Servers   132.233.150.10               │
│           Node Type    Hybrid                       │
│      NetBIOS Scope Id                               │
│   IP Routing Enabled □        WINS Proxy Enabled □  │
│  NetBIOS Resolution Uses DNS  ✓                     │
├─ Ethernet Adapter Information ──────────────────────┤
│                        Intel(R) PRO PCI Adapter   ▼ │
│     Adapter Address    00-A0-C9-E7-3B-65            │
│          IP Address    132.233.150.4                │
│         Subnet Mask    255.255.255.0                │
│      Default Gateway   132.233.150.1                │
│          DHCP Server                                │
│   Primary WINS Server  132.133.150.10               │
│ Secondary WINS Server                               │
│       Lease Obtained                                │
│        Lease Expires                                │
│   ┌────┐ ┌───────┐ ┌──────┐ ┌──────────┐ ┌────────┐ │
│   │ OK │ │Release│ │Renew │ │Release All│ │Renew All│ │
│   └────┘ └───────┘ └──────┘ └──────────┘ └────────┘ │
└─────────────────────────────────────────────────────┘
```

remote computer on a different subnet or network, verify the IP address of the default gateway/router, make sure the remote host/computer is functional, and verify the link between routers. See Figure 5.10.

EXAMPLES: PING   127.0.01
                PING   137.23.34.112

The PING command can also be used to ping a host/computer by the NetBIOS (computer) name or host/DNS name. For example, if you ping by address but not by name, the computer tells you that the TCP/IP is running fine but the name resolution is not working properly. Therefore, you must first check the LMHOSTS file and the WINS server to resolve computer names and then the HOSTS file and the DNS server to resolve domain names.

EXAMPLES: PING   FS1
                PING   WWW.MICROSOFT.COM

A time of 200 milliseconds is considered very good. A time between 200 and 500 milliseconds is considered marginal and a time is over 500 milliseconds is unacceptable. A request timed out indicates total failure, as shown in Figure 5.11.

Another useful command is TRACERT.EXE, which sends out a packet of information to each hop (gateway/router) individually a total of three times. Therefore, the TRACERT command can help determine where the break is in a network. See Figure 5.12 and Table 5.4.

Viewing the current configuration (IPCONFIG), pinging the loopback address (PING 127.0.0.1), and pinging the IP address of your computer will verify that the TCP/IP protocol is properly functioning on the PC. By pinging the IP address of the default gateway or router, as well as other local IP computers, you determine if the computer is communicating on the local network. If it cannot connect to the gateway or any other local computer, either you are not connected properly or the IP protocol is misconfigured (IP address, IP

FIGURE 5.9 IPCONFIG command

```
C:\.ipconfig

Windows 98 IP Configuration

0 Ethernet adapter :

        IP Address. . . . . . . . . : 132.233.150.4
        Subnet Mask . . . . . . . . : 255.255.255.0
        Default Gateway . . . . . . : 132.233.150.1
```

```
C:\.IPCONFIG/ALL

Windows 98 IP Configuration
    Host Name . . . . . . . . . : HOSTPC.DOMAIIN.COM
    DNS Servers . . . . . . . . : 132.233.150.10
    Node Type . . . . . . . . . . Hybrid
    NetBIOS Scope ID. . . . . . :
    IP Routing Enabled. . . . . : No
    WINS Proxy Enabled. . . . . : No
    NetBIOS Resolution Uses DNS : Yes

0 Ethernet adapter :
    Description . . . . . . . . : Intel(R) PRO PCI Adapter
    Physical Address. . . . . . : 00-A0-C9-E7-3B-65
    DHCP Enabled. . . . . . . . : No
    IP Address. . . . . . . . . : 132.233.150.4
    Subnet Mask . . . . . . . . : 255.255.255.0
    Default Gateway . . . . . . : 132.233.150.1
    Primary WINS Server . . . . : 132.133.150.10
    Secondary WINS Server . . . :
    Lease Obtained. . . . . . . :
    Lease Expires . . . . . . . :
```

FIGURE 5.10 PING command

```
C:\.ping 132.233.150.4

Pinging 132.233.150.4 with 32 bytes of data:

Reply from 132.233.150.4: bytes=32 time,<0ms TTL=128
Reply from 132.233.150.4: bytes=32 time,<0ms TTL=128
Reply from 132.233.150.4: bytes=32 time,<0ms TTL=128
Reply from 132.233.150.4: bytes=32 time,<0ms TTL=128

Ping statistics for 132.233.150.4
    Packets: Sent = 4, Received = 4, Lost = 0 (0% loss),
Approximate round trip times in milliseconds:
    Minimum = 0ms, Maximum = 0ms, Average = 0ms
```

subnet mask, or gateway address). If you cannot connect to the gateway but you can connect to other local computers, then check your IP address, IP subnet mask, and gateway address. Also check if the gateway is functioning by using the PING command at the gateway to connect to your computer and other local computers on your network as well as pinging the other network connections on the gateway/router or pinging computers on other networks. If you cannot ping another local computer, but you can ping the gateway, most likely the other computer is having problems and you need to restart this procedure at that computer. If you can ping the gateway, but you cannot ping computers on another gateway, then

FIGURE 5.11 PING command showing total failure

```
C:\.ping 132.233.150.2

Pinging 132.233.150.2 with 32 bytes of data:

Request timed out.
Request timed out.
Request timed out.
Request timed out.

Ping statistics for 132.233.150.2:
    Packets: Sent = 4, Received = 0, Lost = 4 (100% loss),
Approximate round trip times in milli-seconds:
    Minimum = 0ms, Maximum = 0ms, Average = 0ms
```

FIGURE 5.12 TRACERT command

```
C:\.tracert www.novell.com

Tracing route to www.novell.com [137.65.2.11]
over a maximum of 30 hops:

  1    97 ms     92 ms    107 ms   tnt3-e1.scrm01.pbi.net [206.171.130.74]
  2    96 ms     98 ms    118 ms   core1-e3-3.scrm01.pbi.net [206.171.130.77]
  3    96 ms     95 ms    120 ms   edge1-fa0-0-0.scrm01.pbi.net [206.13.31.8]
  4    96 ms    102 ms     96 ms   sfra1sr1-5-0.ca.us.ibm.net [165.87.225.10]
  5   105 ms    108 ms    114 ms   f1-0-0.sjc-bb1.cerf.net [134.24.88.55]
  6   107 ms    112 ms    106 ms   atm8-0-155M.sjc-bb3.cerf.net [134.24.29.38]
  7   106 ms    110 ms    120 ms   pos1-1-155M.sfo-bb3.cerf.net [134.24.32.89]
  8   109 ms    108 ms    110 ms   pos3-0-0-155M.sfo-bb1.cerf.net [134.24.29.202]
  9   122 ms    105 ms    115 ms   atm8-0.sac-bb1.cerf.net [134.24.29.86]
 10   121 ms    120 ms    117 ms   atm3-0.slc-bb1.cerf.net [134.24.29.90]
 11   123 ms    131 ms    130 ms   novell-gw.slc-bb1.cerf.net [134.24.116.54]
 12      *         *         *     Request timed out.
 13   133 ms    139 ms    855 ms   www.novell.com [137.65.2.11]

Trace complete.
```

TABLE 5.4 TRACERT options

Options	Description
–d	In the event a name resolution method is not available for remote hosts, users can specify the –d option to prohibit the utility from trying to resolve host names as it runs. If not using this option, TRACERT will still function, but it will run very slow as it tries to resolve these names.
–h	By specifying the –h option, users can specify the maximum number of hops to which a route can be traced.
Timeout_value	Used to adjust the timeout value. The value determines the amount of time in milliseconds the program will wait for a response before moving on. If you raise this value and the remote devices respond whereas they did not respond previously, then a bandwidth problem may be indicated.
–j	Known as loose source routing, TRACERT –j <router name> <local computer> allows TRACERT to follow the path to the router specified and return to the user's computer.

FIGURE 5.13 IPX protocol

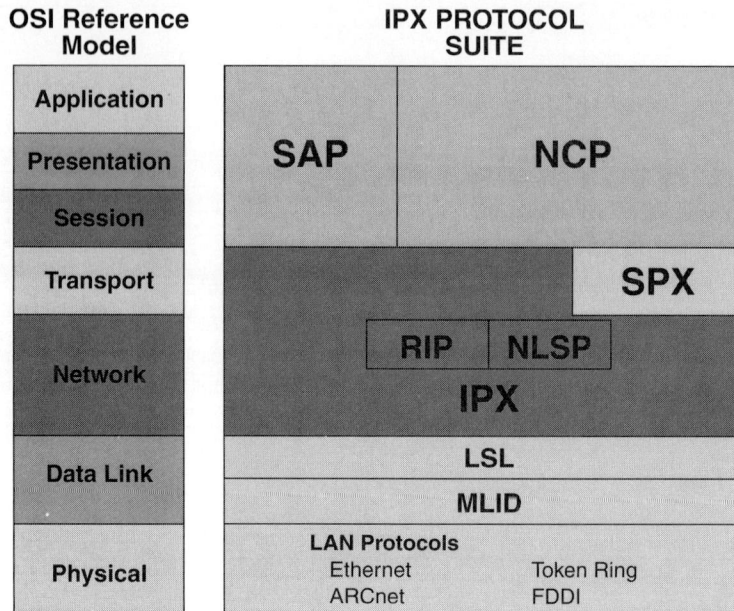

OSI Reference Model	IPX PROTOCOL SUITE

Application / Presentation / Session: **SAP** | **NCP**

Transport: **SPX**

Network: **RIP** | **NLSP** — **IPX**

Data Link: **LSL** / **MLID**

Physical: **LAN Protocols**
Ethernet   Token Ring
ARCnet   FDDI

check the routers and pathways between the two computers by using the PING or TRAC-ERT commands.

You also can use the NETSTAT command to display protocol statistics and current TCP/IP connections. The NETSTAT–A command displays all connections, and NET-STAT–R displays the route table plus active connections. The NETSTAT–E command displays Ethernet statistics, and NETSTAT–S displays per-protocol statistics. If you use NET-STAT–N, addresses and port numbers are not converted to names. For more information, see Windows 2000 Help.

5.3 IPX PROTOCOL SUITE

The **Internetwork packet exchange (IPX)** is a networking protocol suite used by Novell NetWare, Windows NT, and Windows 2000. Windows refers to the IPX protocol as NWLink (NDIS-compliant version of the IPX protocol which is fully compatible with the Novell IPX/SPX protocol). IPX is the fastest routable network protocol suite available. See Figure 5.13.

Similar to the TCP/IP network, users can connect to an IPX network by using one of several interfaces including Ethernet, token ring, or ARCnet. For a Novell NetWare network, the software used to connect to an IPX network first loads a **multiple link interface driver (MLID),** which is basically the driver to control the network card. Different from previous software interfaces used by early versions of IPX and other protocols, the MLID complies with the Novell open data link interface (ODI) architecture that allows a network driver to communicate with multiple protocol stacks such as TCP/IP and IPX.

Since the MLID supports multiple protocols, software is needed to identify the packets that come through the network card and then route to the proper protocol stack. This software is Link Support Layer (LSL).

5.3.1 Frame Types

When installing IPX on Windows 95, Windows 98, Windows NT, or Windows 2000, consider setting the frame type. The frame type defines the way in which the network adapter formats data to be sent over a network. See Table 5.5.

On Ethernet networks, the standard frame type for NetWare 2.2 and NetWare 3.11 is 802.3. Starting with NetWare 3.12, the default frame type was changed to 802.2. Users can choose to automatically detect or manually configure the frame type; however, the frame type is automatically detected when NWLink is loaded. If multiple frame types are detected in addition to the 802.2 frame type, NWLink defaults to the 802.2 frame type. One big prob-

TABLE 5.5   Frame types

Topology	Supported Frame Type
Ethernet	Ethernet II, 802.3, 802.2, and subnetwork access protocol (SNAP), which defaults to 802.2
Token ring	802.5 and SNAP
Fiber distributed data interface (FDDI)	802.2 and 802.3

lem when connecting to an IPX network is using the wrong frame type, so if users have multiple frame types, they must manually configure the operating system to load them simultaneously.

5.3.2   IPX and Sequenced Packet Exchange

As mentioned, IPX is a networking protocol used to interconnect networks. It is a connectionless protocol and therefore does not require a connection to set up before packets are sent to a destination.

Another worthwhile protocol is the **sequenced packet exchange (SPX).** Different from the IPX protocol, the SPX protocol uses packet acknowledgments to guarantee delivery of packets. SPX provides segmentation, reassembly, and segment sequencing of data streams that are too large to fit into the frame size. SPX uses virtual circuits referred to as connections. These connections are assigned specific connection identifiers as defined in the SPX header. Multiple connection IDs can be attached to a single socket.

5.3.3   IPX Addressing

When assigning addresses within the IPX network, there are two types of addresses, the internal and the external IPX network numbers. The **internal IPX network number,** an eight-digit (4-byte) hexadecimal number, is used to identify a server. Unlike TCP/IP, clients are not assigned internal network numbers, but instead use their MAC addresses. The **external IPX address** is used to identify the network. Therefore, all servers and routers on the same physical network must be assigned the same external IPX number. When assigning both the internal and external network numbers, all numbers must be unique. This includes numbers assigned to the network or to the individual servers. See Figure 5.14.

Much like the TCP/IP protocols, IPX also uses socket numbers in order to identify to which network software the packet is sent. Some examples are shown in Table 5.6.

5.3.4   Service Advertising and NetWare Core Protocols

The service advertising protocol (SAP) is used to advertise the services of all known servers on the network, including file servers and print servers. Servers periodically broadcast their service information while listening for SAPs on the network and then storing the information. Clients then access the service information table when they need a network service.

The NetWare core protocol (NCP) consists of the majority of network services offered by a Novell NetWare server. This includes but is not limited to file services, print services, Novell Directory Services (NDS), and message services.

5.4   NETBIOS AND NETBEUI

The **Network Basic Input/Output System (NetBIOS)** is a common program that runs on most Microsoft networks (Windows for Workgroups, Windows 95/98, and Windows NT). Originally created for IBM for its early PC networks, it was adopted by Microsoft and has since become a de facto industry standard.

NetBIOS is a session-level interface used by applications to communicate with NetBIOS-compliant transports such as NetBEUI, IPX, or TCP/IP. It is responsible for establishing logical names (computer names) on the network, establishing a logical connection

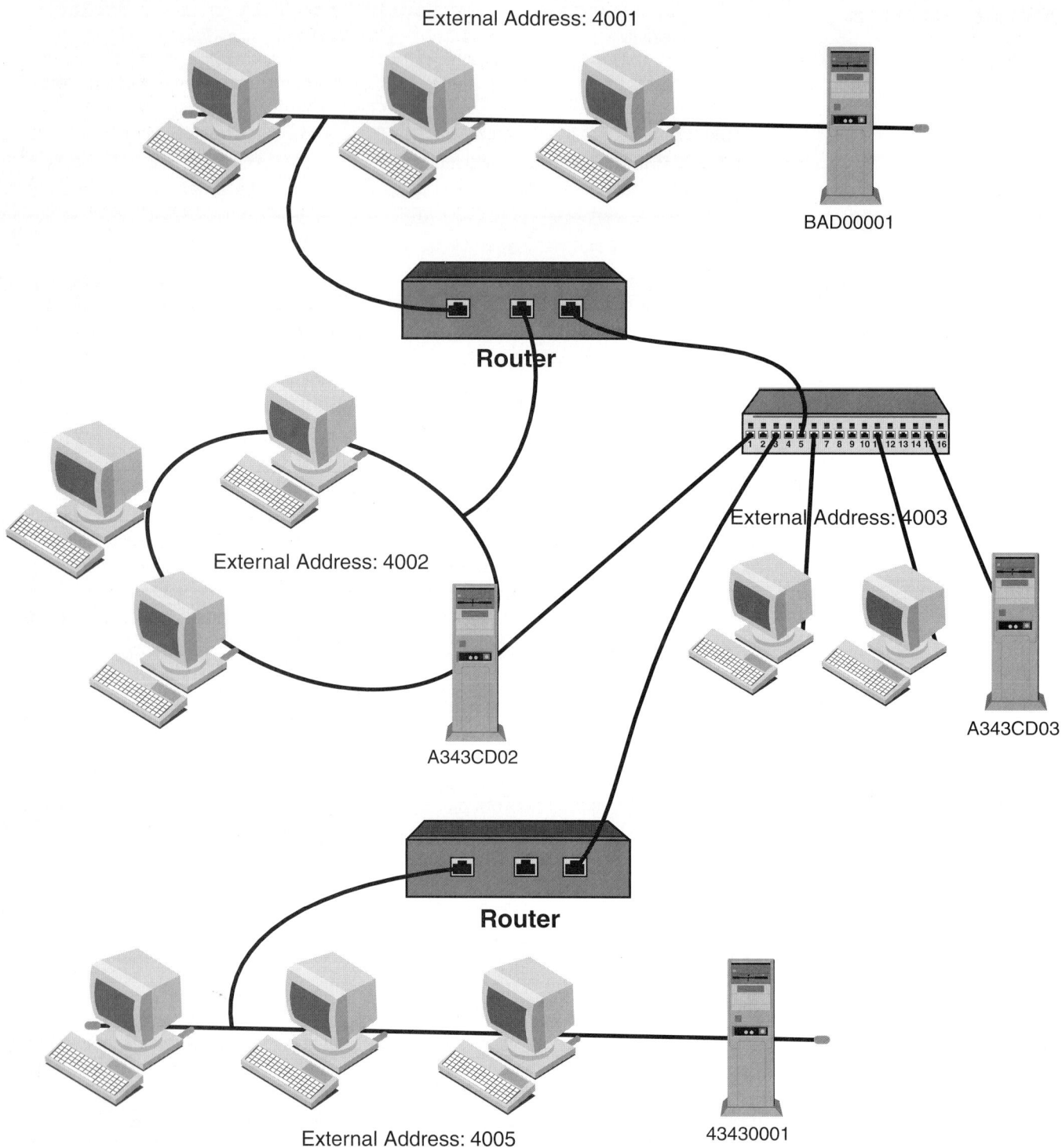

FIGURE 5.14 IPX network

External Address: 4001

BAD00001

Router

External Address: 4003

External Address: 4002

A343CD03

A343CD02

Router

External Address: 4005

43430001

between the two computers, and supporting reliable data transfer between computers that have established a session.

Once a logical connection is established, computers can then exchange data in the form of NetBIOS requests or in the form of a **server message block (SMB).** The SMB protocol, which was jointly developed by Microsoft, Intel, and IBM, defines a series of commands used to pass information between networked computers. Clients connected to a network using NetBIOS over TCP/IP, NetBEUI, or IPX/SPX can send SMB commands. Note that Microsoft refers to NetBIOS over TCP/IP as NBT.

TABLE 5.6 Common IPX network services

Port Number	Network Services
0×0451	NetWare Core Protocol (NCP)
0×0452	Service Advertising Protocol (SAP)
0×0453	Routing Information Protocol (RIP)
0×0455	Novell NetBIOS
0×9001	NetWare Link Services Protocol (NLSP)

Notes:

- If you need to communicate between two networks, you should use TCP/IP or IPX on all servers and client computers.
- You can use the NetBEUI protocol on the servers and then use NetBEUI on client boot disks so that you can perform downloads and installations over the network.

SMB can be broken into four message types: session control, file, printer, and message. The session control consists of commands that start and end a redirector connection to a shared resource at the server. The file SMB messages are used by the redirector to gain access to files at the server. The printer SMB messages are used by the redirector to send data to a print queue at a server and to get status information about the print queue. The message SMB type allows an application to send and receive messages from another workstation.

NetBIOS enhanced user interface (NetBEUI) provides the transport and network layers for the NetBIOS protocol, usually found on Microsoft networks. NetBEUI is a protocol used to transport data packets between two nodes. While NetBEUI is smaller than TCP/IP or IPX and is extremely quick, it will send packets only within the same network. In other words, NetBEUI is not a routable protocol because it cannot send a packet to a computer on another network.

Please note, a recommended method is to install both NetBEUI and TCP/IP on all servers in each computer so that you can perform downloads and installations (particularly using DOS) with local servers and TCP/IP to talk to other networks.

To navigate a Microsoft network, the computers are identified with computer names, no matter what protocol (TCP/IP, IPX and NetBEUI) you are using. The NetBIOS/computer names are up to 15 characters long. To connect the TCP/IP address with the computer name (NetBIOS name), Microsoft networks can use a LMHOSTS file or a WINS server.

A **LMHOSTS** file is a text file similar to a HOSTS file. Each entry should be kept on an individual line. The IP address should be placed in the first column followed by the corresponding computer name. The address and the computer name should be separated by at least one space or tab. Again, like the HOSTS file, the # character is generally used to denote the start of a comment. See figure 5.15. Table 5.7 shows the LMHOST special entries.

A WINS (Windows Internet Naming Service) server determines the IP address with a computer's NetBIOS name. WINS uses a distributed database that is automatically updated with the names of computers that are available and the IP address assigned to each one. Note that UNIX machines do not use NetBIOS names, so, if you want to use name resolution for these machines, you must use HOSTS files/DNS servers.

5.5 OTHER PROTOCOLS

Windows 2000 is designed to work with a wide variety of protocols at the same time. Although Windows 2000 includes the most common protocols, it also has the ability to install other protocols from disks. Some protocols with the Windows 2000 installation CD include the data link control (DLC) and AppleTalk protocols.

TABLE 5.7 LMHOSTS special entries

Predefined Keywords	Description
#PRE	Defines which entries should be initially preloaded as permanent entries in the name cache. Preloaded entries reduce network broadcasts, because names are resolved from cache rather than from broadcast or by querying the LMHOSTS file. Entries with a #PRE tag are loaded automatically when TCP/IP initializes, or manually by typing NBTSTAT–R at a command prompt.
#DOM: [domain_name]	Facilitates domain activity, such as logon validation over a router, account synchronization, and browsing.
#NOFNR	Avoids using NetBIOS-directed name queries for older LAN manager UNIX systems.
#INCLUDE	Loads and searches NetBIOS entries in a separate file from the default LMHOSTS file. Typically the #INCLUDE file is a centrally located shared LMHOSTS file. Refer to the #INCLUDE file by using a universal naming convention (UNC) path. Any computer referred to in the UNC path must also have a name–to–IP address mapping in the LMHOSTS file. This enables Windows 2000 to resolve a computer's NetBIOS name when it reads the #INCLUDE file.
#BEGIN_ALTERNATE #END_ALTERNATE	Defines a redundant list of alternate locations for LMHOSTS files. When you include multiple #INCLUDE lines between #BEGIN_ALTERNATE and #END_ALTERNATE, Windows 2000 includes the first file that it can locate in the list.
#MH	Adds multiple entries for a multihomed computer. A multihomed computer is a computer that contains the IP addresses that belong to different networks.

Note: The NetBIOS name cache and file are always read sequentially. Add the most frequently accessed computers to the top of the list. Add the #PRE-tagged entries near the bottom, because they are loaded when TCP/IP initializes and are not accessed again.

FIGURE 5.15 Sample LMHOSTS file

```
102.54.94.97    rhino        #PRE #DOM:networking #File Server
182.102.93.122  MISSERVER    #PRE                 #MIS Server
122.107.9.10    SalesServer                       #Sales Server
131.107.7.29    DBServer                          #Database Server
191.131.54.73   TrainServ                         #Training Server
```

The **data link control (DLC)** is a special, nonroutable protocol which enables computers that are running Windows 2000 to communicate with computers that are running DLC protocol stacks such as IBM mainframes, IBM AS/400 computers, and Hewlett-Packard network printers that are connected directly to the network using older HP JetDirect network cards. When using the DLC protocol to connect to a network printer, the DLC protocol must be installed and running on the print server for the printer. The protocol does not need to be installed on the computers that send print jobs to the print server.

The **AppleTalk protocol** allows Apple Macintosh computers to communicate with each other. For example, if the AppleTalk protocol is running on the Windows 2000 server, it can share its files and printers with other Apple Macintosh clients.

SUMMARY

1. A protocol is a set of rules or standards that are designed to enable computers to connect with one another, and peripheral devices to exchange information with as little error as possible.

2. A protocol stack, sometimes referred to as a protocol suite, is a set of protocols that works together to provide a full range of network services.

3. The standards for TCP/IP are published in a series of documents called Requests for Comments (RFC).

4. The lowest protocol within the TCP/IP suite is the Internet protocol (IP). It specifies the format of the packets (also called datagrams) and worries about the delivery of the packets.

5. The protocols that work on top of the IP are TCP and UDP. The transport control protocol (TCP) is a reliable, connection-oriented delivery service that breaks data into manageable packets.

6. Each connection on a TCP/IP address is called a host (a computer or other network device that is connected to a TCP/IP network) and is assigned a unique IP address.

7. The format of the IP address is four 8-bit numbers divided by a period (.). Each number can be 0 to 255.

8. Every computer on a TCP/IP network must have the same network number.

9. When connecting to the Internet, network numbers are assigned to a corporation or business.

10. The subnet mask is used to define which bits describe the network number and which bits describe the host address.

11. Network number 127 is used for loopback testing, and the specific host address 127.0.0.1 refers to the localhost or the actual host or computer currently in use.

12. If an individual network is connected to another network and a user must communicate with any number of computers on the other network, the user must also define the default gateway, which specifies the local address of a router.

13. All subnets on a network must use the same network address. The subnet network number and the subnet mask must be the same for all computers.

14. The address resolution protocol (ARP) provides a mechanism so that a host can learn a receiver's MAC address when knowing only the IP address.

15. Fully qualified domain names (FQDN), sometimes referred to as simply *domain names,* are used to identify computers on a TCP/IP network.

16. The FQDN must be translated to a numeric address by either a HOSTS file or by using a DNS server.

17. Every time a TCP/IP host communicates with another TCP/IP host, it will use the IP address and port number to identify the host and service/program running on the host.

18. A TCP/IP port number is a logical connection place by client programs to specify a particular server program that is running on a computer on the network defined at the transport layer.

19. In supernetting, when the address is assigned, the network number and host number are not made along the standard octet boundaries, but instead are made from a specific number of bits from the beginning of the address. A supernet is a group of networks identified by contiguous network addresses. IP service providers can assign to customers blocks of contiguous addresses in order to define supernets as needed.

20. Classless interdomain routing (CIDR) collapses multiple network ID entries into a single entry.

21. The new IP protocol, currently called IP next generation (IPng), consists of 128 bits to allow for more addresses and includes other enhancements to the current 32-bit IP address.

22. The Internetwork packet exchange (IPX) is a fast, routable networking protocol used by Novell NetWare, Windows NT, and Windows 2000.

23. Windows refers to the IPX protocol as NWLINK (NDIS-compliant version of the IPX protocol that is fully compatible with the Novell IPX/SPX protocol).

24. A frame type used in IPX networks defines the way in which the network adapter formats data to be sent over a network.

25. The internal IPX network number, an eight-digit (4-byte) hexadecimal number, is used to identify a server while the external IPX address is used to identify the network.

26. NetBIOS enhanced user interface (NetBEUI) provides the transport and network layers for the NetBIOS protocol, usually found on Microsoft networks.
27. A LMHOSTS file is a text file that is used to resolve NetBIOS names to IP addresses.
28. The data link control (DLC) is a special, nonroutable protocol which enables computers that are running Windows 2000 to communicate with computers that are running DLC protocol stacks, such as IBM mainframes, IBM AS/400 computers, and Hewlett-Packard network printers that are connected directly to the network using older HP Jet-Direct network cards.
29. The AppleTalk protocol allows Apple Macintosh computers to communicate with each other.

QUESTIONS

1. The protocol used by the Internet is _____ .
   a. TCP/IP
   b. IPX
   c. NetBEUI
   d. AppleTalk

2. To connect to a TCP/IP network, you must configure which of the following? (Choose two answers)
   a. TCP/IP address
   b. gateway
   c. IPX address
   d. subnet mask
   e. DNS server address

3. To connect to a TCP/IP network that contains several subnets, which of the following must be configured? (Select all that apply)
   a. TCP/IP address
   b. gateway
   c. IPX address
   d. subnet mask
   e. DNS server address

4. What is the default subnet mask for a class B network?
   a. 255.0.0.0
   b. 255.255.0.0
   c. 255.255.255.0
   d. 255.255.255.255
   e. 127.0.0.1

5. What does the 127.0.01 address represent?
   a. broadcast address
   b. network address
   c. loopback address
   d. subnet address

6. _____ is when you take a large network and divide it into smaller networks.
   a. Subnetting
   b. Broadcasting
   c. Gatewaying
   d. Hosting

7. Which of the following must you consider when deciding which subnet mask to apply to a TCP/IP network? (Choose all that apply)
   a. IP address class
   b. types of computers used on the network
   c. number of subnets
   d. potential for network growth

8. Your company is assigned the network address 150.50.0.0. You need to create seven subnets on the network. A router on one of the subnets will connect the network to the Internet. All computers on the network will need access to the Internet. What is the correct subnet mask for the network?
   a. 0.0.0.0
   b. 255.255.0.0
   c. 255.255.240.0
   d. the subnet mask assigned by InterNIC

9. A company with the network ID 209.168.19.0 occupies four floors of a building. You create a subnet for each floor. You want to allow for the largest possible number of host IDs on each subnet. Which subnet mask should you choose?
   a. 255.255.255.192
   b. 255.255.255.224
   c. 255.255.255.240
   d. 255.255.255.248

10. The Acme Corporation has been assigned the network ID 134.114.0.0. The corporation's eight departments require one subnet each; however, each department may grow to over 2500 hosts. Which subnet mask should you apply?
    a. 255.255.192.0.
    c. 255.255.240.0.
    b. 255.255.224.0.
    d. 255.255.248.0.

11. All IP addresses are eventually resolved to NIC addresses. Which of the following is used to map an IP address to a NIC address?
    a. WINS
    c. DNS
    b. DHCP
    d. ARP

12. The enhanced version of the IP (IPv4) is _____ .
    a. IPng
    c. IPv5
    b. IPX
    d. IPING

13. There are several UNIX computers and mainframes on your Windows NT network. You want to standardize the network protocol used on all computers and provide access to the Internet. Which protocol would you choose for your network?
    a. TCP/IP
    c. NetBEUI
    b. NWLINK
    d. IPX/SPX

14. If you are running Windows NT and want to know the MAC address of the network card in your server, you would type _____ at the command prompt.
    a. IPCONFIG /ALL
    b. WINIPCFG /ALL
    c. MACCFG /ALL
    d. CONFIG /ALL

15. What is the default assigned ports used by a web server browser?
    a. 21
    c. 119
    b. 80
    d. 139

16. What two types of names can be resolved in a Windows TCP/IP platform? There are a total of five answers.
    a. host
    c. network name
    b. computer
    d. NetBIOS name

17. What is used to resolve host names such as Microsoft.com? (Select two answers)
    a. DNS server
    c. LMHOSTS files
    b. HOSTS files
    d. WINS server

18. You are troubleshooting a workstation and find the following information:

IP address:	132.123.140.14
Subnet mask:	255.255.255.0
Default gateway:	131.123.140.200
DNS servers:	130.13.18.3, 140.1.14.240
Router's IP address:	131.123.140.1

    What is the reason why the workstation cannot get to the Internet?

    a. The subnet mask is invalid for this network.
    b. There can only be one IP address for DNS servers.
    c. The IP address of the DNS should match the default gateway address.
    d. The IP address and subnet mask of the workstation are wrong.
    e. The default gateway should match the router's address.

19. Which of the following are transport layer protocols? (Choose all that apply)
    a. TCP
    c. SPX
    b. IP
    d. IPX

20. Which one of the following protocols is an NDIS-compliant version of the Internetwork packet exchange protocol?
    a. IP
    c. NetBEUI
    b. NWLINK
    d. SMB

21. You have static IP addresses assigned to your workstation. If you have two workstations that are assigned the same IP address, what would happen as far as communication with these two workstations?
   a. The second workstation will take over communication when it boots.
   b. The first workstation to boot and log in will communicate.
   c. Both workstations will be okay.
   d. Neither workstation will be able to communicate on the network.

22. One computer on your IPX/SPX network cannot connect to the network. What is the most likely cause of the problem?
   a. incorrect frame type
   b. protocol mismatch
   c. upper memory area conflicts
   d. faulty connectivity devices

23. There is a frame type mismatch on your Novell NetWare network. It is only affecting one computer. Which of the following needs to be reconfigured?
   a. frame type on the client machine
   b. frame type on the server machine
   c. frame binding setting on the client machine
   d. frame binding setting on the server machine

24. Which of the following uses 15-character names to identify computers on a network?
   a. TCP/IP
   b. IPX/SPX
   c. NetBIOS
   d. AppleTalk

25. Which of the following protocols are network layer protocols? (Choose all that apply)
   a. IP
   b. TCP
   c. IPX
   d. SPX

26. Transport protocols provide for communication sessions between computers. Which of the following is a transport protocol?
   a. SNMP
   b. NetBEUI
   c. IP
   d. IPX

27. NetBEUI is the extended user interface of NetBIOS. Which of the following statements is true of NetBEUI? Choose all that apply. There are two possible answers.
   a. NetBEUI is a small, fast, and efficient transport layer protocol used primarily with Microsoft networks.
   b. NetBEUI is a small, fast, and efficient session layer protocol, but it is limited to Novell NetWare networks.
   c. NetBEUI does not support routing.
   d. NetBEUI is a relatively large protocol, which can cause problems in MS-DOS-based clients.

28. Which nonroutable protocol fits in both at the network and transport layers of the OSI model?
   a. TCP/IP
   b. AppleTalk
   c. IPX/SPX
   d. NetBEUI

29. What file is used to resolve a host name to an IP address?
   a. LMHOSTS
   b. HOSTS
   c. HOSTS.SAM
   d. LMHOSTS.SAM

30. Which of the following protocols are routable? (Choose all that apply)
   a. TCP/IP
   b. IPX/SPX
   c. AppleTalk
   d. NetBEUI

31. The purpose of a WINS server in a network is to _____ .
   a. keep a database of host names and corresponding IP addresses
   b. keep communication flowing on a TCP/IP network by assigning IP addresses dynamically to each workstation as it logs in
   c. keep a database of NetBIOS names and corresponding IP addresses
   d. authenticate each workstation as it logs in and determine the MAC address and corresponding IP address of each workstation

32. Routers interconnect networks and provide filtering functions. Which one of the following protocols can be used with a router?
    a. IPX/SPX
    b. TCP/IP
    c. NetBEUI
    d. AppleTalk

33. What is the primary protocol used with Novell NetWare 4.x?
    a. TCP/IP
    b. IPX/SPX
    c. NetBEUI
    d. DLC

34. You are using a computer with NetBEUI, but unfortunately cannot connect to a computer located on another LAN, although you can connect to the computers that are on the same LAN. What is the problem?
    a. You are using the wrong MAC address.
    b. You are using the wrong NetBEUI name.
    c. You are using the wrong gateway address.
    d. NetBEUI is a nonroutable protocol.

35. You have a Windows 95 workstation that is dynamically being assigned an IP address and you want to find out the current IP address. How do you do this?
    a. Click the Network properties and go to the IP address tab.
    b. Run the IPCFG utility.
    c. Run the WINIPCFG utility.
    d. Double-click on the My Computer icon, double-click on the Control Panel icon, and then double-click on the System applet.

36. Which operating systems use NetBIOS names? (Choose all that apply)
    a. UNIX
    b. LINUX
    c. Windows NT Server
    d. Windows 98

37. UNIX workstations resolve addresses on the Internet by using _____ .
    a. WINS
    b. SMTP
    c. SNMP
    d. DNS

38. What protocol provides reliable, connection-based delivery?
    a. TCP
    b. UDP
    c. IP
    d. ARP

39. Several users are complaining that they cannot access one of your Windows NT file servers, which has an IP address that is accessible from the Internet. When you get paged, you are not in the server room but at another company in a friend's office that only has a UNIX workstation available, which also has Internet access. What can you do to see if the file server is still functioning on the network?
    a. Use PING from the UNIX workstation.
    b. Use ARP from the UNIX workstation.
    c. Use WINS from the UNIX workstation.
    d. Use DNS from the UNIX workstation.

40. When configuring a client computer, what information is required for the computer to operate in a TCP/IP environment? (Choose two answers)
    a. IP address
    b. subnet mask
    c. default gateway
    d. DNS server address
    e. WINS server address

41. When configuring a client computer, what information is required for the computer to operate in a TCP/IP WAN environment? (Choose three answers)
    a. IP address
    b. subnet mask
    c. default gateway
    d. DNS server address
    e. WINS server address

42. What tool can you use to see the path taken from a Windows NT system to another network host?
    a. PING
    b. IPCONFIG
    c. WINIPCFG
    d. SNMP
    e. TRACERT

43. Which bits of a subnet mask correspond to the network ID?
    a. ones
    b. zeros
    c. both ones and zeros
    d. neither ones nor zeros

44. You are troubleshooting the network connectivity for one of the client systems at your company, but you do not know the IP address for the system. You are at the client system and you want to make sure that the system can communicate with itself using TCP/IP. How can you accomplish this?
    a. PING 127.0.0.1
    b. PING 255.255.0.0
    c. PING 255.255.255.255
    d. PING 0.0.0.0

45. How are the network ID and the host ID for an IP address determined?
    a. subnet mask
    b. unicast mask
    c. range mask
    d. multicast mask

46. What are the TCP/IP addresses available for multicast transmissions?
    a. 128.0.0.0 to 191.255.0.0
    b. 192.0.0.0 to 223.255.255.0
    c. 224.0.0.0 to 239.255.255.255
    d. 240.0.0.0 to 247.255.255.255

47. At one of your company's remote locations, you have decided to segment your class B address down, since the location has three buildings and each building contains no more than 175 unique hosts. You want to make each building its own subnet and utilize your address space the best way possible. Which subnet mask meets your needs in this situation?
    a. 255.0.0.0
    b. 255.255.0.0
    c. 255.255.255.0
    d. 255.255.255.240

48. You have successfully obtained a class C subnet for your company. What is the default subnet mask?
    a. 255.255.255.0
    b. 255.255.0.0
    c. 255.0.0.0
    d. 255.255.240.0

49. A Windows 2000 Server computer, named SERVER1, resides on a remote subnet. Pat cannot ping SERVER1 using its IP address. He can successfully ping his default gateway address and the addresses of other computers on the remote subnet. What is the most likely cause of the problem?
    a. Pat's computer is set up with an incorrect default gateway address.
    b. Pat's computer is set up with an incorrect subnet mask.
    c. SERVER1 is not WINS enabled.
    d. The IP configuration on SERVER1 is incorrect.
    e. The LMHOSTS file on Pat's computer has no entry for SERVER1.

50. Pat is planning to set up an intranet web server at his company. Employees will be able to access the web server using the server's host name. Which one of the following services should Pat install on his intranet to provide name resolution?
    a. DHCP
    b. FTP
    c. DNS
    d. WINS

51. Pat fails repeatedly to access a Windows 2000 Server computer on a remote subnet. Using Network Monitor to troubleshoot the problem, he finds that every time he tries to connect to the server, his workstation broadcasts an ARP request for the IP address of the remote Windows 2000 Server computer. No other users on the TCP/IP network have trouble accessing the server. What is the most likely cause of Pat's problem?
    a. The workstation is not set up to use DNS.
    b. The workstation is not set up to use WINS.
    c. The workstation is set up with a duplicate IP address.
    d. The workstation is set up with an incorrect subnet mask.

52. Pat connects to a remote UNIX computer using its IP address. He knows TCP/IP is properly installed on this computer. How should he check if the router is working correctly?
    a. Ping 127.0.0.1.
    b. Ping the far side of the router.
    c. Ping the local server.
    d. Ping the near side of the router.

53. A Windows 2000 Professional computer user complains that he cannot connect to any other computers on his network. The network uses both a DHCP and a DNS server. Seated at his workstation, you ping 127.0.0.1 and fail to get a response. What is the most probable cause of the problem?
    a. TCP/IP is not properly installed on the workstation.
    b. The default gateway address on the workstation is incorrect.
    c. The subnet mask on the workstation is incorrect.
    d. The workstation is not set up for DHCP.
    e. The workstation is not set up for DNS.

54. Which switch would you use if you want to prohibit the TRACERT utility from resolving names as it runs?
    a. –r   c. –h
    b. –n   d. –d

55. What value determines the amount of time in milliseconds the program will wait for a response before continuing?
    a. hops   c. wait
    b. timeout   d. time to live

56. You are a network engineer who is implementing a network that consists of approximately 150 Windows 2000 workstations and five Windows 2000 servers. You have multiple sites located in the same city that are connected through a PSTN. Your manager wants you to ensure fast communication by implementing the proper protocol. Each site has a router that is already configured by the telephone company. The workstations are running Windows 98.

    Required result:
    • Use a routable protocol to ensure connectivity between sites.

    Optional results:
    • Ensure communication between the Windows 2000 servers.
    • Log in to all the servers without maintaining multiple logins.

    Proposed solution:
    • Implement NetBEUI as your protocol and use the same usernames on all servers, but different passwords for security purposes.

    a. Achieves the required result and both optional results.
    b. Does not achieve the required result but achieves both optional results.
    c. Does not achieve the required result but does achieve one of the optional results.
    d. Achieves the required result but cannot achieve either of the optional results.
    e. Achieves neither the required result nor either of the optional results.

57. Which of the following entries in a HOSTS file residing on a Windows 2000 Server computer will connect to the UNIX server SERVER1?
    a. 132.132.11.53 #SERVER1 #corporate server
    b. 132.132.11.53 SRV1 #corporate server
    c. 132.132.11.53 SERVER1 #corporate server
    d. 132.132.11 #SERVER1 #corporate server

58. Pat oversees a network with five subnets: Subnet1, Subnet2, Subnet3, Subnet4, and Subnet5. A Windows 2000 computer user on Subnet4 complains that he can connect to local servers, but not to a Windows 2000 Server computer on Subnet1. Other users

on Subnet4 have no trouble connecting to the Subnet1 server. Pat runs IPCONFIG on the workstation and notes the following output:

```
Ethernet Adapter Local Area Connection:

  Connection-specific DNS Suffix:
  IP Address    . . . . : 132.132.223.19
  Subnet Mask   . . . . : 255.255.224.0
  Default Gateway   . . : 132.132.224.1
```

What is the reason why the workstation cannot connect to the remote server?
- a. The default gateway is incorrect.
- b. The IP address must fall between 132.132.224.1 and 132.132.255.254.
- c. The subnet mask must be 255.255.192.0.
- d. The subnet mask must be 255.255.255.224.

59. You are a network engineer who is implementing a network that consists of approximately 150 Windows 2000 workstations and five Windows 2000 servers. You have multiple sites located in the same city that are connected through a PSTN. Your manager wants you to ensure fast communication by implementing the proper protocol. Each site has a router that is already configured by the telephone company. The workstation are running Windows 98.

Required result:
- Use a routable protocol to ensure connectivity between sites.

Optional results:
- Ensure communication between the Windows 2000 server.
- Log in to all the servers without maintaining multiple logins.

Proposed solution:
- Implement TCP/IP as your protocol and use the same usernames on all servers, but different passwords for security purposes.

- a. Achieves the required result and both optional results
- b. Does not achieve the required result but achieves both the optional results
- c. Does not achieve the required result but does achieve one of the optional results
- d. Achieves the required result but cannot achieve either of the optional results
- e. Achieves neither the required result nor either of the optional results

60. You are a network engineer who is implementing a network that consists of approximately 150 Windows NT workstations and five Windows NT servers. You have multiple sites located in the same city that are connected through a PSTN. Your manager wants you to ensure fast communication by implementing the proper protocol. Each site has a router that is already configured by the telephone company. The workstations are running Windows 98.

Required result:
- Use a routable protocol to ensure connectivity between sites.

Optional results:
- Ensure communication between the Windows NT servers.
- Log in to all the servers without maintaining multiple logins.

Proposed solution:
- Implement TCP/IP as your protocol and use the same usernames and passwords on all servers.

- a. Achieves the required result and both optional results.
- b. Does not achieve the required result but achieves both optional results.
- c. Does not achieve the required result but does achieve one of the optional results.
- d. Achieves the required result but cannot achieve either of the optional results.
- e. Achieves neither the required result nor either of the optional results.

61. Look at the following figure. The first number of each host is the IP address, the second number is the subnet mask, and the last number is the default gateway. What is the problem with the TCP/IP network?

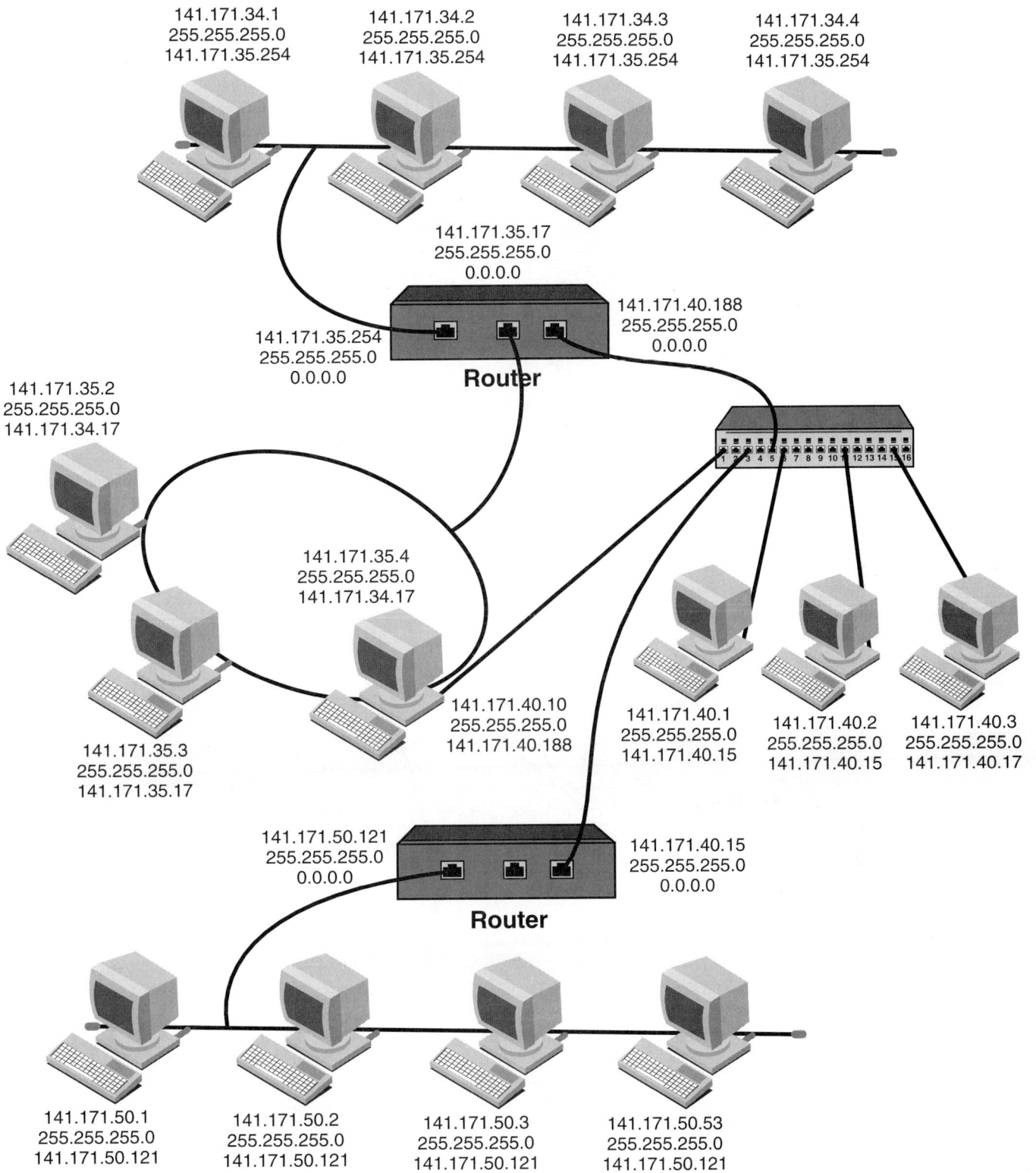

141.171.34.1
255.255.255.0
141.171.35.254

141.171.34.2
255.255.255.0
141.171.35.254

141.171.34.3
255.255.255.0
141.171.35.254

141.171.34.4
255.255.255.0
141.171.35.254

141.171.35.17
255.255.255.0
0.0.0.0

141.171.40.188
255.255.255.0
0.0.0.0

141.171.35.254
255.255.255.0
0.0.0.0

Router

141.171.35.2
255.255.255.0
141.171.34.17

141.171.35.4
255.255.255.0
141.171.34.17

141.171.35.3
255.255.255.0
141.171.35.17

141.171.40.10
255.255.255.0
141.171.40.188

141.171.40.1
255.255.255.0
141.171.40.15

141.171.40.2
255.255.255.0
141.171.40.15

141.171.40.3
255.255.255.0
141.171.40.17

141.171.50.121
255.255.255.0
0.0.0.0

141.171.40.15
255.255.255.0
0.0.0.0

Router

141.171.50.1
255.255.255.0
141.171.50.121

141.171.50.2
255.255.255.0
141.171.50.121

141.171.50.3
255.255.255.0
141.171.50.121

141.171.50.53
255.255.255.0
141.171.50.121

62. Look at the following figure. The first number of each host is the IP address, the second number is the subnet mask, and the last number is the default gateway. What are the two problems with the TCP/IP network?

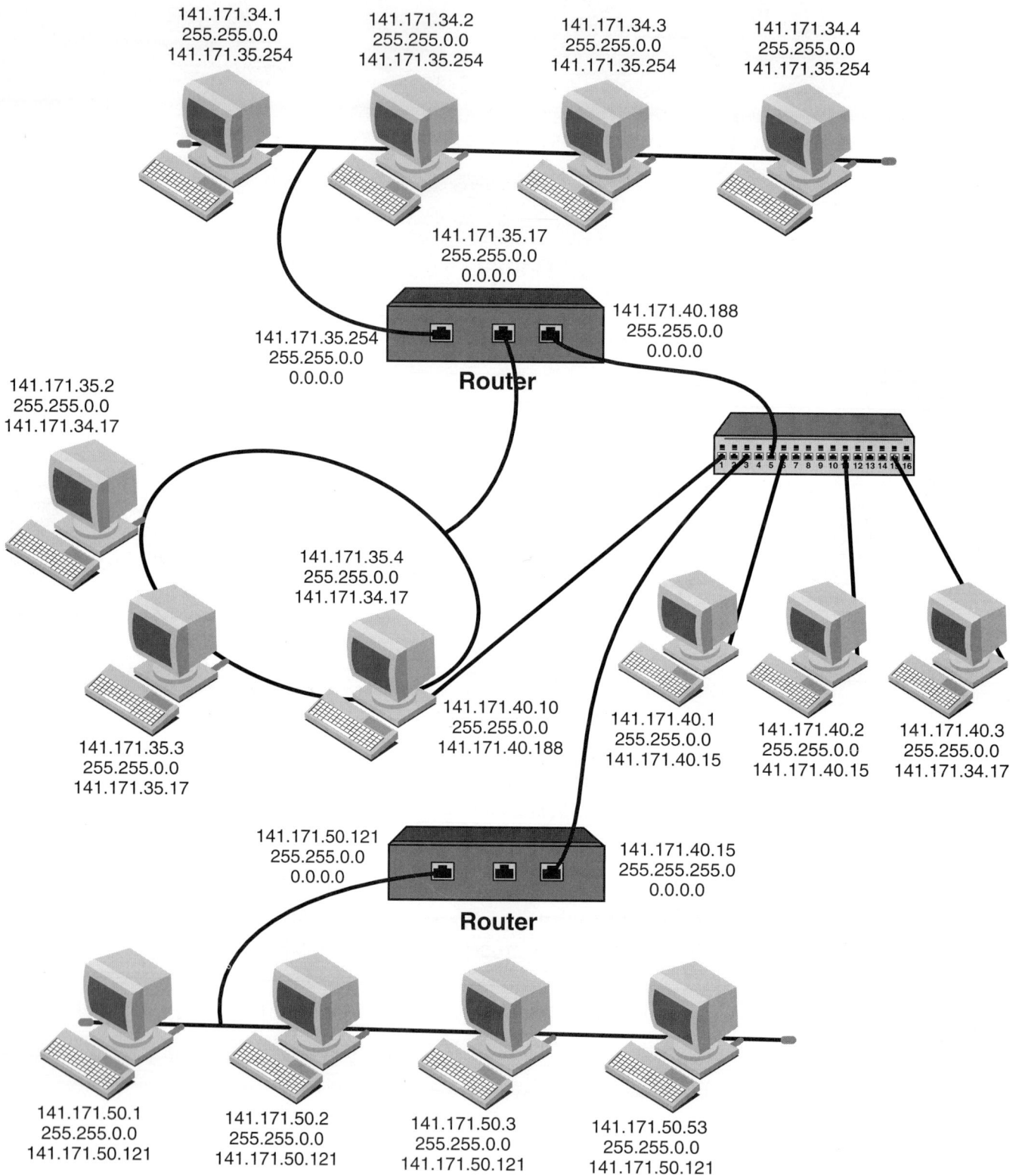

141.171.34.1
255.255.0.0
141.171.35.254

141.171.34.2
255.255.0.0
141.171.35.254

141.171.34.3
255.255.0.0
141.171.35.254

141.171.34.4
255.255.0.0
141.171.35.254

141.171.35.17
255.255.0.0
0.0.0.0

141.171.40.188
255.255.0.0
0.0.0.0

141.171.35.254
255.255.0.0
0.0.0.0

Router

141.171.35.2
255.255.0.0
141.171.34.17

141.171.35.4
255.255.0.0
141.171.34.17

141.171.40.10
255.255.0.0
141.171.40.188

141.171.40.1
255.255.0.0
141.171.40.15

141.171.40.2
255.255.0.0
141.171.40.15

141.171.40.3
255.255.0.0
141.171.34.17

141.171.35.3
255.255.0.0
141.171.35.17

141.171.50.121
255.255.0.0
0.0.0.0

141.171.40.15
255.255.255.0
0.0.0.0

Router

141.171.50.1
255.255.0.0
141.171.50.121

141.171.50.2
255.255.0.0
141.171.50.121

141.171.50.3
255.255.0.0
141.171.50.121

141.171.50.53
255.255.0.0
141.171.50.121

63. Look at the following figure. What is the problem with the IPX network?

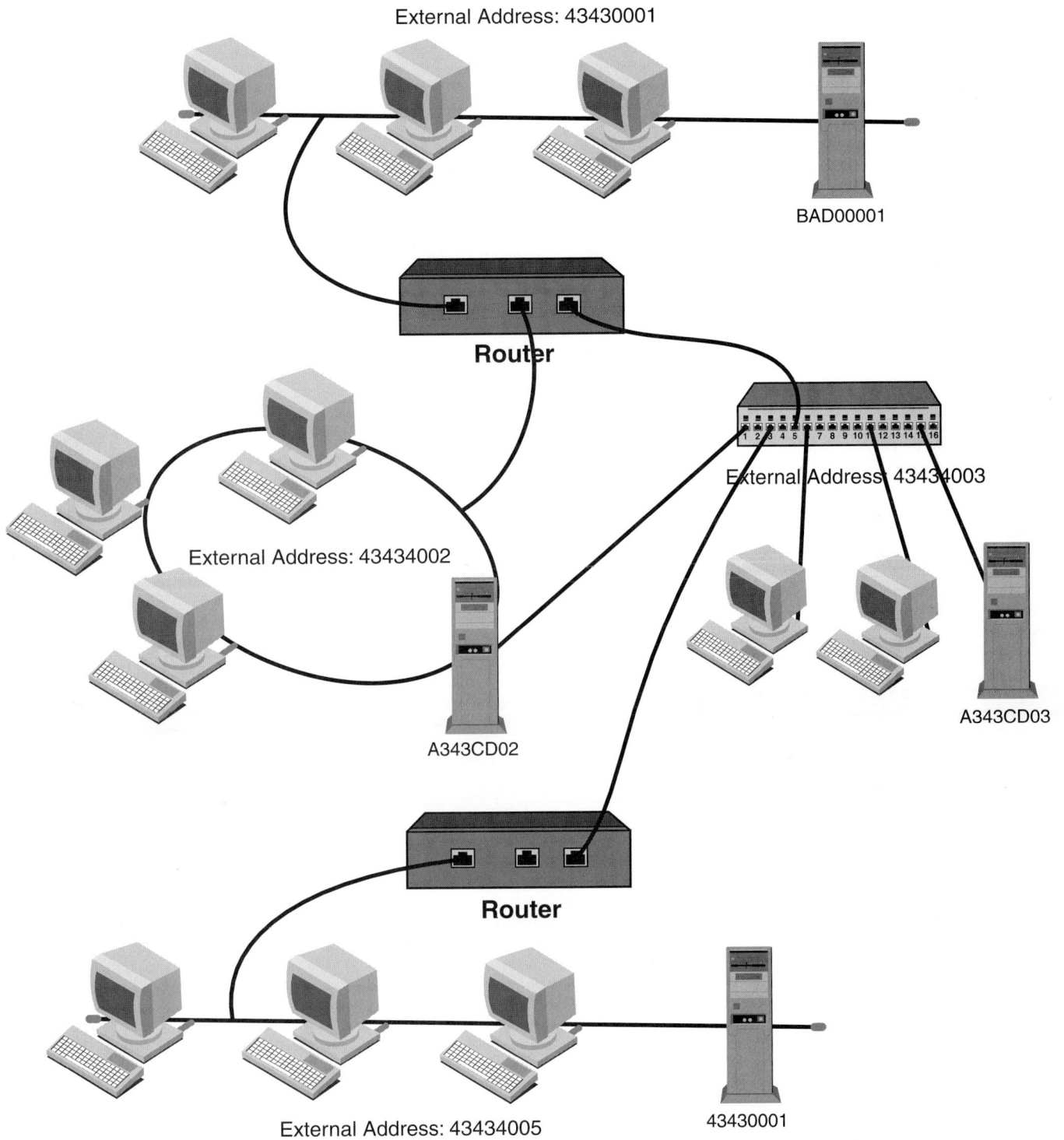

External Address: 43430001

BAD00001

Router

External Address: 43434002

External Address: 43434003

A343CD02

A343CD03

Router

External Address: 43434005

43430001

6

Networking with Windows 95/98

INTRODUCTION

Windows 95 (and Windows 98) is a complete operating system designed to replace DOS and Windows 3.xx. The most obvious difference between Windows 95 and Windows 3.xx is that Windows 95 has a better, more intuitive user interface that supports long file names. Behind the interface, Windows 95 is a preemptive, multithreading, multitasking environment. It is primarily a 32-bit operating system that is not limited by the conventional memory restrictions, yet provides compatibility for most DOS and Windows 3.xx device drivers and applications. It supports plug-and-play technology and has built-in multimedia and network capability.

OBJECTIVES

1. Configure the TCP/IP, IPX, and NetBEUI protocols on a Windows 95 or Windows 98 computer.
2. Share a drive or directory on a Windows 95 or 98 machine.
3. Share a printer on a Windows 95 or 98 machine.
4. Use Network Neighborhood in Windows 95 or 98 to navigate to a network resource.

6.1 LOADING AND CONFIGURING THE NETWORKING SOFTWARE

In Windows 95 and Windows 98, network components are installed and configured using the network dialog box, which is accessed by clicking on the Network applet in the control panel or by accessing the shortcut menu of the Network Neighborhood icon and selecting the Properties option. See Figure 6.1. The Network Neighborhood icon does not show on the desktop unless some network components are installed.

FIGURE 6.1 Windows 95 Network dialog box

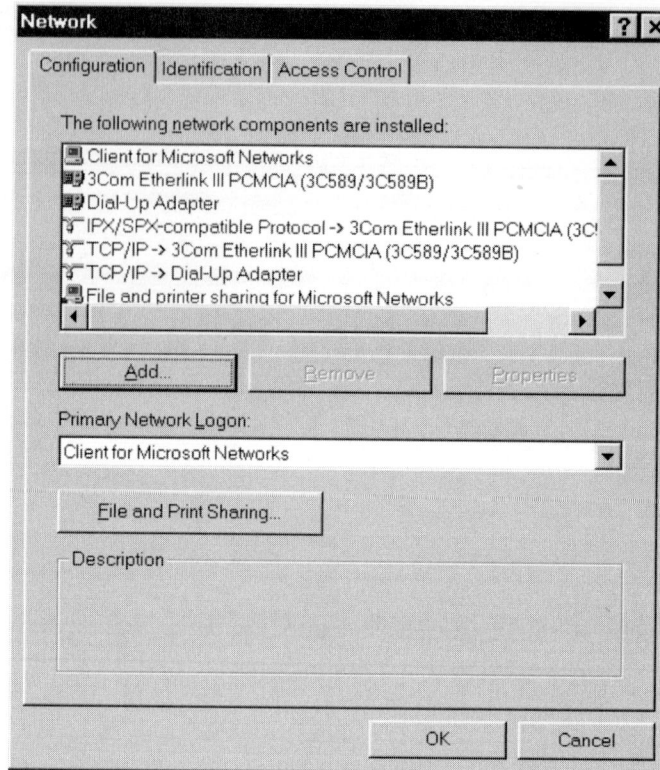

To get full access to the network, install the following network software through the Configuration tab:

1. A client software choice such as Client for Microsoft Networks or Client for NetWare Networks
2. A network adapter driver or dial-adapter driver (such as a modem)
3. A protocol stack such as TCP/IP, IPX/NWLINK, or NetBEUI

6.1.1   Client Software

Client software often uses redirectors and designators. A **redirector** is a small section of code in the network operating system that intercepts requests in the computer and determines if they should be left alone to continue in the local computer's bus or redirected out to the network to another server. For example, when you request a file from the server or send a print job to a network printer, the request is intercepted by the redirector and forwarded out onto the network. A **designator** keeps track of which drive designations are assigned to network resources. For example, when you attach or map to a network drive, you and the software applications may see a drive letter, such as drive M or drive P, as on any other local drive. When these drives are accessed, a redirector will route the request to the network.

The **Client for Microsoft Networks** provides the redirector (VREDIR.VXD) to support all Microsoft networking products that use the server message block (SMB) protocol. This includes support for connecting computers to and accessing files and printers on Windows for Workgroups, Windows 95, Windows 98, Windows NT, and Windows 2000 computers.

If the Client for Microsoft Networks is the primary network logon client, the Microsoft network usually consists of Windows NT and Windows 2000 servers. These could be used to download system policies and user profiles. System policies and user profiles are used to control and manage the computers across the network. For example, you can restrict access to the control panel options and what users can do from the desktop, customize parts of the desktop, or configure network settings.

Under the properties of the Client for Microsoft Networks, users can specify a domain name to log on to and specify the network logon options. See Figure 6.2a. A **domain** is an

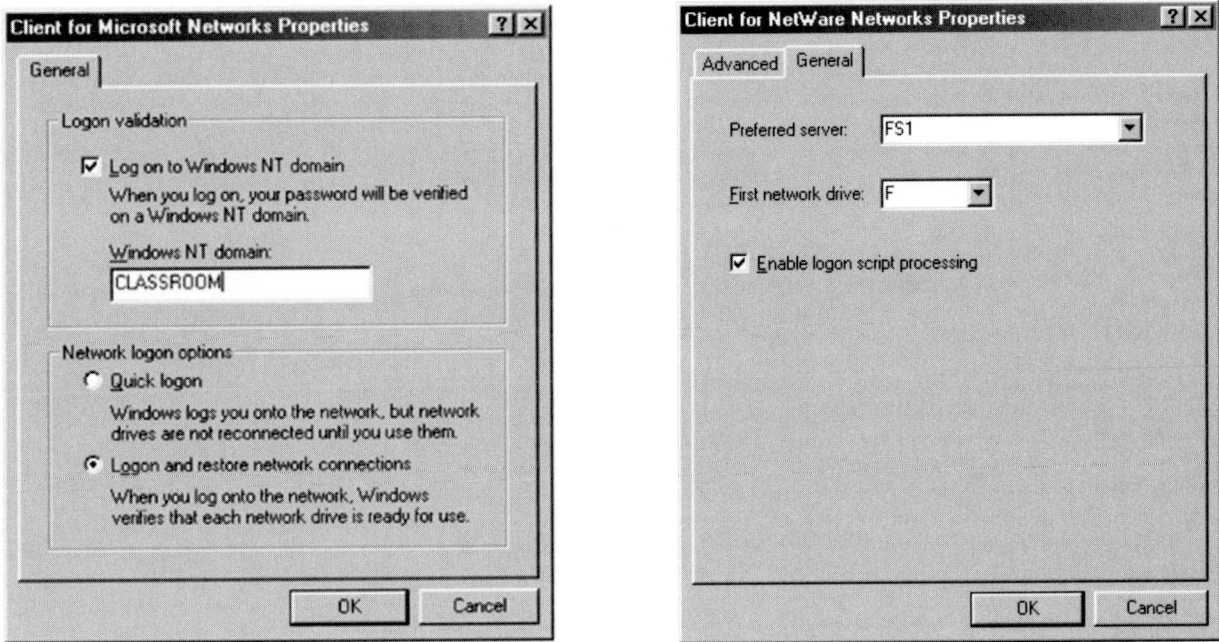

FIGURE 6.2 **Client for Microsoft and NetWare Networks properties**

organization unit consisting of computers, which uses one or more domain controllers to hold the centralized security account information. When a user logs on to a workstation/ client computer on the network, the password is validated by one of the domain controllers and access is granted to the assigned network resources.

Client for NetWare Networks, run with the IPX protocol, is used to process login scripts, support all NetWare 3.xx command line utilities and most 4.x command line utilities, connect and browse to NetWare servers, access printers on the NetWare server, and process login scripts on the NetWare server. See Figure 6.2b. In addition, if you decide to install Client for NetWare, you will need to load a network card driver that supports the ODI protocol. In addition, to get full access to all utilities, you should load the client software provided by Novell.

6.1.2   Protocol Stacks

Since Windows 95/98 can use NDIS- and ODI-compatible network card drivers, it is possible to have one or more protocol stacks on the system. For example, you can load TCP/IP, IPX/SPX-/compatible protocol, NetBEUI, and many others.

If users load the TCP/IP protocol, they must then configure it. If they highlight the TCP/IP protocol and click on the Properties button, the screen shows seven tabs. See Figure 6.3. For example, within the IP Address tab, you either assign the IP address or select to obtain an IP address automatically from a DHCP server. The gateway or address of the routers is set within the Gateway tab. The WINS Configuration and DNS Configuration tabs are used to assign the addresses of the DNS and WINS servers, which are used for name resolution.

The IPX and NetBEUI protocols are not as complicated to configure as the TCP/IP protocol and usually require no configuration. See Figures 6.4 and 6.5. Within the properties of the IPX protocol, select the IPX frame types and enable NetBIOS over IPX/SPX. In either case, TCP/IP, IPX, and NetBEUI all have the option to become the default protocol.

6.1.3   Network Card Drivers

The network card or adapter is the hardware device that allows users to connect to the network, but it requires a software driver before it can be used. If the driver is not listed, then use the Have Disk button and indicate where the installation files are located for the driver that is to be installed. See Figure 6.6.

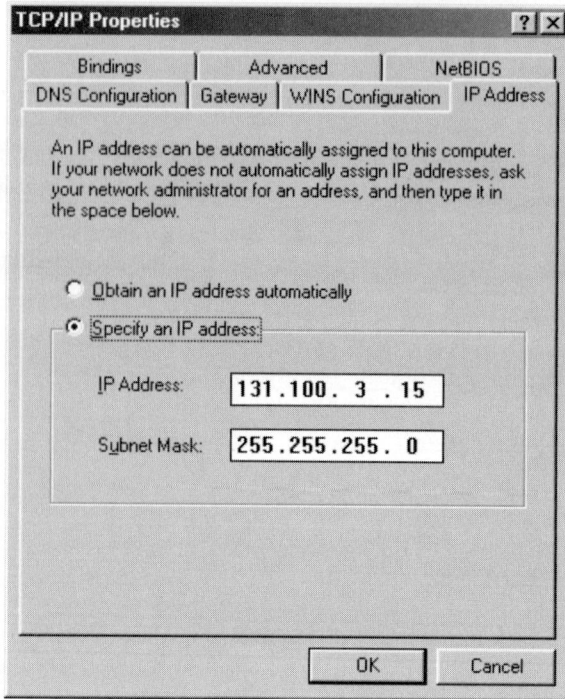

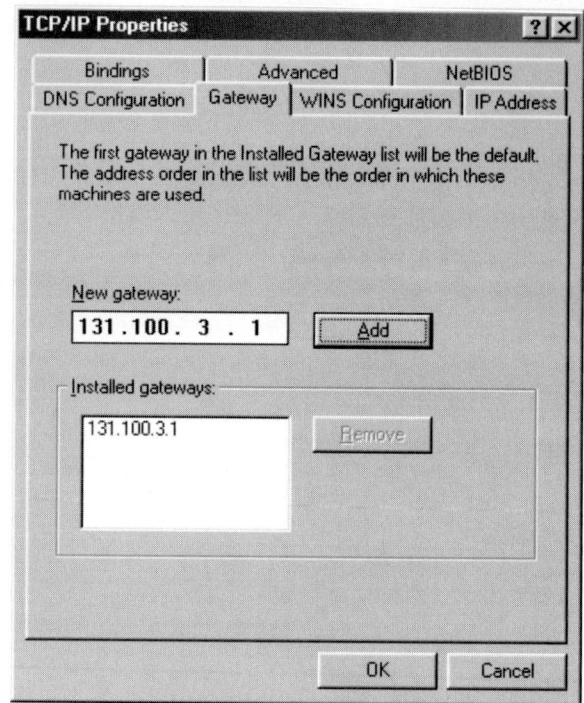

FIGURE 6.3   TCP/IP properties

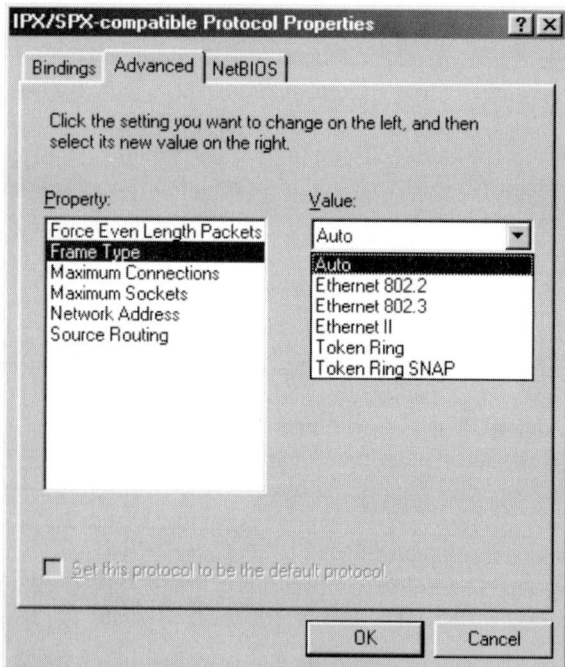

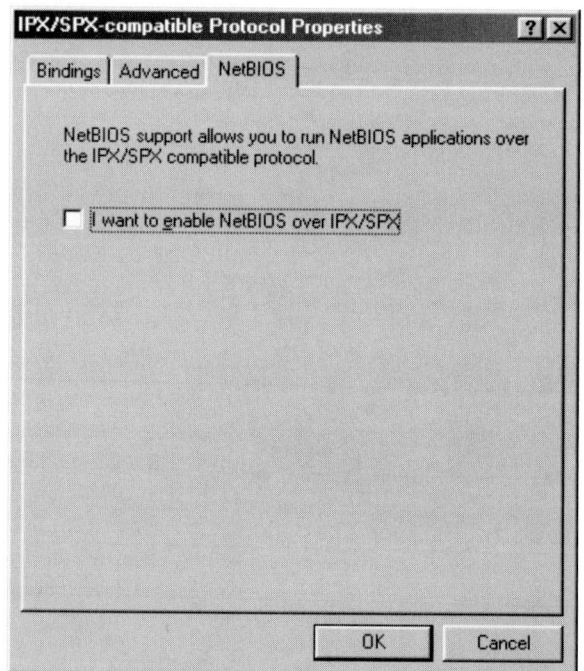

FIGURE 6.4   IPX/SPX protocol properties

If the card is a legacy card, it will go into the properties of the network card and specify the card's resources such as IO address and IRQ. See Figure 6.7a. While most protocols are automatically bonded to the network card, this can be configured in its properties dialog box under the Bindings tab. See Figure 6.7b. In addition, if a card has advanced features, it will usually be configured using the adapter properties dialog box. See Figure 6.8.

162

FIGURE 6.5 NetBEUI protocol properties dialog box

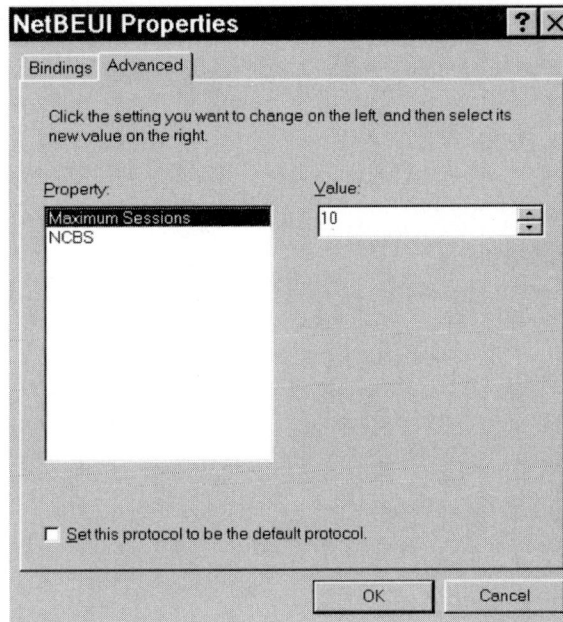

FIGURE 6.6 Adding a network card/adapter driver

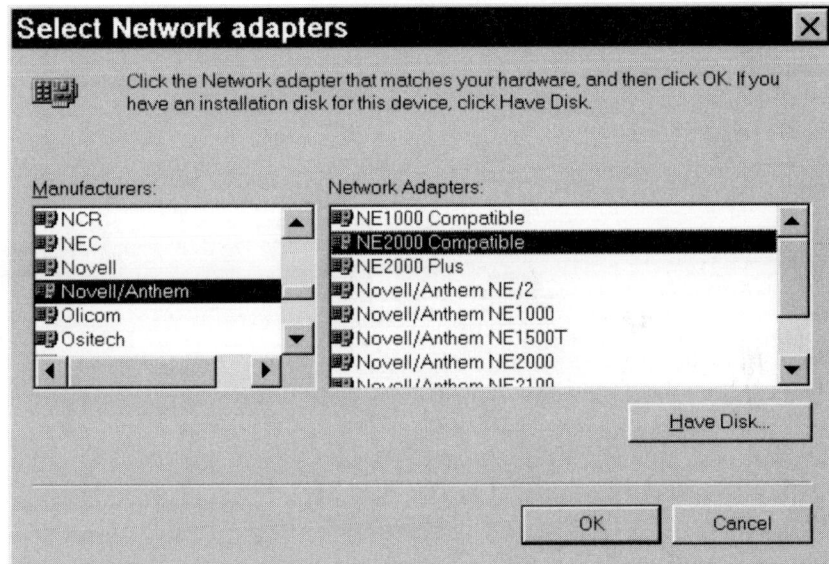

6.1.4 Identification and Access Control

The Identification tab is used to assign the NetBIOS computer name. Much like the MAC addresses, no two computers can have the same computer name on the same workgroup or domain. You can then use utilities such as Network Neighborhood to access the computer and its network resources by using its computer name. The computer name is a maximum of 15 characters. See Figure 6.8.

The last tab in the Network dialog box is Access Control. The share-level security is used in Windows 95/98 to share resources and is secured with a single password for each resource. The user-level security uses the user accounts on a Windows NT server to give access to the network resources. Although there is security when accessing a shared resource from a remote computer (a computer elsewhere on the network), there is no security to files and directories when accessing a Windows 95 locally (when sitting at the actual computer).

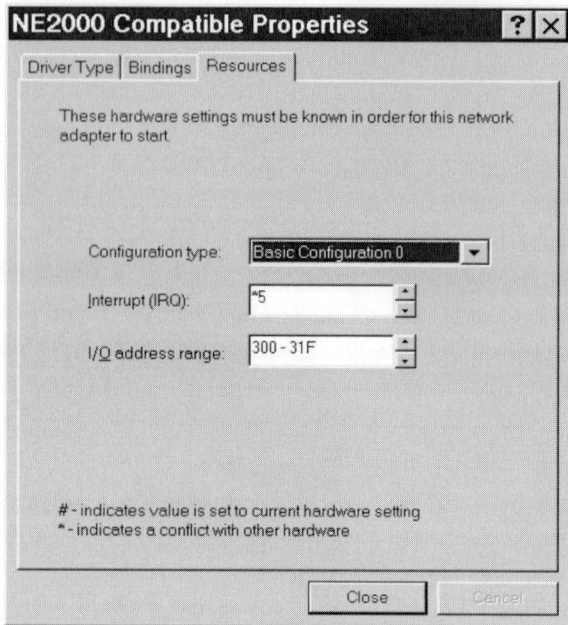

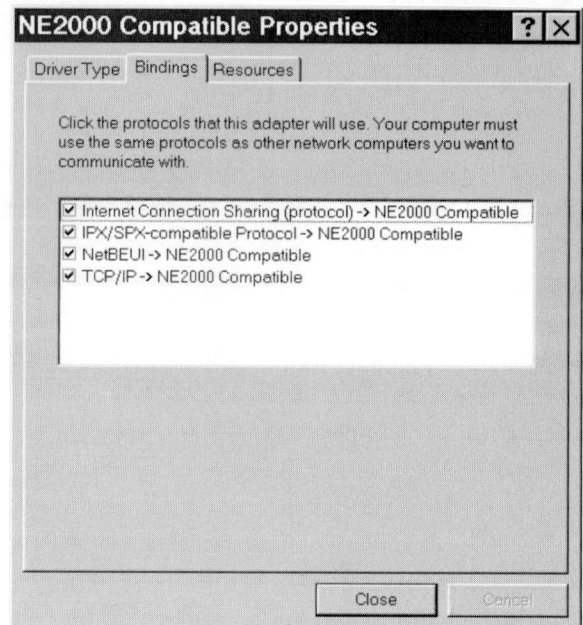

FIGURE 6.7   Network card/Adapter properties dialog box

FIGURE 6.8   Network dialog box

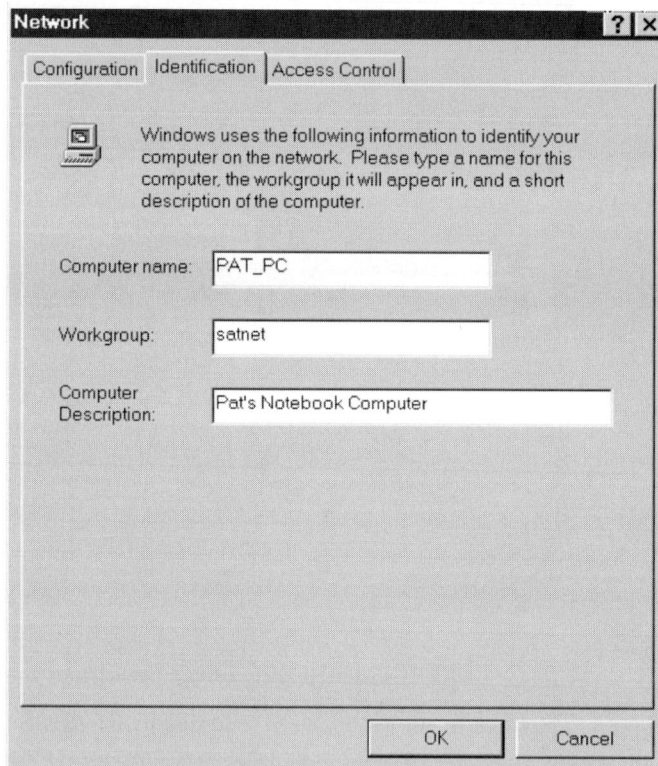

6.2 SHARING DRIVES AND DIRECTORIES

File-and-print sharing for Microsoft networks gives users the ability to share files or printers with Windows for Workgroups, Windows 95, Windows 98, and Windows NT computers. See Figure 6.9. Note that even though you enable file-and-print sharing for an individual computer, you still must share a drive, directory, or printer before it can be accessed through the network.

If users enable file sharing on a Windows 95/98 or Windows NT computer, they can share any drive or directory by right-clicking the drive or directory and selecting the Sharing option from the File menu or by selecting the Sharing option from the shortcut menu of

FIGURE 6.9 File and print sharing dialog box

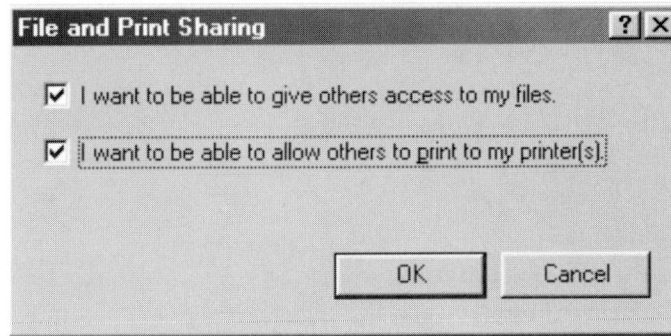

File and Print Sharing   [?] [X]

☑ I want to be able to give others access to my files.

☑ I want to be able to allow others to print to my printer(s).

OK   Cancel

FIGURE 6.10 Sharing drive/directory

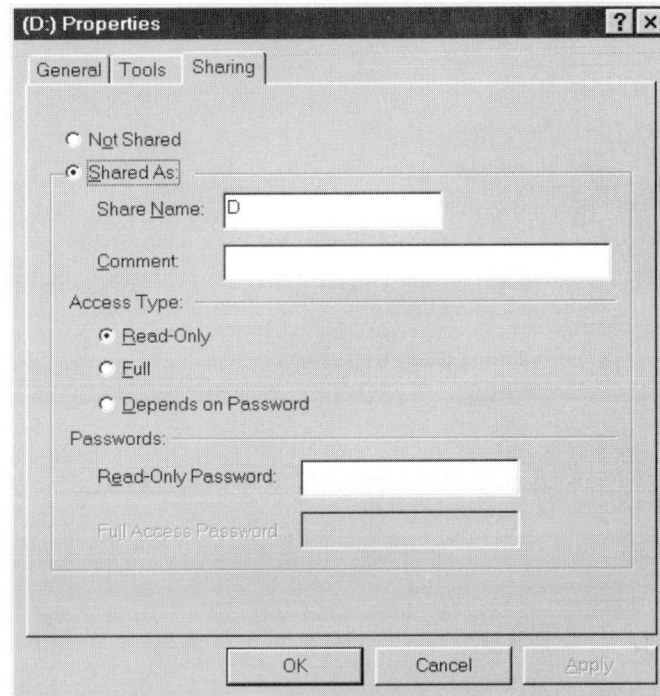

(D:) Properties   [?] [X]

General | Tools | Sharing

○ Not Shared
● Shared As:

Share Name: [D]

Comment: [   ]

Access Type:
● Read-Only
○ Full
○ Depends on Password

Passwords:

Read-Only Password: [   ]

Full Access Password: [   ]

OK   Cancel   Apply

FIGURE 6.11 The hand indicates a shared drive or directory

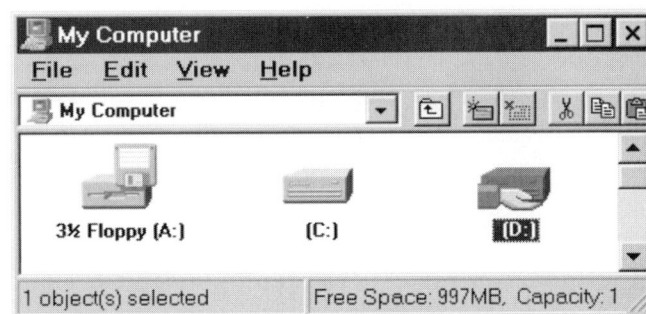

My Computer   [_] [□] [X]

File   Edit   View   Help

My Computer

3½ Floppy (A:)   (C:)   (D:)

1 object(s) selected   Free Space: 997MB, Capacity: 1

the drive or directory. Users would then provide a **share name,** the name seen by other clients and type of access that users can have. See Figure 6.10. A shared drive and directory will be indicated with a hand under the drive or directory. See Figure 6.11.

The shared drive or directory can then be accessed in one of three ways:

1. Using Network Neighborhood and accessing the resources under the server
2. Specifying the **universal naming convention (UNC)** using the Run option under the Start button. The UNC format is specified as \\COMPUTERNAME\SHARENAME.
3. Selecting the Map Drive button on the toolbar of Network Neighborhood or Microsoft Explorer on machines using Active Desktop.

165

FIGURE 6.12 Network Neighborhood showing a computer with network resources

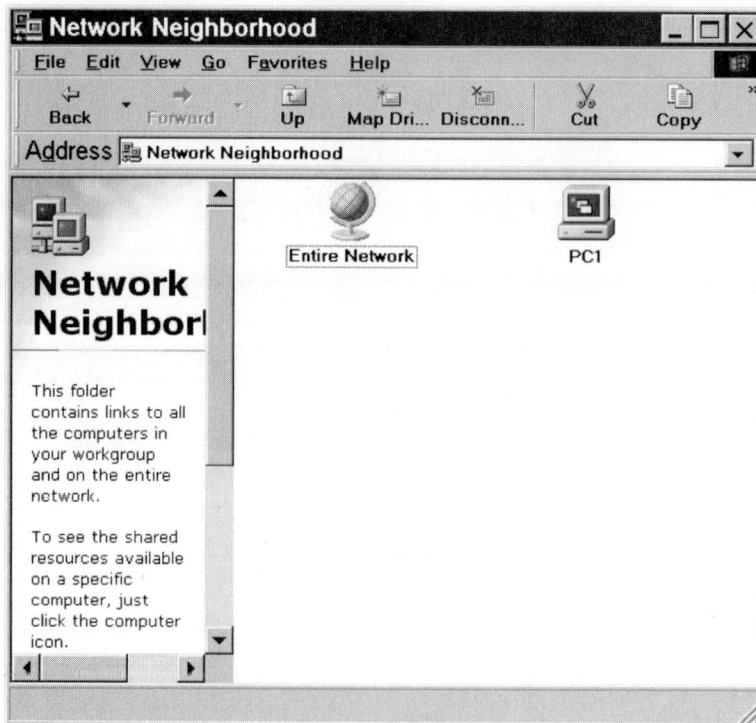

FIGURE 6.13 Network shares shown for a computer on the network

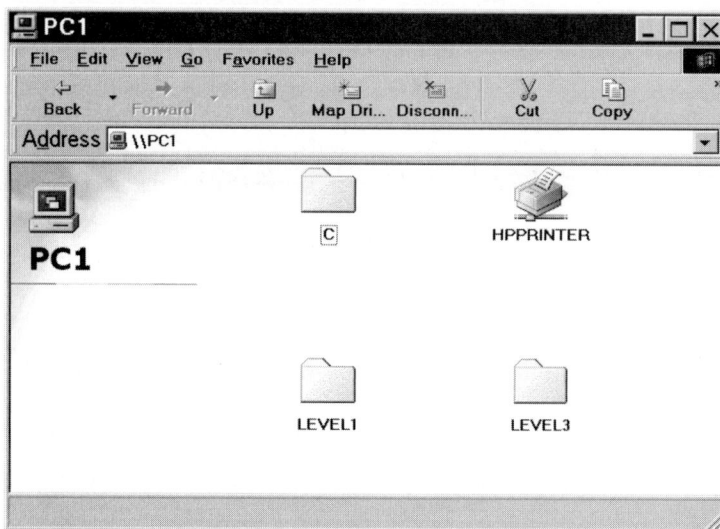

6.3 SHARING PRINTERS

Sharing a printer that is connected to a Windows 95 or 98 machine makes it available to other people. Sharing a printer is similar to sharing files. For example, you first install the print driver using the Add Printer icon in the Printer folder. You then select the printer and choose the Sharing option. After assigning the printer share name and option password, click on the OK button and a hand will appear under the printer to indicate that it is shared. You then go to the client computers and run the Add Printer wizard. Instead of choosing the local printer option, choose the network printer option. You will then be prompted for the UNC name (//COMPUTERNAME/SHARENAME) of the printer. When you print to this printer, the print job will then be automatically redirected to the Windows 95/98 machine and sent to the printer.

6.4 USING NETWORK NEIGHBORHOOD

Network Neighborhood is a **browser service** that shows the available network resources in a graphical environment and allows easy access to them. See Figure 6.12. After installing the appropriate network software (client software, network card driver, and protocol), the Network Neighborhood is automatically placed on the desktop. For example, when you double-click on the icon, it will open a window that displays computers that have network resources available. By double-clicking on one of the computers listed in Network Neighborhood, a new window will open and display the network resources (mostly shared drives, directories, and printers) that are provided by that computer. See Figure 6.13.

SUMMARY

1. In Windows 95 and Windows 98, network components are installed and configured using the Network dialog box, which is accessed by clicking on the Network applet in the control panel or by accessing the shortcut menu of the Network Neighborhood icon and selecting the Properties option.
2. To get full access to the network, install a network client software package, a network adapter driver or dial-adapter driver, and a protocol stack.
3. A redirector is a small section of code in the network operating system that intercepts requests in the computer and determines if they should be left alone to continue in the local computer's bus or redirected out to the network to another server.
4. A designator keeps track of which drive designations are assigned to network resources.
5. The Identification tab in the Network dialog box is used to assign the NetBIOS computer name.
6. No two computers can have the same computer name on the same workgroup or domain.
7. File-and-print sharing for Microsoft networks gives users the ability to share files or printers with Windows for Workgroups, Windows 95, Windows 98, and Windows NT computers.
8. A shared drive and directory will be indicated with a hand under the drive or directory.
9. The shared drive or directory can then be accessed by using Network Neighborhood, by specifying the universal naming convention (UNC) using the Run option under the Start button, or by selecting the Map Drive button on the toolbar of Network Neighborhood or Microsoft Explorer on machines using Active Desktop.
10. The UNC format is specified as \\COMPUTERNAME\SHARENAME.
11. Network Neighborhood is a browser service that shows the available network resources in a graphical environment and allows easy access to them.

QUESTIONS

1. When installing the network components in Windows 95 or Windows 98, which of the following components are required to connect to the network? (Choose all that apply)
   a. network client software
   b. network protocol
   c. network service
   d. network card driver

2. To connect to a Windows NT network, install _____ .
   a. Client Software for Microsoft
   b. Client Software for NetWare
   c. Client Software for UNIX
   d. Client Software for LINUX

3. In Windows 98, to display the network resources that are available on the network, use

   _____ .
   a. Internet Explorer
   b. Network Service Indexer
   c. Network applet in the control panel
   d. Network Neighborhood

4. A hand under a drive or directory indicates the drive or directory is _____ .
   a. password protected
   b. virus protected
   c. shared
   d. invisible to the public

5. A drive letter that is assigned to a directory or drive on a remote computer is known as
   a _____ .
   a. map
   b. pointer
   c. share
   d. ping

6. You are the system administrator for a company. The company's network consists of Windows NT servers, Windows NT workstations, and Windows 95/98 clients. Pat, who you recently hired, needs to configure a new Windows 98 client for a static IP address. You have given her the IP address 132.132.1.1. Pat is not sure where to input the IP address on the system, so you tell Pat to input it in the IP address tab of the _____ .

   a. Network properties
   b. TCP properties
   c. Client properties
   d. TCP/IP properties

7. When installing and configuring networking components, you are asked to provide identification information for each network computer. Which of the following required names does *not* have to be unique?

   a. network name
   b. computer name
   c. workgroup name
   d. IP address

8. Paul wants to share resources with other users on the network. Which service should Paul install?

   a. File and Printer Sharing
   b. Administrator Tools
   c. Microsoft Remote Administration
   d. Peer Resource Sharing

9. You and your friend Jerry are working on a financial report. You open a shared folder on Jerry's computer by using _____ .

   a. System Monitor
   b. Remote Network
   c. Network Neighborhood
   d. Briefcase

10. _____ intercepts requests in the computer and determines if the request should be sent to another server.

    a. Redirector
    b. Designator
    c. Interceptor
    d. Security manager

11. When you map a drive letter to a shared directory, you are using a(n) _____ to point to the shared directory.

    a. redirector
    b. designator
    c. interceptor
    d. security manager

HANDS-ON EXERCISES

EXERCISE 1: PARTITION AND FORMAT THE HARD DRIVE

1. Delete all partitions on your drive.
2. Create a 800-MB primary partition and make it active.
3. Create an extended partition of at least 3.2 GB. Leave at least 200 MB of undesignated space to use in Windows 2000.
4. In the extended partition, create two logical drives, each at least 1.6 GB.
5. Format the C drive and make it bootable.
6. Format the D and E drives.
7. Install the DOS CD-ROM drivers so that the CD drive is active.

EXERCISE 2: INSTALLING WINDOWS 95 OR WINDOWS 98

Insert the Windows 95/98 installation CD in the CD drive and execute the SETUP.EXE file.

- When you get to the Choose Directory dialog box, install Windows in the C:\WINDOWS directory.
- When you get to the Setup Options dialog box, select Typical option.
- When you get to the Analyzing Your Computer dialog box, select Yes (recommended) option so that the setup program can automatically detect hardware devices.
- When you get to the Network Configuration dialog box, do *not* install any network components. This will be done in later exercises.

EXERCISE 3: CREATING AN UNSHIELDED TWISTED-PAIR ETHERNET

1. Determine the type of network card by looking at the card and using its documentation. Include the following:
   - Type of connection
   - Type of expansion slot
   - Speed of the network connection
   - Method of configuration (plug-and-play, software configurable, or jumper/DIP switches)
2. Configure the card for your system.
   a. Plug-and-play card
   - Physically install the card into the system.
   b. Software configurable
   - Start Windows 95 or Windows 98.
   - Using the Windows 95/98 device manager, determine what resources (IRQ, DMA, IO addresses, and memory addresses) are available. The device manager can be accessed by right-clicking on the My Computer icon and selecting the Properties option or double-clicking on the System applet from within the control panel. The resources are shown by highlighting the computer shown at the top of the device manager and clicking on the Properties button.
   - Reboot the computer to the command prompt. This can be done by selecting the Shut Down option from the Start button menu and selecting the Restart in MS-DOS mode.
   - Insert the configuration disk that comes with the network card.
   - Determine the executable file that configures the card and execute it.
   - Within the configuration program, set the resources for the card so that there is no conflicts with any other device.
   - If required, select the card's network speed and connection.
   - Exit the configuration program and shut down the computer.
   - Physically install the card into the computer.
   c. Configuring a legacy card (jumpers/DIP switches)
   - Start Windows 95 or Windows 98.
   - Using the Windows 95/98 device manager, determine what resources (IRQ, DMA, IO addresses, and memory addresses) are available. The device manager can be accessed by right-clicking on the My Computer icon and selecting the Properties option or double-clicking on the System applet from within the control panel. The resources are shown by highlighting the computer shown at the top of the device manager and clicking on the Properties button.
   - Looking at the card's documentation, set the resources for the card so that there is no conflicts with any other device.
   - If required, select the card's network speed and connection.
   - Close Windows and shut down the computer.
   - Physically install the card into the computer.
3. Connect the unshielded twisted-pair cable made in Chapter 4 to the hub and the network card.

EXERCISE 4: INSTALLING THE NETWORK SOFTWARE

To build an actual network, you will need access to two computers, Computer A and Computer B.

1. Open the control panel.
2. Double-click on the Network applet.
3. If there are any network components installed, remove them by highlighting them and clicking on the Properties button.
4. Select the Add button.
5. Select the Client option and click on the Add button.

6. Select the Microsoft option from the Manufacturers list and select Client for Microsoft Networks from the Network Clients list. Click on the OK button.
7. Select the card's manufacturer from the Manufacturers list and select the model of the card from the Models list. Click on the OK button. If the card is not listed, insert the driver disk and use the Have Disk button and browse to the appropriate director on the floppy disk.

> **Note:** It is often recommended to use the newest driver available. These drivers are usually found on the Internet.

8. If the TCP/IP protocol is not listed, click on the Add button and select Protocol from the list and click on the Add button. Then select Microsoft from the Manufacturers list and TCP/IP from Network Protocols. Click on the OK button.
9. From within the Network dialog box, select TCP/IP from the list of installed network components and click on the Properties button.
10. Since we do not have a DHCP server installed that automatically assigns TCP/IP addresses, we manually assign the addresses. Therefore, click on the Specify an IP Address option and enter the following address (unless your instructor specifies differently) and subnet mask.
    • Computer A:
    IP Address: 132.132.1.1
    Subnet Mask: 255.255.255.0
    • Computer B:
    IP Address: 132.132.1.2
    Subnet Mask: 255.255.255.0
11. Click on the OK button.
12. Click on the Identification tab.
13. Unless your instructor specifies differently, enter the following computer name:
    • Computer A: PC1
    • Computer B: PC2
14. Unless your instructor specifies differently, enter "Workgroup" as your name.
15. Click on the OK button.
16. If asked to do so, insert the Windows 95 or 98 installation CD in the CD drive or direct the program to the location of the installation files.
17. If asked to restart the computer, click on the OK button.

EXERCISE 5: TCP/IP UTILITIES

1. Perform WINIPCFG command. Notice the MAC address, the IP address, and the subnet mask.
2. If you are using Windows 98, click on the Start button, select the Programs option, and select the MS-DOS prompt option. Execute the IPCONFIG command. Again, notice the MAC address, the IP address, and the subnet mask.
3. Ping the loopback address (127.0.0.1).
4. Ping the computer's IP address.
5. Ping the IP address of another computer on the same subnet.
6. Ping the computer name of another computer on the same subnet.

EXERCISE 6: FILE-AND-PRINT SHARING

1. From the desktop, select the Network Neighborhood icon and right-click on it. Select the Properties option. You also could open the network dialog box by opening the Network applet in the control panel.

2. Click on the File and Print Sharing button and select the options: I want to be able to give others access to my files and I want to be able to give others access to my printer(s). Click on the OK button.
3. Click on the OK button; and click on the Yes button if asked to reboot the computer.
4. From the desktop, double-click on the My Computer icon.
5. Right-click on the C drive and select the Sharing option.
6. Click on the Shared As: option, and then on the OK button.
7. In the C drive, create a folder called LEVEL1.
8. In the LEVEL1 folder, create a folder called LEVEL2.
9. In the LEVEL2 folder, create a folder called LEVEL3.
10. In the LEVEL1 folder, create a text file called FILE1.TXT. Put your name in the file. You can use EDIT or NOTEPAD.
11. In the LEVEL3 folder, create a text file called FILE3.TXT. Put your name in the file. You can use EDIT or NOTEPAD.
12. From within the C drive, select the LEVEL1 folder. Right-click the folder and select the Sharing option.
13. Select the Shared As: option and leave the Read-Only Access Type option selected. Click on the OK button.
14. From within the LEVEL2 folder, right-click the folder and select the Sharing option.
15. Select the Shared As: option and the Full Access Type option. Then click on the OK button.
16. From another computer on the same subnet and workgroup, double-click on Network Neighborhood.
17. If the computer shows in Network Neighborhood, double-click on it to see the shares. If the computer does not show in Network Neighborhood, click on the Start button and select the Run option. In the Run dialog box, type \\COMPUTERNAME of the other computer and click on the OK button. For example, if you are Computer A, type \\COMPUTERB. If you are Computer B, type \\COMPUTERA.
18. As you can see, the C, LEVEL1, and LEVEL3 shares are displayed, although they are located at different levels on the drive.
19. Open each of the shares.
20. Map a drive letter to the other computer's shared resource by taking the following steps:
    - Start Windows Explorer.
    - Select the Map Network Drive option from the Tools menu.
    - For the Drive pulldown option, select the G drive.
    - For the Path: text box, type \\PCX\LEVEL1, where X is the computer number. For example, if you are using PC2, type \\PC1\LEVEL1 (\\COMPUTERNAME\SHARENAME).
21. Go to the desktop and double-click on the My Computer icon. Double-click on the G drive.

EXERCISE 7: IPX PROTOCOL

1. Display the network dialog box by right-clicking on the Network Neighborhood icon.
2. Click on the Add button, click on the Protocol option from the list, and then click on the Add button.
3. Select Microsoft from the Manufacturers list and IPX/SPX-compatible Protocol from Network Protocols. Click on the OK button.
4. From within the Network dialog box, select IPX/SPX-compatible Protocol from the list of installed network components and click on the Properties button.
5. Click on the Advanced tab and view the Frame Type (choices include Auto, 802.2, 802.3, II, and SNAP) option and the network address (external network address).
6. Click on the OK button.
7. Click on the TCP/IP option, and then click on the Remove button.
8. Click on the OK button.
9. From another computer on the same subnet and workgroup, double-click on Network Neighborhood.

10. If the computer shows in Network Neighborhood, double-click on it to see the shares. If the computer does not show in Network Neighborhood, click on the Start button and select the Run option. In the Run dialog box, type \\COMPUTERNAME, and click on the OK button.
11. As you can see, the C, LEVEL1, and LEVEL3 shares are displayed, although they are located at different levels on the drive.
12. Open each of the shares.

EXERCISE 8: NETBEUI PROTOCOL

1. Display the Network dialog box by right-clicking on the Network Neighborhood icon.
2. Click on the Add button, click on the Protocol option from the list, and then click on the Add button.
3. Select Microsoft from the Manufacturers list and NetBEUI from Network Protocols. Click on the OK button.
4. From within the Network dialog box, select IPX/SPX-compatible Protocol from the list of installed network components and click on the Properties button.
5. Click on the IPX/SPX-compatible Protocol option, and click on the Remove button.
6. Click on the OK button.
7. From another computer on the same subnet and workgroup, double-click on Network Neighborhood.
8. If the computer shows in the Network Neighborhood, double-click on it to see the shares. If the computer does not show in Network Neighborhood, click on the Start button and select the Run option. In the Run dialog box, type \\COMPUTERNAME, and click on the OK button.
9. As you can see, the C, LEVEL1, and LEVEL3 shares are displayed, although they are located at different levels on the drive.
10. Open each of the shares.

EXERCISE 9: TROUBLESHOOTING TCP/IP NETWORK

1. Remove the NetBEUI protocol.
2. Install the TCP/IP network with the following address (unless specified differently by your instructor).
   Computer A:
   IP Address: 132.132.1.1
   Subnet Mask: 255.255.255.0
   Computer B:
   IP Address: 132.132.2.2
   Subnet Mask: 255.255.255.0
3. Try to ping the other computer.
4. Change the address of Computer B to the following addresses:
   IP Address: 132.132.2.201
   Subnet Mask: 255.255.255.0
5. Try to ping the other computer. It should work.

EXERCISE 10: CONNECTING TO THE INTERNET

1. Install the appropriate software to connect to the Internet. You may need to use the Dial-Up Networking program.
2. Install Microsoft Internet Explorer and Netscape Communicator.
3. Connect to the Internet.
4. If you are using a modem to connect to the Internet, execute WINIPCFG or IPCONFIG to view your IP settings.

5. If you are using a modem to connect to the Internet, click on the Details button if using WINIPCFG or execute the IPCONFIG /ALL to see all IP settings.
6. Ping ZDNET.COM.
7. Perform the TRACERT command.
8. Find, download, and install the Adobe Acrobat Reader program and plug in from WWW.ADOBE.COM.
9. Access the IOMEGA.COM site and access one Adobe Acrobat Reader document.

EXERCISE 11: BUILDING A COAXIAL ETHERNET NETWORK

1. Remove the network components such as the Microsoft Client software, TCP/IP protocol, and network card driver.
2. Uninstall the UTP network.
3. Attach a T-connector to the network card.
4. Attach a coaxial segment between two T-connectors.
5. Put T-connectors on the two ends of the coaxial chain.
6. Install the Microsoft Client software, the appropriate network driver, and the TCP/IP protocol and File Sharing.
7. Test the network by trying to access the shared network resources.

EXERCISE 12: CONNECTING MULTIPLE HUBS

1. Disassemble the network.
2. Connect two hubs to a central hub by using either a crossover cable or by using designated ports that allow connection to another hub.
3. If necessary, uninstall the network driver and install the appropriate network card driver. In addition, install any network components that need to be installed.
4. Test the network by trying to access the shared network resources.

EXERCISE 13: GATEWAYS (OPTIONAL)

1. Connect the computer to a larger network.
2. Configure the TCP/IP address and subnet mask that is appropriate to the classroom.
3. Try to ping a computer on another network.
4. Configure the IP gateway (address of the router).
5. Try to ping a computer on another network. It should work.
6. For the IP gateway, put in the address of another computer.
7. Try to ping the other computer.

Unit 2

Windows 2000 Basics

7 Introduction to Windows 2000 Architecture

INTRODUCTION

Previous chapters have focused on connecting a network and getting the network to function. This chapter is the first to specifically discuss Windows 2000. The chapter begins with the various methods of installing Windows 2000, followed by an overview of the OS architecture. Installing the operating system is an important step in learning how to support the OS, but understanding the architecture is not so apparent. In reality, the architecture involves certain terms and concepts, as discussed in this chapter, which will be used throughout the rest of the book.

OBJECTIVES

1. Install Windows 2000 from the installation CD or a network drive.
2. List the system requirements for the various Windows 2000 operating systems.
3. Given a scenario, recommend which Windows 2000 operating system to utilize.
4. Upgrade to Windows 2000.
5. Perform an attended installation of Windows 2000 Professional and Windows 2000 Server.
6. Perform (a) an unattended installation of Windows 2000 Professional and Windows 2000 Server, including use of a server remote installation service (RIS) and a system preparation tool; and (b) an unattended installation of answer files by using Setup Manager to automate the installation of Windows 2000.
7. Upgrade from a previous version of Windows to Windows 2000.
8. Apply update packs to installed software applications.
9. Deploy service packs.
10. Implement, manage, and troubleshoot input and output (I/O) devices.
11. Given a scenario, recommend which of the two types of licensing should be used.
12. Given a common installation problem, determine and fix the cause of that problem.
13. Update drivers.
14. Monitor and configure multiple processing units.
15. Given a scenario, recommend a method to complete automated installations.
16. Compare the two major layers of Windows 2000, user mode and kernel mode.
17. Describe threads and processes and how they relate to each other.
18. Given an application, describe how Windows 2000 will run the program.
19. Describe real memory and virtual memory and how they relate to each other.
20. List and describe the boot sequence.
21. Open and modify the BOOT.INI file.
22. Troubleshoot and fix common boot errors.
23. Demonstrate how to use the Windows 2000 Security dialog box and Task Manager.
24. Set priorities and start and stop processes.

7.1 WHAT IS WINDOWS 2000?

Windows 2000 is an operating system that can function as a desktop operating system or a network file, print server, or application server. Earlier versions of Windows NT used the Windows 3.xx interface while Windows NT 4.0 uses the Windows 95 interface. Windows 2000 uses the Windows 98 active desktop interface (merging of the desktop interface and browser interface), which allows users to put active content from Web pages on to their desktops).

Under these popular interfaces, Windows 2000 uses a different architecture that offers higher performance, greater reliability, and better security than DOS/Windows 3.xx and Windows 95/98. Unfortunately, to achieve these benefits, Windows 2000 is not 100% backward compatible with all hardware and software.

7.2 INSTALLING WINDOWS 2000

To install and run Windows 2000, refer to Table 7.1 for system requirements. Although listed in the Windows 2000 documentation, the requirements in Table 7.1 will offer slow and sometimes unreliable operations. For acceptable performance and reliability, Windows 2000 Professional should have a Intel Pentium II or III processor or better with a minimum of 128 MB of RAM; a Windows 2000 Server should have a Pentium II or III processor or better with a minimum of 256 MB of RAM.

To ensure hardware compatibility with Windows 2000, Microsoft offers users a Hardware Compatibility List (HCL) of devices that have been tested and approved to work with Windows 2000. The list is available on the Windows 2000 compact disk (\SUPPORT\CHL.TXT) and at http://www.microsoft.com/hcl/default.asp. Therefore, users must check this list before installing Windows 2000 or purchasing a system. In addition to ensuring hardware compatibility, users should update to the newest system ROM BIOS for their motherboards.

7.2.1 Installation Process

Windows 2000 can be installed from the installation CD or from a network server. Although most Microsoft programs use SETUP.EXE to install their software, Windows 2000 uses

TABLE 7.1 System requirements for Windows 2000

	Windows 2000 Professional— Intel-Based Machine	Windows 2000 Server— Intel-Based Machine	Windows 2000 Advanced Server— Intel-Based Machine
Processor	• Pentium 133 MHz (or higher) • Supports up to two processors	• Pentium 133 MHz (or higher) • Supports up to four processors	• Pentium 133 MHz (or higher) • Supports up to eight processors
RAM	• A minimum of 64 MB • 128 MB or higher recommended • 4 GB maximum	• A minimum of 128 MB • 256 MB or higher recommended for most network environments • 4 GB maximum	• A minimum of 128 MB • 256 MB or higher recommended for most network environments • 8 GB maximum
Disk space	• 2 GB hard disk with a minimum of 685 MB free space (Additional free hard disk space is required if installing over a network.)	• 2 GB hard disk with a minimum of 1.0 GB free space (Additional free hard disk space is required if installing over a network.)	• 2 GB hard disk with a minimum of 1.0 GB free space (Additional free hard disk space is required if installing over a network.)
Display	• VGA or higher	• VGA or higher	• VGA or higher
Input devices	• Keyboard and pointing device (e.g., mouse, trackball, or glide pad)	• Keyboard and pointing device (e.g., mouse, trackball, or glide pad)	• Keyboard and pointing device (e.g., mouse, trackball, or glide pad)
Networking	• Network card if you need network connectivity	• Network card if you need network connectivity	• Network card if you need network connectivity
Other	• CD-ROM (or network card for installation over the network)	• CD-ROM (or network card for installation over the network)	• CD-ROM (or network card for installation over the network)

Note: A new version called Windows 2000 Data Center Server will support up to 64 GB of memory and 32 processors.

WINNT.EXE file or WINNT32.EXE (both of which are located in the I386 directory on the Windows 2000 installation CD).

WINNT.EXE should be used for a clean installation on a computer that is running Microsoft MS-DOS or Microsoft Windows 3.xx (upgrades of these operating systems are not supported). **WINNT32.EXE** should be used on a clean installation or upgrade on a computer that is running Microsoft Windows NT 4.0, Microsoft Windows 95, or Microsoft Windows 98. WINNT32 can be executed from within the Windows 95, Windows 98, or Windows NT/2000 command prompt. The upgrade paths are shown in Table 7.2.

Before running the installation program or during the installation process, create and size only the partition on which you will install Windows 2000. After installing Windows 2000, you can then use Disk Management to partition and format the remaining hard drive space. While Windows 2000 requires a minimum of 685 MB of disk space, Microsoft recommends

TABLE 7.2 Upgrade paths to Windows 2000

Current OS	Upgrade to
Windows 95	Windows 2000 Professional
Windows 98	Windows 2000 Professional
Windows NT Workstation	Windows 2000 Professional
Windows NT Server	Windows 2000 Server
Windows NT Terminal Service Addition	Windows 2000 Server, Advanced Server
Windows NT Enterprise Edition	Windows 2000 Advanced Server, Data Center
Windows 2000 Advanced Server	Windows 2000 Data Center

Note: There is no upgrade path from Windows 3.xx.

that you install Windows 2000 on a 1-GB partition or larger. A 2-GB partition allows growth if needed. When installing other applications and setup data file storage, then, create them on other partitions.

Windows 2000 supports FAT, FAT32, and NTFS. If you need file and folder security, disk compression, disk quotas (disk space limits for users), or encryption, use NTFS. If you need to dual boot between the first version of Windows 95, use FAT. If you need to dual between the later versions of Windows 95 and Windows 98, use FAT or FAT32.

When executing the WINNT.EXE, the WINNT.EXE programs will copy the installation files from the installation CD or the network drive to a temporary directory (Win_nt.~ls) on the destination hard drive. The system will then reboot and continue with the installation from the installation files in the temporary directory.

The Windows 2000 installation process is broken into four parts: running the setup program, running the Windows 2000 setup wizard, installing Windows 2000 networking components, and completing the installation. The WINNT will allow you to select the partition on which to install Windows 2000, select the file system for the new partition, and give you the option to format the new partition.

When the computer reboots, the setup wizard (Graphical User Interface (GUI)-based program) runs. The setup wizard will set the regional settings, licensing mode, computer name, password, and optional Windows 2000 components such as internet information services (IIS), dynamic host configuration protocol (DHCP), domain name server (DNS), and display settings.

A **software license** is the legal permission for an individual or group to use a piece of software. For Windows 2000, you should have a license to run the Windows 2000 Server and a client access license for each computer that connects to the server. There are two kinds of **client access licenses (CALs)**—per-seat licensing and per-server licensing. The **per-seat licensing** gives a license to access any Windows 2000 server for every specific computer (seat). For example, if you have several servers used by the clients, use the per-seat licensing. The **per-server licensing** gives the maximum number of clients that will connect to that server at any one point in time. For example, if you have multiple servers, ensure that you have sufficient licenses for all connections to all servers. The per-server mode is most economical in a single-server situation or with servers that are only used occasionally. Client access licenses are not required for anonymous or authenticated access to a computer that is running Windows 2000 Server with IIS or some other web server application. In addition, CALs are not needed for Telnet IIS/WWW or FTP connections.

During installation, you need to specify the type of network security group to join: a workgroup or a domain. If you want a computer to join an existing workgroup, assign the name of an existing workgroup. Adding the computer does not give it access to any domain. To join a domain during setup, you must have a computer account in the domain you want to join. If you're upgrading from Windows NT, Setup uses your existing computer account. Otherwise, you'll be asked to provide a new computer account. You would then ask your network administrator to create a computer account during Setup and join the domain. To join a domain during Setup, you would need to provide the user name and password of which has the necessary rights to create a computer account. If you do not have a computer account created and you do not have the rights to add a computer to the domain, finish the installation. Add the computer to the domain later by starting the network and dial-up connections applet in the control panel. Select the network identification hypertext and click the properties button. For more information about workstations and domains, see Chapter 11. For more information about creating computer accounts, see Chapter 12.

During the installation of Windows 2000 networking components, load the Client for Microsoft Networks, detect and load the network card driver, install protocols, and load the file-and-print sharing for Microsoft networks. The setup wizard program will copy the necessary files to the hard disk and configure Windows 2000. During the last phase of installation, it will save the configuration, remove any temporary files, and restart the computer. See Exercise 7.1 at the end of the chapter.

After installing Windows 2000, check if Microsoft has released any fixes or patches and apply them to the Windows 2000 system. For Microsoft Windows NT and Windows 2000 computers, if there are many fixes or patches, Microsoft will release them together as a **service pack.** You should therefore look for the most recent service pack and install that before continuing. In addition, if you make any major changes including adding any services to the computer, then reinstall the service pack. To see if there is a service pack or what the newest service pack is, refer to the http://support.microsoft.com/support/downloads Web site.

7.2.2 Windows 2000 Setup Disks

Users can prepare for the possibility of a system failure on a computer that cannot be started from the CD-ROM drive by creating a set of four Windows 2000 Setup floppy disks so they can start the setup program, start the recovery console, and start the emergency repair process. The best method is to always try safe mode before booting with the floppy disks. For more information on safe mode and repair options, see Chapter 16.

The Windows 2000 Setup floppy disks must match the operating system on the computer that is to be started. Windows 2000 Setup floppy disks that are created from the Windows 2000 Professional setup CD cannot be used to start a computer that is running Windows 2000 Server or vice versa.

To create the four setup disks, perform the following:

1. Insert a blank, formatted, 3.5-inch, 1.44-MB disk into the floppy disk drive.
2. Insert the Windows 2000 CD-ROM into the CD-ROM drive.
3. Click on the Start button and select the Run option.
4. In the Open box, type *d:*\BOOTDISK\MAKEBOOT A: or *d:*\BOOTDISK\ MAKEBT32.EXE A: (where *d:* is the drive letter assigned to the CD-ROM drive) and then click the OK button. MAKEBOOT.EXE runs under DOS and Windows 3.XX, Windows 95, and Windows 98; MAKEBT32.EXE runs under Windows NT and Windows 2000.
5. Eventually, prompts will instruct to insert disks 2, 3, and 4.

7.2.3 WINNT.EXE and WINNT32.EXE Switches

Switches are customization options added to WINNT.EXE or WINNT32.EXE commands to modify the network installation. To use a switch, first type the command, followed by a space, followed by the switch. Some switches require additional information after the command. For example, to force Windows 2000 to install on drive D instead of drive C, type the WINNT /T:D command at the command prompt. If you execute WINNT /? at the prompt, the information displayed in Figure 7.1 will be shown. The /CHECKUP-GRADEONLY switch used with the WINNT32.EXE command will check the system for incompatibilities that will prevent a successful upgrade (same as running chkupgrd.exe utility from Microsoft site) and the /SYSPART:*drive_letter* specifies users to copy the Setup start-up files to a hard disk and mark the disk as active, which can then be installed from the disk onto another computer and will automatically start the next phase of the setup.

7.2.4 Mass Storage Drivers and Hardware Abstraction Layers

As an IT manager, if you have a mass storage controller such as a SCSI or RAID controller for your hard disk, in which the driver does not come with Windows 2000, you must tell Windows where it can locate the third-party driver. Shortly after the first reboot, during the early part of setup, a line at the bottom of the screen will prompt you to press F6 if you need to install a third-party SCSI or RAID driver. When you press the F6 key at this time, it will further prompt you to specify the location of the driver. Before using such a controller, however, check if the controller is on the HCL. If the controller is not supported and you proceed with the installation, an error message will indicate the problem as "inaccessible boot device."

The **Hardware Abstraction Layer (HAL)** is a library of hardware manipulating routines that hides the hardware interface details. It contains the hardware-specific code that handles I/O interfaces, interrupt controllers, and multiprocessor operations so that it can act as the translator between specific hardware architectures and the rest of the Windows 2000 software. As a result, programs written for Windows 2000 can work on other architectures, making those programs portable. Therefore, when loading Windows 2000, be sure to load the appropriate HAL. See Table 7.3.

During setup, typically the correct HAL is loaded. If not, specify which HAL gets installed by pressing the F5 key at the same time that you would press the F6 key as

FIGURE 7.1   WINNT /?

```
Sets up Windows 2000 Server or Windows 2000 Professional.
WINNT  [/s[:sourcepath]]  [/t[:tempdrive]]
        [/u[:answer file]]  [/udf:id[,UDF_file]]
        [/r:folder]  [/r[x]:folder]  [/e:command]  [/a]
/s[:sourcepath]
  Specifies the source location of the Windows 2000 files. The
  location must be a full path of the form x:[path] or
  \Servershare[path].
/t[:tempdrive]
  Directs Setup to place temporary files on the specified Drive
  and to install Windows 2000 on that drive. If you do not
  specify a location, Setup attempts to locate a drive for you.
/u[:answer file]
  Performs an unattended Setup using an answer file (requires
  /s). The answer file provides answers to some or all of the
  prompts that the end user normally responds to during setup.
/udf:id[,UDF_file]
  Indicates an identifier (id) that Setup uses to specify how a
  Uniqueness Database File (UDF) modifies an answer file (see
  /u). The /udf parameter overrides values in the answer file,
  and the identifier determines which values in the UDF file are
  used. If no UDF_file is specified, Setup prompts you to insert
  a disk that contains the $Unique$.udb file.
/r[:folder]
  Specifies an optional folder to be installed. The folder
  Remains after Setup finishes.
/rx[:folder]
  Specifies an optional folder to be copied. The folder is
  Deleted after Setup finishes.
/e Specifies a command to be executed at the end of GUI-mode Setup.
/a Enables accessibility options.
```

TABLE 7.3   Various HALs available in Windows 2000

HAL	Description
ACPI MultiProc	Use for Multiprocessor ACPI systems.
ACPI UniProc	Use for ACPI Multiprocessor board but with a single processor installed.
Advanced Configuration Power Interface PC	Use for single-processor motherboard with single-processor ACPI system.
Compaq Systempro	Use for Compaq Systempro computers.
MPS Uniprocessor PC	Use on non-ACPI systems dual-processor motherboard with a single processor installed.
MPS Multiproc PC	Use on non-ACPI systems with a dual processor running.
Standard PC	Use on any standard PC, non-ACPI, or nonMPS.

mentioned previously. If you select the wrong HAL, unexpected results may occur including an inability to boot the computer.

If Windows 2000 was originally installed on a computer with a single processor and you change your system to use multiple processors, then the HAL on your computer must be updated so that it will recognize and use the multiple processors. To change the HAL, take the following steps:

1. Click on the Start Button. Select Control Panel and then select System.
2. Click the Hardware tab and click on the Device Manager button. (The Device Manager will be explained in more detail in Chapter 8.) See Figure 7.2.
3. Double-click the Computer branch to expand it. Note your type of current support.

FIGURE 7.2 Changing the HAL in Device Manager

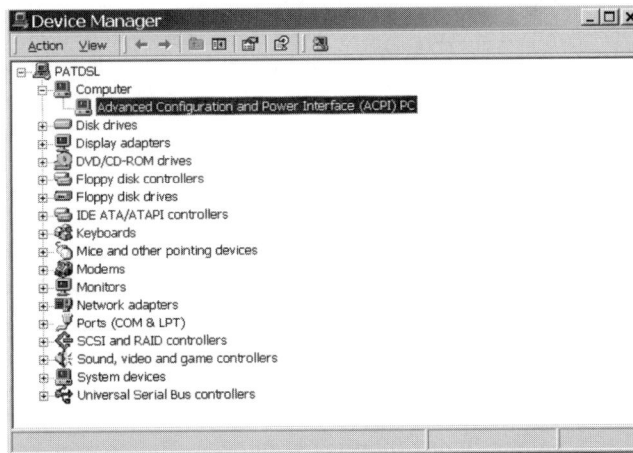

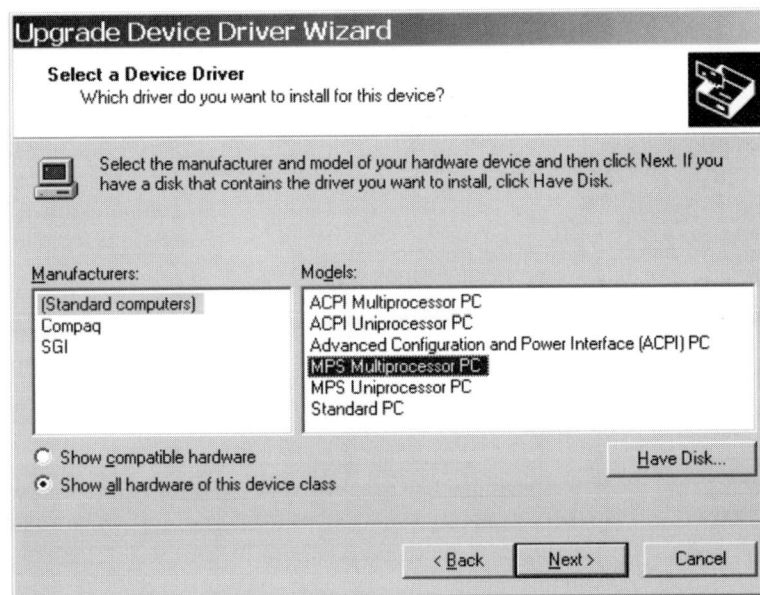

4. Double-click the computer type listed under the Computer branch, click the Drivers tab, click Update Driver, and then click Next.
5. Click "Display a list of known drivers for this device," and then click "Show all hardware of this device class."
6. Click the appropriate computer type (one that matches your current type, except for multiple processors), click Next, and then click Finish.

Use this procedure only to upgrade from a single-processor HAL to a multiple-processor HAL. If you use this procedure to change from a standard HAL to an ACPI HAL (e.g., after a BIOS upgrade) or vice versa, unexpected results may occur including an inability to boot the computer.

7.2.5 Troubleshooting Installation Problems

Some common troubleshooting problems are listed in Table 7.4. For more information, refer to Windows 2000 Server Help files.

7.2.6 Automating Installations by Using Windows 2000 Setup Manager Wizard

When numerous computers will be running Windows 2000 and they all have to be installed, the Windows 2000 Setup Manager is of great help. The **Windows 2000 Setup Manager** wizard allows users to create an **answer file, uniqueness database file (UDF)** and a batch

TABLE 7.4   Common troubleshooting problems

Problem	Troubleshooting Assistance
Media errors	There is an error reading the installation files. If it is a CD, try cleaning the CD and the drive. If this fails to work, try a replacement CD. If installing over the network, check the integrity of the disk and files.
Nonsupported CD-ROM drive	The CD drive is not compatible. Replace the CD-ROM with one that is supported or try installing over the network.
Insufficient disk space	The destination does not have enough space to install Windows NT. Delete some files from the destination drive, reformat the destination drive, or delete and re-create larger partitions.
Failure of dependency service to start	The network services could not start. In the Windows 2000 Setup Wizard, return to the Network Settings dialog box and verify the protocol and network card settings.
Inability to connect to the domain controller	The computer cannot connect to the domain controller. Verify that the domain name is correct and check protocol and network card settings. If you cannot locate a domain controller, install the computer into a workgroup and join the domain after the installation.
Failure of Windows 2000 to install or start	A hardware device on the computer has failed, which prevents Windows 2000 from installing or starting. Verify that Windows 2000 is detecting all hardware and the hardware is on the HCL.

file that launches the Windows 2000 setup program with the answer file as the command-line argument, which can be used as scripts when using the WINNT command. The answer file is used as a generic script for many installations whereas the UDF is used to supply unique information for an installation. The UDF (an ASCII text file created with a text editor such as Notepad) will identify the part of the answer file that data will be merged or replaced in the answer file. The Setup Manager (SETUP.EXE) can be found in the \SUPPORT\RESKIT directory on the Windows 2000 resource kit CD.

EXAMPLE 1   Your job is to install Windows 2000 on 40 desktop computers and 25 notebook computers. To begin, you run the Setup Manager to create two answer files, one for the desktop computers and one for the notebook computers. Only one UDF is needed to supply all of the user-specific or computer-specific information. For example, if you configure all computers to join the SALES domain, you would specify JoinDomain=Sales under the [Network] section in the answer file.

7.2.7   Installing Windows 2000 Over a Network

To install Windows 2000 from a network is very similar to installing it from a CD except the location of the source file is on a distribution server (a server with a shared drive or directory with a copy of the installation files, particularly the contents of the i386 folder). To be able to perform the installation, you will need a FAT partition with at least a 850 MB partition on the target computer. Once the partition is set, install a network client on the client computer. If the client does not have an operating system, you need to boot from a disk that includes a network client, which enables the operating system to connect to the distribution server. After connecting to the distribution server and accessing the shared drive or directory with the installation files, start the WINNT or WINNT32 command. See Appendix B to learn how to install the network client from floppies or to create the boot disk with the network client.

7.2.8 Remote Installation Services

The **remote installation service (RIS)** is used to remotely install Windows 2000 Professional; thus, the topic fits in this chapter that describes various ways of installing Windows 2000. Unfortunately, some of the utilities and services that enable RIS to function are not explained until later in the book, so refer back to this section after learning those utilities and services.

RIS ships as part of the Windows 2000 Server operating system. RIS enables client computers to connect to a server during the initial start-up phase and remotely install Windows 2000 Professional. This reduces the time that administrators spend visiting all computers on a network, thereby reducing the cost of deploying Windows 2000 Professional. Unlike a standard network installation performed by running WINNT.EXE, a remote installation does not require users to know where the installation source files are stored or what information to supply during the setup program. Although any Windows 2000 server can host RIS, there are some prerequisites to be met before using RIS. They are as follows:

- RIS requires DNS, DHCP, and active directory somewhere on the network. See Chapters 10, 11, and 12.
- The drive on the server where RIS is installed must be formatted with the NTFS. See Chapter 13.
- RIS cannot be installed on the same drive or partition as Windows 2000 Server (boot partition).
- Since RIS requires a significant amount of disk space, ensure that the chosen drive contains enough free space to hold the images of files to be installed. Free space is a minimum of 800 MB.

To install RIS:

1. On a Windows 2000 Server, click the Start button, select the Settings option, and select Control Panel.
2. Double-click the Add/Remote Programs applet and double-click the Add/Remove Windows Components option.
3. Scroll down and select Remote Installation Services and click on the Next button.
4. Insert the Windows 2000 Server CD-ROM into the CD drive and click the OK button. The necessary files are copied to the server. After the CD is entered, a dialog box asks if you want to upgrade to the operating system. Click No and exit this screen.
5. Click the Finish button to end the wizard.
6. When prompted to restart your computer, click the Yes button.

To configure the RIS:

1. Click the Start button, select the Run option, and type RISETUP.EXE and click on the OK button. This starts the RIS setup wizard.
2. The Welcome screen appears, indicating some of the requirements to successfully install RIS. Click the Next button.
3. The next screen prompts you to enter the server drive and directory where you would like to install the RIS files. The default drive and directory will be the largest nonsystem, nonboot, NTFS-formatted drive. Click the Next button. The drive on which you choose to install RIS must be formatted with the NTFS file system. RIS requires a significant amount of disk space and cannot be installed on the same drive or partition on which Windows 2000 Server is installed. Ensure that the chosen drive contains enough free disk space for at least one full Windows 2000 Professional compact disc—a minimum of approximately 800 MB to 1 GB.
4. The setup wizard prompts you to either enable RIS at the end of setup or disable the service to allow modification of specific server options before servicing client computers. These options are as follows:
   - *Respond to client computers requesting service.* This option controls whether this RIS server responds to client computers that are requesting service at the end of setup. If this option is checked, the server will respond to clients and provide them with OS installation options. If unchecked, this RIS server will not respond to clients that are requesting service.

- *Do not respond to unknown client computers.* This option controls whether this server responds to unknown client computers that are requesting a remote installation server. A client computer is known if a managed computer account object exists for it within the active directory. This allows the administrator to offer only authorized—that is, prestaged within the active directory—computers the OS installation options from this RIS server. This setting also provides support for multiple remote boot or install servers from different vendors on one physical network. For example, if another vendor's remote install/boot server exists on the same network as the RIS server, you cannot control which server answers the client computer's request. Setting this option and prestaging client computers ensures that this RIS server will service only prestaged client computers.

5. If you select "Respond to client computers requesting service" and click the Next button, the setup wizard prompts you for the location of the Windows 2000 Professional installation files. RIS supports the remote installation of Windows 2000 Professional only. Insert the Windows 2000 Professional CD into the server's CD drive and type the drive letter containing the CD (or browse to a network share that contains the installation source files). Click the Next button.

6. The wizard prompts you to enter the directory name that will contain the workstation files on the RIS server. This directory is created beneath the directory specified in step 3. The directory name should reflect its contents.

7. You are prompted for a friendly description and help text that describes this OS image. The friendly description and help text is displayed to users during the client installation wizard (OSChooser) at initial start-up on a remote client. The help text is displayed when the user selects the description within the client installation wizard. As an IT manager, make sure you provide clear help text to your users, to ensure that they choose the correct OS option at installation time. Click Next to accept the default name of Microsoft Windows 2000 Professional.

8. At this point, you are presented with a summary screen indicating the choices you have made. Click the Finish button to confirm your choices. Once the installation wizard completes, you are ready to either service client computers or additionally configure the RIS settings.

9. Wait while the wizard installs the service and settings you have selected. This takes several minutes.

10. When the installation is done, click on Done.

To administer the majority of the RIS configuration settings, you will use the Active Directory Users and Computers console. To start this console, click Start, select the Programs option, select the Administrative Tools, and then select the Active Directory Users and Computers. On the tree in the left pane, right-click the RIS and select properties. The Remote Install tab determines how this RIS responds to client computers that request service. Then click on the Advanced Settings button to specify the ways in which client computers are installed and to manage the images. See Figure 7.3.

An **image** is a copy of all files of an operating system. It is not a disk image such as those used with Disk Image or Ghost. The Images tab is used for managing the client operating system images that are installed on a RIS server. Its options allow an administrator to add, remove, or modify the properties of an operating system image. There are two types of images that can be displayed on the Images tab:

- A CD image is a copy of all Windows 2000 installation files with answer files. It is not a disk image. When you install RIS, you will begin with the RISTANDARD.SIF answer file. You can then configure the RISTANDARD.SIF file or you can create additional answer files by using the Setup Manager wizard.
- Remote installation preparation (RIPrep) images are created by using the wizard on a source computer that has both Windows 2000 Professional and its applications already configured.

Upon start-up, RIS client computers can connect to a RIS server to install Windows 2000 Professional remotely or to run diagnostic and maintenance utilities. The only client computers that support remote installation are as follows:

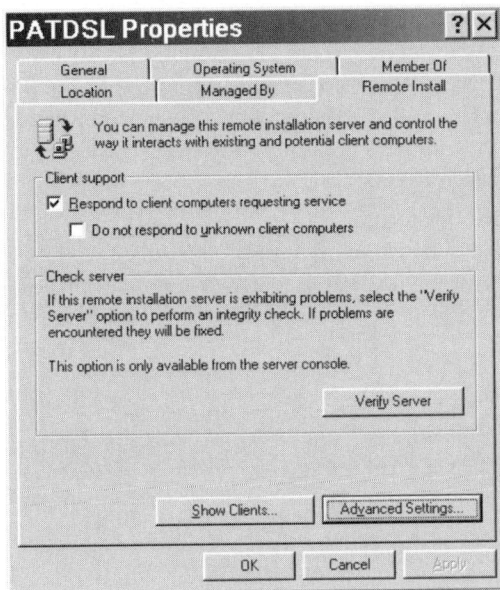

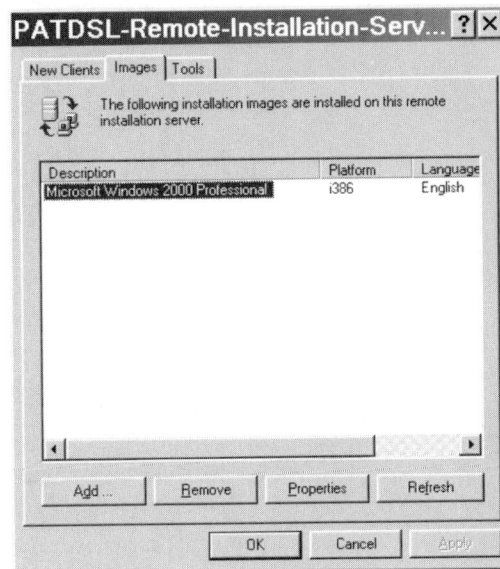

FIGURE 7.3   Configuring Windows 2000 RIS

- Network PC (net PC)
- A network adapter with a PXE-based boot ROM that allows the computer to remotely start the computer with the network card
- A supported network adapter and a remote installation start-up disk

The remote installation boot disk can be used with computers that do not contain a remote boot-enabled ROM on the network card. The boot disk is designed to simulate the PXE boot process for computers that lack support DHCP PXE-based remote boot ROM. The boot disk generator utility is called RBFG.EXE, which is located within the \RemoteInstall\ admin directory on the RIS.

7.2.9   System Difference Tool

Another useful tool for installing many computers is the **system difference tool (SYSDIFF.EXE),** which is also available in the resource kit. The system difference tool enables users to automatically distribute and install applications and limited hardware devices during or after the Windows 2000 setup.

To use the system difference tool, run SYSDIFF.EXE /SNAP SNAP_FILE on a reference computer to take a snapshot of the computer before installation of the applications. After the applications have been installed, then use the SYSDIFF /DIFF SNAP_FILE DIFF_FILE to create the difference file that will contain only the changes that occurred during the installation of the applications. You would then use the SYSDIFF.EXE /APPLY\\ SETUPSERVER\W2K\DIFF_FILE to add or apply the difference file to the other computers.

7.2.10   Disk Image Software

Another way to quickly install Windows 2000 on computers that have identical hardware would be to use **disk copying/disk cloning software,** where the software will create a file containing the disk image of the hard drive. You can then place the image file to a central location and use the disk cloning software to copy the image to other computers. Popular disk cloning software packages include Symantec Ghost Professional and Microhouse ImageCast Deluxe.

Windows 2000 uses a **security identifier (SID)** to uniquely identify the computer, even if the computer name has been changed. Because it can identify a computer, it can stop someone from installing a Windows 2000 computer with the same computer name to impersonate a server. The problem when using a disk image software package is the package also copies the SID to the other computer. Therefore, all the computers have the same SID.

Windows 2000 has the **system preparation tool (SYSPREP.EXE)** that will remove the security identifiers and all other user-specific or computer-specific information from a Windows 2000 prepared system. You would then use the disk image software to create and distribute the disk images. New security identifiers will be created when the new computers are restarted.

The System Preparation Tool has the following syntax:

SYSPREP.EXE [/QUIET] [/NOSIDGEN] [/PNP]

/QUIET—runs SYSPREP in the silent mode where no messages are displayed.

/NOSIDGEN—runs SYSPREP.EXE without generating a SID. Normally each computer must have an individual SID or security identifier. Use this option only to make a clone of a computer as a backup or to allow the end user to customize the computer.

/PNP—Forces plug and play to refresh when the computer reboots to redirect all the hardware devices in the computer.

/REBOOT—reboots the computer after SYSPREP.EXE runs.

7.3 WINDOWS 2000 ARCHITECTURE

Windows 2000 is a modular operating system that has many small, self-contained software components that work together to perform the various operating system tasks. Each component provides a set of functions that acts as an interface to the rest of the system. See Figure 7.4.

7.3.1 User Mode

The Windows 2000 architecture can be divided into two major layers, user mode and kernel mode (also referred to as privileged mode). The **user mode** programs run in ring 3 of the **Intel 386 microprocessor protection model,** and are protected by the operating system. It is

FIGURE 7.4 Windows 2000 architecture

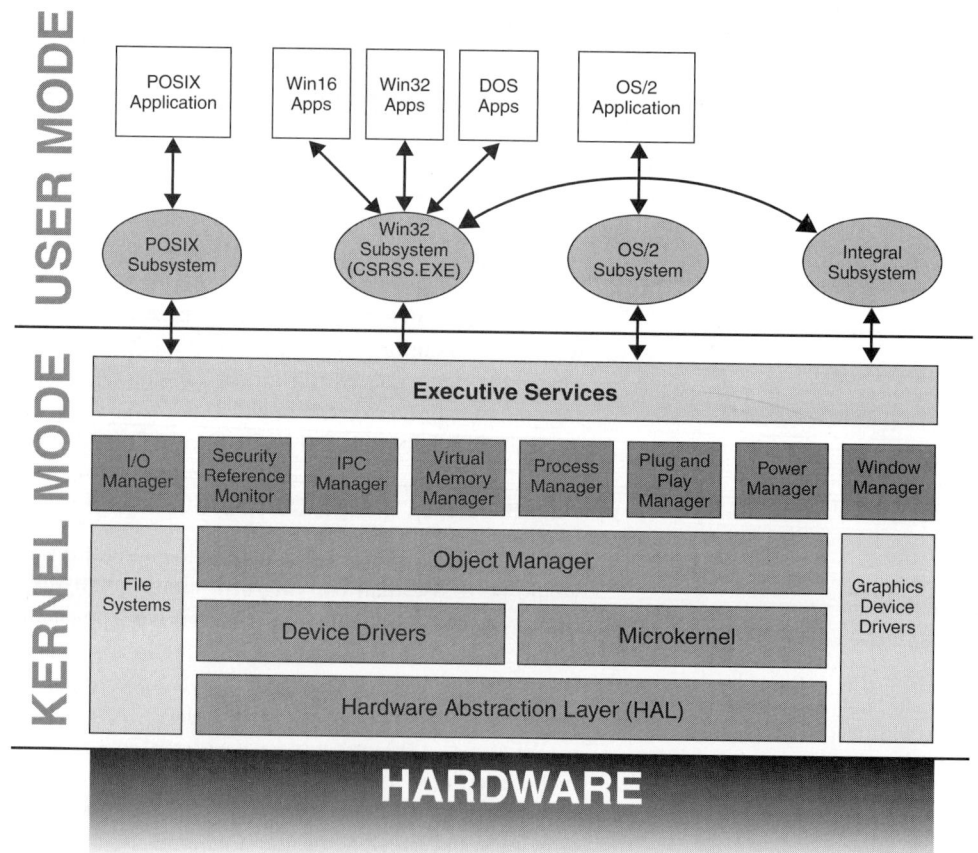

188

a less-privileged processor mode that has no direct access to hardware and can only access its own address space. Since programs running in ring 3 have very little privilege compared with programs running in ring 0 (kernel mode), the programs in user mode should not be able to cause problems with components in kernel mode. Since most of the user's applications run in ring 3, an application can be terminated without causing problems with the other applications that are running on the computer. In addition, programs executed in user mode do not have direct access to hardware. If a program tries to access the hardware directly, the program will be terminated; thus, the system can maintain reliability and security.

The user mode layer contains the environment and integral subsystems. The **environment subsystem** provides the application programming interfaces (APIs) for the programs and converts them to the proper calls for Windows 2000. APIs are a set of routines, protocols, and tools for building software applications. In the Windows environment, APIs provide building blocks, which a program can then use, and a common interface. For example, when a Win32-based application such as Microsoft Word 97 asks the operating system to do something, the Win32 subsystem captures that request and converts it to commands that Windows 2000 can understand. The environment subsystems include the following:

- **32-bit Windows-based system (Client Server Runtime Subsystem—CSRSS.EXE)**—Handles windows and graphic functions for all subsystems. It provides a working environment for Win32, Win16, and DOS-based applications.
- **OS/2 subsystem**—Provides APIs for 16-bit character mode (DOS mode or bounded) OS/2 applications.
- **POSIX subsystem**—Provides API for POSIX-compatible UNIX applications.

Windows 2000 32-bit Windows-based subsystem controls the input/output (I/O) between subsystems so that all applications have a consistent user interface.

The **integral subsystem** performs essential operating system functions such as the security subsystem, the workstation service, and the server service. The security subsystem tracks rights and permissions associated with user accounts, tracks which system resources are audited, and performs login authentication. The workstation services allow access to the network, and the server service provides API to access the network server.

7.3.2 Kernel Mode

The **kernel mode** components run in ring 0 of the Intel 386 (I386) microprocessor protection model. Whereas user mode components are protected by the OS, kernel mode components are protected by the microprocessor. It has direct access to all hardware and all memory including the address space of all user mode processes. It includes the Windows 2000 Executive, Hardware Abstraction Layer (HAL), and the Microkernel.

The **Windows 2000 Executive** services consist of managers and device drivers. They include the I/O Manager, Object Manager, Security Reference Manager, Virtual Memory Manager (VMM), Win32K Window Manager and GDI, local procedure call facility, hardware device drivers, and graphics device drivers. See Table 7.5.

A **kernel** is the central module of an operating system. It is the part of the operating system that loads first and it remains in RAM. Because it stays in memory, a kernel must be as small as possible but still provide all the essential services required by other parts of the operating system and applications. Typically, the kernel is responsible for memory management, process and task management, and disk management.

The **Microkernel** is the central part of Windows 2000, which coordinates all I/O functions and synchronizes the activities of the Executive services. Much like the other kernels in Windows 3.XX and Windows 95/98, the Microkernel determines what is to be performed and when it is to be performed, while handling interrupts and exceptions. Lastly, it is designed to keep the microprocessor(s) busy at all times.

Device drivers are programs that control a device. They each act like a translator between the device and programs that use the device. Each device has its own set of specialized commands that only its driver knows. While most programs access devices by using generic commands, the driver accepts the generic commands from the program and translates them into specialized commands for the device. Device drivers are installed into the

TABLE 7.5 Windows 2000 Executive

Component	Function
I/O Manager	Manages input from and delivers output to the file systems, device drivers, and software cache.
Security Reference Monitor	Enforces security policies on the local computer.
Interprocess Communication (IPC) Manager	Manages communications between clients and servers. It includes the following: • Local procedure call (LPC)—Manages communication when clients and servers exist on the same computer; and • Remote procedure call (RPC)—Manages communications when clients and servers exist on separate computers.
Virtual Memory Manager (VMM)	Implements and controls memory (physical and virtual).
Process Manager	Creates and terminates processes and threads.
Plug-and-Play Manager	Maintains central control of the plug-and-play process. It communicates with device drivers so that drivers can be directed to add and start devices.
Power Manager	Controls power management APIs, coordinates power events, and generates power management requests.
Window Manager and graphical device interface (GDI)	A device driver (WIN32K.SYS) that manages the display system and contains the Window Manager and GDI. The Window Manager will control window displays, manage screen output, and receive input from devices such as the keyboard and mouse. The GDI contains the functions that are required for drawing and manipulating graphics.
Object Manager	Creates, manages, and deletes objects that represent OS system resources such as processes, threads, and data structures.

Executive services section of Windows 2000. Many drivers utilized by Windows 2000 use the **Windows Driver Model (WDM).**

7.3.3 Windows 2000 Processing

A **program,** sometimes referred to as a **process,** is an executable framework that follows a sequence of steps. A program consists of initializing code and data, a private memory address space, system resources (such as files, communication ports, and Windows resources), and one or more threads.

A **thread** is part of a program that can execute independently of other parts. Operating systems that support multithreading enable programmers to design programs whose threaded parts can execute concurrently. This can enhance a program by allowing multitasking, improving throughput, enhancing responsiveness, and aiding background processing. For example, Microsoft Word uses a thread to respond to keys typed on the keyboard by the user to place characters in a document. Other threads are used to check spelling and to paginate the document as the user types. If the user prints, another thread spools or feeds a document to the printer in the background. When a program uses multithreading, the programmer must carefully design the program in such a way that all the threads can run at the same time without interfering with each other.

Windows 2000 is a **preemptive** multitasking operating system, which means the system assigns time slices to threads, tasks, and applications. Threads, tasks, and applications with a higher priority get a larger time slice. At the end of the time slice, the kernel can take back control of the computer without permission from the thread or program. See Figure 7.5.

FIGURE 7.5 Programs under Windows 2000

PREEMPTIVE MULTITASKING

PREEMPTIVE MULTITASKING

Operating System
including
Kernel
GDI
User

Win32 App

Win32 App

Win16 App

Win16 App

Win16 App

Cooperative Multitasking

DOS Virtual Machine/ DOS App

DOS Virtual Machine/ DOS App

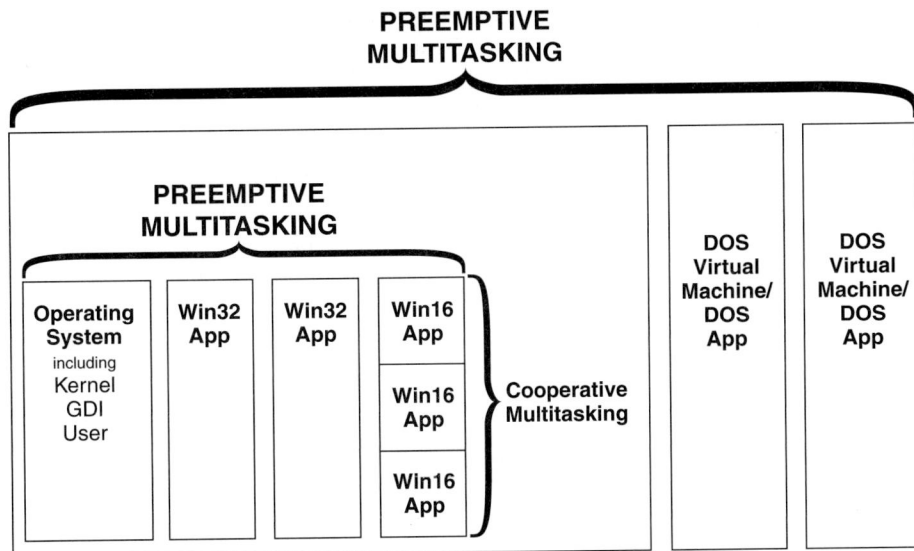

Win16-based applications (used by Windows 3.xx applications) use cooperative multi-tasking. Cooperative multitasking depends on applications that cooperate with each other. Since Windows 3.xx uses cooperative multitasking, the application uses the microprocessor and voluntarily gives control back for other programs. Poorly written software applications that do not give up control can cause problems for other applications.

Win16-based applications in Windows 2000 cannot take advantage of preemptive multitasking. Instead, they share the same memory area or **virtual machine (VM)** and will all be assigned a single time slice. Unfortunately, if a program causes problems, it can cause all Win16-based applications to crash; however, all Win16-based applications are run pre-emptive with Windows 32-based applications and other Windows 95 processes to improve the performance of Win16-based applications.

When the Windows 2000 kernel controls access to the microprocessor, Windows 2000 assigns 32 different priorities to each process, numbered 0 to 31. The levels are divided as follows:

- Levels 0–15 are used for user mode components.
- Levels 16–31 are reserved for kernel mode components.

The Windows 2000 kernel assigns each process a base priority level as well as to each thread that executes within that process. The base priority for a thread can be adjusted by the kernel by two levels below to two levels above the one initially assigned. For example, if a thread were assigned a priority level of 8, the range would be from 6 to 10. If the thread is accepting user input, the kernel might raise the thread's priority level and if the thread requires computations, the kernel might assign it a lower priority.

7.3.4 Windows 2000 Memory Model

Physical memory refers to the actual RAM in the system. Virtual memory is disk pretending to be RAM, allowing for the use of more RAM than what users physically have. Since the disk is much slower than RAM, virtual memory is much slower than physical RAM. The advantage of using virtual memory is it will allow users to run more programs than would have been possible with only the physical memory. The file that makes up virtual memory in Windows NT and Windows 2000 is called a paging file (PAGEFILE.SYS). If a system has multiple hard disks, create a paging file for each disk to increase paging performance.

The Windows NT and Windows 2000 memory model is based on a flat, linear, 32-bit address space, which is managed by the Virtual Memory Manager (VMM). Remember that a 32-bit address (11111111 11111111 11111111 11111111) is the largest binary number that a 32-bit address can form, which equates to 4 GB of memory.

Whatever memory the system has, the VMM groups all physical and virtual memory together and assigns memory addresses to every byte. As processes (programs) request memory resources, the VMM will assign the memory resources to the process and protect them so that another program does not interfere with the memory space of another process.

Recall from Chapter 2 that the Intel microprocessors use demand paging, which swaps data between the RAM and disk only on the demand of the microprocessor. Since the processor does not try to anticipate what data will be needed from the hard drive, the code and data that are stored on disk as virtual memory will remain on disk until they are needed.

In Windows 2000, the physical and virtual memory are grouped into 4-KB blocks called pages. The process of moving data in and out of physical memory is called paging. When physical memory becomes full and a thread needs access to code, or data are not currently in physical memory, the VMM moves some pages from physical memory to a storage disk called a page file. Since this frees RAM, there is now room to load the requested code or data into RAM.

The kernel divides memory into two pools, nonpaged pool and paged pool. Items that are in the nonpaged pool memory must stay in physical memory whereas items in paged pool can be paged to disk if necessary.

When running programs, Windows 2000 uses dynamic memory allocation so that it can use the RAM more efficiently. Dynamic memory allocation gives a program memory when it needs it and takes it away when the program is not using it.

Windows 2000 Advanced Server and Windows 2000 Data Center can recognize 8 GB and 64 GB, respectively. These servers are based on the Intel Physical Address Extension (PAE). The PAE is based on a 36-bit address that supports up to 64 GB of physical memory. Applications using Microsoft Address Windowing Extension (AWE) APIs can create data caches that are 16 times as large as the cache sizes used in the 32-bit model.

Each process has a 4-GB virtual address space. The 4 GB are divided into two areas. The lower addresses (user mode) map to data and code that are private to the process or program. The upper addresses (kernel mode) map to system data and code (OS, device drivers, and file system cache) that are common to all processes. By default, the lower address is 2 GB and the upper address is 2 GB. In Windows 2000, the boundary can be moved with a switch in the BOOT.INI so that the programs have access to 3 GB and the system can access 1 GB.

By default, the system reserves about 50% of physical memory for the file system cache, but the system trims the cache if it is running out of memory. A server running a web server (specialized file server) such as IIS needs to have a large and effective file system cache to maintain efficiency in operation.

7.3.5 Interprocess Communication

Applications that split processing between networked computers are called distributed applications. A client-server application that uses distributed processing has its processes divided between a workstation and a more powerful server (application server). Typically, the client portion formats requests and sends them to the server for processing and the server runs the request and passes the result back to the client.

Interprocess communication (IPC) allows bidirectional communication between clients and servers using distributed applications. IPC is a mechanism used by programs and multiprocesses. IPCs allow concurrently running tasks to communicate between themselves on a local computer or between the local computer and a remote computer. Windows 2000 provides many different IPC mechanisms, including local procedure calls, remote procedure calls, named pipes, and mailslots.

Local procedure calls (LPCs) are used to transfer information between applications on the same computer. **Remote procedure calls (RPCs)** are used to transfer information between applications that are on separate computers. The Microsoft RPC mechanism is unique in that it uses other IPC mechanisms, such as named pipes, NetBIOS, or WinSock to establish communications between the client and the server. Since the RPC are already defined, a programmer can send a message to another server with the appropriate arguments and the server returns a message containing the results of the program that was executed.

Named pipes and mailslots are high-level IPC mechanisms used by network computers. Different from the other mechanisms, both are written as file system drivers. A **named pipe** is connection-oriented messaging via pipes which set up a virtual circuit between the two points to maintain reliable and sequential data transfer. A pipe connects two processes so that the output of one can be used as input to the other. A **mailslot** is connectionless messaging, which means that messaging is not guaranteed. Connectionless messaging is useful for identifying other computers or services on a network, such as the Browser service offered in Windows 2000. Connection to a named pipe and mailslot is done through the **common Internet file system (CIFS)** redirector. A redirector intercepts file I/O requests and directs them to a drive or resource on another computer. The redirector allows a CIFS client to locate, open, read, write, and delete files on another network computer that is running CIFS.

7.3.6 COM, COM+, and DCOM

Component object model (COM) is an object-based programming model designed to promote interoperability by allowing two or more applications or components to easily cooperate with one another, even if they were written by different vendors, at different times, in different programming languages, or if they are running on different computers running different operating systems. COM is the foundation technology upon which broader technologies can be built such as object linking and embedding (OLE) technology and ActiveX.

COM+ is an extension of component object model. COM+ is both an object-oriented programming architecture and a set of operating system services. It adds to COM a new set of system services for application components while they are running, such as notifying them of significant events or ensuring they are authorized to run. COM+ is intended to provide a model that makes it relatively easy to create business applications that work well with the Microsoft Transaction Server (MTS) in a Windows NT system.

Distributed component object model (DCOM) is a set of Microsoft concepts and program interfaces in which client program objects can request services from server program objects in other network computers. Since DCOM uses TCP/IP and HTTP, it can work with a TCP/IP network or the Internet. For example, you can create a page for a Web site that contains a script or program that can be processed before being sent to a requesting user by a more specialized server in the network. Using DCOM interfaces, the web server site program, which is now acting as a client, forwards a RPC to the specialized server. The specialized server provides the necessary processing and returns the result to the web server. It passes the result on to the Web page viewer.

7.4 WINDOWS 2000 BOOT SEQUENCE

During boot-up, the first Windows 2000 file to be read is **NTLDR,** which switches the microprocessor from real mode to protected mode and starts the appropriate minifile system drivers (built into NTLDR) so that it can read the VFAT/FAT16, FAT32, or NTFS file systems.

Next, NTLDR reads the **BOOT.INI** (if one is available) and displays the Boot Loader Operating System Selection menu. See Figure 7.6. NTLDR then loads the operating system (such as Windows NT 2000 Server, Windows 2000 Professional, Windows 95/98, or DOS) as selected from the menu. If you do not select an entry before the timer reaches zero, NTLDR loads the default operating system specified in the BOOT.INI file. If the BOOT.INI file is not present, NTLDR tries to load Windows 2000 from the Windows 2000 directory

FIGURE 7.6 Windows 2000 Boot Loader Operating System Selection menu

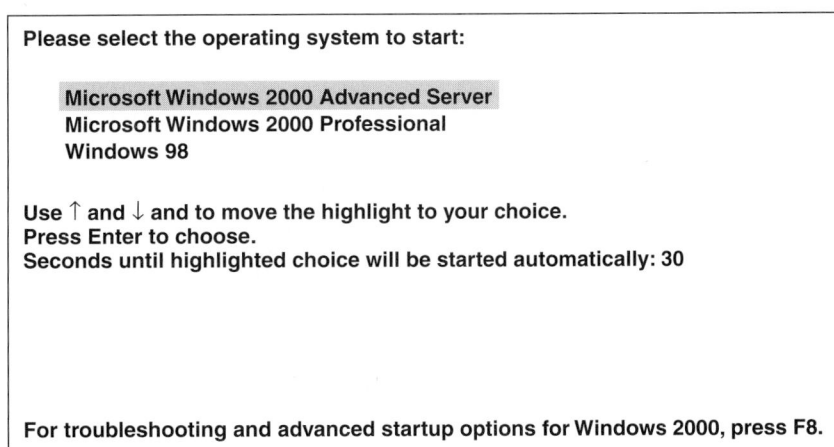

```
Please select the operating system to start:

    Microsoft Windows 2000 Advanced Server
    Microsoft Windows 2000 Professional
    Windows 98


Use ↑ and ↓ and to move the highlight to your choice.
Press Enter to choose.
Seconds until highlighted choice will be started automatically: 30




For troubleshooting and advanced startup options for Windows 2000, press F8.
```

(typically C:\WINNT). In addition, the menu will not be displayed if you only have the one operating system or if you have an operating system that cannot read the NTFS partition and Windows 2000.

NTLDR runs NTDETECT.COM, which attempts to detect the bus/adapter type, serial ports, floating-point coprocessor, floppy disks, keyboard, mouse/pointing device, parallel ports, SCSI adapters, and video adapters. If you have a system that has a SCSI hard disk, for which the BIOS on the SCSI adapter is disabled, NTLDR will load NTBOOTDD.SYS to access the SCSI devices during boot-up.

After NTLDR collects the hardware information, it then gives you an option to press the spacebar to invoke the Hardware Profile/Configuration Recovery menu. This menu will list the hardware profiles that you have saved on your drive and list the Last Known Good Configuration option.

The NTLDR finally loads the NTOSKRNL followed by the hardware abstraction layer (HAL.DLL). It will then load the HKEY_LOCAL_MACHINE\SYSTEM registry key from systemroot\system32\CONFIG\SYSTEM directory. Note that the registry and hive are described in Chapter 8. The SYSTEM hive specifies which device drivers to load during boot-up. When the device drivers are loaded, a hardware list is made and stored in the registry. Lastly, the Session Manager (SMSS.EXE) is loaded, which loads the appropriate services needed for Windows 2000 to function. Lastly, the NTLDR initializes the NT kernel (NTOSKRNL.EXE) and takes control of the boot process.

The active partition that contains the NTLDR and BOOT.INI file is known as the **system partition.** The partition that contains the Windows 2000 operating system files is called the **boot partition.** If a system has only one partition with the initial boot files and the Windows 2000 directory, then the partition is both the system partition and the boot partition.

> **Note:** The *%systemroot%* or *systemroot* indicates the folder into which Windows 2000 is installed, which is located on the boot partition. By default, the Windows 2000 system root directory is C:\WINNT.

DOS, Windows 95, and Windows 98 use a different volume boot sector. In addition, Windows NT and Windows 2000 start with NTLDR whereas DOS, Windows 95, and Windows 98 start with IO.SYS. If DOS, Windows 95, or Windows 98 boot files are already on a partition when you install Windows 2000, then the old boot sector gets copied into a file called **BOOTSECT.DOS.** Therefore, when you select the old operating system from the boot menu, NTLDR loads BOOTSECT.DOS and passes control to it. The operating system then starts up as normal. When dual booting between Windows 2000 and DOS, Windows 95, or Windows 98, neither DOS, Windows 95, nor Windows 98 can read NTFS partitions. See Figure 7.7.

7.4.1 BOOT.INI Files

During boot-up, the BOOT.INI file provides a Boot Loader Operating System Selection menu, which allows the selection between multiple operating systems. The BOOT.INI file

FIGURE 7.7 The BOOT.INI/Boot menu allows two pathways during boot-up

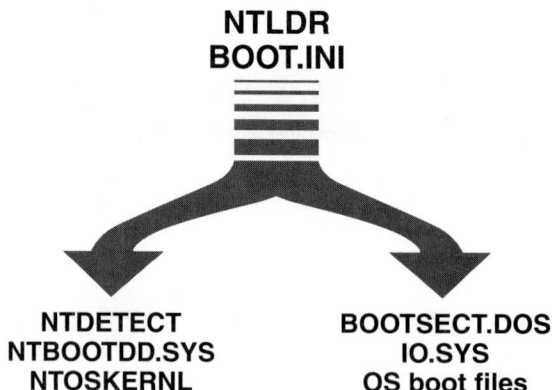

NTLDR
BOOT.INI

NTDETECT
NTBOOTDD.SYS
NTOSKERNL

BOOTSECT.DOS
IO.SYS
OS boot files

194

is a read-only, hidden system text file located in the root directory of the system partition. See Figure 7.8. The BOOT.INI file is divided into two sections, [boot loader] (see Table 7.6) and [operating systems]. The entries in the [boot loader] section configure the time (in seconds) that the Boot Loader Operating System Selection menu appears on the screen and that the default operating system is loaded.

The [operating systems] section contains the list of available operating systems. Each entry includes an ARC path to the boot partition for the operating system, the string to display in the boot loader screen, and optional parameters. The optional parameters are shown in Table 7.7.

An **advanced RISC computing (ARC) path** is used to specify the location (partition) of an operating system. It follows the format:

`multi(x)disk(y)rdisk(z)partition(a)`

or

`scsi(x)disk(y)rdisk(x)partition(a)`

FIGURE 7.8   Typical BOOT.INI file

```
[boot loader]
timeout530
default5multi(0)disk(0)rdisk(0)partition(1)\WINNT
[operating systems]
multi(0)disk(0)rdisk(0)partition(1)\WINNT5"Microsoft Windows 2000
Advanced Server" /fastdetect
multi(0)disk(0)rdisk(1)partition(1)\WINNT5"Microsoft Windows 2000
Professional" /fastdetect
C:\5"Windows 98"
```

TABLE 7.6   Settings in the [boot loader] section of the BOOT.INI file

Entries	Description
Timeout=XX	Specifies the number of seconds the user has to select an operating system from the boot loader screen before NTLDR loads the default operating system. If the value is zero, NTLDR immediately starts the default operating system without displaying the boot loader screen.
Default=	The ARC path to the default operating system.

TABLE 7.7   Optional parameters found in the BOOT.INI file

Entries	Descriptions
/3GB	Changes the standard of allocating 2 GB for each process and 2 GB for the system to 3 GB for each process and 1 GB for the system.
/BASEVIDEO	Specifies that Windows NT uses the standard VGA video driver.
/fastdetect=[comxl comx,y,z]	Disables serial mouse detection. Without a port specification, this switch disables peripherals detection on all COM ports.
/maxmem:n	Specifies the amount of RAM that Windows 2000 uses. Use this switch if you suspect memory is bad.
/noguiboot	Boots the computer without displaying the graphical boot status screen.
/SOS	Displays the device driver names while they are being loaded.

SCSI is used for a SCSI disk with its BIOS disabled. MULTI is used for disks other than SCSI or a SCSI disk with its BIOS enabled. The number after MULTI is the ordinal number of the hardware adapter card starting from zero. The number after DISK is the SCSI bus number and will always be zero for a non-SCSI disk or for a SCSI disk with its BIOS enabled. The number after RDISK is the ordinal number of the disk starting from zero. The number after PARTITION is the ordinal number of the partition. Unlike the other values, it starts at one. When booting from a computer that does not use the correct ARC path, you may get one of the following error messages:

```
BOOT: Couldn't find NTLDR. Please insert another disk.
NT could not start because the following file is missing or
corrupt: \winntroot\system32\ntoskrnl.exe.
NTDETECT V1.0 Checking Hardware. . .
NTDETECT failed/missing
```

7.4.2   Advanced Boot Options

If you press F8 while the Windows 2000 Boot menu is listed, it will display the Advanced Boot menu. The Advanced Boot menu options are listed in Table 7.8 and displayed in Figure 7.9.

After you reboot the computer and log on, the kernel copies the configuration information to the Last Known Good control that is set in the registry. If you install a driver and the driver causes problems during boot, you can then activate the Last Known Good Configuration. This can be executed at the F8 during the boot menu or by pressing the spacebar when it says to open the OS Loader Recovery menu.

Safe mode loads Windows 2000 with the minimum number of device drivers and services necessary to boot successfully. Networking enabled safe mode adds network drivers and services to that of standard safe mode. Safe mode with the command prompt is identical to standard safe mode, except that Windows 2000 runs the command prompt application (CMD.EXE) instead of the Windows Explorer as the shell when the system switches into GUI mode. To keep track of which device drivers and services are part of the standard and networking enabled safe boots, the system uses HKEY_LOCAL_MACHINE\ SYSTEM\CurrentControl\Set\Control\SafeBoot registry key. (Registry is explained in

TABLE 7.8   Advanced Boot menu options

Advanced Boot Option	Description
Safe Mode	Similar to Windows 95/98 safe mode, Window 2000 will load and use basic files and drivers, including the mouse, VGA monitor, keyboard, hard drive, and default system services. However, no network components are loaded.
Enable Boot Logging	Logs the loading and initialization of drivers and services into a NTBTLOG.TXT text file in the Windows NT directory (usually WINNT). The text file can then be used for troubleshooting boot problems.
Enable VGA mode	Starts Windows 2000 with basic VGA drivers (resolution of 640X480 at 16 colors).
Last Known Good Configuration	Starts Windows 2000 using the last saved configuration stored in the registry.
Directory Services Restore Mode	Allows to boot the computer into a mode that lets you restore the active directory (AD) of a domain controller from backup media. This option is not available for Windows 2000 Professional.
Debugging Mode	Turns on debugging. This option is not available for Windows 2000 Professional. See Chapter 16 for more information.

Note: All options except the Last Known Good Configuration creates the NTBTLOG.TXT file and uses the VGA driver (resolution of 640X480 at 16 colors).

FIGURE 7.9   Windows 2000
Advanced Boot menu options

```
Windows 2000 Advanced Options Menu
Please Select an option:

   Safe Mode
   Safe Mode with Networking
   Safe Mode with Command Prompt

   Enable Boot Logging
   Enable VGA Mode
   Last Known Good Configuration
   Directory Services Restore Mode (Windows 2000 domain controllers only)
   Debugging Mode

   Boot Normally
   Return to OS Choices Menu

Use ↑ and ↓ and to move the highlight to your choice.
Press Enter to choose.
```

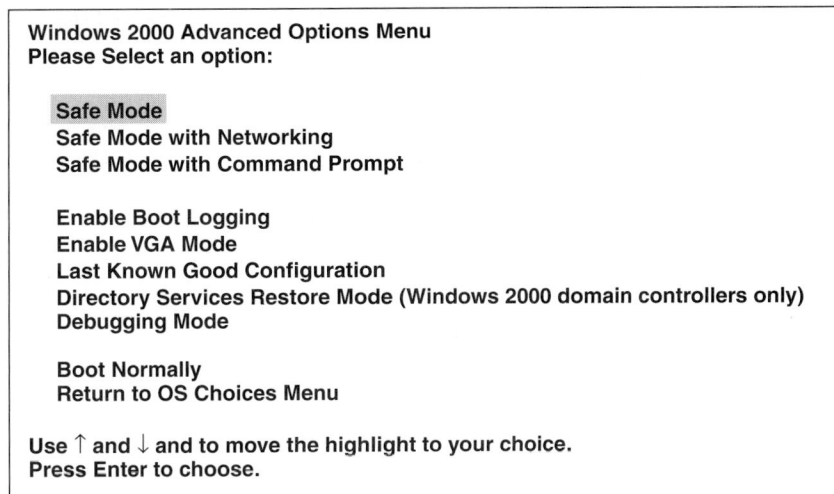

Chapter 8). When any of the safe modes are selected, the NT Loader (NTLDR) passes the /SAFEBOOT parameter to the NTOSKRNL.EXE so that the kernel can specify to only load these drivers and services.

7.5   WINDOWS 2000 START-UP FLOPPY DISK

After installing and configuring Windows 2000, make a Windows 2000 start-up floppy disk. A start-up floppy disk is used to access a drive with a faulty start-up sequence, such as a corrupted boot sector, corrupted MBR, a virus infection, missing or corrupt NTLDR or NTDETECT.COM, or a incorrect NTBOOTDD.SYS file. In addition, the disk can be used to start a computer after the primary disk of a mirrored volume. (A mirrored volume is volumes or partitions located on two different hard drives with duplicate information.) If one hard drive fails, the computer does not go down because the information can be read from the second disk. The start-up disk cannot be used for incorrect or corrupted device drivers or start-up problems that occur after NTLDR.

To create a boot disk, do the following:

1. Format a disk on a computer running Windows 2000. Since DOS and Windows 95/98 use different formats and boot sections, format with Windows 2000 not DOS or Windows 95/98.
2. For an Intel machine, save NTLDR, NTDETECT.COM, NTBOOTDD.SYS (if you have a SCSI controller without a SCSI BIOS), and BOOT.INI file from the system partition to the floppy disk.
3. Test the boot files.

If using a mirrored volume, you should have two start-up floppy disks, one on which the ARC path points to the first disk and the second on which the ARC path points to the second disk. Therefore, after creating the second disk, you will need to modify the BOOT.INI file manually using a text editor such as EDIT or Notepad.

7.6   WINDOWS 2000 LOGON

To log on to a computer that is running Windows 2000, a user provides a user name and password. See Figure 7.10. Windows 2000 authenticates the user during the logon process to verify the identity of the user. Only valid users can gain access to resources and data on a computer or the network. In addition, depending on the situation, you can specify the domain name or computer name in the "Log on to:" text box. Note that the user name is not case sensitive, but the password is case sensitive.

There are two forms of logging on, logging on locally to the computer (sometimes known as interactively) and logging on remotely to another computer. A user can log on locally to a computer that is a member of a workgroup or to a computer that is a member of

FIGURE 7.10   Logon dialog box

a domain, but is not a domain controller, and if it selects the computer name in the "Log on to:" text box. A domain controller does not maintain a local security database; therefore, local user accounts are not available on domain controllers, and a user cannot log on locally to one.

7.7 WINDOWS 2000 SECURITY DIALOG BOX

The Security dialog box provides information such as the user account currently logged on and the domain or computer to which the user is logged on. To access the Windows 2000 Security dialog box, press the Ctrl+Alt+Delete keys. See Figure 7.11. As shown, the screen indicates who you are logged as and when you logged on.

On the dialog box are six buttons. The Log Off button allows you to log off the system without restarting or shutting down the computer. The Shutdown button allows you to shut down or reboot the computer. The Log Off button and the Shutdown button close all running programs. The Lock Workstation button allows you to secure the computer without logging off. All programs remain running. For security reasons, if you are leaving your computer alone, either lock your workstation or log off. The Task Manager button brings up the Task Manager dialog box. See Figures 7.12 and 7.13. **Task Manager** provides a list of the current programs that are running and the overall CPU (including number of threads and processes) and memory usage (including the amount of physical memory, size of the file cache, paged memory, and nonpaged memory). Task Manager also switches between programs and stops a program that is not responding. A new feature of Windows 2000 server, called kill process tree utility, is its ability to end all processes related to the one being terminated by the administrator.

7.8 RUNNING PROGRAMS

As mentioned, Windows 2000 will run WIN32 applications (such as Microsoft Office 97 and Office 2000), WIN16 applications (programs written for Windows 3.xx), and DOS applications.

7.8.1 DOS Applications

DOS programs run in a special WIN32-based application called the NT Virtual DOS machine (NTVDM). The NTVDM provides a simulated MS-DOS environment for MS-DOS-based applications. In addition, since each NTVDM has a single thread and its own address space, if one NTVDM fails, no other NTVDM is affected.

To customize an NTVDM for an MS-DOS-based application, go to the application file name and select Properties from the shortcut menu to create a program information file

FIGURE 7.11 Security dialog box

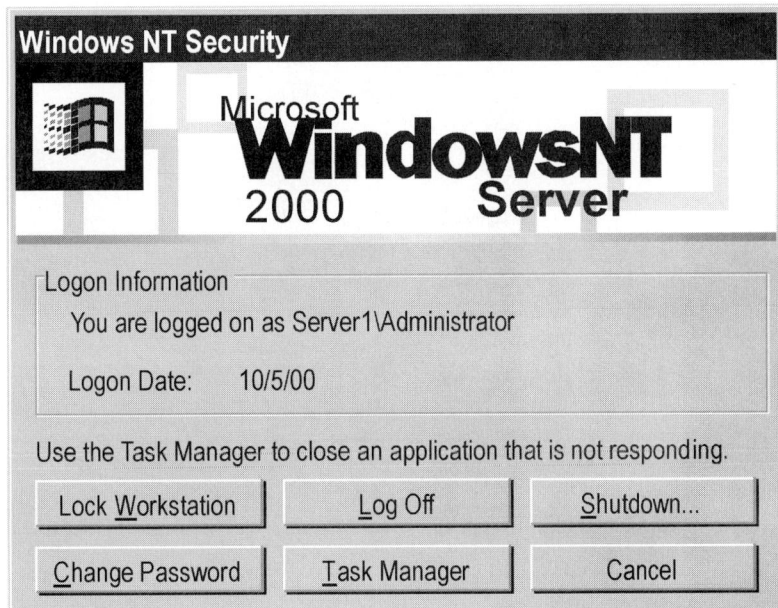

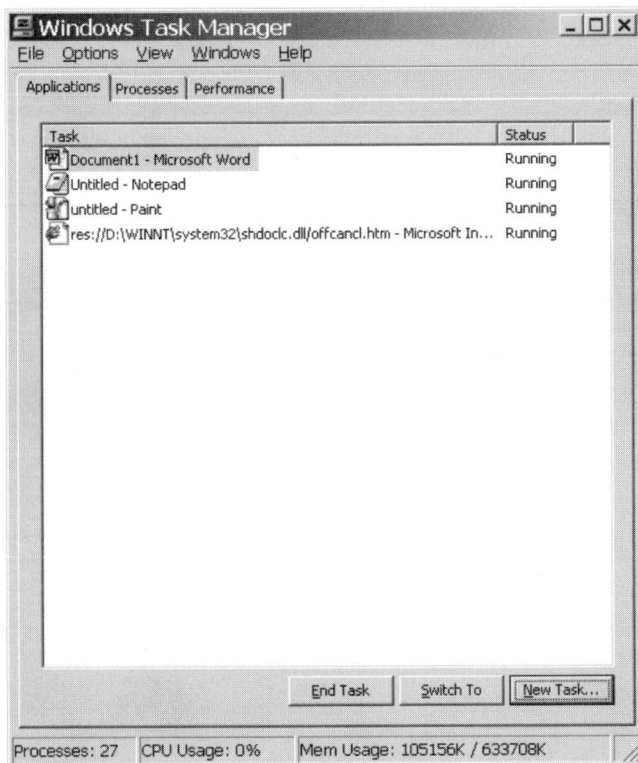

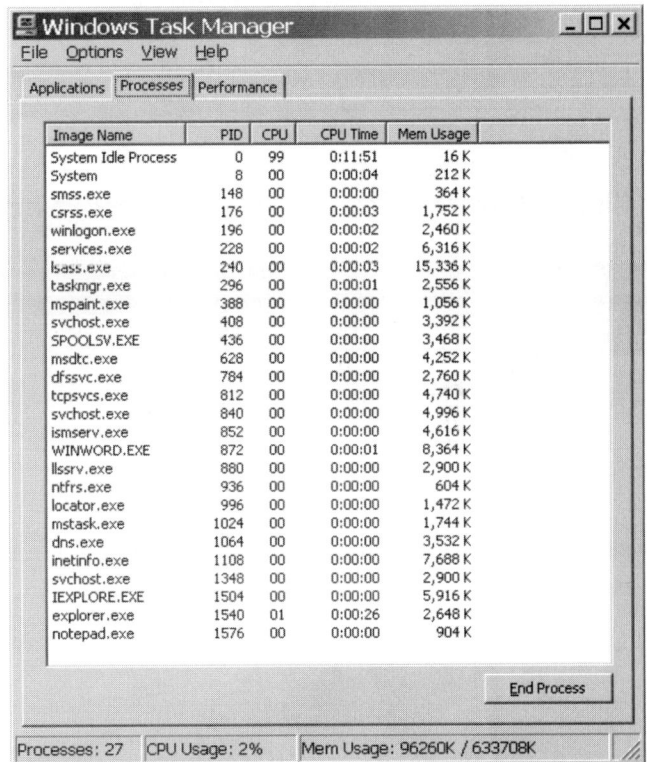

Windows NT Security

Microsoft
WindowsNT
2000 Server

Logon Information
You are logged on as Server1\Administrator

Logon Date: 10/5/00

Use the Task Manager to close an application that is not responding.

Lock Workstation | Log Off | Shutdown...

Change Password | Task Manager | Cancel

Windows Task Manager

File Options View Windows Help

Applications | Processes | Performance

Task	Status
Document1 - Microsoft Word	Running
Untitled - Notepad	Running
untitled - Paint	Running
res://D:\WINNT\system32\shdoclc.dll/offcancl.htm - Microsoft In...	Running

End Task | Switch To | New Task...

Processes: 27 | CPU Usage: 0% | Mem Usage: 105156K / 633708K

Windows Task Manager

File Options View Help

Applications | Processes | Performance

Image Name	PID	CPU	CPU Time	Mem Usage
System Idle Process	0	99	0:11:51	16 K
System	8	00	0:00:04	212 K
smss.exe	148	00	0:00:00	364 K
csrss.exe	176	00	0:00:03	1,752 K
winlogon.exe	196	00	0:00:02	2,460 K
services.exe	228	00	0:00:02	6,316 K
lsass.exe	240	00	0:00:03	15,336 K
taskmgr.exe	296	00	0:00:01	2,556 K
mspaint.exe	388	00	0:00:00	1,056 K
svchost.exe	408	00	0:00:00	3,392 K
SPOOLSV.EXE	436	00	0:00:00	3,468 K
msdtc.exe	628	00	0:00:00	4,252 K
dfssvc.exe	784	00	0:00:00	2,760 K
tcpsvcs.exe	812	00	0:00:00	4,740 K
svchost.exe	840	00	0:00:00	4,996 K
ismserv.exe	852	00	0:00:00	4,616 K
WINWORD.EXE	872	00	0:00:01	8,364 K
llssrv.exe	880	00	0:00:00	2,900 K
ntfrs.exe	936	00	0:00:00	604 K
locator.exe	996	00	0:00:00	1,472 K
mstask.exe	1024	00	0:00:00	1,744 K
dns.exe	1064	00	0:00:00	3,532 K
inetinfo.exe	1108	00	0:00:00	7,688 K
svchost.exe	1348	00	0:00:00	2,900 K
IEXPLORE.EXE	1504	00	0:00:00	5,916 K
explorer.exe	1540	01	0:00:26	2,648 K
notepad.exe	1576	00	0:00:00	904 K

End Process

Processes: 27 | CPU Usage: 2% | Mem Usage: 96260K / 633708K

FIGURE 7.12 Task Manager showing applications and processes running in memory

(PIF) for the MS-DOS application. The PIF describes how the DOS program will run under the Windows 2000 environment. These settings include specifying memory settings, screen settings, and customized CONFIG.NT and AUTOEXEC.NT files for DOS applications. See Figures 7.14, 7.15, and 7.16.

To create customized CONFIG.NT and AUTOEXEC.NT files, do the following:

1. Use a text edit such as Notepad or EDIT to edit the CONFIG.NT or AUTOEXEC.NT files located in the systemroot\SYSTEM32 directory.
2. Save each file with a new name.

199

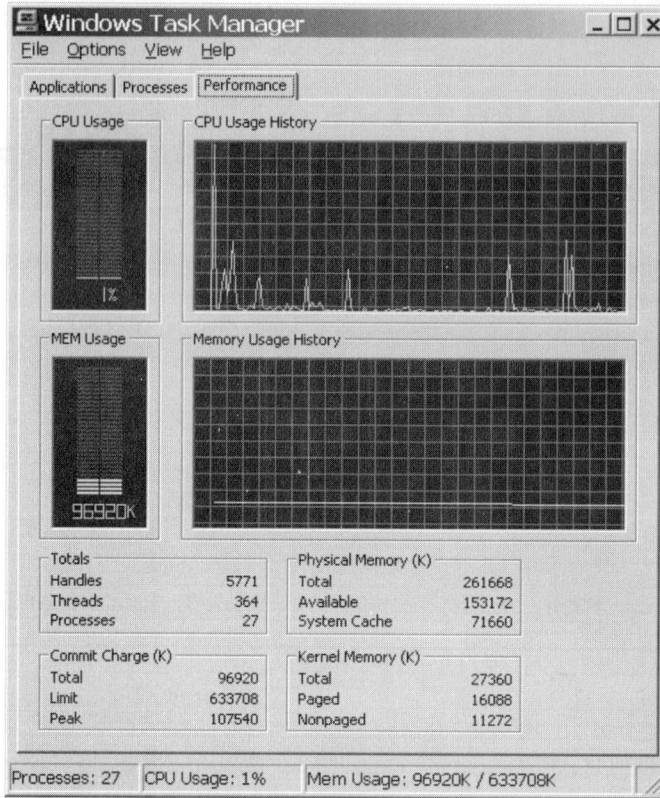

Windows Task Manager

File   Options   View   Help

Applications | Processes | Performance

CPU Usage

CPU Usage History

1%

MEM Usage

Memory Usage History

96920K

Totals
Handles   5771
Threads   364
Processes   27

Physical Memory (K)
Total   261668
Available   153172
System Cache   71660

Commit Charge (K)
Total   96920
Limit   633708
Peak   107540

Kernel Memory (K)
Total   27360
Paged   16088
Nonpaged   11272

Processes: 27 | CPU Usage: 1% | Mem Usage: 96920K / 633708K

FIGURE 7.14   Modifying the PIF settings of a DOS program

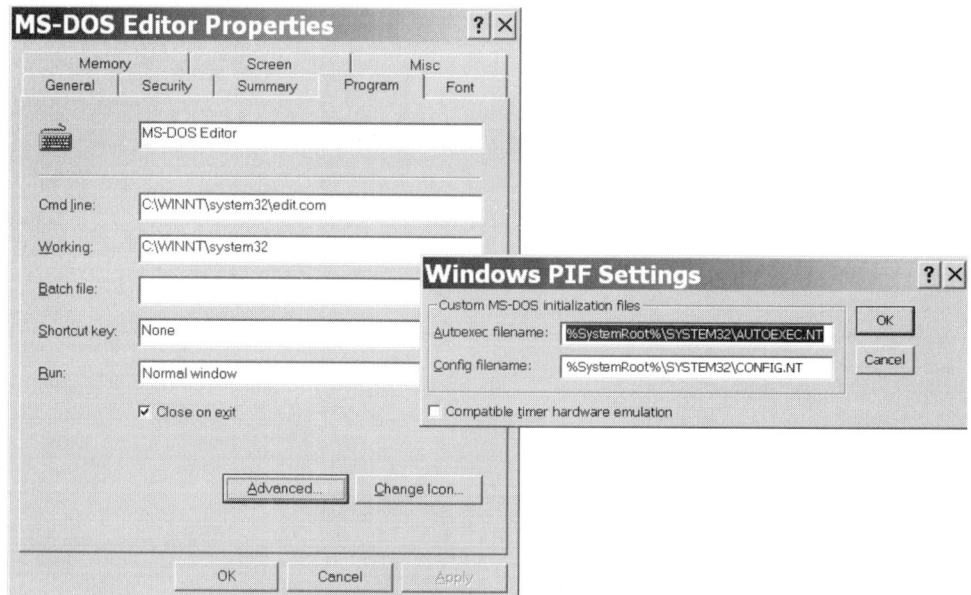

MS-DOS Editor Properties

Memory | Screen | Misc
General | Security | Summary | Program | Font

MS-DOS Editor

Cmd line: C:\WINNT\system32\edit.com

Working: C:\WINNT\system32

Batch file:

Shortcut key: None

Run: Normal window

☑ Close on exit

Windows PIF Settings

Custom MS-DOS initialization files
Autoexec filename: %SystemRoot%\SYSTEM32\AUTOEXEC.NT
Config filename: %SystemRoot%\SYSTEM32\CONFIG.NT

☐ Compatible timer hardware emulation

OK   Cancel

Advanced...   Change Icon...

OK   Cancel   Apply

3. Right-click on the MS-DOS based program shortcut and select the Properties option.
4. Click on the Program tab and select Advanced button.
5. Under Custom MS–DOS–based program initialization files, type the new name for your custom start files.

If Windows displays an error message concerning the CONFIG.NT or AUTOEXEC.NT or if you have problems running MS–DOS based programs, check the CONFIG.NT or AUTOEXEC.NT files for the particular program that you are trying to run.

You can switch between a full screen or DOS window by pressing the ALT+ENTER keys. If an application cannot run in a DOS VM such as the application requires full access

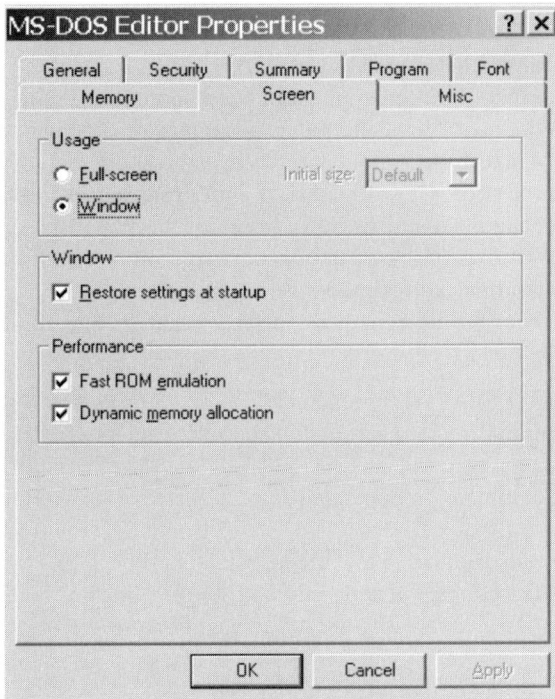

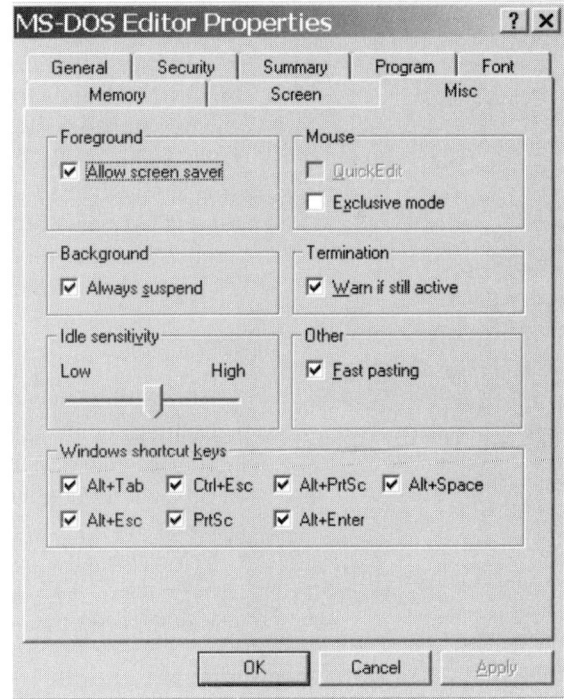

FIGURE 7.15   Screen and Misc tab within the properties of a DOS program in Windows 2000

FIGURE 7.16   Memory settings for a DOS program

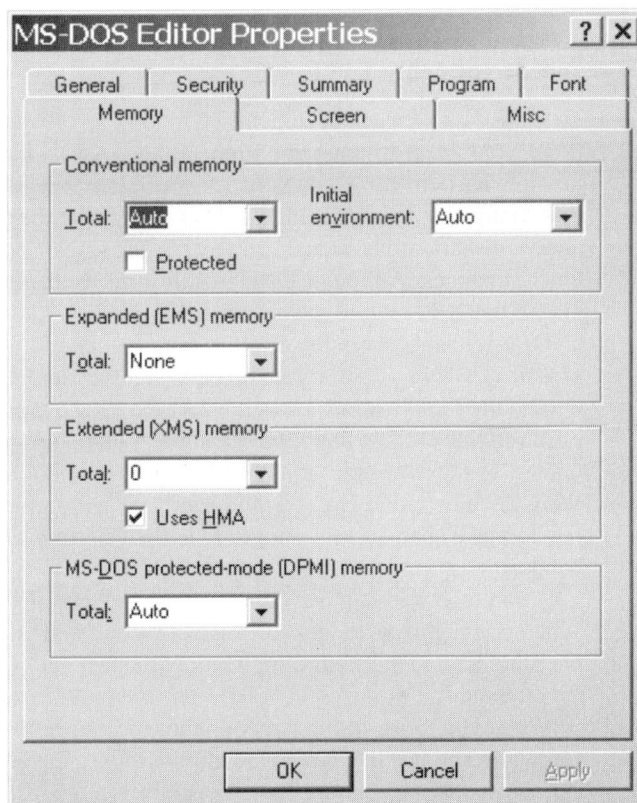

to the computer's resources or the application has video problems, you would run the application in DOS mode.

If you're trying to free up a lot of lower memory for a DOS program, specify different AUTOEXEC.BAT and CONFIG.SYS commands for each application by using the Advanced button in the Program tab.

7.8.2 WIN16-based Applications

Win16-based applications are executed by Win16 on Win32 (WOW) which translates or thunks 16-bit calls to 32-bit calls. Since Win16-based applications use cooperative multitasking, Win16-based applications in Windows 2000 cannot take advantage of preemptive multitasking. Instead, by default, the Win16-based applications share the same memory area and will all be assigned a single time slice. Unfortunately, if a program causes problems, it can cause all Win16-based applications to crash. However, all of the Win16-based applications are run preemptive with Windows 32-based applications and other Windows 2000 processes, which improves the performance of Win16-based applications.

To start a Win16-bit application in its one separate memory space, you can do one of the following:

- At the command prompt, type START /SEPARATE filename.
- On the Start menu, select the Open option from the Start button and select the Run in Separate Memory Space option.
- Create a shortcut and select the Run command in Separate Memory Space option found on the Properties shortcut tab.
- In Windows NT Explorer, select Options from the View menu, select the File Types tab and click on the Win16 applications to be edited. Click on Edit and double-click on Open. Edit the open line to include the separate switch using the following syntax:

```
Cmd /c start /separate path\application_executable %1
```

While running multiple NTVDMs for 16-bit applications, there are some disadvantages. First, starting WIN16 applications in their own memory spaces introduce additional overhead. In addition, if WIN16-based applications do not follow the OLE and DDE specifications, they can not communicate with other NTVDMs.

7.8.3 Sharing Data

To share data between documents, Windows applications can use Dynamic Data Exchange (DDE) or Object Linking and Embedding (OLE). DDE was introduced with Windows 3.0 to allow users to copy data between applications while maintaining a link. Whenever data was changed at the source, it was also changed at the target. For example, you could insert an Excel chart into a Word document. As the spreadsheet data changes, the chart in the document changes. Although DDE is still used by many applications, it is slowly being replaced by OLE.

OLE (enhanced version of DDE) is a compound document standard developed by Microsoft. It allows you to create objects with one application and then link or embed objects into another application. Embedded objects keep their original format and link to the application that created them. For example, you could insert an Excel chart into a Word document. To make changes to the chart, you would double-click on the chart from within the Word document and Excel would run from within the Word document. This would allow you to make changes without going to the Program Manager and starting Excel in its own window.

A common way to move data between applications or documents is to use the clipboard. When you highlight the text or select a picture and you use the Copy or Cut function from the Edit menu in the source application, the text and picture goes into RAM, specifically to the clipboard.

When you move to the target document, choose Paste from the Edit menu. The source application copies the data in the clipboard into the current document using the first available format. If you choose Paste Special from the Edit menu, a dialog box appears, which lists the formats from which you can choose for pasting the data. If you choose Paste Link, the application tries to make an OLE link to the source document. If the ObjectLink format is not available, a DDE link is created instead.

If you press the Print Screen button on your keyboard, you can place a bitmap picture of the entire screen to the clipboard. If you press Alt+Print Screen, the active bitmap picture is placed onto the clipboard.

SUMMARY

1. Windows 2000 is an operating system that can function as a desktop operating system, network file, print server or application server.
2. Windows 2000 uses a different architecture, which offers higher performance, greater reliability and provides better security than DOS/Windows 3.xx and Windows 95/98. Unfortunately, to achieve these benefits Windows 2000 is not 100 percent backward compatible with all hardware and software.
3. To install Windows 2000 Professional, you should have a minimum of 133 MHz, 64 MB of RAM and 650 MB of free disk space running on a Pentium processor.
4. To install Windows 2000 Server, you should have a minimum of 133 MHz, 128 MB of RAM and 1 GB of free disk running on a Pentium processor.
5. Windows 2000 Professional supports up to 2 processors and 4 GB of RAM.
6. Windows 2000 Server supports up to 4 processors and 4 GB of RAM.
7. Advanced Server supports up to 8 processors and 8 GB of RAM.
8. The Windows 2000 data center supports up to 32 processors and 64 GB of RAM.
9. To make sure that you know what hardware is compatible with Windows 2000, you should only use devices that are on Microsoft's hardware compatibility list (HCL).
10. To install or upgrade Windows 2000, use WINNT.EXE file or WINNT32.EXE, both of which are located in the I386 directory on the Windows 2000 Installation CD.
11. WINNT.EXE should be used for a clean installation on a computer running Microsoft DOS or Microsoft Windows 3.xx.
12. WINNT32.EXE should be used on a clean installation or upgrade on a computer running Microsoft Windows NT version 4.0, Microsoft Windows 95, or Microsoft Windows 98.
13. Windows 2000 supports FAT, FAT32 and NTFS.
14. A software license typically allows an individual or a group to use a piece of software.
15. There are two kinds of Client Access Licenses (CAL): per seat licensing and per server licensing.
16. After installing Windows 2000, you should check to see if Microsoft has released any fixes, patches or service packs. Be sure to apply those fixes or patches to the Windows 2000 system.
17. To prepare for the possibility of a system failure on a computer that cannot be started from the CD-ROM drive, you need to create a set of four Windows 2000 setup floppy disks so you can start the setup program, the recovery console and the emergency repair process.
18. To create the Windows 2000 setup floppy disk, use the MAKEBT32.EXE which is located in the BOOTDISK directory on the installation CD.
19. Switches are customization options added to WINNT.EXE or WINNT32.EXE commands to modify the network installation.
20. The Hardware Abstraction Layer (HAL) is a library of hardware manipulating routines that hide the hardware interface details.
21. The Windows 2000 setup manager wizard allows you to create an answer file and uniqueness database file (UDF), which can be used as scripts when using the WINNT command to do automated installations.
22. Remote Installation Services (RIS) enables client computers to connect to a server during the initial startup phase and remotely install Windows 2000 Professional.
23. To install and use RIS, you must have DNS, DHCP and Active Directory on the network. The drive on the server must be formatted with NTFS in which case you cannot install RIS on the same drive or partition as the Windows 2000 Server.
24. The System Difference (SYSDIFF.EXE) tool enables you to automatically distribute and install applications and limited hardware devices during or after the Windows 2000 setup.
25. Windows 2000 uses a security identifier (SID) to uniquely identify the computer, even if the computer name has been changed.
26. Windows 2000 has the System Preparation Tool (SYSPREP.EXE) that will remove the security identifiers and all other user-specific or computer-specific information so that you can install disk images on several systems without them all having the same SID.
27. The User Mode programs run in Ring 3 of the Intel 386 microprocessor protection model are protected by the operating system. It is a less privileged processor mode that has no direct access to hardware and can only access its own address space.

28. The Kernel mode components (sometimes referred to as privileged mode) run in Ring 0 of the Intel 386 microprocessor protection model. The Kernel mode components are protected by the microprocessor.

29. A kernel is the central module of an operating system.

30. The microkernel is the central part of Windows 2000, which coordinates all I/O functions and synchronizes the activities of the executive services.

31. Device drivers are programs that control a device. A device driver acts as a translator between the device and programs that use the device.

32. Most drivers used by Windows 2000 use the Windows Driver Model (WDM), which is also compatible with Windows 98.

33. A program, sometimes referred to as a process, is an executable program that follows a sequence of steps.

34. A thread is part of a program that can execute independently of other parts.

35. Windows 2000 is a preemptive multitasking operating system.

36. Preemptive multitasking is when the operating system assigns time slices to threads, tasks and applications. Threads, tasks and applications with a higher priority get a larger time slice.

37. The file that makes up virtual memory in Windows NT and Windows 2000 is called a paging file. (PAGEFILE.SYS).

38. The Windows NT and Windows 2000 memory model is based on a flat, linear 32-bit address space, which is managed by the Virtual Memory Manager (VMM).

39. The process of moving data in and out of physical memory is called paging.

40. Interprocess Communication (IPC) allows bidirectional communication between clients and servers using distributed applications. IPC is a mechanism used by programs and multi-processes.

41. During boot up, the first Windows 2000 file read is NTLDR, which switches the microprocessor from real mode to protected mode and starts the appropriate minifile system drivers.

42. NTLDR reads the BOOT.INI (if one is available) and displays the Boot Loader Operating System Selection menu.

43. NTLDR runs NTDETECT.COM, which attempts to detect the bus/adapter type, serial ports, floating-point co–processor, floppy disks, keyboard, mouse/pointing device, parallel ports, SCSI adapters and video adapters.

44. The active partition that contains the NTLDR and BOOT.INI file is known as the system partition.

45. The partition that contains the Windows 2000 operating system files is called the boot partition.

46. The default directory for Windows 2000 is C:\WINNT.

47. An ARC (Advanced RISC Computing) path is used to specify the location (partition) of an operating system.

48. If you press the F8 key while the Windows 2000 boot menu is listed, the advanced boot menu displays.

49. Windows 2000 authenticates the user during the logon process to verify the identity of the user. Only valid users can gain access to resources and data on a computer or the network.

50. There are two forms of logging on: logging on locally to the computer (sometimes known as interactively) and logging on remotely to another computer.

51. Windows 2000 can run Win16, Win32, DOS, OS/2 and POSIX applications.

52. To share data between documents, Windows applications can use Dynamic Data Exchange (DDE) or Object Linking and Embedding (OLE).

53. A Windows 2000 startup floppy disk is used to access a drive with a faulty startup sequence.

QUESTIONS

1. What are the minimum requirements to run Windows 2000 Professional?
   a. A Pentium processor—133 MHz, 32 MB of RAM and 650 of free disk space.
   b. A Pentium processor—133 MHz, 64 MB of RAM and 1 GB free disk space.

   c. A Pentium processor—166 MHz, 64 MB of RAM and 650 MB of free disk space.

   d. A Pentium processor—233 MHz, 128 MB of RAM and 1 GB of free disk space.

   e. A Pentium II processor—233 MHz, 128 MB of RAM and 2 GB of free disk space.

2. What are the minimum requirements to run Windows 2000 Server?

   a. A Pentium processor—133 MHz, 64 MB of RAM and 650 of free disk space.

   b. A Pentium processor—133 MHz, 128 MB of RAM and 1 GB free disk space.

   c. A Pentium processor—166 MHz, 64 MB of RAM and 650 MB of free disk space.

   d. A Pentium processor—233 MHz, 128 MB of RAM and 1 GB of free disk space.

   e. A Pentium II processor—233 MHz, 256 MB of RAM and 2 GB of free disk space.

3. Which of the following statements are not true? (Choose all that apply)

   a. Windows 2000 Professional can support only 1 processor.

   b. Windows 2000 Server can support up to 4 processors.

   c. Windows 2000 Advanced Server can support up to 8 processors

   d. Windows 2000 Data Center can support up to 16 processors.

4. You just installed Windows 2000 on a computer on which Windows NT was installed. Unfortunately, some of the components are not being detected by Windows 2000. What should you do?

   a. Try to install it again.

   b. Make sure that the hardware devices are on the compatibility list.

   c. Reformat the system and reinstall.

   d. Make sure that you have the newest BIOS.

5. To install Windows 2000, you would use the _____ command.

   a. install   c. WINNT

   b. sctup   d. start

6. Which of the following statements are true. (Select all that apply)

   a. If you have Windows 98, you can upgrade to Windows 2000 Professional.

   b. If you have Windows 98, you can upgrade to Windows 2000 Server.

   c. If you have Windows NT Server Enterprise edition, you can upgrade to Windows 2000 Advanced Server.

   d. If you have Windows NT Server, you can upgrade to Windows 2000 Advanced Server.

   e. If you have Windows 2000 Server, you can upgrade to Windows 2000 Advanced Server.

7. If you have many servers for your workplace with over 200 users, you should choose which type of license?

   a. per server   c. single

   b. per seat licensing   d. multiple

8. True or False—You just bought a copy of Windows 2000 Server and you cannot wait to get it installed. Since it was brand new, you don't have to worry about fixes, patches or service packs.

9. To repair your computer, you sometimes need to use the four Windows 2000 Setup disks. How do you make the setup disks?

   a. Use the /B command with the WINNT command

   b. Use the /OX command with the WINNT command

   c. Use the BOOTDISK.EXE command

   d. Use the MAKEBOOT.EXE command

   e. Use the Add/Remove software applet in the Control Panel

10. Which of the following procedures will allow you to install Windows 2000 Professional over the network onto an MS-DOS computer?

   a. Booting the MS-DOS computer from the network installation startup disk and running SETUP.EXE from the command prompt

   b. Connecting to the network directory that contains the Windows NT installation files and running WINNT.EXE

c. Connecting to the network directory that contains the Windows 2000 installation files and running SETUP.EXE

d. Connecting to the network directory that contains the Windows 2000 installation files and running WINNT32.EXE

11. When using the WINNT32 command, which switch would you use to check your system for incompatibilities that might prevent a proper installation?
    a. /HELP
    b. /CHECKUPGRADEONLY
    c. /C
    d. /MAKELOCALSOURCE

12. What acts as a translator between our hardware and the Windows 2000 OS specific to the computer architecture?
    a. device drivers
    b. TSRs
    c. ACPI
    d. HAL
    e. kernel
    f. microkernel

13. If you get an Inaccessible Boot Device error, what is most likely the cause of the problem?
    a. Your controller is not supported because you do not have the proper driver specified.
    b. There is a read error from the installation CD.
    c. There is a write error on the installation target.
    d. You do not have enough disk space in your installation target.

14. If you get a media error during installation of Windows 2000, what is most likely the cause of the problem?
    a. Your controller is not supported because you do not have the proper driver specified.
    b. There is a read error from the installation CD.
    c. There is a write error on the installation target.
    d. You do not have enough disk space for your installation target.

15. What do you use to create the answer files and UDFs so that you can do automated installations?
    a. WINNT /U:namefile1 /UDF:id
    b. Use the Add/Remove software applet in the Control Panel.
    c. Use the Windows 2000 Setup Manager Wizard from the Windows NT installation CD.
    d. Use the Windows 2000 Setup Manager Wizard from the Resource Kit.

16. You would like to use RIS to install Windows 2000 Professional on a client computer. How would you create a network boot disk?
    a. WINNT/B
    b. WINNT32/R
    c. RBFG.EXE
    d. RISDISK.EXE
    e. RISETUP.EXE

17. Which of the following services are needed to use RIS? (Choose all that apply)
    a. DHCP
    b. WINS
    c. DNS
    d. active directory
    e. NetBIOS

18. You want to automate the installation of Windows NT workstation on 200 identical laptop computers and 50 identical desktop computers. What is the minimum number of answer files and UDF files required to complete this installation?
    a. 1 answer file and 2 UDF files
    b. 2 answer files and 1 UDF file
    c. 2 answer files and 250 UDF files
    d. 250 answer files and 2 UDF files

19. Suppose you are responsible for installing Windows 2000 Professional on 75 identical desktop computers and 25 identical laptop computers. Your company's network consists of a single domain, Corp.
    Required result:
    You must automate the installation of Windows 2000 Professional on the 100 computers.
    Optional desired results:
    You want to automate the installation of a standard suite of Win32-based applications. You want to configure all computers to join the Acme domain during the installation of Windows 2000 Workstation.

Proposed solution:
Create two answer files, one for each hardware platform. Add "OemPreinstall=Yes" to the [Unattended] section. Add "JoinDomain=ACME" to the [Network] section. Create a UDF with a section for each computer that specifies the user name and computer name. Install Windows 2000 Professional on a reference computer. Run SYSDIFF.EXE /SNAP to take a snapshot of the reference computer. Install the applications on the reference computer, then run SYSDIFF.EXE /DIFF to create the difference file. Add the command SYSDIFF.EXE /APPLY to the OEM\Cmdlines.Txt file. Start installation of Windows NT Workstation on the computers by specifying the appropriate answer file and the UDF in the WINNT.EXE command.

Which result does the proposed solution produce?

    a. The proposed solution produces the required result and produces both of the optional desired results.

    b. The proposed solution produces the required result and produces only one of the optional desired results.

    c. The proposed solution produces the required result but does not produce either of the optional desired results.

    d. The proposed solution does not produce the required result.

20. You want to use a disk imaging software package to create several Windows 2000 computers with the same configuration. What command will allow you to do that?

    a. SYSDIFF   c. WINN32 /WALKER

    b. WINDIFF   d. SYSPREP

21. Which of the following types of applications does Windows 2000 *not* support?

    a. bounded OS/2 applications   d. DOS applications

    b. Macintosh applications   e. POSIX applications

    c. Win32 applications

22. _____ is a less-privileged mode that has no direct access to hardware and can only access its own address space.

    a. User mode   c. Isolated mode

    b. Protected mode   d. Kernel mode

23. What is the central part of the operating system?

    a. kernel   c. device drivers

    b. boot files   d. compilers

24. Another name for a process is a _____ .

    a. thread   c. task

    b. program   d. suite of programs

25. Which of the following best describes the relationship between processes and threads?

    a. Each process consists of at least a single thread.

    b. Each process consists of only one thread.

    c. Each process consists of two or more threads.

    d. Each process consists of two threads—one for input, the other for output.

26. What is the smallest unit that can be scheduled with the Microkernel?

    a. thread   c. queue

    b. process   d. segment

27. All memory in Windows 2000 is managed by the:

    a. kernel   c. VMM

    b. Microkernel   d. process manager

28. Windows 2000 primarily uses _____ multitasking.

    a. cooperating   c. task switching

    b. preemptive   d. context switching

29. Windows 2000 Server supports only _____ of RAM

    a. 2 GB   d. 16 GB

    b. 4 GB   e. 32 GB

    c. 8 GB

30. To speed disk access, the _____ , which is a subset of the physical memory, is set aside to retain recently used information.
    a. virtual memory
    b. system cache
    c. L2 cache
    d. IPC

31. Which statement best describes how 16-bit Windows applications are run by default on a Windows NT Workstation computer?
    a. 16-bit Windows applications are run as a single thread in a single NTVDM with a shared address space.
    b. 16-bit Windows applications are run as a single thread in a single NTVDM with separate address spaces.
    c. 16-bit Windows applications are run as separate threads in a single NTVDM with a shared address space.
    d. 16-bit Windows applications are run as separate threads in a single NTVDM with separate address spaces.
    e. 16-bit Windows applications are run as separate threads in separate NTVDMs with separate address spaces.

32. What function do the AUTOEXEC.NT and CONFIG.NT files serve?
    a. They contain the set of commands that need to be executed to start Windows NT.
    b. They contain the set of commands that need to be executed to start the Win16 NTVDM.
    c. They provide a location to load files required to run Windows 16-bit applications.
    d. They act as backup files for the AUTOEXEC.BAT and CONFIG.SYS files used to boot Windows NT.

33. Your friend Pat has a computer with Windows 2000 Professional that will not start. You format a blank disk using Windows 2000 and copy NTLDR, NTDETECT.COM, NTBOOTDD.SYS, and BOOT.INI from your computer that has Windows 2000 Professional to the floppy disk. Your computer has Windows 2000 load from your D drive. When you try to boot Pat's computer using this disk, you receive the following error message:

```
Windows NT could not start because the following file is
missing or corrupt: \Winnt Root\System32\Ntoskrnl.Exe.
Please reinstall a copy of the above file.
```

What should you do to correct this problem?
    a. Use Windows 2000 Backup program to restore the registry on Pat's computer.
    b. Copy the NTOSKRNL.EXE file from the bootable floppy to the root directory on Pat's computer.
    c. Copy NTLDR, NTDETECT.COM, NTBOOTDD.SYS, and BOOT.INI from the bootable floppy to the root directory of the primary partition on Charlie's computer.
    d. Edit the BOOT.INI file on the bootable floppy to specify the correct ARC path name to the Windows 2000 boot partition on Pat's computer.

34. Which one of the following statements describes limitations of MS-DOS applications running on Windows NT?
    a. They cannot share memory space.
    b. They cannot be preemptively multitasked.
    c. They cannot be started with different priorities other than NORMAL.
    d. They cannot be selectively terminated using Task Manager.

35. Your computer is configured to dual boot between Windows 2000 Professional and Windows 98. You are trying to boot to Windows 98, and you receive the following error message:

```
Non-System disk or disk error
Replace and press any key when ready.
```

Which one of the following is the most likely cause of the problem?
    a. The NTLDR file is missing.
    b. The NTDETECT.COM file is missing.
    c. The BOOTSECT.DOS file is missing.

    d. The BOOT.INI file is missing.

    e. The BOOT.INI file is using the wrong ARC path.

36. Your Windows NT Server computer contains two IDE disks on the same controller card. Each disk contains only one primary partition. The Windows NT Server operating system is located on the first hard disk, and the boot files are located on the second hard disk. Which of the following is the correct ARC name for the system partition?

    a. multi(0)disk(1)rdisk(0)partition(1)

    b. multi(0)disk(0)rdisk(1)partition(1)

    c. multi(0)disk(0)rdisk(0)partition(0)

    d. multi(1)disk(1)rdisk(1)partition(1)

    e. multi(1)disk(1)rdisk(0)partition(0)

37. Which of the following ARC names correctly indicates that the Windows NT boot files are located on the third partition on a SCSI disk with BIOS disabled, whose target ID is zero on the first SCSI controller?

    a. scsi(0)disk(0)rdisk(0)partition(2)

    b. scsi(0)disk(0)rdisk(0)partition(3)

    c. scsi(0)card(0)disk(0)partition(2)

    d. scsi(1)card(1)disk(1)partition(3)

38. The system and boot partitions have failed on the PDC. Fortunately, you are using disk mirroring on the PDC. How do you now modify the BOOT.INI file on your boot floppy disk so that the system will be started from the mirror partition?

    a. Add the /2 switch to the end of the appropriate line in the [Boot Loader] section.

    b. Modify the NTFS drive path and file name specifications in the [Boot Loader] section.

    c. Modify the ARC path name specifications in the [Operating Systems] section.

    d. The BOOT.INI file does not need to be modified as NTDETECT.COM will detect the change.

39. Your computer is configured to dual boot between Windows 98 and Windows 2000 Professional. However, your computer is only able to boot to Windows 98 and the boot menu never appears on the screen when you start the computer. What is the most likely cause of the problem?

    a. The NTLDR file is corrupt.

    b. The master boot record on the disk has been overwritten.

    c. The timeout value in the BOOT.INI file is set to zero.

    d. An entry has been added to the CONFIG.NT file to force the system to boot to Windows 98.

40. What allows us to use specialized programs to create a document by using objects from one program and inserting them into another program?

    a. OLE   c. networks

    b. multitasking   d. IFSHLP.SYS

41. Within the Windows 2000 Architecture, user mode is made of which two components? (Select two answers)

    a. environment subsystems   c. integral subsystem

    b. internal subsystems   d. VMM

42. Windows 2000 uses what type of multitasking?

    a. symmetric   c. asymmetric

    b. 32 bit   d. 64 bit

43. Which WINNT32 switch would you use if you wanted to mark the installation partition as active after the copying has finished?

    a. /ACTIVE   c. BOOTSECTOR

    b. /SYSPART   d. /SYSTEM

44. Which of the following components are installed by default on Windows 2000 Server? (Choose all that apply)

    a. Internet information server (IIS)   c. certificate services

    b. MicroSoft Script Debugger   d. terminal services

EXERCISE 1: INSTALLING WINDOWS 2000 PROFESSIONAL

1. You will need a 4-GB hard drive or larger. Create a 2-GB (2048-MB) active primary partition on the hard drive. When using the FDISK command and it asks you to support large disks, answer No.
2. Create a 1-GB (1024-MB) extended partition. From the extended partition, create a 1-GB (1024-MB) logical drive. Leave the rest of the disk unused.
3. Format the 2-GB primary partition (drive C) and make it bootable.
4. Format the 1-GB partition (drive D).
5. Reboot the computer with the proper DOS CD-ROM drivers. You can also boot with a bootable network disk and connect to a network drive to load the Windows 2000 files.
6. Load the Windows 2000 Professional CD or go to the network drive/directory where the Windows 2000 installation files are located.
7. Change into the I386 directory by using the CD I386 command.
8. Type in WINNT and press enter.
9. When Windows 2000 Setup needs to know where the Windows 2000 files are located, press Enter. Windows 2000 will copy the installation files over the C drive. Be patient, this will take a few minutes.
10. When the MS-DOS-based portion of setup is complete, remove any floppy disks from the A drive and press Enter to restart the computer.
11. Windows 2000 will welcome you to Setup. To set up Windows now, press Enter.
12. When the license agreement appears, press the F8 key to continue.
13. When Windows shows the partitions, select the D drive (1024-MB partition). This way, the system partition will be the C drive and the boot partition for Windows 2000 Professional will be the D drive. Press the Enter key to install.
14. Setup will install Windows 2000 on partition D. Select "Leave the current file system intact (no changes)" and press Enter. Windows 2000 will copy more files. The system will reboot again.
15. When the system reboots and starts a graphical interface, it will then automatically detect and install the hardware devices. Next it will ask for the regional settings. Select the appropriate regional settings and click on the Next button.
16. To personalize your software, enter your name and your company's name. Click on the Next button.

> **Note:** To get the most out of this book, you will need to work with a partner or a second computer. The computer on the left will be designated as Computer A and the computer on the right will be Computer B.

17. The next screen shows a random computer name. Change the computer name to WS2000-xxy, where *xx* represents your two-digit partner number in the class and *y* represents A if you are Computer A or B if you are Computer B. If you are not doing this in class, use 01. For example, if you are the first set of partners and are using the computer on the left, use WS2000-01A. If you are the first set of partners and are using the computer on the right, use WS2000-01B. Lastly, for the password, enter "password" in the Administrator password and confirm password text boxes. Click on the Next button.
18. If your computer has a modem, a modem dialing information box will appear. Enter your area code and type in the appropriate options for your computer. Click on the Next button.
19. For the date and time settings, enter the proper information and click on the Next button.
20. Windows 2000 will configure the networking settings. Select typical settings and click on the Next button.
21. The next window asks if you want to be a member of a workgroup or a computer domain. For now, select No and click on the Next button. It will then copy some files and perform final tasks.
22. When the Windows 2000 setup wizard is complete, click on the Finish button and the computer will reboot.
23. The Welcome to the Network Identification wizard begins. Click on the Next button.
24. It will ask, "Who can log on to this computer?" Since we will assume that this is going to be a secure environment, select the users who then must enter a user name and password to use this computer. Click on the Next button, then on the Finish button.

25. Log on as the administrator.
26. If you are connected to the Internet, start Internet Explorer, open the Tools menu, and select the Windows Update option to go Microsoft's Web site to search for updates. Download and install any updates that are needed.

EXERCISE 2: INSTALLING WINDOWS 2000 SERVER (OR WINDOWS 2000 ADVANCED SERVER)

1. Reboot the computer with the proper DOS CD-ROM drivers. You can also boot with a bootable network disk and connect to a network drive to load the Windows 2000 files.
2. Load the Windows 2000 Server CD or go to the network drive/directory where the Windows 2000 installation files are located.
3. Change into the I386 directory by using the CD I386 command.
4. Type in WINNT and press Enter.
5. When Windows 2000 Setup needs to know where the Windows 2000 files are located, press Enter. Windows 2000 will copy the installation files over the C drive. Be patient, this will take a few minutes.
6. When the MS-DOS-based portion of setup is complete, remove any floppy disks from the A drive and press Enter to restart the computer.
7. Windows 2000 will welcome you to Setup. To setup up Windows now, press Enter.
8. When the license agreement appears, press the F8 key to continue.
9. When Windows shows the partitions, select the C drive (2047-MB partition). This way, the system partition will be the C drive and the boot partition for Windows 2000 Server will be the C drive. Press the Enter key to install.
10. Setup will install Windows 2000 on partition C. Select the "Convert the partition to NTFS" option and press Enter. Windows 2000 will copy more files. The system will reboot again.
11. When the system reboots and starts a graphical interface, it will then automatically detect and install the hardware devices. Next it will ask for the regional settings. Select the appropriate regional settings and click on the Next button.
12. To personalize your software, enter your name and your company's name. Click on the Next button.
13. In the next screen, select the per-server licensing mode and enter 50 connections. Click on the Next button.

> **Note:** To get the most out of this book, you will need to work with a partner or a second computer. The computer on the left will be designated as Computer A and the computer on the right as Computer B.

14. The next screen shows a random computer name. Change the computer name to Server2000-xxy, where *xx* represents your two-digit partner number in the class and *y* represents A if you are Computer A or B if you are Computer B. If you are not doing this in class, use 01. For example, if you are the first set of partners and are using the computer on the left, use Server2000-01A. If you are the first set of partners and are using the computer on the right, use Server2000-01B. Lastly, for the password, enter "password" in the Administrator password and confirm password text boxes. Click on the Next button.
15. The installation wizard will ask you to add or remove components of Windows 2000. Since there is nothing that we want to add at this time, click on the OK button.
16. If your computer has a modem, a modem dialing information box will appear. Enter your area code and type in the appropriate options for your computer. Click on the Next button.
17. For the date and time settings, enter the proper information and click on the Next button.
18. Windows 2000 will configure the networking settings. Select Custom Settings and click on the Next button.
19. The network components chosen by default are Client for Microsoft Networks, file-and-print sharing for Microsoft networks, and Internet protocol (TCP/IP). Click on the Internet protocol and click on the Properties button.

211

20. In the Internet protocol (TCP/IP) properties dialog box, click on the "Use the following IP address" option and input the following:

IP address: 140.100.1.1xx
Subnet mask: 255.255.255.0

where *xx* is the computer number assigned. If you are working at home, use 01. Therefore, the address would be 140.100.1.101. In addition, if your network has a gateway, specify the gateway address. Click on the OK button and then on the Next button.

21. The next window asks if you want to be a member of a workgroup or a computer domain. For now, select No and click on the Next button. It will then copy some files and perform final tasks.

22. When the Windows 2000 setup wizard is complete, click on the Finish button and the computer will reboot.

23. When the boot menu appears, select the Windows 2000 Server.

24. Log on as the administrator.

25. If you are connected to the Internet, start Internet Explorer, open the Tools menu, and select the Windows Update option to go Microsoft's Web site to search for updates. Download and install any updates that are needed.

26. Go to the WWW.ADOBE.COM Web site and download the newest version of Adobe Acrobat reader.

27. Go to the WWW.WINZIP.COM Web site and download the evaluation copy of the newest version of WinZip.

28. Install both Acrobat reader and WinZip on your server.

EXERCISE 3: BOOTING WINDOWS 2000

1. Press the CTRL+ALT+DEL keys and click on the Shut Down button. Select Restart from the pulldown menu and click on the OK button.

2. During the reboot, study what is displayed during boot-up, specifically the boot menu.

3. Start a Command Prompt window by clicking on the Start button, clicking on the Accessories option, and clicking on the Command Prompt option.

4. The C:\> prompt should be showing. Type in the DIR command and press Enter.

5. The directory listing shows the WINNT, Documents and Settings, Program Files, and Inetpub folder. Type in the DIR /AH and press Enter.

6. The directory listing shows the BOOTSECT.DOS, NTLDR, NTDETECT.COM, BOOT.INI files. Type EDIT BOOT.INI and press Enter.

7. The BOOT.INI displays the boot menu that appears during boot-up.

8. Type EXIT and press Enter.

9. Reboot the computer again. When the boot menu appears, press a left arrow key.

10. Note at the bottom of the screen the "For troubleshooting and advanced startup options for Windows 2000, press F8" message. Press the F8 key.

11. The Windows Advanced Options menu appears. Select Safe Mode and press Enter. Make sure the Windows 2000 Server option is highlighted and press Enter.

12. Press the CTRL+ALT+DEL keys to log on as the administrator. Then perform another reboot. Again, go into the Windows 2000 Advanced Options menu. Select Safe Mode with Command Prompt. Select the Windows 2000 Server and press Enter.

13. Log on as administrator. Close the Command Prompt window. Press the CTRL+ALT+DEL keys and perform another reboot.

14. Boot the Windows 2000 Server and log on as the administrator.

15. Open the Command Prompt window.

16. Type CD WINNT and press Enter.

17. Type EDIT NTBTLOG.TXT and press Enter.

18. Using the arrow keys, study the NTBTLOG.TXT file.

19. Press the Alt+F to open the file menu and select the Exit option. If it asks you to save the changes, choose No.

20. Close the Command Prompt window.

EXERCISE 4: EXPLORING THE WINDOWS 2000 SECURITY DIALOG BOX AND WINDOWS TASK MANAGER

1. Press the CTRL+ALT+DEL keys to open the Windows 2000 Security dialog box.
2. Click on the Lock Computer option.
3. Unlock the computer.
4. Open the Windows 2000 Security dialog box and click on the Change Password button. Change the password to today.
5. Change the password back to password.
6. Open the Windows 2000 Security dialog box and select Task Manager. Click on the Process tab. After you study the Process tab, click the Performance tab.
7. At the bottom of the Windows Task Manager window, notice the number of processors and the CPU usage. In addition, notice the amount of system cache, which is located beneath the graphs.
8. Go back to the Processes tab and check for a NOTEPAD.EXE listing. Don't spend too much time on this.
9. Open the WordPad program. It can be opened by clicking on the Start button, selecting the Accessories option, and selecting the WordPad option.
10. Notice that there is a WORDPAD.EXE listed in the Processes tab and a Document-WordPad listed in the Application tab.
11. In the Application tab, click on the Document-WordPad listing and click on the End Task button.
12. Close the Task Manager.

EXERCISE 5: CREATING WINDOWS 2000 START-UP FLOPPY DISKS

1. Insert a blank floppy disk in drive A.
2. Double-click My Computer to open it.
3. Right-click the A drive and select the Format option.
4. When the Format A:\ dialog box appears, click on the Start button. Click on the OK button to format the disk.
5. When the format is complete, click on the OK button and close the Format A:\ window.
6. Double-click on the C: icon to open it.
7. To show the hidden files, open the Tools menu and select the Folder Options option. Click on the View tab and click on the "Show hidden files and folders." In addition, click on the "Hide protected operating system files (Recommended)" option and the "Hide file extensions for known file types" option to take off the check mark. Click on the Yes button when "Deleting or editing the protected operating systems files can make your computer inoperable" warning appears. Click on the OK button.
8. While pressing the CTRL key down, click on the following files.

    NTLDR.EXE
    NTDETECT.COM
    NTBOOTDD.SYS (if available)
    BOOT.INI

9. While these files are still highlighted, right-click on one of them and select the Send To option. Select the 3 1/2 floppy (A) option.
10. Close all windows.
11. Another way to do a shutdown is to click on the Start button and select the Shut Down option.
12. With the disk still in drive A, select the Restart option and click on the OK button.
13. Although the menu is identical to the previous boot menu, the difference is it is being executed from the floppy disk. Select the Windows 2000 Server option and press Enter.
14. Open a window for the A drive.
15. Delete the NTLDR.EXE file from the A drive. You can delete the file by right-clicking on the file and selecting the Delete option. If it asks if you are sure, click Yes. Make sure it is the A drive and not the C drive.
16. Close all windows and reboot the computer again with the disk in drive A.

17. You can see that a NTLDR is missing. Normally, when this happens you would expect something has happened to the NTLDR file in the C drive (boot drive). Of course, we are using the floppy disk instead.
18. Remove the disk in drive A. Use the CTRL+ALT+DEL keys to reboot the computer into the server.
19. Log on as administrator.
20. Double-click on the My Computer icon, then double-click on the C: drive icon.
21. Insert the disk. To open a second window, double-click on My Computer, then double-click on the A drive.
22. Position the windows to view both of them.
23. Drag the NTLDR file from the C drive window to the A drive window. Because the windows represent two different drive letters, the file is copied (a duplicate is made and placed in the A drive). If the two windows represented the same drive letter, but different directories/folders, then the file would be moved. In this particular case, if this file is moved, then the system would not boot, which is one reason why we are making the Windows 2000 boot disk.
24. Remove the disk and label the outside of the disk as the Windows 2000 Start-up floppy disk. If you modify the BOOT.INI file, save it to the disk after the changes have been made and tested.
25. To test the start-up disk, insert the disk in drive A and reboot the computer from the floppy disk.

EXERCISE 6: LOOKING AT THE BOOT.INI FILE

1. Double-click on My Computer to open a My Computer window.
2. Double-click on the C drive to open a C:\ window.
3. If you cannot see the BOOT.INI file and the other boot files (NTLDR and NTDE-TECT.COM), open the Tools menu and select the Folder option. Select the View tab.
4. If the option is not already enabled, under Advanced Settings, select the "Show hidden files and folders" option.
5. Under the Advanced Settings, deselect the "Hide file extension for known file types and Hide protected operating system files (Recommended)." Click on the OK button.
6. When it asks if you are sure you want to display these files, click on the Yes button.
7. Right-click the BOOT.INI file and select the Properties option. Deselect the "Read-only in the Attributes section" and click on the OK button.
8. Right-click the BOOT.INI file and select the Open With option. Select Notepad and click on the OK button.
9. As you can see, the default timeout is 30 seconds. Change the 30 to a 15.
10. Open the file menu and save the file.
11. Click on the Start button and reboot Windows 2000. Notice that the timer is now 15 seconds instead of 30 seconds. Select the Windows 2000 Server.
12. Open the BOOT.INI file with Notepad again. For the default entry, change the partition from 1 to 2.
13. Save the changes for the BOOT.INI file. Again, click on the Start button and reboot Windows 2000.
14. When the timer begins, let the time run out so that it will pick the Default menu option. Notice that while it selected the correct menu option, Windows Professional loaded because the ARC path indicated partition 2 and not partition 1.
15. From Windows 2000 Professional, we could use notepad. Instead, click on the Start button, select the Accessories option, and select the Command prompt.
16. Switch to the C:\. If you do not know how, type in C: followed by pressing the Enter key and typing in CD\ followed by the Enter key.
17. At the C:\> prompt, type in EDIT BOOT.INI and press Enter.
18. Change the partition back to 1 for the ARC path for the default entry.
19. Press the Alt key and hold it down. While the Alt key is down, press the F key. Use the arrow keys to select the Save option.
20. Open the File menu again by using the Alt+F keys and select the Exit option.
21. Type EXIT at the prompt and then press Enter.
22. Restart Windows to test if the Default menu option brings up Server.

8

Windows 2000 Basics

INTRODUCTION

Once you have Windows 2000 installed, then it is time to start using it. This chapter discusses how to operate and configure Windows 2000 as an operating system, not as a network operating system. When creating and configuring Windows 2000, you will find it quite similar to Windows 98 and Windows NT, but when you start getting into more of the advanced features of Windows 2000, you will start seeing some major differences in this new version.

OBJECTIVES

1. Demonstrate how to configure Windows 2000 taskbar.
2. Demonstrate common file and directory management tasks.
3. Demonstrate how to configure Windows 2000 using the Control Panel.
4. Given common hardware problems, troubleshoot and fix such problems.
5. Implement, manage, and troubleshoot display devices.
6. Configure multiple-display support.
7. Install, configure, and troubleshoot a video adapter.
8. Manage and troubleshoot driver signing.
9. Given a computer, optimize the performance of the paging files.
10. Manage hardware profiles.
11. Configure advanced power management (APM).
12. Configure, manage, and troubleshoot the Task Scheduler.
13. Given a UPS, configure Windows 2000 to communicate with the UPS.
14. Starting with an empty MMC, create a custom console and save it in one of four MMC modes.
15. Compare the four MMC modes.
16. Use the Services console to enable, stop, or pause various services.
17. Demonstrate using the Task Scheduler to schedule a maintenance task such as disk defragmenter or ScanDisk.
18. Given a key and its location in the registry, find and make the correct changes.

8.1 WINDOWS 2000 INTERFACE

The Windows 2000 interface is based on the desktop, which is the graphical space on the screen from which all control and display presentations begin. On the desktop, you will find the My Computer icon, the Recycle Bin, a My Documents shortcut, and the taskbar (with the Start button). In addition, you may have a My Network Place icon to access network resources (files, applications, and printing) and an Inbox to receive electronic mail and various folders (directories) and shortcuts to programs and documents. See Figure 8.1.

The **My Computer** icon represents your computer. It includes all disk drives and the control panel. From within the My Computer icon, you can access and manage all files on your drives.

The **Recycle Bin** is used as a safe delete. When you delete a file using the GUI interface, Windows will store the file in the Recycle Bin. It will remain there until you empty it or you

FIGURE 8.1 Windows 2000 interface

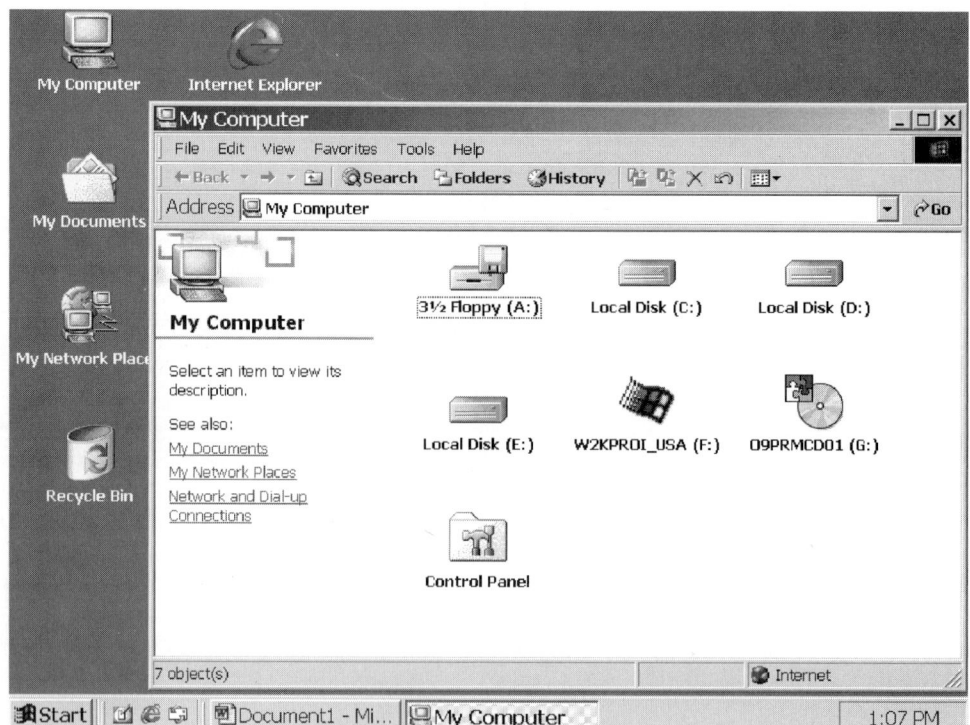

start running out of disk space (Windows will recycle the oldest files first). If you delete a file and decide that you want to undelete it, then open the Recycle Bin and drag it to any location or choose the Restore option from the File menu.

The **My Documents** shortcut points to a My Documents folder. By default, the My Documents folder is stored in the C:\Documents and Settings*user_login_name* folder. Users can change the target folder location of their My Documents folder by right-clicking the My Documents icon on the desktop, selecting the Properties option, and specifying a new location on the Target tab. See Figure 8.2.

The **taskbar** is located at the bottom of the screen. See Figure 8.3. You can configure the taskbar to be always visible or to auto hide, so that it appears only when you move the mouse pointer to the bottom of the screen. Auto hide can be used to give you more free area on the desktop.

The taskbar is divided into four areas—the Start button, the notification area, the active program buttons, and quick launch programs. The **Start button** is used to start programs. In addition, you can open recently accessed documents, access the Control Panel and printer folder and find files, and get help for Windows 2000. The Start button is a series of shortcuts to programs.

The **notification area,** located on the right side of the taskbar, is used for the clock and for any programs running in the background such as printers and modems. The rest of the taskbar is blank or holds the active program buttons, which can be used to switch between open programs.

When a window is open, it will usually contain a Minimize button, a Maximize/Restore button, and a Close button in the top right corner and a Control Menu icon in the top left corner. In addition, most program windows will include a menu and many will include one or more toolbars.

Like Windows 95, Windows 98, and Windows NT, Windows 2000 uses **shortcut menus,** which are accessed by clicking the secondary mouse button (usually the right mouse button). It contains common commands that you can use on the item that you clicked. For example, if you click a file with the right mouse button, then you can choose to open the file, copy the file, delete the file, or show the properties of the file. If you click the disk drive in

FIGURE 8.2   Properties of My Documents

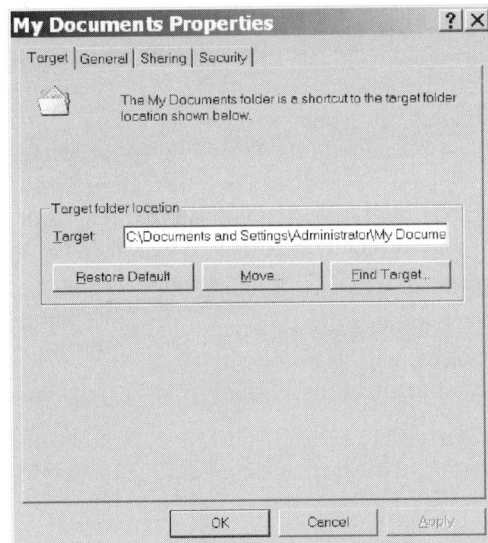

FIGURE 8.3   The taskbar

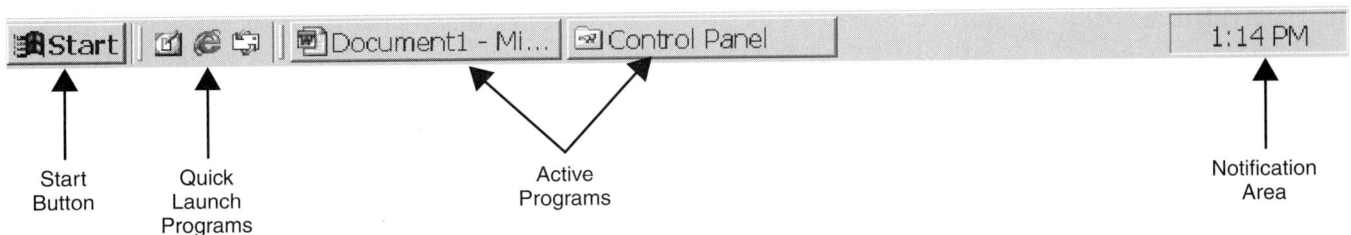

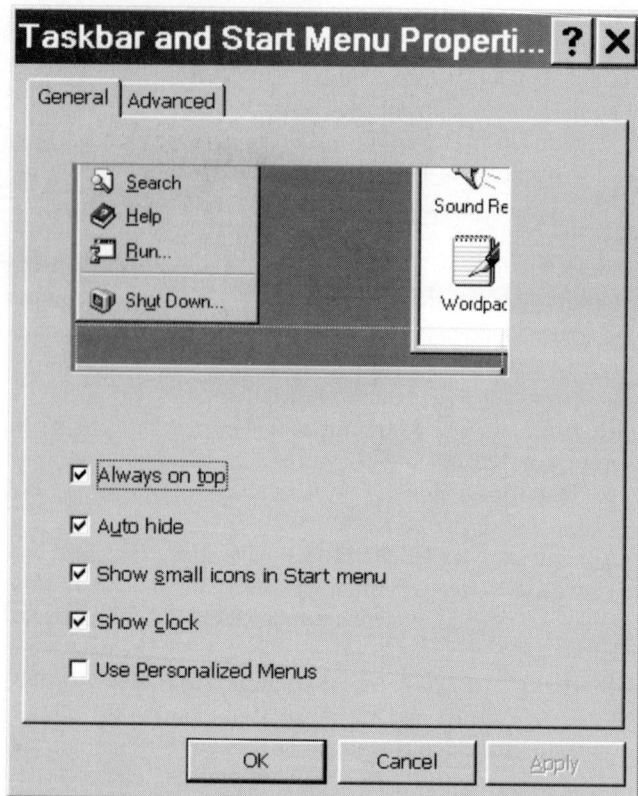

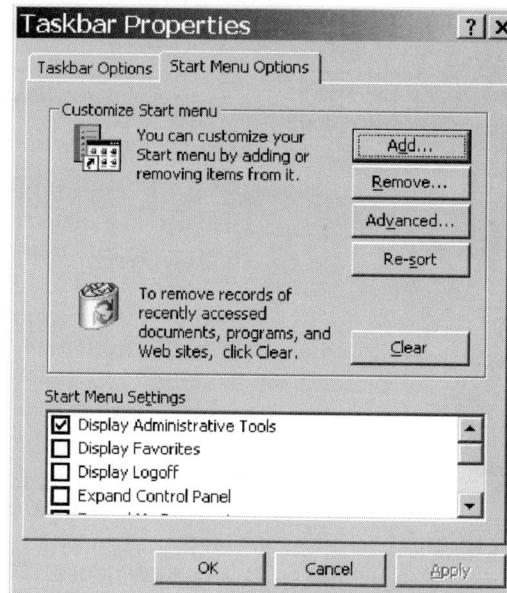

FIGURE 8.4 Taskbar properties

My Computer with the right mouse button, you can open the disk, format the disk, or show the properties of the disk. If you highlight some text in a word processor such as Microsoft Word and click on the text with the right mouse button, you can choose the Cut, Copy, or Paste options or change the formatting of the text.

As stated, the desktop contains the My Computer and Recycle Bin icons. In addition, it can hold folders, files, and shortcuts. To create a folder or a shortcut, simply use the desktop's shortcut menu and select the New option. Note that content visible on the desktop is kept in C:\Documents and Settings*user_login_name*\Desktop folders.

The taskbar can be configured by using the Properties button (Taskbar Options) on the taskbar's shortcut menu or by using Taskbar in the Settings option under the Start button. In the Taskbar Options tab, specify the following:

1. If the taskbar auto hides when not in use
2. If the taskbar will show the time
3. If the Start menu will use small icons or large icons

From the Start Menu Programs tab, add, remove, or reorganize the programs listed under the Start button or clear the Documents menu. See Figure 8.4.

8.2 FILE MANAGEMENT

Since Windows 2000 is an operating system, become familiar with how to accomplish common tasks such as managing files and disks, running programs, and configuring the Windows NT environment. Disk and file management is usually done with My Computer or **Windows Explorer.** See Figure 8.5. You can start Windows Explorer by clicking on the Start button, select the Programs option, select the Accessories option, and select Windows Explorer. The most common actions are shown in Table 8.1.

If you want to move, delete, or copy several files or folders at the same time, use the CTRL and SHIFT keys to select multiple files. For example, if you press and hold down the CTRL key, you can select any file or folder within the same drive or folder. If you press the SHIFT key and keep it down, you can select two files or folders and everything listed be-

FIGURE 8.5 Windows Explorer

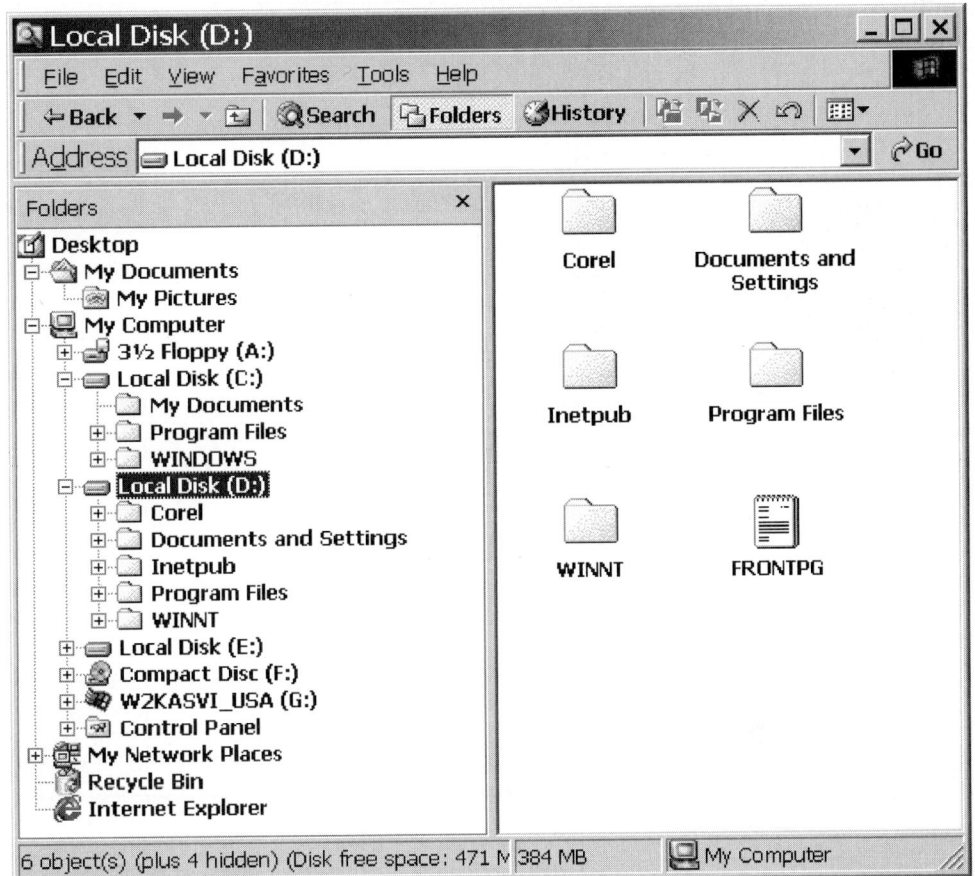

tween the two will be selected. After the file, or directories, have been selected, you can then copy, delete, or move them as one.

To start a DOS session to perform DOS commands at a prompt, click on the Start button, select the Programs option, select the Accessories option, and select the Command Prompt option.

8.3 RUNNING PROGRAMS

Windows 2000 will run Win16-based applications (Windows 3.xx), Win32-based applications (Windows 95/98 and Windows NT), DOS applications, and OS/2 bounded applications. To run a program, select the program by using the Program option in the Start button. In addition, you can start an executable file by way of the following:

1. Click on the file (or data document) using Windows Explorer or My Computer.
2. Create a shortcut on the desktop or in any folder.
3. Use the Run option in the Start button. See Figure 8.6.
4. Type the name of the executable file in an MS-DOS command session.

Since Windows 2000 is a multitasking environment, it allows you to start several programs simultaneously. You can then change between different programs as follows:

• Press and hold down the ALT key and repeatedly press the TAB key until you get the desired application.
• Use the ALT+ESC keys to move to the next application.
• Select the Application button on the taskbar.
• Use the Task Manager.

8.3.1 Removing a Windows Application

During installation, a Windows program often adds or modifies information to the registry. To remove these entries and program files on the hard drive, many programs include an

TABLE 8.1   Common disk and file management actions

Task	Action
To create a folder (directory) on the desktop, in the root directory, or in another folder	Select New from the File menu or New from the shortcut menu (right mouse button) and select the Folder option.
To delete a file or directory	Select the file or folder by clicking on it and dragging it to the Recycle Bin. Pick the Delete option from the shortcut or file menu or press the DEL key on the keyboard.
To format a disk	Select the Format option from the drive's shortcut menu or select the Format option from the My Computer's file menu.
To copy a floppy disk	Select the Copy Disk option from the drive's shortcut menu or from the My Computer file menu.
To copy a file or directory from one drive to another	Drag the file (click on the file and keep the left mouse button clicked while moving the mouse) to its new destination.
	Another way is to first select the file or directory and select the Copy option from the shortcut menu or the Edit menu (Explorer or any disk or folder window). You then go to your destination and select the Paste option from the shortcut menu or Edit menu.
To move a file or directory to a different folder within the same drive	Press the Ctrl key while dragging the file to its new destination. If you do not press the Ctrl key, it will make a shortcut to the file or directory instead of moving the file.
	Another way is to first select the file or directory and select the Cut option from the shortcut menu or the Edit menu (Explorer or any disk or folder window). Then go to your destination and select the Paste option from the shortcut menu or Edit menu.
To rename a file	Select the file and choose the Rename option in the file menu or in the shortcut menu.
	Another way is to click once on the file name.
To view or change the directory or file attributes (read-only, hidden, or system)	Select the folder or file and select the Properties from the File menu or shortcut menu.

Note: Many 32-bit Windows applications also allow you to cut, copy, delete, and rename by using the shortcut menu from within some dialog boxes including the Open, Save, and Save As dialog boxes.

FIGURE 8.6   Run option under the Start button

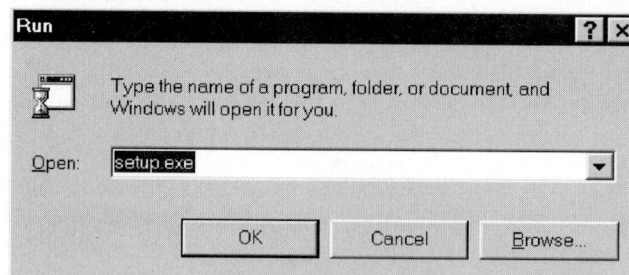

FIGURE 8.7   Registered files
shown in the Folder options
dialog box

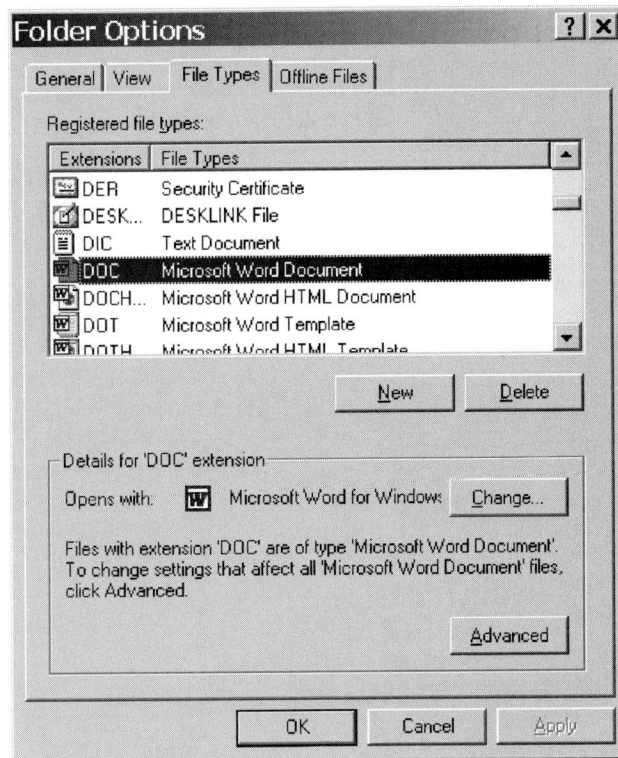

additional uninstall program or have an uninstall option within the installation program. In addition, if the program follows the Windows 2000 API standards, it can be removed by using the Add/Remove Programs applet in the Control Panel. Even when you uninstall a program, some of its files may remain on the hard drive.

8.3.2   Windows Associations

When you double-click on a document in Windows 2000, it will recognize the file name extension and start the appropriate program to open the file. Associating an extension with a program is known as registering a file. The information identifying the file name extensions is located in the registry and can be accessed and modified by selecting Options under the View menu in the My Computer, disk, or folder window and clicking on the File Type tab or by opening the Folder Option applet in the Control Panel. See Figure 8.7.

8.4   CONTROL PANEL

There are several utilities used to configure Windows 2000. The main utilities are the Control Panel and Registry Editor. The Windows 2000 Control Panel is a graphical tool to configure the Windows environment and hardware devices. See Figure 8.8. It can be accessed from the Settings option in the Start button under My Computer. In addition, there are various shortcuts to directly access certain Control Panel applets (icons). The most commonly used applets are the System applet, the Display applet, the Add/Remove Programs applet, and the Add New Hardware applet.

8.4.1   System Applet

The System applet has five tabs: General, Network Identification, Hardware, User Profiles, and Advanced. See Figure 8.9. The General tab shows the user name and company entered during installation, the type of processor, and the amount of RAM in the system, as shown in Figure 8.9a. The Network Identification tab shows the full computer name. This tab will be discussed in Chapter 9.

FIGURE 8.8 Control panel

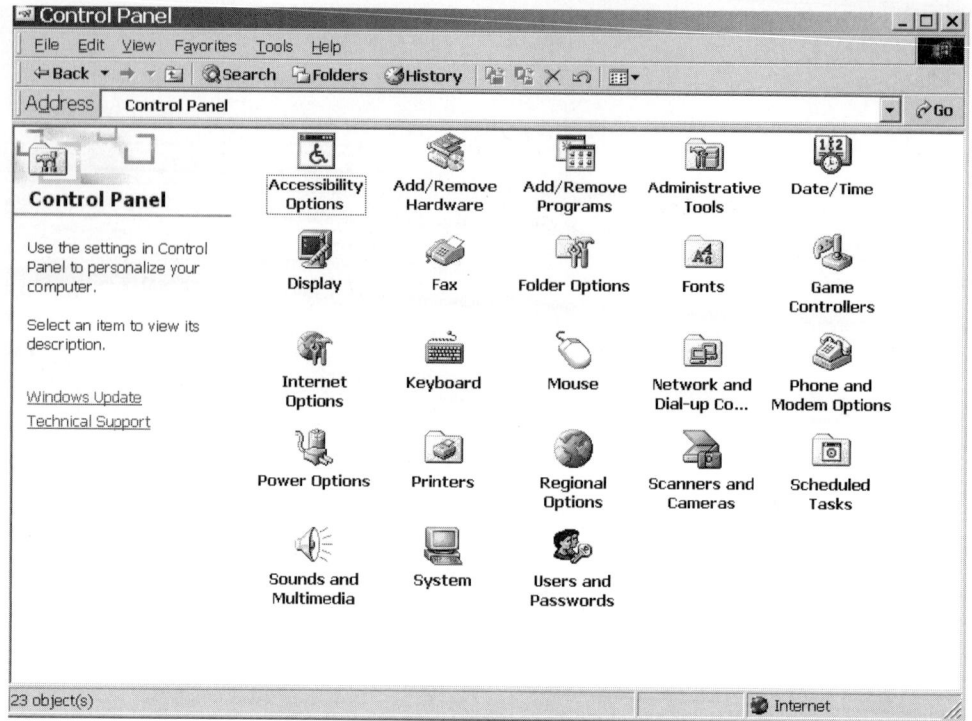

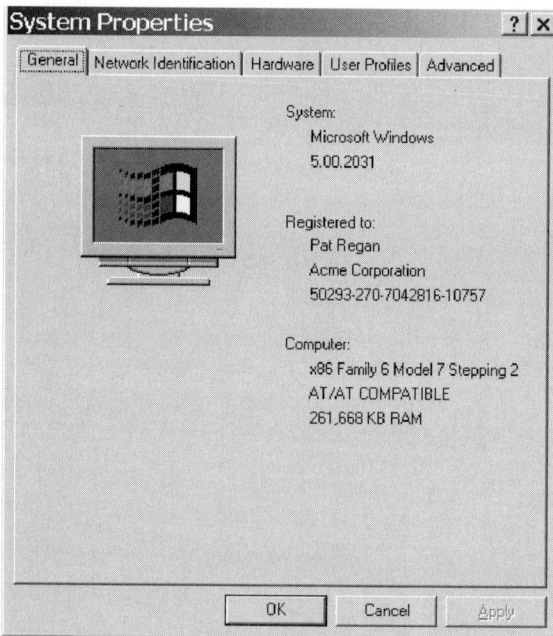

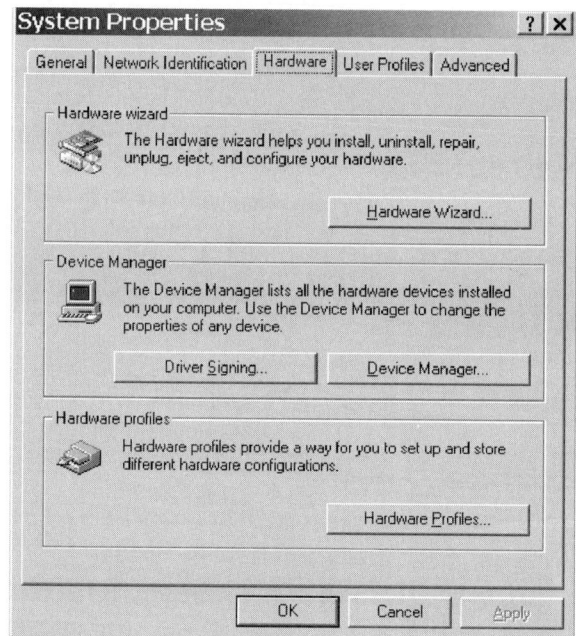

FIGURE 8.9 Opening the System applet in the Control Panel will display the System Properties dialog box. Within the dialog box are five tabs: General, Network Identification, Hardware, User Profiles, and Advanced

The Hardware tabs allow access to the Hardware wizard, the Device Manager, and Hardware profiles. The Hardware wizard helps you install, uninstall, repair, eject, and configure your hardware. See Figure 8.9b.

The **Device Manager** lists all hardware devices on your computer and allows you to change the properties of any device. If you open the View menu and select Resources by Type, you will see the IRQ, DMA, I/O address, and memory areas used by the different hardware components. See Figure 8.10.

If you open the View menu and select Devices by Type or Devices by Connection, you see all hardware devices organized in a tree structure. When a red X appears through an icon

FIGURE 8.10 Device manager
showing the resources used by
the various devices within the
computer

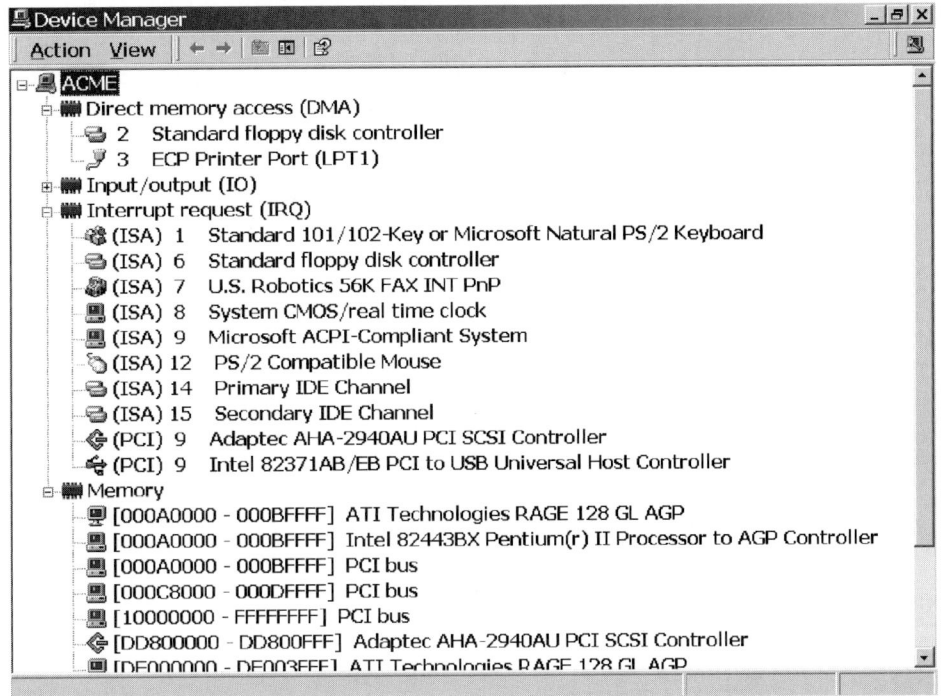

FIGURE 8.10 Device manager showing the resources used by the various devices within the computer

FIGURE 8.11 Device Manager
showing all devices on the
computer

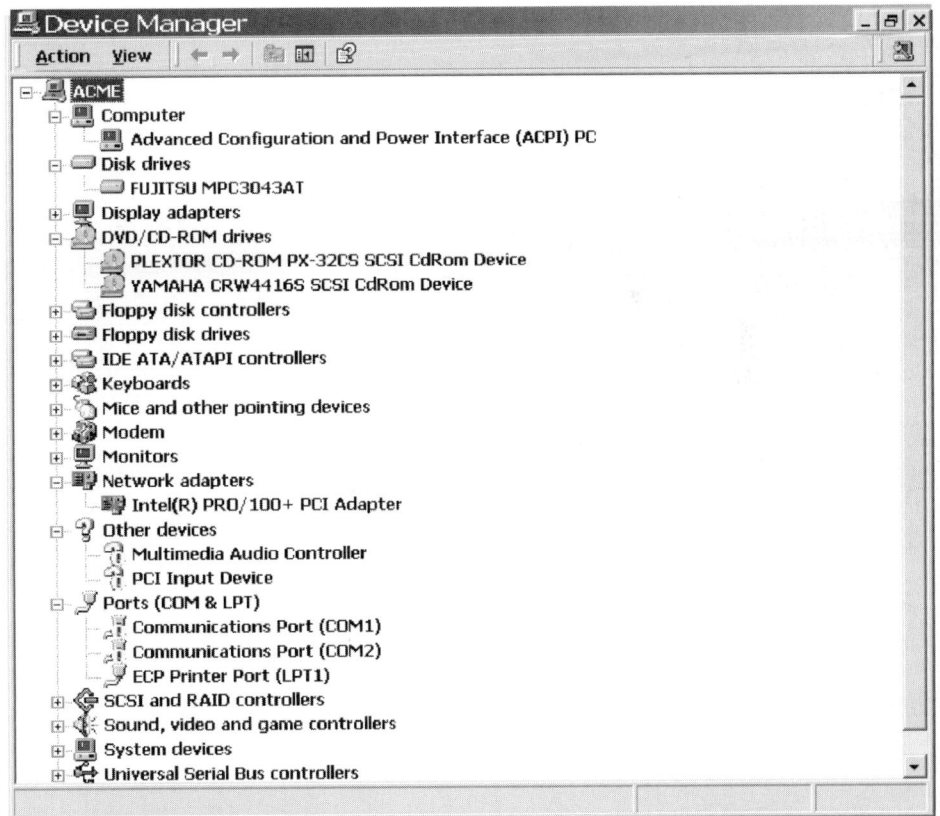

FIGURE 8.11 Device Manager showing all devices on the computer

this means the hardware device has been disabled. A yellow circled exclamation point through the icon means the hardware device has a problem. The problem could be a resource conflict (IRQ, DMA, I/O address, or memory address) or the drivers are not loaded properly. See Figures 8.11 through 8.14.

When viewing device information on your Windows 2000–based computer using Device Manager, you may see an unknown device listed next to a yellow question mark. Determining the cause of this unknown device can be difficult, because there are few indications

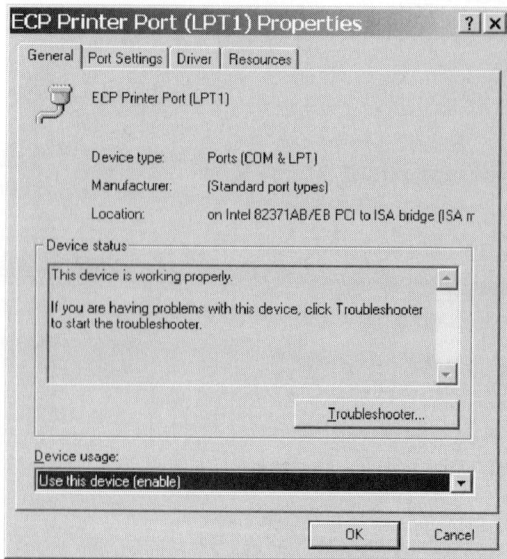

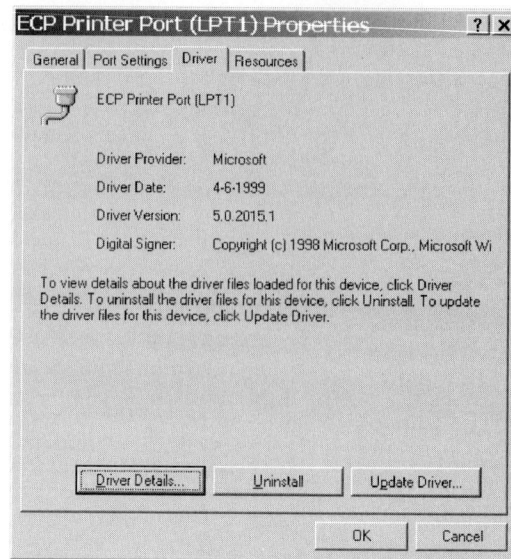

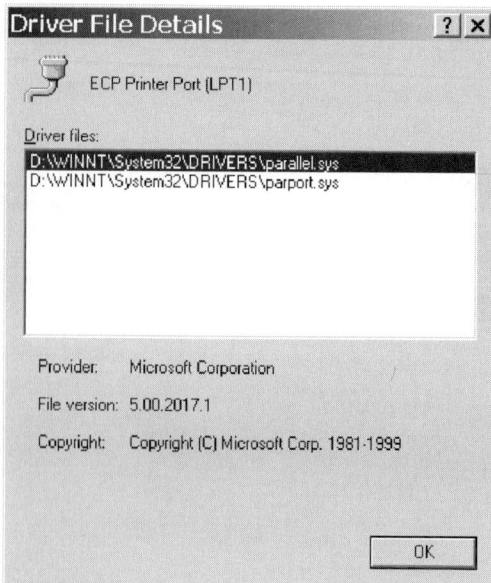

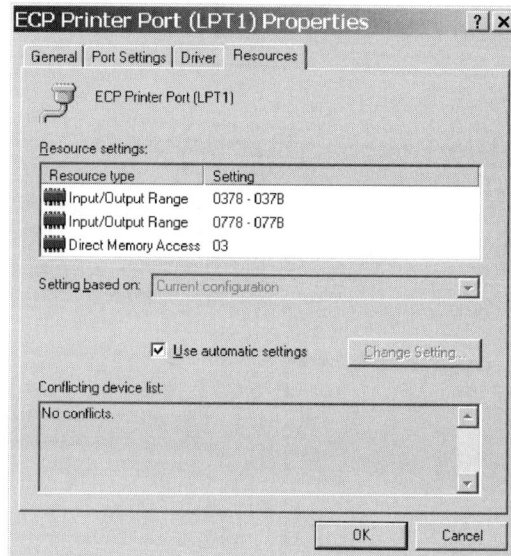

FIGURE 8.12   ECP Printer Port properties as accessed through the Device Manager, including drivers and resources used

of what could be creating it. The unknown device may also cause a conflict when trying to install a driver for another device.

The most common reasons Device Manager may list a device as unknown are as follows:

- The Device Manager does not have a device driver.
- You are using a Windows 95/98 device driver.
- A device ID is not present.
- Windows 2000 does not recognize the device ID.
- The device is created by software.
- The system has faulty hardware or firmware.

When a device driver for a device is not available, Device Manager displays the device as unknown, and places it in the Other Devices folder. This commonly occurs with universal serial bus (USB) and IEEE 1394 composite devices. A status of "Error Code 1" or "Error Code 10" also may be displayed when you view the properties of the device in Device Manager.

You cannot use virtual device driver (.vxd) files common to Windows 95/98 drivers with Windows 2000, and attempting to install them on your Windows 2000–based computer may cause the device to be listed as unknown in Device Manager. This usually occurs when the

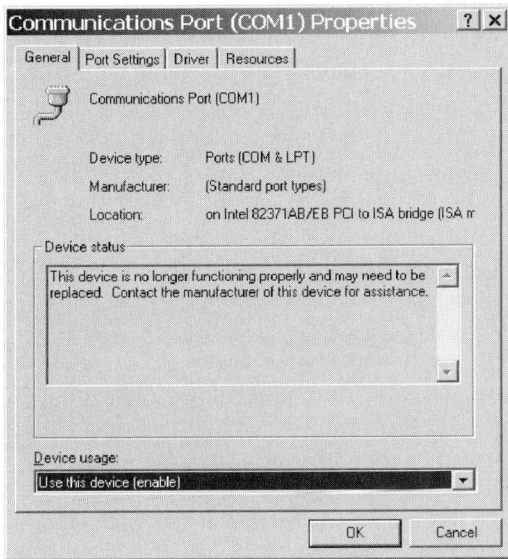

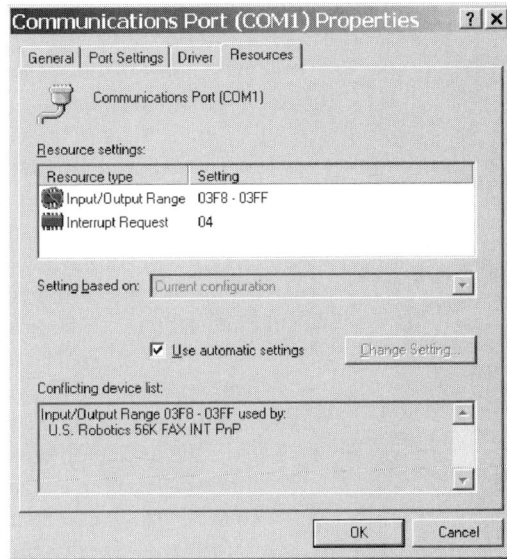

FIGURE 8.13 If a device is not working properly, you can find more information about the problem by showing the properties of the device in the Device Manager

FIGURE 8.14 You can enable or disable a device by using the properties for the device in the Device Manager

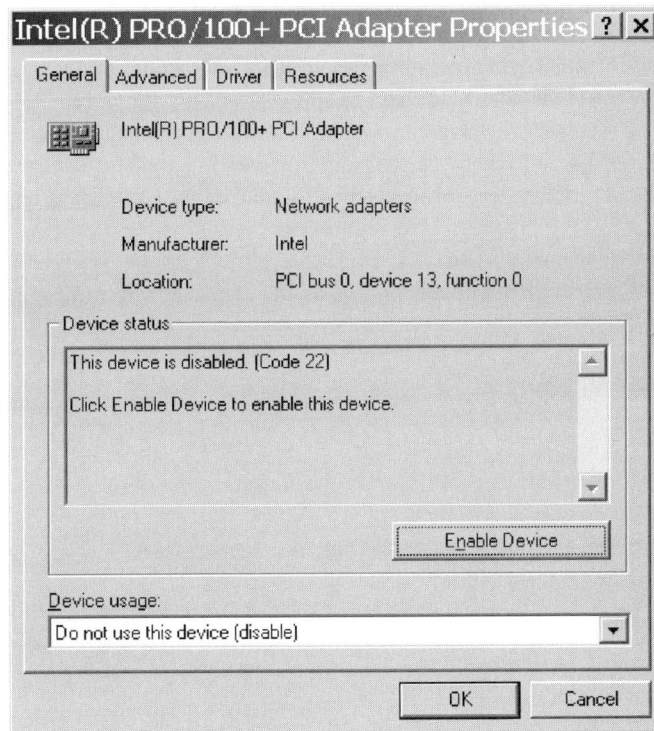

device driver manufacturer does not properly differentiate between the two drivers, or assumes that Windows 2000 is capable of using Windows 98/95 .vxd files.

Every hardware device has a special identifier used by plug-and-play (PNP). This identifier can consist of several different types, such as vendor ID, device ID, subsystem ID, subsystem vendor ID, or revision ID. If a device ID is not present, or your Windows 2000–based computer does not recognize the device ID, Device Manager may list the device as unknown.

Virtual devices that are created with software also would be considered as unknown when a driver tries to use the InstallShield installation program or a similar method and the device was removed in Device Manager, but still may have entries left over in the registry. To determine if an unknown device is being created by software, try starting Windows 2000 in safe mode. In addition, click on the Start button, select the Programs option, and select the Administrative Tools and then select Computer Management. In the Computer Management console, click the System Information folder, double-click the Software Environment folder, and double-click the Startup Programs folder. You should then check the event log for errors relating to any of these programs to see if one is not working properly. If you find a related event, uninstall the associated program. In addition, look at the Components folder in the System Information folder. The Components folder will list the common name for the device and the name of the device driver associated with it. In the PNP Device ID column, it will list the device IDs such as the PCI ID, ISA ID, an ID for some other bus type or unknown type. The Error Code column lists the error code associated with a specific problem such as a bad or incompatible device driver.

The unknown device can also be caused by faulty hardware or firmware. To isolate the correct hardware device, remove the hardware devices from the Windows 2000 computer one at a time until the unknown device is no longer listed in Device Manager. In addition, check if the device driver is digitally signed. If Windows 2000 detects that a device driver is not digitally signed, a Not Digitally Signed message is displayed.

To see hardware in a category, click on the plus (+) sign next to the hardware type. To see information about a piece of hardware, double-click on the hardware device icon or select the hardware device and click on the Properties button. The Properties will usually contain several tabs indicating if the device is working properly and what drivers are loaded, and it gives you an option to update them and to view and change the device resources (if it is a plug-and-play device).

Within the Advanced tab of the System Properties dialog box, you will find the Performance Options, Environment Variables, and Startup and Recovery buttons. See Figure 8.15.

The Performance Options allows you to adjust how the microprocessor resources are distributed between running programs. Selecting Applications assigns more resources to the foreground applications, whereas the Background Services assigns an equal amount of resources to all programs. You should select the Background Services if it is a server with network resources/services or Applications if you want the server to act more like a desktop machine. See Figure 8.16.

The Virtual Memory area allows you to configure the paging file size (virtual memory) and the maximum size of the registry. The minimum paging file is 2 MB. For Windows

FIGURE 8.15   Advanced tab in the System Properties dialog box

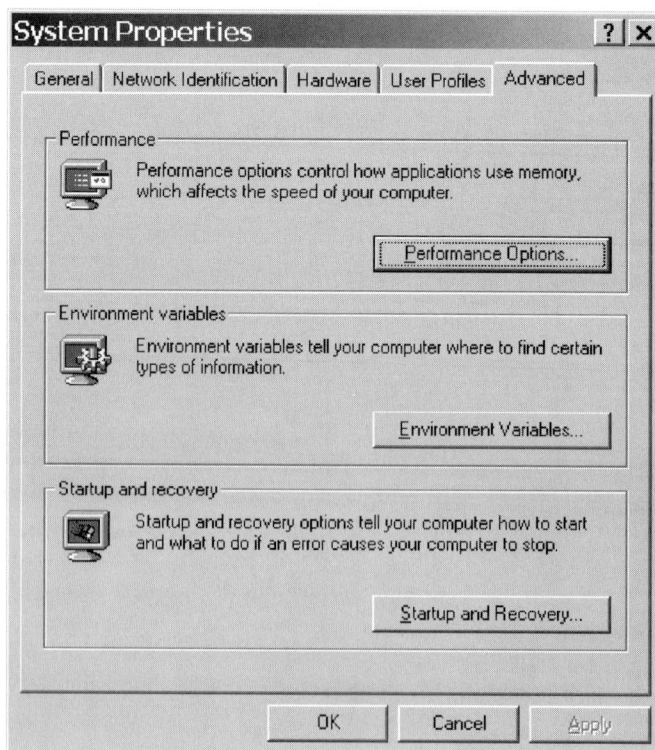

FIGURE 8.16   Performance options within the System Properties dialog box

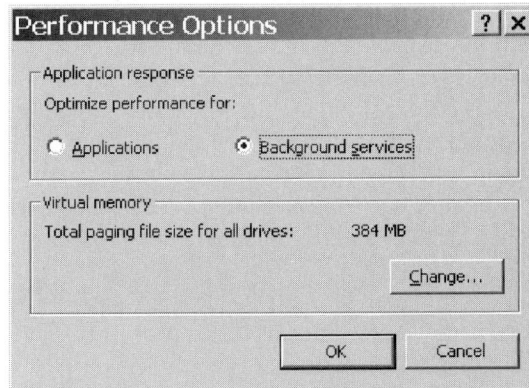

FIGURE 8.17   Changing the Virtual Memory settings.tif

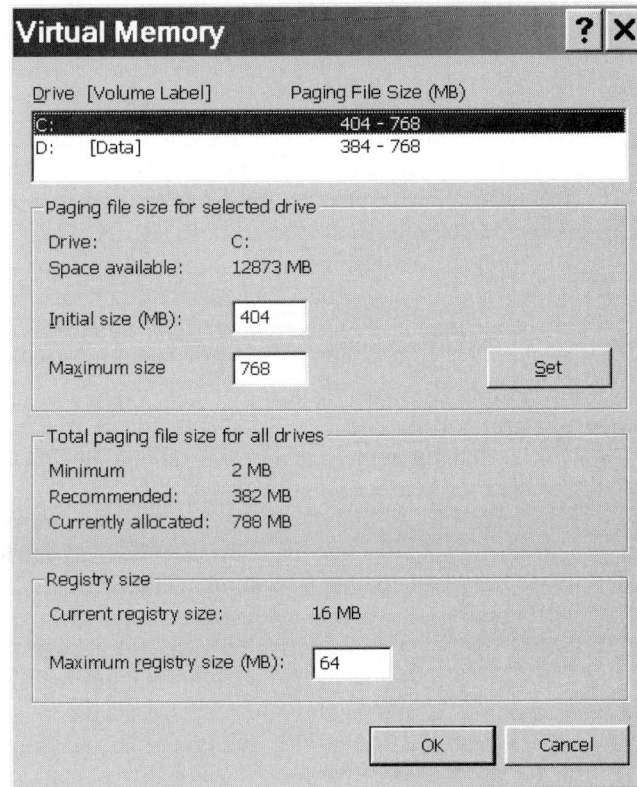

2000 Professional, the default size of the file is equal to the total amount of RAM plus 12 MB, not to exceed the amount of available disk space. Usually the size of the paging file can stay at the default value assigned during installation, but the recommended size for the paging file should be 1.5 times the amount of RAM available on your system and it should not exceed 2.5 times the amount of RAM installed. On computers with 4GB or more of RAM, this is an ineffective use of disk space. A paging file size of 2060 MB is the recommended size for computers with 4 GB or more of physical memory. When changing the virtual memory settings, be sure to click on the OK button and not the Cancel button. See Figure 8.17.

As mentioned in earlier chapters, virtual memory is much slower than physical RAM, because the memory is on the hard drive, which is a mechanical device. If you want to enhance system performance and you have several physical hard drives (not necessary partitions/logical drives), you can create a paging file on each disk by moving the paging file off the boot partition (the partition with the WINNT directory). This step is possible because the hard disk controller can read and write to multiple hard disks simultaneously and the Virtual Memory Manager (VMM) tries to write the page data to the paging file on the disk that is the least busy.

Lastly, you can enhance system performance by setting the initial size of the paging file to the value displayed in the Maximum Size box in the Virtual Memory dialog box. This occurs because when the system needs more RAM beyond what it already has, it will increase the size of the paging file. If the file is already set to its maximum size, then it eliminates the time required to enlarge the file from the initial size to the maximum size. The maximum size of the file is determined by the largest contiguous (continuous) space of disk space. The disk defragmenter utility in the Accessories system tools also is important because it ensures that a maximum amount of the drive is contiguous and available for page file use.

The Environment Variables dialog box allows you to change system and user environment variables such as the TEMP and TMP variables. System environment variables are used for all users whereas user environment variables are different for each user. The environment variables are set by first using the AUTOEXEC.BAT file, then by using the system environment followed by the user environment variables. You can prevent Windows 2000 from searching the AUTOEXEC.BAT file by editing the \HKEY_CURRENT_USER\ SOFTWARE\Microsoft\WindowsNT\CurrentVersion\Winlogon\Parseautoexec and setting it to zero.

The last option under the system properties dialog box is Startup and Recovery settings. The System Startup section allows you to determine the default operating system listed in the BOOT.INI file and the time in seconds that the boot menu is displayed on the screen before automatically selecting the default operating system. Note that when these values are changed using this dialog box, the BOOT.INI file is changed. The Recovery section provides options to assist in troubleshooting Stop errors. You must be logged on as a member of the Administrator group to change these options. See Table 8.2.

To help diagnose Stop errors, you can create a memory dump containing debugging information. The debugging information can then be used by support engineers to resolve these errors. To write the debugging information, a paging file must be on the system partition (the partition with the WINNT directory), the paging file must be at least 1 MB larger than the amount of physical RAM in your computer, and you must have enough disk space to write the file in the location you specify.

A **hardware profile** is a set of instructions that tells Windows 2000 which devices to start upon computer start-up or what settings to use for each device. For example, if you have a notebook computer, you might create two different hardware profiles, one for when the computer is docked in a docking station and one for when it is not or if you work from two locations.

To create or modify a hardware profile, use the System Properties dialog box (double-click on the System applet of the Control Panel or use the secondary mouse button on the My Computer icon and select the Properties applet), select the Hardware tab and click on the Hardware Profile button.

To create a new profile, select one of the profiles listed and click on the Copy button to create a copy of the selected profile. The first profile listed is the default profile. If you want to change the order of the profiles, you can select a profile and use the up and down arrows on the sides to move it (or them) up or down on the list.

TABLE 8.2   Recover options

Option	Description
Write an Event to the System Log	If enabled, it writes an event to the system log when the system stops unexpectedly
Send an Administrative Alert	If enabled, it sends an administrative alert to administrators when the system stops unexpectedly.
Write Debugging Information To	If enabled, it will write debugging information to the specified file name. This file can then be used by support engineers to diagnose problems.
Automatically Reboot	If enabled, it allows Windows 2000 to reboot whenever the system stops unexpectedly.

One of the leading causes of instability in Windows NT is buggy device drivers. While you should only use hardware that is on the HCL, there is no guarantee that the driver is bug-free. If the drivers are on the HCL, however, there is a better chance that they have no bugs.

To keep your system loaded exclusively with drivers from the HCL, Microsoft introduced **driver-signing technology** into Windows 2000. To configure the device driver signature verification system, click on the Driver Signing button in the hardware tab of the System applet. In the Driver Signing Options dialog box, you can disable this feature by selecting the Ignore option, which displays a warning if a driver is being loaded that does not have the signature or blocks the driver with the signature all together.

8.4.2 Display Applet

The Display applet allows you to configure the Windows environment including menus, windows, icons, the desktop, and the screen saver. See Figure 8.18. In addition, it allows you to configure your video system, install or upgrade video drivers, and choose the video resolution and number of colors. See Figure 8.19. Access is through the Control Panel or the Properties option on the shortcut menu of the desktop.

Originally, screen savers were used to prevent burn-in to a monitor. Burn-in occurred when a monitor displayed the same screen for such a long period of time that the screen image remained even after that screen had changed. Although they are no longer needed to prevent burn-in, screen savers have grown into a form of entertainment.

Although most people who are using Windows 95 and 98 are already familiar with screen savers, some may not be aware of the password-protected feature. When you enable password protection, you add a level of security to your system if you leave the system and forget to log off. When the screen saver activates and you press any key or move the mouse, you must specify the login password to be able to continue.

In addition, if you wish to use a screen saver, be sure to select a simple one, not a 3-D variety, especially OpenGL screensavers. 3-D screen savers have a tendency to take up too much processing.

If you have video problems, there are several tools available for troubleshooting. First, if you choose the wrong driver or you select the wrong resolution or refresh rate and thus cannot see anything on the screen, you can reboot the computer, press F8 during the boot menu, and select the VGA mode. Another tool is the troubleshooting tool tab which reduces

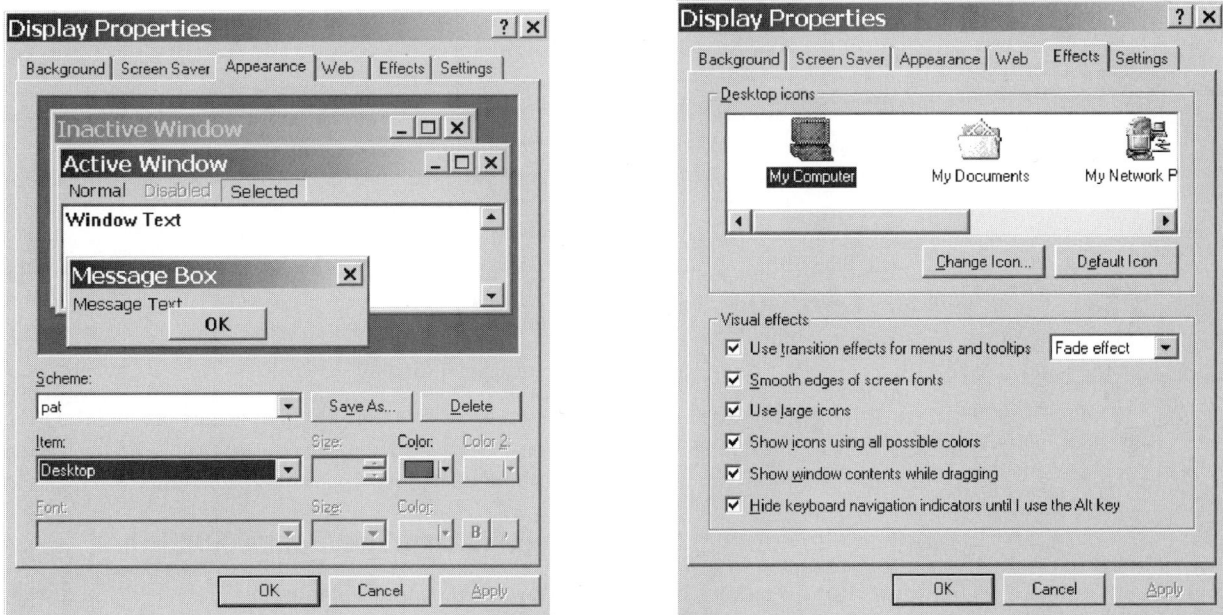

FIGURE 8.18   The Appearance and Effects tabs within the Display Properties allows you to change the appearance of Windows

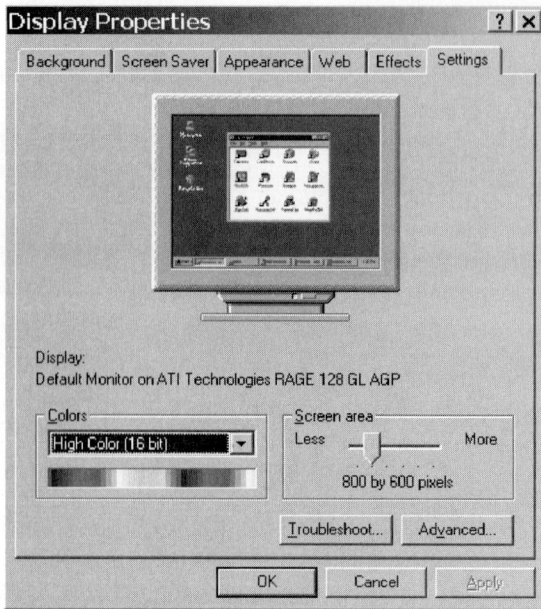

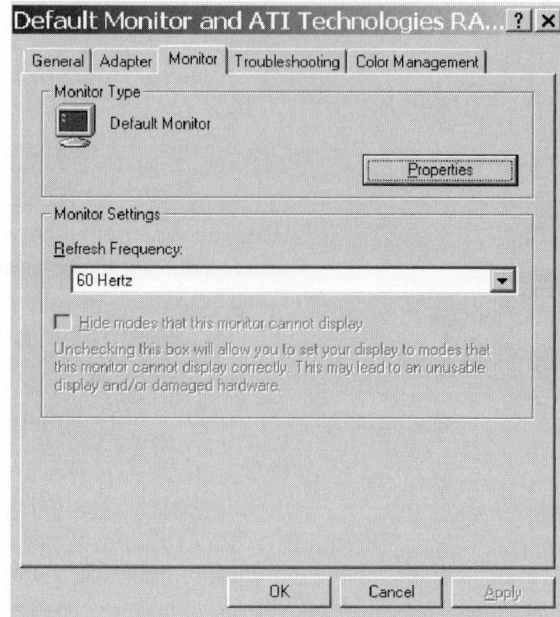

FIGURE 8.19   The Settings tab of the Display Properties dialog box can be used to adjust the number of colors and the resolution rate of the display system. By clicking on the Advanced button you can adjust the monitor refresh rate

the hardware acceleration of the video card. If you suspect a video problem, reduce the hardware acceleration to see if the problem goes away.

Different from Windows 95—original version of Windows 98 and Windows NT 4.0—Windows 2000 supports the use of up to 10 monitors. To support multiple monitors, use peripheral component interconnect (PCI) or accelerated graphics port (AGP) devices.

If one of the video cards is built into the motherboard, it must be a multiple-display-compatible card. In addition, the motherboard adapter becomes the secondary adapter. Lastly, if your motherboard has a built-in video card, you must install Windows 2000 before installing the additional video cards. If not, Windows 2000 Setup will disable the motherboard adapter.

To configure your display in a multiple environment, do the following:

1. Open the Display Properties dialog box by double-clicking on the Display applet of the Control Panel or by using the shortcut menu of the desktop and selecting the Properties option.
2. Click on the Settings tab.
3. Click the monitor icon for the primary display device.
4. Select the display adapter for the primary display and select the color depth and resolution.
5. Click the monitor icon for the secondary display device.
6. Select the display adapter for the secondary display and then select the Extend My Windows Desktop Onto This Monitor check box.
7. Select the color depth and resolution for the secondary display.
8. Repeat the last three steps for each additional display.

The positions of the display can then be changed by clicking the Settings tab and dragging the displays to their desired position. See Figure 8.20.

If you have problems with the multiple displays, review Table 8.3.

8.4.3   Add/Remove Hardware Applets and Add/Remove Software

Different from Windows NT 4.0, Windows 2000 supports plug-and-play (PNP) hardware. This means that by physically installing or connecting a PNP device, Windows 2000 will automatically configure the device (including the system resources such as I/O addresses,

FIGURE 8.20   Configuring
multiple monitors

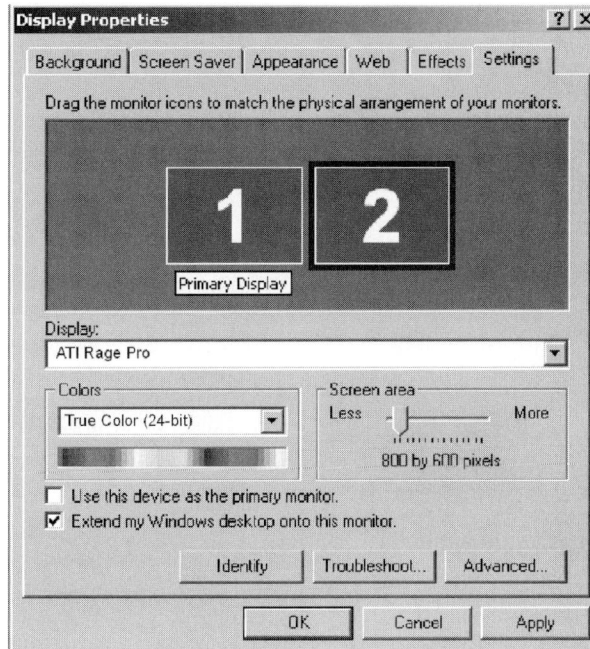

TABLE 8.3   Multiple display
problems

Problem	Solution
The output on the secondary displays is not visible.	Check if the correct video driver is installed and check the status of the video adapter in the Device Manager. You can also try to switch the order of the adapters in the slots.
The Extend My Windows Desktop Onto This Monitor check box is unavailable.	Select the secondary display instead of the primary display in the Display Properties dialog box. In addition, make sure that the hardware supports multiple hardware and that the Windows 2000 can detect the secondary display.
An application fails to display on the secondary display.	Run the application on the primary display. If it is a DOS program, switch the application to full-screen mode and if it is a Windows application, maximize the application. Lastly, disable the secondary display to determine whether the problem is specific to multiple display support.

IRQs, and DMAs) and load the appropriate driver. Occasionally, you will need to use the Add/Remove Hardware wizard to discover a PNP device. If the plug-and-play device causes a resource conflict, you can often use Device Manager to resolve such conflicts.

For non-PNP devices, configure the device by using jumpers or dip switches or by running a setup utility that comes with the device. To find out what the free resources are (I/O addresses, IRQ, DMA, and memory addresses), use Device Manager. After the card is configured and physically installed, Windows 2000 can often identify the hardware and install the appropriate drivers to activate the driver. If not, you will have to use the Add/Remove Hardware wizard. You may need to provide configuration information such as the resources of the device. See Figure 8.21.

The Add/Remove Programs applet helps you manage programs on your computer. It prompts you through the necessary steps to add a new program, change or remove an existing program, or add or remove Windows 2000 components that were not installed during the original installation. See Figures 8.22 and 8.23.

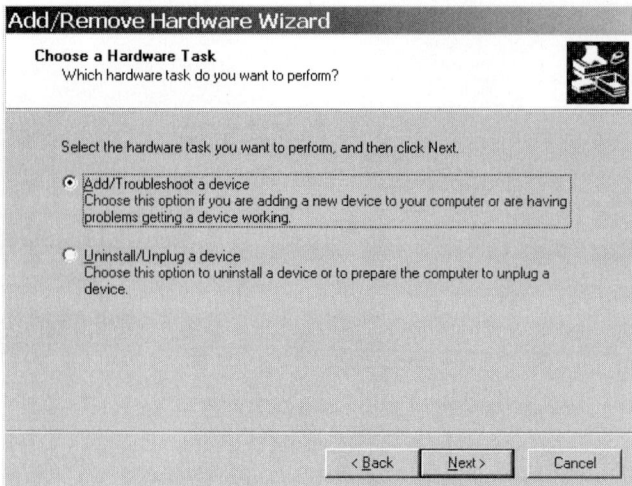

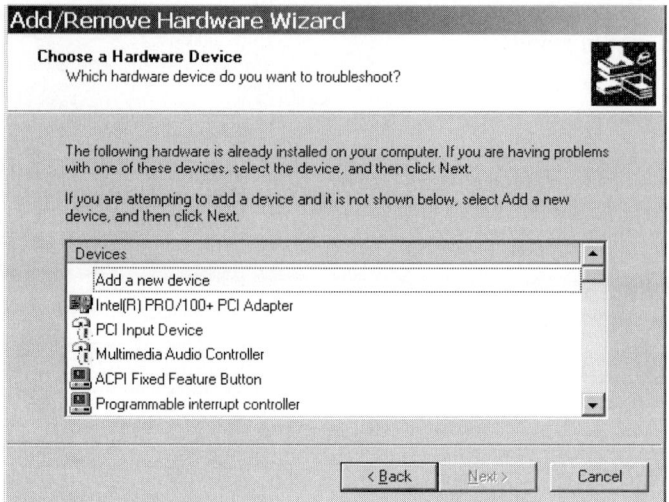

FIGURE 8.21   Using Add/Remove Hardware applet

FIGURE 8.22   The Add/Remove Programs applet can be used to change or remove programs that are already installed on your program

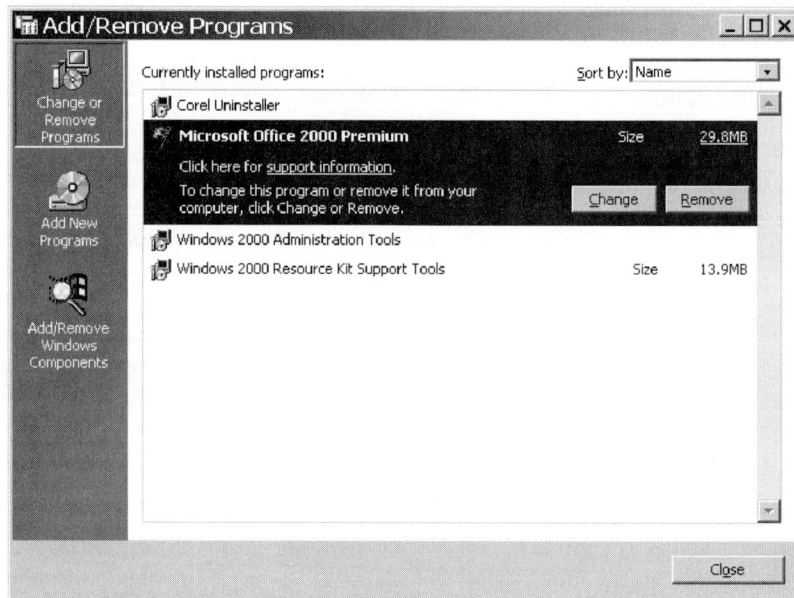

FIGURE 8.23   The Add/Remove applet can be used to add or remove Windows Components without reinstalling Windows

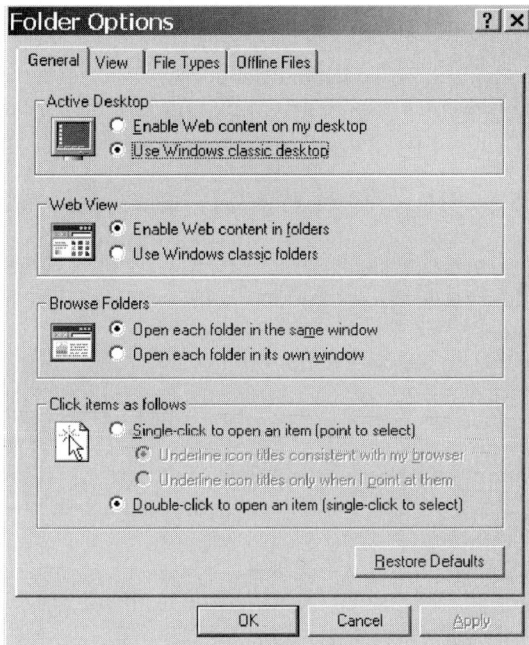

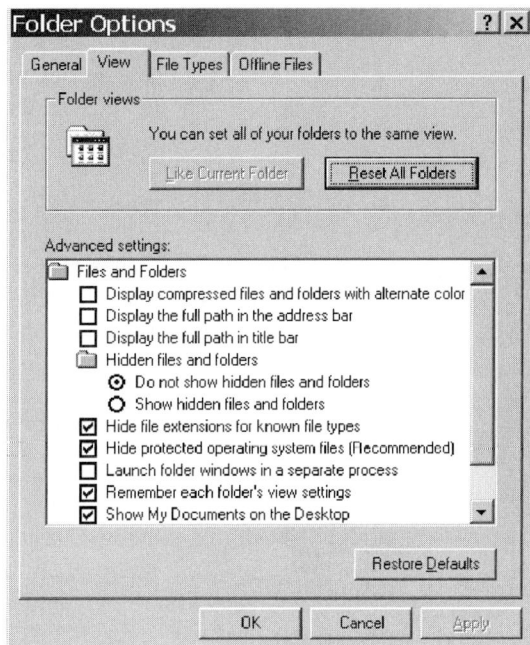

FIGURE 8.24   Folder Options

8.4.4   Folder Options

The Folder Options dialog box is used to configure if the Windows will display hidden files in a window. See Figure 8.24. The File Types tab will show the file name extensions for known file types and the View tab can be used to hide the protected operating system files.

8.4.5   Power Options and the UPS

Another Control Panel item is the Power Option applet. This option allows us to configure Windows 2000 to conserve power when not in use. For example, you configure the computer to automatically shut off the monitor or hard drive after a set time limit of nonuse. For servers, it is recommended that you should disable power-saving features so that your server can be working at full capacity at all times. Instead, use a simple screen saver.

For notebook computers, Windows 2000 supports Hibernate mode and Standby mode to extend battery life. When a computer hibernates, it stores whatever it has in memory on your hard disk and then shuts down. When your computer comes out of hibernation, it returns the content saved to the hard disk back to RAM. In other words, it maintains the state of open programs and connected hardware while it hibernates.

Standby mode, also known as Suspend mode, is when the computer switches to a low power state where devices, such as the monitor and hard disks, turn off while there is enough power to maintain the content of the RAM. When you want to use the computer again, it comes out of standby quickly and your desktop is restored exactly as you left it. Since standby does not save your desktop state to disk, a power failure while in Standby mode can cause you to lose unsaved information. Of the two modes, Standby mode returns to a ready state faster than Hibernate mode. Of course, it does need some power to maintain itself.

To enable Standby mode, you would use the System Standby option in the Power Schemes tab where you would specify the minutes of inactivity before going into Standby mode. To enable Hibernate mode, select the Hibernate option in the Hibernate tab. You can then go into Hibernate mode or Standby mode by clicking on the Start button, selecting the Shutdown option and selecting the appropriate mode.

An uninterruptible power supply (UPS) provides power from internal batteries when the AC power fails. It is usually rated to provide a specific amount of power for a specified time. Many UPS devices can communicate with the operating system of the pending shutdown when the power from the UPS is spent. The UPS service within Windows 2000 pauses the

233

Server service to prevent any new connections and sends a message to notify users of the power failure. The UPS service then waits a specified time before notifying users to save their work and quit their sessions. The server itself will eventually shut down to protect the integrity of the information stored there. If power is restored before shutdown of the server, another message is sent to inform users that power has been restored and normal operations have resumed.

The UPS will typically connect through a serial port to communicate, when a power failure occurs, that the battery power is low and a remote shutdown will occur by the UPS device. You can set options for the operations of the UPS using Power Options in the Control Panel. After installing the UPS and configuring Windows 2000 to communicate with the UPS, perform a test of the UPS by disconnecting the power. Remember: To configure the UPS settings, you must be an Administrator or a member in the Administrators group.

8.4.6 Accessibility Tools

The accessibility tools that ship with Windows 2000 are intended to provide a minimum level of functionality for users with special needs. You can adjust the appearance and behavior of Windows 2000 to enhance accessibility for some vision–impaired, hearing–impaired, and mobility–impaired users without requiring additional software or hardware. Windows 2000 includes the following programs to enhance accessibility even if you choose not to install accessibility services:

- Magnifier—enlarges a portion of the screen for ease of viewing.
- Narrator—uses text–to–speech technology to read the contents of the screen aloud. This is useful for people who are blind or who have low vision.
- On–screen Keyboard—provides users with limited mobility the ability to type on–screen using a pointing device.
- Utility Manager—enables users with administrator–level access to check an Accessibility program's status, start or stop an Accessibility program, and designate to have the program start when Windows 2000 starts.

The accessibility tools available in Accessibility Options in Control Panel perform various functions:

- Sticky Keys—enables simultaneous keystrokes while pressing one key at a time.
- Filter Keys—adjusts the response of your keyboard.
- Toggle Keys—emits sounds when certain locking keys are pressed.
- Sound Sentry—provides visual warnings for system sounds.
- Show Sounds—instructs programs to display captions for program speech and sounds.
- High Contrast—improves screen contrast with alternative colors and font sizes.
- Mouse Keys—enables the keyboard to perform mouse functions.
- Serial Keys—allows the use of alternative input devices instead of a keyboard and mouse.

A wide variety of hardware and software products are available to make personal computers easier to use for people with disabilities. Among the different types of products available for MS-DOS and the Microsoft Windows operating systems are:

- programs that enlarge or alter the color of information on the screen for people with visual impairments.
- programs that describe information on the screen in Braille or synthesized speech for people who are blind or have difficulty reading.
- hardware and software utilities that modify the behavior of the mouse and keyboard.
- programs that enable the user to type using a mouse or his or her voice.
- word or phrase prediction software that allow users to type more quickly and with fewer keystrokes.
- alternate input devices, such as single switch or puff-and-sip devices, for people who cannot use a mouse or a keyboard.

Most users with disabilities will need utility programs with higher functionality for daily use. For a list of Windows-based accessibility utilities, see Microsoft Accessibility on the Microsoft Web site.

8.5 ADMINISTRATIVE TOOLS

Several utilities that are used to control and configure the many network services are grouped into the Administrative Tools folder, which can be accessed from within a Control Panel applet or by clicking on the Start button and clicking on the Programs option. While this book will discuss these utilities throughout, this chapter will introduce the Microsoft Management Console (MMC) and the Event Viewer.

8.5.1 Microsoft Management Console

The **Microsoft Management Console (MMC)** is a primary administrative tool that is used to manage Windows 2000. See Figure 8.25. It provides a standardized method to create, save, and open the various administrative tools provided by Windows 2000, BackOffice products, and third-party services. Some of the Windows 2000 snap-ins include the following:

- System tools—container for managing local users and groups
- System information—container for managing hardware resources, monitoring the software environment, and controlling system services
- Storage—container for a disk defrag tool, a removable storage control panel, and an optional archive manager
- Server applications and services—DHCP, DNS, RAS, and IIS services
- Links to shared directories, web links, and resource kits

Note that most snap-ins can be used on remote computers.

To start the Microsoft Management Console,

1. Click on the Start button;
2. Click on the Programs option;
3. Click on the Administrative Tools option; then
4. Select the Computer Management option.

To start an empty MMC, go to the command prompt and type MMC.EXE. Every MMC has a console tree, which displays the hierarchical organization of snap-ins (plugable modules) and extensions (a snap-in that requires a parent snap-in). The adding and deleting of the snap-ins and extensions allows each user to customize the console. The console settings can then be saved to a MMC file (file with a .MSC extension) and can be used later, shared by others and even e-mailed on other computers. By default, MMC stores snap-ins in the

FIGURE 8.25 Computer Management console is a prebuilt MMC

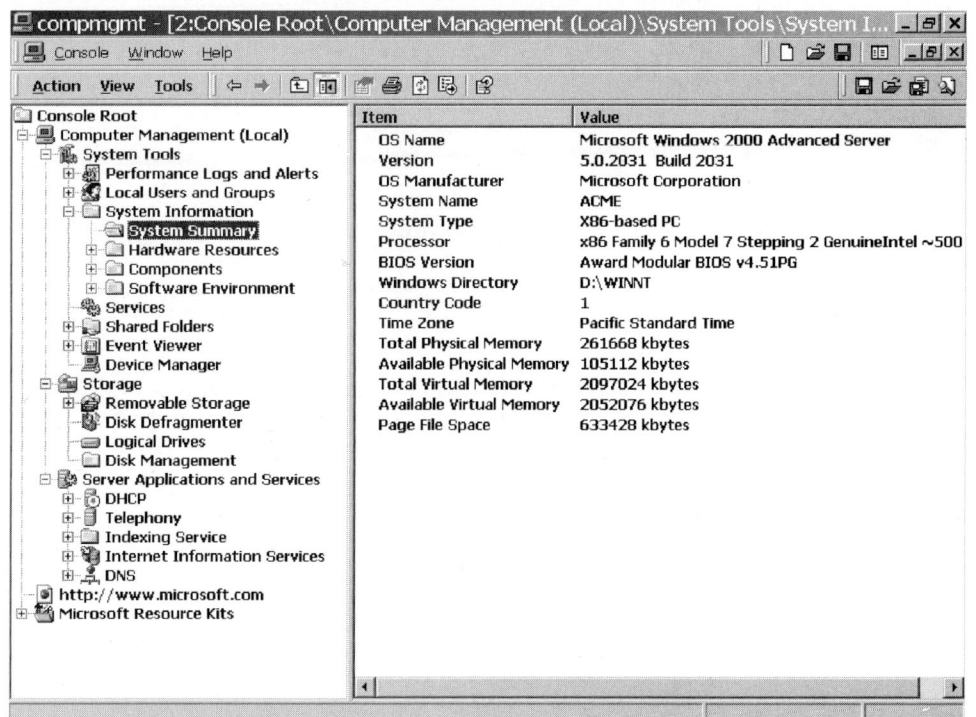

235

%systemroot%\system32 folder, the folder into which Windows 2000 is installed. By default, the Windows 2000 directory is C:\WINNT.

When building a custom console, assign various console modes (access options), as listed in Table 8.4. By default, all new MMCs are saved in author mode. To change the console mode, open the Console menu, select the Option option, and click on the Console tab. See Figure 8.26.

To add snap-ins, do the following:

1. Open the MMC.
2. Select the Add/Remove Snap-in from the Console menu.
3. Click on the Add button from within the Add Standalone Snap-in dialog box.
4. Click the snap-in that you want and click on the Add button.
5. Repeat steps 3 and 4 as needed.

TABLE 8.4   MMC modes

Console Mode	Description
Author mode	Used to add or remove snap-ins, create new windows, view all portions of the console tree, and save the MMCs.
User mode—full access	Primarily used to distribute an MMC console to other administrators. Users cannot add snap-ins, remove snap-ins, or save the MMC console.
User mode—limited access, multiple windows	Users cannot add snap-ins, remove snap-ins, or save the MMC console. While users can view the console in multiple windows, they cannot open new windows or gain access to a portion of the console tree.
User mode—Limited access, single window	Users cannot add snap-ins, remove snap-ins, or save the MMC. Users cannot view the console in multiple windows, open new windows, or gain access to a portion of the console tree.

FIGURE 8.26   Changing the MMC mode

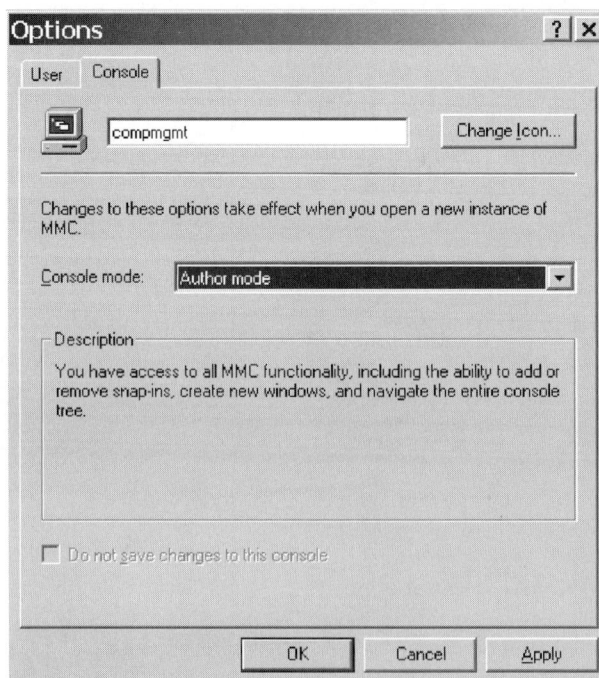

To remove a snap-in, do the following:

1. Open the MMC.
2. Select the Add/Remove Snap-in from the Console menu.
3. Click on the snap-in that you want to remove.
4. Click on the Remove button.
5. Repeat steps 3 and 4 as needed.

8.5.2 Event Viewer

Event Viewer is a useful utility that is used to view and manage logs of system, program, and security events on a computer. Event Viewer gathers information about hardware and software problems and monitors Windows 2000 security events. Event Viewer can be executed by clicking on the Start button, clicking on Programs, clicking on Administrative Tools, and then clicking on Event Viewer or by adding it to the MMC. See Figure 8.27.

Windows 2000 Event Viewer starts with three kinds of logs:

Application log—The application log contains events logged by programs. For example, a database program might record a file error in the programs log. Program developers decide which events to monitor. The application log can be viewed by all users.

Security log—The security log contains valid and invalid logon attempts as well as events related to resource use such as creating, opening, or deleting files or other objects. For example, if you have enabled logon and logoff auditing, attempts to log on to the system are recorded in the security log. By default, security logging is turned off. To enable security logging, use Group Policy to set the audit policy or change the registry. To audit files and folders, you must be logged on as a member of the Administrators group or have been granted the Manage auditing and security log right in Group Policy. See Chapter 12 for policies and rights.

System log—The system log contains events that are logged by the Windows 2000 system components. For example, the failure of a driver or other system component to load during start-up is recorded in the system log. The event types logged by system components are predetermined by Windows 2000. The application log can be viewed by all users.

FIGURE 8.27 Event Viewer

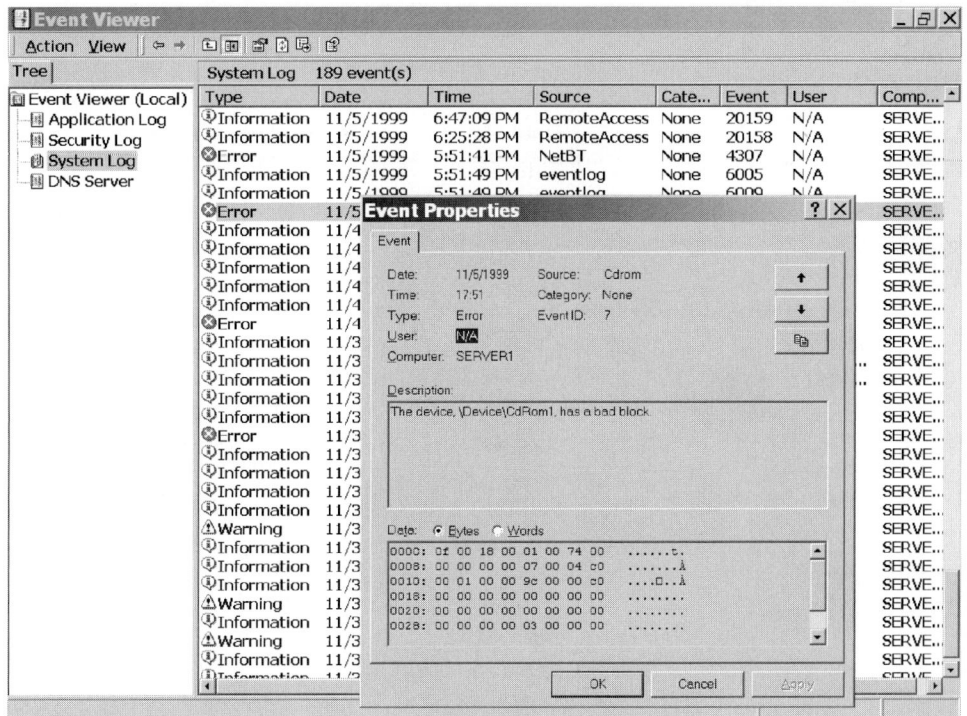

237

All BackOffice family applications (Microsoft applications designed to work on Windows NT or Windows 2000 servers) can post security events to the Windows 2000 event log. Several servers also make their own logs, which can be viewed by the Windows 2000 Event Viewer as well.

The five types of events are error, warning, information, success audit, and failure audit. They are shown Table 8.5.

When you double-click on an event, the Event Properties window will appear. The Event Properties can be divided into two parts, event header and event description. The event header information is shown in Table 8.6. The description of the event is the most important information within the Event Properties window and will usually indicate what happened or the significance of the event.

TABLE 8.5   Event Viewer event types

Event Type	Description
Error	A significant problem occurs, such as loss of data or loss of functionality, for example, when a service fails during start-up.
Warning	An event that is not necessarily significant, but may indicate a possible future problem. For example, when disk space is low, a warning will be logged.
Information	An event that describes the successful operation of an application, driver, or service. For example, when a network driver loads successfully, an information event will be logged.
Success audit	An audited security access attempt that succeeds. For example, a user's successful attempt to log on the system will be logged as a success audit event.
Failure audit	An audited security access attempt that fails. For example, if a user tries to access a network drive and fails, the attempt will be logged as a failure audit event.

TABLE 8.6   Event header information

Information	Meaning
Date	Date the event occurred.
Time	Local time the event occurred.
User	User name on whose behalf the event occurred.
Computer	Name of the computer where the event occurred. The computer name is usually your own, unless you are viewing an event log on another Windows 2000 computer.
Event ID	Number identifying the particular event type. The first line of the description usually contains the name of the event type.
Source	Software that logged the event, which can be either a program name, such as SQL Server, or a component of the system or of a large program, such as a driver name.
Type	Classification of the event severity: error, information, or warning in the system and application logs; and success audit or failure audit in the security log. In Event Viewer's normal list view, these are represented by a symbol.
Category	Classification of the event by the event source. This information is primarily used in the security log. For example, for security audits, this corresponds to one of the event types for which success or failure auditing can be enabled in Group Policy.

8.5.3 Services

A **service** is a program, routine, or process that performs a specific system function to support other programs. To manage the services, use the Services console (located under Administrative Tools) or the MMC with the Services snap-in. To start, stop, pause, resume, or restart services, right-click on the service and click on the desired option. Figure 8.28 shows the many services. On the left of the service name is a description.

To configure a service, right-click the service and click on the Properties option. On the General tab, under the start-up type pulldown option, set the following options:

- Automatic—specifies that the service should start automatically when the system starts.
- Manual—specifies that a user or a dependent service can start the service. Services with manual start-up do not start automatically when the system starts.
- Disable—prevents the service from being started by the system, a user, or any dependent service.

The Workstation service allows a user who is sitting at the computer to access the network to then access network resources such as shared folders. The workstation is also known as the redirector.

The Server allows a computer to provide network resources. When you pause the Server service, only users in the computer's Administrators and Server Operators groups will be able to make new connections to the computer. When you stop the Server service, all users who are connected over the network to the computer will be disconnected. If you stop a Server service, the affected computer can no longer be administered remotely and you must start the Server service locally; therefore, it is a good idea to warn connected users before stopping the Server service.

8.5.4 Task Scheduler

Task Scheduler is a program that comes with Windows 2000 which will schedule programs, batch files, and documents to run once, at regular intervals or at specific times. Task Scheduler starts each time you start Windows 2000 and runs in the background.

To use the scheduling service, double-click on the Scheduled Tasks applet in the Control Panel. You can schedule new tasks by double-clicking Add Scheduled Tasks, which starts the Scheduled Task wizard. From the wizard, you can then choose when you want

FIGURE 8.28   Services console

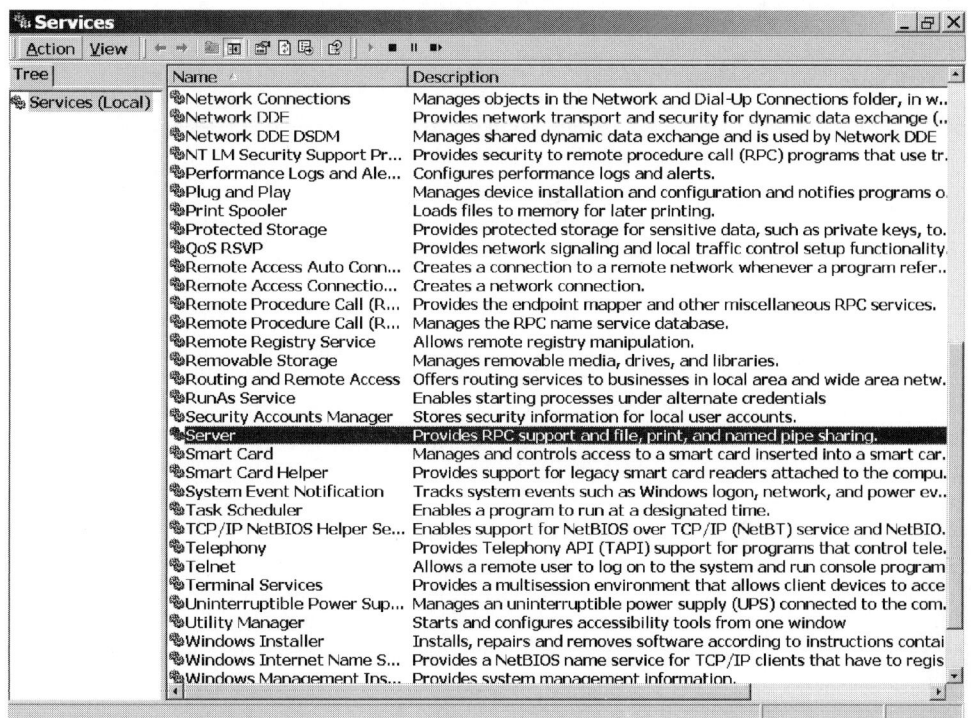

the program to run and which program you want to run. After an item has been scheduled, you can then manage the item by opening the item listed within the Add Scheduled Tasks applet.

In Windows 2000, tasks are scheduled and performed based on standard Windows 2000 security permissions. You can set permissions for which users or groups can view, delete, modify, or use a task. In addition, the items that make up the task (the scripts, programs, and documents) are controlled by whatever ACLs are present for those individual items.

To configure security for a scheduled task, add or remove a user on the Security tab. Once a user name is added to a scheduled task, you can set permissions for that user's involvement with the task.

8.6 THE REGISTRY AND REGISTRY EDITOR

The **registry** is a central, secure database in which Windows 2000 stores all hardware configuration information, software configuration information, and system security policies. Components that use the registry include the Windows NT Kernel, device drivers, setup programs, NTDETECT.COM, hardware profiles, and user profiles. These items are discussed throughout the book, so do not worry if some sound unfamiliar.

The data kept in the registry are added and modified by a variety of system modules that start during boot-up and are modified by various configuration tools such as the Control Panel, Windows 2000 Setup program, User Manager, adding or removing a hardware device, adding or removing a printer, and other administrative utilities.

In addition, data are added and modified when installing software that uses a single application programming interface (API). You can think of an API as having a set of standard commands that can be used by any software package to access the registry. For example, a software package can find out what type of hardware is installed (including type of processor, resolution, and number of colors of the video system and the IRQ and DMA settings for a device) and the version of drivers and other software modules.

Since Windows 2000 is a multiuser system, it records and preserves security and graphical desktop information on an individual basis. Therefore, it contains a permanent record of per-user, per-application, per-machine configuration information. To keep the entire system secure, the key or setting within the registry is protected with an Access Control List (ACL), which allows selected users to modify the contents of the registry and grants to others read-only access to that data.

8.6.1 Registry Editor

If the need arises to view or change the registry, you can use the **Registry Editor (REGEDIT.EXE** or **REGEDT32.EXE)** utilities. REGEDT32.EXE is automatically installed in the WINNT\SYSTEM32 folder. REGEDIT.EXE is automatically installed in the WINNT folder. REGEDIT.EXE is easier to use because of its Explorer-style interface. See Figure 8.29. The REGEDT32.EXE is the Windows NT 4.0 version of the Registry Editor that provides access to Windows 2000 security permissions and auditing. See Figure 8.30.

Occasionally you may have the need to view or edit the registry to add or change a value that cannot be changed in the Control Panel or other utility or to add, view, and change hardware settings that cannot be done with Device Manager. You should only make changes when following directions in a magazine article, book, manual, or with a support person. If you make the wrong changes, Windows 2000 may not run properly or may not boot at all.

8.6.2 Structure of the Registry

The registry is organized in a hierarchical structure. It is first divided into five **subtrees.** See Table 8.7. Subtrees have names that begin with the string HKEY, which stands for "**handle to a key.**" A subtree is similar to a root directory of a disk.

Of the different subtrees, the two main ones are the HKEY_LOCAL_MACHINE and the HKEY_USERS. The HKEY_LOCAL_MACHINE contains information about the type of hardware installed, drivers, and other system settings. The HKEY_USERS contains information about all users who log on to the computer including the DEFAULT generic user settings. The DEFAULT user serves as a template for any new users.

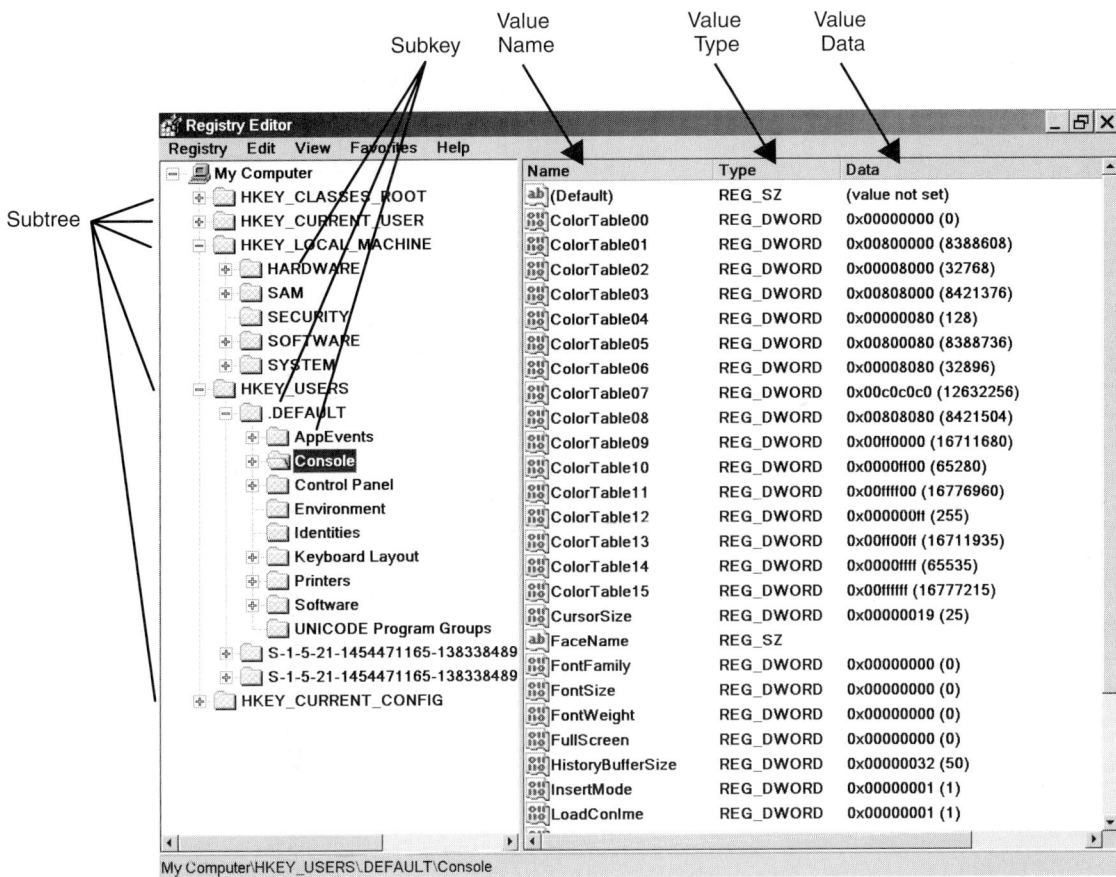

FIGURE 8.29   REGEDIT.EXE

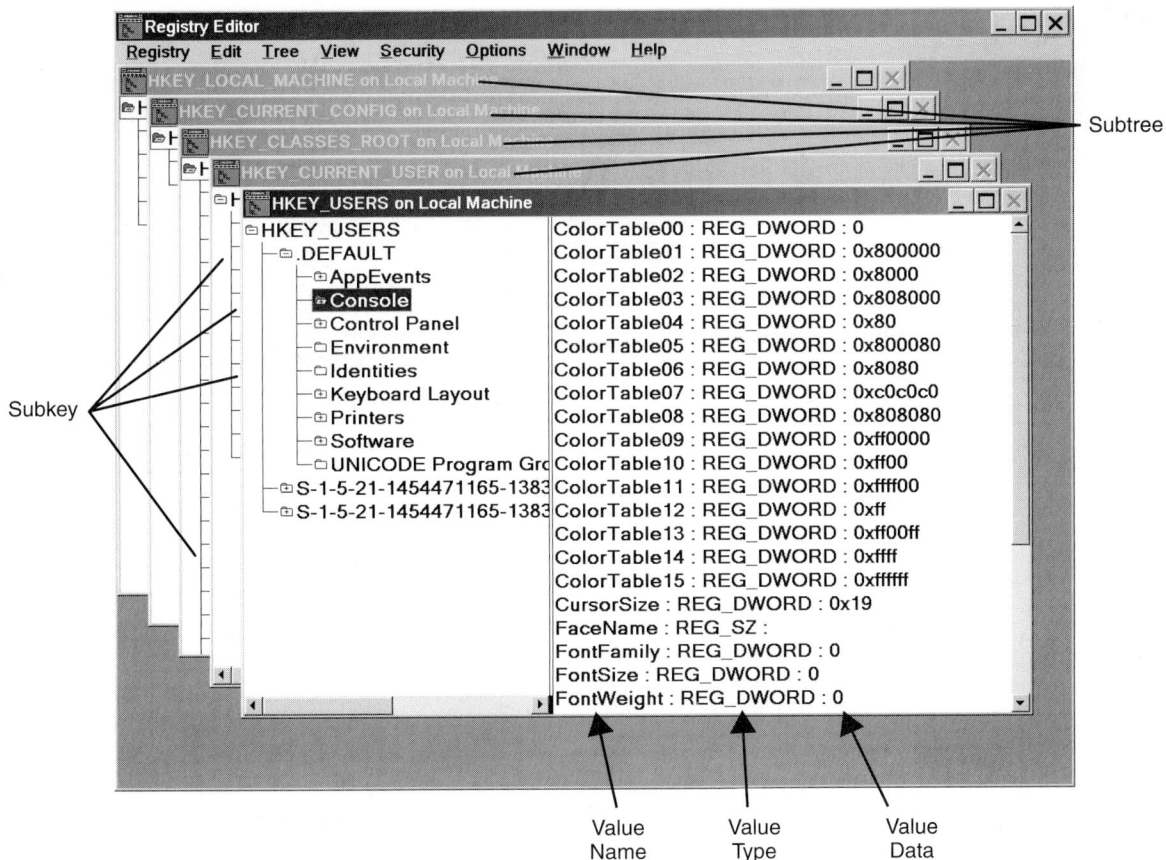

FIGURE 8.30   REGEDT32.EXE

TABLE 8.7   Windows 2000 Registry subtrees

Subtrees	Description
HKEY_CLASSES_ROOT	Contains file associations and OLE information.
HKEY_CURRENT_USER	Contains settings for applications, desktop configurations, and user preferences for the user currently logged on. The information retrieves a copy of each user account that is used to log on to the computer from the NTUSER.DAT file and stores it in the *systemroot*\PROFILES\username key. This subkey points to the same data contained in HKEY_USERS\SID_currently_logged_on_user. Note: This subtree takes precedence over HKEY_LOCAL_MACHINE for duplicated values.
HKEY_LOCAL_MACHINE	Contains information about the type of hardware installed, drivers, and other system settings. Information includes the bus type, system memory, device drivers, and start-up control data. The data in this subtree remain constant regardless of the user.
HKEY_USERS	Contains information about all users who log on to the computer including the DEFAULT generic user settings. The DEFAULT user serves as a template for any new users.
HKEY_CURRENT_CONFIG	Contains information about the current running hardware configuration. This information is used to configure settings such as the device drivers to load and the display resolution to use. This subtree is part of the HKEY_LOCAL_MACHINE subtree and maps to HKEY_LOCAL_MACHINE\SYSTEM\CurrentControlSet\Hardware Profiles\Current.

The root keys are then divided into **subkeys,** which may contain other subkeys. Think of the subkeys as folders within the subtree. Within the subkey is a value entry, which consists of three parts: the name of the value, the data type of the value, and the value itself. Data types describe the format of the data. The data types of the values are shown in Table 8.8.

8.6.3   HKEY_LOCAL_MACHINE Subtree

As mentioned, the **HKEY_LOCAL_MACHINE** subtree is one of the two main trees from which the other trees are made. Its structure is similar to the others. It has five subkeys, which are shown in Table 8.9. The HARDWARE key contains a detailed description of the installed hardware such as the type of the motherboard, video adapter, SCSI adapters, serial ports, parallel ports, sound cards, and network adapters. Data in the HARDWARE key are volatile and computed at boot time. When your hardware configuration changes, the changes are reflected in the HARDWARE key at the next boot.

The next two keys, SAM, which stands for Security Accounts Manager, and SECURITY have no visible information, as they point to security policies and user authentication information. The values in these two keys are created, modified, and removed with either the User Manager or the User Manager for Domains. To keep the Windows 2000 secure, the Registry Editor does not display these data.

The SOFTWARE key contains a list of file extensions and associated applications—one key for each application that follows the configuration database registration procedure and one key for each loaded network driver. The SYSTEM key describes bootable and non-bootable configurations in a group of ControlSet, where each ControlSet represents a unique configuration. Within each ControlSet, two keys describe operating system components and service data for that configuration. This key also records the configuration used to boot the running system (CurrentControlSet), along with any failed configurations and the Last Known Good Configuration. Finally, the Setup key records the command used to

Data Type	Description
REG_DWORD	Data represented by a number that is 4 bytes (double word) long. Many parameters for device drivers and services are this type and are displayed in Registry Editor in binary, hexadecimal, or decimal format.
REG_SZ	A fixed-length text string.
REG_EXPAND_SZ	A variable-length data string. This data type includes variables that are resolved when a program or service uses the data.
REG_BINARY	Raw binary data represented as a string of hexadecimal digits. Windows 2000 interprets every two hexadecimal digits as a byte value.
REG_MULTI_SZ	A multiple string. Values that contain lists or multiple values in a form that people can read are usually this type. Entries are separated by spaces, commas, or other marks.
REG_FUL_RESOURCE _DESCRIPTOR	Stores a resource list for hardware components or drivers. You cannot add or modify entries with this data type.

Subkey	Description
HARDWARE	Contains type and state of physical devices attached to the computer. Much of this information is collected and built during boot-up. Therefore, it does not point to a specific file on a disk.
SAM	The directory database for the computer. The SAM subkey points to the SAM and SAM.LOG files in the systemroot\SYSTEM32\CONFIG directory.
SECURITY	The security information for the local computer. It maps to the SECURITY and SECURITY.LOG files in the systemroot\SYSTEM32\CONFIG directory. Applications query information by using security APIs.
SOFTWARE	Information about the local computer software that covers all users. IT points to the SOFTWARE and SOFTWARE.LOG files in the systemroot\SYSTEM32\CONFIG directory. It also contains file association and OLE information.
SYSTEM	Information about system devices and services. When you install or configure device drivers or services, they add or modify information under this hive. The SYSTEM hive maps to the SYSTEM and SYSTEM.LOG files in the systemroot\SYSTEM32\CONFIG directory. The registry keeps a backup of the data in the SYSTEM hive in the SYSTEM.ALT file.

install Windows 2000 and the boot disk and provides a list of the OEMSETUP files required to install hardware components.

EXAMPLE 1 Consider the following value:

HKEY_LOCAL_MACHINE\HARDWARE\DESCRIPTION\System\ CentralProcessor\0\~MHz

See Figure 8.31. Within the HKEY_LOCAL_MACHINE, you will find the HARD-WARE subkey. By clicking on the plus (+) sign next to the HARDWARE subkey, you will open it to show the DESCRIPTION subkey. In the System subkey, you will find the CentralProcessor subkey and under that you will find the 0 subkey. In the 0

FIGURE 8.31
HKEY_LOCAL_MACHINE values

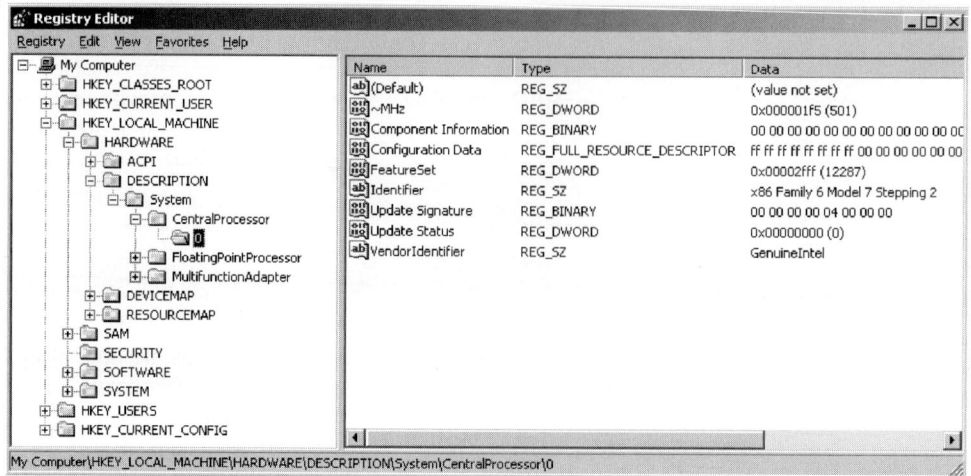

FIGURE 8.31
HKEY_LOCAL_MACHINE values

subkey you will find the ~MHz value, which is shown on the left pane. Next to the ~MHz value, you will find the data type of REG_DWORD, which means that we expected eight hexadecimal digits (4 bytes of information). Lastly, the value is set to 0x000001F5 (hexadecimal), which is equivalent to 501 MHz (decimal).

8.6.4 Hives

The registry stores most of its information in sets of files called hives. A **hive** is a discrete body of keys, subkeys, and values. Each hive has a corresponding registry file and .LOG file located in the WINNT\SYSTEM32\CONFIG folder. Windows 2000 uses the .LOG file to record changes and ensure the integrity of the registry. When you work with the registry, you view and edit subtrees and their contents; when you back up and restore the registry, you work with hives.

Files without extensions contain a copy of the hive. Files with the .LOG extension contain a record of any changes to the hive. The files with the .SAV files contain copies of the hives at the end of Setup's text mode step. The .ALT file contains the Last Known Good information, in case anything happens to the system hive. The .DAT files contain user profile information. Table 8.10 lists the Windows 2000 hives.

Under normal circumstances, your system will reference the files without extensions. When writing changes to a hive file, the system will note the changes in the hive's log, so that it can cancel any unfinished changes in case the hive's updating process is interrupted. If something happens to the critical system hive, the .ALT version replaces it during the

TABLE 8.10   Windows 2000 hives

Hive	Hive Support Files
HKEY_LOCAL_MACHINE\SAM	SAM, SAM.LOG, and SAM.SAV
HKEY_LOCAL_MACHINE\Security	SECURITY, SECURITY.LOG, and SECURITY.SAV
HKEY_LOCAL_MACHINE\Software	SOFTWARE, SOFTWARE.LOG, and SOFTWARE.SAV
HKEY_LOCAL_MACHINE\System	SYSTEM, SYSTEM.ALT, SYSTEM.LOG, and SYSTEM.SAV
HEY_CURRENT_CONFIG	SYSTEM, SYSTEM.ALT, SYSTEM.LOG, and SYSTEM.SAV
HKEY_USERS\DEFAULT	DEFAULT, DEFAULT.LOG, and DEFAULT.SAV
HKEY_CURRENT_USER	NTUSER.DAT and NTUSER.DAT.LOG

boot process. The .ALT then makes itself the new system hive and creates another backup .ALT file.

By default, most hive files are stored in the *%systemroot%*\SYSTEM32\CONFIG folder. The *%systemroot%* indicates the name of the directory that is holding the Windows 2000 files. By default, Windows 2000 installs to the WINNT folder. The location of user profile information for each user of a computer, including the NTUSER.DAT and NTUSER.DAT.LOG, may depend on whether the installation of the operating system was a fresh installation or was installed as an upgrade from Windows 95, Windows 98, or Windows NT. In fresh installations and upgrades from Windows 95 and 98, the NTUSER.DAT and NTUSER.DAT.LOG files are stored in the *%systemdrive%*:\DOCUMENT AND SETTINGS*Username* folder. In installations that are upgrades from Windows NT, the NTUSER.DAT/NTUSER.DAT.LOG files are stored in the *%systemroot%*\PROFILES*Username* folder.

8.6.5 Register Size Limits

When Windows 2000 is running, the registry is stored in the paged pool, which is a portion of virtual memory. A value called the *registry size limits* prevents programs from completely filling the paged pool with registry data.

You can view and set the size of the registry by using System in the Control Panel. On the Advanced tab, click Performance Options and then click Change. For more information about setting the registry size, see "To change the maximum size of the computer's registry" and "To change the size of the virtual memory paging file."

By default, the registry size limit is 33% of the size of the paged pool. A registry size limit of up to 80% of the size of the paged pool is allowed. The minimum registry size is 16 MB. If you attempt to set the registry size to anything smaller, the system resets this value to 16 MB.

You should change the registry size only if the computer is a domain controller for a large network or if you receive an error message warning that the registry is too small. A large value for the registry size limit does not cause the system to use that much space unless it is actually needed by the registry. In addition, a large value does not guarantee that the maximum space is actually available for use by the registry.

8.6.6 Best Practices of the Registry

Before you make changes to the registry, make a backup copy by using a program such as Windows Backup. After you make changes to the registry, update your emergency repair disk (ERD) and automated system recovery (ASR) disk and tapes. For troubleshooting purposes, keep a list of the changes you make to the registry. For more information, see "Backing up, restoring, and recovering the Windows 2000 Registry." Do not replace the Windows 2000 Registry with the registry of another version of the Windows or Windows NT operating systems or from any computer that is not identical in every way.

Incorrectly editing the registry may severely damage your system. When possible, use the Control Panel, tools, and programs other than Registry Editor to edit the registry. Limit the number of people who have access to the registry. For example, because members of the Administrators group have full access to the registry, add only users who need Administrator rights to this group. You also can use Group Policy to restrict the use of Registry Editor (both REGEDT32.EXE and REGEDIT.EXE) for users who do not need access to the registry, or you can simply remove Registry Editor from the computers of these users.

If you use REGEDT32.EXE, make sure Read Only Mode on the Options menu is checked until you are ready to make changes. When you are done making changes and have saved your edits, recheck Read Only Mode. Alternatively, be sure to check Confirm on Delete on the Options menu. Never leave Registry Editor running unattended.

8.6.7 Troubleshooting Registry Problems

A corrupt or damaged registry can manifest itself in a number of ways: You may notice that files are missing or that components are not behaving as expected, you may receive an error message or a Stop error, or you may even be unable to start the operating system.

If you suspect a problem with the registry, try the following suggestions:

- If you have kept a record of the changes made to the registry, try undoing the most recent change.
- If you know which components have been affected, determine what the correct registry settings are for those components.
- Try using the Last Known Good Configuration for your computer.
- If you have a backup copy of the registry (as part of your backup of System State data) and can open Windows Backup, try to restore your registry settings. If you cannot start Windows 2000, try repairing the operating system with your ERD. For more information, see Chapter 16.

In a worst-case scenario, reinstall Windows 2000 and all of its applications. If you have backed up your entire system, this job will be much easier and quicker.

8.7 WINDOWS FILE PROTECTION

Earlier versions of the Windows operating system do not prevent shared system files from being overwritten by program installations. After these changes are made, the user often experiences unpredictable performance results, which range from program errors to an unstable operating system. This problem affects several types of files, most commonly Dynamic Link Libraries (DLLs) and Executable (EXE) files.

Windows 2000 includes a new feature called Windows File Protection (WFP), which prevents the replacement of certain monitored system files. By using this feature, file version mismatches can be avoided. The WFP feature uses the file signatures and catalog files generated by code signing to verify if protected system files are the correct Microsoft versions. The WFP feature does not generate signatures of any type.

The WFP feature provides protection for system files using two mechanisms. The first mechanism runs in the background. The WFP feature is implemented when it is notified that a file in a protected folder is modified. Once this notification is received, the WFP feature determines which file was changed. If the file is protected, the WFP feature looks up the file signature in a catalog file to determine if the new file is the correct Microsoft version. If it is not, the file is replaced from the Dllcache folder (if it is in the Dllcache folder) or the distribution media. By default, the WFP feature displays the following dialog box to an administrator:

```
A file replacement was attempted on the Protected Sys-
tem File file name. To maintain system stability, the
file has been restored to the correct Microsoft ver-
sion. If problems occur with your application, please
contact the application vendor for support.
```

The second protection mechanism provided by the WFP feature is the System File Checker (SFC.EXE) tool. At the end of GUI-mode setup, the System File Checker tool scans all protected files to ensure they are not modified by programs installed using an unattended installation. The System File Checker tool also checks all catalog files used to track correct file versions. If any catalog files are missing or damaged, the WFP feature renames the affected catalog file and retrieves a cached version of that file from the Dllcache folder. If a cached copy of the catalog file is not available in the Dllcache folder, the WFP feature requests the appropriate media to retrieve a new copy of the catalog file.

The System File Checker tool gives an administrator the ability to scan all protected files to verify their versions. The System File Checker tool also checks and repopulates the %systemroot%\System32\Dllcache folder. If the Dllcache folder becomes damaged or unusable, you can use the sfc/scanonce or sfc/scanboot command to repair its contents. All SYS, DLL, EXE, TTF, FON and OCX files included on the Windows 2000 CD-ROM are protected. However, due to disk space considerations, maintaining cached versions of all these files in the Dllcache folder is not desirable on all computers.

Depending on the size of the SFCQuota value in the HKEY_LOCAL_MACHINE\ SOFTWARE\Microsoft\Windows NT\CurrentVersion\Winlogon registry key (the default size is 0xFFFFFFFF, or 400 MB), the WFP feature keeps verified file versions cached in the Dllcache folder on the hard disk. The SFCQuota setting can be made as large or small as needed by the system administrator. Setting the SFCQuota value to 0xFFFFFFFF causes the WFP feature to cache all protected system files (approximately 2,700 files). After the registry values are set and saved, you then execute sfc/scannow at a command prompt. This causes the System File Checker tool to verify all protected file versions and add all missing protected files to the Dllcache folder.

If a file change is detected by the WFP feature, the affected file is not in the Dllcache folder, and the corresponding file in use by the operating system is the correct version, the WFP feature copies that version of the file to the Dllcache folder. If the affected file in use by the operating system is not the correct version or the file is not cached in the Dllcache folder, the WFP feature attempts to locate the installation media. If the installation media is not found, the WFP feature prompts an administrator to insert the appropriate media to replace the file or the Dllcache file version.

8.8 WINDOWS SECONDARY LOGON SERVICE

Windows 2000 secondary logon allows administrators to log on with a non-administrative account and still be able to perform administrative tasks (without logging off) by running trusted administrative programs in administrative contexts. In this scenario, system administrators require two user accounts: a regular account with basic privileges, and an administrative account (this can be a different administrative account for each administrator or a single administrative account shared among administrators).

Secondary logons address the security problems presented by administrators running programs that may be susceptible to "Trojan Horse" attacks (such as running Microsoft Internet Explorer in the administrative context while accessing a non-trusted Web site). Even though secondary logon is primarily intended for system administrators, it can be used by any user with multiple accounts to start programs under different account contexts without the need to log off.

To start a command shell in local computer administrative context while logged on as a normal user:

1. Click the Start button and select the Run option.
2. Type runas /user:*machine_name*\administrator cmd, where the *machine_name* is the name of your computer, and click on OK.
3. When a console window appears and prompts you for a password, type the password for the administrator account and press the Enter key.
4. A new console will appear running in the administrative context with the title of the console stating running as machine_name\administrator.

Any command-based administrative programs from this console window.
To start a Control Panel tool in Administrative Context while logged on as a normal user

1. Click the Start button, select the Settings option and select the Control Panel option.
2. Select the particular tool you want to run in administrative context (for example: Add/Remove Hardware).
3. Highlight the selected tool by using a single left-click on the icon.
4. Hold down the Shift key and right-click on the icon. You will notice the Run As command that appears in the command list.
5. Select the Run As command. You will be prompted with a dialog box titled, "Run program as other user".
6. Type the administrator account name and password in the appropriate fields. Please note that the domain name can also be changed.
7. After entering the credential for the administrator account, click on OK and the program associated with the tool will start in the administrative context.

The following example uses a shortcut to the Computer Management program, but this method will work on shortcuts of .EXE files and shortcuts of registered file types like .TXT, .DOC and .MSC files.

While logged on as a normal user:

1. Use Windows Explorer to create a shortcut on your desktop for the file COMP-MGMT.MSC. COMPMGMT.MSC can be found in the \%WINDIR%\SYSTEM32 directory. By default, this is the \WINNT\SYSTEM32 directory, located on the boot partition.
2. Highlight the Shortcut to Compmgmt icon on your desktop using a single left-click.
3. Hold down the Shift key while right-clicking on the Shortcut to Compmgmt icon on the desktop.
4. Select the Run As command. You will be prompted with the "Run program as other user" dialog box.
5. Type the name and password for the administrator account in the appropriate fields. Click on OK.
6. This will start an MMC console with the Computer Management snap-in loaded. This snap-in is now running in administrative context.

You can also configure a shortcut to always run using alternate credentials when opened by configuring the Properties for the shortcut as follows:

1. Close and open MMC consoles and highlight the Shortcut to Compmgmt icon on your desktop using a single left-click. Right-click on the icon and select Properties.
2. In the center area of the Properties dialog box find the checkbox labeled, "Run as different user". Select the checkbox and click on OK to close the Properties dialog box.
3. Double-click on the Shortcut to Compmgmt icon to start the console.
4. You will be prompted with the "Run program as other user" dialog box. Enter the credentials in the appropriate fields and click on OK.

To start an MMC in Administrative Context Using a Saved MSC file while logged on as a normal user:

1. Use Windows Explorer to copy the file MSC file to your desktop.
2. Highlight the MSC icon on your desktop by using a single left-click.
3. Hold down the Shift key and right-click on the Compmgmt icon on the desktop.
4. Select the Run As command. You will be prompted with the "Run program as other user" dialog box.
5. Type the name and password for the administrator account in the appropriate fields. Click on OK.

A new MMC console will now appear with the Computer Management snap-in loaded. This snap-in is now running in the administrative context. In a similar fashion, a system administrator can create custom Microsoft Management Consoles containing frequently used administrative snap-ins and run them in administrative context using secondary logon.

To run the Windows Explorer shell in administrative security context while logged on as a normal user:

1. To start Task Manager, right-click on the Task bar and select Task Manager.
2. Select the Processes tab.
3. Select EXPLORER.EXE. Click on End Process. Click Yes on the warning pop-up message. The entire desktop will disappear. You will still have any programs that you started including Task Manager.
4. Select the Programs tab.
5. Click on New Task.
6. Type runas/user:machine/domain name\administrator explorer.exe. Click on OK.
7. A Console window will appear and prompt for the password. Minimize Task Manager, type the password and press Enter. The desktop will return including the task bar, shortcuts, Startup folder items, etc.
8. Perform necessary administrative tasks. For example: clicking on Start, Settings and Control Panel will bring up control panel in administrative context.
9. When finished, log off Administrator. A new shell will automatically start, running in the originating user context.

SUMMARY

1. The Windows 2000 interface is based on the desktop, where you do all of your work. On the desktop you will find the My Computer icon, the Recycle Bin, the My Documents shortcut, and the task bar (with the Start button).

2. The My Computer icon represents your computer. It includes all disk drives and the Control Panel.

3. The Recycle Bin is used as a safe delete.

4. By default, the My Documents folder is stored in the C:\Documents and Settings\ *user_login_name* folder.

5. Disk and file management is usually done with My Computer or Windows Explorer.

6. There are several utilities used to configure the Windows 2000. The main ones are the Control Panel and Registry Editor.

7. The Device Manager lists all hardware devices on your computer and allows you to change the properties of any device.

8. The default size of the paging file (virtual memory) is equal to the total amount of RAM plus 12 MB, not to exceed the amount of available disk space.

9. The recommended size for the paging file should be 1.5 times the amount of RAM available on your system.

10. A hardware profile is a set of instructions that tells Windows 2000 which devices to start upon computer start-up or what settings to use for each device.

11. To keep your system loaded exclusively with drivers from the Hardware Compatibility List, Microsoft introduced driver-signing technology into Windows 2000.

12. The Display applet allows you to configure the Windows environment including menus, windows, icons, the desktop, and the screen saver.

13. If you enable password protection, when the screen saver activates, then you must specify the login password to be able to continue.

14. Windows 2000 supports the use of up to 10 monitors.

15. Windows 2000 supports plug-and-play (PNP) hardware.

16. Several of the utilities used to control and configure the many network services are grouped into the Administrative Tools folder, which can be accessed from within a Control Panel applet or by clicking on the Start button and clicking on the Programs option.

17. The Microsoft Management Console (MMC) is a primary administrative tool used to manage Windows 2000.

18. The Event Viewer is a useful utility that is used to view and manage logs of system, program, and security events on a computer.

19. A service is a program, routine, or process that performs a specific system function to support other programs.

20. The registry is a central, secure database in which Windows 2000 stores all hardware configuration information, software configuration information, and system security policies.

21. The data kept in the registry are added and modified by a variety of system modules that start during boot-up and are modified by various configuration tools such as the Control Panel, Windows 2000 Setup program, User Manager, adding or removing a hardware device, adding or removing a printer, and other administrative utilities.

22. If the need arises to view or change the registry, use the Registry Editor (REGEDIT.EXE or REGEDT32.EXE) utilities.

23. The registry is first divided into five subtrees. The subtrees have names that begin with the string HKEY, which stands for "handle to a key."

24. The root keys are then divided into subkeys, which may contain other subkeys.

QUESTIONS

1. You just installed a Windows application in Windows 2000. Typically to start the program, you would usually _____ .
   a. find the executable file using My Computer and double-click on it
   b. find the executable file using Windows Explorer and double-click on it
   c. click on the Start button, click on the Programs options, and click on the program in the appropriate folder
   d. click on right mouse on the desktop and select the program from the shortcut menu

2. When you delete a file by mistake, the file can be undeleted using _____ .
   a. Recycle Bin
   b. Trash Can
   c. Swap File
   d. Paging File
   e. C drive

3. Which of the following will you *not* find in the My Computer icon?
   a. drive
   b. C drive
   c. Control Panel
   d. Recycle Bin
   e. printer folder

4. You have a program that follows the API specification. While removing the program, you want to make sure that the registry settings and program files are removed. Which applet would you use in the Control Panel?
   a. System applet
   b. Add/Remove Programs applet
   c. Add New Hardware applet
   d. Administrative Tools

5. How do you modify the appearance of the Windows 2000 desktop? (Choose all that apply)
   a. Double-click on the Display applet in the Control Panel.
   b. Double-click on the System applet in the Control Panel.
   c. Double-click on the Add/Remove Hardware applet in the Control Panel.
   d. Right-click the My Picture folder.
   e. Right-click the desktop and choose the Properties option.

6. To change the video card driver, you would _____ .
   a. use the Add/Remove applet in the Control Panel and select the Video tab
   b. use the Display applet in the Control Panel and select the Settings tab
   c. modify the registry
   d. modify the SYSTEM.INI file using a text editor

7. Which tool(s) can be used to install an updated driver for a Windows 2000 device? (Choose all that apply.)
   a. Registry Editor
   b. SETUP.EXE program
   c. Add/Remove Program applet
   d. Add/Remove Hardware applet
   e. Device Manager

8. What happens when you use your mouse to right-click on an object in Windows 2000?
   a. The properties sheet for that object will open.
   b. A special Windows 95 property sheet will open that allows you to configure the Windows 95 environment.
   c. A shortcut menu pertaining to that object will open.
   d. You copy the object into the clipboard.
   e. You open the Explorer for that object.

9. All of the devices are shown in the Device Manager. What does it mean when one of the devices has a circled exclamation point?
   a. The hardware has a problem.
   b. The hardware is disabled.
   c. The device is a high-priority device.
   d. The device is currently being used and cannot be removed at this time.
   e. The hardware was never installed.

10. You have an MS-DOS-based game that has to be run in a full screen. How do you make sure the program will always start in a full screen?
    a. Use the PIF editor to create a PIF file for the game.
    b. Using the shortcut menu of the DOS program, select the Properties shortcut.
    c. Press the Alt+Enter keys.
    d. Right-click the icon for the program and choose Full screen.

11. You want to make sure deleted items are removed from the Recycle Bin immediately. How do you configure the Recycle Bin?
    a. Select the Display applet in the Control Panel.
    b. Select the System applet in the Control Panel.
    c. Right-click the Recycle Bin and select the Properties options.
    d. Delete the Recycle Bin from the desktop.

12. What files make up the registry?
    a. REGISTRY.DAT
    b. BOOT.INI
    c. SYSTEM.DAT
    d. SYSTEM.INI
    e. WIN.INI
    f. hives

13. How can you copy a file to a floppy disk from Windows Explorer? (Select two answers)
    a. Right-click the file name and choose Send To.
    b. Drag the file to the floppy disk drive icon on the taskbar.
    c. Drag the file to the desktop.
    d. Right-click the file name and choose Copy To.
    e. Drag the file to the A drive icon within My Computer.

14. You installed a modem on your computer, but the system does not recognize the modem. What do you do next?
    a. Run the Add/Remove Hardware applet in the Control Panel.
    b. Reinstall Windows 2000 and let Setup automatically install all necessary drivers.
    c. Copy the required driver from the installation CD to the hard drive.
    d. Use the Add/Remove Programs applet in the Control Panel.

15. The Device Manager can be found by _____ .
    a. double-clicking the System applet in the Control Panel
    b. clicking on the Start button and selecting the Settings option
    c. double-clicking on the DEVMAN.EXE file
    d. double-clicking on the Device Manager icon on the taskbar

16. What utility is used to perform common file and disk management functions?
    a. Explorer
    b. File Manager
    c. System Editor
    d. PIF Editor

17. Most of the Windows 2000 configuration information is kept in _____ .
    a. initialization files (*.INI)
    b. system files (*.SYS)
    c. Windows 2000 Registry
    d. configuration files (*.CFG)
    e. dynamic link files (*.DLL)

18. All of the devices are shown in the Device Manager. What does it mean when one of the devices has a red X?
    a. The hardware has a problem.
    b. The hardware is disabled.
    c. The device is currently being used and cannot be removed at this time.
    d. The hardware was never installed.
    e. The device is not on or does not have power.

19. You want to increase the initial paging file size. How do you configure the paging file?
    a. Edit the registry.
    b. Click the Change button on the Performance tab in System Properties.
    c. Click the Virtual Memory button on the General tab in System Properties.
    d. Click the Paging File button on the Hardware Profiles tab in System Properties.

20. Pat, a member of the users group, would like to view the application, security, and system logs on his Windows 2000 Professional computer using Event Viewer. What must he do to accomplish this?
    a. Nothing, provided Pat has a functioning account on the workstation, because he will be able to view all of these logs.
    b. Pat must ask the Administrator of the computer to grant the proper NTFS permissions for viewing the logs to him.

    c. Pat must ask the Administrator of the computer to add his account to the Administrators local group.

    d. Pat must ask the Administrator of the workstation to grant him access to Event Viewer.

21. Pat uses his portable computer both at home and in his office. He has a docking station at the office so that he can connect to the company's network. However, Pat does not want to start any network services while he is using his computer at home. How would Pat configure his computer?

    a. Create two different user profiles.

    b. Create two different hardware profiles.

    c. Create two different roaming user profiles.

    d. Create two different BOOT.INI files.

22. Pat wants to optimize the paging file on her computer. Her computer has three physical disks and each has only one partition. The boot partition is located on the first physical disk. How would Pat optimize the paging file on her computer?

    a. Create one paging file and put it on the disk that contains the boot partition.

    b. Create one paging file and put it on any disk other than the one that contains the boot partition.

    c. Create three paging files, one for each of the three physical disks.

    d. Create two paging files, one for each physical disk except for the physical disk that contains the boot partition.

23. You want to connect an uninterruptible power supply (UPS) device to your Windows NT Workstation computer. How would you install the UPS device? (Select all that apply)

    a. Connect the UPS device to a serial port.

    b. Connect the UPS device to a parallel port.

    c. Configure the UPS using the Devices option in Control Panel.

    d. Configure the UPS using the UPS option in Control Panel.

24. When Windows NT encounters data loss or failure of major functions, it produces kernel Stop errors. Which program would you use to view Stop errors and the time that they occurred?

    a. Event Viewer   c. Disk Administrator

    b. Server Manager   d. Kernel Debugger

25. Which one of the following log files on a Windows 2000 computer can only be viewed by members of the Administrators local group?

    a. application log file

    b. Device.Log file

    c. security log file

    d. system log file

26. How would you ensure that your workstation is locked automatically if you leave for more than 15 minutes?

    a. Select a password-protected screen saver.

    b. Enforce a system policy that locks the workstation if there is no activity for 15 minutes.

    c. Schedule periodical shutdowns of the workstation.

    d. Enable locking of the workstation on the Startup/Shutdown tab of System Properties.

27. Your Windows NT workstation is continually logged on to the Internet. You are also connected to a departmental domain. You want to ensure that sensitive information on your computer cannot be accessed by unauthorized users who are browsing the Internet. Which one of the following security precautions should you implement on your computer?

    a. Enable IP forwarding.

    b. Disable the Workstation service on your machine.

    c. Disable the Server service on your machine.

    d. Enable the Computer Browser service on your machine.

28. Which utility would you use to search for a particular value in the registry?

    a. REGEDIT.EXE   c. RDISK.EXE

    b. REGEDT32.EXE   d. POLEDIT.EXE

29. Which tool would you use to run maintenance utilities at a specific time?
    a. Task Scheduler
    b. Task Manager
    c. Job Manager
    d. VMM

30. The primary Windows 2000 Administrative Tool is the _____ .
    a. MAT—Microsoft Admin Tool
    b. MMC—Microsoft Management Console
    c. MZA—Microsoft Zero Administration
    d. IEA—Internet Explorer Administration

31. Which registry subtree contains data for the computer including hardware information?
    a. HKEY_CURRENT_USER
    b. HKEY_HARD_CONFIG
    c. HKEY_LOCAL_MACHINE
    d. HKEY_DEFAULT

32. Which one of the following user mode MMC types allow the creation of a new MMC window?
    a. delegated access, single window
    b. full access
    c. delegated access, multiple windows
    d. author

33. You have just opened an MMC tool in author mode. Which of the following actions can you perform in this MMC? (Select all that apply)
    a. Save MMC console settings.
    b. Open additional windows.
    c. Remove snap-ins.
    d. Add snap-ins.

34. Which of the following are user mode MMC types? (Choose all that apply)
    a. limited access, single window
    b. delegated access, double windows
    c. full access
    d. delegated access, multiple windows

35. You have just opened an MMC tool in user mode with full access. Which of the following actions can you perform in this MMC? (Choose all that apply)
    a. Save MMC console setting.
    b. Add snap-ins.
    c. Create windows.
    d. Remove snap-ins.

36. What are the valid MMC modes? (Choose the two applicable answers.)
    a. read-only
    b. author
    c. user
    d. admin

37. Which service allows you to access network resources?
    a. PNP
    b. Workstation
    c. Server
    d. Print Spooler

38. To stop new users from logging in but still allow the current users to stay connected, you would _____ .
    a. stop the Server service
    b. pause the Server service
    c. stop the Workstation service
    d. pause the Workstation service

HANDS-ON EXERCISES EXERCISE 1: INTRODUCTION TO THE WINDOWS 2000 INTERFACE

1. Start Windows 2000 and log in.
2. If any windows are open, close them.
3. Start the Calculator by clicking on the Start button, clicking on Programs, clicking on Accessories, and selecting the Calculator. Close the Calculator by clicking on the Close button on the top right corner of the Calculator.
4. Start the Calculator by double-clicking on My Computer, double-clicking on the C: drive, double-clicking on the WINNT, double-clicking the SYSTEM32 folder, and then double-clicking on the CALC file. In the WINNT and SYSTEM32 folder, you will probably have to click on the Show All Files option on the left of the window.
5. Start the Calculator by using Windows Explorer. Close the Calculator by double-clicking on the small calculator icon at the top left of the Calculator.
6. Start the Calculator by using the Run option under the Start button and specifying the path of C:\WINNT\SYSTEM32\CALC.EXE. Close the Calculator.

7. Start the Calculator by using the Run option under the Start button and using the Browse button to find the CALC.EXE. Close the Calculator.

8. Open a window for the C:\Documents and Settings*user_login_name* folder (where *user_login_name* is the name used during the login) and notice the contents of the desktop directory.

9. Create a shortcut for the Calculator (C:\WINNT\SYSTEM32\CALC.EXE) program. You can do this by using the shortcut menu (right mouse button on the desktop) and select the New option followed by the Shortcut option. Try to determine what distinguishes a shortcut.

10. Test the Calculator shortcut. Close the Calculator.

11. Right-click the desktop, select the New option, select the Text Document option. Use Notepad to create a text file called NAME.TXT located in the C:\ folder. Include your first and last name in the file. Close Notepad.

12. Double-click on the NAME file under the C drive. Notice that Notepad automatically started. Close Notepad.

13. On the desktop, make a shortcut to the NAME file. This can be done by accessing the shortcut menu (clicking on the desktop with the right mouse button—secondary mouse button), clicking on the New option, and clicking on the Shortcut option.

14. Test the shortcut to name.txt file. Close Notepad.

15. Create a new folder called COMMON APPS.

16. In the COMMON APPS group, create icons for the Calculator (CALC.EXE) and Notepad (NOTEPAD.EXE). CALC.EXE is located in the C:\WINNT\SYSTEM32 directory and Notepad is located the C:\WINNT\.folder.

17. Test the newly created shortcuts.

18. Open the desktop folder again and notice the contents of the DESKTOP folder. It should be easy to see the purpose of the Desktop folder.

19. Delete the COMMON APPS folder, the Calculator shortcut, and Names shortcut.

20. Open a window for the C:\WINNT directory.

21. Select the Large Icons option under the View menu.

22. Select the Small Icons option under the View menu.

23. Select the List option under the View menu.

24. Select the Details option under the View menu.

25. Select the Toolbars option in the View option.

26. Find and click on the Large Icons button. You may need to increase the window size to show all buttons on the toolbar.

27. Open the View menu, click on the Arrange Icons option under the By Type option. "By type" means that it will sort by file name extension.

28. In the WINNT window, open the shortcut menu by clicking on right mouse button without clicking on an icon click on the Arrange Icons option and click on the By Name option.

29. Make sure the Windows window does not auto arrange its icons. This can be done by opening the View menu and clicking on the Arrange Option menu. If there is a check mark on the Auto Arrange option, click on it to shut it off. If there is not a check mark on the Auto Arrange option, click somewhere on the desktop to close the menu.

30. Move some of the files around without putting one on top of another.

31. Enable the Auto Arrange feature for the Windows window.

32. Close all windows.

33. Double-click on the My Computer icon.

34. Double-click on the C: drive.

35. Open Windows for the C:\WINNT, C:\WINNT\FONTS, and C:\WINNT\SYSTEM32 folders.

36. Close all windows.

37. Double-click on My Computer icon.

38. Select the Folder Options option under the Tools menu.

39. Under the General tab, select the "Open each folder in the same window" and click on OK.

40. Double-click on the C drive, double-click on the WINNT folder, and double-click on the SYSTEM32 folder. Notice how many windows are open.

41. Click on the Up One Level button. Keep clicking on this button until you get back to the My Computer window.
42. Select the Folder Options option under the Tools menu. Under the General tab, select the "Open each folder in its own window" option and click on OK.
43. Close all windows.

EXERCISE 2: FILE MANAGEMENT

1. Insert a disk in drive A. Under My Computer, use the shortcut menu of the A drive and format the disk.
2. Open a window for the A drive.
3. Create a folder called COMMAND on the A drive using the shortcut menu.
4. Open the A:\COMMAND folder.
5. Open a window for the C:\WINNT\SYSTEM32.
6. Close all windows except for the two COMMAND and SYSTEM32 windows.
7. Copy the FORMAT.COM, ATTRIB.EXE, and DISKCOPY.COM files from the C:\WINNT\SYSTEM32 folder to the A:\COMMAND folder.
8. Create a DATA folder in the A:\ folder.
9. Move the FORMAT.COM file from the A:\COMMAND folder to the A:\DATA folder.
10. Copy the FORMAT.COM file from the A:\DATA folder to the A:\COMMAND directory.
11. Using the shortcut menu of the A drive, perform a disk copy to another disk. A disk must be in the disk drive before the Copy Disk option will appear.
12. Rename the FORMAT.COM file to DESTROY.COM in the A:\DATA folder.
13. Hide the DESTROY.COM file by using the shortcut menu of the DESTROY.COM file and selecting Properties.
14. If you cannot see hidden files, select Options under the View menu, click on the View menu, select Show all Files, and click on the OK button.
15. Use the COPY option from the shortcut menu of XCOPY.EXE file in the C:\WINNT\SYSTEM32 directory.
16. Make the A:\ window active. On the A:\ windows, select the Paste option from the shortcut menu.
17. Delete the DESTROY.COM file.
18. Using the A:\ window, highlight both folders at the same time and delete them.
19. Start Windows Explorer.
20. Using the shortcut menu of the A drive shown in Windows Explorer, format the disk.
21. Create a folder called COMMAND on the C drive using the shortcut menu of the C drive window.
22. Click on the DOS COMMAND folder. Notice the right side of the windows should show no files.
23. Click on the plus (+) next to the C:\WINNT folder in Explorer to show the subdirectories under WINNT.
24. Click on the SYSTEM32 folder under the WINNT directory.
25. Copy FORMAT.COM, ATTRIB.EXE, and DISKCOPY.COM files from the C:\WINNT\SYSTEM32 folder to the C:\COMMAND folder.
26. Delete the FORMAT.COM and ATTRIB.EXE file from the C:\COMMAND directory.
27. Double-click on the Recycle Bin. These are the files that you deleted in the GUI interface.
28. Drag the FORMAT.COM file to the desktop.
29. Highlight ATTRIB.EXE file and select Restore from the Shortcut menu or click on the Restore button on the left side of the Recycle Bin window.
30. Look in the COMMAND folder and notice that the ATTRIB.EXE file is back.
31. Select the Empty Recycle Bin option from the File menu.
32. Close the Recycle Bin window.
33. Access the shortcut menu of the Recycle Bin and select Properties. Notice the maximize size of the Recycle Bin.
34. Click on the OK button.

EXERCISE 3: DOS APPLICATIONS

1. Open a window for the C:\WINNT\SYSTEM32 folder.
2. Right-click inside the window (not a file or folder), select the New option followed by the Text File option.
3. Type in the following:
   MEM
   PAUSE
4. Save the file as GO.BAT and close Notepad. Be sure to save it in the C:\WINNT\ SYSTEM32 folder.
5. Double-click the GO.BAT file to run it. Record the numbers for extended memory.
6. Select the Properties option in the shortcut menu of the SHOW MEMORY shortcut. Notice the extra tabs for a DOS program compared with a Windows program.
7. Change the properties so that you have 4096 KB of extended memory and change the working directory to C:\WINNT\SYSTEM32. In addition, configure the shortcut to run in Windows. Hint: Use the memory and screen tabs.
8. Run the GO.BAT file again and compare the amounts of extended memory with those recorded.
9. Delete the GO.BAT file.

EXERCISE 4: CONTROL PANEL

System Applet

1. Open the Control Panel.
2. Double-click on the System applet. Record the Windows 2000 version, the processor ID listed under Computer, and the amount of physical RAM.
3. Click on the Hardware tab followed by the Device Manager button.
4. Open the View menu and select the Resources by Type option.
   a. Find what device is using IRQ 5.
   b. Find which IRQ is being used for COM2 and the hard drive.
   c. Find which IRQs are free.
   d. Find what device is using I/O address 300.
   e. Find what device is using DMA 2.
   f. Find what DMAs are being used by the sound card.
5. Open the View menu and select the Devices by Type option.
6. On the device tree, find and double-click on COM2.
   a. Find and record the device status.
   b. Find and record the bits per second, the number of data bits and stop bits, the type of parity, and the type of flow control.
   c. Find and record the resources used by COM2.
   d. Find and record the entries in the Conflicting Device List.
7. Disable the COM2 by using the Device Usage option in the General tab.
8. If you have a sound card loaded, find it and delete the sound card from the device tree.
9. Close the Device Manager.
10. Click on the Hardware wizard button. You also can use the Add Remove Hardware applet in the Control Panel. Click on the Next button.
11. Select the Add/Troubleshoot a Device option and click the Next button. The sound card should automatically be detected and the drivers loaded. If not, manually load the drivers by either specifying the type of card or the location of the drivers. When done, click on the Finish button.
12. Click on the Advanced tab followed by the Performance option. If the computer is going to be used as a production server by many people, optimize for Background services. If it is used primarily by you to run applications, optimize for Applications.
13. Click on the Change button under Virtual Memory.
14. Notice the initial size and maximize size. Add 20 MB to the initial size by typing the number and clicking on the Set button.

15. Select the D drive. Type in 384 MB for the initial size and 768 MB for the maximize size. Click on the Set button. Remember, you should create two paging files on the same physical driver. It is better to put them on two or more physical drivers, not logical drives. In addition, do not use the drive that has the boot volume.
16. Remove the entry for the D drive by putting in zero for the initial size and maximize size. Click on the Set button. Click on OK to close the Virtual Memory dialog box and then click on OK to close the Performance Option dialog box.
17. Click on the Startup and Recovery button. Change the time to display the operating systems to 20 seconds. Click on OK.
18. Open the BOOT.INI file and see that the new time is 20 seconds.

Display Applet

19. Using the Display applet, change the background pattern to brick.
20. Enable the Flying Window screen saver and have it activate after 2 minutes of inactivity.
21. Load the High-Contrast Black (Large) appearance schemes.
22. Load the Lilac appearance scheme.
23. Increase the horizontal icon spacing to 75 pixels.
24. Change the icon text to bold, 15-point Arial.
25. Load the Windows default appearance scheme.
26. Click on the Settings tab.
27. Record the current resolution and number of colors.
28. Change the screen resolution to 640x480 and the number of colors to 16. Click on OK.
29. Change the screen resolution to 800x600, number of colors to 256 colors, and small fonts. Click on OK.
30. Change the screen resolution to 800x600, number of colors to high color, and large fonts. Click on OK.
31. Change the display back to the recorded resolution and number of colors.
32. Click on the Advanced Properties button.
33. Record the manufacturer and type of video card and monitor.
34. Close the Display dialog box.

Add/Remove Programs

35. Go to the WWW.ADOBE.COM site and download the Adobe Acrobat Reader.
36. Double-click the Add/Remove Programs applet in the Control Panel.
37. Under Change or Remove Programs, find WinZip and click on it. Then click on the Change/Remove button to remove WinZip.
38. Click on the Add/Remove Windows Components, click on the Other Network File and Print Services option, click on the Details button and select the Print services for Unix. Click on OK.
39. Click on the Next button. Use the wizard to finish the installation.
40. Close the Add/Remove Programs dialog box.

EXERCISE 5: MMC CONSOLE

Computer Management Console

1. Click on the Start button, select the Programs option, select the Administrative Tools option, and select the Computer Management option.
2. Open the Event Viewer. Click on the System option under Event Viewer. Look for any errors. If you find one, double-click to view it.
3. Go back to the Event Viewer and click on the Application option under the Event Viewer. Look for any errors. If you find one, double-click to view it.
4. Close the Event Viewer branch.
5. Click on the System Summary under System Information. Record the amount of virtual memory currently being used.

6. Double-click on the Hardware Resources and click on the IRQs to show which devices are using which devices.

7. Click on the Services option located under Services and Applications. This will display all of the services loaded on your system.

8. To get a better view of the services, open the View menu and select the Detail option. Record the status of the service, workstation, and print spooler service.

9. Right-click the Server option and select the Stop option.

10. Right-click the Server option and select the Start option.

11. Right-click the Server option and select the Properties option.

12. Record the start-up type.

13. Close the Computer Management console.

Adding and Removing Components to the MMC

14. Click on the Start button and select the Run option. Execute the MMC command.

15. Open the Console menu and select the Add/Remove Snap-in option. Click on the Add button.

16. Double-click on the Disk Defragmenter and the Event Viewer. If the Select Computer dialog box comes up, select local computer. The Select Computer dialog box allows you to remotely administer to other computers. Click on the Close button. Click on OK.

17. Click on the Defragmenter. Double-click on Event Viewer.

18. Open the Console menu and select the Save option. Save the MMC console as MISC.MSC on the desktop. Close the Console.

19. From the Desktop, double-click on the MISC.MSC file. Close the console.

20. Right-click the MISC.MSC file and select the Copy option.

21. Right-click the taskbar and select Properties option. Select the Advanced tab, followed by the Advanced button.

22. The Start menu should already be highlighted. In the right pane, right-click and select the Paste option.

23. Close the Start menu dialog box. Close the Taskbar dialog box by clicking on the OK button.

24. Click on the Start button and run the MISC.MSC program.

25. Right-click on the System Event Viewer and select the New Window from Here option. Open the View menu and select the Tile Horizontally option.

26. Open the Console menu and select the Options option. Change the Console mode to "User mode—limited Access, single window." Click on the OK button.

27. Close the console and save the settings. When a warning appears asking if you are sure that you want to display a single window, click Yes.

28. Start the MISC.MSC file. If you right-click on any of the components, you will notice the New Window from Here option.

29. Close the console.

Exercise 6: Working with the Registry

1. Start the Display applet in the Control Panel.

2. Set the Wallpaper to Water Color and tile it.

3. Set the screen saver to flying windows and set the wait time to 3 minutes.

4. Start REGEDIT.EXE.

5. Find the following value:

   HKEY_CURRENT_USER\Control Panel\Desktop\ScreenSaveTimeOut

   Notice the value and compare it with the wait time of the screen saver. (Remember 1 minute equals 60 seconds.)

6. Change the ScreenSaveTimeOut value to 120 seconds.

7. Find the following value:

   HKEY_CURRENT_USER\Control Panel\Desktop\TileWallPaper

   Although the TileWallPaper value is a string, it only uses a 1 or 0 value. 1 indicates to tile the wallpaper and 0 indicates not to tile the wallpaper.

8. Find the following value:

    `HKEY_CURRENT_USER\Control Panel\Desktop\WallPaper`

9. Find the following value:

    `HKEY_LOCAL_MACHINE\hardware\description\system.`

    Record the BIOS and BIOS date.

10. Find the following value:

    `HKEY_LOCAL_MACHINE\hardware\description\system\`
    `centralprocessing\0`

    Record the processor identifier and the speed.

11. Click on My Computer listed at the top of the tree.

12. Open the Edit menu and select the Find option. Search for the processor until you get to HKEY_LOCAL_MACHINE\hardware\description\system\centralprocessor\ 0 parameters.

13. Exit REGEDIT.EXE

9 Networking with Windows 2000

INTRODUCTION

Before going into the advance network features of the TCP/IP protocol and Windows 2000, we must also discuss the basics of network connectivity. Although some of this chapter may seem like a review from previous chapters, such as connecting Windows 95 or Windows 98 to a network, the material has enough differences to warrant further discussion.

OBJECTIVES

1. Install, configure, and troubleshoot network adapters.
2. Given a Windows 2000 computer, install and configure common network clients, network services, and network protocols.
3. Given a Windows 2000 computer, optimize the computer for data throughput depending on its function.
4. When a Windows 2000 computer is connected to the network, test its connection by using common network commands such as PING, TRACERT, and ARP.

9.1 WINDOWS 2000 NETWORK ARCHITECTURE

As discussed in Chapter 3, the OSI model is a modular approach to networking. The OSI model is made of layers, each with specific tasks to perform. Although the OSI model is a common learning tool in network, it does not correspond to any common network architecture. Windows 2000 network architecture is arranged in a similar fashion, in that its components are arranged in layers and each has specific tasks to perform within its assigned layer. See Figure 9.1.

The **network driver interface specification (NDIS)** layer provides a communication path between two network devices. It is similar to the network layer in the OSI model as

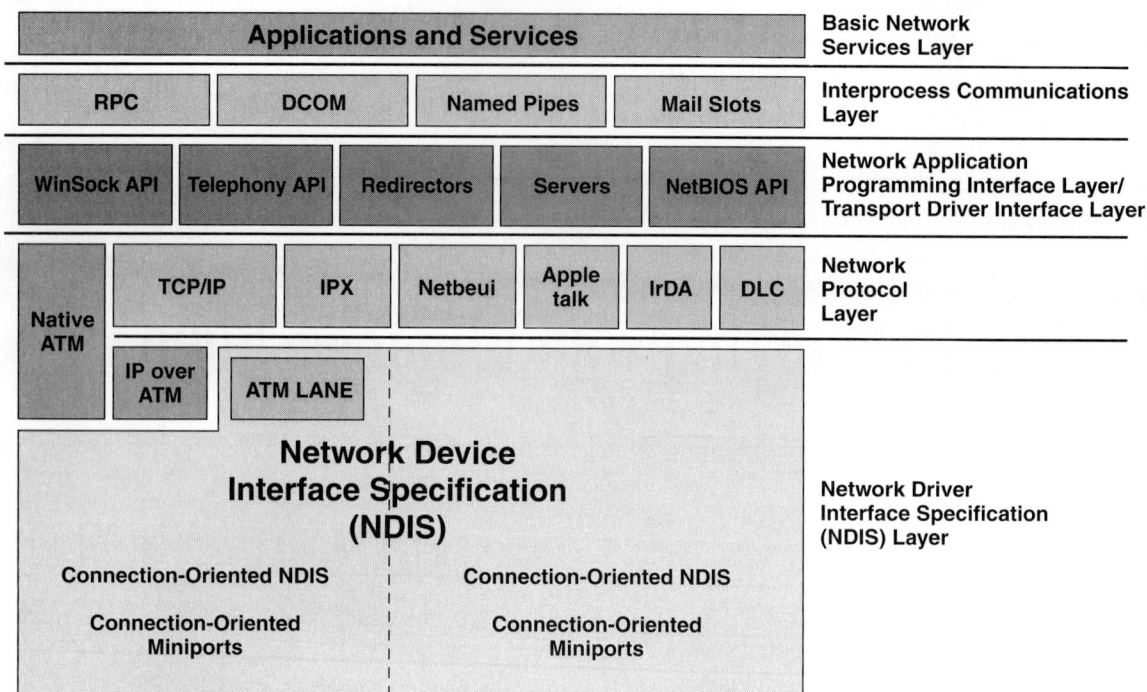

FIGURE 9.1 Windows 2000 network architecture

it acts as the link between the network adapters and the network protocols above it and manages the binding of network interface connections to the protocol. When a packet gets to the card, the NDIS will route it to the appropriate protocol. These protocols could be connection-oriented protocols such as ATM or TCP/IP, or it could be connectionless protocol such as IPX. This layer contains the network media including Integrated Services Digital Network (ISDN) lines, Ethernet, token ring, FDDI, and other fiber-optic pathways. It also includes the **miniport drivers** that control the network interface adapter and connect the hardware device to the protocol stack.

Windows 2000 NDIS is implemented by a file called NDIS.SYS. It is also referred to as the NDIS wrapper because the code surrounds all NDIS device drivers. When a protocol is bound to a network component **(binding),** it is linked to a protocol stack. NDIS allows an unlimited number of network adapters in a computer and an unlimited number of protocols bound to one or more network adapters.

The **network protocol layer,** similar to the data and network layer of the OSI model, includes the protocols that allow clients and applications to send data over the network. It includes TCP/IP, NWLink (IPX/SPX), NetBEUI, infrared data association (IrDA), AppleTalk, and data link control (DLC).

The **transport driver interface layer** provides a standard interface between network protocols and the network services (network applications, network redirectors, and network application programming interfaces (APIs). The network API provides standard programming interfaces for network applications and services including WinSock, NetBIOS, telephony API (TAPI), messaging API (MAPI), and others. The **interprocess communications layer** is the protocol that manages communications between clients and servers, including RPC, DCOM, named pipes, and mailslots. See Chapter 7.

The top layer is the **basic network services layer.** It supports network user applications by providing needed services such as file services, print services, network address management, and name services.

9.2 NETWORK COMPONENTS

You can add, remove, or configure the network components of Windows 2000 using the Network and Dial-up Connections applet in the Control Panel or by using the My Network Places icon on the desktop. When opening the Network and Dial-up Connections

FIGURE 9.2   Network and Dial-up Connections dialog box

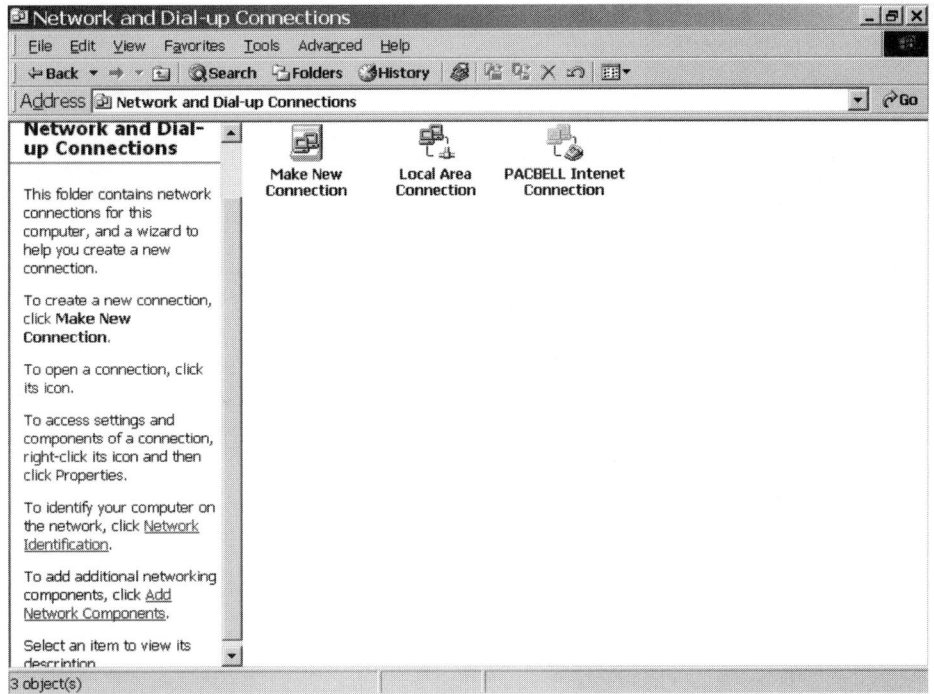

FIGURE 9.3   Network Identification tab

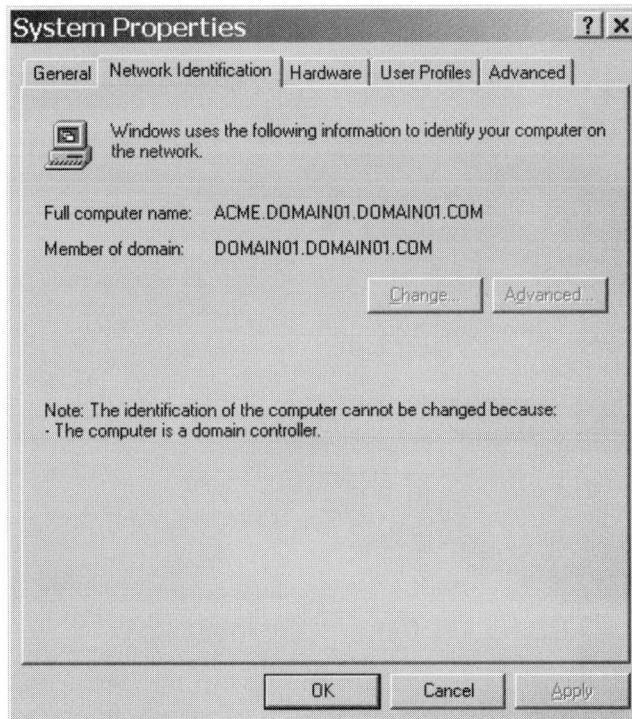

applet, you will find the Make New Connection and Local Area Connection icons. See Figure 9.2.

On the left side of the window are Network Identification and Add Network Components hyperlinks. The Network Identification hyperlink opens the Network Identification tab of the System Properties dialog box. From there, you can change the full computer/DNS name and the domain name/NetBIOS name. If the computer is a domain controller, however, you cannot change the computer and domain names. See Figure 9.3.

FIGURE 9.4   Windows Network Components

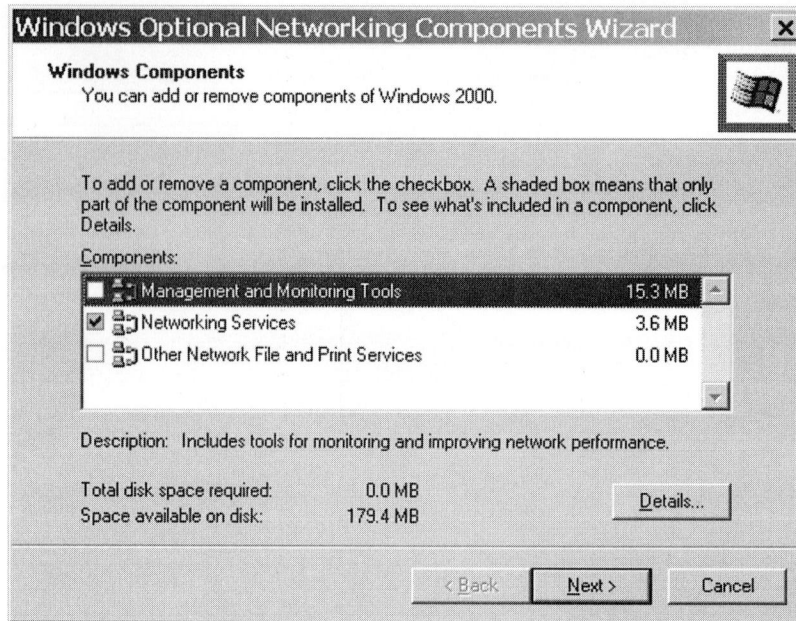

FIGURE 9.5   Management and Monitoring Tools dialog box

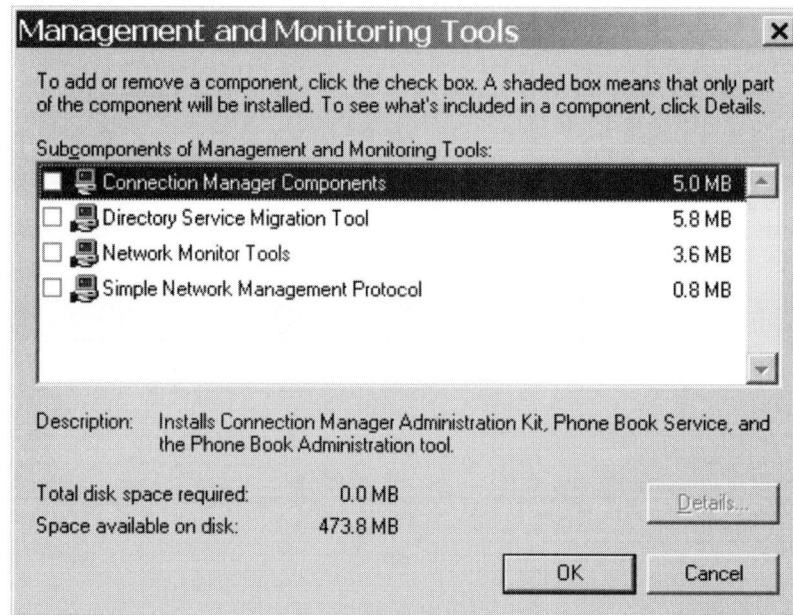

9.2.1   Add Network Components

The Add Network Components hyperlink opens the Windows Optional Networking Components Wizard dialog box. You can then add or remove components within the following three categories:

- Management and Monitoring Tools—includes the Connection Manager components, Directory Service Migration Tool, Network Monitor Tools, and Simple Network Management Protocol (SNMP) utilities.
- Networking Services—includes DNS, DHCP, and WINS services.
- Other Network File and Print Services—includes file and print services for Macintosh and print services for UNIX.

The individual utilities are listed by clicking on the Details button. See Figures 9.4, 9.5, and 9.6.

FIGURE 9.6 Networking
Services dialog box

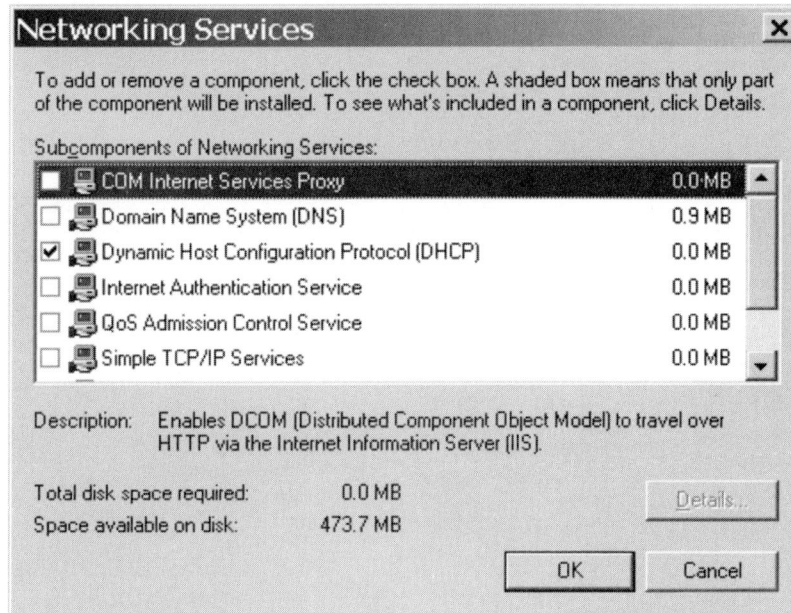

FIGURE 9.7 Network
connection types

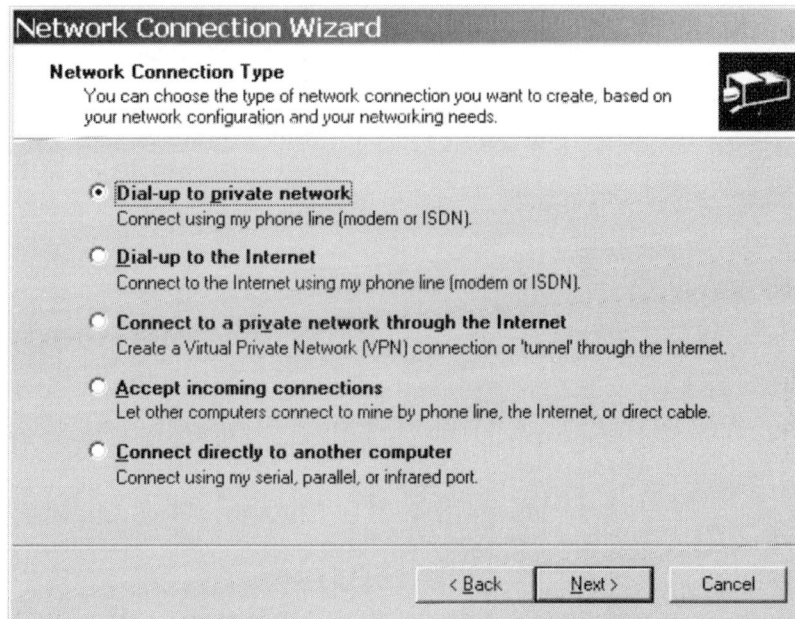

9.2.2 Make New Connection

The Make New Connection icon starts the Network Connection Wizard so that you can establish a network connection. The network connections include connecting to a private network or the Internet using a phone line and connecting to a private network using the Internet (creating a tunnel). In addition, you can set up to accept incoming connections using the phone line or the Internet and to connect directly to another computer using a serial port, parallel port, or infrared port. See Figure 9.7.

9.2.3 Local Area Connection

A double-click on the Local Area Connection icon will show the computer's status on the network. This includes how long the computer has been connected, the speed of the connection, the number of packets that have been sent and received, and an option to disconnect from the network. See Figure 9.8. For a single-link connection and for individual links

FIGURE 9.8   Local Area Connection Status dialog box

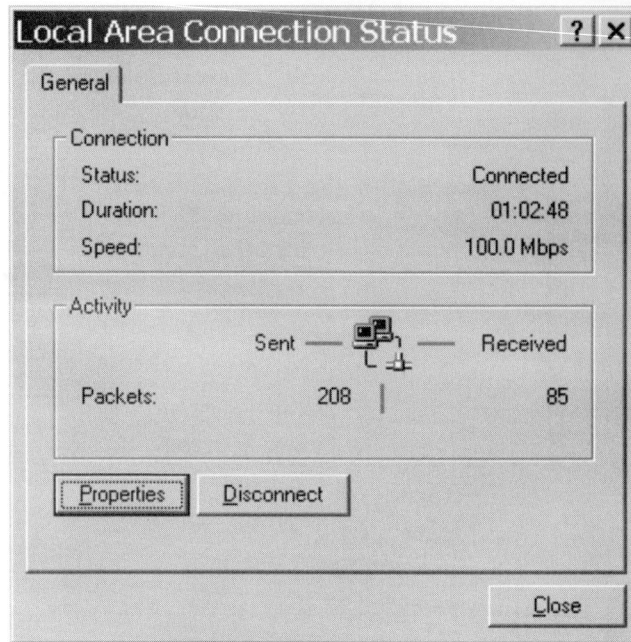

FIGURE 9.9   Local Area Connection Properties dialog box

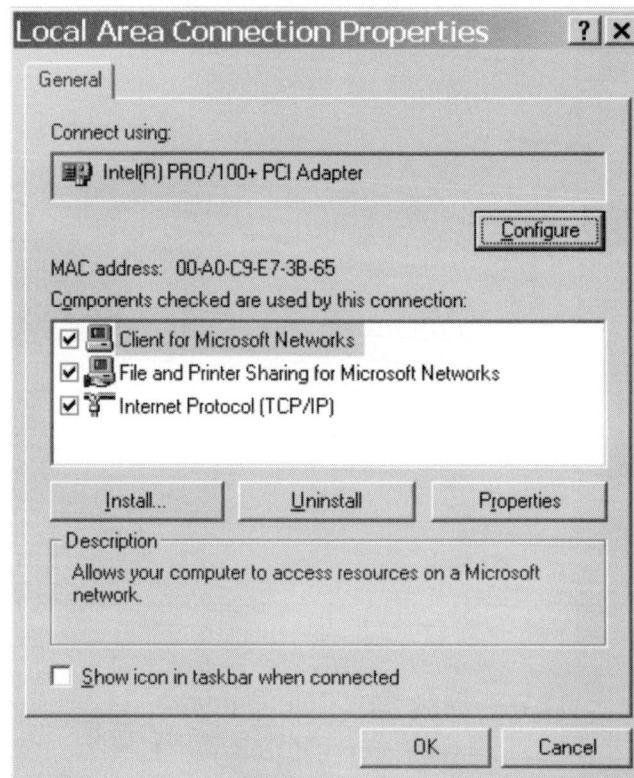

in a multilink connection, the speed is negotiated and fixed at the time the connection or link is established. For multilink connections, this speed is equal to the sum of the speeds of the individual links. These speeds will vary as links are added or deleted. A multilink connection is a combination of two or more physical communications links' bandwidth into a single logical link used to increase the remote access bandwidth and throughput.

The Properties button is used to install and configure the network card, client software, network services, and protocols. The default client is Client for Microsoft Networks with an option to add Gateway (and Client) Services for NetWare. The default service loaded is the File and Print Sharing for Microsoft Networks, which allows other users to connect to

FIGURE 9.10   Network Adapter Properties

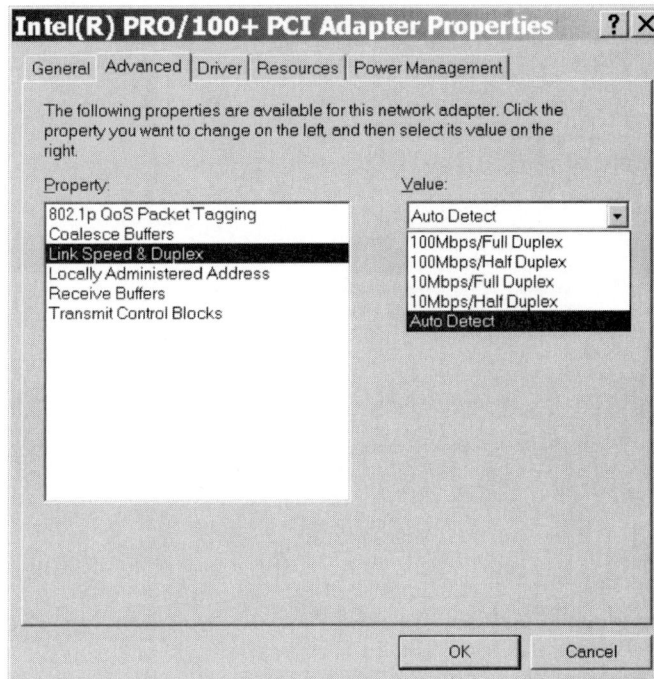

Intel(R) PRO/100+ PCI Adapter Properties

General | Advanced | Driver | Resources | Power Management

The following properties are available for this network adapter. Click the property you want to change on the left, and then select its value on the right.

Property:
- 802.1p QoS Packet Tagging
- Coalesce Buffers
- Link Speed & Duplex
- Locally Administered Address
- Receive Buffers
- Transmit Control Blocks

Value:
Auto Detect
- 100Mbps/Full Duplex
- 100Mbps/Half Duplex
- 10Mbps/Full Duplex
- 10Mbps/Half Duplex
- Auto Detect

OK   Cancel

TABLE 9.1   Server optimization

Options	Environment
Minimize memory used	Choose this if using the server mainly as a workstation and will have fewer than 10 connections to the server.
Balance	Choose this if using the server as a workstation and a server will support between 10 and 64 connections.
Maximize throughput for file sharing	Choose this if using the server as a domain controller or a file-and-print server. If you have at least 128 MB of RAM, this option will result in a good-size system cache.
Maximize throughput for network applications	Choose this if using the server as an application server such as a SQL server exchange, or IIS. This is typically best used by distributed applications that do their own memory caching.

the shared drives, directories, and printers on the computer. See Figure 9.9. Clicking the configure button will bring up the Properties dialog box for the network card adapter. The parameters located within the dialog box will vary depending on the network adapter manufacturer and model. See Figure 9.10. If the card and driver are ACPI compatible, then there may be a tab for Power Management.

If you click to highlight the File and Printer Sharing Microsoft Networks and click on the Properties button, you can optimize the server depending on its function. For example, if the server is to be a domain controller or a file/print server, you would select "Maximize throughput for file sharing." See Table 9.1 and Figure 9.11. If it is an application server such as a SQL server, choose "Maximize throughput for network applications." The default is "Maximize throughput for file sharing."

9.2.4   TCP/IP Networks

The default protocol is the Internet Protocol (TCP/IP). If you access the properties of the TCP/IP protocol, you can select Obtain an IP address automatically which is handed to the computer using a DHCP server or you can manually set the IP address, the subnet mask and the default gateway.

FIGURE 9.11 Server
optimization dialog box

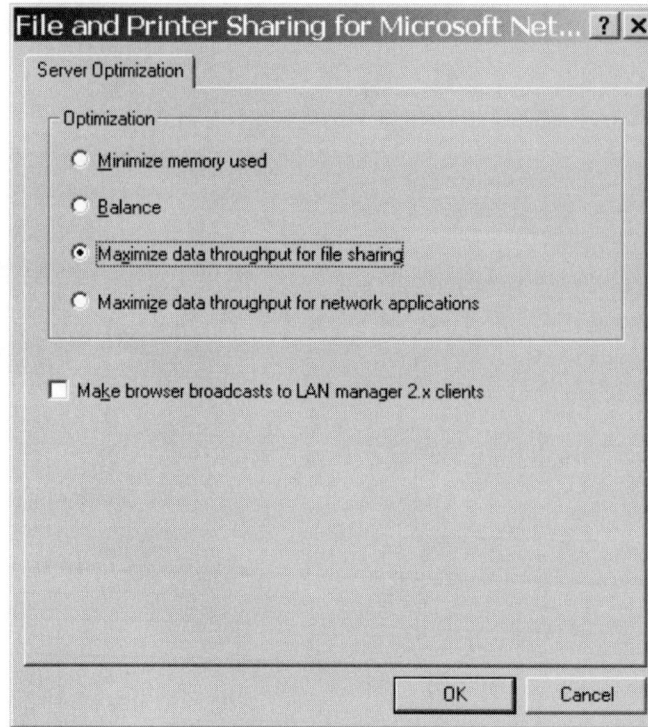

The IP address identifies the computer on the IP network while the subnet masks specify which bits of the IP address is used as the network address and which bits are used for the host address. Of course, to communicate on a TCP/IP network, you must use an IP address and a subnet mask. The default gateway is the gateway or router address that connects your local area network to another network. Therefore, if you need to connect to another subnet or network, use the default gateway. At the bottom of the Internet Protocol (TCP/IP) Properties dialog box, you can specify the DNS server address. The DNS server is for name resolution to resolve DNS names such as MICROSOFT.COM or MIT.EDU to an IP address.

If the computer is set to accept a DHCP server and one does not respond, the computer will use Automatic Private IP Addressing (also available in Windows 98), which generates an IP address in the form of 169.254.x.y (where x.y is the unique host address) and the subnet mask as 255.255.0.0. After the computer generates the address, it broadcasts this address until it can find a DHCP server. When you have an Automatic Private IP address, you can only communicate with computers on the same network/subnet that have an Automatic Private IP address. By default, the Automatic Private IP Addressing feature is enabled. To disable this feature, change the HKEY_LOCAL_MACHINE\SYSTEM\CurrentControlSet\Services\TCPIP\ Parameters\Interfaces\Adapter\ IPAutoconfigurationEnabled to zero in the registry.

By clicking on the Advanced button you can add additional gateways, DNS servers, and WINS servers. These additional gateway addresses, DNS servers, and WINS servers are only accessed when the first address is inaccessible, however. See Figure 9.12.

Like most TCP/IP implementations, the TCP/IP connection can be tested with the IPCONFIG, PING, ARP, and TRACERT utilities. The IPCONFIG will display addressing and settings of the TCP/IP protocol. The PING utility can be used to verify a TCP/IP connection while the TRACERT will show all router addresses as it connects to another computer. Lastly, the ARP protocol can be used to display the MAC address. For more information on the IPCONFIG, PING, and TRACERT commands, see Chapter 5.

The PATHPING command is a route tracing tool that combines features of the PING and TRACERT commands with additional information that neither of those tools provides. The PATHPING command sends packets to each router on the way to a final destination over a period of time, and then computes results based on the packets that are returned from each hop. Since the command shows the degree of packet loss at any given router or link, it is easy to determine which routers or links might be causing network problems. A number of switches are available, as shown in Table 9.2 and Figure 9.13.

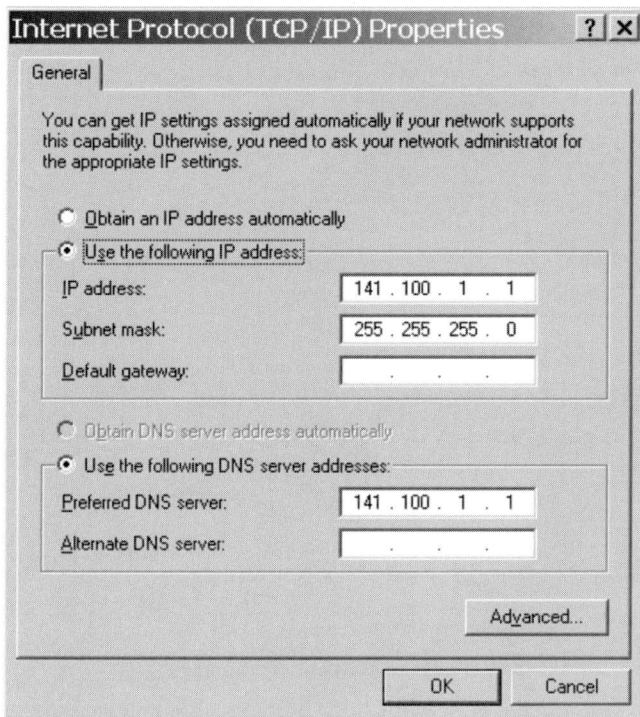

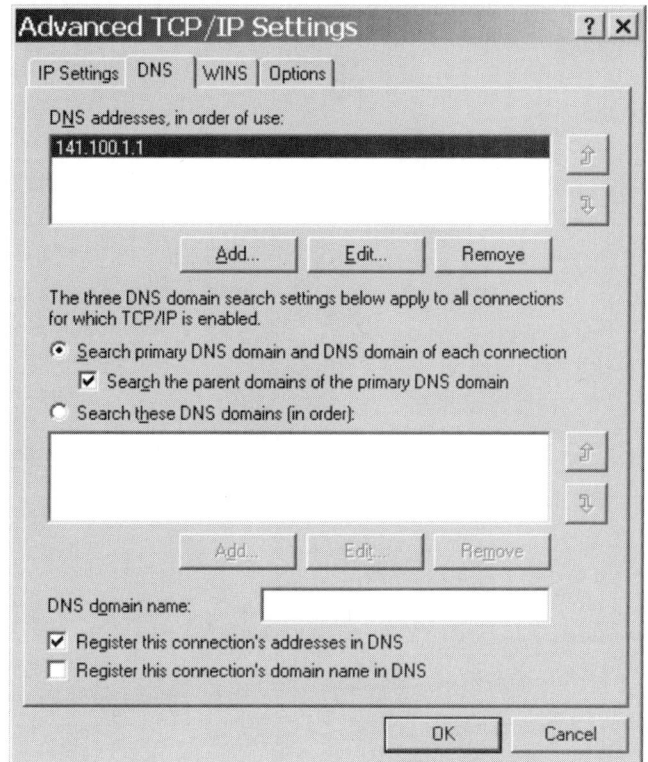

FIGURE 9.12   TCP/IP properties

TABLE 9.2   Options of the PATHPING command

Switch	Name Function
-n	Does not resolve addresses to host names.
-h *Maximum hops*	Maximum number of hops to search for target. The default number of hops is 30, and the default wait time before a timeout is 3 seconds.
-g *Host-list*	Loose source route along host list.
-p *Period*	Number of milliseconds to wait between pings. The default period is 250 milliseconds
-q *Num_queries*	Number of queries per hop. The default number of queries to each router along the path is 100.
-w *Timeout*	Waits this many milliseconds for each reply. Default is 100.
-T	Attaches a layer-2 priority tag (e.g., for IEEE 802.1p) to the packets and sends it to each of the network devices in the path. This helps in identifying the network devices that do not have layer-2 priority configured properly. The -T switch is used to test for quality of service (QoS) connectivity.
-R	Checks to determine whether each router in the path supports the resource reservation protocol (RSVP), which allows the host computer to reserve a certain amount of bandwidth for a data stream. The -R switch is used to test for quality of service (QoS) connectivity.

The **NetDiag** command is a command line, diagnostic tool that helps isolate networking and connectivity problems by performing a series of tests to determine the state of the network client. NetDiag diagnoses network problems by checking all aspects of a host computer's network configuration and connections. In addition to troubleshooting TCP/IP issues, it examines a host computer's internetwork packet exchange (IPX) and NetWare

269

FIGURE 9.13   PATHPING command

```
c:\>pathping -n acme

Tracing route to acme [7.54.1.196]
over a maximum of 30 hops:
  0  172.16.87.35
  1  172.16.87.218
  2  192.68.52.1
  3  192.68.80.1
  4  7.54.247.14
  5  7.54.1.196


Computing statistics for 125 seconds...
                Source to Here     This Node/Link
Hop   RTT     Lost/Sent = Pct    Lost/Sent = Pct   Address
  0                                                 172.16.87.35
                                    0/ 100 =  0%    |
  1   41ms      0/ 100 =  0%       0/ 100 =  0%    172.16.87.218
                                   13/ 100 = 13%    |
  2   22ms     16/ 100 = 16%       3/ 100 =  3%    192.68.52.1
                                    0/ 100 =  0%    |
  3   24ms     13/ 100 = 13%       0/ 100 =  0%    192.68.80.1
                                    0/ 100 =  0%    |
  4   21ms     14/ 100 = 14%       1/ 100 =  1%    7.54.247.14
                                    0/ 100 =  0%    |
  5   24ms     13/ 100 = 13%       0/ 100 =  0%    7.54.1.196

Trace complete.
```

configurations. NetDiag is included in the Support Tools folder on the Windows 2000 Server compact disk. For information about installing and using the Windows 2000 Support Tools and Support Tools Help, see the file Sreadme.doc in the Support Tools folder on the Windows 2000 Server CD.

9.2.5   Other Network Protocols

Two common network protocols are NWLink (IPX/SPX) and NetBEUI, both of which are much simpler to configure than the TCP/IP network protocol. For the NetBEUI protocol, nothing has to be done because it is nonroutable.

For the NWLink, specify the internal network number (eight-digit hexadecimal number) that identifies the server. In addition, specify the frame type and external number that identifies the subnet. If you select the auto frame type detection, it will automatically detect the frame type used by the network adapter to which it is bound. If NWLink detects no network traffic or if multiple frame types are detected in addition to the 802.2 frame type, NWLink sets the frame type to 802.2. See Figure 9.14.

To help troubleshoot the IPX protocol in Windows 2000, use the **IPXROUTE** command to determine computer settings and perform diagnostic tests to resolve communication problems. See Table 9.3 and Figure 9.15.

9.2.6   Bindings

There are several ways to enhance network performance. First, when choosing which protocols to install, only install and enable the network protocols that you need. This will also reduce network traffic. Second, you can open the Network and Dial-up Connections folder,

FIGURE 9.14   NWLink IPX/SPX
NetBIOS dialog box

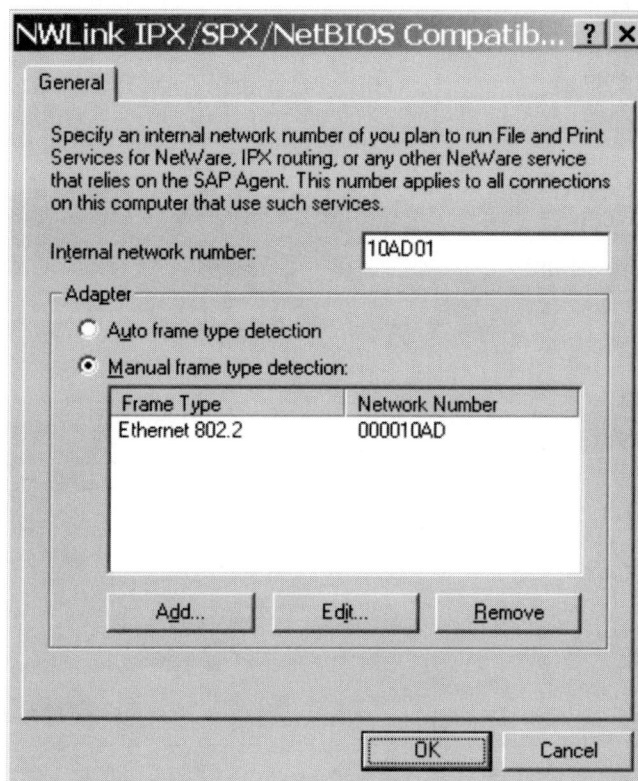

**FIGURE 9.14   NWLink IPX/SPX
NetBIOS dialog box**

NWLink IPX/SPX/NetBIOS Compatib... ? X

General

Specify an internal network number of you plan to run File and Print
Services for NetWare, IPX routing, or any other NetWare service
that relies on the SAP Agent. This number applies to all connections
on this computer that use such services.

Internal network number: 10AD01

Adapter
○ Auto frame type detection
● Manual frame type detection:

Frame Type	Network Number
Ethernet 802.2	000010AD

Add...   Edit...   Remove

OK   Cancel

**TABLE 9.3   IPXROUTE
command**

Command	Description
IPXROUTE CONFIG	Displays the current IPX status including the network number, media access control (MAC) address, interface name, and frame type. See Figure 9.14.
The IPXROUTE RIPOUT *network_number*	Uses RIP to determine if there is connectivity to a specific network.

**FIGURE 9.15   IPXROUTE
CONFIG command**

```
NWLink IPX Routing and Source Routing Control Program v2.00

Num   Name                      Network    Node          Frame
================================================================
1.    IpxLoopbackAdapter        1234cdef   000000000002  [802.2]
2.    Local Area Connection     00000000   00c0f056bb98  [802.2]
3.    Local Area Connection 2   00000000   00a0c9e73b65  [802.2]
4.    NDISWANIPX                00000000   7ab920524153  [EthII]-

Legend
======
- down wan line
```

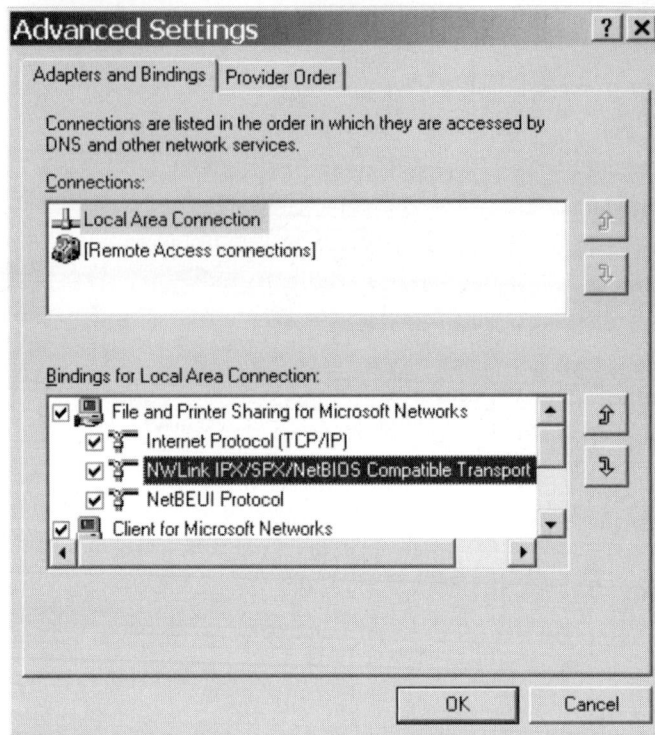

open the Advanced menu and then select the Advanced Settings option. This process enables you to use multiple protocols and to select the order of the protocols that are used. For example, if you are connecting to both the Microsoft Windows network and NetWare network using both TCP/IP and IPX protocols, you can set the TCP/IP protocol as the primary protocol for the Microsoft Windows network and IPX for the NetWare protocol by changing the binding order. Simply click on the protocol and use the arrows on the right side to adjust the position of the protocol. See Figure 9.16.

9.2.7 My Network Places

If the **My Network Places** is on the desktop, bring up the shortcut menu by using the secondary mouse button on the icon and clicking on the Properties option to bring up the Network and Dial-up Connections dialog box. If you double-click the My Network Places icon, you will find Add Network Place, Computers Near Me, and Entire Network icons. The Add Network Place helps you add a link to the network where you can store your documents such as a shared folder, a web folder, or an FTP site. The Computers Near Me icon will show the computers in your workgroup or network domain. The Entire Network shows all domains and workgroups.

9.2.8 ARP Utility

The address resolution protocol (ARP) is the TCP/IP protocol that provides IP address–to–MAC address resolution. To help manage the ARP protocol, you can use the ARP utility. See Figure 9.17.

   To minimize the ARP broadcasts on the network, Windows 2000 maintains a cache of hardware-to-software addresses or MAC addresses to TCP/IP addresses for future use. This cache contains the dynamic ARP cache entries and static ARP cache entries. Dynamic ARP cache entries are added and deleted automatically during normal use of TCP/IP sessions with remote computers. Dynamic entries age and expire from the cache if not reused within 2 minutes. If a dynamic entry is reused within 2 minutes, it may remain in the cache and age up to a maximum cache life of 10 minutes before being removed or requiring cache renewal by using the ARP broadcast process. Static entries are added manually by using

FIGURE 9.17 ARP utility

```
ARP -a [inet_addr] [-N if_addr]
-a              Displays current ARP entries by interrogating the current protocol data.
                If inet_addr is specified, the IP and Physical addresses for only the
                specified computer are displayed. If more than one network interface uses
                ARP, entries for each ARP table are displayed.
-g              Same as -a.
inet_addr       Specifies an internet address.
-N if_addr      Displays the ARP entries for the network interface specified by if_addr.
-d              Deletes the host specified by inet_addr.
-s              Adds the host and associates the Internet address inet_addr with the
                Physical address eth_addr. The Physical address is given as 6 hexadecimal
                bytes separated by hyphens. The entry is permanent.
eth_addr        Specifies a physical address.
if_addr         If present, this specifies the Internet address of the interface whose
                address translation table should be modified. If not present, the first
                applicable interface will be used.
Example:
  > arp -s 157.55.85.212   00-aa-00-62-c6-09  .... Adds a static entry.
  > arp -a                                    .... Displays the arp table.
```

FIGURE 9.18 Display the contents of the ARP cache

```
C:\>ARP -A

    Interface: 141.3.12.4 on Interface 0x1
      Internet Address        Physical Address       Type
        141.3.12.4              00-a0-04-67-7d-a4    dynamic
```

the ARP command with the –s option and will remain in the ARP cache until the computer is restarted.

To view the address resolution protocol cache, type ARP –A at the command prompt. See Figure 9.18. In this example, the cache entry indicates that the remote host computer at 141.3.12.4 resolved to a MAC address of 00-a0-04-67-7d-a4 assigned in the remote computer's network adapter hardware. Remember that you can use IPCONFIG/ALL or WINIPCFG on the machine they want, to identify the MAC address. If you recently used the ping command to a host computer, you can then access the ARP cache to view the MAC address.

To establish a static address, use the –S option. For example, to assign the MAC address of 00-a0-04-67-7d-a4 to the TCP/IP address of 141.3.12.4, type in the following command:

```
arp -s 157.55.85.212 00-aa-00-62-c6-09
```

SUMMARY

1. Windows 2000 network architecture is similar to the OSI model, in that the Windows 2000 network components are arranged in layers and each has specific tasks to perform within its assigned layer.
2. The network driver interface specification (NDIS) layer provides a communication path between two network devices.
3. The miniport drivers control the network interface adapter and connect the hardware device to the protocol stack.
4. When a protocol is bound to a network component (binding), the protocol is linked to a protocol stack.
5. The default client is Client for Microsoft Networks with an option to add Gateway (and Client) Services for NetWare.

6. The default service loaded is the File and Print Sharing for Microsoft Networks, which allows other users to connect to the shared drives, directories, and printers on the computer.

7. Like most TCP/IP implementations, the TCP/IP connection in Windows 2000 can be tested with the IPCONFIG, PING, ARP, and TRACERT utilities.

QUESTIONS

1. To add or configure the network components, use _____ .
   a. Network and Dial-up Connections
   b. Device Manager
   c. Add/Remove Hardware
   d. System
   e. Network Neighborhood

2. If you have a domain controller, you cannot change the _____ .
   a. IP address
   b. network card driver
   c. computer name
   d. subnet mask

3. Doing which one of the following will show the connection speed of an interface?
   a. right-click My Network Places
   b. double-click My Network Places
   c. right-click the Local Area Connection
   d. double-click the Local Area Connection

4. Which is the default client for Windows 2000?
   a. Client Services for Microsoft
   b. Client Services for NetWare
   c. File and Print Sharing for Microsoft Networks
   d. File and Print Sharing for NetWare

5. Which services are loaded by default in Windows 2000? (Choose all that apply)
   a. Client for Microsoft Networks
   b. Client Services for NetWare
   c. File and Print Sharing for Microsoft Networks
   d. File and Print Sharing for NetWare

6. To specify "Maximize data throughput for file sharing" or "Maximize data throughput for network applications," you would click to highlight _____ .
   a. TCP/IP and click on the Properties button
   b. File and Printer Sharing Networks and click on the Properties button
   c. My Network Place and click on the Properties button
   d. File and Printer Sharing for Microsoft Networks

7. Which one of the following would you choose if your server is a domain controller?
   a. Minimize memory used
   b. Balance
   c. Maximize data throughput for file sharing
   d. Maximize data throughout for network applications

8. What is the default protocol or protocols loaded in Windows 2000? (Select all that apply)
   a. TCP/IP
   b. IPX/NWLink
   c. NetBEUI
   d. DLC
   e. AppleTalk

9. Automatic Private IP Addressing can provide an IP address in which range?
   a. 169.254.0.0 to 169.254.255.255
   b. 150.101.0.0 to 150.101.255.255
   c. 130.0.0.0 to 130.255.255.255
   d. 127.0.0.0 to 127.255.255.255

10. If you load NWLink, you need to specify what? (Select two answers)
    a. gateway
    b. subnet mask
    c. internal network number
    d. frame type

11. How do the binding orders of client machines affect system performance?
    a. Client binding order has no effect on system performance.
    b. You should move slower protocols to the top of the binding order on client machines to improve performance.
    c. You should move frequently used protocols to the top of the binding order on client machines to improve performance.
    d. Client binding order has no effect on system performance, but it is usually possible to improve system performance by reordering server binding orders.

12. In Windows 2000, to see all domains, you would use _____ .
    a. Add Network Place
    b. Computers Near Me
    c. Entire Network
    d. Network Neighborhood

13. The Add Network Place _____ .
    a. helps you add a link to the network where you can store your documents such as a shared folder, a web folder, or an FTP site
    b. creates a network share
    c. creates a web share
    d. sets up a private network connection

14. You want to examine the IP address–to–MAC address resolution of outgoing packets on your Windows 2000 Server computer. Which of the following should you use?
    a. TRACERT
    b. ARP
    c. PING
    d. NBTSTAT

HANDS-ON EXERCISES

EXERCISE 1: USING THE TCP/IP NETWORK

1. Start a DOS prompt by clicking on the Start button, selecting the Programs option, selecting the Accessories option and then selecting the Command prompt option.
2. Execute the IPCONFIG command and record the following settings:
   IP address
   subnet mask
   default gateway
3. Use the IPCONFIG /ALL command and record the following settings:
   MAC address
   WINS server (if any)
   DNS server (if any)
   DHCP enabled?
4. Ping the loopback address of 127.0.0.1.
5. Ping your IP address.
6. Ping your instructor's computer.
7. Ping your partners's computer.
8. If you have a router on your network, ping your gateway or local router connection.
9. If your network has a DHCP network, right-click My Network Places and select the Properties option.
10. Click on the Internet Protocol (TCP/IP) and then click on the Properties button.
11. In the Internet Protocol (TCP/IP) Properties dialog box, select "Obtain an IP address automatically." Click on the OK button.
12. Execute the IPCONFIG command at the Command prompt and record the following settings:
    IP address
    subnet mask
    default gateway
    WINS server (if any)
    DNS server (if any)
    DHCP enabled?
13. Ping the loopback address of 127.0.0.1.
14. Ping your IP address.

15. Ping your instructor's computer.
16. Ping your partners's computer.
17. If you have a router on your network, ping your gateway or local router connection.
18. At the Command prompt, execute the IPCONFIG /RELEASE to remove your values specified by a DHCP server.
19. At the Command prompt, execute the IPCONFIG command and compare the recorded values from number 4.
20. At the Command prompt, execute the IPCONFIG /RENEW command.
21. At the Command prompt, execute the IPCONFIG command and compare the recorded values from number 4.
22. Have your instructor stop the DHCP server (or DCHP service).
23. At the Command prompt, execute the IPCONFIG /RELEASE command followed by the IPCONFIG /RENEW. (There will be a pause while Windows 2000 attempts to locate a DHCP server.)
24. At the Command prompt, execute the IPCONFIG prompt. Record the address and try to determine from where this address originated.
25. Try to ping your instructor's computer and the local gateway. You should not be able to ping the IP addresses (and subnet mask) to indicate that the two computers are on two different networks.
26. After your partner has acquired an Automatic Private IP address, try to ping each other. Since these addresses are on the same network (physically and logically), it should work.
27. Go back into the TCP/IP dialog box and enter the static addresses that you recorded in steps 3 and 4.
28. Test your network by pinging your partner, instructor's computer, and gateway.
29. Disconnect the network cable from the back of the computer.
30. Look at the taskbar in the notification area (near the clock) and find the red X.
31. From the Command prompt, type in IPCONFIG.
32. Go into the Network and Dial-up Connection dialog box by right-clicking My Network Places and selecting Properties. Note the red X.
33. Connect the cable back into the network card. Note that the red X in both places goes away. In addition, note that the icon disappears altogether from the notification area.
34. Right-click Local Area Connection and select Properties.
35. Select the "Show icon in taskbar when connected." Click on the OK button. Close the Local Area Connection dialog box.
36. Go to the notification area and note the new icon representing the network connection.
37. Without clicking on the new icon, move the mouse pointer onto the icon. Without moving the mouse, note the information given.
38. Double-click the icon to bring up the Local Area Connection dialog box.
39. From the Command prompt, execute the ARP–A
40. Close all windows.

EXERCISE 2: NAVIGATING THE NETWORK

1. Right-click the desktop and create a folder called Shared.
2. Right-click the Shared directory and select the Sharing option.
3. Select the "Share this folder option" and click the OK button. A hand should appear underneath the Shared directory.
4. Open the Shared directory and create a text file with your name. To create a text file, right-click the Open folder, select the New option and then select the Text Document option.
5. Close all windows.
6. Double-click My Network Places to open it.
7. Double-click Computers Near Me.
8. Double-click your computer to see all shared resources.
9. Double-click the Shared directory to see the text file that you created.
10. Click the Up button twice on the toolbar to go back to the Computers Near Me window.
11. Double-click your partner's computer. Double-click the Shared directory to see your partner's text file.

12. Click the Up button three times on the toolbar to go back to the Computers Near Me dialog box.
13. Double-click the Entire Network icon, then click on the entire contents hyperlink on the left side of the window.
14. Double-click the Microsoft Windows Network icon. You should see your workgroup and any domains that might be out there.
15. Double-click your workgroup to see all computers assigned to the workgroup.
16. Use the Up arrow button on the toolbar to go back to the My Network Places window.
17. Double-click the Add Network Place icon.
18. In the "Type the location of the Network Place" text box, type the following: *partnercomputername*\SHARED
19. Click on the Next button, then click on the Finish button.
20. Close the Shared window.
21. Note the new icon in the My Network Places. Double-click on it.
22. Close all windows.

EXERCISE 3: USING IPX PROTOCOL

1. Bring up the Network and Dial-up Connection window again.
2. Double-click on the Local Area Connection and click on the Properties button.
3. Click on the Install button. In the Select Network Component Type dialog box, click the Protocol option and click on the Add button. In the Select Network Protocol dialog box, click "NWLink IPX/SPX/NetBIOS Compatible Transport Protocol," and then click OK.
4. In the Local Area Connection Properties dialog box, click the "NWLinkIPX/SPX/ NetBIOS Compatible Transport Protocol" and then click Properties. Note which type of frame detection is selected by default.
5. Close the Local Area Connection Properties dialog box.
6. In the Network and Dial-up Connections window, open the Advanced menu and select the Advanced Settings option.
7. In the Advanced Settings dialog box, under Client for Microsoft Networks, unbind TCP/IP by clearing the Internet Protocol (TCP/IP) check box. Click the OK button and close the Network and Dial-up Connections window.
8. Restart the computer.
9. Open a Command prompt window.
10. At the prompt, type in IPXROUTE CONFIG and press Enter. Note the frame type, network address, and MAC address.
11. Close the Command prompt window.
12. Try to access the share that you set up in Exercise 1.
13. Open the Network and Dial-up Connections window.
14. Double-click the Local Area Connection and click on the Properties button.
15. In the Local Area Connection Properties dialog box, click "NWLinkIPX/SPX/ NetBIOS Compatible Transport Protocol" and click the Uninstall button. Click on the Yes button to uninstall NWLink.
16. When it attempts to restart your computer, click the No button. Close all windows and reboot the computer.

Unit 3

Windows 2000 Server Basics

10

Advanced TCP/IP Topics

INTRODUCTION

Before users can start navigating your network, you will need some type of name resolution. In Chapter 5 we defined both a DNS and a WINS server which are used to resolve names and return IP addresses. This chapter deals with installing and configuring these types of servers. In addition, we will learn to install and configure DHCP servers so that you can automatically assign IP addresses and other IP parameters to your clients.

OBJECTIVES

1. Given a name, specify which method of name resolution would be used to resolve the name to an IP address.
2. Implement a DNS and NetBIOS naming scheme for all computers on a given network.
3. Given a fully qualified domain name, identify each component of the name.
4. Explain how a DNS zone relates to a domain and subdomain.
5. List and describe each type of zone.
6. Configure a root name server.
7. Install and configure a DNS server in Windows 2000.
8. Configure a DNS client.
9. Configure zones for dynamic updates.
10. Test the DNS Server service.
11. Implement a delegated zone for DNS.
12. Manually create DNS resource records.
13. Manage and monitor DNS.
14. Create a zone or subzone in a DNS name space using Windows 2000.
15. Install and configure a WINS server.
16. Install and configure a WINS proxy server.
17. Given a problem in name resolution, troubleshoot and fix the problem.
18. Install and configure a DHCP server in Windows 2000.
19. Configure a scope and common DHCP parameters on a Windows 2000 DHCP Server.
20. Install and configure a DHCP relay agent.
21. Configure a superscope and multicast scopes on a Windows 2000 DHCP Server.
22. Configure DHCP for DNS integration.
23. Authorize a DHCP server in active directory.
24. Explain how SNMP agents and managers communicate with each other.
25. Install and configure Windows 2000 as a SNMP agent or manager.

10.1 INTRODUCTION TO DOMAIN NAME SYSTEM NAME RESOLUTION

Two services can be used to find and provide IP addresses, the WINS service and the domain name system (DNS) service. If a computer knows the destination's NetBIOS name, it can send the NetBIOS name to the WINS service and the WINS service will send back the corresponding IP address of the destination. Since only Microsoft OS uses NetBIOS names, WINS only works with Microsoft products. If a computer knows the fully qualified domain names (FQDNs), it can send that to a DNS service and the DNS service will return the corresponding IP address. Most modern OSs and programs can use FQDNs.

Domain name system (DNS) is a hierarchical client-server-based distributed database management system that translates Internet domain names such as MICROSOFT.COM to an IP address. As mentioned in previous chapters, it is used because domain names are easier to remember than IP addresses. The DNS clients are called **resolvers** and the DNS servers are called **name servers.**

DNS can be thought of as its own little network. If one DNS server cannot translate a particular domain name, then another one tries, and so on, until the correct IP address is returned. DNS is most commonly associated with the Internet, but private networks can also use DNS to resolve computer names and to locate computers within their local networks without being connected to the Internet.

The first TCP/IP networks used HOSTS files to translate from domain names (such as MICROSOFT.COM, ACME.COM, or MIT.EDU) to IP addresses. As some networks grew so quickly, manually updating and distributing the HOSTS file was not as effective. For the Internet (the largest network that uses DNS), no single organization is responsible for keeping the DNS system updated. Instead, a distributed database exists on many different name

TABLE 10.1 RFC documents that describe DNS

RFC	Title
974	Mail Routing and the Domain System
1034	Domain Names—Concepts and Facilities
1035	Domain Names—Implementation and Specification
1123	Requirements for Internet Hosts—Application and Support
1886	DNS Extensions to Support IP Version 6
1912	Common DNS Operational and Configuration Errors
1995	Incremental Zone Transfer in DNS
1996	A Mechanism for Prompt DNS Notification of Zone Changes
2136	Dynamic Updates in the Domain Name System (DNS UPDATE)
2181	Clarifications to the DNS Specification
2182	Selection and Operation of Secondary DNS Servers
2219	Use of DNS Aliases for Network Services
2308	Negative Caching of DNS Queries (DNS NCACHE)

servers around the world, with no one server storing all the information; thus, DNS allows for almost unlimited growth.

The most popular implementation of the DNS protocol is the Berkeley Internet Name Domain (BIND), which was developed for the University of California Berkeley's BSD UNIX operating system. The primary specifications for DNS are defined in Requests for Comments (RFC) 1034 and 1035. DNS uses either UDP port 53 or TCP port 53 as the underlying protocol. See Table 10.1.

10.1.1 Domain Name Space

The **DNS name space** describes the hierarchical structure of the DNS database as an inverted logical tree structure. Each node on the tree is a partition of the name space called a **domain.** Domains can be further partitioned at node points within the domain into subdomains. Names of the domains and subdomains can be up to 63 characters long. See Figure 10.1.

The top of the tree is known as the **root domain.** It is sometimes shown as a period (.) or as empty quotation marks (" "), indicating a null value. Immediately below the root domain are the **top-level domains.** The top-level domains indicate a country, region, or type of organization. Three-letter codes indicate the type of organization. See Table 10.2. For example, COM indicates commercial (business) and EDU stands for educational institution.

Two-letter codes indicate countries, which follow the International Standard 3166. For example, CA stands for Canada, AU for Australia, FR for France, and UK for United Kingdom. For a list of two-letter codes, go to gopher://gopher.scit.wlv.ac.uk/00/pub/info-magic/standards/iso/3166/codes.

The **second-level domain names** are variable-length names that are registered to an individual or organization for use on the Internet. These names are almost always based on the appropriate top-level domain, depending on the type of organization or geographic location where a name is used.

EXAMPLE 1

Domain Name	Second-Level Domain Name
MICROSOFT.COM	Microsoft Corporation
CISCO.COM	Cisco Corporation
MTI.EDU	MTI University
ED.GOV	United States Department of Education
ARMY.MIL	United States Army
W3.ORG	World Wide Web Consortium
NATO.INT	North Atlantic Treaty Organization
PM.GOV.AU	Prime Minister of Australia

FIGURE 10.1   DNS name space

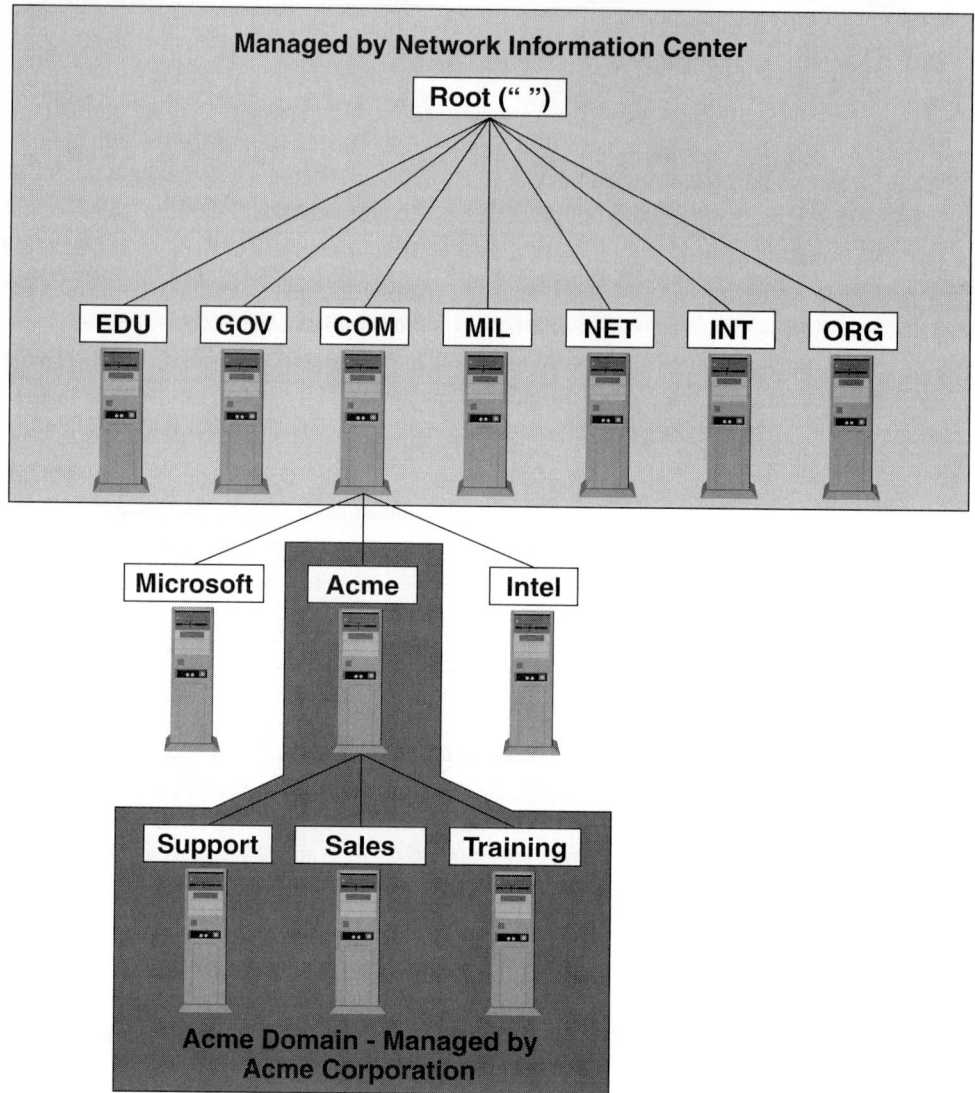

TABLE 10.2   Three-letter codes indicating the type of organization

Code	Meaning
com	Commercial
edu	Educational
gov	Government
int	International organization
mil	Military
net	Network related
org	Miscellaneous organization

Second-level domain names must be registered by the authorized party. For example, for years, Network Solutions Inc. (http://www.networksolutions.com/) ran a government-sanctioned monopoly on registrations for .COM, .NET, and .ORG domain names. But as the U.S. government handed the control of the Internet to an international body, several companies including Network Solutions Inc. now handle the registration of these three-letter codes. Since most of the common top-level domains names are already taken, some countries such as Tonga (TO) are selling their domain names. Therefore, some commercial and user sites may be using one of these two-letter codes. Another worth mentioning is the domain name ".TV" given by the Island nation, Tuvalu. The .TV web address symbolizes rich media, streaming video and broadband content. To register .TV domain names as well as .COM, .NET and .ORG domain names, you could use www.register.com.

Subdomain names are additional names that an organization can create which are derived from the registered second-level domain name. The subdomain allows an organization to divide a domain into a department or geographical location, allowing the partitions of the domain name space to be more manageable. A subdomain must have a contiguous domain name space, which means that the domain name of a zone (child domain) is the name of that zone added to the name of the domain or parent domain.

A **host name** is a name assigned to a specific computer within a domain or subdomain by an administrator to identify the TCP/IP host. Multiple host names can be associated with the same IP address, although only one host name can be assigned to a computer using the Windows Control Panel/Network applet. If the DNS is seen as a tree, it represents the leaf or object of the tree. Much like a subdomain, it is the left-most label of the DNS domain name. The host name can then be used in place of an IP address such as PING or other TCP/IP utilities. Total length of an FQDN cannot exceed 255 characters. The host name does not have to be the same as the NetBIOS (computer) name. By default, TCP/IP setup uses the NetBIOS name for the host name, replacing illegal characters, such as the underscore, with a hyphen (-).

A **fully qualified domain name (FQDN)** describes the exact position of a host (computer) within the domain hierarchy and it is considered to be complete. When used in a DNS domain name, it is stated by a trailing period (.) to designate that the name of the host is located off the root or highest level of the domain hierarchy.

EXAMPLE 2

SERVER1.SALES.ACME.COM

COM indicates a commercial business
ACME is the domain name
SALES is the subdomain name
SERVER1 is the name of the server located within the SALES subdomain

---

REGISTERING YOUR DOMAIN

The quickest and easiest way to register your domain name is through your ISP. If that method is not possible, use the following steps:

1. Find the company to register the domain name. For domain names with the .COM, .ORG, and .NET domain suffixes, go to WWW.NETWORKSOLUTIONS.COM. or www.REGISTER.com.
2. Pick a name and ensure its availability. At the website you can search for a name without any obligation. Try to select a name that indicates your company or business, although that task may be difficult because most easily identifiable domain names are already taken.
3. Arrange your domain name, either through your ISP or by setting up your own DNS server.
4. Complete the registration form at the company where you are registering the domain name. Follow the company's instructions. During the entire registration period, keep checking for small errors and omissions, as these can delay the processing of the registration form. The registration form will usually need three contacts—admin, technical, and billing. You will usually receive an acknowledgment, which will contain a tracking number. Be sure to keep that tracking number for future inquiries.
5. You will receive a message telling you the date your domain will be activated in the root name servers. It may take a few hours to a few days after the activation date because of the time required for the distributed database to be replicated throughout the Internet with the added address. Note that there will be a fee to activate the domain name.

When you create a domain name space, consider the following domain guidelines and standard naming conventions:

- To minimize the level of administrative tasks, limit the number of domain levels.
- If possible, each subdomain should have a unique name throughout the entire domain to ensure that the subdomain name is unique.
- Use simple, yet meaningful domain names that are easy to remember and navigate.
- Use standard DNS characters including A–Z, a–z, 0–9, hyphen, and unicode characters, which include the additional characters needed for foreign languages such as French, German, and Spanish.
- Keep names of the domain and subdomains to 63 characters long.
- Keep total length of an FQDN at or below 255 characters.

10.1.2 DNS Zones

A **DNS zone** is a portion of the DNS name space whose database records exist and are managed in a particular DNS database file. Each zone is based on a specific domain node, or root domain, which is the authority source for that node. Zone files do not necessarily contain the complete DNS name since subdomains may be its own zone. If subdomains are added below the domain, they can be part of the same zone or belong to another zone.

The computer that maintains the master list for a zone is the primary name server for that zone and is considered the authority there. A DNS server might be configured to manage one or more zones.

EXAMPLE 3

Question: Why would you take a domain and partition it into several subdomains (zones), so that they can be controlled by separate DNS name servers?

Answer: By breaking up domains across multiple zone files, the zone files can be stored, distributed, and replicated to other DNS servers. This enables the system to do the following:

- To delegate management of a domain/subdomain as it relates to its location or department within the organization
- To divide one large zone into smaller zones so that the traffic load can be distributed among multiple servers for better DNS name resolution performance and for fault tolerance

When choosing how to structure zones, use a DNS structure that reflects the structure of your organization.

In Windows 2000, there are three types of zones to configure. They are the standard primary zone, the standard secondary zone, and the active directory integrated zone. See Table 10.3.

The **primary name server** stores and maintains the zone file locally. Changes to a zone, such as adding domains or hosts, are done by changing files at the primary name server. A **secondary name server** gets the data from the zone of another name server—either a primary name server or another secondary name server—for that zone across the network. The process of obtaining this zone information across the network is referred to as a **zone**

TABLE 10.3 Zone types

Zone Type	Description
Standard primary	The master copy of a new zone.
Standard secondary	A replica of an existing zone. Standard secondary zones are read only.
Active directory integrated	A zone that is stored in active directory (Windows 2000 only). Updates of the zone are performed during active directory replication.

transfer. Zone transfers occur over TCP port 53. The **active directory integrated zone** is only found in Windows 2000 and has the zone defined using the active directory, not the zone files.

The source of the zone information for a secondary name server is referred to as a master name server. A master name server can be either a primary or secondary name server for the requested zone.

Question: Why do you want to have a secondary name server?

Answer: There are three reasons to have a secondary name server. The first reason is for fault tolerance. You should have at least two DNS name servers serving each zone. In each client's configuration, both name servers would be listed. If the first server listed cannot be contacted, the client will then contact the second name server. The second reason would be to divide the load between different name servers so that the performance for name resolution will be increased. Lastly, the DNS servers could be used to service computers that are located in remote locations so that they would not have to use a slow WAN link.

10.1.3 Active Directory Integrated Zones

Chapter 11 of this book discusses the active directory. Basically, it is a directory service that stores all information about the network resources and services such as user data, printers, servers, databases, groups, computers, and security policies. In addition, it identifies all resources on a network and makes them accessible to users and applications. Rather than maintaining two separate naming services (active directory and a DNS-based name space), active directory integrated zones have the zone data stored as an active directory object and replicated as part of the domain replication. In addition, since active directory integrated zones are stored in the active directory, there is no zone database file on a primary server. When you store a zone in active directory, the zone database file is copied into the active directory and deleted on the primary server for the zone. One advantage of using the active directory integrated zone is that it has the ability to have more than one DNS server to update a DNS zone. Active directory integrated zones can only be created on servers that are configured as domain controllers and configured to run the DNS dynamic update protocol. If you want to use active directory, you must have a DNS server, but not necessarily an active directory integrated zone.

10.1.4 Zone Files

A **zone file** is a file that defines a zone. All files are kept on the DNS name server. While most DNS systems such as UNIX must be configured by editing these text files, Microsoft DNS uses a user-friendly interface (DNS console). The zone database files are traditionally maintained on computers that are running the DNS Server service. Remember that in Windows 2000 a zone can be stored in the active directory rather than in a zone database file.

The primary file, known as the zone database file (*%systemroot%*\SYSTEM32\DNS\ *domain_name*.DNS), contains the resource records (RRs) for that part of the domain for which the zone is responsible. See Figure 10.2 for a sample zone file. There are various resource record types that are defined for the DNS database. Table 10.4 lists some of the more common types of resource records. For more information on resource records, refer to RFC 1034, RFC 2052, and RFC 2065.

10.1.5 Name Resolution

When a client computer needs an IP address, it will send a name query to the DNS Client service (resolver) located on the client. The DNS Client service will then check the locally cached information and local HOSTS file (if present). Be aware that the cache area is not the cache inside the processor, but rather an area of memory in RAM set aside to hold DNS entries. The entries in the cache come from the preloaded entries of the HOSTS file or from previous answered responses. When previous queries are cached, data are kept for a preset time, known as the time to live (TTL). If the query does not match an entry in the

TABLE 10.4 Common resource record types

Resource Record	Purpose
SOA (Start of Authority)	Identifies the name server that is the authoritative source of information or data within a domain. An SOA record is created automatically when you create a new zone. A primary server for a given zone lists itself in an SOA record to show that it is the source for this zone. The first record in the zone database file must be the SOA record.
NS (Name Server)	Provides a list of name servers that are assigned to a domain.
A (Host Address)	Provides a host name to an Internet protocol (IP) version 4, 32-bit address. For more information, see RFC 1035.
PTR (Pointer)	Resolves an IP address to a host name (reverse mappings). For more information, see RFC 1035.
CNAME (Canonical Name)—Alias	Creates an alias or alternate DNS domain name for a specified host name. The most common or popular use of an alias is to provide a permanent DNS-aliased domain name for generic name resolution of a server-based name such as www.acme.com and ftp.acme.com to more than one computer or IP address used in a web server. In this way, you can assign acme.com to one server, www.acme.com to a second server, and ftp.acme.com to a third server. If you do use the same server for all three entries and you decide it to split the service, simply change the CNAME resource record to point to the new server.
SRV (Service)	Locates servers that are hosting a particular service. Note: SRV records are new in Windows 2000 DNS Server services. This record enables you to maintain a list of servers for a well-known server port and transport protocol type ordered by preference for a DNS domain name. For more information, see the Internet draft "A DNS RR for specifying the location of services (DNS SRV)."
MX (Mail Exchanger)	Identifies which mail exchanger to contact for a specified domain and in what order to use each mail host.

FIGURE 10.2 Sample zone database file

```
;
;   Database file acme.com.dns for acme.com zone.
;       Zone version:   6
;
@                       IN   SOA server1.acme.com.
administrator.acme.com. (
                        6               ; serial number
                        900             ; refresh
                        600             ; retry
                        86400           ; expire
                        3600            ) ; minimum TTL
;
;   Zone NS records
;
@                       NS   server1.acme.com.
;
;   Zone records
;
server2000-01a      A   132.132.60.45
server1             A   132.132.20.1
testserver          A   132.132.20.20
```

cache, then the resolution process continues with the client querying a DNS server to resolve the name.

When the resolver queries a DNS server, it will perform a recursive query. A **recursive query** asks the DNS server to respond with the requested data or with an error stating that either the requested data or the specified domain name does not exist. Note that the name server does not refer the query to another name server unless it is configured as a forwarder, in which it will forward the DNS request as a recursive query.

When the DNS server receives a request, it will check if the host name is located in its own zone database file, in which it is an authority. If it is not listed in the zone file, it will then check the cache area. From then on, the DNS server will use iterative queries to resolve the name. An **iterative query** gives the best current answer back as a response. The best answer will be the address being sought or an address of a server that would have a better idea of its address.

EXAMPLE 4 A program needs to find the address of SERVER.SUPPORT.ACME.COM. The program will send the request to the resolver. The resolver will look in its cache area and its HOSTS file. If no address is found, the resolver will forward the query to its preferred DNS name server.

The DNS server will check if it knows the address of SERVER.SUPPORT. ACME.COM. If not, it will then ask the root server if it knows the address of SERVER.SUPPORT.ACME.COM. The root server will not know the address, so it will respond with the best answer by providing the address of the COM root server. The DNS server will then ask the COM root server for the address of SERVER.SUPPORT. ACME.COM. The COM root server will respond back with its known best answer of the ACME.COM name server address. The DNS server will query the ACME.COM name server for the address of SERVER.SUPPORT.ACME.COM. The DNS server will then respond back to the resolver with the address of the SUPPORT.ACME.COM. The DNS server will query the SUPPORT.ACME.COM name server for the address of SERVER.SUPPORT.ACME.COM and will respond back with the address.

The preferred server will then respond back to the client with the correct address. The program will then use the IP address to communicate its needs.

As shown by this example, the resolver queried its DNS server using a recursive query while the DNS server queried other DNS servers using iterative queries. See Figure 10.3.

Although the recursive query process can be resource intensive, it has some performance advantages for the DNS server. When the recursion process is complete, the information is cached by the server. The cache information will be used again to help speed the answering of subsequent queries. Over time, the cache information can grow to occupy a significant portion of the server memory resources.

Typically, the process of domain name resolution occurs very quickly. Occasionally, it may be delayed. If the delay is too long, the browser will come back and say that the domain does not exist, even though you know that is true, because your computer got tired of waiting and timed out. Yet, when you try again, there is a good chance it will work because the authoritative server has had enough time to reply and the name server has stored the information in its cache.

DNS servers can be configured to send all recursive queries to a selected list of servers, known as **forwarders.** Servers used in the list of forwarders provide recursive lookup for any queries that a DNS server receives which it cannot answer based on its own zone records. During the forwarding process, a DNS server that is configured to use forwarders essentially behaves as a DNS client to them. Typically, forwarders are used on remote DNS servers that use a slow link to access the Internet.

DNS servers use a mechanism called **round-robin,** or **load sharing,** to share and distribute loads for network resources. Round-robin rotates the order of resource records data that are returned in a query answer, in which multiple resource records exist of the same type for a queried DNS domain name. Since the client is required to try the first IP address listed, a DNS server that is configured to perform round-robin will rotate the order of the A resource records when answering client requests.

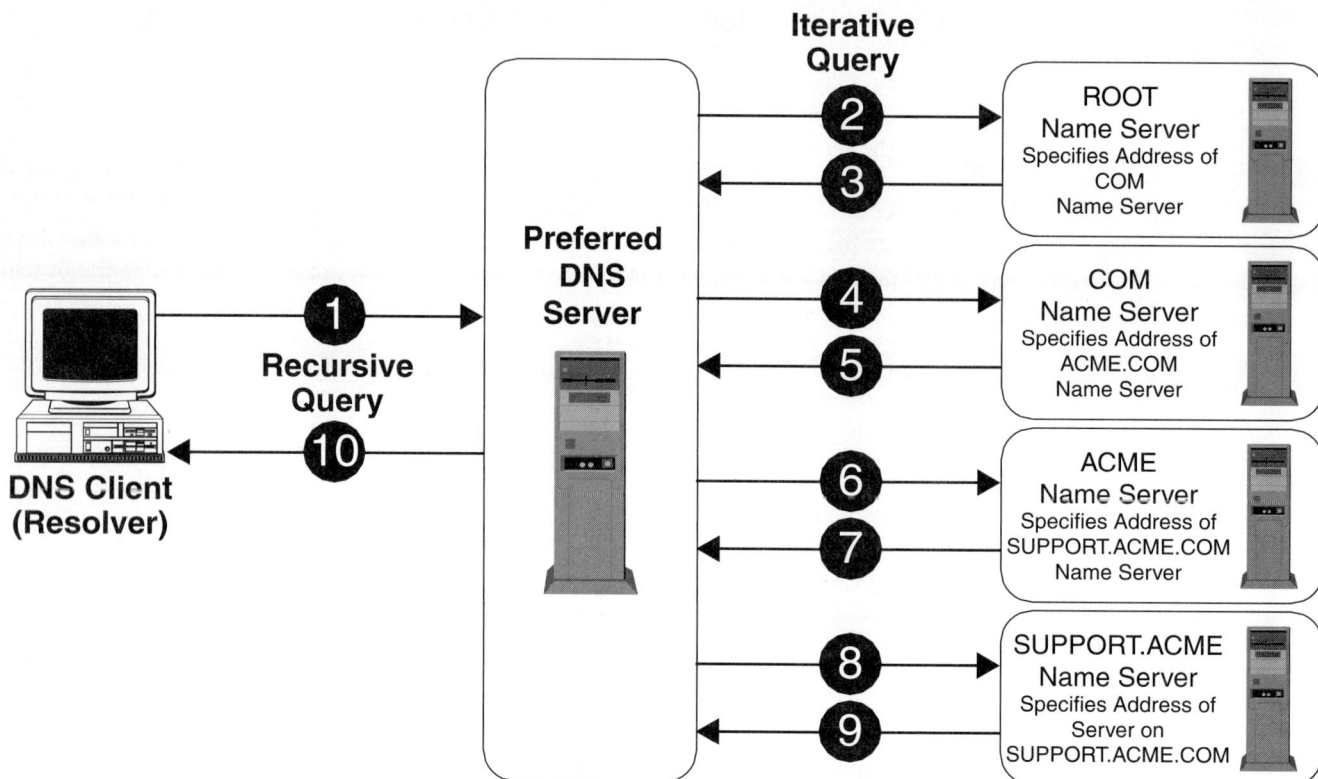

Iterative Query

ROOT
Name Server
Specifies Address of
COM
Name Server

COM
Name Server
Specifies Address of
ACME.COM
Name Server

ACME
Name Server
Specifies Address of
SUPPORT.ACME.COM
Name Server

SUPPORT.ACME
Name Server
Specifies Address of
Server on
SUPPORT.ACME.COM

Preferred
DNS
Server

DNS Client
(Resolver)

Recursive
Query

FIGURE 10.3   A DNS client will use recursive query with the preferred server to find out an IP address, whereas the preferred server will typically use iterative query to discover the IP address

10.1.6   Reverse Queries

Occasionally, the resolver may perform a **reverse query,** in which a resolver knows an IP address and wants to know the host name. To prevent the searching of all domains for an inverse query, a special domain called "*in-addr.arpa*" was created. Nodes in the in-addr.arpa domain are named after the numbers in the IP address. Since IP addresses get more specific from left to right, the order of IP addresses is reversed when building the in-addr.arpa domain. Once this domain is built, special resource records called pointer records (PTR) are added to associate the IP addresses to the corresponding host names. To find a host name for an IP address, the resolver would query the DNS server for a PTR record for the address.in-addr.arpa. For example, to find the IP address of 1.2.3.4, the resolver would query the DNS server for a PTR record of 4.3.2.1.in-addr.arpa. See Figure 10.4 for a sample reverse zone database file.

In addition to reverse lookups, some DNS servers support an **inverse query.** Like the reverse lookup, a client making an inverse query provides the IP address and requests a FQDN. Instead of using the in-addr.arpa domain to find the answer, it will check its own zone for the answer. If the answer is not in the zone, it will return an error message. Inverse queries are not as thorough as reverse queries and so are not used often.

10.2   WINDOWS 2000 DNS SERVICES

Microsoft 2000 includes an RFC-compliant DNS name server that is implemented as in Windows NT. Because the Microsoft DNS is an RFC-compliant DNS server, it creates and uses standard DNS zone files and supports all standard resource record types. While being RFC compliant, it offers integration with other Windows 2000 services including active directory, WINS, and DHCP. Windows 2000 also offers DNS dynamic update that enables DNS client computers to register and dynamically update their resource records with a DNS server whenever changes occur. This reduces the need for manual administration of zone

FIGURE 10.4   Sample reverse zone database file

```
20.132.132.in-addr.arpa - Notepad                              _ □ X
File   Edit   Format   Help
;                                                                    ▲
;  Database file 20.132.132.in-addr.arpa.dns for 20.132.132.in-addr.arpa z
;      Zone version:  2
;
@                         IN  SOA server1.acme.com.  administrator.acme.com.
                              2           ; serial number
                              900         ; refresh
                              600         ; retry
                              86400       ; expire
                              3600      ) ; minimum TTL

;|
;  Zone NS records
;

@                         NS      server1.acme.com.

;
;  Zone records
;

20                        PTR     whitehouse.acme.com.

                                                                    ▼
◄                                                                  ► //
```

records, especially for clients who frequently move or change locations and use DHCP to obtain an IP address.

10.2.1   Installing Windows 2000 DNS Services

Before installing a DNS server, be sure that the TCP/IP protocol is installed, a static IP address is assigned, and the appropriate DNS domain name is specified. During the DNS Server service installation process, the Windows 2000 server does the following:

- Installs the DNS Server service and starts the service automatically without restarting the computer.
- Installs the DNS snap-ins and adds the DNS Management shortcut to the Administrative Tools menu.
- Adds the following key for the DNS Server service to the registry: HKEY_LOCAL_MACHINE\System\CurrentControlSet\Services\Dns
- Creates the *%systemroot%*\SYSTEM32\DNS folder, which contains the DNS database files including the zone database file (*domain_name*.DNS), the reverse lookup file (*z.y.w.x*.IN-ADDR.ARPA), the cache file (CACHE.DNS), and an optional boot file (a BIND-specific implementation that controls how the DNS Server service starts).

To install a DNS server, do the following:

1. Open the Add/Remove Programs applet in the Control Panel.
2. Click on the Add/Remove Windows Components.
3. Click on the Next button.
4. Click to highlight the Networking Services option.
5. Click on the Details button.
6. In the Subcomponents of Networking Services, make sure that there is a checkmark in the box next to the Domain Name System (DNS). See Figure 10.5.
7. Click on the OK button.
8. Click on the Next button.
9. If the Insert Disk dialog box appears, insert the Windows 2000 installation CD, ensure that the path to the source files is correct, and click on the OK button.
10. Click on the Finish button.

The primary tool used to manage local and remote Windows 2000 DNS servers is the DNS console, which can be accessed from within the Administrative Tools folder or by using the Microsoft Management Console (MMC) snap-in. See Figure 10.6.

291

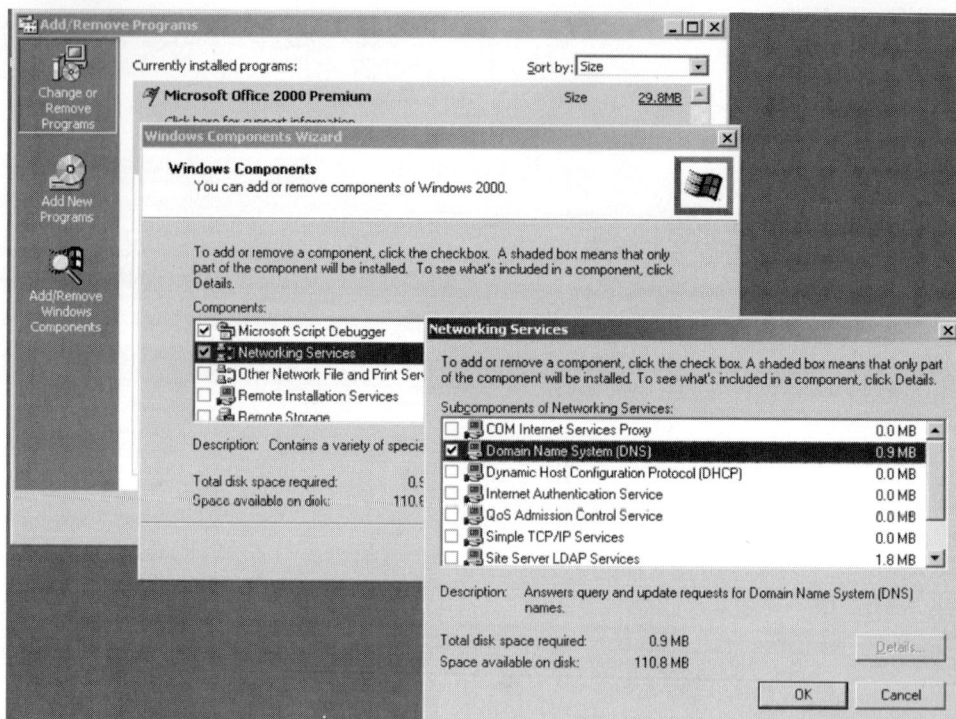

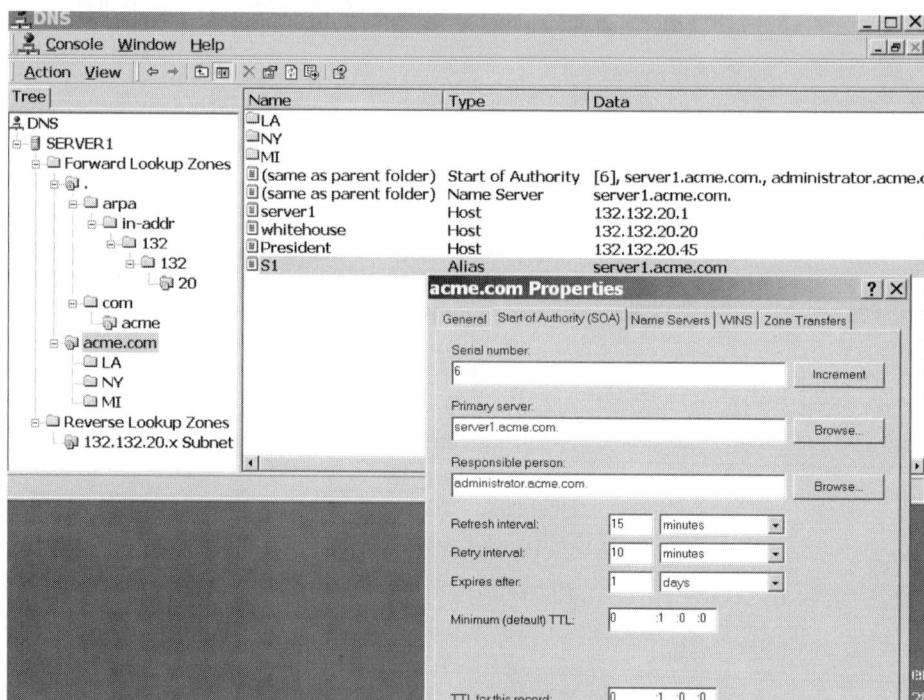

When you open the DNS console for the first time, the right pane will state that the DNS server has not been configured yet and to configure the server you need to open the Action menu and select the Configure the Server option. When this occurs, the DNS Server Configuration wizard will guide you through the process to configure the name server.

10.2.2 Root Name Servers and Resource Records

When starting the DNS server for the first time, do the following:

1. Configure the root name server.
2. Configure the forward lookup zone.

FIGURE 10.7   Root hints of a
DNS server

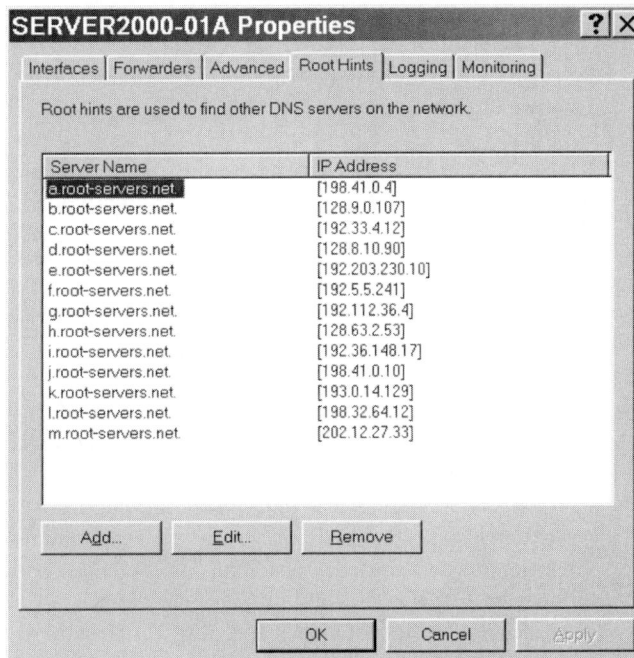

FIGURE 10.8   For a private
network, create a root domain
(standard primary forward
lookup zone) represented by a
period (.)

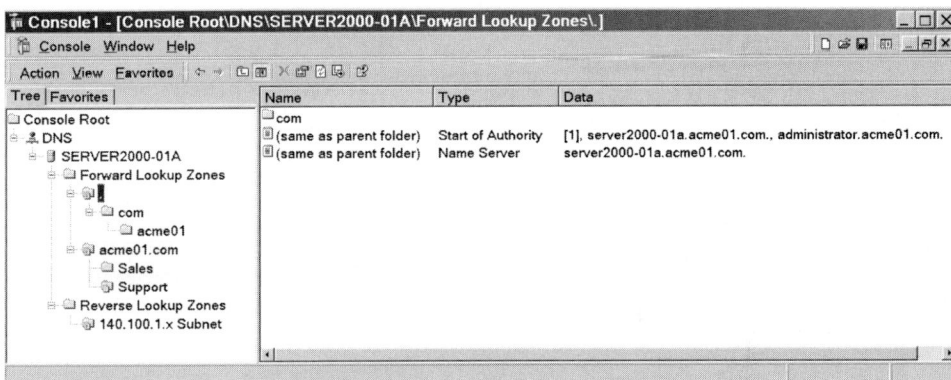

3. Configure the reverse lookup zone.
4. Create resources records (RRs).

The **root name server** contains the resource records for all top-level name servers in the domain name space, such as the COM and EDU domain. By default, the DNS service implements **root hints** using a file, CACHE.DNS, stored in the *%systemroot%*\SYSTEM32\DNS folder on the server computer. This file normally contains the NS and A resource records for the Internet root servers. See Figure 10.7.

If you are using a DNS service on a private network that is not connected to the Internet, you can edit or replace the CACHE.DNS file with similar records that point to your own internal root DNS servers. In addition, the root name server would use a root domain, which is a standard primary forward lookup zone represented by a period (.). You should then delete the CACHE.DNS files for the root servers. See Figure 10.8.

A **forward lookup zone** is that part of the DNS system that allows you to perform name-to-address resolution (forward lookup queries). On name servers, configure at least one forward lookup zone in order for the DNS service to work. The forward lookup zone can be created by using the DNS Server wizard or starting the Create New Zone wizard by right-clicking the Forward Lookup Zone folder and selecting the Create a New Zone.

A **reverse lookup zone** allows you to perform address-to-name resolution (reverse lookup queries). The reverse lookup zone differs from forward lookup zones in that it is required only to run some troubleshooting tools such as NSLOOKUP and to record a name instead of an IP address in Internet information services (IIS) log files. The reverse lookup zone can be created by using the DNS Server wizard or starting the Create New Zone wizard by right-clicking the Reverse Lookup Zone folder and selecting the Create a New Zone.

Once you create your zones, you can then create resource records (entries in the database file). To add a resource record, right-click the zone to which you want to add the record, and select the type of new record that you want to create.

10.2.3   Creating Zones and Zone Transfers

To configure the standard primary zone or to create a standard secondary zone, right-click the appropriate DNS server and click Create a New Zone option to start the New Zone wizard. The New Zone wizard will then create a standard primary, a secondary primary, and an active directory integrated zone. A fourth type of server, a **caching-only name server,** performs name resolution for the clients and then caches or stores the results, allowing for fast name resolution and reduced WAN traffic. Different from the other servers, it does not contain any zone files and is therefore not an authority for any zone. Of course, the cache is populated with the most frequently requested names and their associated IP addresses. To configure a caching-only server, install the DNS service on a computer that is running Windows 2000 Server and do not configure any forward or reverse lookup zones.

If you select the standard secondary zone, you must specify a name server from which to obtain the zone information. The name server from which the information is being copied is known as the master server, which can be a standard primary server or another standard secondary server. To add a secondary server for an existing zone, you need to have network access to the server acting as the master server, since it is the source of the zone data. In addition, if you are accessing the DNS server remotely, you may need to add the applicable DNS server to the console first before you can manage it.

To distribute the load of the DNS database, improve name resolution, improve performance, and provide fault tolerance, additional servers are used to host a zone file. To replicate and synchronize all copies of the zone files, a zone transfer is required between the various DNS servers. When a new DNS server is added to the network and is configured as a new secondary server for an existing zone, it must perform a **full initial transfer (AXFR)** to obtain a full copy of the zone files.

After the Windows 2000 DNS Servers are replicated and synchronized, they will keep the zone files synchronized by performing an **incremental transfer (IXFR),** in which the secondary server pulls only those zone changes it needs to synchronize its copy of the zone with its source. To keep track of the changes, each version of the zone file includes a serial number in the start of authority (SOA) resource record. If the serial number is the same between two zone files, no transfer is made. If the serial number at the source is greater than at the requesting secondary server, a transfer is made of only those changes to RRs for each incremental version of the zone.

The zone transfer is always initiated at the secondary server for a zone by sending a SOA query to the configured master server. The query includes the following:

- The refresh interval expires for the zone as specified in the SOA resource record (default 900 seconds/15 minutes).
- A secondary server is notified of zone changes by its master server.
- A DNS Server service is started at a secondary server for the zone.
- The DNS console is used at a secondary server for the zone to manually initiate a transfer from its master server.

The source server answers the query for its SOA record. The destination server checks the serial number of the SOA record in the response.

Windows DNS servers support DNS Notify (RFC 1996), which permits a means of initiating notification to secondary servers when zone changes occur. DNS notification implements a push mechanism for notifying a select set of secondary servers for a zone when it is updated. Servers that are notified can then initiate a zone transfer. For secondary servers to be notified by the master DNS server, each secondary server must first have its IP address in the notify list of the master server. This list can be accessed when you right-click on the zone and choose the Properties option, select the Zone Transfer tab, and select the Notify button.

When a secondary name server starts up, it contacts the master name server and initiates a zone transfer for each zone for which it is acting as a secondary name server. If the data on the master name server have changed, zone transfers can occur periodically as set by the refresh parameters in the SOA record of the zone file.

When a zone is updated at the master or source server, the serial number field in the SOA RR is also updated, indicating a new local version of the zone. The master server then sends a DNS notify message to the other servers that are part of its configured notify list so that it knows that a zone transfer has to be completed.

For Windows 2000, use DNS notification only to notify servers that are operating as secondary servers for a zone. For replication of directory-integrated zones, DNS notification is not needed, because any DNS servers that load a zone from active directory automatically poll the directory approximately once every 15 minutes to update and refresh the zone. If you configure a notify list, you can actually degrade system performance by causing unnecessary additional transfer requests for the update zone.

To change a zone type, left-click on any zone to highlight it and select the Properties option from the shortcut menu. On the General tab, click on the Change button, select the zone type, and click on the OK button.

10.2.4 Creating a Subdomain

To create a subdomain from within the DNS console, click to highlight the name of the zone in which you want to create the subdomain. Right-click the zone name to bring up the shortcut menu and click on the New Domain option. Type the name of the subdomain in the New Domain dialog box and click on the OK button.

After creating a subdomain, you can delegate authority of it to another DNS server/zone. This allows you to distribute the load of the DNS database and improve name resolution performance. When a query is being done on a zone, it will refer to the NS (name server) resource records to find the name server for the target zone being queried.

To delegate authority for a subdomain, click to highlight the name of the domain that you want to delegate authority (typically the domain, not the subdomain), right-click the domain name to bring up the shortcut menu, and select the New Delegation option. You would then follow the Add New Delegation wizard to guide you through the rest of the process, including specifying the name of the domain to which you are delegating authority and adding the names and IP address of the server(s) that will host the delegated zone.

10.2.5 Dynamic DNS Updates

Since DNS has become the primary naming resolution tool for Windows 2000, every computer that holds a network service such as file or print sharing has to be registered. A large network that has many computers using DHCP to obtain a new IP address needs to dynamically register and update the DNS server's resource records. Windows 2000 provides client and server support for the use of dynamic updates, as described in RFC 2136.

By default, client computers running Windows 2000 dynamically update their host (A) and pointer (PTR) resource records (RRs) in DNS with a FQDN and their IP addresses. The FQDN uses the domain name appended to the computer name. For example, if the computer name is HOST3 and the DNS suffix is ACME.COM, the registered name would be HOST3.ACME.COM.

Dynamic updates can be sent for any of the following reasons or events:

- At start-up time, when the computer is turned on
- When an IP address is added, removed, or modified in the TCP/IP properties configuration for any one of the installed network connections
- When an IP address lease changes or renews with the DHCP server any one of the installed network connections. For example, when the computer is started or if the IPCONFIG/RENEW command is used
- When the IPCONFIG/REGISTERDNS command is used to manually force a refresh of the client name registration in DNS

Note that the DHCP client service performs the updates, not the DNS client services.

Dynamic updates are sent or refreshed periodically. By default, Windows 2000 sends a refresh once every 24 hours. Unfortunately, the names are not removed from DNS zones if they become inactive or are not updated within the refresh interval (24 hours). DNS does not use a mechanism to release or tombstone names, although DNS clients attempt to delete or update old name records when a new name or address change is applied. The Windows 2000 DNS service allows dynamic update to be enabled or disabled on a per-zone basis at each server configured to load either a standard primary or directory-integrated zone. Standard secondary zones are not supported because they get the information from the standard primary zone through zone transfers.

To enable DNS dynamic updates for clients, open the DHCP console and right-click on the applicable DHCP server. In the shortcut menu, pick the Properties option and click on the DNS tab. Then put a check in the "Automatically update DHCP client information in DNS" check box.

If you are using active directory integrated zones, use secure dynamic update to automatically update a zone file. Secure dynamic update allows only the computers that you specify to make new entries or modify existing entries in a zone. By default, all authenticated computers in a forest (multiple Windows 2000 domains connected together) can make new entries in a zone, and only the computer that created a name is allowed to modify the data associated with that name.

To allow only secure dynamic updates, right-click on the applicable zone and select the Properties option. On the General tab, verify that the zone type is active directory integrated and that the "Allow dynamic update?" dropdown list is set to "Only secure updates." By default, the Windows 2000 clients will try to use unsecured dynamic updates first. If that request fails, then they use secure updates. Unless the DNS server prohibits the clients, the clients try to overcome their previous registered resource records.

DNS service also provides the ability to use WINS servers to look up names not found in the DNS domain name space by checking the NetBIOS name space managed by WINS. See Figure 10.9. To use WINS lookup integration, two special resource record types are enabled and added to the zone, WINS and WINS-R. A good example to use WINS lookup is when you are using a mixed-mode client of Windows and UNIX clients, whereas the UNIX clients and some early-version Microsoft clients can locate your WINS clients by extending DNS host name resolution into the WINS-managed NetBIOS name space. If you use a mixture of Windows DNS servers and other DNS servers to host a zone, enable the "Do not replicate this record" option for any primary zones when using the WINS lookup record.

FIGURE 10.9   DNS service also provides the ability to use WINS servers to look up names not found in the DNS domain name space by checking the NetBIOS name space managed by WINS

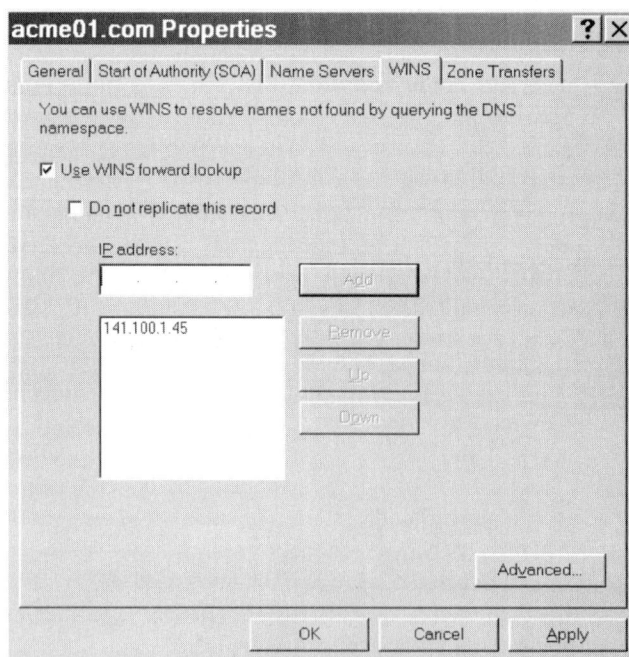

This prevents the WINS lookup record from being included in zone transfers to other DNS servers that do not support or recognize this record. If not, it would cause data errors and failed zone transfers.

10.2.6 DNS Best Practices

To get the best results from a WINS server, do the following:

- Enter the correct e-mail addresses of those who are responsible for each zone so that these people can be notified for query errors and security problems. Most Internet e-mail addresses contain the at (@) sign, but this sign must be replaced with a period (.).
- Use secondary or caching-only servers for your zones to assist in off-loading DNS query traffic.
- If planning a large DNS design, such as for a large Internet service provider (ISP) that supports the DNS name service, read the most current RFCs.

10.2.7 Monitoring and Troubleshooting DNS Services

In Windows 2000, DNS services are essential to the network. Therefore, Windows 2000 has several utilities for monitoring and troubleshooting DNS servers. These include the following:

- DNS administrative tool is used to test DNS servers and monitor their ability to process and resolve queries. See Figure 10.10.
- Command line utilities, such as NSLOOKUP, are used to verify resource records and troubleshoot DNS problems.
- Logging features, such as DNS server log, can be viewed using Event Viewer.

You can configure the DNS Server service to perform queries on a scheduled basis to ensure that the service is operating correctly. In the DNS console, right-click the server that you want to monitor, select the Properties option, and click on the Monitoring tab. From here, you can perform a simple query, a recursive query, or both. If the recursive query fails, the DNS server is unable to contact any of the servers listed in Root Hints. This problem can result if the configuration of the root name server is invalid or the root name server cannot be contacted. For more suggestions, see the Windows 2000 help files and resource kit.

NSLOOKUP is a diagnostic tool that displays information from the DNS servers. It is only available if the TCP/IP protocol has been installed. NSLOOKUP has two modes, interactive and noninteractive. Use interactive mode when you require more than one piece of data. To run interactive mode, at the Command prompt, type NSLOOKUP. To exit interactive mode, type Exit. Use noninteractive mode when you require a single piece of data.

FIGURE 10.10 The DNS administrative tool is used to test DNS servers and monitor their ability to process and resolve queries

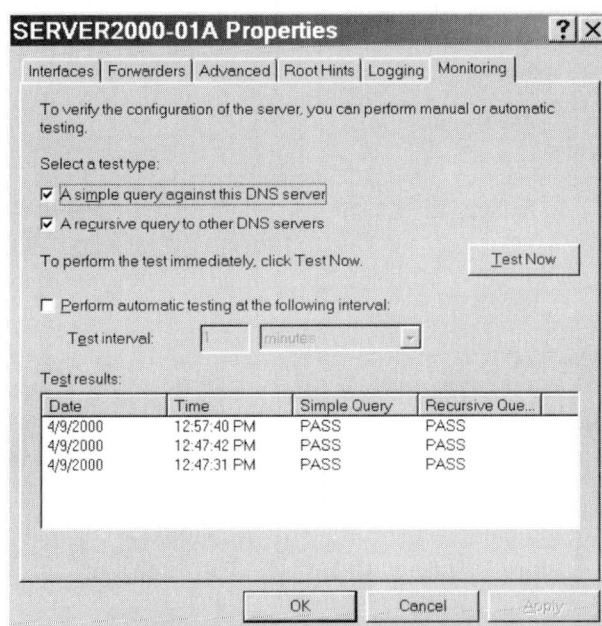

Type the NSLOOKUP command at the Command prompt with the proper parameters, and the data then return.

Name resolution problems can occur if a client computer is resolving names incorrectly or if the client computer name is not registered with the DNS servers on your network. If you determine that the client computer is resolving names incorrectly, use the IPCONFIG /FLUSHDNS from the Command prompt to flush and reset the cache on the client computer. If a client's name records are missing from the servers, use the IPCONFIG / REGISTERDNS to force the client computer to renew its registration.

10.3 INTRODUCTION TO WINDOWS INTERNET NAME SERVICE

If you try to access a computer using its NetBIOS (computer) name, such as using a UNC name to specify a network resource, the computer will need to determine the IP address. Initially, it will broadcast onto the network to ask for the IP address of the computer. Unfortunately, because the broadcast usually does not go across routers, computers on other subnets do not get resolved. In addition, if you have several computers doing these types of broadcasts, then the broadcasts will slow the performance of the network.

Other methods to resolve NetBIOS names are to use either LMHOSTS files or to access a WINS server. Like HOSTS files, it would be extremely difficult to maintain and distribute the LMHOSTS files to all computers in a large network, especially since addresses may change frequently under a network using DHCP servers.

10.3.1 Using a WINS Server

A **Windows Internet Name Service (WINS) server** contains a database of IP addresses and NetBIOS (computer names) that update dynamically. For clients to access the WINS server, they must know its address. Therefore, the WINS server requires a static address—one that does not change. When the client accesses the WINS server, the client sends a message directly to the WINS server versus doing a broadcast. When the WINS server gets the requests, it knows which computer sent the request and the WINS can reply directly to the originating IP address. The WINS database stores the information and makes it available to the other WINS clients. (The WINS database is located at *systemroot*System32\WINS\Wins.mdb.) The WINS registration generates little excessive network traffic because it does not use broadcast. The WINS protocol is based on and is compatible with the protocols defined for NetBIOS Name Server (NBNS) in RFCs 1001/1002.

When a WINS client starts up, it registers its name, IP address, and type of services within the WINS server's database. The type of service is designated by a hexadecimal value, which is placed at the end of the name. For example, when starting a Windows 2000 computer called Server2, it will register three mappings, which are Server2[00h] (workstation), Server2[03h] (messenger), and Server2[20h] (file server). The NetBIOS can be a maximum 15 characters, not counting the hexadecimal value which represents the service. See Table 10.5 for the list of NetBIOS network services.

TABLE 10.5 NetBIOS network services

NetBIOS Name Suffix	Network Service/Resource Identifier
\\computer_name[00h]	Workstation service
\\computer_name[03h]	Messenger service
\\computer_name[06h]	RAS
\\computer_name[20h]	Server service
\\computer_name[21h]	RAS client service (on a RAS client)
\\computer_name[BEh]	Network monitoring agent service
\domain_name[1Bh]	The PDC in its role as the domain master browser
\domain_name[1Dh]	The master browser for each subnet
\domain_name[1Ch]	The domain controllers (up to 25 IP addresses) within the domain

Since WINS was made only for Windows operating systems, other network devices and services (such as a network printer and UNIX machines) cannot register with a WINS service. Such addresses, then, would have to be added manually.

Names held in the WINS database are given a time to live (TTL) or renewal interval during name registration. A name must be refreshed before this interval ends or it will be released from the database. Names are refreshed by sending a name refresh request to the WINS server by the WINS client. Windows clients will attempt a refresh at one-half the renewal interval and will keep trying to contact the WINS server until the time expires. NetBIOS names are explicitly released when the client performs a proper shutdown or silent when the name is not refreshed within the renewal interval.

When a client node registers a name that already exists in the WINS database and has a different IP address, the WINS server must determine if the name with the old IP address still exists. Therefore, the WINS server will send a name query request to the old IP address. If the old address responds with a positive name query response, then the WINS server will reject the new registration with a negative name registration response. If the old address does not respond to the name query, the new registration is accepted.

10.3.2 WINS Clients

A WINS client can be configured to use one of four NetBIOS name resolution methods. They include B (broadcast) node, P (point-to-point) node, M (mixed) node, and H (hybrid) node. See Table 10.6. In either case, when trying to resolve a computer name, it will always check its own local NetBIOS name cache. Like the DNS name cache, it remembers names and addresses of computers that it recently communicated with. By default, when a system is configured to use WINS for its name resolution, it adheres to H node for name registration. By using DHCP servers or the registry, you can force the client into one of these modes.

TABLE 10.6 NetBIOS resolution modes

Node Type	Registry Value	Description
B (broadcast) node	1	A computer doing B-node name resolution relies on broadcasts to convert names into IP addresses. B-node name resolution is not the best option on larger networks, because the broadcast will load the network and will usually not go through routers. Microsoft really uses B node, which will check the LMHOSTS file after a broadcast.
P (point-to-point) node	2	A computer doing P-node name resolution uses a NetBIOS Name Server (NBNS)/WINS server to look up NetBIOS names to get IP addresses. All systems must know the IP address of the NBNS. The main drawback of P-node name resolution is if the NBNS cannot be accessed, there will be no way to resolve names and thus no way to access other systems on the network by using NetBIOS names.
M (mixed) node	4	An M-node computer first tries a broadcast to resolve a name. It that attempt fails, the computer looks up the name in a NetBIOS Name Server. In other words, an M-node computer first acts as a B node, and if that fails, it tries to act as a P node. M node has the advantage over P node in that if the NBNS is unavailable, systems on the local subnet can still be accessed through B-node resolution. M node is typically not the best choice for larger networks because it uses B-node and thus results in broadcasts. However, when you have a large network that consists of smaller subnetworks connected via slow wide area network (WAN) links, M node is a preferred method as it will reduce the amount of communication across the slow links.
H (hybrid) node (Default)	8	Finally, an H-node computer first does a P-node lookup; if that fails, the computer does a broadcast. In either case, the NetBIOS name resolution will try the LMHOSTS file after trying a broadcast and/or WINS server.

EXAMPLE 4 The H (hybrid) node will first check if the name is the local machine name. It will then check the NetBIOS cache area for remote names. Resolved names will remain in the cache area for 10 minutes. If the name has not yet been resolved, it will try the WINS server followed by a broadcast. Lastly, it will look in the LMHOSTS file, if the system has one, followed by the HOSTS file and DNS server if it was configured.

10.3.3 WINS Server Replication

To provide fault tolerance, it is recommended to have more than one WINS server with the same WINS database. By having more than one WINS server, a WINS client can go to the second WINS server when the first one is unavailable. To make sure that the WINS server has the same information, replicate the information from one WINS server to the other. These servers are known as replication partners. Windows 2000 supports up to 12 replication partners, in which the first and second WINS servers are the primary and secondary servers, and any remaining servers are backup WINS servers.

A **WINS replication partner** can be added and configured as either a pull partner, a push partner, or a push/pull partner. The push/pull partner is the default configuration and is the type recommended for use in most cases. A **pull partner** is a WINS server that requests new database entries from its partner. The pull occurs at configured time intervals or in response to an update notification from a push partner. A **push partner** is a WINS server that sends update notification messages. The update notification occurs after a configurable number of changes to the WINS database.

Pull partners configure at certain time intervals, so you should use a pull partner across slow links; for example, you could have it replicate every 24 hours beginning at 12:00 midnight, when traffic is at a minimum. A push partner should be used with servers connected across fast links, because push replication occurs when a particular number of updated WINS entries are reached. When you need immediate replication, force the WINS servers to replicate using WINS console.

You can configure a WINS server to automatically configure other WINS server computers as its replication partners, using periodic multicasts to announce their presence. These announcements are sent as Internet Group Management Protocol (IGMP) messages for the multicast group address of 224.0.1.24 (the well-known multicast IP address reserved for WINS server use). With this automatic partner configuration, other WINS servers are discovered when they join the network and are added as replication partners.

10.3.4 WINS Proxy Agent

A **WINS proxy agent** is a WINS-enabled computer that is configured to act on behalf of other host computers that cannot directly use WINS. WINS proxies help resolve NetBIOS name queries for computers that are located on a subnet where there is not a WINS server, by hearing broadcast on the subnet of the proxy agent and forwarding those responses directly to a WINS server. This keeps the broadcast local, yet gets responses from a WINS server without using the P node. Most WINS proxies are only useful or necessary on networks that include NetBIOS broadcast-only (B-node) clients, and therefore, are typically not needed.

10.3.5 Installing and Managing WINS Server

Before installing a WINS server, ensure that the TCP/IP protocol is installed and a static IP address is assigned. The WINS server is included in the Windows 2000 Server installation files and is enabled during installation or by using the Control Panel's Add/Remove Software applet.

The primary tool used to manage local and remote WINS servers is the WINS console, which can be accessed from within the Administrative Tools folder or by using the Microsoft Management Console (MMC) snap-in. See Figure 10.11.

10.3.6 NBTSTAT Command

To troubleshoot NetBIOS name resolution over TCP/IP, you can use the NBTSTAT command. Its options are shown in Table 10.7.

```
┌─────────────────────────────────────────────────────────────────────────────────┐
│ ⬚WINS                                                                    _ □ ✕  │
├─────────────────────────────────────────────────────────────────────────────────┤
│ ‖ Action  View  ‖ ⇦ ⇨ ‖ ⬚ ⊞ ‖ ⬚                                                  │
├──────────────────────────┬──────────────────────────────────────────────────────┤
│ Tree │                    │ Active Registrations   Items found by name: 5        │
│                           ├──────────────┬──────────────┬───────────┬────────┬───┤
│ ⬚ WINS                    │ Record Name  │ Type         │ IP Address│ State  │ Static │ Owner │
│  ─ ⬚ Server Status        ├──────────────┼──────────────┼───────────┼────────┼───┤
│  ⊟ ⬚ SERVER2000-01A [63.197.142.129] │ ⬚SERVER2000-01A │ [00h] WorkStation │ 63.197.142.129 │ Active │ │ 63.197.142.129 │
│    ─⬚ Active Registrations │ ⬚SERVER2000-01A │ [03h] Messenger │ 63.197.142.129 │ Active │ │ 63.197.142.129 │
│    └⬚ Replication Partners │ ⬚SERVER2000-01A │ [20h] File Server │ 63.197.142.129 │ Active │ │ 63.197.142.129 │
│                           │ ⬚SERVER2000-01A │ [87h] Other │ 140.100.1.101 │ Released │ │ 63.197.142.129 │
│                           │ ⬚SERVER2000-01A │ [6Ah] Other │ 140.100.1.101 │ Released │ │ 63.197.142.129 │
│                           ├──◀──────────────────────────────────────────────▶──┤
└───────────────────────────┴─────────────────────────────────────────────────────┘
```

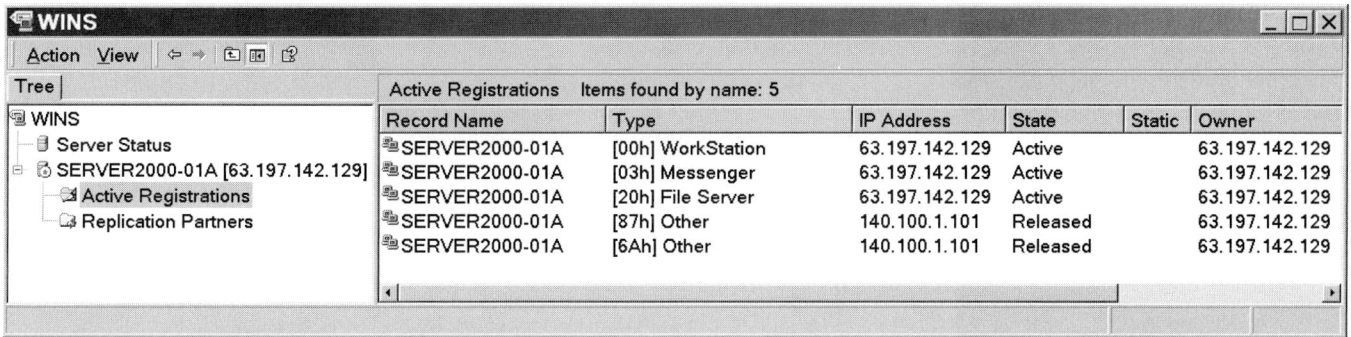

FIGURE 10.11   WINS console

TABLE 10.7   NBTSTAT command options

Option	Description
nbtstat –n	Displays the names that were registered locally on the system by programs such as the server and redirector.
nbtstat –c	Shows the NetBIOS name cache, which contains name-to-address mappings for other computers.
nbtstat –R	Purges the name cache and reloads it from the LMHOSTS file.
nbtstat –RR	Releases NetBIOS names registered with a WINS server and then renews their registration.
nbtstat –a *name*	Performs a NetBIOS adapter status command against the computer specified by *name*. The adapter status command returns the local NetBIOS name table for that computer plus the media access control address of the adapter.
nbtstat –S	Lists the current NetBIOS sessions and their status, including statistics.

10.3.7   WINS Best Practices

To get the best results from a WINS server, do the following:

- Use the default settings to configure WINS servers. The preconfigured WINS settings provide the optimal configuration for most conditions and should be used in most WINS network installations.
- Schedule consistency checking for an off-peak time such as at night during the weekends. This can be done by clicking on the server using the WINS console and clicking and selecting Verify Database Consistency from the Action menu.
- To provide for the best performance, consider a fast disk system including using RAID to improve disk-access time.
- For disk defragmentation and improved disk performance, use off-line WINS compaction on a regular basis.
- Perform regular backups of the WINS database. This can be done by clicking on the server using the WINS console and clicking and selecting Backup Database from the Action menu.
- Configure clients with more than one WINS server IP address. Thus, when one server is not available, it will then contact the second WINS server.
- To refresh the client entries in WINS and replicate them to other replication partners, use the NBTSTAT –RR command.

10.4   INTRODUCTION TO DHCP SERVICES

A **dynamic host configuration protocol (DHCP)** server maintains a list of IP addresses called a pool. When a user needs an IP address, the server removes the address from the pool and issues it to the user for a limited time. Issuing an address is called leasing. Using

TABLE 10.8 Common RFC for DHCP

RFC Number	Description
RFC 2131	Dynamic host configuration protocol
RFC 2132	DHCP options and BOOTP vendor extensions
RFC 951	Bootstrap protocol (BOOTP)
RFC 1534	Interoperation between DHCP and BOOTP
RFC 1542	Clarifications and extensions for the bootstrap protocol
RFC 2136	Dynamic updates in the domain name system (DNS UPDATE)
RFC 2241	DHCP options for Novell Directory Services
RFC 2242	NetWare/IP domain name and information

a DHCP server to issue an address is more reliable and requires less labor than setting every computer manually, and you can get by with fewer IP addresses, because computers not on the network are not using IP addresses.

Created for diskless workstations, the **bootstrap protocol (BOOTP)** enabled a booting host to configure itself dynamically. See Table 10.8. DHCP is an extension of BOOTP. It is used to automatically configure a host during boot-up on a TCP/IP network and to change settings while the host is attached. There are many parameters that you can automatically set with the DHCP server. Some of the more common parameters are as follows:

- IP address
- Subnet mask
- Gateway (router) address
- Address of DNS servers
- Address of WINS servers
- WINS client mode

10.4.1 DHCP Requests

A host computer that is configured to get a DHCP address sends a DHCPDISCOVER message on the local IP subnet to find the DHCP server(s). The client does not know the address(es) of the DHCP server, so it uses an IP broadcast address for the DHCPDISCOVER message. All available DHCP servers respond with a DHCPOFFER message. If more than one server is available, the client usually selects the first server to respond, but no rule specifies which server the client has to use. Regardless of how many servers respond, the client broadcasts a DHCPREQUEST message that identifies which server the client will use and implicitly informs all other servers that the client will not use them. The selected server responds to the client with a DHCPPACK message that contains the assigned IP address, any other network parameter assignments, and the lease or amount of time for which the DHCP server assigns the client the IP address. The client sends messages to UDP port 67 on the DHCP server and the server sends messages to UDP port 68 on the DHCP client.

Computers running any of the following operating systems can be DHCP clients.

- Windows 2000 Professional or Windows 2000 Server
- Microsoft Windows NT version 3.51
- Microsoft Windows 95 or Windows 98
- Microsoft Windows for Workgroups version 3.11 with Microsoft TCP/IP-32 installed
- Microsoft MS-DOS with the Microsoft Network Client version 3.0 for Microsoft MS-DOS installed, and using the real-mode TCP/IP driver
- Microsoft LAN Manager version 2.2c (LAN Manager 2.2c for OS/2 is not supported)

A Windows 2000 Server that is acting as a DHCP server needs to use a static IP address, but does not need to be a domain controller. To install a DNS server, do the following:

1. Open the Windows Component wizard by clicking on the Start button, select the Settings option, and select the Control Panel.
2. Double-click Add/Remove Programs and then click Add/Remove Windows Components.

3. Under Components, scroll to and click Networking Services. Click Details.
4. Under Subcomponents of Networking Services, click Dynamic Host Configuration Protocol (DHCP), and then click OK.
5. If prompted, type the full path to the Windows 2000 distribution files and click Continue.
6. After the required files are copied to the hard disk and the system is rebooted, manage the DHCP server using the DHCP console. The DHCP console can be opened by clicking on the Start button, selecting the Programs option, selecting the Administrative Tools and then the DHCP option.

10.4.2 DHCP Scope and Options

Traditionally, a DHCP server can have only one scope per subnet per DHCP server. Therefore, you can exclude addresses within the scope. For example, you determine the scope will be between 132.132.20.10 and 132.132.20.110. Let us say that you have a network printer assigned a static IP address of 132.132.20.50 and a UNIX server using a static IP address of 132.132.20.51. You would then have to exclude these two addresses. You cannot use the scopes of 132.132.20.10 to 132.132.20.49 and 132.132.20.52 to 132.132.20.110, because each subnet can have only one scope. When creating a scope, you also provide the scope's subnet mask and the duration of the lease (default is 8 days). See Table 10.9. The DHCP database is stored in the *systemroot*\SYSTEM32\DHCP\DHCP.MDB directory.

To provide fault tolerance, assign two or more DHCP servers for every subnet. With two DHCP servers, if one server is unavailable, the other server can take its place and continue to lease new addresses or renew existing clients. When setting up the two DHCP servers, the scopes of the two servers should include scopes with different addresses to avoid both servers assigning the same address to two different computers.

Before installing a DHCP server, be sure the TCP/IP protocol is installed and a static IP address is assigned to the DHCP server. The DHCP server is included in the Windows 2000 Server installation files and is enabled during installation or by using the Control Panel's Add/Remove Software applet.

The primary tool used to manage local and remote DHCP servers is the DHCP console, which can be accessed from within the Administrative Tools folder or by using the Microsoft Management Console (MMC) snap-in. See Figure 10.12.

In Windows 2000, you must authorize a DHCP server in active directory before the server can issue leases to DHCP clients. If this is not done, the DHCP service on a Windows 2000 Server in the forest will not initialize. This step will ensure that another user does not install another DHCP server, which can cause havoc on the network. To authorize a DHCP server for a domain, a user must be a member of the Enterprise Admins group, which exists in the root domain of the active directory forest. Active directory will be explained in Chapter 11.

TABLE 10.9   DHCP scope parameters

Parameter	Description
Name	The name of the scope (optional)
Comment	The comment for the scope, used primarily to describe it (optional)
IP Address Range from Address	The starting IP address of scope or address pool
IP Address Range to Address	The ending IP address of scope or address pool
Mask	The subnet mask assigned to DHCP clients
Exclusion Range Start Address	The starting IP address of the range to exclude within the scope or address pool (optional)
Exclusion Range End Address	The ending IP address of the range to exclude within the scope or address pool (optional)
Lease Duration Unlimited	A parameter that indicates that DHCP leases assigned to clients never expire
Lease Duration Limited to	The number of days, hours, and minutes that a DHCP client lease is available before it must be renewed

FIGURE 10.12 DHCP console

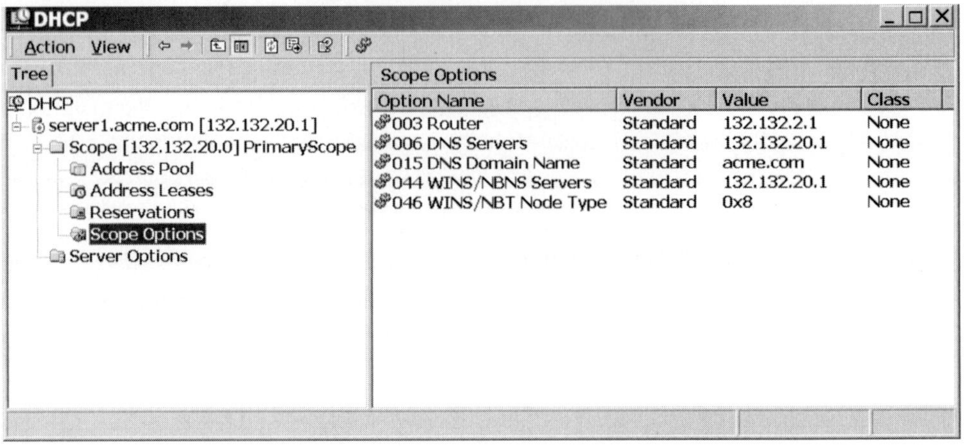
DHCP
Action View
Tree
DHCP
server1.acme.com [132.132.20.1]
Scope [132.132.20.0] PrimaryScope
Address Pool
Address Leases
Reservations
Scope Options
Server Options
Scope Options
Option Name
Vendor
Value
Class
003 Router Standard 132.132.2.1 None
006 DNS Servers Standard 132.132.20.1 None
015 DNS Domain Name Standard acme.com None
044 WINS/NBNS Servers Standard 132.132.20.1 None
046 WINS/NBT Node Type Standard 0x8 None

To authorize a DHCP server in active directory, perform the following steps:

1. From the DHCP console, right-click DHCP and select the Browse Authorized Servers option.
2. In the Authorized Servers in the Directory dialog box, click the Add button.
3. In the Authorized DHCP Server dialog box, enter the name or IP address of the DHCP server to authorize, and then click OK.
4. In the DHCP dialog box, click Yes to confirm the authorization.

There are three levels of DHCP scopes that can be configured for DHCP clients. They are global, scope, and client. Global options are available to all DHCP clients. Use global options (known as Server Options in the Windows 2000 DHCP console) when all clients on all subnets require the same configuration information. For example, if you want all clients to use the same DNS or the same WINS server, then use the global options to specify the DNS or WINS server that will preconfigure all scopes with the DNS or WINS address. Any settings used in the scope option or client option will overwrite the global option.

The scope options are available only to those clients within the specific scope. The options are only overwritten by client options such as those set in the TCP/IP properties of Windows 95 or 98.

When you create a scope using the DHCP console, you will be using the New Scope wizard. After inputting the scope using the wizard, you will then specify the subnet mask and exclusion addresses. After you create a scope, you cannot change the IP address range or subnet mask that is assigned by the scope. Instead, you must delete the scope and create a new scope with the correct information.

After the scope has been defined using the Create Scope wizard, the wizard will then ask to configure the DHCP options (default gateway/router, domain name and DNS server, WINS server name and address, and scope activation). If you decide to add DHCP options or change the current DHCP options, then use the DHCP console. See Table 10.10.

After you create a scope, you must activate it to make it available for lease assignments. To activate a scope, in DHCP, right-click the entry for the scope, point to Task, and then click Activate. It is recommended that you finish configuring the scope before activating it so that clients will receive complete configuration information.

If you have a client that must always use the same address, then reserve an address by using client reservation. The DHCP server will then assign the reserved address to the computer with the specified MAC address. If multiple DHCP servers are configured with a scope that covers the range of the reserved IP address, then the client reservation must be made and duplicated at each of these DHCP servers. Otherwise, the reserved client computer can receive a different IP address, depending on which DHCP responds.

10.4.3 DHCP Relay Agent

A **DHCP relay agent** is a computer that relays DHCP and BOOTP messages between clients and servers on different subnets. Thus, a single DHCP server can handle several subnets without the DHCP server being connected directly to those subnets.

TABLE 10.10   Common DHCP options

DHCP Options	Description
003 Router	The IP address of a router or default gateway address
006 DNS Server	The IP address of a DNS server
015 DNS Domain Name	The DNS domain name for client resolution
044 WINS/NBNS Servers	The IP address of a WINS server available to clients
046 WINS/NBT Node Type	The type of NetBIOS over TCP/IP name resolution to be used by the client. Options are: 1 = B node (broadcast) 2 = P node (peer) 4 = M node (mixed) 8 = H node (hybrid)
047 NetBIOS Scope ID	The local NetBIOS scope ID

FIGURE 10.13   Installing a DHCP relay agent

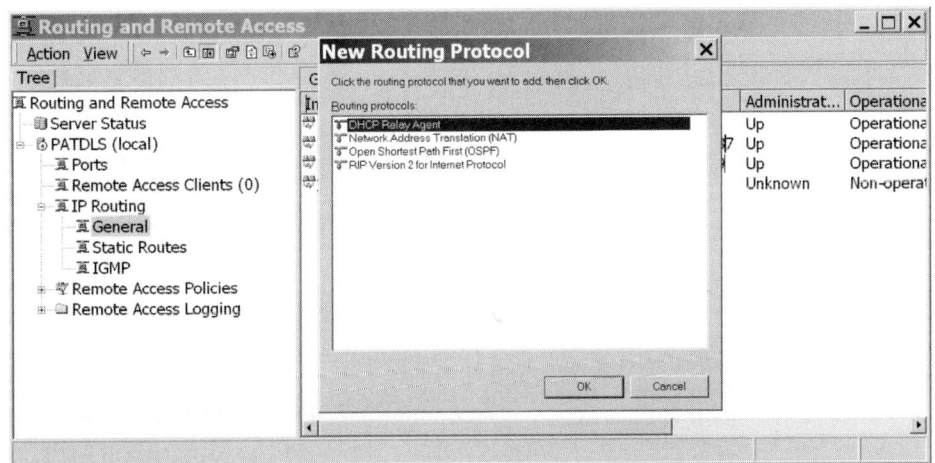

To add the DHCP relay agent, do the following:

1. Click on the Start button, select the Programs option, select the Administrative Tools, and select the Routing and Remote Access option.
2. Right-click on the General option and select the New Routing Protocol option. The General option can be found by opening the Server Name option followed by opening the IP Routing option. See Figure 10.13.
3. Select the DHCP relay agent routing protocol and select the OK button.
4. Right-click DHCP Relay Agent, and select the New Interface option.
5. Click the interface you want to add and click on the OK button.
6. If needed, in Hop-Count Threshold and Boot Threshold (seconds), click the arrows to modify the thresholds. The Hop-Count Threshold provides a space for you to type a value for the maximum number of DHCP relay agents that will handle DHCP relayed traffic. You can also click the arrows to select a new setting. The default value is 4 hops. The maximum value is 16 hops. The Boot Threshold (seconds) provides a space for you to type the number of seconds the relay agent waits before forwarding DHCP messages. You can also click the arrows to select a new setting. The default value is 4 seconds. This option is useful when you want a local DHCP server to respond first, but if the local DHCP server does not respond, you want to forward messages to a remote DHCP server. Click on the OK button.

10.4.4   Superscopes

The Windows 2000 implementation of DHCP supports superscopes. A **superscope** allows you to group multiple scopes (child scopes) as a single administrative entity. Using a superscope, the DHCP server computer can activate and provide leases from more than one scope to clients on a single physical network. Superscopes serve the following functions:

- When running out of addresses assigned to an address pool and more computers need to be added to the network, you can use a superscope to extend the address space for the same physical network segment by adding another range of addresses (scope).
- If an IP network is renumbered, you can use a superscope to migrate a subnet from one scope to another.
- You can use two or more DHCP servers on the same subnet to provide scope redundancy and fault tolerance.
- You can allow two DHCP servers on the same subnet to manage separate logical IP networks.

To create a superscope, click the applicable DHCP server and click New Superscope from the Action menu. The Superscope option is only available when at least one scope is available—when one is not currently part of a superscope. You would then use the New Superscope wizard. A superscope can have scopes added to it during or after its creation.

10.4.5   Option Classes

Traditionally, all DHCP clients are treated equally, and the server is unaware of the specific type of clients; thus, the configuration information issued by the DHCP server is the same for all DHCP clients. Starting with Windows 2000, you can use option classes to provide unique configurations to specific types of client computers.

There are two types of option classes, vendor-defined classes and user-defined classes. Vendor-defined classes identify a DHCP client's vendor type and configuration. For example, you can configure a vendor-defined class to provide a custom configuration for computers that are running a specific operating system. User-defined classes allow DHCP clients to differentiate themselves by specifying what type of client they are, such as a desktop or laptop. For example, an administrator can configure the DHCP server to assign different options depending on the type of client receiving them. Since a notebook computer is constantly moved and reconnected to the network, users specify a shorter lease to the notebook computer compared with the desktop clients. Note that a client computer needs to be configured with a user-defined class identifier before it sends this identifier to a DHCP server.

To create a new user or vendor class, do the following:

1. In the DHCP console, click the applicable DHCP server.
2. On the Action menu, choose either Define User Classes or Define Vendor Classes and then click on the Add button.
3. In New Class, type the required information.

For Windows 2000 DHCP client computers, the IPCONFIG /SETCLASSID *class* command (where *class* is the unique identity of the user class) can be used to set the specified DHCP class ID string and the IPCONFIG /SHOWCLASSID to confirm that the user class was configured correctly.

10.4.6   Multicast Scopes

IP multicast allows a host to communicate with several hosts simultaneously by transmitting information via one data stream. Multicast can greatly reduce the network traffic that bandwidth-hungry applications such as videoconferencing, software distribution, and Webcast create.

You can create a **multicast scope** so that the DHCP server will issue an IP address to an individual client and a shared multicast address to several clients. For multicasting to work correctly, all routers between the server that is sending packets to the multicast address and the receiving client computers must be configured to recognize the multicast address. See Chapter 19.

To create a multicast scope, do the following:

1. In the DHCP console, click the name of the DHCP server for which you want to create a multicast scope, and then wait for the server status to update.
2. Right-click the name of the DHCP server, and then click New Multicast Scope.
3. In the New Multicast Scope wizard, specify the name and description of the multicast scope, the multicast IP address range, and the number of routers through which multicast traffic can pass.
4. Specify any excluded IP addresses and the lease duration.
5. Activate the multicast scope when prompted.

10.4.7 DHCP Services Best Practices

To get the best results from a DHCP server, do the following:

- Since the DHCP is disk intensive, purchase hardware with optimal disk performance characteristics such as a RAID disk system.
- If routing and remote access service is used on your network to support dial-up clients, reduce the lease time on scopes that service these clients.
- If you have a stable, fixed network where available addresses are plentiful, increase the lease time to reduce DHCP-related broadcast traffic.
- Integrate DHCP services with other services such as WINS and DNS.
- Use the appropriate number of DHCP servers, which depends on the number of DHCP-enabled clients on your network, the transmission speed between network segments, the speed of network links, and if DHCP packets can communicate through routers.
- For dynamic updates such as DNS dynamic updates performed by the DHCP service, use the default client preference settings.

10.4.8 Monitoring and Troubleshooting DHCP Services

Windows 2000 stores the DHCP database in the *systemroot*\SYSTEM32\DHCP directory. By default, the database is automatically backed up every 15 minutes to the *systemroot*\SYSTEM32\DHCP\BACKUP\JET\NEW directory. When DHCP Server service starts, DHCP will perform consistency check of its database and attempt to fix any errors it encounters.

The DHCP Server service records service start-up and shutdown events and critical errors in the Windows system log, which can be viewed with the Event Viewer. You can monitor the details of DHCP operations by enabling detailed event logging. If the detailed event logging is enabled, it will create detailed log files of its activities in files called DHCP-SRVLOG.*xxx* where *xxx* is the first three letters of the day of the week, which are placed in the DHCP database directory.

To enable logging in DHCP, perform the following steps:

1. From the DHCP console, right-click the server you are configuring, and then select the Properties option.
2. In the DHCP Properties dialog box, on the General tab, click Enable DHCP Audit Logging.

If the Event Log contains Jet database messages that indicate a corruption of the DHCP database, first look for disk problems and back up the DHCP database. Then repair the database by using the **JetPack** program from the Windows 2000 Server installation CD. To run the JetPack program, take the following steps:

1. Stop the DHCP server.
2. At the Command prompt, change to the directory where the DHCP database is located (by default *systemroot*\system32\dhcp).
3. Type JETPACK DHCP.MDB *temp* (where *temp* is a file name for a temporary database location that is used during repair) and then press Enter.
4. Start the DHCP server.

If JetPack cannot correct the database, you will then have to restore it from a backup. The JETPACK DHCP.MDB temp command will also compact the database.

10.5 BROWSING THE NETWORK

Windows uses a computer browser service to easily find network computers and their services. Examples of the computer browser service include Network Neighborhood, My Network Places, the NET VIEW command, and Windows Explorer.

The computer browser service used by Windows-based computers operates on each subnet to find network computers and their services. Computers that act as master browsers collect, maintain, and provide the browser list. If a domain is large, then a domain master browser will collect and distribute the master browser lists.

The computer browser service relies mostly on broadcast communications within each subnet in a network. Older Windows versions typically use NetBIOS-based browsing services. Windows 2000 Active Directory can provide computer browser service by using global catalogs.

10.6 SIMPLE NETWORK MANAGEMENT PROTOCOL

The **Simple Network Management Protocol (SNMP)** has become the de facto standard for internetwork management. It allows the system to configure remote devices, monitor network performance, detect network faults, detect inappropriate access, and audit network usage. Remote devices include hubs, bridges, routers, and servers.

SNMP contains two primary elements, a manager and agents. The **SNMP manager** is the console through which the network administrator performs network management functions. The SNMP works by sending messages, called protocol data units (PDUs), to different parts of a network. Agents store data about themselves in **management information bases (MIBs).** The manager, which is the console through which the network administrator performs network management functions, will request the information from the MIB. The **SNMP agent** returns the appropriate information to the manager.

A **trap** is an unsolicited message sent by a SNMP agent to a SNMP management system when the agent detects that a certain type of event has occurred locally on the managed host. The SNMP manage console receives a trap message known as a trap destination. For example, a trap message might be sent when a system restarts or when a router link goes down.

Each SNMP management host and agent belongs to an SNMP community. An **SNMP community** is a collection of hosts that are grouped for administrative purposes. Deciding what computers should belong to the same community is generally but not always determined by the physical proximity of the computers. Communities are identified by the names assigned to them.

Although most networks utilize a user name and a password for authentication, SNMP messages are only authenticated by the community name. A host can belong to several communities at the same time, but a SNMP agent does not accept requests from a management system in a community that is not on its list of acceptable community names. If the community name is incorrect, the agent sends an "authentication failure" trap to its trap destination. Therefore, it is the responsibility of the administrator to set hard-to-guess community names.

Community names are transmitted as clear text, that is, without encryption. Because unencrypted transmissions are vulnerable to attacks by hackers with network analysis software, the use of SNMP community names represents a potential security risk. However, Windows 2000 IP Security can be configured to help protect SNMP messages from these attacks. For more information about IP security, see Chapter 18.

The SNMP service requires the configuration of at least one default community name. The name Public is generally used as the community name because it is a common name that is universally accepted in all SNMP implementations. You can delete or change the default community name or add multiple community names. If no community names are defined, the SNMP agent will deny all incoming SNMP requests.

When an SNMP agent receives a message, the community name contained in the packet is verified against the agent's list of acceptable community names. After the name is determined to be acceptable, the request is evaluated against the agent's list of access permissions for that community. The types of permissions that can be granted to a community are shown in Table 10.11.

TABLE 10.11 Community permissions

Permission	Description
None	The SNMP agent does not process the request. When the agent receives an SNMP message from a management system in this community, it discards the request and generates an authentication trap.
Notify	This is currently identical to the permission of None.
Read Only	The agent does not process SET requests from this community. It processes only GET, GET-NEXT, and GET-BULK requests. The agent discards SET requests from manager systems in this community and generates an authentication trap.
Read Create	The SNMP agent processes or creates all requests from this community. It processes SET, GET, GET-NEXT, and GET-BULK requests, including SET requests that require the addition of a new object to a MIB table.
Read Write	Currently identical to Read Create.

There are two versions of the Simple Network Management Protocol. SNMP 1 reports only whether a device is functioning properly. The industry has attempted to define a new set of protocols, called SNMP 2, that would provide additional information, but the standardization efforts have not been successful. Instead, network managers have turned to a related technology called RMON that provides more detailed information about network usage.

To install the SNMP service, you must be logged on as an administrator or as a member of the Administrators group. To install the SNMP service, do the following:

1. In the Control Panel, double-click the Add/Remove Programs.
2. Click the Add/Remove Windows Components to start the Windows Components wizard.
3. Click to highlight Management and Monitoring Tools and then click Details.
4. Select Simple Network Management Protocol check box, and click OK.
5. Click the Next button and finish the wizard.

To configure agent properties, do the following:

1. In the Control Panel, double-click Administrative Tools and then double-click Computer Management.
2. In the console tree, click Services, which is located under Services and Applications.
3. In the Details pane, click to highlight SNMP Service.
4. Right-click the SNMP Service in the Details pane and select the Properties option.
5. On the Agent tab, in Contact, type the name of the user or administrator for this computer. See Figure 10.14.
6. In Location, type the physical location of the computer or the contact.
7. Under Service, select the appropriate check boxes for this computer, and then click OK. See Table 10.12

To configure traps, do the following:

1. In the Control Panel, double-click Administrative Tools and then double-click Computer Management.
2. In the console tree, click Services, which is located under Services and Applications.
3. In the Details pane, click to highlight SNMP Service.
4. Right-click the SNMP Service in the Details pane and select the Properties option.

309

FIGURE 10.14   SNMP Agent properties

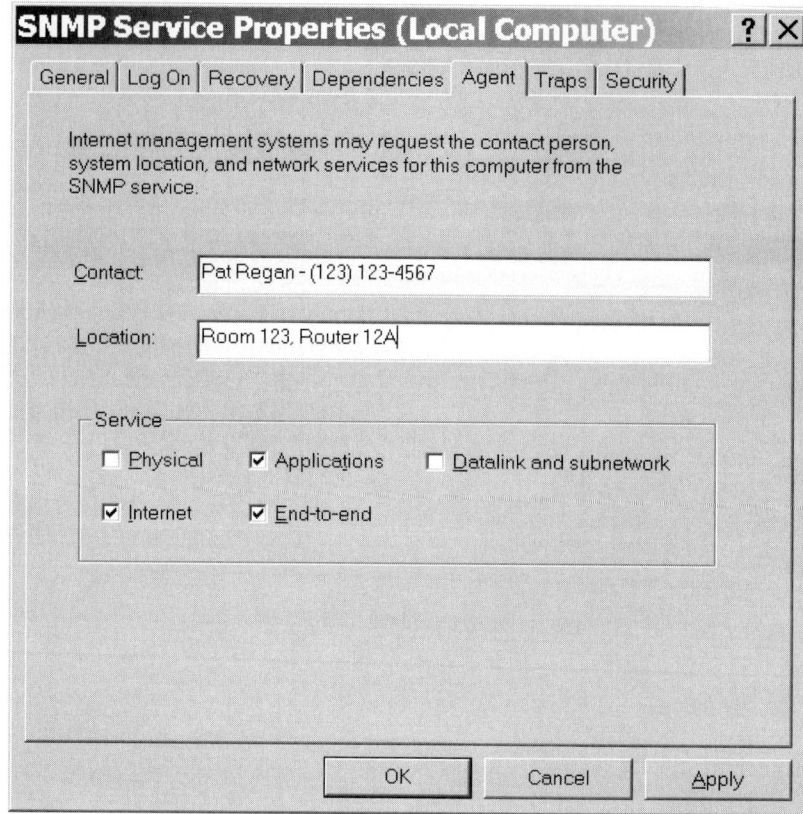

SNMP Service Properties (Local Computer)   ? X

General | Log On | Recovery | Dependencies | Agent | Traps | Security

Internet management systems may request the contact person, system location, and network services for this computer from the SNMP service.

Contact: Pat Regan - (123) 123-4567

Location: Room 123, Router 12A

Service
☐ Physical   ☑ Applications   ☐ Datalink and subnetwork
☑ Internet   ☑ End-to-end

[OK]   [Cancel]   [Apply]

TABLE 10.12   SNMP service types

Agent Service	Select if this computer:
Physical	Manages physical devices, such as a hard disk partition.
Applications	Uses any applications that send data using the TCP/IP protocol suite. This service should always be enabled.
Datalink and Subnet	Manages a bridge.
Internet	Is an IP gateway (router).
End-to-end	Is an IP host. This service should always be enabled.

5. On the Traps tab, under Community Name, type the case-sensitive community name to which this computer will send trap messages, and then click Add to List. See Figure 10.15.
6. In Trap Destinations, click Add.
7. In Host Name, IP, or IPX Address, type information for the host, and click Add.
8. Repeat steps 5 through 7 until you have added all the communities and trap destinations you want.

To configure security, do the following:

1. In the Control Panel, double-click Administrative Tools and then double-click Computer Management.
2. In the console tree, click Services, which is located under Services and Applications.
3. In the Details pane, click to highlight SNMP Service.
4. Right-click the SNMP Service in the Details pane and select the Properties option.
5. On the Security tab, select Send Authentication Trap if you want a trap message sent whenever authentication fails. See Figure 10.16.
6. Under Accepted Community Names, click Add.

FIGURE 10.15 SNMP Trap properties

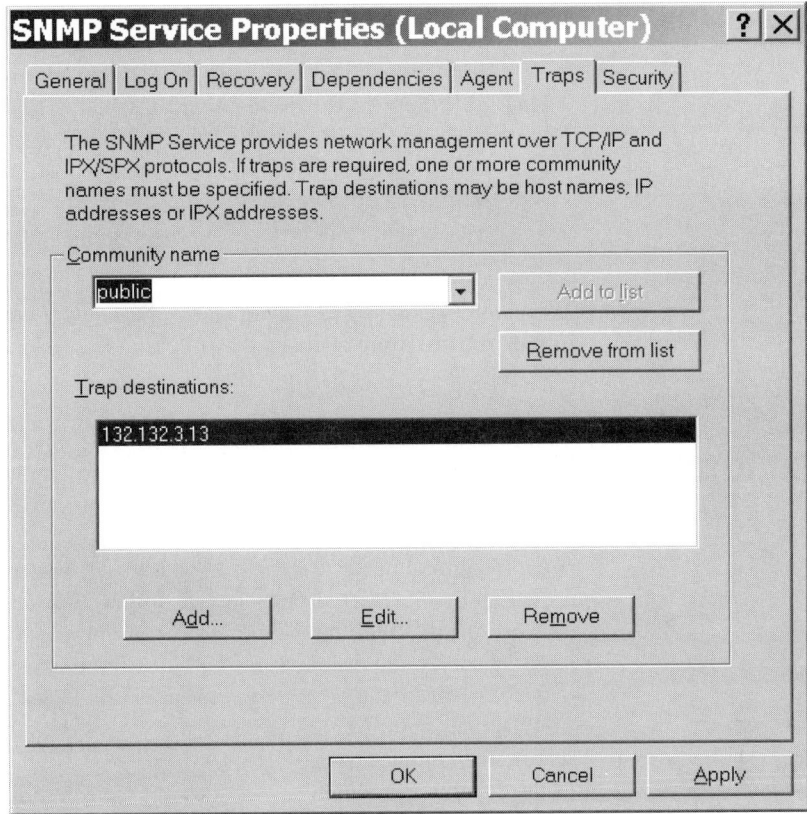

SNMP Service Properties (Local Computer)

General | Log On | Recovery | Dependencies | Agent | Traps | Security

The SNMP Service provides network management over TCP/IP and IPX/SPX protocols. If traps are required, one or more community names must be specified. Trap destinations may be host names, IP addresses or IPX addresses.

Community name

public ▼   Add to list

Remove from list

Trap destinations:

132.132.3.13

Add...   Edit...   Remove

OK   Cancel   Apply

FIGURE 10.16 SNMP Security properties

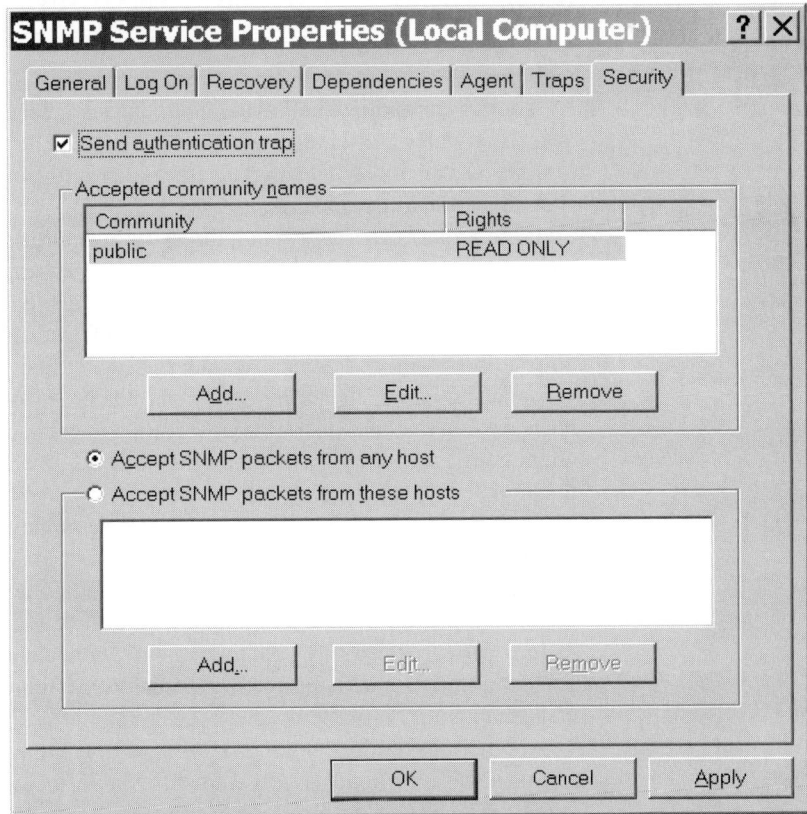

SNMP Service Properties (Local Computer)

General | Log On | Recovery | Dependencies | Agent | Traps | Security

☑ Send authentication trap

Accepted community names

Community	Rights
public	READ ONLY

Add...   Edit...   Remove

◉ Accept SNMP packets from any host
○ Accept SNMP packets from these hosts

Add...   Edit...   Remove

OK   Cancel   Apply

7. Under Community Rights, select a permission level for this host to process SNMP requests from the selected community.
8. In Community Name, type a case-sensitive community name, and then click Add.
9. In SNMP Service Properties, specify whether to accept SNMP packets from a host:
   - To accept SNMP requests from any host on the network, regardless of identity, click "Accept SNMP packets from any host."
   - To limit acceptance of SNMP packets, click "Accept SNMP packets from these hosts," click Add, type the appropriate host name, IP, or IPX address, and then click Add again.

You can make changes to an entry by clicking the entry, and then clicking the Edit button. You can delete a selected entry by clicking the Remove button.

10.7 TROUBLESHOOTING ADVANCED TCP/IP SERVICES

Problem: A DNS client received a *Name not found* error message.

Solution: First ensure that the DNS client computer has a valid IP configuration for the network. Verify that the TCP/IP configuration settings for the client computer are correct, including the IP address, subnet mask, gateway, and those settings that are used for DNS name resolution using WINIPCFG or IPCONFIG /ALL. If you are unsure what the IP address is for the preferred DNS server, observe it by using the IPCONFIG command.

Next ensure that the DNS server is running properly by trying to ping it. If you cannot ping the server, the problem lies with the client, the DNS server, or some link between the two. Remember that it can be caused by the client's IP configuration, the DNS server's TCP/IP configuration, the DNS server's network cards, the cable connected to the DNS server, or the routes to the DNS server.

The client of the DNS server does not have authority for the failed name and cannot locate the authoritative server for this name. If the server is the authority, make sure that the record is there.

Lastly, ensure that the DNS server is still running, by bringing up the DNS console, clicking on the zone, and selecting the Properties option from the Action button.

Problem: The DNS client appears to have received a response with stale or incorrect information in it.

Solution: The client of the DNS server does not have authority for the failed name and is using stale information from its local DNS database. Determine whether the DNS server is authoritative for the name and proceed accordingly.

Another option in which dynamic updates are enabled is to force registration and update at the computer that is targeted by the query. If the target computer is running Windows 2000, you can force it to update the registration of its RRs by using the IPCONFIG / REGISTERDNS command.

If the preferred server is not a direct authority for the queried name, it likely already answered the query based on information obtained and cached during an earlier recursive lookup. In this case, you might consider clearing the server name's cache. This compels the server to use new recursive queries for this RR data and rebuild its cache contents based on current information. If it is a secondary server, then you can initiate a zone transfer at the secondary server to its master server to update the zone.

The name that was queried was specified in error, either through user input or in a stored client configuration. Verify that the name was correctly specified in the application where the name query originated.

For directory-integrated zones, it is also possible that the affected records for the erred query have been updated in active directory, but not replicated to all DNS servers loading the zone. By default, all DNS servers that load zones from active directory poll it at a set interval (typically every 15 minutes), and update the zone for any incremental changes to it. In most cases, a DNS update takes no more than 20 minutes to replicate to all DNS servers used in an active directory domain environment using default replication settings and reliable high-speed links.

Another possibility for the erred data is in whether WINS lookup integration is enabled and used with the zone. If you are using WINS lookup with your zones, verify that WINS is not the source of the erred data.

SUMMARY

1. In Windows 2000, there are two methods to resolve names, domain name system (DNS) and Windows Internet Naming Service (WINS).
2. DNS is the primary method for name resolution for Windows 2000.
3. Domain name system is a hierarchical client-server-based distributed database management system that translates Internet domain names to an IP address.
4. Immediately below the root domain are the top-level domains, which indicate a country, region, or type of organization.
5. Subdomain names are additional names that an organization can create which are derived from the registered second-level domain name.
6. A zone is a portion of the DNS name space whose database records exist and are managed in a particular DNS database file, which is based on a specific domain node.
7. When the resolver queries a DNS server, it will perform a recursive query, where the query asks the DNS server to respond with the requested data or with an error stating that either the requested data or the specified domain name does not exist.
8. To prevent the searching of all domains for an inverse query, a special domain called "in-addr.arpa" was created.
9. NSLOOKUP is a diagnostic tool that displays information from the DNS servers.
10. A WINS server contains a database of IP addresses and NetBIOS (computer names) that update dynamically.
11. A WINS client can be configured to use one of four NetBIOS name resolution methods. They include B (broadcast) node, P (point-to-point) node, M (mixed) node, and H (hybrid) node.
12. A WINS replication partner can be added and configured as either a pull partner, a push partner, or a push/pull partner. The push/pull partner is the default configuration and is the type recommended for use in most cases.
13. A WINS proxy agent is a WINS-enabled computer that is configured to act on behalf of other host computers that cannot directly use WINS.
14. Dynamic host configuration protocol (DHCP) is used to automatically configure a host during boot-up on a TCP/IP network and to change settings while the host is attached.
15. A DHCP relay agent is a computer that relays DHCP and BOOTP messages between clients and servers on different subnets.
16. A superscope allows you to group multiple scopes (child scopes) as a single administrative entity.
17. Using a superscope, the DHCP server computer can activate and provide leases from more than one scope to clients on a single physical network.
18. IP multicast allows a host to communicate with several hosts simultaneously by transmitting information via one data stream.
19. You can create a multicast scope so that the DHCP server will issue an IP address to an individual client and a shared multicast address to several clients.
20. The Simple Network Management Protocol (SNMP) has become the de facto standard for internetwork management. It allows the system to configure remote devices, monitor network performance, detect network faults, detect inappropriate access, and audit network usage. Remote devices include hubs, bridges, routers, and servers.
21. A trap is an unsolicited message sent by an SNMP agent to a SNMP management system when the agent detects that a certain type of event has occurred locally on the managed host.
22. Each SNMP management host and agent belongs to an SNMP community. An SNMP community is a collection of hosts that are grouped for administrative purposes.

QUESTIONS

1. Pat installs and configures an Internet Information Server (IIS) web server on a Windows 2000 Server. Pat configures IIS for both WWW and FTP. He wants users to access the same IIS server as either WWW.DATA.ACME.COM or FTP.DATA.ACME.COM. Which resource record type must Pat add?
   a. AFSDB
   b. CNAME
   c. MB
   d. MG
   e. MINFO

2. Pat's network contains four Windows 2000 Server computers that set up as web servers. Pat wants to enable Windows 98 and Apple Macintosh computers running Internet Explorer to access each of these web servers by using a host name. What should Pat use?
   a. DHCP
   b. DNS
   c. FTP
   d. WINS

3. Julie has just installed the DNS service on a Windows 2000 Server computer. Julie needs to add a resource record for her domain's mail server. Which resource record must she add?
   a. CNAME
   b. MX
   c. PTR
   d. WKS

4. A _____ is a portion of the DNS name space whose database records exist and are managed in a particular DNS database file.
   a. FQDN
   b. zone
   c. subdomain
   d. host name

5. The top of a DNS name space is referred to as the _____ .
   a. root directory
   b. root container
   c. root domain
   d. master domain

6. Which of the following are *not* a second-level domain name? (Choose all that apply)
   a. COM
   b. " "
   c. EDU
   d. MIL
   e. WWW
   f. FTP

7. _____ names are additional names that an organization can create which are derived from the registered second-level domain name so that they can divide their domain.
   a. Second-level domain
   b. Zone
   c. Subzone
   d. Subdomain

8. What is the maximum length of a FQDN?
   a. 11 characters
   b. 63 characters
   c. 128 characters
   d. 255 characters

9. Windows 2000 has three types of zones. Which is *not* one of these?
   a. self-replicating
   b. standard primary
   c. active directory integrated
   d. standard secondary

10. When Windows 2000 uses dynamic update for DNS names, what is the default time before it sends a refresh update?
    a. 1 hour
    b. 12 hours
    c. 24 hours
    d. 8 days

11. When a client does not know the address of a host, it will send what kind of query to its preferred DNS server to find out the address?
    a. interactive
    b. iterative
    c. recursive
    d. broadcast

12. When trying to resolve a host name from its IP address, DNS uses a special domain called a _____ .
    a. in-addr.arpa
    b. caching only
    c. PTR
    d. conical

13. If a client's name records are missing from a DNS server, what command would you use to force the client to renew its registration?
    a. IPCONFIG /RENEW
    b. IPCONFIG /REGISTERDNS
    c. NSLOOKUP
    d. IPCONFIG /FLUSHDNS

14. From his Windows 98 computer, Pat tries to connect to a Windows 2000 Server using http://support.acme.com. He receives the message: "Internet Explorer cannot open the Internet site http://support.acme.com. A connection with the server could not be established." Pat has no problems when connecting to other remote servers using host

names. His computer does not use a HOSTS file for name resolution. Which of the following are the most likely causes of the problem? (Choose all that apply).

    a. The ARP cache on the Windows 98 computer does not have an entry for support.acme.com.

    b. The DNS server has no entry for support.acme.com.

    c. The IP address on the Windows 2000 computer for the default gateway is incorrect.

    d. The Windows 98 computer was provided with an incorrect IP address for support.acme.com from the DNS server.

15. Pat has installed a DNS server on his network. To provide for DNS database redundancy, he wants to implement a backup DNS server. Pat should do this by setting up a

    _____ .

    a. caching-only server

    b. forwarding

    c. replication server

    d. secondary server

16. Which of the following are true of DNS in Windows 2000? (Choose all that apply)

    a. DNS on Windows 2000 can be integrated into the active directory.

    b. DNS on Windows 2000 supports dynamic updates.

    c. DNS on Windows 2000 supports both full zone transfer (AXFR) and incremental zone transfer (IXFR).

    d. A DNS server must be available before active directory can be installed.

17. Which file should the DNS root server use to connect to the Internet?

    a. CACHE.DNS

    b. DOMAIN.DNS

    c. HOSTS

    d. HOSTS.DNS

    e. REVERSE-NETID.in-ADDR.ARPA.DNS

18. Pat's Windows 2000 Professional computer resides on a WINS-enabled network. In which order will Pat's computer perform name resolution if his computer is configured to use an LMHOSTS file?

    a. local cache, broadcasting, LMHOSTS file, WINS server

    b. local cache, WINS server, broadcasting, LMHOSTS file

    c. WINS server, local cache, broadcasting, LMHOSTS file

    d. WINS server, local cache, LMHOSTS file, broadcasting

19. Pat manages a TCP/IP network of both Windows-based and UNIX computers with three subnets—Subnet1, Subnet2, and Subnet3—connected by a single router. After installing a WINS server on Subnet1, he configures each Windows-based computer to recognize the WINS server. However, he soon discovers that the UNIX computers on Subnet2 and Subnet3 are not able to use the WINS server for name resolution. What can Pat do to correct this problem?

    a. Install a second router.

    b. On Subnet2 and Subnet3, install BOOTP relay agents.

    c. On Subnet2 and Subnet3, install WINS proxies.

    d. On the WINS server, install DHCP.

    e. On the WINS server, resolve the names of the clients on Subnet2 and Subnet3.

20. Pat wants to see a list of all NetBIOS names currently cached on this Windows 2000 Server. What command should he use?

    a. ARP

    b. NBTSTAT

    c. NETSTAT

    d. Network Monitor

    e. PING

    f. TRACERT

21. At each of his company's two offices, Pat has installed a WINS server. A T1 line connects both offices. He wants both servers to replicate their WINS databases with one another. How does Pat do this?

    a. Make each WINS server a WINS client of the other WINS server.

    b. Set the Directory Replicator Server service to start automatically.

c. Set up each WINS server as both a push partner and a pull partner of the other WINS server.

d. Set up each WINS server as the secondary WINS server of the other server.

22. Pat's company is headquartered in Sacramento with a second office in New York. Both offices use TCP/IP as the networking protocol. The Sacramento office has 10 Windows 2000 Servers and 800 Windows 2000 Professional computers. The Chicago office has 2 Windows 2000 Server computers and 150 Windows 2000 Professional computers. The WINS server in Sacramento is called SERVER1. The WINS server in New York is called SERVER2. Pat wants to set up WINS database replication between the two WINS server.

Required result:

He must replicate the Sacramento WINS server database to the New York WINS server.

Optional desired results:

1. He must replicate the New York WINS server database to the Sacramento WINS server.
2. He wants to be sure that the Sacramento WINS database is replicated to New York at least once a day.

Proposed solution:

Configure Sacramento to push its WINS database update information to New York once every 1000 updates. Configure New York to pull Sacramento's WINS database update information once every 24 hours.

Which result does the proposed solution produce?

a. Achieves the required result and both optional results.

b. Does not achieve the required result but achieves both the optional results.

c. Does not achieve the required result but does achieve one of the optional results.

d. Achieves the required result but cannot achieve either of the optional results.

e. Achieves neither the required result nor either of the optional results.

23. You discover that errors in the LMHOSTS file of your Windows 2000 Server computer are creating problems on your network. After correcting the LMHOSTS file, which command will issue to purge the server's NetBIOS name cache?

a. ARP   d. PING

b. NBTSTAT   e. ROUTE

c. NETSTAT

24. Pat moves a Windows 2000 Professional computer originally configured at a DHCP client from one subnet to another on a DHCP network. He discovers that the workstation cannot connect to the local Windows 2000 Server. What is the most likely cause of the problem?

a. The Windows 2000 Professional computer's default gateway address has been manually set.

b. The Windows 2000 Professional computer's default gateway address is set as a local-level option on the DHCP server.

c. The Windows 2000 Professional computer's IP address has been manually set.

d. The workstation's IP address is set as a scope-level option on the DHCP server.

25. You administer a network of 1000 client computers on five subnets. You use DHCP to assign IP addresses to all client computers. You install two DNS servers on the network. Now you want to specify the IP addresses of both DNS servers on each client computer across all five subnets. How do you configure the DHCP option?

a. As a client option   c. As a scope option

b. As a global option   d. As an Internet option

26. Pat uses DHCP to assign IP addresses to all client computers on his network. Pat sets up the DHCP server to assign the IP address of the WINS server to all client computers. What else should Pat do to allow client computers to use WINS?

a. In the DHCP console, specify the NetBIOS resolution mode.

b. On each client computer, enter the IP address of at least one DNS server.

c. On each client computer, enter the IP address of the local gateway.

d. One each client computer, specify the NetBIOS scope ID.

27. Pat has just changed the DHCP configuration on his Windows 2000 Server. He wants so see if the client computer has registered the changes. Which command should Pat execute?
    a. ARP –S
    b. IPCONFIG /ALL
    c. NBSTART –S
    d. NETSTART –E
    e. PING –S

28. You want DHCP to assign IP addresses to all Windows-based computers on your network. Each Windows 2000 Server is to be assigned that same unique IP address each time that server is booted up. How should you proceed?
    a. Exclude a range of IP addresses that will be assigned to the servers.
    b. For each server, create a separate scope that contains that server's IP address.
    c. For each server, implement a client reservation.
    d. For each server, specify an unlimited lease period.

29. Pat administers several hundred computers on a TCP/IP network with six subnets. Many users have notebook computers that run Windows 98 and Windows 2000. Pat wants to automatically assign IP addresses to these laptop computers each time they connect. Which service must Pat configure?
    a. DHCP
    b. DNS
    c. FTP
    d. SNMP
    e. WINS

30. On a network with six subnets, all client computers use only one DHCP server. A global option on the DHCP server specifies the IP address of the DNS server that the client computers are to use. A user complains that hc cannot connect to other computers using their host names. Running the IPCONFIG utility on the user's computer, you ascertain that all IP settings are correct except for the IP address of the DNS server. What could account for the difference between the DNS server's IP address as assigned by the DHCP server and its IP address as found on the user's computer? (Choose all that apply)
    a. The client computer is on a subnet other than the one for which the DHCP option was defined.
    b. The DHCP scope is not activated for the client's subnet.
    c. The IP address for the DNS server has been created for the client computer as a client option.
    d. The user's computer is no longer a DHCP client, and the IP address for the DNS server has been manually set at the user's computer.

31. Pat uses DHCP, WINS, and DNS to administer a large TCP/IP network. He wants to reserve the IP addresses of his three domain controllers and his DNS server. What information must Pat have when reserving these servers in DHCP console?
    a. MAC addresses
    b. IP addresses
    c. lease durations
    d. subnet masks

32. Pat is designing a TCP/IP network with two subnets: Subnet1 and Subnet2. Pat configures a Windows 2000 Server computer as a static router between the two subnets. He installs a second Windows 2000 Server computer as a DHCP server on Subnet1 and a third Windows 2000 Server computer on Subnet2. Pat wants client computers on Subnet2 to obtain their IP configurations from the DHCP server on Subnet1. Which service must Pat install on the third Windows 2000 Server computer on Subnet2?
    a. LLC sublayer
    b. proxy service
    c. RAS service
    d. DHCP relay agent service
    e. RIP for IP service

33. Suppose the following situation exists:
    Pat administers a TCP/IP network of Microsoft-based computers on five subnets. Pat wants to install two DHCP servers, each on a separate subnet that will automatically assign IP addresses to the host computers.
    Required result:
    Each DHCP server must act as a backup server if the other DHCP server is down.

Optional desired results:

1. DHCP should provide the same unique IP addresses to each Windows 2000 Server computer when that server initializes.
2. DHCP should assign the IP addresses of the WINS servers and DNS servers to all DHCP clients.

Proposed solutions:

1. On each subnet, install the DHCP relay agent.
2. On one of the DHCP servers, define a DHCP scope for each subnet. In each scope, define the IP address range available to that subnet. Assign half of the available IP addresses on each subnet to each scope.
3. On the other DHCP server, define a DHCP scope for each subnet. In each scope, define the IP address range available to that subnet. Assign the remaining half of the available IP addresses on each subnet to each scope.
4. On the DHCP servers, enable and configure the 044 WINS/NBNS servers, the 046 WINS/NBT node type options, and the 006 DNS server options.

Which result does the proposed solution produce?

a. Achieves the required result and both optional results.
b. Does not achieve the required result but achieves both optional results.
c. Does not achieve the required result but does achieve one of the optional results.
d. Achieves the required result but cannot achieve either of the optional results.
e. Achieves neither the required result nor either of the optional results.

34. You manage a network with seven subnets. The network has two DHCP servers, two WINS servers, and three DNS servers. All client computers on the network are Windows based and use DHCP. One user reports receiving a message that he has a duplicate IP address. What is the cause of the problem?

a. DHCP lookup is not enabled on one of the DNS servers.
b. The user's computer is registered with more than one WINS server.
c. DHCP servers have overlapping scopes.
d. WINS lookup is not enabled on one of the DHCP servers.

35. You are setting up a DHCP relay agent on a Windows 2000 Server. What information do you need?

a. DHCP node type
b. DHCP relay agent lease duration
c. IP address of the DHCP server
d. NetBIOS name of the DHCP server

36. Which service must be started on a Windows 2000 Professional computer so that trap messages can be forwarded to various host names on a network?

a. Alerter service
b. EventLog service
c. SAP service
d. SNMP service

37. Pat wants to send TCP/IP protocol statistics from his Windows 2000 computer to a UNIX computer. Both computers are on the same TCP/IP network. What must Pat install on each computer.

a. NETSTAT.EXE on the UNIX computer and Network Monitor Agent on the server
b. Performance Monitor on the UNIX computer and SNMP service on the server
c. Protocol Analyzer on the UNIX computer and Network Monitor Agent on the server
d. SNMP management software on the UNIX computer and SNMP service on the server

38. Pat wants the SNMP service on his Windows 2000 Server computer to send trap messages to an SNMP trap destination. Which of the following must be supplied? (Select two answers)

a. SNMP management station's community
b. SNMP management station's IP address
c. SNMP management station's scope ID
d. SNMP management station's subnet mask

39. You want to manage and monitor the hosts on your Windows-based network using SNMP. Which of the following do you need to install or configure on your network? (Choose all that apply)
    a. Performance Monitor
    b. SNMP agent
    c. SNMP manager software
    d. TCP/IP

40. Which service automatically registers the domain names to the Windows DNS server using dynamic update?
    a. DNS service
    b. WINS service
    c. DHCP service
    d. SNMP service

LAB EXERCISES

EXERCISE 1: INSTALLING AND CONFIGURING DNS SERVICES

One partner will configure the primary zone, while the other will configure a secondary zone and designate his or her partner's computer as the master server. Record the type of zone you will configure in this exercise.

Configuring the DNS Domain Name of Your Computer

Computers A and B should do the following:

1. Right-click on the My Computer icon and select the Properties option.
2. Click on the Network Identification tab and click on the Properties button. Click on the More button.
3. For the primary DNS suffix, enter ACMExx.COM, where *xx* is your partner number. If you are the first group or are working at home, enter ACME01.COM. Click on the OK button.
4. Notice the full computer name. Click on the OK button.
5. Close the System Properties window and reboot the system.

Installing the DNS services

Computer A should do the following:

6. Open the Add/Remove Programs applet in the Control Panel.
7. Click on the Add/Remove Windows Components.
8. Click on the Next button.
9. Click to highlight the Networking Services option.
10. Click on the Details button.
11. In the Subcomponents of Networking Services, ensure the check box is marked next to the Domain Name System (DNS).
12. Click on the OK button.
13. Click on the Next button.
14. If the Insert Disk dialog box appears, insert the Windows 2000 installation CD, ensure that the path to the source files is correct, and click on the OK button.
15. Click on the Finish button.

Configuring the DNS Server

Computer A should do the following:

16. Open the DNS console by clicking on the Start button, clicking the Programs option, clicking the Administrative Tools option, and selecting the DNS option.
17. The DNS tree is shown in the left pane. Click on the + symbol next to the SERVER2000-01A server. The tree will expand to show Forward Lookup and Reverse Lookup.
18. Right-click the server and select the Configure the Server option. The Configure New DNS Server wizard will begin. Click on the Next button.
19. On the Forward Lookup Zone page, select Yes; create a forward lookup zone and click on the Next button.
20. On the Zone Type page, select the Standard Primary Zone. Click on the Next button.

21. On the Zone Name page, enter the ACMExx.COM, where *xx* represents your partner number. This should be the same domain name that was used in the Identification tab specified previously. Click on the Next button.
22. On the Zone File page, click on the Next button to use the default file name.
23. On the Reverse Lookup Zone page, select Yes; create a reverse lookup zone option and click on the Next button.
24. On the Zone Type page, select the Standard Primary option and click on the Next button.
25. In the Reverse Lookup Zone, enter the first three octets address of the TCP/IP address (140.100.1) and click on the Next button.
26. On the Zone File page, select the Create a New file with this file name and click on the Next button.
27. Click on the Finish button.
28. In the DNS console, expand the entire tree. See Figure 10.17.
29. Right-click the domain name for the forward lookup zone and select the Properties option. If you are the first group, right-click on ACME01.COM.
30. Change the Allow Dynamic Update option to Yes. Click on the OK button.
31. Right-click the 140.100.1x subnet for the reverse lookup zone and select the Properties option.
32. Change the Allow Dynamic Update option to Yes. Click on the OK button.

Computer B should do the following:

33. Start a Command prompt.
34. At the Command prompt, execute PING *address_of_serverA* command, where the *address_of_serverA* is the address of Server A.
35. At the Command prompt, execute PING *computernameA* command, where the *computernameA* is the NetBIOS name of Server A. This command should work because it is finding the computer using broadcast, not DNS.
36. At the Command prompt, execute PING *DNS_name of serverA* command, where the *DNS_name_of_serverA* is the full computer name including the domain name. For example, if you are in group 1, it should be Server2000-01A.acme01.com.
37. Using the properties of the TCP/IP of your Local Area Connection in My Network Place, set the preferred DNS server to the address of Server A.
38. At the Command prompt, execute PING *DNS_name of serverA* command, where the *DNS_name_of_serverA* is the full computer name including the domain name. For example, if you are in group 1, it should be Server2000-01A.acme01.com. It should work now because it is getting the address from the DNS server. See Figure 10.18.

Adding Resource Records

Computer B should do the following:

39. At the Command prompt, execute PING PC.ACMExx.COM command, where *xx* is your partner number.

Computer A should do the following:

40. From the DNS console, right-click the domain name for the forward lookup zone and select the New Host option. If you are the first group, right-click on ACME01.COM.

FIGURE 10.17

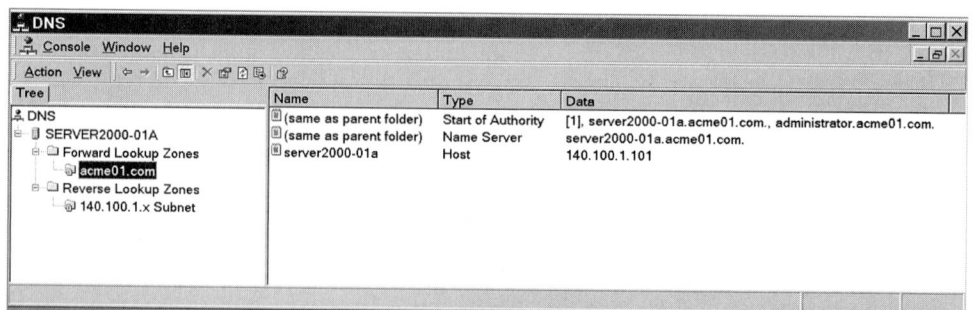

41. In the Name text box, type in PC. Note that it will automatically attach ACMExx.COM.
42. Specify the address of 140.100.1.100.
43. Select the "Create associated pointer (PTR) record to automatically create a reverse lookup" entry and click on the Add Host button. Click on the Done button.

Computer B should do the following:

44. From the Command prompt, ping PC.ACMExx.COM. Notice that it translated the address with no problem. Because we do not have a computer called "PC" as part of our domain with this specific address, the request will time out.

Computer A should do the following:

45. Right-click the PC Host entry and select the Delete option.
46. From the DNS console, right-click the domain name for the forward lookup zone and select the New Alias option. If you are the first group, right-click on ACME01.COM.
47. Leave the Alias Name text box blank. Note that it will assume ACMExx.COM.
48. In the Full Qualified Name text box, specify the full qualified domain name of your server (SERVER2000-01A.ACMExx.COM).

Computer B should do the following:

49. From the Command prompt, ping your domain name (ACMExx.COM).
50. Click the Start button and select the Run option. Type in MMC and click the OK button.
51. Open the Console menu and select the Add/Remove Snap-in option.
52. Click on the Add button. Select the DNS option and click the Add button. Click on the OK button.
53. At the top of the console tree, right-click DNS and select the Connect to Computer option. Select the "The following computer" option, specify SERVER2000-01A. ACMExx.COM. Click on the OK button.
54. Right-click the domain name under Forward Lookup Zone and select the New Alias option.
55. Specify WWW for the alias name and specify the full qualified domain name for the DNS server. Click on the OK button.
56. Create an alias for FTP, which should also point to the domain server.
57. Add a host record for your instructor's computer.
58. Close the MMC console.
59. Ping WWW.ACMExx.COM and FTP.ACME.xx.COM.

Test the DNS Server

Computer A should do the following:

60. Right-click the DNS server in the DNS console and select the Properties option.
61. From the Monitoring Tab, select "A simple query against this DNS server" and "A recursive query to other DNS servers."
62. Click on the Test button and note the test results.
63. Click on the OK button.
64. At the Command prompt, execute the NSLOOKUP command.
65. At the NSLOOKUP Command prompt, type in the name of your domain (ACMExx. COM) and press Enter.
66. To set the default server, execute SERVER SERVER2000-xxA.ACMExx.COM at the prompt. For the first group, type in NSLOOKUP SERVER200001A.ACME01.COM.
67. To list the A records, execute ls–t a ACMExx.COM command at the Command prompt. The command is case sensitive.
68. To look at the name servers, execute the type=ns command at the NSLOOKUP Command prompt, then type in the domain name, and then press Enter.
69. To look at the SOA records, execute the type=soa command at the NSLOOKUP Command prompt, then type in the domain name, and then press Enter.
70. Type in Exit, to go back to the Command prompt.
71. Close the Command prompt window.

72. Double-click My Computer, double-click the C drive, double-click the WINNT folder, double-click the SYSTEM32 folder, and then double-click the DNS folder.
73. Right-click the Domain Name file and select the Open With option. Select Notepad. View the contents of the file and close it when done.

Create a Secondary DNS Server

Computer B should do the following:

74. Open the Add/Remove Programs applet in the Control Panel.
75. Click on the Add/Remove Windows Components.
76. Click on the Next button.
77. Click to highlight the Networking Services option.
78. Click on the Details button.
79. In the Subcomponents of Networking Services, ensure the check box is marked next to the Domain Name System (DNS).
80. Click on the OK button.
81. Click on the Next button.
82. If the Insert Disk dialog box appears, insert the Windows 2000 installation CD, ensure that the path to the source files is correct, and click on the OK button.
83. Click on the Finish button
84. Open the DNS console from the Administrative Tools menu.
85. Right-click the server and select Configure the Server option.
86. On the Forward Lookup Zone page, select Yes; the Create a Forward Lookup Zone option is selected. Click the Next button.
87. On the Zone Type page, select the Standard Secondary option and click the Next button.
88. On the Zone Name page, type in your domain name (ACMExx.COM). If you are the first group, it will be ACME01.COM. Click on the Next button.
89. On the Master DNS Servers page, in the IP Address box, type the IP address of your partner's computer, click the Add button and then the Next button.
90. On the Reverse Lookup Zone page, select Yes; the Create a Reverse Lookup Zone option is selected. Click the Next button.
91. On the Zone Type page, select the Standard Secondary option, and then click the Next button.
92. On the Reverse Lookup Zone page, type the first three octets of your IP address (140.100.1) in the IP Address box and click the Next button.
93. On the Master DNS Servers page, type the IP address of your partner's computer. Click the Add button, then the Next button.
94. Click the Finish button.

Creating a Subdomain

Computer A should do the following:

95. Start the DNS console.
96. Right-click the Domain Name and select the New Domain option.
97. For the new domain type, type in "Sales" and click the OK button.

Deleting a Secondary Domain

Computer B should do the following:

98. Right-click the Secondary Domain in your DNS console and select the Delete option. Click on the OK button.

Declaring Delegation of Authority

Computer B should do the following:

99. Click on the Start button. Select the Programs option and select Administrative Tools. Then right-click DNS and select the Run As option.

100. In the Run As Other User dialog box, specify Administrator as the user and Password as the password. Delete the contents of the Domain box. Click on the OK button.
101. Right-click DNS in the console and click Connect to Computer.
102. In the Select Target Computer dialog box, click "The following computer in the text box," type in the name of your partner's computer (SERVER2000-xxA), and click OK.
103. In the console tree, expand the server, expand the forward lookup zones, and expand the domain.
104. Right-click the Domain and select the New Delegation option. Click the Next button.
105. On the Delegated Domain Name page, type the name of your domain in the Domain box and click the Next button.
106. On the Name Servers page, click the Add page.
107. In the New Resource Record dialog box, type the full qualified domain name for your computer and click the Resolve button. Click on the OK button.
108. On the Name Servers page, click the Next button.
109. On the Completing the New Delegation Wizard page, click the Finish button.

EXERCISE 2: INSTALL AND CONFIGURE A WINS SERVICES

Installing the WINS services

Computer A should do the following:

1. Open the Add/Remove Programs applet in the Control Panel.
2. Click on the Add/Remove Windows Components.
3. Click on the Next button.
4. Click to highlight the Networking Services option.
5. Click on the Details button.
6. In the Subcomponents of Networking Services, ensure the check box is marked next to the Windows Internet Name Services (WINS).
7. Click on the OK button.
8. Click on the Next button.
9. If the Insert Disk dialog box appears, insert the Windows 2000 installation CD, ensure that the path to the source files is correct, and click on the OK button.
10. Click on the Finish button.

Configuring the WINS Client

Computer A should do the following:

11. To make the WINS server register itself in its WINS database, make the WINS server a client. To make a WNS server a client, right-click My Network Places and select the Properties option. Right-click the Local Area Connection and select Properties.
12. Click Internet Protocol (TCP/IP) and click on the Properties button.
13. In the Internet Protocol (TCP/IP) Properties dialog box, click the Advanced button.
14. In the WINS tab, click the Add button.
15. In the TCP/IP WINS Server dialog box, type the IP address of the WINS server, and click the Add button.
16. Click the OK button to close the Advanced TCP/IP Settings dialog box and click OK to close the Internet Protocol (TCP/IP) Properties dialog box.
17. Click OK to close the Local Area Connection Properties dialog box.

Computer B should do the following:

18. To configure the client to use the WINS server, right-click My Network Places and select the Properties option. Right-click the Local Area Connection and select Properties.
19. Click Internet Protocol (TCP/IP) and click on the Properties button.
20. In the Internet Protocol (TCP/IP) Properties dialog box, click the Advanced button.
21. In the WINS tab, click the Add button.

22. In the TCP/IP WINS Server dialog box, type the IP address of the WINS server, and click the Add button.
23. Click the OK button to close the Advanced TCP/IP Settings dialog box, and click OK to close the Internet Protocol (TCP/IP) Properties dialog box.
24. Click OK to close the Local Area Connection Properties dialog box.

Managing the WINS Server

Computer A should do the following:

25. From the Administrative Tools, start the WINS console.
26. Right-click Active Registration and select the Find by Name option.
27. In the "Find names beginning with" text box, type in an asterisk (*) and click on the Find Now button. Notice all of your entries.
28. To add a static WINS record, right-click Active Registration and select the New Static option.
29. In the Computer Name text box, type in TESTPC. In the IP address, type in 100.100.100.100. Click on the OK button.
30. Use the Find Names option to show all entries in the WINS table. Notice the entries in the WINS table, especially the Static column.

Computer B should do the following:

31. From the Command prompt, ping TESTPC. Since the TESTPC does not exist, PING will respond, "Destination host unreachable." Most importantly, note the address resolution.

Computer A should do the following:

32. Right-click the three entries for TESTPC and select the Delete option.

Setting Up Replication

Computer B should do the following:

33. Open the Add/Remove Programs applet in the Control Panel.
34. Click on the Add/Remove Windows Components.
35. Click on the Next button.
36. Click to highlight the Networking Services option.
37. Click on the Details button.
38. In the Subcomponents of Networking Services, ensure the check box is marked next to the Windows Internet Name Services (WINS).
39. Click on the OK button.
40. Click on the Next button.
41. If the Insert Disk dialog box appears, insert the Windows 2000 installation CD, ensure that the path to the source files is correct, and click on the OK button.
42. Click on the Finish button.

Both Computers should do the following:

43. From the WINS console, right-click Replication Partners and select the New Replication Partner option.
44. In the WINS Server text box, type the IP address of your partner's computer and click on the OK button.
45. In the Details pane, right-click your partner's computer name, and then click the Properties option.
46. In the Advanced tab, verify that the replication partner type is push/pull and that persistent connections are used for both push and pull replications.

Computer A should do the following:

47. Set the start time for pull replication to 18:00 (6 PM) and the replication interval to 12 hours.

Computer B should do the following:

48. Set the start time for pull replication to 21:00 (9 PM) and the replication interval to 12 hours.

Both Computers should do the following:

49. In the "Number of changes in version IS before replication" box, type 250.
50. Click the OK button to close the Properties dialog box for the replication partner.

Computer A should do the following:

51. Right-click Replication Partners and select the Replicate Now option.
52. Click the Yes button to be sure and click on the OK button.

Computer B should do the following:

53. From the WINS console, right-click Active Registration and list all owners. The entries should be the same as your partners. If not, wait another minute and try again.

EXERCISE 3: INSTALL AND CONFIGURE A DHCP SERVER

Installing a DHCP Server

Computer A should do the following:

1. Open the Add/Remove Programs applet in the Control Panel.
2. Click on the Add/Remove Windows Components.
3. Click on the Next button.
4. Click to highlight the Networking Services option.
5. Click on the Details button.
6. In the Subcomponents of Networking Services, ensure the check box is marked next to the Windows Internet Name Services (WINS).
7. Click on the OK button.
8. Click on the Next button.
9. If the Insert Disk dialog box appears, insert the Windows 2000 installation CD, ensure that the path to the source files is correct, and click on the OK button.
10. Click on the Finish button.

Creating a Scope

Computer A should do the following:

11. Open the DHCP console from the Administrative Tools.
12. Right-click the server and click the New Scope option. Click on the Next button.
13. On the Scope Name page, type the name of your server in the Name text box and click on the Next button.
14. Use 140.100.1.XX for the start address and 140.100.1.XX+1 for the end address, where *XX* is your computer number. Change the subnet mask to 255.255.255.0. Click on the Next button.
15. Since we have no exclusions, click the Next button.
16. On the Lease Duration, click the Next button to keep the default of 8 days.
17. On the Configure DHCP Options page, select the "Yes, I want to configure these options now" option and click the Next button.
18. On the Router (Default Gateway) page, type in the address of your gateway or local router. If you do not have one, for now type in 140.100.1.1. Click the Add button and then the Next button.
19. On the Domain Name and DNS Servers page, type in ACME.COM for the parent domain and type in the address of your DNS server. Click the Add button and then the Next button.
20. On the WINS Servers page, type in the address of your primary WINS server address. Click the Add button and then the Next button.

21. Select the "Yes, I want to active this scope now" option and click the Next button. Click the Finish button.
22. Click on the Scope option and look at the various options that were configured with the wizard.

Making a Client Reservation

Computer A should do the following:

23. From the DHCP console, right-click Reservations and select the New Reservation option.
24. Put in the MAC address of your partner's network card. You can find this out by executing the IPCONFIG /ALL command at your partner's computer.

Testing the DHCP Server

Computer B should do the following:

25. At the Command prompt, execute the IPCONFIG /ALL command.
26. Right-click My Network Places and select the Properties option.
27. Right-click Local Area Connection and select the Properties option.
28. Click the Internet Protocol (TCP/IP) and click on the Properties button.
29. In the Internet Protocol (TCP/IP) dialog box, click the "Obtain an IP address automatically" option.
30. Click the "Obtain DNS server address automatically" option.
31. Click on the Advanced button.
32. In the WINS tab, click the WINS address and click the Remove button.
33. Click the OK button to close the Advanced TCP/IP Settings, click OK to close the Internet Protocol (TCP/IP) Properties dialog box, and click OK to close the Local Area Connection Properties dialog box. Remember, any settings set manually will override settings given by a DHCP server.
34. At the Command prompt, execute the IPCONFIG /ALL command and study the current setting.

11

Introduction to the Active Directory

INTRODUCTION

One of the biggest changes from Windows NT to Windows 2000 was the introduction of Active Directory, a true directory service. Before we continue on how to create user and computer accounts and to the everyday management activities of a network, it is important to understand what Active Directory is and how it works.

OBJECTIVES

1. Compare and contrast workgroups and domains.
2. Describe the relationships between Active Directory tree, domain, and organizational units (OUs).
3. Install the Active Directory.
4. Create sites and site links.
5. Create global catalog servers.
6. Transfer operations master roles.
7. Implement an OU structure.
8. Explain the purpose of the global catalog.
9. Describe the roles of the domain controller, and given a domain controller, identify its role.
10. Describe the relationship between sites and domains.
11. Move Active Directory objects.
12. Control access to Active Directory objects.
13. Delegate administrative control of objects in Active Directory.
14. Manage and troubleshoot Active Directory replication.
15. Compare and contrast private and public-key encryption.
16. Compare and contrast digital envelope, digital signature, and digital certificate.
17. List, compare, and contrast the three types of authentication used in Windows 2000.
18. Install a time sync with a time server.

11.1 INTRODUCTION TO DIRECTORY SERVICES

Directory services is a network service that identifies all resources on a network and makes those resources accessible to users and applications. Resources can include e-mail addresses, computers, and peripheral devices (e.g., printers). Ideally, the directory service should make the physical network, including topology and protocols, transparent to the users on the network. They should be able to access any resources without knowing where or how they are physically connected.

11.2 X.500 DIRECTORY SERVICES AND LIGHTWEIGHT DIRECTORY ACCESS PROTOCOL

There are a number of directory services that are widely used. The most common is the **X.500,** which uses a hierarchical approach, in which objects are organized in a similar way to the files and folders on a hard drive. At the top of the structure is the root container with "children" organized under it. X.500 is part of the OSI model; however, it does not translate well into a TCP/IP protocol environment. Therefore, many of the protocols that are based on the X.500 do not fully comply with it.

Windows 2000 Active Directory uses **lightweight directory access protocol (LDAP)** to communicate and perform basic read, write, and modify operations; query an object in the Active Directory; and allow Windows 2000 Servers to communicate with Windows 3.51 and Windows 4.0 servers.

11.3 WORKGROUP

Computers and devices on a peer-to-peer network are usually organized into logical subgroups called **workgroups.** In a workgroup, each computer has a local security database so that it tracks its own user and group account information. The user information is not shared with other workgroup computers; therefore, you must create users on each workgroup computer if you want to use the other computer's network resources. A workgroup is used more as a basic grouping of the computers and is only intended to help users find objects such as printers and shared folders within that group.

Workgroup Advantages
- A workgroup does not require a computer that is running Windows 2000 Server to hold centralized security information.
- A workgroup is simple to design and implement.

- A workgroup is convenient for a limited number of computers in close physical proximity and should not consist of more than 10 computers.
- Each user must manage his or her own computer resources.

Workgroup Disadvantages

- A user must have a user account of each computer to which he or she wants to gain access.
- Any changes to user accounts, such as changing a user's password or adding a new user account, must be made on each computer.
- Workgroups require each user to administer his or her own computer, thus, each user must be relatively proficient for the workgroup to work well.

11.4 ACTIVE DIRECTORY

Resources on a network are useless if users cannot find or manage them. Active Directory provides a treelike summary of resources to users and allows users to manage resources from anywhere on the network. It maintains the network information in files that are distributed around the network. This provides fault tolerance and higher performance.

Active Directory is a network directory service that combines domains, X.500 naming services, DNS, X.509 digital certificates, and Kerberos authentication. It stores all information about the network resources and services such as user data, printers, servers, databases, groups, computers, and security policies. Active Directory supports several name formats so that it can be compatible with several standards, as shown in Table 11.1.

The Active Directory database uses a data store to hold all information about the objects. The data store is sometimes referred to as the directory. It is stored on domain controllers and is accessed by network applications and services. The Active Directory uses the **extensible storage engine (ESE),** which allows the Active Directory object database to grow to 17 terabytes, giving it the ability to hold up to 10 million objects, although it is best to have no more than 1 million objects.

Active Directory is made of the following files:

- NTDS.DIT is a single file that makes up the Active Directory database. It stores the Active Directory objects on the domain controller. The default location is the *%systemroot%*\NTDS folder.
- The EDB*.LOG files are the transaction log files with the EDB.LOG being the default log file. Each transaction log file is 10 MB. When the EDB.LOG is full, it is renamed to EDB*nnnnnn*.log, where *nnnn* is an increasing number starting from one.
- The EDB.CHK is a checkpoint file used by the database engine to track the data not yet written to the Active Directory database file. The checkpoint file is a pointer that maintains the status between memory and the database file on disk.
- RES1.LOG and RES2.LOG are reserved transaction log files to reserve disk space to provide sufficient room to shut down if all other disk space is being used.

TABLE 11.1 Name format supported by Windows 2000 Active Directory

Format	Description
RFC 822	Internet e-mail addresses that follow a user_name@domain.xxx format
HTTP uniform resource locators (URL)	Hypertext transfer protocol web page addresses that use DNS and follow the http://domain.xxx/path_to_page
Universal naming convention (UNC)	A NetBIOS address that follows the \\domainname\sharename
LDAP URL	An address that specifies the server or resource on the Active Directory services tree that follows the LDAP://Domain.xxx/CN=Commonname_of_Resource, OU=Organization_Name, DC=DomainComponentName name

During the installation of Active Directory, a folder called **System Volume (SYSVOL)** is installed onto the hard drive. In the SYSVOL folder is another folder called SYSVOL, which is shared. The shared folder stores the server copy of the domain's public files, such as the Group Policy and Net Login (logon scripts), which are replicated among all domain controllers in the domain. The contents of SYSVOL are replicated to all domain controllers in the domain. Because the creation of SYSVOL requires a volume to be formatted with NTFS, if an NTFS-formatted volume cannot be found or if there is not sufficient free disk space, the installation cannot proceed.

A handy method for checking SYSVOL replication is to create a file named after the originating computer, such as %systemroot%\Sysvol\Sysvol\domain_name\file_name_equals_server_name. Then observe which other domain controllers receive the new file. Updates to servers in remote sites are governed by a schedule.

The Active Directory, similar to a file system found on your hard drive or DNS, is a structured hierarchy consisting of domains and organizational units that are connected to form a tree of objects (computers, users, and network resources). The top of the Active Directory is called the root domain. Domains that are under the root domain are child domains to the root domain, and the root domain is the parent domain to its child domains. In addition, a child domain could have another domain under it. Therefore, the child domain of the root domain is the parent domain to its child domains.

11.4.1   Windows 2000 Domain

The Windows 2000 **domain** is a logical unit of computers and network resources that define a security boundary. It is typically found on medium or large networks or networks that require a secure environment. Different from a workgroup, a domain uses one database to share its common security and user account information for all computers within the domain. Therefore, it allows centralized network administration of all users, groups, and resources on the network.

DOMAIN BENEFITS
- A domain provides centralized administration because all user information is stored centrally.
- A domain provides a single logon process for users to gain access to network resources and only needs to log on once to gain those resources.
- A domain provides scalability so that it can create very large security databases.

When the Active Directory tree is partitioned into several domains, the Active Directory database is broken into smaller parts and stored on different domain controllers. See Figure 11.1. Since the parts are smaller and more manageable, there is a lesser load on the individual domain controllers. Therefore, the Active Directory database can be massive.

One advantage to the administrator and the user is since all computers and network resources are tracked in a single central database (stored in a domain controller), the administrator only has to create one user account and password for each user. Therefore, the user only has to log on once to the domain and is allowed to use any available resource located within the domain, assuming that the user has the necessary permissions to use the resource. When the user logs in, the user is logging onto the domain, not a specific server within the domain. Thus, any server that does authentication within the domain can do the authentication for the user.

The domain is not limited to a single network or a specific type of network configuration. For example, a domain can consist of computers that are located within a single LAN, on a subsection of a LAN, or on computers that span several LANs connected with any number of physical connections including Ethernet, token ring, dial-up lines, ISDN, fiber optics, and wireless technology.

11.4.2   Trust Relationships

Whereas small networks can store accounts and resources in a single domain, large organizations typically establish multiple domains. Windows 2000 Server Directory Services

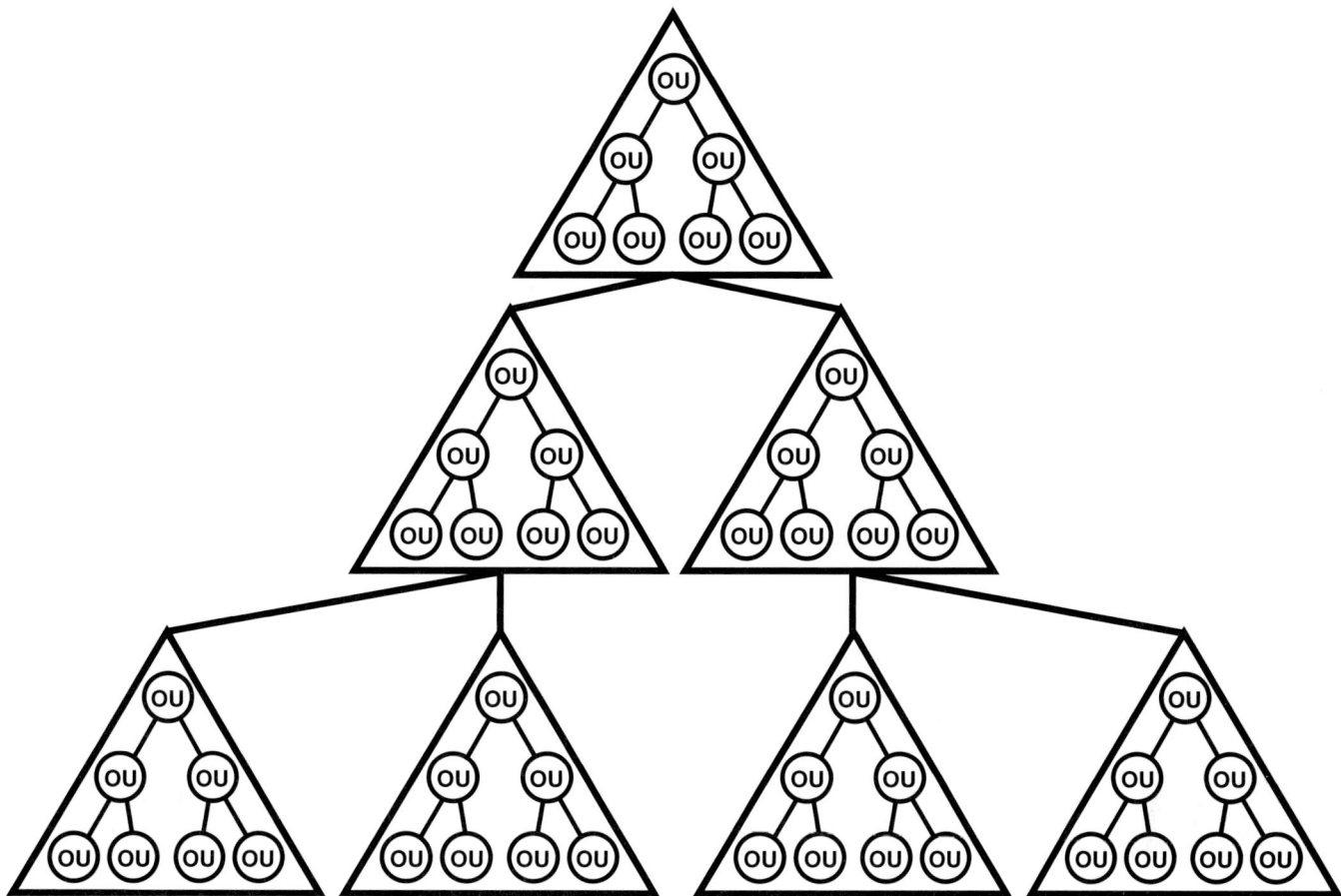

FIGURE 11.1 Active Directory tree

provides security across multiple domains through trust relationships. A **trust relationship** is a link that combines two domains into one administrative unit that can authorize access to resources on both domains.

The two types of trust relationships in Windows 2000 domains are as follows:

- **One-way trust relationship**—One domain trusts the users in the other domain to use its resources. More specifically, one domain trusts the domain controllers in the other domain to validate user accounts to use its resources. The resources that become available are in the trusting domain, as are the accounts that can use them. The one-way trusts are nontransitive, which means that if domain A trusts domain B and domain B trusts domain C, then domain A does not automatically trust C unless another trust is set up between domains A and C. One-way nontransitive trust relationships are typically made between existing domains. Since these types of trust relationships must be created manually, a one-way nontransitive trust relationship requires a large amount of administrative overhead for large networks.
- **Two-way trust relationship**—Each domain trusts user accounts in the other domain. Users can log on from computers in either domain to the domain that contains their account. Two-way trust relationships are transitive, which means that if domain A trusts domain B and domain B trusts domain C, then domain A trusts C. Two-way transitive trust relationships are the default in Windows 2000. When you create a new child domain, a two-way transitive trust relationship is made between the parent domain and the child domain.

If you want to set up a two-way trust, but you want it to be nontransitive, then you will have to set up two one-way trusts in which Domain A trusts Domain B and vice versa.

To create an explicit domain trust:

1. On a system that has Active Directory, click on the Start button, select the Programs option, select the Administrative Tools option, and select the Active Directory Domains and Trusts.

2. In the console tree, right-click the domain node for the domain you want to administer, and then select the Properties option.
3. Select the Trusts tab.
4. Depending on your requirements, click either "Domains trusted by this domain" or "Domains that trust this domain," and then click the Add button.
5. If the domain to be added is a Windows 2000 domain, type the full DNS name of the domain; or if the domain is running an earlier version of Windows, type the domain name.
6. Type the password for this trust and confirm the password. The password must be accepted in both the trusting and trusted domains.

11.4.3   Organizational Units

To help organize the objects within the domain and minimize the number of domains created to organize objects, the domain can contain organizational units. **Organizational units (OUs)** are used to hold users, groups, computers, and other organizational units. See Figure 11.2. An organizational unit can only contain objects that are located in the domain to which the organizational unit belongs. There are no restrictions on the depth (layers) of the OU hierarchy; however, a shallow hierarchy performs better than a deep one. Therefore, create an OU hierarchy no deeper than necessary, and that mirrors the organization's function or business structure.

Since domains and organizational units are used to contain objects (users, network resources, and other domains), they are sometimes referred to as containers. Containers can be assigned Group Policy settings or be used to delegate administrative authority. Group

FIGURE 11.2   Organization units used in the Active Directory tree

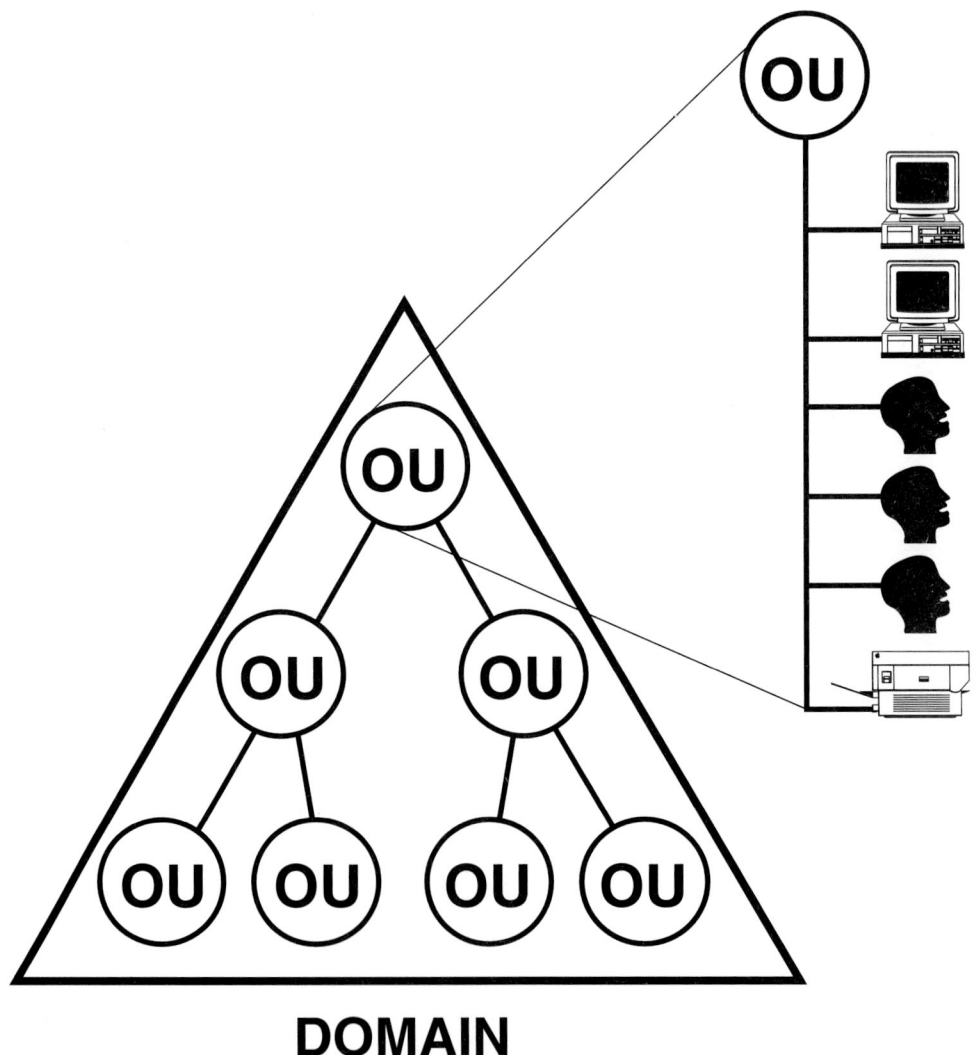

DOMAIN

Policy settings control the work environments for users in the domain or organizational unit. Delegated administrative authority allows assigned users to manage objects within the container.

11.4.4 Objects

An **object** is a distinct, named set of attributes or characteristics that represents a network resource, including computers, people, groups, and printers. Attributes have values that define specific objects. For example, the attributes of a user account might include the user's first and last names, department, and e-mail address while printers might include the printer name and location. The most common object types include user account, computer, domain controllers, groups, shared folder, and printers. See Table 11.2.

As new objects are created in Active Directory, they are assigned a 128-bit unique number called a **globally unique identifier (GUID),** sometimes referred to as a security identifier (SID). While objects have several names including a common name, a relative name, or some other identifier and can be moved to another container when so desired, the GUID stays the same. Therefore, it can be used to locate an object regardless of name or location.

The **schema** of the Active Directory contains a formal definition and set of rules for all objects and attributes of those objects. The default schema contains definitions of commonly used objects, such as user accounts, computers, printers, and groups.

To enable the Active Directory to be flexible, the schema allows you to define new objects and attributes for existing objects. For example, you could add badge numbers to the user objects. By default, the Active Directory servers do not allow the schema to be edited without changing the \HKLM\SYSTEM\CurrentControlSet\Services\DTDS\Parameters\ Schema Update Allows value within the registry of the Active Directory to one. You can never delete a class or attribute within the Active Directory schema.

A **name space** is a set of unique names for resources or items used in a shared computing environment. A **distinguished name (DN)** for an object uniquely identifies that object by using the actual name of the object plus the names of container objects and domains that contain the object. The distinguished name identifies the object and its location in a tree.

A DNS distinguished name would have a format such as the following:

```
CN=<Object Name>,OU=<Organizational Unit(s)>,
O=<Organization>,C=<CountryCode>
```

For example:

```
CN=pregan,OU=users,OU=support,O=Acme.com,C=US
```

TABLE 11.2   Object classes

Object Type	Description
User account	The information that allows a user to log on to Windows 2000 including the user logon name
Computer account	The information about a computer that is a member of the domain
Domain controllers	The information about a domain controller including its DNS name, the version of the OS loaded on the controller, the location, and who is responsible for managing the domain controller
Groups	A collection of user accounts, groups, or computers that you can create and use to simplify administration
Organizational unit (OU)	A container of other objects, including other OUs
Shared folders	A pointer to a shared folder on a computer (A pointer contains the address of certain data, not the data itself.)
Printers	A pointer to a printer on a computer

A LDAP distinguished name would have a format such as the following:

```
/O=<Organization>/DC=<DomainComponent(s)>/CN=<Object
Name(s)>
```

For example:

```
/O=Internet/DC=Com/DC=Acme/DC=Support/CN=Users/CN=pregan
```

11.4.5 Forests

If you have two or more existing trees and you need to merge them, then combine them into a grouping called a forest. A **forest** consists of one or more trees that are connected by two-way, transitive trust relationships. In a forest, each tree still has its own unique name space. Therefore, a forest is useful in organizations that need to maintain organizational structures, such as a company that needs distinct public identities for its subsidiaries. By default, the name of the root tree, or the first tree that is created in the forest, is used to refer to a given forest.

Active Directory Tree
- Hierarchy of domains
- Contiguous name space
- Transitive Kerberos trust relations between domains
- Common schema
- Common global catalog

Active Directory Forest
- One or more sets of trees
- Disjointed name spaces between these trees
- Transitive Kerberos trust relationships between the trees
- Common schema
- Common global catalog

A domain controller has a replica (copy) of every object in its own domain. If you have more than one domain controller for a domain, ideally, all objects listed in the Active Directory are stored in both servers. Therefore, if one server goes down, you can still access the objects in the database; and if multiple servers are running, performance will be increased because both servers will provide access to those objects. If you make changes to an object on one controller, those changes will be duplicated to the other servers in a timely manner.

When you search for an object or retrieve information about an object within a domain, it can retrieve all information from any domain controller. If you need to search for an object or retrieve information about an object in another domain that is part of the forest, access a replica of the global catalog within the current domain to find the object.

The **global catalog** holds replica information of every object in the Active Directory. Instead of storing the entire object, however, it stores those attributes most frequently used in search operations (such as a user's first and last names). It can even be configured to store additional properties as needed. In addition, the global catalog has sufficient information about each object to locate a full replica of the object at its respective domain controller. Since all objects in the forest are listed in the global catalog, all domains within a forest share a single global catalog. The domain controller that contains the global catalog is called the global catalog server. By default, a global catalog is created automatically on the initial domain controller in the forest.

Clients must have access to a global catalog to log on. In addition, a global catalog is necessary to determine group memberships during the logon process. If users are members of the Domain Admins group, they are able to log on to the network even when a global catalog is not available. If your network has any slow or unreliable links, then enable at least one global catalog on each side of the link for maximum availability and fault tolerance.

To enable or disable a global catalog, do the following:

1. Click the Start button, select the Program option, select the Administrative Tools option, and select the Active Directory Sites and Services option.
2. In the console tree, double-click the domain controller that is hosting the global catalog.
3. Right-click NTDS Settings, and then click Properties.
4. Select the Global Catalog check box.

11.4.6 Planning Your Domain Structure

When deciding between a single domain with organizational units or a forest, it is easy to see why a single domain is the easiest to administer—all users and network resources are located within this domain. If you have a small network or a simple medium network, you should probably use the single domain model.

If you determine that the single domain model no longer meets your needs, then consider using the tree structure based on your organizational structure. The organizational structure could be based on geography or function.

As a general rule, when a network is a far-reaching WAN and you decide to use multiple domains to organize your network resources, organize your domain by geography. Your domain should not span the WAN, because of the added traffic that must go back and forth to keep the domain controllers in sync. It is always a good idea to have a local domain controller for all domains at each site so that the users can access the Active Directory database quickly. The exception to this rule is if you have fast links that connect the different sites (256 KB or faster), then organize your domain by function.

Another consideration when designing your domain structure is basing your design on organizational units that allow you to delegate administrative control over smaller groups of users, groups, and resources (decentralized administration). If you are using domains for the various WAN sites, either assign site administrators to each site who would have full authority over the site or limit the authority. In any case, those site administrators would have no control over the other sights. If you are using a domain design by function, then assign administrator by function. For example, you can have an administrator for all sales people and an administrator for all research people.

Lastly, when choosing your directory structure, try to look for future growth. This way, you can plan for future organization units if needed.

11.5 PHYSICAL STRUCTURE

Forests, domains, and organizational units are considered logical structure because none follow any subnet or network boundary. The physical structure of the Active Directory, which uses subnet/network boundaries, consists of domain controllers and sites.

11.5.1 Domain Controllers

The computer that stores a replica (copy) of the account and security information of the domain and defines the domain is the **domain controller.** A Windows 2000 domain controller is a Windows 2000 Server with an NTFS partition running Active Directory services. The directory data (account and security information) is stored in the **NTDS.DIT** file on an NTFS partition on the domain controller. Access to domain objects is controlled by **Access Control Lists (ACLs).** ACLs contain the permissions associated with objects that control which users can gain access to an object and what type of access users can gain to the objects. The domain controller also manages user-domain interactions including user logon processes, authentication, and directory searches.

Active Directory uses **multimaster replication,** which means that there is no master domain controller/primary domain controller as in Windows NT. Instead, all domain controllers store writable copies of the directory. When a change is made to one of the domain controllers, it is the job of the domain controller to replicate those changes to other domain controllers within the same domain and within a short time. By adding a domain controller to a domain, the server is automatically configured for replication.

Another type of server worth mentioning on the domain is a member server. A **member server** is used to run applications that are dedicated to specific tasks, such as managing print servers, managing file servers, running database applications, or running as a web server. It does not store copies of the directory database, and therefore does not authenticate accounts or receive synchronized copies of the directory database.

For each domain, it is recommended to have more than one domain controller. If you only have one domain controller and it goes down for any reason, the users could not use any of the network resources because there is no server to authenticate them and to give the user permissions to use the network resources. The extra domain controllers do not provide fault tolerance for the different network services. They only provide fault tolerance for the Active Directory that contains the account and security information. Whereas Windows NT has primary and secondary domain controllers, in Windows 2000, domain controllers are all equal to one another. Additional domain controllers also would divide the workload of authentication between the different controllers, allowing for faster network response time.

> Because domain controllers store all user account information for a domain, each should be locked in a secure room. Also, only administrators should be allowed to log on interactively to the console of a domain controller. This process will be explained in the next chapter.

To track all changes to an Active Directory, each update is assigned its own 64-bit **unique sequence number (USN).** This number is based on a counter that is incremented whenever a change is made and on what system the changes occurred. When a server replicates an update to other Active Directory servers, it sends the USN with the change. Each domain controller maintains an internal list of replication partners and the highest USN received from them. The domain controller receiving the update requests only those changes with USNs higher than previously received. If two changes occur on two different controllers and are replicated at the same time, a replication collision is detected and is resolved by using the update with the later time stamp. Therefore, it is very important to keep systems time synchronized between Windows 2000 Servers.

11.5.2 Master Operations

Since Active Directory supports multimaster replication of directory data between all domain controllers in the domain, it is impractical to have certain tasks handled by multiple domain controllers. Therefore, those tasks are assigned to individual domain controllers called operations masters. In an Active Directory, there are at least five different operations master roles that are assigned to one or more domain controllers. If you have only one domain controller, then it performs all five roles. See Table 11.3.

To identify the role of a server, do the following:

1. Click on the Start button, select the Programs option, select Administrative Tools, and select Active Directory Users and Computers.
2. Right-click Active Directory Users and Computers and select the Operations Masters option.
3. The roles in Operations Master appear in the appropriate tabs. See Figure 11.3.

To transfer a role, do the following:

1. Click on the Start button, select the Programs option, select Administrative Tools, and select Active Directory Users and Computers.
2. In the console tree, right-click the domain controller node that will become the new PDC emulator master, and then click Connect to Domain.
3. Type the domain name or click Browse and select the domain from the list.
4. In the console tree, right-click Active Directory Users and Computers, and then click Operations Master.
5. On the appropriate tab, click the Change button.

11.5.3 Sites and Active Directory Replication

A **site** is one or more IP subnets connected by a high-speed link (128 Kbps or higher), typically defined by geographical locations. Sites are based on IP subnets and all subnets can only belong to one site. Multiple subnets can be assigned to a single site. When a user logs on, Active Directory clients locate an Active Directory server in the same site as the user.

Active Directory will automatically generate a simple bidirectional ring topology for replication in the same domain and site with up to seven domain controllers. The ring ensures that if one domain controller goes down, it still has an available path to replicate its information to other domain controllers. When the number of servers grows beyond seven,

TABLE 11.3   Operating master roles

Operation Master Roles	Scope of Operation Master	Description
Schema master	One per forest	The schema master domain controller controls all updates and modifications to the schema. To update the schema of a forest, you must have access to the schema master.
Domain naming master	One per forest	The domain controller holding the domain naming master role controls the addition or removal of domains in the forest.
Relative ID (RID) master	One Per domain	When a domain controller creates a user, group, or computer object, it assigns the object a unique security ID. The security ID consists of a domain security ID (identifies the domain to which the object belongs) and a relative ID that identifies the object within the domain. The relative ID master allocates sequences of relative IDs to each of the various domain controllers in its domain. To move an object between domains, initiate a move on the domain controllers acting as the relative ID master of the domain that currently contains the object.
PDC emulator	One Per domain	If the domain contains computers operating without Windows 2000 client software or if it contains Windows NT backup domain controllers (BDCs), the PDC emulator acts as a Windows NT primary domain controller. It processes password changes from clients and replicates updates to the BDCs.
		In a Windows 2000 domain that is operating in native mode, the PDC emulator receives preferential replication of password changes performed by other domain controllers in the domain. If a password was recently changed, that change takes time to replicate to every domain controller in the domain. If a logon authentication fails at another domain controller due to a bad password, that domain controller will forward the authentication request to the PDC emulator before rejecting the logon attempt.
Infrastructure master	One Per domain	When you rename or move a member of a group (and that member resides in a different domain from the group), the infrastructure master of the group's domain is responsible for updating the group so it knows the new name or location of the member. The infrastructure master distributes the update via multimaster replication. The administrator, when renaming or moving a member of a group, may see that the member of a group notices a temporary inconsistency such as the member not showing up in the group.

FIGURE 11.3   Displaying the domain's Operations Master

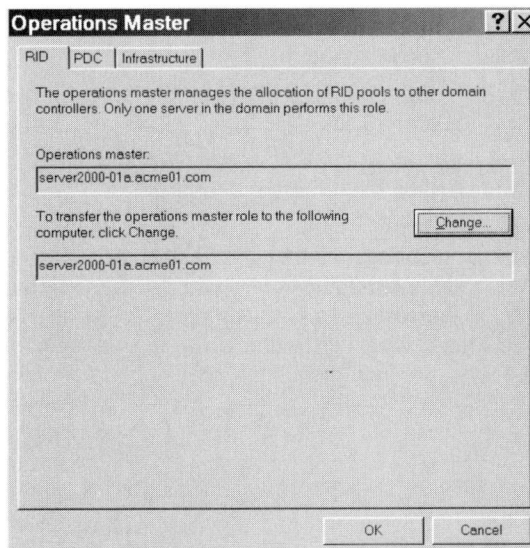

the number of connections that are needed are estimated so that if a change occurs at any one domain controller, there are as many replication partners as needed to ensure that no domain controller is more than three replication hops from another domain controller. These optimizing connections are created at random and are not necessarily created on every third domain controller. Note: The replication path, or site topology, within a site is automatically managed by a service called the Knowledge Consistency Checker (KCC).

When an update occurs on a domain controller, the replication engine waits for a configurable interval, which is 5 minutes by default. It then sends a notification message to the first replication partner, informing it of the change. Each additional direct partner is notified after a configurable delay, which is 30 seconds by default. Items that are security-sensitive are immediately replicated and partners are notified immediately. If no changes occur during a configuration period, which is 1 hour by default, a domain controller initiates replication with its replication partners to ensure that no changes from the originating domain controller were missed.

One reason to use sites is to control replication traffic. When a domain controller replicates to another domain controller in a site, replication information is done without being scheduled. Urgent changes such as password changes, account lockout policy changes, freshly locked accounts, and domain password policies are replicated immediately. In addition, the replication data are sent uncompressed, which keeps the domain controller processing down to a minimum. See Table 11.4.

The default replication pattern used by the Active Directory is optimized for a single location or site with high-speed connectivity. If your network includes multiple locations or sites, however, the replication pattern would have to be configured so that a domain controller will replicate all changes to the domain controllers within a site and have a single domain controller to replicate across a slower WAN link to the other sites.

A **bridgehead server** is a single server located in each site that is designated to perform site-to-site replication. Bridgehead servers are designated automatically or they can be assigned manually by an administrator. The links between bridgehead servers are assigned time schedules to indicate what times of day they are available to carry replication traffic. The replication interval indicates how often the bridgehead servers poll the other side of a link for replication changes. The first site in Active Directory is called "Default-First-Site-Name," which is created automatically for the administrator. This site is a member of the default site link called "DEFAULTIPSITELINK," which also is created automatically for

TABLE 11.4 Comparing site and intersite replication

Site Replication	Intersite Replication
Replication traffic is not compressed to save processor processing.	Replication traffic is compressed to save bandwidth.
Replication partners notify each other when changes need to be replicated, to reduce replication latency.	Replication partners do not notify each other when changes need to be replicated, to save bandwidth.
Replication partners poll each other for changes on a periodic basis.	Replication partners poll each other for changes on a specified polling interval, during scheduled periods only.
Replication uses the remote procedure call (RPC) transport.	Replication uses the TCP/IP or SMTP transport.
Replication connections can be created between any two domain controllers located in the same site. Connections can be made with multiple domain controllers to reduce replication latency.	Replication connections are only created between bridgehead servers (servers that handle all intersite replication for that domain). Connections between bridgehead servers use the lowest-cost route, according to the site link cost. Connections will only be created over higher-cost routes if all domain controllers in lower-cost routes are unreachable.

the administrator. If the administrator creates two additional sites (Site1 and Site2, for example), then the administrator must define a site link, of which each site is going to be a member, before the sites can be written to Active Directory.

By default, all site links are bridged or transitive; that is, all site links for a specific transport implicitly belong to a single site link bridge for that transport. If your IP network is not a fully routed IP network, you do not need to configure any site link bridges.

Replication within sites requires little or no planning because it is fully automatic. When you have multiple sites, however, you should use the following steps to optimize Active Directory synchronization traffic.

1. Identify sites that are well connected through backbones, and create low-cost site links between them.
2. Identify sites that are interconnected with a comparable transport, and create medium-cost site links between them—for example, full mesh links (remote sites that are connected over telecommunication links), frame relay cloud links (a point-to-point system that uses a private virtual circuit), and medium area network links with T1 connections.
3. Identify remaining WAN links.
4. Create a site link for each pair of sites that cross a WAN link.
5. Create a schedule that meets user needs.
6. Avoid high-frequency times.

EXAMPLE 1

Question: You have three sites. Site A is connected to Site B via a 256-K WAN link. Site A is also connected to Site C via a 256-K WAN link. You wish to create site links to optimize Active Directory synchronization traffic. How many site links should you create?

Answer: The answer is one link. All WAN links are the same speed.

EXAMPLE 2   You have four sites. Site A, Site B, and Site C are connected to a central ATM switch via 256-K WAN links. Site D is connected to the ATM switch via a 128-K WAN link. You wish to create site links to optimize Active Directory synchronization traffic. How many site links should you create?

Again, the answer is two links. Site A, Site B, and Site C use one link because they are connected with a comparable transport. Site D is connected by a slower WAN link, to the other site through switch, which makes up the second link.

The most useful tool for monitoring and troubleshooting replication is the Active Directory Replication Monitor, or REPLMON. This tool is installed by running the Setup program in the SUPPORT/TOOLS directory on the Windows 2000 CD. Another useful tool is the DSASTAT, which allows you to compare the directories on Windows 2000 domain controllers. Lastly, besides Replmon, you can also use Repadmin (command line utility) to force replication events.

There are a number of tasks that should be performed on a regular basis to ensure that the Active Directory structure is up-to-date and accurately addresses the needs of the users. The best utility for this is NTDSUTIL.EXE, command line utility provided in the Windows 2000 Server Resource kit that can be used to perform several administrative tasks including compacting, repairing and checking the integrity of the Active Directory database and clean up directory metadata. For example, the NTDSUTIL can be used to analyze and clean up core directory structure if you removed a domain controller from the network. It can also be used to do an authoritative restore, which is discussed in Chapter 16.

Avoid confusing trees and forests with sites. Trees and forests are used to manage administration and security in an organization; sites reflect geographical boundaries. You may choose to arrange a site's trees and forests using a geographical or an organizational approach, but doing so does not affect changes to the sites of the domains. In addition, Active Directory does not impact the name space in any way.

As mentioned, you should have more than one controller for both fault tolerance and faster authentication. Placing domain controllers and DNS servers at the different physical sites greatly improves the performance of the directory in a wide area network (WAN), since authentication will be done locally and not through a slow WAN link. Having multiple servers at a site will offer fault tolerance and load balancing for requests. In addition, having multiple global catalog servers is a necessity.

To create a site, do the following:

1. Click the Start button, select the Programs option, select the Administrative Tools option, and select the Active Directory Sites and Services option.
2. Right-click the Sites folder, and select the New Site option.
3. In Name, type the name of the new site.
4. Click a Site Link Object, and then click OK.
5. Associate a subnet with a site for this newly created site.
6. Move a domain controller from an existing site into this new site, or install a new domain controller.
7. If desired, choose a different licensing computer other than the one automatically selected.
8. Delegate control of the site.

To delete a site, do the following:

1. Open Active Directory Sites and Services.
2. In the console tree, double-click on Sites.
3. Right-click the Site Container, and then click Delete.

To bridge all site links, do the following:

1. Click the Start button, select the Programs option, select the Administrative Tools option, and select the Active Directory Sites and Services option.
2. Double-click the Sites folder, and then double-click the Inter-Site Transports folder.
3. Right-click the appropriate intersite transport folder (such as IP or SMTP), and then click Properties.
4. Select the "Bridge all site links" check box.

To create a site link, do the following:

1. Click the Start button, select the Programs option, select the Administrative Tools option, and select the Active Directory Sites and Services option.
2. In the console tree, right-click the intersite transport protocol you want the site link to use, and then click New Site Link.
3. In Name, type the name to be given to the link.
4. Click two or more sites to connect, and then click Add.
5. Configure the site link's cost, schedule, and replication frequency by right-clicking the site link that you want to configure and select the Properties option.

If you create a site link that uses SMTP, you must have an Enterprise Certification Authority (Enterprise CA) available and SMTP must be installed on all domain controllers that will use the site link.

To create a site link bridge, do the following:

1. Click the Start button, select the Programs option, select the Administrative Tools option, and select the Active Directory Sites and Services option.
2. In the console tree, right-click the intersite transport folder for which you want to create a new site link bridge, and then click New Site Link Bridge.
3. In Name, type a name for the site link bridge.
4. Click two or more site links to be bridged, and then click Add.

If you have enabled "Bridge all site links," this procedure is redundant and will have no effect.

11.6 INSTALLING AND CONFIGURING DOMAIN CONTROLLERS AND ACTIVE DIRECTORIES

When you install Windows 2000 Server, it defaults to being a member server of the domain. To become a domain controller, install the server and then promote it.

Before installing a domain controller, identify the DNS name space and verify that the computer has a NTFS partition and that the TCP/IP protocol is installed and configured correctly. In addition, make sure that a DNS server is available on the network or you will have to install and configure the DNS service on the domain controller computer.

To install a domain controller, you can do the following:

1. Click Start, point to Programs, point to Administrative Tools, and then click Configure Your Server.
2. Click Active Directory, and click the Start Hypertext to start the Active Directory Installation wizard. You must scroll down to get to the Start hypertext.
3. Follow the instructions in the Active Directory Installation wizard.

You can start the Active Directory Installation wizard by clicking on the Start button, then on the Run option, and then executing the DCPROMO command. See Figure 11.4.

The Active Directory Installation wizard will prompt you when the domain controller is used to create the following:

- A new domain or an additional domain controller for an existing domain
- A new tree or to be part of a child domain in an existing tree
- A new forest or to join an existing forest

If you are joining an existing domain, tree, or forest, then the wizard will prompt you for network credentials of a user account with sufficient rights to create objects in the Active Directory.

After the Active Directory has already been installed, it can then be configured by using the Active Directory Domains and Trusts console, the Active Directory Sites and Services console, and the Active Directory Users and Computers console or by using the MMC with the respective snap-ins. See Figure 11.5.

You can also administer Active Directory remotely, from a computer that is not a domain controller. To use the Active Directory Administrative Tools remotely, from a computer that is not a domain controller such as one running Windows 2000 Professional, install the Windows 2000 Administrative Tools.

When first created, the Active Directory is configured for **mixed mode,** which allows the Windows 2000 to communicate with Windows 3.51 and 4.0. Unfortunately, when running in mixed mode, Windows 2000 cannot use universal or nested groups. In addition, clients that use operating systems other than Windows 2000 cannot use the transitive domains that would normally be available through transitive trust relationships.

The default domain mode setting on Windows 2000 is domain controllers. Mixed mode allows Windows NT and Windows 2000 backup domain controllers to coexist in a domain. Mixed mode does not support the universal and nested group enhancements of Windows

FIGURE 11.4 Configure Your Server wizard can be used to promote a server into a domain controller

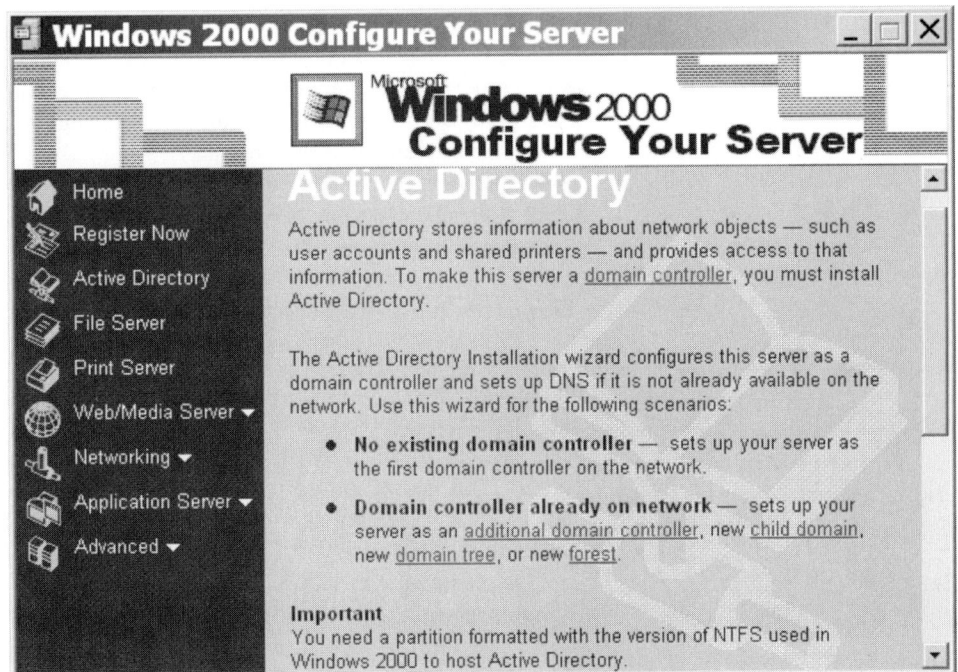

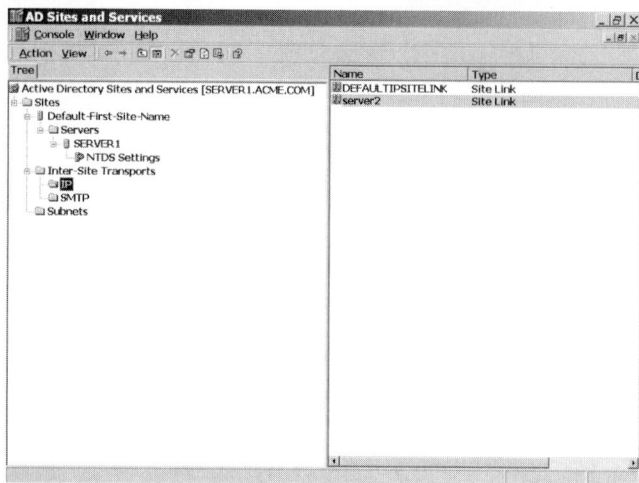

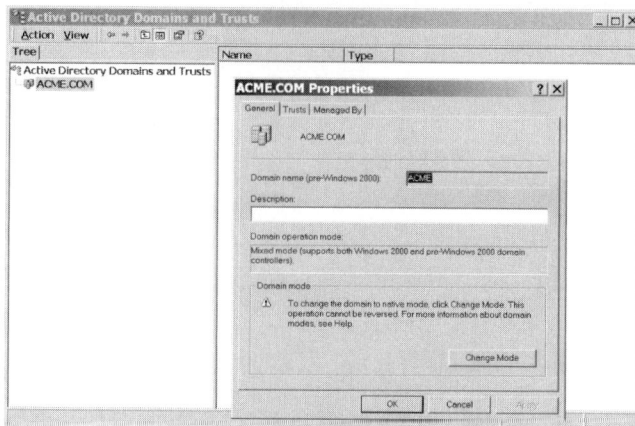

FIGURE 11.5 (a) Active Directory Sites and Services and (b) Active Directory Domains and Trusts consoles

2000. The domain mode setting can be changed to Windows 2000 native mode when all Windows NT domain controllers are removed from a domain.

11.7 ENCRYPTION AND DECRYPTION

In networks, sensitive data are transmitted across the network, sent as e-mail messages, and stored on files on disk drives. **Encryption** is the process of disguising a message or data in what appears to be meaningless jumble (cipher text) to hide and protect the sensitive data from unauthorized access. **Decryption** is the process of converting data from encrypted format to its original format. **Cryptography** is the art of protecting information by transforming it (encrypting it) into cipher text.

To encrypt and decrypt a file, you must use a key. A **key** is a string of bits that is used to map text into a code and a code back to text. You can think of the key as a super-decoder ring that is used to translate text messages to a code and back to text. There are two types of keys, public keys and private keys.

11.7.1 Private-Key Encryption

The most basic form of encryption is the **private-key encryption**, also known as the symmetric algorithm. This encryption system requires each individual to possess a copy of the key. For this system to work as intended, you must have a secure way to transport the key to other people. Keeping multiple keys for each person can be quite cumbersome, so private-key algorithms are generally used for bulk data encryption, because they are fast and easily implemented in hardware.

The two general categories of private-key algorithms are block and stream cipher. A block cipher encrypts one block of data at a time. A stream cipher encrypts each byte of the data stream individually.

11.7.2 Public-Key Encryption

Public-key encryption, also known as asymmetric algorithm, uses two distinct but mathematically related keys, public and private. The public key is the nonsecret key that is made available to anyone you choose, or made available to everyone by posting them in a public place. Availability is often through a digital certificate. The private key is kept in a secure location and is used only by you. When data need to be sent, they are protected with a secret-key encryption that was encrypted with the public key of the recipient of the data. The encrypted secret key is then transmitted to the recipient along with the encrypted data. The recipient will use the private key to decrypt the secret key. The secret key will then be used to decrypt the message itself.

For example, say you want to send data to someone. You would retrieve his or her public key and encrypt the data. You encrypt the data and the secret key, and send both the data and the secret key. Since the recipient's private key is the only thing that can decrypt the secret key—which is the only thing that can decrypt the message—the data can be sent over an insecure communications channel.

11.7.3 Digital Envelopes, Signatures, and Certificates

A **digital envelope** is a type of security that encrypts the message using symmetric encryption and encrypts the key to decode the message using public-key encryption. This technique overcomes one problem of public-key encryption—being slower than symmetric encryption because only the key is protected with public-key encryption, thus providing little overhead.

A **digital signature** is a digital code that can be attached to an electronically transmitted message that uniquely identifies the sender. Like a written signature, the purpose of a digital signature is to guarantee that the individual who is sending the message really is who he or she claims to be.

A **digital certificate** is an attachment to an electronic message that is used for security purposes such as for authentication. It is also used to verify that a user who is sending a message is who he or she claims to be and to provide the receiver with a means to encode a reply. An individual wishing to send an encrypted message applies for a digital certificate from a **certificate authority (CA).** The CA issues an encrypted digital certificate that contains the applicant's public key and a variety of other identifying information. The CA makes its own public key readily available through print publicity or the Internet. The recipient of an encrypted message uses the CA's public key to decode the digital certificate attached to the message, verifies it as issued by the CA, and then obtains the sender's public key and identification information held within the certificate. With this information, the recipient can send an encrypted reply. See Figure 11.6.

To install CA services for Windows 2000, do the following:

1. Double-click the Add/Remove applet in the Control Panel.
2. Click on Add/Remove Windows Components.
3. Select the Certificate Services.
4. If asked to continue because the computer cannot be renamed or the computer cannot join or be removed from a domain, click on the Yes button.

FIGURE 11.6 Certification Authority console

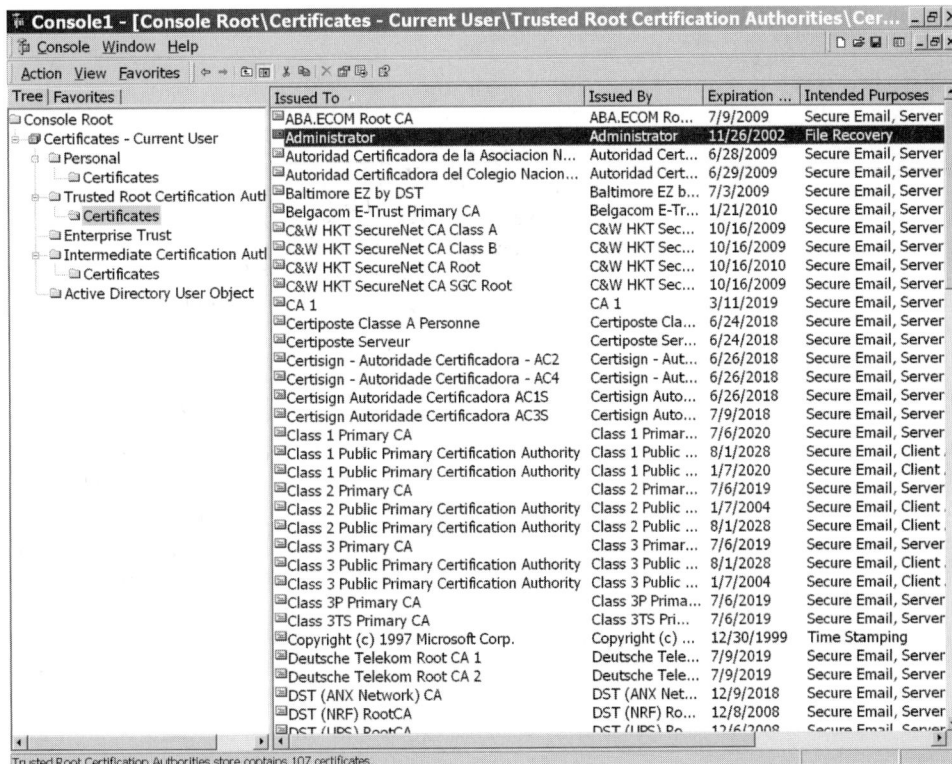

5. Click on the Next button.
6. Select the type of certification authority and click on the Next button.
7. Enter the information to identify the CA and click on the Next button.
8. For the Data Storage Location, keep the default locations and click on the Next button.
9. If you have an IIS service running, it will ask you to stop the service. Click on the OK button to stop the service.
10. Click on the Finish button.

11.7.4   RSA and DES Standards

The **RSA standard** (created by Ron **R**ivst, Adi **S**hamir, and Leonard **A**dleman) defines the mathematical properties used in public-key encryption and digital signatures. The key length for this algorithm can range from 512 to 2048, making it a very secure encryption algorithm. This algorithm uses a number known as the public modulus to produce the public and private keys. This number is formed by multiplying two prime numbers. The security of this algorithm is found in the fact that while finding large prime numbers is relatively easy, factoring the result of multiplying two large prime numbers is not as easy. If the prime numbers used are large enough, the problem approaches being computationally impossible. The RSA algorithm has become the de facto standard for industrial-strength encryption, especially for data sent over the Internet.

The **Data Encryption Standard (DES)** was developed in 1975 and was standardized by ANSI in 1981 as ANSI X.3.92. It is a popular symmetric-key encryption method that uses block cipher. The key used in DES is based on a 56-bit binary number, which allows for 72,057,594,037,927,936 encryption keys. Of these 72 quadrillion encryption keys, a key is chosen at random. If you want to send an encrypted file from one person (source person) to another person (target person), the source person will encrypt the secret key with the target person's public key, which was obtained from his or her certificate. Because the target person's key was used to encrypt the secret key, only the target person using his or her private key will be able to decrypt the DES secret key and decrypt the DES-encrypted data. (Note: The U.S. government has banned the export of DES outside of the United States.)

Triple DES is a stronger alternative to regular DES, and is used extensively in conjunction with Virtual Private Network implementations. Triple DES encrypts a block of data as a DES secret key. The encrypted data are encrypted again using a second DES secret key. Finally, the encrypted data are encrypted a third time using yet another secret key. Triple DES is of particular importance as the DES algorithm keeps being broken into shorter and shorter times.

11.7.5   X.509 Digital Certificates

X.509 certificates are the most widely used digital certificates. X.509—Version 3 (X.509v3)—is the standard certificate format used by Windows 2000 certificate-based processes. Note that X.509 is actually an ITU recommendation, which means that it has not been officially defined or approved.

The X.509 describes two levels of authentication: simple authentication based on a password to verify user identity, and strong authentication using credentials formed by using cryptographic techniques. It is recommended that you use the strong authentication to provide secure services. Windows 2000 users certificates are stored in the Active Directory as attributes and can be freely communicated with the system and obtained by users of the Directory in the same way as other information. See Figure 11.7.

11.8 AUTHENTICATION

Authentication, the layer of network security, is the process by which the system validates the user's logon information. Authentication is crucial to secure communication. Users must be able to prove their identity to those with whom they communicate and must be able to verify the identity of others. Typically in Windows 2000, when a user logs on, that person is then authenticated for all network resources which the user has permission to use. The user has to log on only once, even though the network resources may be located throughout several computers.

FIGURE 11.7 X.509 certificate

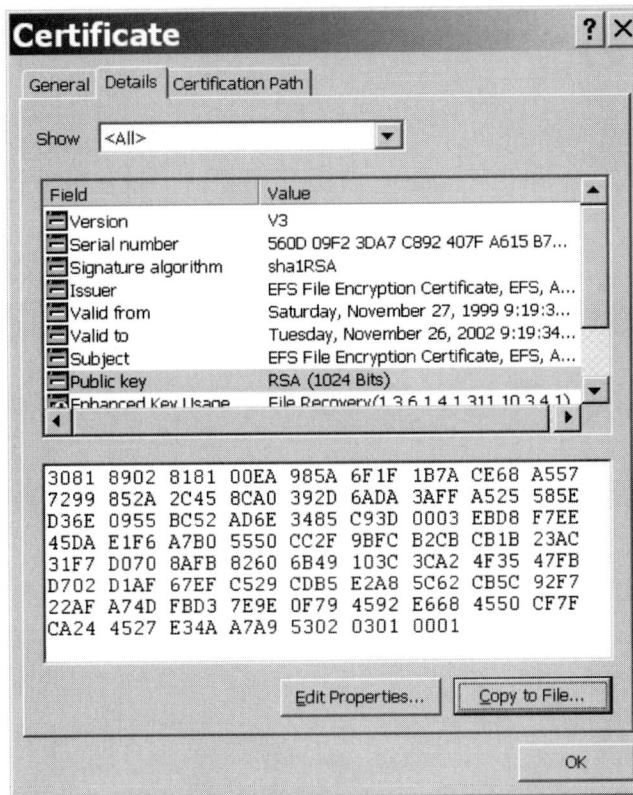

Type of Authentication	Description
Kerberos V5 authentication	Kerberos is a primary security protocol for authentication within a domain. It verifies both the identity of the user and network services by issuing tickets for accessing network services.
NTLM authentication	NTLM stands for NT LAN Manager. NTLM authentication is used for transactions between two computers in a domain, where one or both computers are running Windows NT 4.0 or earlier or when not participating in a domain, such as a stand-alone server or workgroup.
Secure Sockets Layer/ Transport Layer Security (SSL/TLS) authentication	SSL and TLS are protocols used for secure network communications using a combination of public- and secret-key technologies. It is typically used to access a secure web server.
IP Security (IPSec)	IPSec is used on a Virtual Private Network (VPN) to ensure private, secure communications over IP networks. It utilizes public-key cryptography in conjunction with digital signatures to provide secure tunnels. It can also use DES and triple DES encryption algorithms to encrypt the tunneled data.

11.8.1 Interactive Authentication

Windows 2000 has several forms of authentication. To perform a logical interactive logon, sit at a computer that is not a domain controller and press the Ctrl+Alt+Del to log in, enter a user name and password, and specify the individual computer/workgroup. During a local interactive logon, the steps in authentication are as follows:

1. The user provides the user name and password. The computer then forwards the information to the security subsystem of the local computer.

2. The domain controller will compare the user name and password with the local Security Accounts Manager (SAM) database.
3. If the information matches and the user account and password are valid, Windows 2000 creates an access token for the user.
4. Any time the user makes a connection to a computer, the user name and password is forwarded to the computer. The computer then authenticates the user and returns an access token.

When you log on as a local user, the local user account is not available to the domain/domain controllers. Note also that a user cannot log on locally to a domain controller.

11.8.2 NTLM Authentication

Windows 2000 has several forms of authentication. To perform a network authentication process, sit at a computer that is not a domain controller and press the Ctrl+Alt+Del to log in, enter a user name and password, and specify the domain name. During a network authentication process, various protocols can be used including NTLM and Kerberos.

In NTLM authentication, the client selects a string of bytes, uses the password to perform a one-way encryption of the string, and sends both the original string and the encrypted string to the server. The server receives the original string and uses the password from the account database to perform the same one-way encryption. If the result matches the encrypted string sent by the client, the server concludes that the client knows the user name/password pair. NTLM does not send the password in any form, so it is more secure than Basic authentication.

When a user starts a computer running Windows 2000, the user is prompted to press Ctrl+Alt+Delete to log on. Windows 2000 then prompts to provide a user name, password, and domain name. The steps in Windows 2000 authentication process are as follows:

1. Windows 2000 forwards the user name and password in an encrypted format to a domain controller. Windows 2000 compares the logon information with the user information that is stored in the appropriate database.
2. If the information matches and the user account is valid, Windows 2000 creates an access token for the user. An access token is the user's identification for the computers in the domain or for that local computer, and it contains the user's security settings. These security settings allow the user to gain access to the appropriate resources and to perform specific system tasks.
3. Any time a user makes a connection to a computer, that computer authenticates the user and returns an access token.

11.8.3 Kerberos Authentication

Kerberos is an authentication service developed at MIT for the TCP/IP network, which is the default authentication method for Windows 2000. Its purpose is to allow users and services to authenticate themselves to each other without allowing other users to capture the network packet on the network and resending it so that they can be authenticated over an unsecure media such as the Internet. Therefore, Kerberos provides strong authentication for client-server applications by using symmetric secret-key cryptography. After a client and server have used Kerberos to prove their identities, they can also encrypt all of their communications to assure privacy and data integrity as they go about their business.

When requesting a service, a user's identity must be established. To do this, a ticket is presented to the server, with proof that the ticket was originally issued to the user, not a captured network packet containing the password or ticket that is resent. There are three phases to authentication using Kerberos. In the first phase, the user obtains credentials to be used to request access to other services. In the second phase, the user requests authentication for a specific service. In the final phase, the user presents those credentials to the end server.

To get the initial Kerberos ticket under Windows 2000, the user is prompted for the user name and password. The client computer will first send the user name to the Windows 2000 Server. The server checks if the user is listed in the security database. If the user name is valid, a random session key and a ticket consisting of the client's name, the name of the ticket-granting server, the current time, a lifetime of the ticket, and the client's IP address are generated. The ticket and random session key are encrypted with the user's password and sent back to the client.

Once the response has been received by the client, the password is converted to a key and is used to decrypt the response from the authentication server. The ticket and the session key are stored for future use and the user's password and key are erased from memory.

To gain access to the server, the client builds an authenticator for the desired server. The application builds an authenticator containing the client's name, the IP address, and current time, which is encrypted with the session key and the ticket for the server and sent to the server. Once the authenticator and ticket have been received by the server, the server decrypts the ticket, uses the session key included in the ticket to decrypt the authenticator, and compares the information in the ticket with that in the authenticator, the IP address from which the request was received, and the present time. If everything matches, it allows the request to proceed.

If the client specifies that it wants the server to also prove its identity, the server adds a time stamp that the client sent in the authenticator, encrypts the result in the session key, and sends the result back to the client. At the end of the exchange, the server is certain of the client's identity. If mutual authentication occurs, the client is also convinced that the server is authentic. Moreover, the client and server share a key which no one else knows, and can safely assume that a reasonable recent message encrypted in that key originated with the other party.

When a program requires a ticket to access a network service, it sends a request to the ticket-granting server. The request contains the name of the server for which a ticket is requested, along with the ticket-granting ticket and the authenticator. The ticket-granting server then checks the authenticator and ticket-granting ticket. If valid, the ticket-granting server generates a new random session key to be used between the client and the new server. It then builds a ticket for the new server containing the client's name, the server name, the current time, the client's IP address, and the new session key just generated.

The ticket-granting server then sends the ticket, with the session key and other information, back to the client. This time, however, the reply is encrypted in the session key that was part of the ticket-granting ticket. Of course, this ticket is only good for that service at that server.

As you can see throughout the entire process of getting authentication, the password was never sent on the wire. Therefore, a packet that contains the password cannot be compromised. In addition, many of these tickets and requests are time stamped, to keep track of when the packet was made so that packets are not captured and re-sent at a later time. Lastly, the encryption algorithm is different for every client.

FOR MORE INFORMATION

Kerberos: The Network Authentication Protocol—
http://web.mit.edu/kerberos/www/

The Moron's Guide to Kerberos—http://gost.isi.edu/brian/security/kerberos.html

The Kerberos Network Authentication Service—http://gost.isi.edu/info/kerberos/

Windows 2000 Kerberos Authentication—
http://technet.microsoft.com/cdonline/content/complete/windows/win2000/
win2ksrv/technote/kerberos.htm

11.9 ACTIVE DIRECTORY CLIENTS

Computers that are running Windows 2000 Professional, Windows 2000 Server, Windows NT Workstation, Windows NT Server, Windows 95, Windows 98, and Windows for Workgroup can be configured to participate in either a domain or a workgroup. When setting up one of these computers for networking in a workgroup, specify a computer name and a workgroup name. If the workgroup name matches a domain name, the computer name appears in the Browse list for that domain and can browse computers having shared resources, whether participating in a domain or a workgroup. To determine whether the computer participates in a domain or a workgroup, during setup specify that the computer logs on to either a Windows NT Server domain or a workgroup.

For a computer to connect to and make full use of an Active Directory network, it needs to have Active Directory Network Client software installed. A computer configured with the Active Directory Client can locate a domain controller and log on to the network. Computers that are running Windows 2000 or Windows 2000 Professional already have Active Directory Client. Computers running Windows 95 or Windows 98 need to have add-on Active Directory Client software installed. Computers not running Active Directory Client software will appear just like a Windows NT directory. The Active Directory Client is provided in the CLIENTS folder on the Windows 2000 Server CD.

To log on to an Active Directory network, an Active Directory Client must first locate an Active Directory domain controller for its domain. To locate a domain controller for a specified domain, an Active Directory Client sends a DNS name query with a SRV Query type and ldap._tcp.*domain_name* Query name. For example, to log on to the domain ACME.COM, an Active Directory Client sends a DNS name query of the type SRV for the name_ldap._tcp.acme.com. A DNS server will respond back with the list of DNS names of the domain controllers and their IP addresses. From the list of domain controllers' IP addresses, the client attempts to contact each domain controller. The first domain controller that responds is used for the logon process.

Question: A small company uses a single LAN. How many domain controllers would you recommend?

Answer: Recommend two domain controllers. Although the network could get by with one domain controller, recommend two to allow for fault tolerance and faster network response time.

Question: A small company uses several LANs that are connected to form a WAN through various WAN links. The company's corporate office is located in Sacramento. Other offices in Cleveland, Los Angeles, and Houston need to be part of the network. How many domain controllers would you recommend?

Answer: Recommend two domain controllers located at the corporate office to provide fault tolerance for the corporate office and allow for fast authentication for the corporate users. Next, recommend having a domain controller at each of the different sites to also provide fault tolerance to the network, but allow the users at the remote sites to authenticate locally without sending requests back to the corporate office. This setup would significantly increase the users' response time.

Question: A large company uses several LANs that are connected to form a WAN through various WAN links. The company's corporate office is located in Sacramento. Other offices in Cleveland, Los Angeles, and Houston need to be part of the network. How many domain controllers would you recommend?

Answer: Since this is a large network, it would probably be best to use multiple domains to divide the network resources into smaller, more manageable units. Until you know more about the organization and the WAN links that connect the individual LANs, recommend that the corporate office represent the root domain and the remote sites have their own child domains. Each domain would have two domain controllers to provide fault tolerance.

11.10 TIME SYNCHRONIZATION

Windows 2000 uses a new time synchronization service to synchronize the date and time of computers running on a Windows 2000-based network. As mentioned earlier in the chapter, Kerberos authentication uses the workstation time as part of the authentication ticket generation process.

When a client boots, it contacts a domain controller for authentication. As the two computers exchange authentication packets, the client adjusts its local time based on the time of the domain controller. If the domain controller is ahead of the client's time by less than 2 minutes, then the client immediately adjusts its time to match the target time. If the domain controller is behind the client's time by less than 2 minutes, then the client slows its clock over a period of 20 minutes until the two times are in synch. If the local time is off by more than 2 minutes, then the client immediately sets its time to match the target time. To keep the clients in synch with the time server, the client performs these checks every 8 hours.

To keep the domain controllers in synch with each other, Windows 2000 uses a **Windows Time Synchronization Service.** The first domain controller in the forest, which acts as the

PDC emulator, becomes the authoritative timekeeper for the entire enterprise. Therefore, the PDC emulator should gather its time from an external source such as an observatory or atomic clock. To configure the PDC emulator, recognize an external SNTP time server as the authoritative timekeeper using the NET TIME /SETSNTP:*server* command from the Command prompt, where the *server* is the IP address or server name that uses the **Simple Network Time Protocol (SNTP).** SNTP defaults to using UDP port 123. If this port is not open to the Internet, you cannot synchronize your server to Internet SNTP servers.

There are many SNTP time servers throughout the world that are satisfactory for this function. Some are as follows:

- ntp2.usno.navy.mil at 192.5.41.209
- tick.usno.navy.mil at 192.4.41.40
- tock.usno.navy.mil at 192.5.41.41
- bitsy.mit.edu at 18.72.0.3
- bonehed.lcs.mit.edu at 18.26.4.105
- clock.isc.org at 192.5.5.250
- clock.nc.fukuoka-u.ac.jp at 133.100.9.2
- ntp.nml.csiro.au at 130.155.98.1
- ntp-p1.obspm.fr at 145.238.110.49
- ntp2.ja.net at 193.63.94.26

Therefore, to set up the PDC emulator to point to the first time server listed, type in:

```
NET TIME /SETSNTP:192.5.41.209
```

or

```
NET TIME /SETSNTP:NTP2.USNO.NAVY.MIL
```

For the system to update from a time source, you must have the Windows Time service started.

SUMMARY

1. A common directory service is the X.500, which uses a hierarchical approach, in which objects are organized in a similar way to your files on a hard drive.
2. The lightweight directory access protocol (LDAP), based on X.500, allows Windows 2000 Active Directory to communicate and perform basic read, write, and modify operations; queries an object in the Active Directory; and allows Windows 2000 Servers to communicate with Windows 3.51 and 4.0 servers.
3. A workgroup is an organizational unit of computers that is typically found in small peer-to-peer networks. In a workgroup, each computer has a local security database so that it tracks its own user and group account information.
4. Active Directory is a directory service that combines domains, X.500 naming services, DNS, X.509 digital certificates, and Kerberos authentication.
5. NTDS.DIT is a single file located on the domain controller that makes up the Active Directory database.
6. The System Volume (SYSVOL) stores the server copy of the domain's public files, such as the Group Policy and Net Login (logon scripts), which are replicated among all domain controllers in the domain.
7. The Windows 2000 domain is a logical unit of computers and network resources that define a security boundary.
8. A trust relationship is a link that combines two domains into one administrative unit that can authorize access to resources on both domains.
9. Organizational units are used to hold and organize users, groups, computers, and other organizational units.
10. An object is a distinct, named set of attributes or characteristics that represents a network resource, including computers, people, groups, and printers.
11. The schema of the Active Directory contains a formal definition and set of rules for all objects and attributes of those objects.
12. A forest consists of one or more trees that are connected by two-way, transitive trust relationships.

13. The global catalog holds a replica of every object in the Active Directory.
14. The computer that stores a replica (copy) of the account and security information of the domain and defines the domain is known as the domain controller.
15. Active Directory uses multimaster replication.
16. The schema master domain controller controls all updates and modifications to the schema.
17. The relative ID master allocates sequences of relative IDs to each of the various domain controllers in its domain.
18. The PDC emulator acts as a Windows NT primary domain controller.
19. When you rename or move a member of a group (and that member resides in a different domain from the group), the infrastructure master of the group's domain is responsible for updating the group so it knows the new name or location of the member.
20. A site is one or more IP subnets connected by a high-speed link (128 Kbps or higher), typically defined by geographical locations.
21. A bridgehead server is a single server located in each site that is designated to perform site-to-site replication.
22. Before installing a domain controller, identify the DNS name space and verify that the computer has a NTFS partition and that the TCP/IP protocol is installed and configured correctly.
23. To start the Active Directory Installation wizard, run the DCPROMO command.
24. Encryption is the process of disguising a message or data in what appears to be meaningless jumble (cipher text) to hide and protect the sensitive data from unauthorized access.
25. Kerberos is an authentication service developed at MIT for the TCP/IP network, which is the default authentication method for Windows 2000.
26. For a computer to connect to and make full use of an Active Directory network, it requires network client software.
27. Windows 2000 uses a new time synchronization service to synchronize the date and time of computers running on a Windows 2000–based network.

QUESTIONS

1. If you combine domain trees, you form a domain _____ .
   a. forest
   b. jungle
   c. collective
   d. group

2. _____ is a network service that identifies all resources on a network and makes those resources accessible to users and applications.
   a. Security services
   b. Directory services
   c. Application services
   d. Communication services

3. Which of the following is commonly referred as a peer-to-peer network that has each computer keep its own security database?
   a. forest
   b. workgroup
   c. domain
   d. replication partner

4. Which of the following uses a centralized security database to authenticate users on the network?
   a. client
   b. workgroup
   c. domain
   d. replication partner

5. If you have performed a default installation of Windows 2000 Server with Active Directory, which folder will be shared as the system volume?
   a. *%systemroot%\sysvol\sysvol*
   b. *%systemroot%\system32*
   c. *%systemroot%\system32\system*
   d. *%systemroot%\sysvol*

6. Which of the following are true of a Windows 2000 domain controller? (Choose all that apply)
   a. A domain controller can be demoted to be a member server and vice versa.
   b. Changes to the Active Directory can be made on any domain controller.

c. All domain controllers in Windows 2000 are equal status with no primary or master server.

d. Both Windows 2000 Server and Windows 2000 Professional can be used as a domain controller.

7. Which of the following are true of a global catalog in Windows 2000?
   a. A global catalog contains only the commonly queried objects and all attributes for a single domain only.
   b. A global catalog contains only the commonly queried objects and all attributes for a forest.
   c. A global catalog contains all objects and all attributes for a forest.
   d. A global catalog contains all objects and common attributes for a forest.
   e. A global catalog contains all objects and common attributes for a single domain only.

8. You wish to install Active Directory on a Windows 2000 Server but currently have no DNS server. Which of the following options would allow Active Directory to be installed? (Choose all that apply)
   a. Install DNS server on a separate computer after installing Active Directory.
   b. Install DNS server on a separate computer before installing Active Directory.
   c. Install both DNS server and Active Directory at the same time on the same server.
   d. Install a WINS server before installing Active Directory.

9. Which server will become the global catalog server in a forest by default?
   a. first domain controller in the first domain in the forest
   b. first domain controller in each domain
   c. every domain controller in the forest
   d. every Windows 2000 Server in the forest

10. The _____ of the Active Directory contains a formal definition and set of rules for all objects and attributes of those objects.
    a. RID master
    b. GUID
    c. global catalog
    d. schema

11. Which of the following statements are true about trust relationships between domains? (Choose all that apply)
    a. One-way trust relationships are nontransitive.
    b. Two-way trust relationships are nontransitive.
    c. One-way trust relationships are transitive.
    d. Two-way trust relationships are transitive.

12. Which of the following are containers? (Choose all that apply)
    a. organizational units
    b. domains
    c. groups
    d. domain controllers

13. The computer that stores a replica (copy) of the account and security information of the domain and defines the domain is known as the _____ .
    a. member server
    b. domain controller
    c. ACL
    d. Windows 2000 Advanced Server

14. The Active Directory data (account and security information) is stored in the _____ file.
    a. NTFS.DIT
    b. NTDS.DIT
    c. NTFS.DIR
    d. NTDS.DIR

15. Which of the following are true of a site with respect to Windows 2000 domains? (Choose all that apply)
    a. A site must be contained within a single domain.
    b. A site contains one or more IP subnets on a LAN.
    c. A site must contain at least one whole domain.
    d. A site is configured to optimize Active Directory synchronization traffic.

16. Which options are available for site link replication methods? (Choose all that apply)
   a. IP
   b. IPX
   c. SMTP
   d. SMB

17. What name is given to the site link created by default on Windows 2000?
   a. DEFAULTIPSITELINK
   b. DEFLINK
   c. IPLINK
   d. DEFAULTSITE

18. Which of the following statements are true of organizational units (OUs) in a Windows 2000 domain? (Choose all that apply)
   a. Administrative control of OUs can be delegated to other users.
   b. An OU can contain other OUs.
   c. Administrators have permission to create OUs by default.
   d. Users and computers are two of the built-in OUs in a domain.

19. You have installed Windows 2000 Member Server and now wish to promote it to a domain controller. What should you do?
   a. Run WINNT /U.
   b. Run DCPROMO.
   c. Run UPDATEAD.EXE.
   d. Reinstall Windows 2000.

20. Which of the following statements are true of a mixed-mode domain in Windows 2000? (Choose all that apply)
   a. A mixed-mode domain supports group nesting.
   b. A mixed-mode domain supports universal groups.
   c. A mixed-mode domain can contain Windows 2000 domain controllers.
   d. A mixed-mode domain can contain Windows NT 4.0 domain controllers.

21. Consider the following site link scenario: You have three sites (Site A, Site B, and Site C). Site A is connected to Site B via a 256-K WAN link. Site A is also connected to Site C via a 256-K WAN link. Site B is connected to Site C via a 56-K WAN link. You wish to create site links to optimize Active Directory synchronization traffic. How many site links should you create?
   a. 2
   b. 1
   c. 3
   d. 4

EXERCISES

EXERCISE 1: INSTALLING ACTIVE DIRECTORY

Computer A should do the following:

1. To start the Active Directory Installation wizard and to make the stand-alone server into a domain controller, execute the DCPROMO.EXE file. Click on the Start button and select the Run option. Enter DCPROMO in the Open text box and click on the OK button. Click on the Next button.
2. For the Domain Controller Type page, select the "Domain controller for a new domain" option and click on the Next button.
3. For the Create a Tree or Child Domain page, select the "Create a new domain tree" option and click on the Next button.
4. For the Create or Join Forest page, select the "Create a new forest of domain trees" option and click on the Next button.
5. For the New Domain Name page, enter ACMExx.COM in the "Full DNS name for new domain" text box. Click on the Next button.
6. On the NetBIOS Domain Name page, leave the default Domain NetBIOS name and click on the Next button.
7. On the Database and Log Locations page, leave the default file locations and click on the Next button.
8. On the Shared System Volume page, leave the default location for the SYSVOL folder and click on the Next button.
9. On the Configure DNS page, click on the "Yes, install and configure DNS on this computer (recommended)" option and click on the Next button.

10. On the Permissions page, select the "Permissions compatible only with Windows 2000 servers" option and click on the Next button.
11. On the Directory Services Restore Mode Administrator Password page, enter the administrator password and click on the Next button.
12. On the Summary page, click on the Next button.
13. Click on the Finish button.

EXERCISE 2: ADDING A SECOND DOMAIN CONTROLLER

Computer B should do the following:

1. To start the Active Directory Installation wizard and to make the stand-alone server into a domain controller, execute the DCPROMO.EXE file. Click on the Start button and select the Run option. Enter DCPROMO in the Open text box and click on the OK button. Click on the Next button.
2. For the Domain Controller Type page, select the "Additional domain controller for an existing domain" option and click on the Next button.
3. For the Create a Tree or Child Domain page, select the "Create a new domain tree" option and click on the Next button.
4. For the Create or Join Forest page, select the "Create a new forest of domain trees" option and click on the Next button.
5. For the New Domain Name page, enter ACMExx.COM in the "Full DNS name for new domain" text box. Click on the Next button.
6. On the NetBIOS Domain Name page, leave the default Domain NetBIOS name and click on the Next button.
7. On the Database and Log Locations page, leave the default file locations and click on the Next button.
8. On the Shared System Volume page, leave the default location for the SYSVOL folder and click on the Next button.
9. On the Configure DNS page, click on the "Yes, install and configure DNS on this computer (recommended)" option and click on the Next button.
10. On the Permissions page, select the "Permissions compatible only with Windows 2000 servers" option and click on the Next button.
11. On the Directory Services Restore Mode Administrator Password page, enter the administrator password and click on the Next button.
12. On the Summary page, click on the Next button.
13. Click on the Finish button.

If you have a DHCP server still running, it will need to be authorized for the domain. To authorize a domain, do the following:

14. From the DHCP console, right-click DHCP and select the "Browse authorized servers" option.
15. In the Authorized Servers in the Directory dialog box, click the Add button.
16. In the Authorized DHCP Server dialog box, enter the name or IP address of the DHCP server to authorize, and then click OK.
17. In the DHCP dialog box, click Yes to confirm the authorization.

EXERCISE 3: SYNCHRONIZATION WITH A TIME SERVER

This exercise can only be performed if you are connected to the Internet.

1. Change the date and time to a different value.
2. Start a Command Prompt window and execute the following command:

```
NET TIME /SETSNTP:192.5.41.209
```

3. In the Administrative Tools, start the Services console.
4. Right-click the Windows Time service and select the Start option. Notice the date and time of the system.

12 Managing Computers, Users, and Groups

INTRODUCTION

In Chapter 11 we installed the Active Directory. Now you are ready to learn how to manage the network by creating and managing users, computers, and groups. In addition, you will learn the first steps to take in securing the network, including setting up passwords and specifying logon times, locations, and Group Policies.

355

OBJECTIVES

1. Create a template and use it to create new accounts.
2. For a user account, create, configure, and manage user profile and login scripts.
3. List and describe the two types of groups.
4. List and describe the three types of group scopes.
5. Delegate control of an organizational unit to a person or group.
6. Move an object from one container to another.
7. Implement, configure, manage, and troubleshoot local Group Policy, account settings, and account policies.
8. Create and manage local users and groups.
9. Implement, configure, manage, and troubleshoot user rights.
10. List and describe, in order of priority, the types of group policies.
11. Compare and contrast rights and permissions.
12. Install applications by using Windows Installer packages.
13. Implement, configure, manage, and troubleshoot local user authentication and security configuration.

12.1 USER ACCOUNTS

A **user account** enables a user to log on to a computer and domain with an identity that can be authenticated and authorized for access to domain resources. All users who log on to the network should have their own unique user account and password.

In Windows 2000, there are two different types of user accounts. The **domain user account** can log on to a domain to gain access to the network resources. A domain controller must authenticate the domain user account. See Figure 12.1. A **local user account** allows users to log on at and gain resources on only the computer where they create such an account. The local user account is stored in the local security database on the computer that the user logs on to. When using a local user account, you will be unable to access any of the network resources except as a peer on a peer-to-peer network. Remember that a local user account is not available to the domain/domain controllers, so, a user cannot log on locally to a domain controller.

12.1.1 Built-in User Accounts

Like Windows NT, Windows 2000 starts with two built-in accounts, the administrator and guest accounts. The administrator account is used to manage the overall computer and domain configuration; create and maintain computers, user accounts, and groups; manage security policies; assign rights and permissions; and create and maintain printers.

Since the Administrator account is the most important account within the domain, there should be certain precautions and procedures to follow.

FIGURE 12.1   Active Directory Users and Computers console

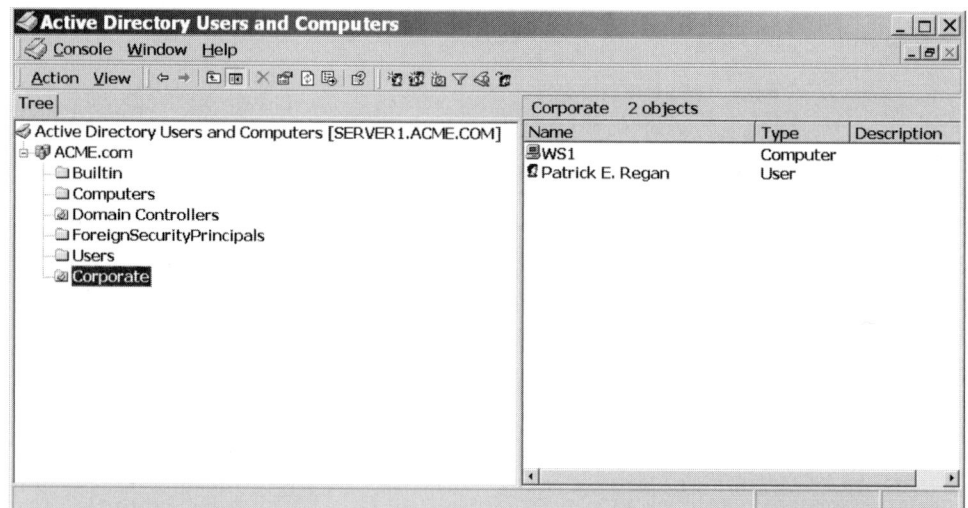

- Since every Windows NT and Windows 2000 computer begins with having an Administrator account, it is a good idea to rename that account. Then, if someone is trying to get into the system, the hacker will have to guess both the name of the administrator and the password. You can only rename the Administrator account; you cannot delete it.
- You should not use the administrator as a general user. Instead, create a second account and use it as a general user and the administrator account for administrative tasks only.

The guest account should be accessible for the occasional users who need to log on and to gain access to resources. The guest account and any other account that performs the same function as the guest account should only be used in low-security networks. Unlike the administrator account, the guest account is disabled by default. If you enable it, you should assign it a password. Similar to the administrator account, you cannot delete it, but you can rename it.

12.1.2 Creating a User Account

To create a domain user account, open Active Directory Users and Computers console and right-click the OU in which you want to create the user account. Click on the New option and click the User option. To create a local user account on a computer, open Computer Management and use Local Users and Groups in System Tools.

In the first dialog box (see Figure 12.2), find the following fields:

> **First name**—User's first name (required)
> **Initial**—User's middle initial
> **Last name**—User's last name (required)
> **Full Name**—User's full name created by combining the first name, initial, and last name. The full name must be unique within the container where you create the user account.
> **User logon name**—The user's unique logon name. The logon name must be unique within the directory.
> **User logon name (pre-Windows 2000)**—The name used in Windows NT domains.

When you click on the Next button, you get to a second dialog box, which defines password requirements. See Figures 12.3 and 12.4. In this dialog box are the following fields:

> **Password**—Authenticates the user. Although a password is not required, one should be assigned for greater security. Passwords are case sensitive.
> **Confirm Password**—Confirms the password by typing it a second time.
> **User must change password at next logon**—Forces the user to change the password when logging in to authenticate logger use.
> **User cannot change password**—Prohibits any person from a group of people who are using the same user account to change the password. Also allows the administrators to control passwords.

FIGURE 12.2   New Object dialog box

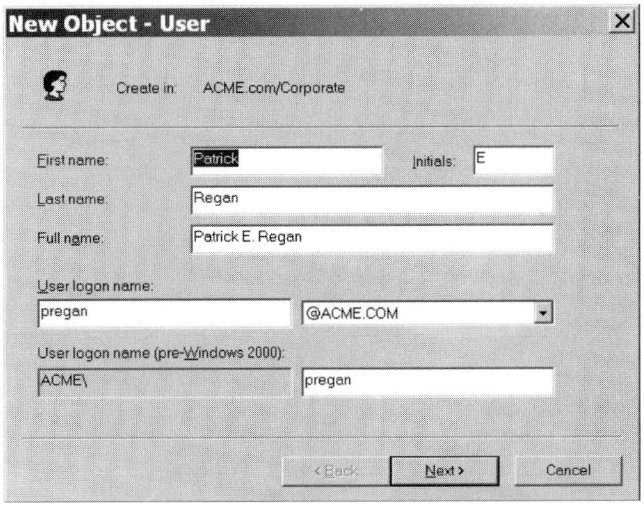

New Object - User

Create in: ACME.com/Corporate

First name: Patrick   Initials: E
Last name: Regan
Full name: Patrick E. Regan

User logon name:
pregan   @ACME.COM

User logon name (pre-Windows 2000):
ACME\   pregan

< Back   Next >   Cancel

Password never expires—Prohibits a password from being changed, even if a user account is configured to periodically force the user to change the password.

Account disabled—Prevents the use of this account in situations such as when an employee has not yet started or has left the company and no replacement has been assigned.

Under properties of the user are several tabs which can be used for future searches. Use of these fields is recommended. These tabs include the General tab (see Figure 12.5), Address tab, Telephones tab, and Organization tab.

The Account tab is similar to the second dialog box shown during the creation of the user object. From here you can change any of the password requirements and make a password automatically expire on a certain day, for example, a user's last day of work. The logon hours are used to specify when the user can log on during the week. For example, if a user never comes in the middle of the night, then stop the account from logging in during that time. This situation is ideal for temporary workers. The Log On To option is used to specify from which computer the person can log on. See Figure 12.6.

FIGURE 12.3   New User's Password dialog box

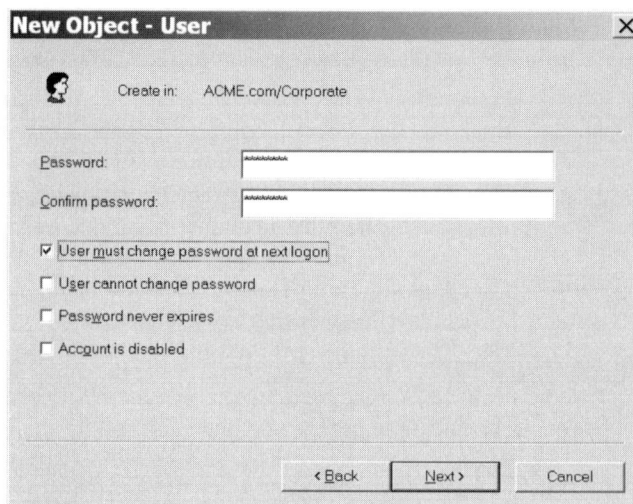

FIGURE 12.4   Account tab in the User's Properties dialog box

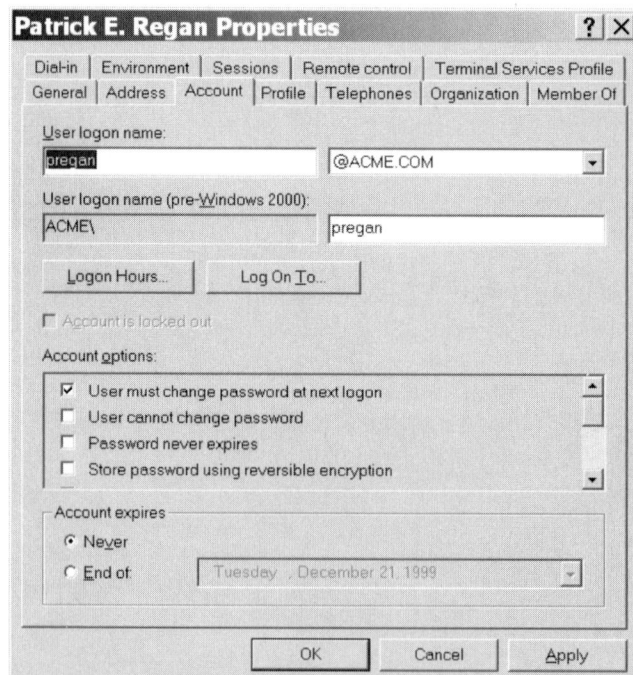

Question: The safety manager of your company was just fired. What should you do?

Answer: Most novice administrator's first reaction would be to immediately delete the account. Although this will prevent the person from accessing the network and causing all kind of problems, it is probably not the best solution.

Prior to making your decision, consider how much time and work goes into maintaining a user account. First, you have to create the account and fill-in addresses, phone numbers, and titles. Second, the manager has been assigned to groups and has been given rights and permissions to files, printers, and other network resources. It would be a better decision to disable the account, so when a person is hired to replace

FIGURE 12.5 General tab in the User's Properties dialog box

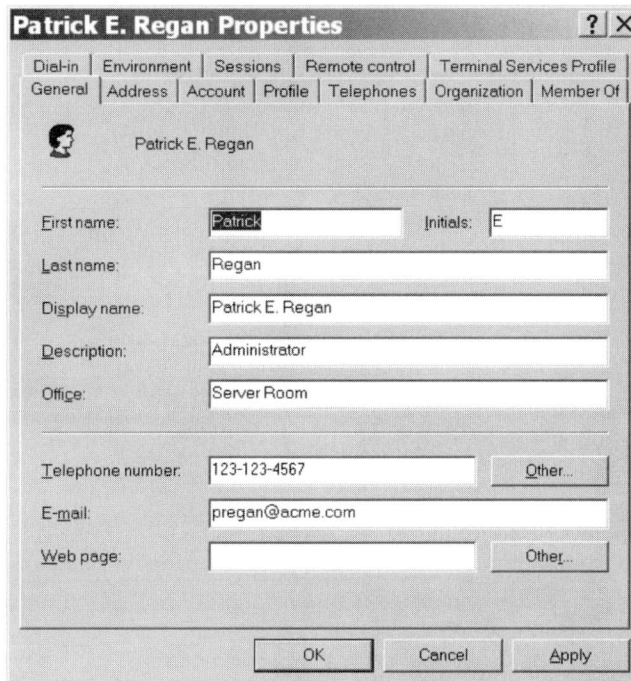

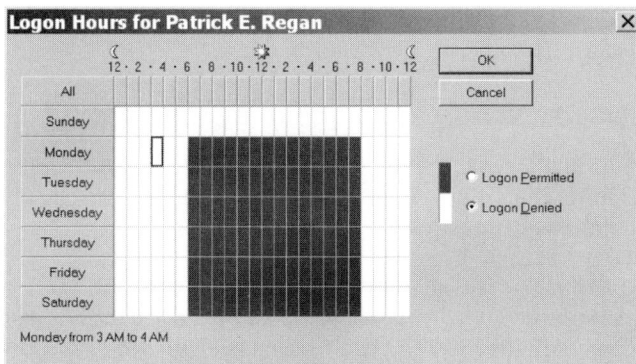

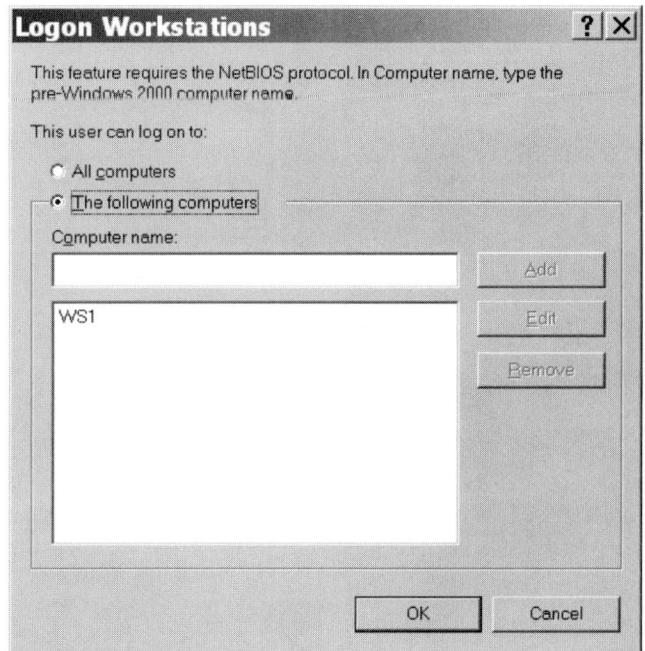

FIGURE 12.6 (a) User's Logon Hours and (b) Logon Workstations restrictions

359

USER NAME GUIDELINES	PASSWORD GUIDELINES
• User logon domains can contain up to 20 characters (lowercase characters, uppercase characters, digits, and valid nonalphanumeric characters). • Invalid characters include ″ / \ [ ] : ; | = , + * ? < and >. • The user logon name field can accept more than 20 characters, but only the first 20 will be recognized. • The user logon name is not case sensitive. • If two employees have duplicate names, add additional letters or numbers to the name to differentiate the two accounts. • Use a consistent naming conventional for all users so that it will be easier to find names that follow a consistent pattern. For example, you can use the first initial, middle initial, and last name or first name and last initial. If you have 100+ employees, however, then perhaps the first name, middle initial, and last name will differentiate all of the users. Create a document stating the naming scheme of user accounts and provide this document to all administrators who create accounts. • To identify the type of employee, add additional letters at the beginning or end of each name. For example, add a *T* before the name or an *X* at the end to indicate the employee is under contract or is a temporary.	• Passwords can be up to 128 characters. • Set a minimum length of 8 characters. • Always use passwords that are not obvious. As part of the responsibilities of the administrator, train the employees of your company to not use obvious names. • Use passwords that combine uppercase characters, lowercase characters, numeral digits, and valid nonalphanumeric characters. • Invalid characters include " / \ [ ] : ; | = , + * ? < and >. • Change passwords regularly.

the safety manager, you can rename the user, reset the password, and reactivate the account. This practice will save time and work, assure managers that they did not miss anything when assigning rights and permissions, and enable the person to access all old files.

12.1.3 Creating User Templates

Creating multiple user accounts simultaneously can be a long and tedious task. The best method is to create a template and then use it to create the needed accounts. You can only create templates in domains with Active Directory.

EXAMPLE 1   Your job is to create 10 users that work for the finance department of your company. The best way is to create a user with the first name of Template and the last name of FinanceUser. You can then preconfigure the Template for FinanceUser with the proper rights, groups, and settings. After the template is finished, you can then use the Copy option to create new accounts. Disable the template account when you are finished copying it.

FIGURE 12.7   Profile tab in the User's Properties dialog box

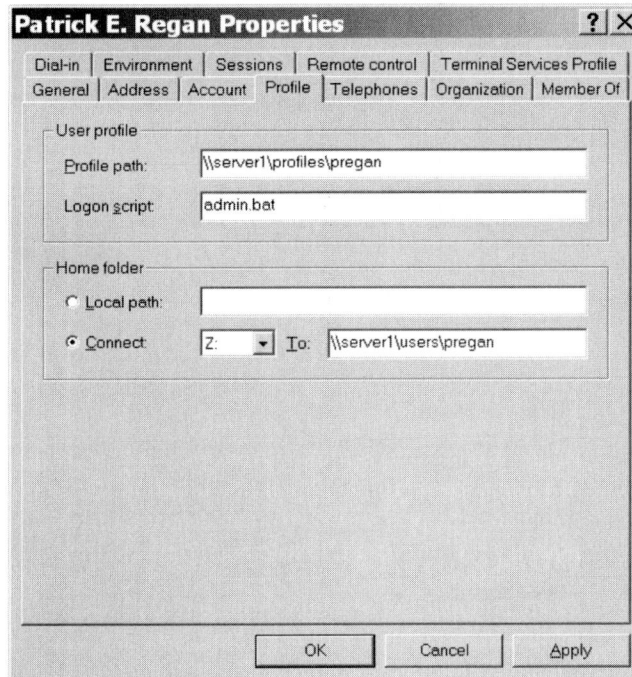

12.1.4   Home Directories

The **home directory** is a folder used to hold or store a user's personal documents. It can be accessed from any Microsoft operating system including DOS, Windows for Workgroups 3.11, Windows 95, Windows 98, and Windows 2000. Although a home directory is typically assigned to an individual user, it can be made available to many users if needed.

The home directory can be stored on the user's computer or in a shared folder on the file server. By having home directories on a file server, users can gain access to their own home directory no matter which computer they are logging in from. In addition, if all home directories are located in one location, it is easy to back up all of the user's data files. Lastly, if the home files are on an NTFS partition, you can further restrict access to the user's personal files. See Figure 12.7.

From the Active Directory Users and Computers console, right-click the user and click Properties. If the home directory is to be on a file server, then click on the Connect option, specify a drive letter, and enter the UNC path (\\servername\sharename\directory). If the home directory is on the user's computer, however, then click on the Local Path option and enter the drive letter and directory path of the home directory. Note that if you use *%username%* for the last subdirectory in the path, the login user name will be substituted.

To set up a home directory on a network file server, create and share a folder in which to store all home directories on a network server. Typically, this folder is called USERS, but it can be any name that the administrators choose. For the shared folder, remove the default permission Full Control from the Everyone group and assign Full Control to the Users group. This ensures that only users in the Domain Users group can gain access to the shared folder. When you specify the location of the home directory of the user and you specify *%username%* to name a folder on an NTFS volume, the user directory is created with the user's login name, the NTFS Full Control permission is assigned to the user, and all other permissions are removed from the folder including those for the administrator account. For example, if you have just created an account for JGordon and you assign her a home directory of \\SERVER\USERS\%USERNAME%, then a folder\USERS\JGordan will be created on the server disk. (Please refer to Chapter 14 for a discussion on NTFS rights.)

12.1.5   User Profiles

A **user profile** is a collection of folders and data that stores the user's current desktop environment and application settings. See Table 12.1. A user profile also records all network

TABLE 12.1   Profile folders

Source	Parameters Saved
Accessories	All user-specific program settings affecting the user's Windows environment, including Calculator, Clock, Notepad, and Paint.
Application data and registry hive	Application data and user-defined configuration settings such as a custom dictionary. The program manufacturer decides what data to store in the User profile folder.
Cookies	Cookies store user information and preferences. Cookies are mostly generated when using a browser such as by Microsoft Internet Explorer.
Control Panel	All user-defined settings made in Control Panel.
Desktop contents	Items stored on the desktop including files, shortcuts, and folders.
Favorites	Shortcuts to favorite locations on the Internet.
Mapped network drive	Any user-created mapped network drives.
My Documents	User-stored documents. The My Documents shortcut can be redirected to the home directory.
My Network Places	The NetHood folder stores shortcuts to other computers on the network.
My Pictures	User-stored picture items.
Online user education bookmarks	Any bookmarks placed in the Windows 2000 Help system.
Printer settings	Printhood folder stores shortcuts to defined network printer connections.
Screen colors and fonts	All user-definable computer screen colors and display text settings.
Start Menu	Shortcuts to program items stored in the Start menu. There are several parts to the Start menu (user start menu, all user start menu, and default start menu), which are merged into the Start menu that is shown to the user.
Templates	User template items.
Windows Explorer	All user-definable settings for Windows Explorer.
Windows 2000–based programs	Any program written specifically for Windows 2000 can be designed so that it tracks program settings on a per-user basis. If this information exists, it is saved in the User profile folder.

connections that are established when a user logs on to a computer such as mapped drives to shared folders on a network server. When users log on to a system, then, they will get the same desktop environment that they had previously on the computer. On computers running Windows 2000, user profiles are automatically created and maintained on the local computer when users log on to a computer for the first time. See Figure 12.8.

The three types of user profiles are as follows:

- **Local user profile**—Created the first time you log on to a computer and is stored on a computer's local hard disk (C:\Documents and Settings-user_logon_name folder). Any changes made to your local user profile are specific to the computer on which you made the changes.
- **Roaming user profile**—Created and stored on a server. Since the profile is on a server, it can be accessed from any computer. Therefore, this profile is available every time you log on to any computer on the network.

FIGURE 12.8 Microsoft Explorer showing a user's profile

FIGURE 12.8 Microsoft Explorer showing a user's profile

- **Mandatory user profile**—Used as a roaming profile to specify particular settings for individuals or an entire group of users. When the user logs off, Windows 2000 does not save any changes that the user made during the session. Mandatory user profile can only be changed by the administrator.

Another important user profile is the default user profile. It serves as the basis for all user profiles. Every user profile begins as a copy of the default user profile, which is stored on each computer running Windows 2000 Professional or Windows 2000 Server.

To set up a roaming profile, do the following:

1. On a server, create a shared folder to hold the roaming profile. Typically, the folder name is Profile, although it can be named anything. For share, create a Profiles folder if it does not already exist, and share the folder with authenticated users allowing read-only permissions.
2. Log on by using the template account and configure the desktop environment such as desktop appearance, network connections, applications, shortcuts, and Start menu items.
3. Log off so that Windows 2000 creates a user profile on the computer in the login user name profile folder (C:\Documents and Settings\user_logon_name). Use this user profile for your template.
4. Log on as administrator and copy the user profile template (C:\Documents and Settings\user_logon_name folder) to the user's profile folder (\\servername\ sharedname\username) on the server. You can copy the profile by right-clicking on the My Computer icon, selecting the Properties option, selecting the User Profiles tab, and use the Copy To button. Do not assign a roaming user profile to multiple accounts unless it is a mandatory user profile. See Figure 12.9.
5. Using the Active Directory Users and Computers console, right-click the user, and click on the Profile tab.
6. In the Profile Path text box, if the profile is to be used only by the Windows 2000 clients, enter the user profile path to the folder that contains the profile. If the profile is to be used by a combination of Windows NT 3.X, Windows NT 4.0, and Windows 2000, then the user profile path must contain a file name that has a .USR file name extension.

If users will log on only to computers running Windows 2000, the user profile path should be to a folder name and should not include an extension of .usr or .man. If the folder

FIGURE 12.9   Copying a User Profile

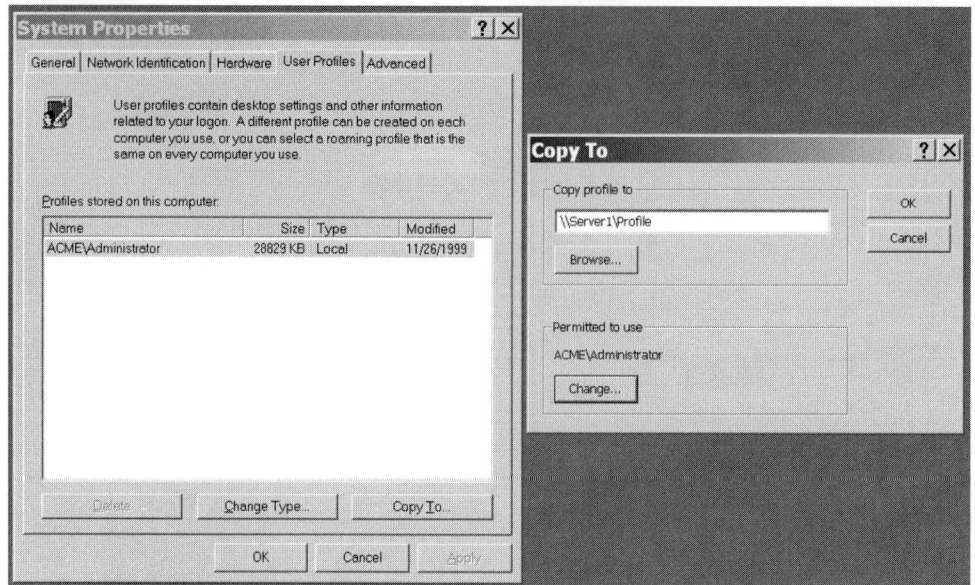

specified in the user profile path does not exist, it is automatically created the first time the user logs on.

To assign the same preconfigured roaming user profile to multiple user accounts, first enter the user profile path separately for each user account (see step 4 above). Then, in Control Panel, double-click System and then click the User Profiles tab. This will copy the preconfigured user profile to the server location for each user.

In the profile directory you will find a hidden file called NTUSER.DAT. This file is the registry portion of the user profile that contains the user environment settings, such as the desktop appearance. This file can be made read-only if you change the file name to NTUSER.MAN, which turns this into a mandatory profile. In fresh installations, the NTUSER.DAT and NTUSER.DAT.LOG are stored in the *systemdrive*\Documents and Settings*username* folder. In installations that are upgrades from Windows NT, the NTUSER.DAT and NTUSER.DAT.LOG files are stored in the *systemroot*\Profiles*username* folder.

You can also create a profile for Windows 95 and Windows 98 machines. Instead of NTUSER.DAT, use USER.DAT file for a roaming profile and USER.MAN for the mandatory profile. To create the profile on a Windows 95 or Windows 98 client, copy the profile folder to the network server. To create a mandatory user profile, do the following:

- Open Active Directory Users and Computers.
- In the Details pane, right-click the applicable user account and select the Properties option.
- In the Profile tab, type the path information ending with the .MAN file name extension.

12.2   GROUPS

A **group** is a collection of user accounts. Groups are tools that help simplify administration. When you assign users, and rights and permissions, to a group, those rights and permissions apply to everyone in the group. Users can be members of multiple groups and groups can belong to other groups. They are, however, not containers of users and other objects. Instead, a group is an object that lists the members. For example, if you delete the group, the members of the group are not deleted, but their rights and permissions that have been assigned to the group are lost.

In Windows 2000, there are two group types, security and distribution. The **security group** is used to assign permissions and gain access to resources. This group can also be used for nonsecurity purposes such as those that you would find in a distribution group. The **distribution group** is used only for nonsecurity functions such as those used to distribute e-mail messages to many people. Typically, distribution groups are used only for specific applications such as the Microsoft Exchange Mail Server.

The security group can be divided into **group scopes,** which define how the permissions and rights are assigned to the group. In Windows 2000, there are three scopes—global group, domain local group, and universal group.

12.2.1 Global and Domain Local Group Scopes

Global groups are primarily used to group people within a domain. Thus, user accounts and global groups are listed from their domain. The global group can be assigned access to resources in any domain. Although rights and permissions assignments are valid only within the domain in which they are assigned, by applying groups with global scope uniformly across the appropriate domains, you can consolidate references to accounts with similar purposes.

Using global groups helps to manage directory objects that require daily maintenance, such as user and computer accounts. Because groups with global scope are not replicated outside of their own domain, accounts in a group having global scope can be changed frequently without generating replication traffic to the global catalog.

The **domain local group** is primarily used to assign rights and permissions to network resources that are in the domain of the local group. Different from global groups, domain local groups can list user accounts, universal groups, and global groups from any domain and local groups from the same domain.

EXAMPLE 2 Say, for example, that we have a printer and some files to be accessed in the SALES domain. We have users in the FINANCE and SALES domains that need to access these resources.

- First, create two global groups called ANALYSTS and MANAGERS in the FINANCE domain and create a global group called SALESPEOPLE in the SALES domain. Then assign the respective users from the FINANCE domain to the ANALYSTS group and MANAGERS group, and the users from the SALES domain to the SALESPEOPLE group.
- Next, create a domain local group called SALES_RESOURCES in the SALES domain. Assign rights and permissions to the SALES_RESOURCES group to the printer and directories in the SALES domain.
- You can then assign users from either domain and assign the ANALYSTS group, MANAGER group, and SALESPEOPLE group. Since groups consist of many people, it is best to assign the groups first. Then if you still have certain individuals that need different or special access, you can assign them as needed.

As you can see, the global groups are used to group people with common purposes or that have the need for the same rights and permissions. The advantage of using this scheme is that when you add members to the two global groups they will automatically get access to the resources.

12.2.2 Universal Groups

The universal security group is new to Windows 2000, and is only available in native mode (not mixed mode), which means there are no Windows NT domain controllers, only Windows 2000 domain controllers. In a mixed-mode network, the Universal Group option will be grayed out. It can contain users, universal groups, and global groups from any domain and it can be assigned rights and permissions to any network resource in any domain in the domain tree or forest. Thus, the universal group is much easier to use and is more flexible.

Universal groups are best used to consolidate groups that span domains. To do this, add the accounts to groups with global scope and nest these groups within groups having universal scope. Using this strategy, any membership changes in the groups having global scope do not affect the groups with universal scope. The membership of a group with universal scope should not change frequently, since any changes to these group memberships cause the entire membership of the group to be replicated to every global catalog in the forest.

You can change a global group into a universal group if the global group is not a member of another global group. You can change a domain local group to a universal group if it does not contain any another domain local group.

FIGURE 12.10   Creating a
Group

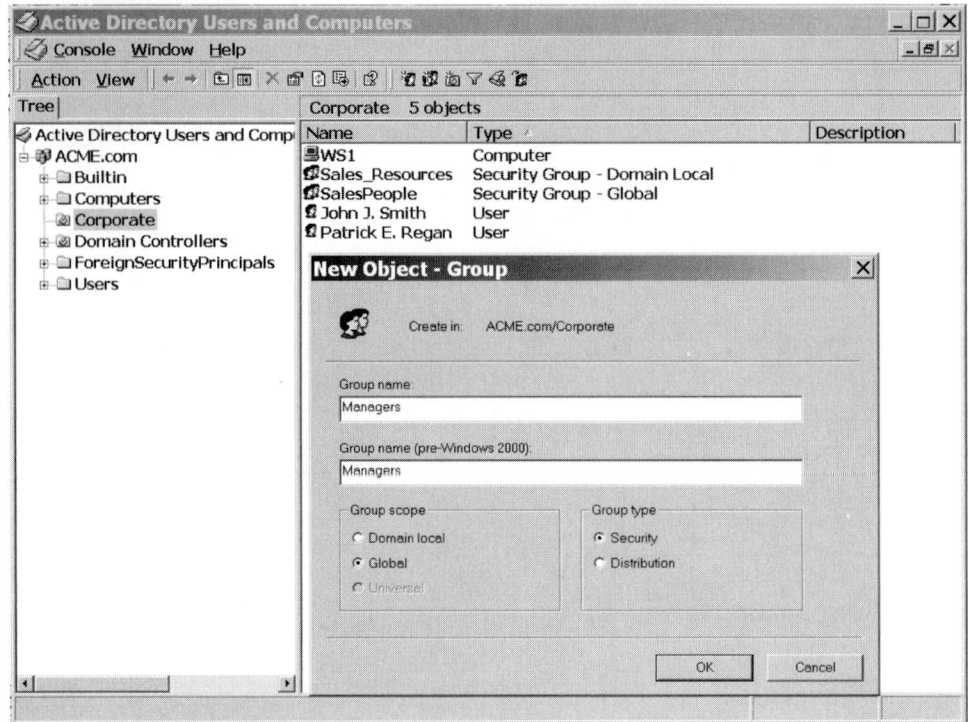

12.2.3   Creating Groups

To create a group, use the Active Directory Users and Computers console or the MMC by using the Active Directory Users and Computers snap-in. To create a group, do the following:

1. Open the Active Directory Users and Computers console.
2. Right-click on the organizational unit in which you want to create the group, and click on the New option and then on the Group option. See Figure 12.10.
3. Enter the group name, enter the pre-Windows 2000 group type, and select the group scope and group type. Click on the OK button.
4. To add people to the list or to assign users and groups to a group, right-click on the group and select the Properties option. See Figure 12.11. You can then use the assign rights and permissions to groups. See Figure 12.12.

When choosing a group name, use intuitive names. It can contain up to 256 uppercase or lowercase characters except for the following:

$$ " \ / \ \backslash \ [\ ] \ : \ ; \ | \ = \ , \ + \ * \ ? \ < \ > $$

A group name cannot consist solely of periods (.) or spaces. A group name cannot be identical to any other group or user name on the domain. Thus, if you need the domains to have the same name, you must utilize a consistent naming scheme instead. For example, if you need a Managers group for each domain, create the names with the domain name built in, for example, Managers LA and Managers NY.

To help identify groups, they are assigned unique security identifiers (SIDs), much like SIDs assigned to domain controllers. If you delete a group and re-create a group with the same name, the second group would have a different SID. Since these are different groups, the users, rights, and permissions will have to be reassigned.

12.2.4   Built-in Groups

Similar to the administrator and guest accounts, Windows 2000 has default groups called built-in groups. These default groups have been granted a useful collection of rights and permissions. Using built-in groups can save the administrator both time and effort. They have been carefully designed to help users perform common duties such as performing

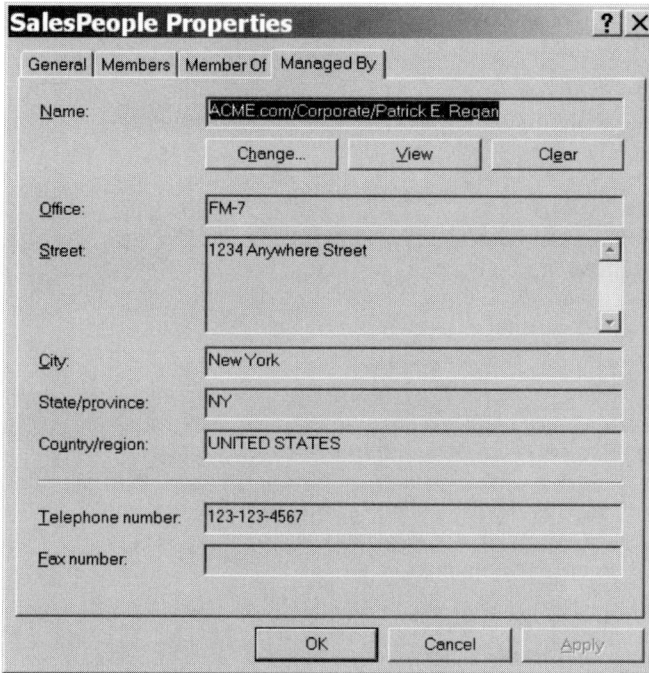

FIGURE 12.11   Group properties

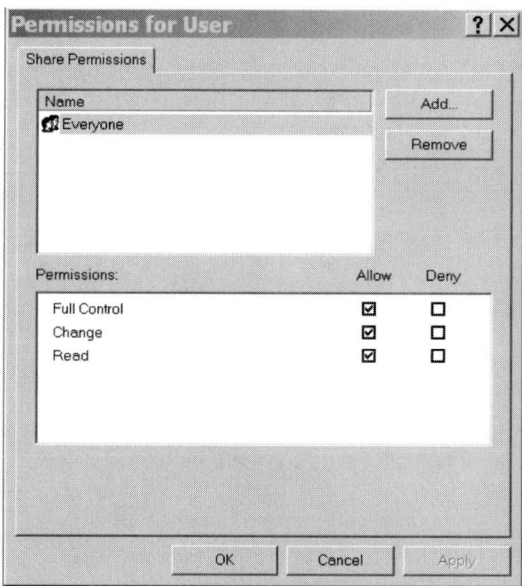

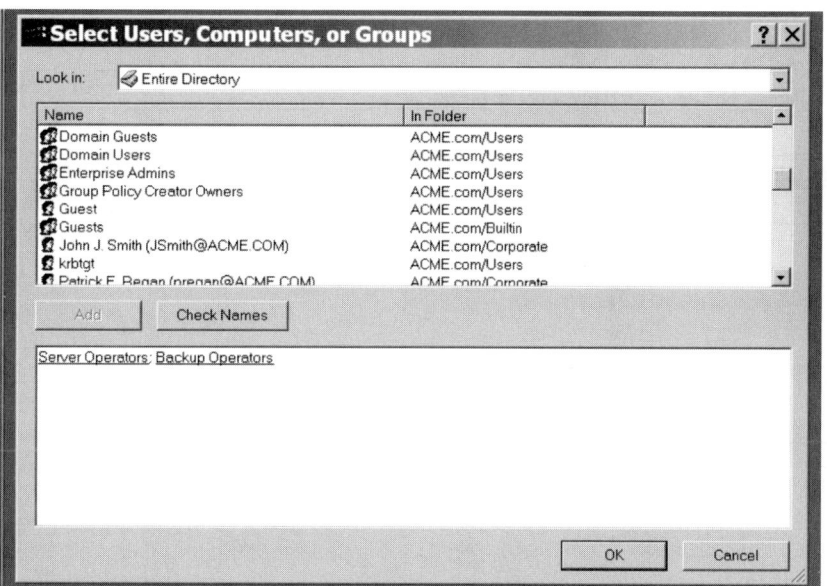

FIGURE 12.12   Assign permissions to users and groups

backups or managing printers. There are four categories of built-in groups: global, domain local, local, and system. See boxed material for more information.

Like a global group that you create, a built-in global group is used to group common types of users. In addition, the domain local group is used to assign rights and permissions to various network resources. The only difference is that by default, Windows 2000 automatically assigns members to some of these groups and has already assigned rights and permissions to these groups. Therefore, any member of these groups will automatically have these rights and permissions.

Built-in local groups exist for any computer that is running Windows 2000 (including stand-alone servers, member servers, domain controllers, and workstations). Built-in local groups have rights to perform system tasks on a single computer.

BUILT-IN GLOBAL GROUP

Domain Admins—Admins can perform administrative tasks on any computer within the domain. By default, the Administrator account is a member.
Domain Guests—Windows 2000 automatically adds Domain Guests to the Guests domain local group. By default, the Guest account is a member.
Domain Users—Windows 2000 automatically adds each new domain user account to the Domain Users group.
Enterprise Admins—This is used to group administrators that need to have administrative control over the entire network (all domains). For this to happen, the Enterprise Admins must be added to the Administrators domain local groups. By default, the Administrator account is a member.

BUILT-IN DOMAIN LOCAL GROUPS

Account Operators—Members can create, delete, and modify user accounts and groups. Note that members cannot modify the Administrators group or any of the operators groups.
Administrators—Members can perform all administrative tasks on all domain controls and the domain itself. By default, the Administrative user account and the Domain Admins global group are members.
Backup Operators—Members can back up and restore all domain controllers by using Windows Backup.
Guests—Members can perform only tasks for which they have been granted rights. By default, the Guest user account and the Domain Guests global group are members.
Print Operator—Members can set up and manage network printers on domain controllers.
Server Operators—Members can share disk resources and back up and restore files on a domain controller.
Users—Members can perform only tasks for which you have granted rights and gain access only to resources for which you have assigned permissions. By default, the Domain Users group is a member. Use this group to assign permissions and rights that every user with a user account in your domain should have.

BUILT-IN LOCAL GROUPS

Administrators—Members of the Administrator group have full control over the computer. It is the only built-in group that is automatically granted every built-in right and ability in the system so that members can install, upgrade, and configure the operating system.
Backup Operator—Members of the Backup Operators group can back up and restore files on the computer, regardless of any permissions that protect those files. They can also log on to the computer and shut it down, but they cannot change security settings.
Power Users—Members of the Power Users group can create user accounts, but can modify and delete only those accounts they create. They can create local groups and remove users from local groups they have created. They can also remove users from the Power Users, Users, and Guests groups. They can install and remove applications except those reserved for administrators and they can configure printers, Date/Time options, power options, and Control Panel resources. They cannot modify the Administrators or Backup Operators groups, nor can they take ownership of files, back up or restore directories, load or unload device drivers, or manage the security and auditing logs.
Users—Members of the Users group can perform most common tasks, such as running applications, using local and network printers, and shutting down and locking the workstation. Users can create local groups, but can modify only the local groups that they created. Users cannot share directories or create local printers.
Guests—The Guests group allows occasional or one-time users to log on to a workstation's built-in Guest account and be granted limited abilities. Members of the Guests group can also shut down the system.

The built-in system groups exist on all computers running Windows 2000. Different from the other groups, the membership of the built-in system groups cannot be modified. Therefore, you do not see system groups when you administer groups unless you are adding that group to another group.

12.3 COMPUTER ACCOUNTS

Computers that are running Windows NT or Windows 2000 are, by design, much more secure than Windows 95 or Windows 98. For example, to use the computer, you must log on to the computer with a user name and password. If you do not log on, you cannot bypass the logon screen by pressing the Escape key and so access the computer's files and programs, or access any network resources.

To use a Windows NT or Windows 2000 computer that is not a domain controller, create a computer account, which is used to uniquely identify the computer on the domain. By having the computer account, you can then audit the computer to ensure that only authorized people who have access to the computer can assign permissions to gain access to network resources.

A computer account, which matches the name of the computer, is an account created by an administrator to uniquely identify the computer on the domain. By being able to identify the computer, a secure communications channel can be created between the client computer and the domain controller. The computer account also allows the administrator to remotely manage the computer user environment and manage the computer user and group accounts.

To create a computer account, do the following:

1. Start the Active Directory User and Computers console or snap-in.
2. Right-click on the container in which you want to create the computer account.
3. Enter the name of the computer.
4. Click on the OK button.

FIGURE 12.13   Creating a computer object

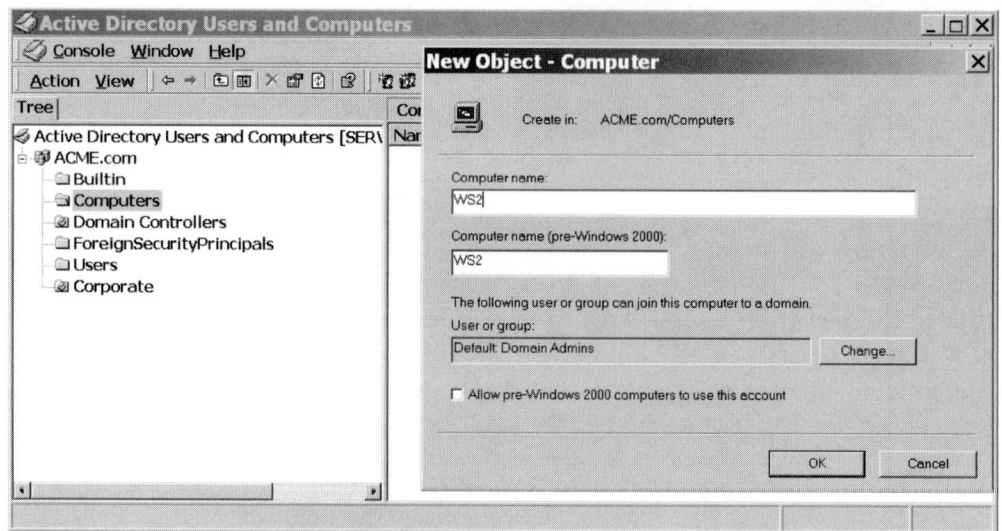

Before you click on the OK button, there is one other option to consider. In Figure 12.13, the following user or group can join this computer to a domain text box which specifies the person(s) who can actually join the computer to the domain. This person will then go to the computer and add the computer to the domain, and would at that time provide a user name and password to make sure that only authorized people can add the computer. See Figure 12.13.

12.4   MOVING OBJECTS

To move an object—including an organizational unit with the same domain, user, or computer—right-click the object and select the Move option. Then select the container where you want to move it and click on the OK button. To move organizational units between domains, use MOVETREE, which comes with the Resource Kit. To move an object between two containers, the person performing the move must have the ability to create the object in the destination container and to delete the object from the source container. Lastly, you can move multiple objects at the same time.

When you move the object, permissions that are assigned directly to the object remain the same. Since the object is now in a new OU, it inherits the new permissions from the parent container. In addition, moving a user object, computer object, or a subtree containing user or computer objects can change the Group Policy that is applied to those objects. Note: You must have the proper permissions to move an object. Group policies and permissions are explained later in this chapter.

12.5   MANAGING SECURITY SETTINGS

One important feature of any network is security. The network administrator is ultimately responsible for the network security. Windows 2000 provides many levels of security, including:

- User name and passwords
- Account restrictions
- Mandatory user profiles
- Group Policies
- Rights and permissions
- Auditing

The user name and password are the first line of defense, as every user requires these to get access to the network and its resources. For the password to be effective, the administrator sets guidelines and policies on how the user should use the network, provides training of these guidelines, and provides frequent reminders. These guidelines include the following:

- Do not give your password to anyone.
- Change your password frequently.

- Do not use obvious or easy passwords.
- Keep people from seeing you type in your password.
- When you leave your computer unattended, log off from the computer.
- Use a password-protected screen saver.
- If you see a security problem or a potential security problem, report it to the network administrator.

Account restrictions, which specify where a user can log on to and which computer a user can log on from, was discussed in Section 12.1.2. Mandatory user profiles, which limit permanently modifying the desktop properties, were discussed in Section 12.1.5. Auditing will be discussed in Chapter 16.

12.6 RIGHTS

A **right** authorizes a user to perform certain actions on a computer, such as logging on to a system interactively or backing up files and directories. Administrators can assign specific rights to individual user accounts or group accounts. Rights are managed with the User Rights policy. Some of the popular user rights are shown in Table 12.2. User rights can be found by opening the Group Policy, opening Computer Configuration, opening Windows Settings, opening Security Settings, opening Local Policies, and then opening User Rights Assignment.

To simplify the administration of rights, user rights are best administered by using groups. If a user is a member of multiple groups, the user's rights are cumulative, which means that the user has more than one set of rights. The only time that rights assigned to one group may conflict with those assigned to another is in the case of certain logon rights. In general, however, user rights assigned to one group do not conflict with those assigned to another group. To remove rights from a user, the administrator simply removes the user from the group. In this case, the user no longer has the rights assigned to that group.

12.7 PERMISSIONS

A **permission** defines the type of access that is granted to an object or object attribute. The permissions that are available for an object depend on the type of object. For example, a user has different permissions than a printer. When a user or service tries to access an object, its access will be granted or denied by an object manager. Common object managers are shown in Table 12.3.

When a computer, user, or group is assigned rights or permissions, each is also assigned a security identifier (SID). Similar to the SID that is assigned to a domain, a SID assigned to a computer, user, or group is a unique alphanumeric structure. The first part of the SID identifies the domain in which the SID was issued and the second part identifies an account object within the domain. Therefore, when a computer, user, or group accesses an object, all three are identified by their SID and not their user name.

Each object uses an Access Control List (ACL) to list users and groups. The ACL is divided into Discretionary Access Control Lists (DACLs) and System Access Control Lists (SACLs). The DACL contains the access control permissions for an object and its attributes, and the SIDs, which can use the object. The permissions and rights that a user has are referred to as access control entries (ACEs). The SACL contains events that can be audited for an object. The ACEs include Deny Access and Grant Access.

Every object in Active Directory has an owner who controls how permissions are set on an object and to whom permissions are assigned. The person who creates the object automatically becomes the owner and by default, has full control over the object, even if the ACL does not grant that owner access. If a member of the Administrators group creates an object or takes ownership of an object, then the Administrators group becomes the object group. A member of the Domain Administrator group has the ability to take ownership of any object in the domain and then change permissions.

Standard permissions are the most common and frequently assigned permissions that apply to the entire object. Assigning standard permissions is sufficient for most administrative tasks. The standard permissions are divided into special permissions, which provide a finer degree of control. See Table 12.4.

TABLE 12.2   Windows 2000 user rights

User Rights	Description	Groups Assigned This Right by Default
Access this computer from a network	Allows the user to connect to the computer over the network.	Administrators, Everyone, and Power Users
Add workstations to domain	Allows the user to add a computer to a specific domain. The user specifies the domain through an administrative user interface on the computer being added, creating an object in the Computer container of Active Directory. The behavior of this privilege is duplicated in Windows 2000 by another access control mechanism (permissions attached to the Computer container or organizational unit).	Authenticated Users
Back up files and directories	Allows the user to circumvent file and directory permissions to back up the system. Specifically, the privilege is similar to granting the following permissions on all files and folders on the local computer: Traverse Folder/Execute File, List Folder/Read Data, Read Attributes, Read Extended Attributes, and Read Permissions. See also *Restore files and directories*.	Administrators, Backup Operators
Change the system time	Allows the user to set the time for the internal clock of the computer.	Administrators, Power Users
Create a page file	Allows the user to create and change the size of a page file. This is done by specifying a paging file size for a given drive in the System Properties Performance options.	Administrators
Debug programs	Allows the user to attach a debugger to any process. This privilege provides powerful access to sensitive and critical system operating components.	Administrators
Enable Trusted for Delegation on user and computer accounts	Allows the user to set the Trusted for Delegation setting on a user or computer object. The user or object that is granted this privilege must have write access to the account control flags on the user or computer object. A server process either running on a computer that is trusted for delegation or running by a user that is trusted for delegation can access resources on another computer. This uses a client's delegated credentials, as long as the client account does not have the Account Cannot Be Delegated account control flag set. Misuse of this privilege or of the Trusted for Delegation settings could make the network vulnerable to sophisticated attacks using Trojan horse programs that impersonate incoming clients and use their credentials to gain access to network resources.	Administrators
Force shutdown from a remote system	Allows a user to shut down a computer from a remote location on the network. See also the *Shut Down the System* privilege.	Administrators
Increase quotas	Allows a process with write property access to another process to increase the processor quota assigned to that other process. This privilege is useful for system tuning, but can be abused, as in a denial-of-service attack.	Administrators
Load and unload device drivers	Allows a user to install and uninstall plug-and-play device drivers. Device drivers that are not plug-and-play are not affected by this privilege and can only be installed by administrators. Since device drivers run as trusted (highly privileged) programs, this privilege could be misused to install hostile programs and give these programs destructive access to resources.	Administrators
Log on locally	Allows a user to log on at the computer's keyboard. Since most protection can be bypassed by being able to log on directly to a machine without going through the network, this right should only be given to a few people.	Administrators, Account Operators, Backup Operators, Print Operators, and Server Operators

TABLE 12.2 *(Continued)*

User Rights	Description	Groups Assigned This Right by Default
Manage auditing and security log	Allows a user to specify the object access auditing options for individual resources such as files, Active Directory objects, and registry keys. Object access auditing is not actually performed unless you have enabled it in the computerwide audit policy settings under Group Policy or under Group Policy defined in Active Directory. This privilege does not grant access to the computerwide audit policy. A user with this privilege can also view and clear the security log from the Event Viewer.	Administrators
Modify firmware environment values	Allows modification of the system environment variables, either by a user through the System Properties or by a process.	Administrators
Profile a single process	Allows a user to use Windows NT and Windows 2000 performance-monitoring tools to monitor the performance of nonsystem processes.	Administrators, Power Users
Profile system performance	Allows a user to use Windows NT and Windows 2000 performance-monitoring tools to monitor the performance of system processes.	Administrators
Restore files and directories	Allows a user to circumvent file and directory permissions when restoring backed up files and directories, and to set any valid security principal as the owner of an object. See also the *Back up files and directories* privilege.	Administrators, Backup Operators
Shut down the system	Allows a user to shut down the local computer.	Administrators, Backup Operators, Everyone, Power Users, and Users
Take ownership of files or other objects	Allows a user to take ownership of any securable object in the system, including Active Directory objects, files and folders, printers, registry keys, processes, and threads.	Administrators

TABLE 12.3   Common object types

Object Type	Object Manager	Management Tool
Files and folders	NTFS	Windows Explorer
Shares	Server Service	Windows Explorer
Active Directory objects	Active Directory	Active Directory Users and Computers console or snap-in
Registry keys	The registry	Registry Editor (REGEDT32.EXE)
Services	Service controllers	Security Templates, Security Configuration and Analysis
Printer	Print spooler	Printer folder

TABLE 12.4   Standard permissions

Object Permission	Description
Full Control	Contains all permissions for the object, including take ownership.
Read	Users can view objects and object attributes, including the object owner and the Active Directory permissions.
Write	Enables users to change all object attributes.
Create All Child Objects	Enables to add any child object to an OU.
Delete All Child Objects	Enables users to delete any child object from an OU.

FIGURE 12.14   Add or change
the permission of the user
object

Edward J. Morell Properties   ? ✕

Environment	Sessions	Remote control	Terminal Services Profile
General	Address	Account	Profile	Telephones	Organization
Published Certificates	Member Of	Dial-in	Object	Security

Name

Administrators (ACME01\Administrators)
Authenticated Users
Cert Publishers (ACME01\Cert Publishers)
Charles L. Gee (CLGee@acme01.com)
Domain Admins (ACME01\Domain Admins)
Enterprise Admins (ACME01\Enterprise Admins)
Everyone

Add...

Remove

Permissions: Allow   Deny

Full Control	☐	☐
Read	☑	☐
Write	☑	☐
Create All Child Objects	☑	☑
Delete All Child Objects	☑	☐
Change Password	☑	☐

Advanced...   Additional permissions are present but not viewable
here. Press Advanced to see them.

☑ Allow inheritable permissions from parent to propagate to this object

OK   Cancel   Apply

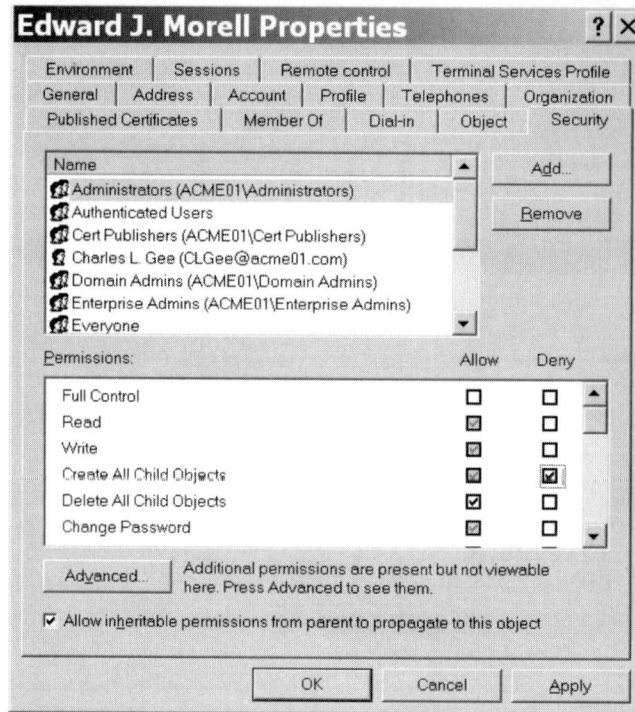

Explicit permissions are those specifically given to the object when the object is created
or assigned by another user. Inherited permissions are those rights that are given to a container and that apply to all child objects under the container. By using inherited permissions,
you can manage them more easily and can ensure consistency of permissions among all objects within a given container.

To add or change the permission for an object, do the following:

1. Using the Active Directory Users and Computers console, enable the Advanced Features
   option.
2. Right-click the object, select the Properties option, and click on the Security tab.
3. To add a new permission, click the Add button and click the user account or group to
   which you want to assign permissions, click Add, then click on the OK button.
4. In the Permissions box, click on the Allow or Deny check boxes for each permission that
   you want to add or remove. See Figure 12.14.

The permissions can be allowed or denied for each user or group. To explicitly allow or
deny the permission, click the appropriate check box. If a check box is shaded, the permission was granted to the user or group for a container that the object is in and the permission
was inherited. If the allow or deny box is not checked for a permission, then the permission
may still be granted from a group. You would then have to check which groups the user or
group is a member of to determine if the rights are granted or denied.

To view special permissions, do the following:

1. Right-click the object, click on the Security tab, and click on the Advanced button.
2. In the Access Control Settings dialog box on the Permission tab, click the entries that
   you want to view, and click on the View/Edit button.
3. In the Apply Onto: box, select where the permissions are to be applied.
4. Select or deselect the desired permissions. See Figure 12.15.

Every object, whether in the Active Directory or an NTFS volume, has an owner. The
owner controls how permissions are set on the object and to whom permissions are granted.
When an object is created, the person creating the object automatically becomes the owner.
Administrators will create and own most objects in Active Directory and on network servers
(when installing programs on the server). Users will create and own data files in their home
directories, as well as on network servers.

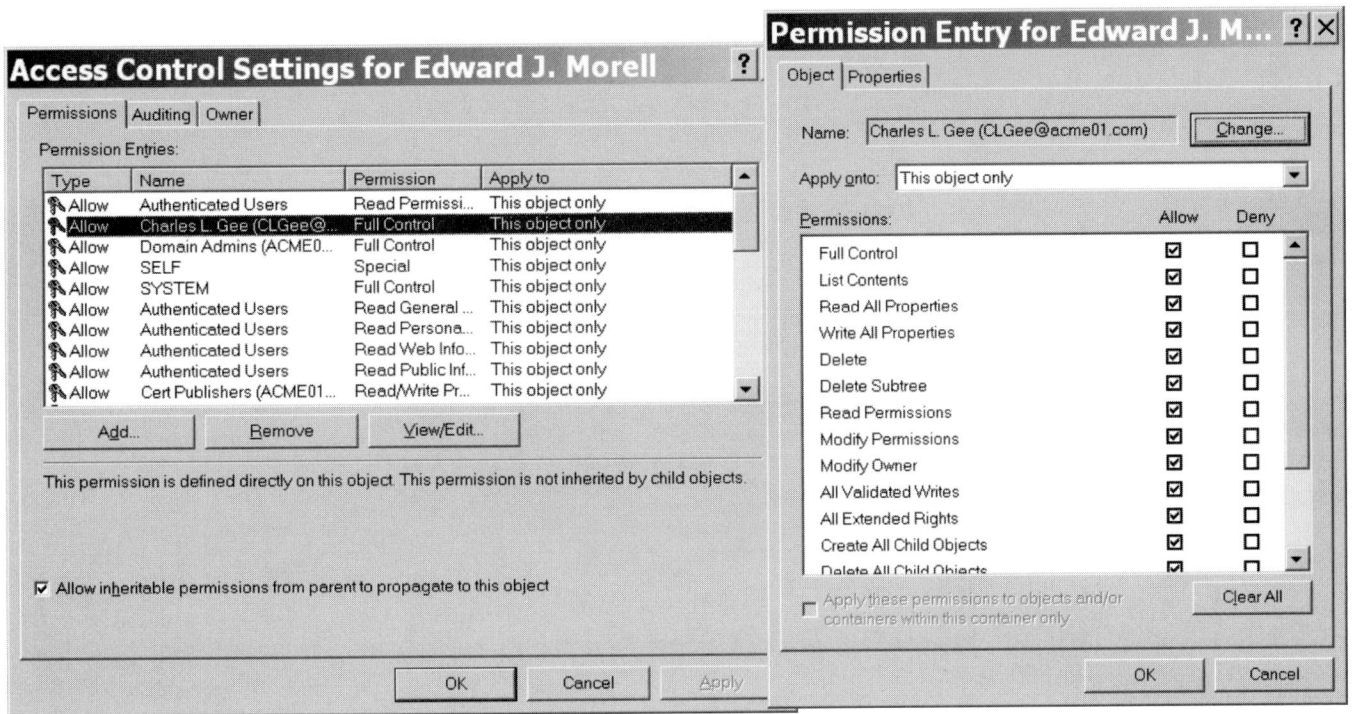

FIGURE 12.15 Changing the Access Control List for an object

Ownership can be transferred in the following ways:

- The current owner can grant the take ownership permission to other users, allowing those users to take ownership at any time.
- An administrator can take ownership of any object under his or her administrative control. For example, if an employee leaves the company suddenly, then the administrator can take control of the employee's files.

Although administrators can take ownership, they cannot transfer ownership to others. This restriction keeps administrators accountable for their actions.

To take ownership of an object, do the following:

1. Right-click the object, select the Properties option, and click on the Security tab.
2. Click on the Advanced button, click the Owner tab, and click your user account. See Figure 12.16.
3. Click on the OK button to close the Access Control Settings dialog box, and click on the OK button to close the object window.

The best way to give sufficient permissions to an organizational unit is to delegate administrative control to the container (decentralized administration), so that the user or group will have administrative control for the OU and the objects in it. To delegate control to an OU, run the Delegation of Control wizard. To start the wizard, right-click the desired OU and select the Delegate Control. You can then select the user or group to which you want to delegate control, the organizational units and objects you want to grant those users the right to control, and the permissions to access and modify objects. For example, a user can be given the right to modify the Owner of Accounts property, without being granted the right to delete accounts in that OU. See Figure 12.17.

The most obvious way to stop inherited permissions is to change the original permission granted at the container. Other ways are as follows:

1. Select a conflict permission (Allow Deny) at the lower level or object, which will override the inherited permission.
2. Clear the "Allow inheritable permission from parent to propagate to this object" check box. By clearing the check box, you will then have the opportunity to either copy previously inherited permissions to the object or remove them from the object. See Figure 12.18.

FIGURE 12.16   Taking ownership of an object

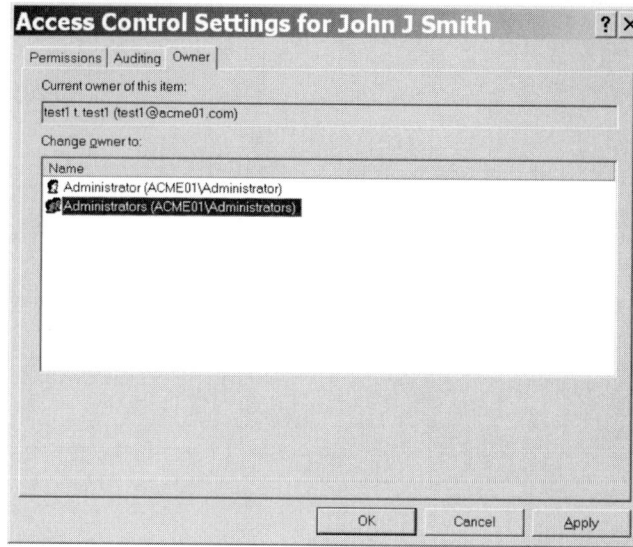

Access Control Settings for John J Smith   ? ×

Permissions | Auditing | Owner |

Current owner of this item:

test1 t. test1 (test1@acme01.com)

Change owner to:

Name
👤 Administrator (ACME01\Administrator)
👥 Administrators (ACME01\Administrators)

[OK]   [Cancel]   [Apply]

FIGURE 12.17   Using the Delegation of Control wizard to delegate permissions over a container

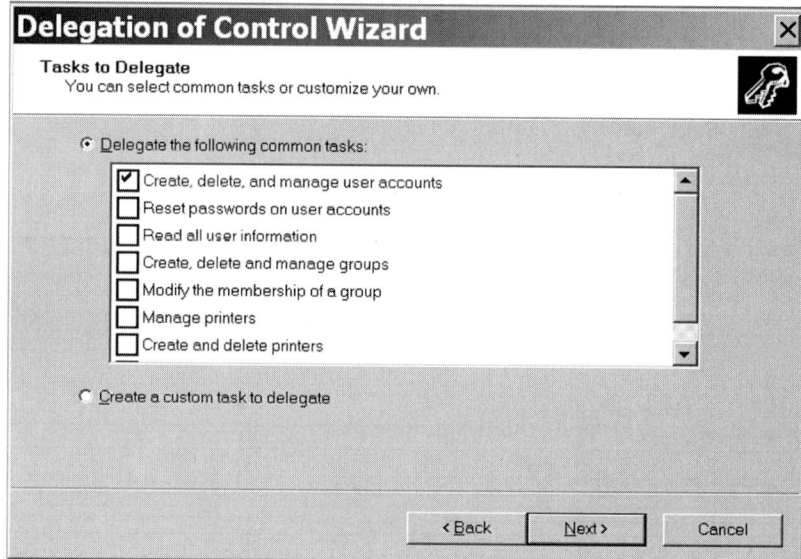

Delegation of Control Wizard   ×

Tasks to Delegate
You can select common tasks or customize your own.

○ Delegate the following common tasks:

☑ Create, delete, and manage user accounts
☐ Reset passwords on user accounts
☐ Read all user information
☐ Create, delete and manage groups
☐ Modify the membership of a group
☐ Manage printers
☐ Create and delete printers

○ Create a custom task to delegate

[< Back]   [Next >]   [Cancel]

FIGURE 12.18   Preventing inherited permissions for a container

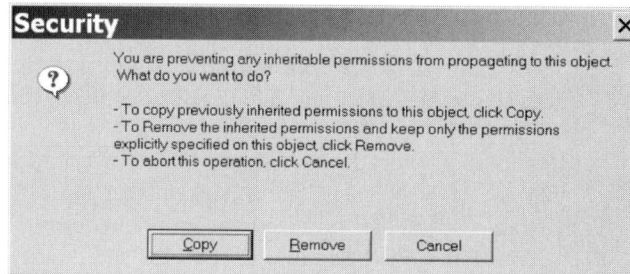

Security   ×

⚠ You are preventing any inheritable permissions from propagating to this object. What do you want to do?

- To copy previously inherited permissions to this object, click Copy.
- To Remove the inherited permissions and keep only the permissions explicitly specified on this object, click Remove.
- To abort this operation, click Cancel.

[Copy]   [Remove]   [Cancel]

Some privileges can override permissions that are set on an object. For example, a user who is logged on to a domain account as a member of the Backup Operators group has the right to perform backup operations for all domain servers. However, this requires the ability to read all files on those servers, even files on which the owners have set permissions that explicitly deny access to all users, including members of the Backup Operators group. A user right—in this case, the right to perform a backup—takes precedence over all file and directory permissions.

12.8 POLICIES

The **IntelliMirror** management technologies are a set of powerful features built into Windows 2000 that allows users' data, software, and settings to follow them. The features of IntelliMirror increase the availability of a users' data, personal computer settings, and computing environment by intelligently managing information, settings, and software. Based on policy definitions, IntelliMirror is able to deploy, recover, restore or replace users' data, software, and personal settings in a Windows 2000-based environment.

At the core of IntelliMirror are three features:

- User settings management
- User data management
- Software installation and maintenance

IntelliMirror is not a particular service or technology, but several services and technologies including group policies, offline folder synchronization, folder redirection, Windows Installer, roaming user profiles, and Remote Installation Services.

Policies serve as a tool that is used by administrators to define and control how programs, network resources, and the operating system behave for users and computers in the Active Directory structure. These settings include the following:

System settings—Application settings, desktop appearance, and behavior of system services

Security settings—Local computer, domain, and network security settings

Software Installation settings—Management of software installation, updates, and removal

Scripts settings—Scripts for when a computer starts and shuts down and when a user logs on and off

Folder redirection settings—Storage for users' folders on the network

The Active Directory is a structured hierarchy, where different levels of policies enable you to customize your configuration.

The different levels of policies are applied in the following order:

1. Windows NT 4.0–style policies
2. Unique local Group Policy object
3. Site Group Policy objects, in administratively specified order
4. Domain Group Policy objects, in administratively specified order
5. Organizational unit Group Policy objects, from the highest to lowest organizational unit and in administratively specified order

Group Policy settings are inherited, cumulative, and affect all computers and user accounts in the Active Directory container with which the Group Policy is associated. If you have settings that are in conflict of each other, then the later policies will overwrite the earlier policies. Therefore, settings in the OU Group Policy will overwrite any settings in the other policies.

12.8.1 Windows NT 4.0–Style Policies

Windows NT 4.0–style policies consist of a set of system registry settings that you create and place on workstations as they log on to the network. These registry entries affect machine and user-specific settings on the workstation. The Windows NT 4.0–style policies control what resources are available to users including what applications appear on the desktop, which applications appear on the Start menu, what accesses there are to basic system areas including the Control Panel, and what accesses there are to the Command prompt.

Windows 95 and Windows 98 system policies are contained in the file CONFIG.POL, whereas Windows NT system policies are contained in the NTCONFIG.POL file. These files are created with the appropriate System Policy Editor such as Poledit.exe and stored in the server's NETLOGON shared folder (\SYSTEM32\REPL\IMPORT\SCRIPTS). The system policies can be defined on a user. Although System Policy Editor (Poledit.cxe) has been largely replaced by Group Policy, it is still useful under some special circumstances.

FIGURE 12.19   Selecting
Group Policies

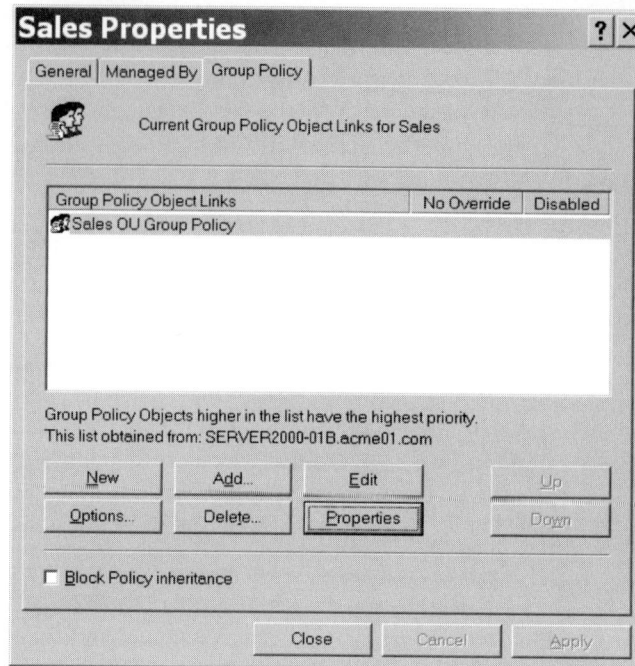

12.8.2   Group Policies

Each computer that is running Windows 2000 has one local Group Policy object, which can be stored on individual computers regardless if they are part of an Active Directory environment or a networked environment. These settings can be overwritten by any nonlocal Group Policies (sites, domains, and organizational units). The local group policy can be added as a snap-in to the MMC or can be executed as a program in the Administrative Tools. The local Group Policy object resides in SystemRoot\System32\GroupPolicy. Computers that are running Windows NT 4.0 or earlier do not have a local Group Policy object. See Figure 12.19.

The nonlocal group policies are stored on the domain controller as Group Policy templates in the SYSVOL volume and are only available in an Active Directory environment. Different from the local group policies, the nonlocal group policies are stored in Group Policy objects (GPOs). As objects, the GPOs can be assigned to multiple sites, domains, or organizational units; and sites, domains, and organizational units can have multiple Group Policy objects.

There are five major categories under which Group Policies can be configured. They are as follows:

- **Folder Redirection**—Stores users' folders (My Documents, Start menu, Application Data, Desktop, and My Pictures) on the network.
- **Security**—Similar to account policies under User Manager in NT 4.0, includes settings for the local computer, the domain, and network security.
- **Administrative Templates**—NT 4.0 administrators will recognize this section as system policies, in a more convenient and flexible configuration. Includes desktop, application, and system settings.
- **Software Installation**—Completely new, this enables an administrator to have software installed automatically at the client machine, or removed automatically.
- **Scripts**—Similar to logon scripts in NT 4.0, but we can now specify a **start-up** and a **shutdown** script for the computer as well as a logon and a logoff script for the user.

Group Policy objects other than the local Group Policy object consist of two parts that are stored separately, the **Group Policy container (GPC)** and the **Group Policy template (GPT).** An administrator can create several GPOs in a given GPC and assign the appropriate GPO to the computers or users that need the settings contained in that GPO. Information that is small and infrequently changed resides in the GPT, while information that is

FIGURE 12.20 Excluding the GPO assigned to the site, domain, or OU

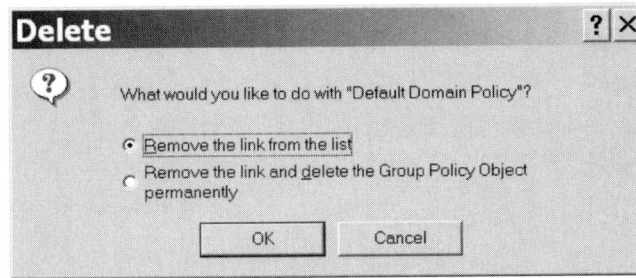

large or frequently changed is kept in the GPC. The Group Policy user interface does not expose them separately.

If you want to exclude certain users or computers from processing the GPO assigned to the Site/Domain/OU to which they belong, simply remove the users' or groups' "Apply group policy" permissions. This effectively creates a filter. You can also delegate control over GPOs so that a manager can change what a GPO does for his or her department, but cannot create any new GPOs or change the scope of a GPO. See Figure 12.20.

It is also possible to disable Group Policy objects without deleting them. If you do this (from Group Policy—Options) it will only disable it for that container and any subcontainers that inherit the settings. If another administrator is "linked" to that GPO from another container, then the GPO is still active in that container.

To create a nonlocal group policy, do the following:

1. Open the Active Directory Sites and Services console or the MMC with the Active Directory Sites and Services snap-in.
2. Right-click the site (Domain Controllers folder), the domain, or the organizational unit and select the Properties option.
3. Click on the Group Policy tab.
4. Click on the New button and name the profile.

If you have multiple group policies assigned to a site, domain, or organizational unit, then specify the order by clicking on the GPO and using the Up and Down buttons. See Figure 12.19.

As mentioned earlier when a GPO is created and associated with an Active Directory container, the settings from the parent container flow down into the child containers. In other words, the child container inherits those settings from above. It was also mentioned that those settings could be overwritten by Group Policies that are executed later.

The inheritance or flowdown of rights can be stopped or blocked by using the Block Policy Inheritance checkbox which is located with the properties of the container. This means that when the box is checked, the container does not inherit any policy settings from the parent-level Group Policies.

Sometimes, you may want to block some of the earlier policies from being overwritten by the later executed policies or block the policy inheritance. To maintain these settings, open the Properties of the GPO and select No Override. You would typically use this in one of the higher levels such as the site or domain GPO to make sure that the administrators of the OUs do not overwrite settings that you want to assign to everyone. See Figure 12.21.

The "Disabled: the Group Policy Object is not applied to this container" option is a troubleshooting tool to help isolate a container's origin. For example, if an object is inheriting a setting that is not the desired setting, you may want to see where the setting originated. Since there can be many GPOs, the disable feature allows you to temporarily disable the GPO without removing its link to the container or allows you to delete the GPO. If the settings that are causing the problem go away when you disable it, you know that the setting is involved in the GPO that you disabled.

If you choose to delete a Group Policy object, you will be given two options. You can either remove the association with the GPO to the container or you can remove the link and delete the GPO. If other containers were using the GPO, however, then those containers would also lose these settings.

Because Group Policy can apply settings from more than one Group Policy object to a site, domain, or organizational unit, you can add Group Policy objects that are associated

FIGURE 12.21 To prevent
overriding a GPO or to disable a
GPO for a container

Sales OU Group Policy Options [?][X]

Link Options:

☐ No Override: prevents other Group Policy Objects from overriding policy set in this one

☐ Disabled: the Group Policy Object is not applied to this container

[OK] [Cancel]

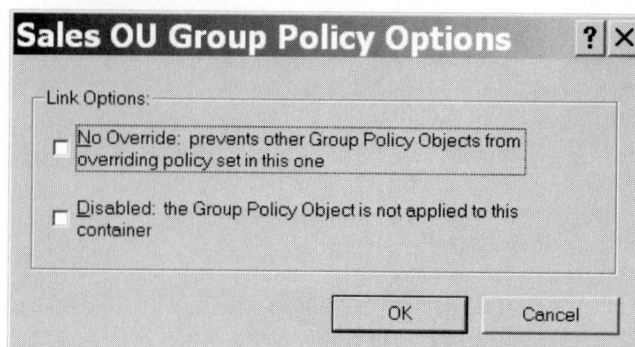

with other directory objects. You can also prioritize how these Group Policy objects affect the directory object to which they are applied.

In Windows 2000, computers can belong to security groups. Administrators can use security groups to further refine which computers and users a Group Policy object influences. For any Group Policy object, administrators can filter the Group Policy object's effect on computers that are members of specified security groups. This filtering occurs using the standard **Access Control List (ACL) editor.** To use the ACL editor, click a Group Policy object's property sheet, and then click Security. The ACL editor can also be used by administrators to delegate who can modify the Group Policy object.

12.8.3   Modifying Policy Settings

To edit a nonlocal group policy, do the following:

1. Open the Active Directory Sites and Services console or the MMC with the Active Directory Sites and Services snap-in.
2. Right-click the site (Domain Controllers folder), the domain, or the organizational unit and select the Properties option.
3. Click on the Group Policy tab.
4. Select the desired GPO and click on the Edit button.

You also can create a MMC with the appropriate snap-ins showing the various Group Profile objects such as that shown in Figure 12.22.

Windows 2000 periodically refreshes Group Policies throughout the network. By default, the client computers are updated every 90 minutes with an offset of $+/-30$ minutes. The domain controllers are updated every 5 minutes. The settings that belong to the Computer Configuration are applied when the computer boots. The settings that belong to the User are applied during login.

To refresh the Group Policy immediately, do the following:

1. Click on the Start button and select the Run option.
2. To refresh the User Configuration settings, in the Run dialog box, input SECEDIT / REFRESHPOLICY USER_POLICY. To refresh the Computer User Configuration settings, in the Run dialog box, input SECEDIT /REFRESHPOLICY MACHINE_POLICY.

12.8.4   Computer Security Settings

The computer security settings are used to combat unauthorized access to resources, viruses, theft of data, or disruption of workflow. The computer security settings are subdivided into the following categories:

Account Policies—Used to configure password policies, account lockout policies, and Kerberos protocol policies for the domain.

Local Policies—Include auditing policies, the assignment of user rights and privileges, and various security options.

Event Log—Used to configure the size, access, and retention parameters for the application logs, the system logs, and security logs.

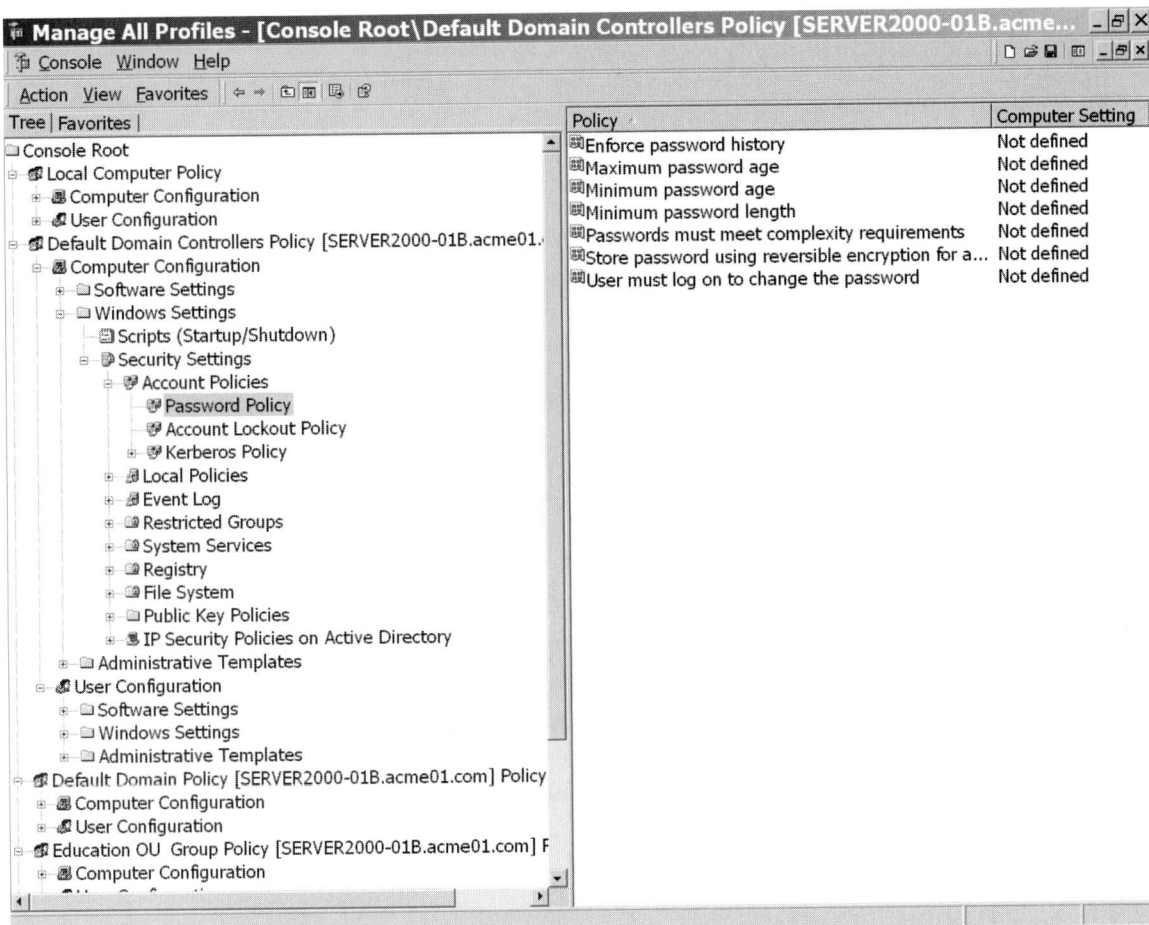

FIGURE 12.22   A MMC showing several policies on the same console

System Services—Used to configure security and start-up settings for services running on a computer.

Registry—Used to configure security on registry key.

File System—Used to configure security on specific file paths.

Public Key Policies—Used to configure encrypted data recovery agents, domain roots, trusted certificate authorities, and so forth. You also can configure public key policies in User Configuration.

IP Security Policies on Active Directory—Used to configure network Internet protocol (IPI) security.

As you open a profile and the various levels of a profile, you will see that the profile stores hundreds of settings. This following section will discuss some of the more important settings. Since the GPO only modifies user settings and no computer settings are altered, you can reduce the time it takes for a client to process the GPO if you disable the computer configuration settings for the GPO.

12.8.5   Password Policy

Account policies should not be configured for organizational units that do not contain any computers, since organizational units that contain only users will always receive account policy from the domain.

When setting account policies in Active Directory, keep in mind that Windows 2000 only allows one domain account policy—the account policy applied to the root domain of the domain tree. The domain account policy will become the default account policy of any Windows 2000 workstation or server that is a member of the domain. The only exception to this rule is when another account policy is defined for an organizational unit. The account policy settings for the organizational unit will affect the local policy on any computers contained in the organizational unit. For more information, see Active Directory overview found in Chapter 11.

The password policy determines the password settings for domain and local user accounts. It can be found by opening the Computer Configuration, opening the Windows Settings, opening Account Policies, and clicking on Password Policies. The popular password policy settings are as follows:

Enforce password history—Remembers a specified number of passwords. Therefore, when a user changes a password, you cannot use the same password. For example, if the Enforce password history is set to 3, users would have to change the password three times before they can use the same password again.

Maximum password age—Specifies how often the password must be changed.

Minimum password age—Specifies how long a user would have to wait before changing a password.

Minimum password length—Specifies the minimum number of characters for a password.

Passwords must meet complexity requirements—When enabled, the password contains at least six characters and must contain characters from at least three of the following four classes:
English uppercase letters (A, B, C, ..., Z)
English lowercase letters (a, b, c, ..., z)
Westernized Arabic numerals (0, 1, 2, ..., 9)
Nonalphanumeric ("special characters") such as punctuation symbols

For a secure network, try setting the Enforce password history to 5 or more, set the maximum password age to between 30 and 45 days, and set the minimum password length to eight characters. Enable passwords to meet complexity requirements.

12.8.6   Account Lockout Policy

The account lockout policy determines when and for whom an account will be locked out of the system. It can be found by opening the Computer Configuration, opening the Windows Settings, opening Account Policies, and clicking on Account Lockout Policies. The account lockout policy has the following settings:

Account Lockout Duration—When an account is locked out, it specifies the duration of the lockout. If you want the account to be unlocked/reset by an administrator, set the value to 99999.

Account Lockout Threshold—The number of invalid logins within the time specified in the Reset Account Lockout Counter After before the account is locked. This setting will eliminate hackers from trying passwords until one works.

Reset Account Lockout Counter After—The time that the number of invalid logins are counted before the invalid login counter is reset.

For a secure network, set the Account Lockout Duration to either 60 minutes or 99999 depending on your situation, set the Account Lockout Threshold to 3, and reset the Account Lockout Counter After to 30 minutes.

12.8.7   Logon Script

Logon scripts are batch files that run every time the user logs on. Batch files are text files that have a BAT file name extension. They can contain commands, which can be executed at the prompt. You can create logon scripts in Notepad, VBscript, and JScript.

A logon script can be assigned to individual users or groups of users. To assign an individual user account, right-click the User account, select the Properties option, and click the Profile tab. When using group profiles, you can assign logon scripts to an entire site, domain, or organizational unit and also be activated during shutdown.

With the use of profiles under Windows, logon scripts are not as important as they once were. Yet, logon scripts still come in handy to ensure certain printer and network connections, set environment settings, and auto load applications. Individual logon scripts can be created and assigned to multiple users or containers.

By default, logon scripts are stored and executed from a domain controller's SYSVOL share. If you remember, the SYSVOL directory is replicated among all the domain controllers so that it will be available no matter which domain controller authenticates the user and provides fault tolerance of the login scripts.

To assign user logon scripts, do the following:

1. Open the Group Policy snap-in.
2. In the console tree, click the Scripts node [located in policy_name Policy, User Configuration, Windows Settings, Scripts (Logon/Logoff)].
3. In the Details pane, double-click the Logon icon.
4. In the Logon Properties page, click Add.
5. In the Add a Script dialog box, set the options you want to use, and then click OK. The options are as follows:
   - Script Name: Type the path to the script, or click Browse to search for the script file in the Netlogon share of the domain controller.
   - Script Parameters: Type any parameters you want to use as you would type them on the command line. For example, if your script included parameters called //logo (display banner) and //I (interactive mode), then type //logo //I
6. In the Logon Properties page, specify any of the following options:
   - Logon Scripts For: Lists all scripts currently assigned to the selected Group Policy object. If you assign multiple scripts, they are processed according to the order specified. To move a script up in the list, click it and then click Up; to move it down, click Down.
   - Add: Opens the Add a Script dialog box, where you can specify any additional scripts to use.
   - Edit: Opens the Edit Script dialog box, where you can modify script information such as name and parameters.
   - Remove: Removes the selected script from the Logon Scripts list.
   - Show Files: Select to view the script files stored in the selected Group Policy object.

If the logon script is stored in a subdirectory of the domain controller's logon script path (sysvol\domainname\scripts), precede the file name with that relative path.

The most common command used in login scripts is the NET command. See Figure 12.23. The NET USE command is used to connect to a shared resource such as a drive, directory, or printer that has been shared or designated for access from clients. This process is called drive mapping.

To access a shared network drive or directory, use the following syntax:

```
NET USE d: \\servername\sharename /Y
```

FIGURE 12.23   Sample Logon Script

```
@echo off
NET USE H: %HOMESHARE%
NET USE G: \\SERVER1\%USERNAME%
NET USE H: \\SERVER1\FINANCE
NET USE I: \\SERVER3\REPORTS
NET USE LPT1: \\SERVER2\HP4-143
IF %USERDOMAIN% = = DOMAIN1 NET USE LPT2: \\SERVER2\HP4-150
IF %USERDOMAIN% = = DOMAIN2 NET USE LPT2: \\TSERVER2\ACCTLASR
NET TIME
```

The *d:* is the local drive letter so assigned to point to the shared directory. The server name (NetBIOS name) locates where the shared drive or directory resides and the share name is the actual name given by the system administrator when creating the share. The \\servername\ sharename is known as the universal naming convention (UNC) name. If the command asks for your confirmation to access, the /Y will automatically respond Yes. For example, to access the shared directory SOFTWARE, which is located on the CLASSROOM server, use the following:

```
NET USE G: \\CLASSROOM\SOFTWARE
```

After this command is executed, as far as the client is concerned, he or she has a G drive. The G drive is actually a pointer that points to the SOFTWARE shared directory on the CLASSROOM server. Therefore, for all practical purposes, the G drive acts like any other local drive. Assuming that you have the proper rights or permissions, you can execute files, access data files, and copy files from or to the G drive. If there already is a G drive, then the computer will ask if you want to overwrite the drive mapping.

Since Windows 2000 remembers its previous drive mappings using profiles and you try to execute a NET USE command to the same drive letter, the NET.EXE will respond, "System error 85 has occurred. The local drive name is already in use." Therefore, you could delete any drive mappings when they are no longer being used by using the disconnection option (/D):

```
NET USE d: /D /Y
```

where the *d:* is the local drive letter assigned to the shared drive or directory. If the command asks for your confirmation to access, the /Y will automatically respond Yes. Therefore, to disconnect from the G drive that was pointing to the SOFTWARE share, execute the following command at the prompt:

```
NET USE G: /D /Y
```

The NET USE command can also be used to connect to a network printer. The proper syntax then is as follows:

```
NET USE LPTx: \\servername\printername
```

where the LPTx is the local printer port assigned to the network printer, the servername is the name of the print server, and the printername is the name of the shared printer. For example, to connect to the IPRINTER network printer that is connected to the INSTRUCT print server using your local LPT2, use the following command:

```
NET USE LPT2 \\INSTRUCT\IPRINTER
```

Thus, all print jobs sent to the LPT2 printer port will be redirected to the IPRINTER, even if you do not really have an LPT2 port.

The NET TIME command can be used to synchronize each client workstation to the server clock. Synchronizing all clients to the same clock is necessary for security and reliability, as you can rely on the time and date stamps of files. For example, to synchronize the client workstation to the server called INSTRUCT, use the following command:

```
NET TIME \\INSTRUCT
```

Variables can be used to customize the logon script by substituting the relevant information during execution. For example, if you use %USERNAME%, the user's name will be substituted, whereas %USERDOMAIN% will substitute the name of the domain where the user is logging on. The variables are listed in Table 12.5. The %USERNAME% will point to an individual home directory; the %HOMESHARE% will point to the location of the home directory.

Therefore, using

```
NET USE H: %HOMESHARE%
```

will create an H drive that points to the home directory.

TABLE 12.5 Variables used in login scripts

Variable	Description
%HOMEDRIVE%	The user's local workstation drive letter connected to the user's home directory
%HOMEPATH%	The full path of the user's home directory
%HOMESHARE%	The share name containing the user's home directory
%OS%	The operating system running on the user's computer
%PROCESSOR_ARCHITECTURE%	The processor type (such as 80386) of the user's workstation
%PROCESSOR_LEVEL%	The processor level of the user's workstation
%USERDOMAIN%	The domain name
%USERNAME%	The user name of the user logging on

Therefore, using

```
NET USE H: %HOMESHARE%
```

will create an H drive that points to the home directory.

To specify a logon script, do the following:

1. Open the Active Directory Users and Computers console.
2. Right-click the user to which you want to assign a logon script.
3. Click on the Profile tab.
4. In the Logon Script text box, enter the name of the logon script you want to assign to the user.

12.8.8 Security Templates

In Windows 2000, User Rights Assignment has been integrated with Group Policy. While it is possible to change security settings for a local machine in a Group Policy Object, a better approach is to use a security template. A security template provides a single place where all system security can be viewed, analyzed, changed, and applied to a single machine or to a Group Policy Object as shown in Table 12.6.

Security templates do not introduce new security parameters, it simply organizes all existing security attributes (account policies, local policies, restricted group, registry settings, file system settings and system services) into one place to ease security administration. Security templates can also be used as a base configurations for security analysis, when used with the Security Configuration and Analysis snap-in.

Windows 2000 comes with pre-configured security templates for common machine configurations, like workstations, secure servers, and domain controllers. The security templates simplify security administration and helps to eliminate gaps in security. For servers deployed on the Internet, rather than laboriously going through a checklist to make sure the server was secure, it is now only necessary to apply a security template. This results in a substantial savings of administrators' time.

To import a security template to a Group Policy object, in a console from which you manage Group Policy settings, click the Group Policy object to which you want to import the security template. It can be found by opening the Policy Object Name, Computer Configuration, Windows Settings, Import Policy and selecting the security template that you want to import. The security settings are applied when the computer starts or as the Group Policy settings dictate.

To customize a predefined security template:

1. In the Security Templates snap-in, double-click Security Templates.
2. Double-click the default path folder (Systemroot\Security\Templates), and right-click the predefined template you want to modify.
3. Click Save As and specify a file name for the security template.
4. Double-click the new security template to display the security policies (such as Account Policies) and double-click the security policy you want to modify.

TABLE 12.6 Windows 2000 predefined security template

Security Template	Description
Basic (basic*.inf)	The basic configurations apply the Windows 2000 default security settings to all security areas except those pertaining to user rights.
Compatible (compat*.inf)	The default Windows 2000 security configuration gives members of the local Users group strict security settings, while members of the local Power Users group have security settings that are compatible with Windows NT 4.0 user assignments so that local users group can use legacy programs. It is not considered a secure environment.
Secure (secure*.inf)	The secure templates implement recommended security settings for all security areas except files, folders, and registry keys. Besides increasing security settings for account policy and auditing, it also removes all members from power users group. These are not modified because file system and registry permissions are configured securely by default.
Highly Secure (hisec*.inf)	The highly secure templates define security settings for Windows 2000 network communications. The security areas are set to require maximum protection for network traffic and protocols used between computer running Windows 2000. As a result, such computers configured with a highly secure template can only communicate with other Windows 2000 computers. They will not be able to communicate with computers running Windows 95 or 98 or Windows NT.

5. Click the security area you want to customize (such as Password Policy), then double-click the security attribute to modify (such as Minimum Password Length).
6. Check the Define this policy setting in the template check box in order to allow editing.

12.9 FOLDER REDIRECTION AND OFFLINE FOLDERS SYNCHRONIZATION

Folder Policy, configured using group policies, redirects certain Windows 2000 special folders located under Documents and Settings such as My Documents and My Pictures to network locations. To individualize a user's redirected folder, it is recommended to incorporate %username% into the path; for example, \\SERVER\SHARE\%USERNAME%\ MY DOCUMENTS.

The benefits of using folder redirection is if a user logs on to various computers on the network, his or her documents are always available. Data stored on a shared network server can be backed up as part of routine system administration.

Offline Files offers users to continue their work with network files and program even when they are not connected to the network. If you lose your connection to the network or disconnect your notebook computer, you still can view shared network items that have been made available offline as it was when you were connected. You can continue to work with them as you normally would. You have the same access permissions to those files and folders as you would have if you were connected to the network. When the network connection is restored, any changes that you made while working offline are updated to the network. When you or someone else on the network have made changes to the same file, you have the option of saving your version of the file to the network, keeping the other version, or saving both. Note: When the status of your connection changes, an Offline Files icon appears in the status area of the taskbar, and an informational balloon is displayed over the status area to notify you of the change.

To set up your computer to use Offline Files:

1. Open My Computer.
2. On the Tools menu, click Folder Options.
3. On the Offline Files tab, make sure that the Enable Offline Files check box is selected. By default, the Enable Offline Files check box is selected in Windows 2000 Professional but is cleared in Windows 2000 Server.
4. Select Synchronize all offline files before logging off to get a full synchronization. Leave it unselected for a quick synchronization.

To make a file or folder available offline:

1. In My Computer or My Network Places, click the shared network file or folder that you want to make available offline.
2. On the File menu, click on the Make Available Offline option. Note: The Make Available Offline option will appear on the File menu only if your computer was set up to use Offline Files.

To undo making a file or folder available offline, right-click the item and click Make Available Offline again to remove the check mark. To view a list of all of the network files that are available offline, on the Offline Files tab, click View Files.

Synchronization can be initiated manually or Synchronization Manager can be set to control when offline files are synchronized with the network. Synchronization Manager also controls whether a full or a quick synchronization is performed. A full synchronization ensures that you have the most current version of every network file that has been made available offline. A quick synchronization is much faster than a full synchronization, but may not provide the most current version of every network file that has been made available offline.

12.10 WINDOWS INSTALLER

Windows 2000 contains a set of new change and configuration management technology features known as IntelliMirror. Using IntelliMirror, you can configure data and applications to follow clients around a network, centrally define client desktop settings for computers and users, and centrally deploy Windows 2000 Professional on desktops using remote installation services (RISs), as discussed in Chapter 7.

One important part of IntelliMirror is the Windows Installer. Using Windows 2000, administrators can control user file systems and registry entries and approve the applications that users can install on their computers. Windows Installer is a service that is responsible for managing installation processes under Windows 2000 to Windows clients (Windows 95, Windows 98, Windows NT 4.0, and Windows 2000).

Windows Installer uses the .msi file format, which replaces the setup program (typically setup.exe) that you have used for years to install software. Windows Installer technology consists of two parts:

- The .MSI package file, which contains a relational database of information that stores instructions for installing and removing applications. When you make modifications to a database table, the changes automatically propagate across the database because of its relational nature.
- The MSIEXEC.EXE installer program that runs on the client side, which uses a dynamic link library (DLL) called MSI.DLL to read the .MSI package file. The program copies files to the hard disk, creates shortcuts, modifies registry entries, and performs other tasks. When you install Windows Installer on your computer, Windows 2000 automatically configures the file association so that .MSI files associate with Windows Installer and the system runs MSIEXEC.EXE.

With Windows Installer, you can install and uninstall applications without affecting the system state or harming other applications. The tool is smart in that it tracks all modifications made to the system during software installation. If your installation crashes, Windows Installer will roll your system back to its original state.

In addition to installing applications, you can also upgrade applications and apply patches. If you want a user to always use a newer version of an application, you can specify for the upgrade to be mandatory. You must alter the software order listed in the GPO so that the newer version is higher than the older version.

Windows Installer uninstalls applications properly, removing registry entries and deleting application files without harming the shared files that other applications use. You can right-click a Windows Installer package (an .MSI file) in Windows Explorer and select Install, Repair, or Uninstall. Alternatively, you can use the Control Panel's Add/Remove Programs applet to change or remove a Windows Installer package. In addition, when setting up the removal options, specify either an optional removal or a forced removal.

Applications can ask Windows Installer to determine whether any program files are damaged or missing. If so, the tool can repair a program by replacing corrupted or missing files. To repair missing files and correct registry entry problems, choose Detect and Repair from an application's help menu.

You can also have Windows Installer restore program shortcuts to the Start menu by selecting "Restore my shortcuts while repairing." When you click Start, Windows Installer detects and repairs damaged Microsoft Office 2000 installations. Windows Installer cannot repair your personal data files, but it can repair software that it installed.

If you have installed Microsoft Office 2000, you are already familiar with another feature of Windows Installer—Advertising. Advertising is a feature that lets you install a small subset of applications initially and have additional components install automatically when you use them for the first time. Windows Installer can also remove components that you have not used for a while.

Software Installation works in conjunction with Group Policy and Active Directory. When you deploy software, you can choose to either assign it or publish it. Assigned software can be targeted at users or computers. If you assign an application to a user, the icons show up on the desktop and/or Start menu, but the program is only installed when the user runs it for the first time. The advantage of assigning an application to a user is that it is available wherever the user logs on. If assigned to the computer, the applications are available to all users on that computer and are installed the next time the system is restarted. If you publish the software, the user can install it through Add/Remove Programs, or through a file that requires that particular program (a file association). Published programs cannot self-repair, cannot be published to computers, and are not advertised on the user's desktop or Start menu, but only through Add/Remove Programs.

Assigned apps require a Windows Installer file (.msi); published apps can use Windows Installer files or .ZAP files (an administrator-created text file that specifies the parameters of the program to be installed and the file extensions associated with it). The .ZAP installations cannot self-repair, cannot install with higher privileges, and will typically require user intervention to completely install.

To use a GPO to deploy software, perform the following steps:

1. Create or edit a GPO in either User Configuration or Computer Configuration, depending on whether the software is to be assigned to users or computers.
2. Expand Software Settings. Right-click Software Installation, point to New, and then click Package. When the File Open dialog box appears, select the package file, and then click Open.
3. In the Deploy Software dialog box, select a deployment method, and then click OK.

To change the software deployment options for a software package, perform the following steps:

1. In Software Installation, right-click the deployed package, and then click Properties.
2. In the Properties dialog box of the application, click the Deployment tab, and set any of the options.

SUMMARY

1. A user account enables a user to log on to a computer and domain with an identity that can be authenticated and authorized for access to domain resources.
2. All users who log on to the network should have their own unique user account and password.
3. The domain user account can log on to a domain to gain access to the network resources.

4. A local user account allows users to log on at and gain resources on only the computer where they create such an account.

5. Like Windows NT, Windows 2000 starts with two built-in accounts, the administrator and guest accounts.

6. To create multiple user accounts simultaneously, it is best to create a template and then use it to create the needed accounts.

7. The home directory is a folder used to hold or store a user's personal documents.

8. A user profile is a collection of folders and data that stores the user's current desktop environment and application settings.

9. The three types of user profiles are local user profile, roaming user profile, and mandatory user profile.

10. A group is a collection of user accounts that is used to simplify administration by assigning rights and permissions to the group. Everyone who is listed in the group will be assigned these rights and permissions.

11. In Windows 2000, there are two group types, security and distribution.

12. The security group is used to assign permissions and to gain access to resources.

13. The distribution group is used only for nonsecurity functions, such as those used to distribute e-mail messages to many people.

14. The security group can be divided into group scopes (global, domain local group, and universal), which define how the permissions and rights are assigned to the group.

15. Windows 2000 has default groups, which have been granted a useful collection of rights and permissions.

16. Policies serve as a tool that is used by administrators to define and control how programs, network resources, and the operating system behave for users and computers in the Active Directory structure.

17. A right authorizes a user to perform certain actions on a computer, such as logging on to a system interactively or backing up files and directories.

18. A permission defines the type of access that is granted to an object or object attribute.

19. To delegate control to an organizational unit, run the Delegation of Control wizard.

20. Windows 2000 contains a set of new change and configuration management technology features known as IntelliMirror.

21. Windows Installer uses the .msi file format, which replaces the setup program.

22. To move an object, including an organizational unit within the same domain, user, or computer, right-click the object and select the Move option.

QUESTIONS

1. You have recently created some user account objects in an OU in your Active Directory. Later you decide to move these accounts to another OU. Which of the following are true of moving objects in Active Directory? (choose all that apply)
   a. Any permissions assigned directly to that object will be moved with the object.
   b. All current permissions on the object will be lost and replaced with the inherited permissions from the new container.
   c. Multiple objects can be moved simultaneously.
   d. Any currently inherited permissions on the object will be lost and replaced with inherited permissions from the new container.

2. Your company has a forest that consists of two domain trees. Each domain tree contains a root domain and two subdomains. You have just created a domain local group and would like to assign permissions to this group of resources within your company. To which resources can permissions be given for this domain local group?
   a. only resources on the same computer as the domain local group
   b. only resources in the same domain as the domain local group
   c. only resources in the same domain tree as the domain local group
   d. any resource in the forest

3. You have just created a global group in a native-mode Windows 2000 Domain in a forest. Which of the following can you add to this global group? (choose all that apply)
   a. any other global group from the domain
   b. any user account from the domain

c. any user account from the forest

d. any global group from the forest

4. Your company has a forest that consists of two domain trees. Each domain tree contains a root domain and two subdomains. You have just created a universal group and would like to assign permissions to this group of resources within your company. To which resources can permissions be given for this universal group?

a. only resources on the same computer as the universal group

b. any resource in the forest

c. only resources in the same domain as the universal group

d. only resources in the same domain tree as the universal group

5. Where in the Group Policy settings would you configure computer start-up and shut-down scripts?

a. Client Configuration   c. User Configuration

b. Start-up Configuration   d. Computer Configuration

6. You have created a GPO for both a parent and a child OU. The parent OU GPO has the "no override" setting enabled. The child OU has the "block inheritance" setting enabled. Which GPO settings will be applied to the child OU?

a. parent GPO settings   c. both GPO settings

b. child GPO settings   d. neither GPO setting

7. Which feature of Windows 2000 allows an administrator to enforce desktop settings for users?

a. System Policy   c. User Policy

b. Group Policy   d. Universal Policy

8. Which two parts comprise the contents of a GPO?

a. GPS   c. GPC

b. GPT   d. GPA

9. Which of the following objects can have an associated GPO? (choose all that apply)

a. global group   c. domain

b. site   d. OU

10. You have associated a GPO with both a parent OU and one of its child OUs. Which of the following statements are true of how the GPO settings will be applied? (choose all that apply)

a. If the GPO settings are not compatible, then the parent OU GPO settings take precedence.

b. If the GPO settings are compatible with each other, then *both* GPOs will be applied.

c. If the GPO settings are not compatible, then the child OU GPO settings take precedence.

d. The child OU GPO settings will always take precedence.

11. Which one of the following group scopes has its membership listed in the global catalog?

a. domain local group   c. universal group

b. global group   d. all groups

12. You have just created a GPO. Where will the Group Policy container (GPC) for this GPO be stored?

a. DNS database   c. SYSVOL share

b. DHCP server   d. Active Directory

13. You have just created a GPO. Where will the Group Policy template (GPT) for this GPO be stored?

a. DNS database   c. DHCP server

b. SYSVOL share   d. Active Directory

14. Which of the following statements are true of groups in Windows 2000? (choose all that apply)

a. Computer accounts and user accounts can be added to groups.

b. A distribution group can be changed to a security group in native mode.

c. A security group can be changed to a distribution group in native mode.

d. A security group can be changed to a distribution group in mixed mode.

15. By default, which of the following profile folders can be redirected to an alternate location using the Folder Redirection setting in Group Policy? (choose all that apply)

a. My Documents

b. Control Panel

c. Start menu

d. Application Data

e. My Pictures

f. Desktop

16. You have just created a GPO for an OU in your enterprise. The GPO only modifies user settings; no computer settings are altered. How can you reduce the time it takes for a client to process this GPO to determine what settings need be applied?

a. Disable the user configuration settings for the GPO.

b. Copy the GPO to every local computer.

c. Split the settings into several GPOs and apply them all.

d. Disable the computer configuration settings for the GPO.

17. Where in the Group Policy settings would you configure user logon and logoff scripts?

a. Client Configuration

b. Computer Configuration

c. User Configuration

d. Start-up Configuration

18. You would like to redirect a user's desktop folders to an alternate location on the network using Group Policy. You do not, however, want all desktop folders to be redirected to the same location. Which folder redirection option would you choose to enable folders to be redirected to alternate locations?

a. Advanced

b. Custom

c. Basic

d. Manual

19. You have associated GPOs with your site, domain, and OUs. In which order are GPOs processed?

a. site, domain, OU

b. domain, OU, site

c. domain, site, OU

d. OU, site, domain

20. You have created a GPO and allocated it to a particular OU in the Active Directory. You do not want this GPO to be applied to all user objects in this OU. How can you apply the OU GPO settings to only some of the user objects in the OU? (choose all that apply)

a. Create a security group that contains only the users that will utilize the GPO settings. Give only this security group both READ and APPLY GROUP POLICY permissions on the GPO.

b. This procedure cannot be done. All GPO settings for an OU will apply to all objects in the OU.

c. Move the user objects that do not require the GPO settings to a sub-OU within the parent OU. Use the Block Inheritance setting on the child OU to stop the parent GPO settings from being applied.

d. Create a security group that contains all the users that will not utilize the GPO settings. DENY the APPLY GROUP POLICY permission on the GPO for this security group.

21. By default, how often will a client computer refresh its Group Policy settings?

a. at user logon only

b. at computer restart only

c. every 90 minutes

d. never

22. You wish to change the scope of a group that you have recently created. Which of the following scope changes can be performed? (choose all that apply)

a. A global group can be changed to a universal group.

b. A universal group can be changed to a global group.

c. A domain local group can be changed to a universal group.

d. No changes can be made.

23. Which two of the following are types of user groups in Windows 2000?

a. distribution groups

b. security groups

c. logon groups

d. access groups

24. You have just created a global group in a native-mode Windows 2000 Domain in a forest. Which of the following can you add to this global group? (choose all that apply)
   a. any other global group from the domain
   b. any user account from the domain
   c. any user account from the forest
   d. any global group from the forest

25. Windows Installer packages can be used to install software on which of the following clients? (choose all that apply)
   a. Windows 95
   c. Windows NT 4.0
   b. Windows 98
   d. Windows 2000

26. You are investigating ways to automatically remove software on a user's computer using Group Policy. Which two removal options are available?
   a. optional removal
   c. temporary removal
   b. permanent removal
   d. forced removal

27. What are the differences between assigning an application to a user and assigning an application to a computer? (choose all that apply)
   a. Applications cannot be assigned to a computer as some user interaction is required before the software can be installed.
   b. Applications assigned to the user will be available where the user logs on, whereas applications assigned to the computer are available to all users of that computer.
   c. Applications assigned to the computer cannot be used interactively as can applications assigned to a user.
   d. Applications assigned to a user require the user to either double-click an associated document or click the shortcut to the program, whereas applications assigned to a computer are automatically installed the next time the computer restarts.

28. You have recently published a software upgrade to your users via Group Policy. You have noticed, however, that when a user invokes an associated document, the older version is installed. How can you install the newer version using document invocation?
   a. Change the upgrade to be mandatory.
   b. Publish the application to the computers rather than the users.
   c. Alter the software order listed in the GPO so that the newer version is higher than the older version.
   d. Run the SETUP.MSI program with the Run option.

29. When looking at the files on your Windows 2000 server you notice some with an .MSI extension. What are these files?
   a. Multimedia System files
   c. Microsoft International Keyboard files
   b. Group Policy files
   d. Windows Installer files

30. Your job is to deploy a non-Windows Installer package to your users by using a .zap file. Which of the following are true of deploying non-Windows Installer packages? (choose all that apply)
   a. They can only be assigned.
   c. They are resilient.
   b. They can only be published.
   d. They are not resilient.

31. Which of the following statements are true of assigning an application using Group Policy? (choose all that apply)
   a. Assigned applications are advertised on the user's desktop.
   b. Assigned applications are resilient and can automatically repair themselves.
   c. Assigned applications are listed in the Add/Remove Programs applet.
   d. Assigned applications can be installed by document invocation.

32. You would like to deploy two different word-processing applications to separate OUs in your company. Users in the first OU require Word 2000 for word processing and those in the second OU require WordPerfect. You have assigned the separate packages to each OU but are concerned that all users would install Word 2000 if they invoked a .DOC file. How can you ensure that document invocation in each OU will install the correct word-processing package?

a. Create two OUs for each department containing only the computers. Publish the correct application to each of these OUs.
b. For each OU, remove the file associations for the unrequired software package.
c. For each OU, modify the file association order so that the required package is higher in the list than the unrequired package.
d. Do not worry, because only the required application is assigned to each OU and installed by document invocation.

33. What are the benefits of using Windows Installer packages to deploy software on your network? (choose all that apply)
a. They can be used to deploy software to non-Windows 2000 clients.
b. They can be removed cleanly.
c. Packages installed using Windows Installer are resilient, and can automatically replace missing or corrupt files.
d. They allow custom installations of the software.

34. You accidentally delete a user's account on a domain controller and you must re-create it with the same permissions and rights. What is the best way to accomplish this task?
a. Assuming that the domain controller has not yet performed synchronization, highlight the deleted user account in Active Directory Users and Computers console and select Undelete from the Console menu.
b. Re-create a new user account with the same name and assign it the security identifier (SID) of the deleted account.
c. Re-create a new user account with the same name and assign it all the permissions and rights of the deleted account.
d. Re-create a new user account with the same name; permissions and rights will be retained since they are contained in the ACLs for domain objects (files, directories, etc.).

EXERCISES

For these labs, X will represent either A or B depending on if you are Computer A or Computer B. This will allow both users to perform the exercise.

EXERCISE 1: CREATING ORGANIZATIONAL UNITS

1. Start the Active Directory User and Computers console.
2. To create an organizational unit called SALES, right-click the ACMExx.COM domain, select the New option, and select Organizational Unit. In the Name text box, input SALES and click on the OK button.
3. Using the same method, create the following organizational units under ACMExx.COM:

   RESEARCHX
   EDUCATIONX
   MANUFACTURINGX

EXERCISE 2: CREATING USERS

1. To create a user in the ACMExx.COM domain, right-click the ACMExx.COM domain, select the New option, and select User. In the New Object–User dialog box, input your first name, middle initial, and last name. For your user logon name, use your first initial, middle initial, and last name without spaces. For example, if your name is Paul G. Rogers, your login name is PGRogers. Click on the Next button.
2. Enter the password of PW (uppercase) and enable the "Password must change password on next logon" option. Click on the Finish button.
3. After your account has been created, right-click on your account and select the Properties option. Input your description as domain administrator and your office as the server room. Input your telephone number, e-mail address, and web page URL if you have one.

Click on the Address tab and type in your address. Click on the Telephones tab and input your phone numbers. Click on the Organization tab and input Administrator for the title, IT for the department, and Acme Corporation for the company. Click on the OK button.

4. Create the following users in the appropriate organizational unit.

First Name	Middle Initial	Last Name	User Logon Name	Title	Department	Organizational Unit
Charles	L	Gee	CLGeeX	Sales Mgr.	Sales	Sales
Frank	J	Biggs	FJBiggsX	Sales Rep.	Sales	Sales
Herold	W	Jones	HWJonesX	Sales Rep.	Sales	Sales
Paul	L	Ray	PLRayX	Sales Rep.	Sales	Sales
Juan	O	Hermes	JOHermesX	Sales Rep.	Sales	Sales
Jill	K	Knight	JKKnightX	Sales Admin. Asst.	Sales	Sales
Jean	A	Mao	JAMaoX	Training Mgr.	Education	Education
Edward	J	Morell	EJMorellX	Trainer	Education	Education
Donna	L	Starr	DLStarrX	Manufact. Mgr.	Manufacturing	Manufacturing
Eric	O	Skow	EOSkowX	Manufact. Technician	Manufacturing	Manufacturing
Victor	N	Sloan	VNSloanX	Manufact. Technician	Manufacturing	Manufacturing
Sonny	K	Wong	SKWongX	Research Engineer	Manufacturing	Research
Gina	J	Smith	GJSmithX	Research Engineer	Manufacturing	Research

5. Right-click on the SALES organizational unit, select the New option, and select the Computer option. Enter the name of your partner's Windows 2000 Professional workstation. It should be WS2000-xxy, where *xx* represents your two-digit partner number and *y* represents either A or B. Click on the OK button.

6. Right-click each of these users and input their appropriate departments.

EXERCISE 3: HOME DIRECTORIES

1. Open My Computer and open the C drive. In the C drive, right-click on empty space within the C drive window, select the New, and select the Folder option. Name the folder USERS and press the Enter key. In a real working environment, do not store data files on the system and/or boot partition.

2. Right-click the USERS folder and select the Sharing option. Select the Share This Folder option. Click on the Permissions button. Since Everyone is already highlighted, click on the Remove button.

3. Click on the Add button. Select the USERS group and click on the OK button. In the Permissions section, make sure the check box for Allow Full Control is marked. Click on the OK button to close the Sharing window and click OK to close the Properties windows.

4. Within the Active Directory Users and Computers console, open the Properties window of Charles L. Gee. Click on the Profile tab. Select the Connect To option under Login Scripts, select the Z drive, and in the To: text box enter:

```
\\SERVER2000-xxy.acmeXX.com\Users\%USERNAME%
```

5. Open the Users folder and note the CLGeeX folder.

6. Do the same for Frank J. Biggs, Herold W. Jones, and Paul L. Ray.

EXERCISE 4: FINDING AN OBJECT

1. Right-click the ACMExx.COM domain and select the Find option.

2. Under the Users, Contacts and Groups tab, enter Jill in the Name: text box and click on the Find Now button. Notice that it had no problem finding Jill Knight.

3. Double-click on Jill K. Knight at the bottom of the window. Click on the OK button.
4. Click on the Advanced tab. Click on the Field button, select the User option, and then select the Department field. Click on the Condition pulldown menu and look at the various conditions available. For the value, input Sales. Click on the Add button and click on the Find Now button.

EXERCISE 5: CREATING GROUPS

1. Right-click the ACMExx.COM domain, select the New option, and select the Group option. Click on the Global Group scope and Security Group type. Enter the Managers in the Group Name text box and click on the OK button.
2. Right-click the Managers group and select the Properties option. Click on the Members tab, click on the Add button, click on Charles L. Gee, and then click on the Add button. Click on and add Jean A. Mao and Donna L. Starr. Click on the OK button to close the Select Users, Contacts, or Computers dialog box. Click on the OK button to close the Managers Property window.
3. Right-click the ACMExx.COM domain, select the New option, and then select the Group option. Click on the Global Group scope and Security Group type. Enter the ManagersX in the Group Name text box and click on the OK button.
4. Right-click the ManagersX group and select the Properties option. Click on the Members tab, click on the Add button, click on Charles L. Gee, and then click on the Add button. Click on and add Jean A. Mao and Donna L. Starr. Click on the OK button to close the Select Users, Contacts, or Computers dialog box. Click on the OK button to close the Managers Property window.
5. Right-click the ACMExx.COM domain, select the New option, and select the Group option. Click on the Global Group scope and Security Group type. Enter the ProductStaffX in the Group Name text box and click on the OK button.
6. Right-click the ProductStaffX group and select the Properties option. Add all staff members from the manufacturing department.
7. Right-click the ACMExx.COM domain, select the New option, and select the Group option. Click on the Domain Local Group scope and Security Group type. Enter the ProductResourcesX in the Group Name text box and click on the OK button.
8. Right-click the ProductResourcesX group and click on the Properties option. Add the Managers group, ProductStaffX group, and Jill K. Knight to the group.
9. Create a universal group called SalesX.

The only step remaining is to assign rights and permissions to the Product Resources group. When new employees are added, they only have to be added to the appropriate group and they will automatically inherit these rights.

10. Open My Computer and open the C drive. In the C drive, right-click on empty space within the C drive window, select the New and select the Folder options. Name the folder TestShare and press the Enter key.
11. Right-click the TestShare folder and select the Sharing option. Select the Share This Folder. Click on the Permissions button. Since Everyone is already highlighted, click on the Remove button.
12. Click on the Add button. Select the ProductResourcesX and click on the OK button. Click on the OK button to close the Sharing window.

In this case, we did not need to use the local and global groups because both users and resources are in the same domain. Remember that global groups are intended to group people within a specific domain, whereas domain local groups are to give rights and permissions to objects.

EXERCISE 6: DISABLE, RENAME, AND DELETE AN ACCOUNT

1. Paul Ray just got fired. To disable the account, right-click on the Paul Ray account and select the Disable Account option. Note the small X that appears.

2. Right-click the Paul L. Ray account and select the Rename option. While the entire name is highlighted, press the Delete key and then press Enter.
3. Change the name to Tom J. Landers.

   Full name: Tom J. Landers
   First name: Tom
   Last name: Landers
   Display name: Tom J. Landers
   User logon name: TJLanders
   User logon name: (pre-Windows2000): TJLanders

4. Click on the OK button.
5. Right-click on the Tom J. Landers account and select the Reset Password option. Change the password to TEST. Click on the OK button to close the Windows and click on the OK button to close the Confirmation dialog box.
6. Right-click on the Tom J. Landers account and select the Enable Account option.
7. Right-click on the Tom J. Landers account and select the Properties option. Select the Profile tab and note the location of the home folder. Although the account has been changed, Tom Landers was able to get access to everything that Paul Ray had including his home directory, even though his home directory is still called Paul Ray. Eventually, rename the directory and be sure that Tom's profile tab indicates the new name of the home folder.
8. Right-click on the Tom Landers folder and select the Delete option. Say Yes to the Are You Sure? dialog box.
9. Paul Ray got rehired into his original position. Unfortunately, since his account was deleted, a new one will have to be created. Therefore, in the Sales organizational unit, create a new Paul Ray account.
10. Using the Profile tab located within Paul Ray's properties, re-create his home folder.
11. When the "Home directory was not created" warning appears, click on the OK button. Because Paul's old account was created, Paul does not automatically get the same rights that he had before when the new account is re-created. All of Paul's previous rights, permissions, and group memberships will have to be given again.

EXAMPLE 7: USING TEMPLATES

1. Right-click the Sales organizational unit and select New User. Specify the following:

   First name: Sales
   Last name: Template
   User logon name: _SALES_TEMPLATE

   Click on the Next button. The reason why you would use the first underscore (_) is it will show first on the list if they are alphabetized by user name.
2. Specify the password of "password" and enable the "User must change password at next logon." Click on the Next button.
3. Click on the Finish button.
4. Right-click the Sales Template and select the Properties option.
5. Put an address and phone number in the Address and Telephones tab.
6. Using the Member Of tab, assign the template to the Sales group.
7. Right-click the Sales template again and select the Copy command to create a new user called Charlie Brown. Use the password of password for Charlie Brown.
8. Right-click the Charlie Brown account and select the Properties option. View the various tabs that you modified when creating the _SALES_TEMPLATE.

EXERCISE 8: USER RIGHTS

1. Log out as the Administrator.
2. Try to log in as Charlie Brown. It should not work, because Charlie Brown has not been given the right to log on locally to a domain controller.

3. Log in as the Administrator.
4. In the Administrative Tools, open the Domain Controller Security Policy console.
5. If you open Security Settings, Local Policies, User Rights Assignment, there you will find Logon Locally. Double-click on it.
6. Click the Add button, click on the Browse button, select the Charlie Brown account, click on the Add button, and then click on the OK button twice.
7. Log out as the Administrator and log in as Charlie Brown. When you log in, change the password to PW.
8. As Charlie Brown, try to disable the Frank Biggs account.
9. Log out as Charlie Brown and log in as Administrator.
10. Right-click the Sales organizational unit and select the Delegate Control option. Click the Next button.
11. Click the Add button, select Charlie Brown, and click on the Add button. Click the OK button. Click on the Next button.
12. Select both the Create, Delete, and Manage User Accounts and Modify the Membership of a Group options. Click on the Next button. Click the Finish button.
13. Log out as the Administrator and log in as Charlie Brown.
14. Disable the Frank Biggs account.
15. Log out as Charlie Brown.

EXAMPLE 9: GROUP POLICIES

1. Log in as the Administrator.
2. Using the Run option, start a Microsoft Management Console.
3. In the MMC, open the Console menu and select the Add/Remove Snap-in option.
4. In the Add/Remove Snap-in dialog box, click on the Add button.
5. Click on the Group Policy snap-in and then click on the Add button. In the Select Group Policy Object dialog box, click on the Browse button.
6. In the Browse for a Group Policy Object dialog box, click the Domain Controllers. AcmeXX.com. Click the OK button.
7. Click the Default Domain Controllers Policy and click the OK button. Click the Finish button.
8. Click on the Group Policy snap-in again and click on the Add button. In the Select Group Policy Object dialog box, click on the Browse button.
9. In the Browse for a Group Policy Object dialog box, click the Sales.AcmeXX.com. Click the OK button.
10. Click the New Group Policy Object button (next to the Up Folder button). Since the New Group Object is highlighted, rename it to the Sales Group Policy and click the OK button. Click the Finish button.
11. Click on the Group Policy snap-in and click on the Add button. With the Local Computer selected for the Group Policy Object, click on the Finish button.
12. Click on the Close button. Click on the OK button.
13. Under the Default Domain Controllers Policy, open Computer Configuration, Windows Settings, Security Policy, Account Policies, and click on Password Policy.
14. In the Details pane, double-click the Minimum Password Length option.
15. In the Template Security Policy Setting dialog box, enable "Define this policy setting in the template." Specify at least eight characters. Click the OK button.
16. Under Password Policy, click on Account Policy option.
17. In the Details pane, double-click the Account Lockout Duration.
18. Enable "Define this policy setting in the template." Keep the default of 30 minutes and click on the OK button. Click on the OK button.
19. Log out as the Administrator.
20. Log in as Charlie Brown.
21. Log in six times with the password TEST. This should lock out the account.
22. Log in as the Administrator.
23. Using the Active Directory Users and Computers console, right-click Charlie Brown's user account and select the Properties option.

24. In the Account tab, remove the X in the Accounts Locked Out box.
25. Try to change Charlie Brown's password to LETMEIN. Then change the password to PASSWORD.
26. In the Active Directory Users and Computers console, right-click the Sales department and select the Properties option.
27. In the Group Policy tab, click the Add button. Click the Up Folder button and double-click the Domain Controllers.amcexx.com folder. Select the Default Domain Controllers Policy. Click the OK button.
28. Click the Default Domain Controllers Policy and click on the Up button to make the Default Domain Controller Policy have a higher priority than the Sales Policy. Although we will not use the option, note the Block Policy Inheritance. Click the OK button.

EXAMPLE 10: USER PROFILES

1. Log in as Charlie Brown. Change the background to Solar Eclipse. (The background is changed using the Display applet in the Control Panel.)
2. Log in as the Administrator.
3. Create a PROFILE folder on the C drive.
4. Right-click the PROFILE folder and select the Sharing option.
5. Click on the Share This Folder and click on the OK button.
6. Right-click the My Computer applet and select the Properties option.
7. On the User Profile tab, click the Charlie Brown profile and click the Copy To button.
8. In the Copy Profile to text box, specify \\server2000-XXx\Profile\CBrownx.
9. When it asks if you want to continue, click on the Yes button.
10. Click the OK button.
11. Using the Active Directory Users and Computers console, right-click Charlie Brown and then click the Profile tab.
12. In the Profile Path text box, specify the \\server2000-XXx\profile\CBrownx. Click on the OK button.
13. From your partner's computer, log on as Charlie Brown. Note that the desktop background should be the solar eclipse.

13

Introduction to Disk Management

INTRODUCTION

Since all of the user's data are kept on disk, it is easy to see why disk management is essential to managing the server. Having a good understanding of how the disk systems work and how they can be utilized effectively improves the server's performance and reliability. In this chapter you will learn about the file system, about forms of RAID that are available through Windows 2000, and how to set up disks.

OBJECTIVES

1. Convert from one file system to another file system.
2. Configure file systems by using NTFS, FAT32, or FAT.
3. Monitor and configure disks.
4. Monitor, configure, and troubleshoot volumes.
5. Monitor and configure disk quotas.

13.1 DISK STRUCTURE AND FILE SYSTEMS

The **disk structure** does not describe how a hard drive or floppy disk physically works, but how it stores files on the disk. In other words, it describes the formatting of the disk (file system, partitions, the root directory, and the directories). A **file system** is the overall structure in which files are named, stored, and organized. Examples of file systems include FAT, FAT32, and NTFS.

Disks are divided into a circular grid consisting of tracks and sectors. **Tracks** are concentric circles in that they share the same center much like rings in a tree. Tracks are numbered starting with the outside track as track 0. The tracks are then further divided into **sectors,** each typically 512 bytes of useable data. Sectors are numbered on a track, beginning with track 1. See Figure 13.1.

In Windows 2000, a **volume** is a logical unit of disk space that functions as though it was a physical disk. It can be a portion of a disk, an entire disk, several disks combined, or several portions of a disk combined. Often the volume is assigned a drive letter.

13.1.1 File Allocation Table

When you save a file, it can be stored at the beginning, the middle, or the end of the disk. When you retrieve a file, do not worry about its physical location, because when you specify a name (and sometimes the path), the operating system finds the file. The directory structure, the methods for organizing a volume (partition or floppy disk), and the method of file storage and retrieval are called the file system.

For operating systems, the most basic storage unit is not a sector, but a **cluster.** Another name for cluster is **allocation unit.** It consists of one or more sectors (usually more). The size of the cluster depends on the operating system, the version of the operating system, the file system that the operating system is using, and the size of the volume (partition or floppy disks). See Table 13.1.

The **file allocation table (FAT)** is an index used to remember which file is located in which cluster. It lists each cluster, if the cluster is being used by a file, the name of the file using the cluster, and the next cluster if the file does not fit within the cluster.

EXAMPLE 1 Your partition is using clusters of eight sectors (4096 bytes). If you have a file that is 2048 bytes, the file allocation table will have an entry for the cluster and the file belonging to the cluster. Although the file is only half as large as the cluster, the file allocation table can only list one file as part of the allocation unit. Therefore, the 2048-byte file will use the entire 4096 bytes.

FIGURE 13.1 Tracks and sectors

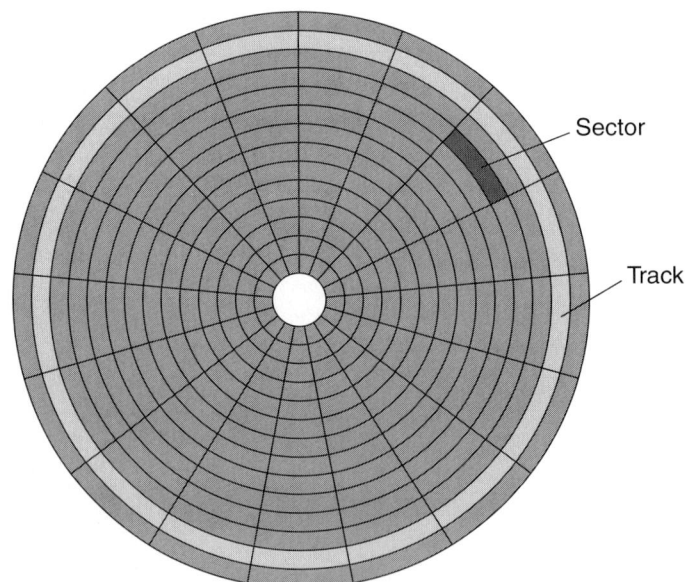

Sector

Track

TABLE 13.1   File system default cluster size

Drive Size	FAT	FAT32	NTFS
0 MB–15 MB	8 Sectors (4 KB)		1 Sector (512 bytes)
16 MB–31 MB	1 Sector (512 bytes)		1 Sector (512 bytes)
32 MB–63 MB	2 Sectors (1 KB)		1 Sector (512 bytes)
64 MB–127 MB	4 Sectors (2 KB)		1 Sector (512 bytes)
128 MB–255 MB	8 Sectors (4 KB)		1 Sector (512 bytes)
256 MB–511 MB	16 Sectors (8 KB)		1 Sector (512 bytes)
512 MB–1023 MB	32 Sectors (16 KB)	8 Sectors (4 KB)	2 Sectors (1 KB)
1024 MB–2047 MB	64 Sectors (32 KB)	8 Sectors (4 KB)	4 Sectors (2 KB)
2048 MB–4095 MB	128 Sectors (64 KB)	8 Sectors (4 KB)	4 Sectors (2 KB)
4096 MB–8191 MB		8 Sectors (4 KB)	4 Sectors (2 KB)
8192 MB–16,384 MB		16 Sectors (8 KB)	4 Sectors (2 KB)
16,385–32,767 MB		32 Sectors (16 KB)	4 Sectors (2 KB)
32,768 MB+		64 Sectors (32 KB)	4 Sectors (2 KB)

Note: Windows NT and Windows 2000 using NTFS can vary the cluster size from 512 bytes to 4 KB.

EXAMPLE 2   Your partition is using clusters of eight sectors (4096 bytes). If you have a file, which is 8193 bytes, the file allocation table would have three entries. The first part of the file would be stored in a cluster and listed in the file allocation table. After the first cluster is filled, there are still 4097 bytes remaining, so the entry of the first cluster will list the location of the next cluster. (The cluster may or may not be the next cluster.) The second cluster would then be filled up, leaving 1 byte. Again, the second cluster will list a third cluster, where the last byte would reside. Since the file allocation table can only list one file per cluster, the single byte would take up the entire 4096 bytes of disk space.

13.1.2   Directories, Files, and Attributes

The storage model of the PC is based on an inverted directory tree structure. Every volume has a starting point called the **root directory,** which is located at the top of the tree structure. The root directory will hold files and/or directories. The **directories** under the root directory can also be referred to as **subdirectories.** Newer operating systems may even refer to a directory as a **folder.** Each directory can also hold files and/or more subdirectories.

A **file** is a collection of related information that is referenced by name. Each file or directory on a volume can be uniquely identified by using the file name (including the file extension, if any) and the path (or location on the tree). Every file on the system is stored in a directory. The directory itself is a specially marked file or table that contains information about files and subdirectories. Every directory, including the root directory, consists of a small database which stores the file name and extension, the file attributes, and dates and times of when the file was created or last modified.

Files are stored in clusters. Since files are often larger than a cluster, a file will be spread among many clusters. On the last cluster, to indicate the end of the file, a special end-of-file (EOF) character is used. This way, the operating system knows to ignore the rest of the cluster since it does not have any valid data.

The file allocation table and the directory work closely together. When you open a file, you specify the name. To find the file, the operating system looks in the directory specified. If no directory was specified, it looks in the current directory. After it finds the file name, it looks at the starting cluster. The operating system reads the data in the starting cluster. The operating system then jumps to the file allocation table, specifically the starting cluster to read the next cluster for the file. It then jumps to the next cluster and retrieves the data. It

TABLE 13.2   Common Microsoft file systems

	FAT	FAT32	NTFS
Operating systems	Used by DOS, Windows 95, Windows 98, Windows NT, and Windows 2000	Used by Windows 95B/OSR2, Windows 98 and Windows 2000	Used by Windows NT and Windows 2000
Volume size	Floppy disk to 4 GB	512 MB to 32 GB	10 MB to 2 TB
Maximum file size	2 GB	4 GB	Limited by size of volume
Limit on entries in root directory	Yes	No	No
File and directory security	No	No	Yes
Directory and file compression	No	No	Yes
Cluster remapping	No	No	Yes

TABLE 13.3   The maximum number of root directory entries under FAT

Volume Type	Maximum Number of Root Directory Entries
360 KB 5¼″ Floppy Disk	112
720 KB 3½″ Floppy Disk	112
1.2 MB 5¼″ Floppy Disk	224
1.44 MB 3½″ Floppy Disk	224
2.88 MB 3⅓″ Floppy Disk	448
Hard Disk	512

will continue jumping between the file allocation table to find the next cluster and the clusters to retrieve the data until it gets to the EOF character.

13.2   FILE SYSTEM

Throughout the history of the PC, there have been a few different file systems. Today, the most common include FAT, FAT32, and NTFS. See Table 13.2.

13.2.1   FAT

The most common file system used is the file allocation table (FAT). FAT is a simple and reliable file system that uses minimal memory. It supports file names of 11 characters, which include the 8 characters for the file name and 3 characters for the file extension.

Early versions of DOS used FAT12, which used a 12-bit binary number for each cluster. Therefore, the biggest cluster number that FAT12 could see was 4086. The number would have been 4096, but some clusters were reserved. FAT12 is still used on floppy disks and hard disk partitions that are smaller than 16 MB.

Newer DOS machines use FAT16. Instead of a 12-bit number for the clusters, it uses a 16-bit number, allowing for 65,526 clusters. It is used for volumes up to 2 GB.

The FAT root directory is fixed in size and so is limited to a specific number of entries. See Table 13.3. Therefore, when you reach the maximum number of files and subdirectories in the root directory, you cannot add more, although there may still be plenty of disk space.

For FAT, a directory entry (see Table 13.4) will use 11 bytes for the entire file name—8 bytes for the file name and 3 bytes for the extension. The file name is meant to identify the file and, thus, should describe the contents of the file. The extension is used to identify the

TABLE 13.4   FAT/VFAT
directory format

Content	Byte Size
File name	8 bytes
File extension	3 bytes
File attributes	1 byte
Time of creation	3 bytes
Date of creation	2 bytes
Date of last access	2 bytes
Time of last modified	2 bytes
Date of last modified	2 bytes
Starting cluster	2 bytes
Size in bytes	4 bytes

type of file. For example, files that have a DOC extension are usually Microsoft Word documents, and files that have an EXE extension are executable files. To indicate that a file is deleted, the operating system will change the first character to a lowercase sigma (σ). This way, assuming all clusters are intact and have not been used by other files, a file can be undeleted by providing a first character. If the entire eight characters for the file name and the entire three characters for the file extension are not used, then the remaining characters become spaces. Note that the dot separating the file name and the extension is not stored as part of the file allocation table or the directory. It is shown by the operating system to indicate a separation between the two.

VFAT is an enhanced version of the FAT structure, which allows Windows 95, Windows 98, Windows NT, and Windows 2000 to support long file names (LFNs) up to 255 characters. If users refer to FAT, they probably mean VFAT. Since it is built on ordinary FAT, each file must have an 8-character name and a 3-character extension to be backward compatible for DOS and Windows 3.XX applications. Therefore, programs running in DOS and Windows 3.XX will not see the LFNs. When running a WIN32 program (programs made for Windows 95, Windows 98, Windows NT, or Windows 2000), they can see and make use of the longer names.

To accomplish the DOS file names and the LFNs, VFAT uses additional directory entries to store the LFNs. Therefore, when an LFN is saved, the first entry will need a truncated name (known as an alias) of the file. It will then use additional directory entries to hold the rest of the LFN. Each entry is 32 bytes in length. Since the root directory has a set number of entries, it is not wise to place too many LFNs in the root directory. To make sure that DOS does not use the directory entries, which hold the LFNs, the entries have the Read-only, Hidden, System, and Volume Label attributes enabled. Having all of these active causes older software to basically ignore the extra directory entries that are being used by VFAT.

A single LFN can use many directory entries (since each entry is only 32 bytes in length), and for this reason it is recommended that LFNs not be placed in the root directory, where the total number of directory entries is limited.

EXAMPLE 3   Under Windows 2000, you save a file called *The Budget for 1999 and 2000.DOC*. The first entry will take the first six characters—ignoring the spaces and adding a tilde (~) followed by a number in the directory would be *THEBUD~1.DOC*. Therefore, under a DOS or Windows 3.XX program, you would see *THEBUD~1.DOC*, whereas under Windows 95/98 and Windows NT programs you would see *The Budget for 1999 and 2000.DOC*. It would take several directory entries to store the alias and even more to save the long file name. If you try to save a second file with the name of *The Budget for 2000 and 2001.DOC,* the alias would be THEBUD~2.DOC.

After the file name and file extension, 1 byte is used for attributes. The **file attribute** field stores a number of characteristics about each file. See Table 13.5. Attributes can be either on or off. The most common attributes include Read-only, Hidden, System, and Archive. One of these attributes indicates if the file is a real file or a directory. Since DOS reserves 1 byte for attributes, it can keep track of up to eight attributes. Remember, 1 byte

TABLE 13.5 DOS file attributes

Attribute	Abbreviation	Description
Read-only	R or RO	When a file is marked as read-only, it cannot be deleted or modified. The opposite of read-only is read-write.
Hidden	H	When a file is marked as hidden, it cannot be seen during normal directory listings.
System	S or Sy	When a file is marked as system, it should not be moved. In addition, it usually cannot be seen during normal directory listings.
Volume Label		The name of the volume.
Subdirectory		A table that contains information about files and subdirectories.
Archive	A	When a file is marked as archive, the file has not been backed up. Anytime a file is new or has been changed, the archive attribute comes on automatically. When the archive attribute is off, the file has been backed up.
Directory	D	An attribute to indicate a directory listing instead of a file.

FIGURE 13.2 Properties of a file, including its attributes

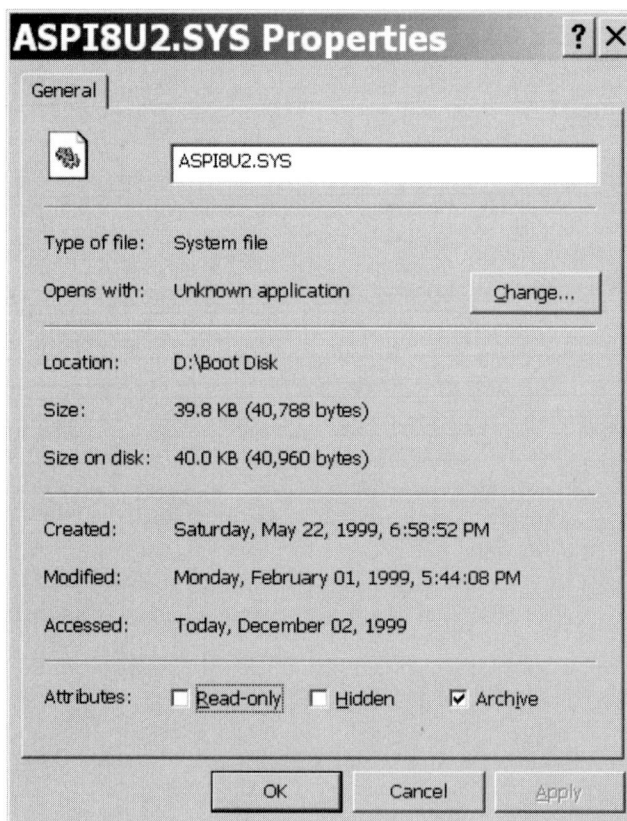

equals 8 bits (on–off switch). To change the attributes, use the ATTRIB command or Explorer (right-click the file and select the Properties option). See Figure 13.2.

Next, operating systems record the date and time that the file or directory was created, modified, and accessed and the size of the file in bytes. The files are stamped with the date and time from the real-time clock (RTC) on the motherboard. If the RTC clock has an incorrect date and time, then so will the file.

Lastly, because files are often larger than one cluster, the cluster that starts the file or subdirectory will be numbered and stored. Then when a file is retrieved from a subdirectory, it reads the directory first to find the starting point of the file. It then goes to the file allocation table to find the location of the second cluster, the third cluster, and so on, until the file has been retrieved into RAM.

13.2.2 FAT32

FAT32, which uses 32-bit FAT entries, was introduced in the second major release of Windows 95 (OSR2/Windows 95B) and is an enhancement of the FAT/VFAT file system. It supports hard drives up to 2 TB, although Windows 2000 supports volumes up to 32 GB. It uses space more efficiently, such as 4-KB clusters for drives up to 8 GB which results in 15% better use of disk space relative to large FAT drives.

The root directory is an ordinary cluster chain and thus can be located anywhere in the drive. It allows dynamic resizing of FAT32 partitions (without loosing data) and allows the FAT mirroring to be disabled, which allows a copy of the FAT other than the first to be active. Consequently, FAT32 drives are less prone to failure in critical data areas such as the FAT.

To install FAT32 on a volume, use the Disk Manager or the FDISK utility from the Windows 95—OSR2/Windows 95B or Windows 98 on a hard drive over 512 MB. When running FDISK, it will ask whether to enable large disk support. If you answer yes, any partition you create that is greater than 512 MB will be marked as a FAT32 partition.

13.2.3 NTFS

NTFS is a file system for Windows NT and Windows 2000, designed for both the server and the workstation. It provides a combination of performance, reliability, and compatibility. It supports long file names, yet maintain an 8.3 name for DOS and Windows 3.XX programs. Because it is a 64-bit architecture, NTFS is designed to support up to 2^{64} bytes = 18,446,744,073,709,551,616 bytes = 16 exabytes.

When you create (format) a NTFS volume, the high-level format program creates a set of files that contains the metadata used to implement the file system structure. The NTFS file system uses approximately 1 MB for the metadata files. See Table 13.6.

TABLE 13.6 NTFS file system

System File	File Name	Description
Master File Table	$Mft	The file allocation table for the NTFS volume that lists the contents of the NTFS volume
Master File Table2	$MftMirr	A mirror or copy of the Master File Table, used when a Master File Table becomes unreadable
Log File	$LogFile	A list of transaction steps, used by the Log File System for recoverability
Volume	$Volume	The name, version, and other information about the volume
Attribute Definitions	$AttrDef	A table of attribute names, numbers, and descriptions
Root Filename Index	$.	Root directory
Cluster Bitmap	$Bitmap	A representation of the volume showing which allocation units are in use
Partition Boot Sector	$Boot	Includes the bootstrap for the volume, if this is a bootable volume
Bad Cluster File	$BadClus	A location where all the bad clusters in the volume are located
Quota Table	$Quota	Disk quota usage for each user on a volume
Uppercase Table	$Upcase	Used for converting lowercase characters to the matching Unicode uppercase characters

Since Windows NT and Windows 2000 include enhanced security, NTFS supports a variety of multiuser security models and allows computers running other operating systems to save files to the NTFS volume on a server. This includes DOS, Windows 3.XX, Windows 95, Windows 98, Windows NT, Windows 2000, UNIX, POSIX, and even Macintosh computers. It does not allow DOS direct access to an NTFS volume directory except through the network (assuming you have the proper permissions or rights to access the volume). It can compress individual files or directories, including infrequently used files or directories.

To make its own volume more resistant to failure, NTFS writes updates to a log area and supports remapped clusters. If a system crash occurs, the log area can be used to quickly clean up problems. If a cluster is found to be bad, it can move the data to another cluster and mark the cluster as bad so that the operating system does not use it.

FAT is simpler and smaller than NTFS and uses an unsorted directory structure. NTFS is generally faster for complex file structures, because it uses a B-tree directory structure, which minimizes the number of disk accesses required to find a file, thus making access to the file faster, especially if it is a larger folder.

NTFS supports volume set and directory/file compression. A volume set combines several hard drives (or parts of hard drives) to be combined into a single volume. If the volume needs to be expanded again, simply add another hard drive and expand the volume. NTFS allows an individual file or directory to be compressed without compressing the entire drive.

13.3 DISK MANAGEMENT

In Windows 2000, to configure and manage disks, use the **Disk Management** console or the Computer Management console (click on the Start button, click on the Programs option, click on the Administrative Tools, and then select the Computer Management option). See Figure 13.3. To use the Disk Management console to configure and manage disks, you must be a member of the Administrators or Server Operators group.

> Data on the system are probably most important on the network and server. Therefore, before making any kinds of changes to the disk systems, back up all important data. Actually, you should back up important data on a regular basis.

After you make changes to your disk configuration, right-click Disk Management or open the Action menu, and use both the Refresh and Rescan Disks options. The Refresh option updates drive letters, file system information, volume information, and removable media information. The Rescan Disks option updates hardware information by scanning all attached disks for disk configuration changes. Be aware that disk rescanning may take several minutes depending on the number of drives.

When you connect a hard drive from another computer, the disks are imported automatically. If the status of the disk appears as Foreign, however, right-click the new disk and use the Import Foreign Disk option. If the status of an imported volume appears as "Failed: Incomplete Volume," you have imported one or more disks that has either a spanned or striped volume and RAID-5 volumes, but is missing one or more disks.

13.3.1 Drive Letters and Paths

Windows 2000 allows an unlimited amount of volumes, but unfortunately does not allow the user to assign more than 26 drive letters (A–Z). Typically, letters A and B are reserved for the floppy disk. If you do not have a floppy disk drive B, you can use the letter B as a network drive. These drive letters only apply to the local machine, not to network machines that connect to a Windows 2000 Server.

For many operating systems, drive letters are dynamic, which means that every time you turn on the machine, a drive letter is assigned to each volume. Unfortunately, when you add a drive or create partitions, sometimes the drive letters are remapped. The problem occurs when a program is set to look for or access a particular drive letter only to find

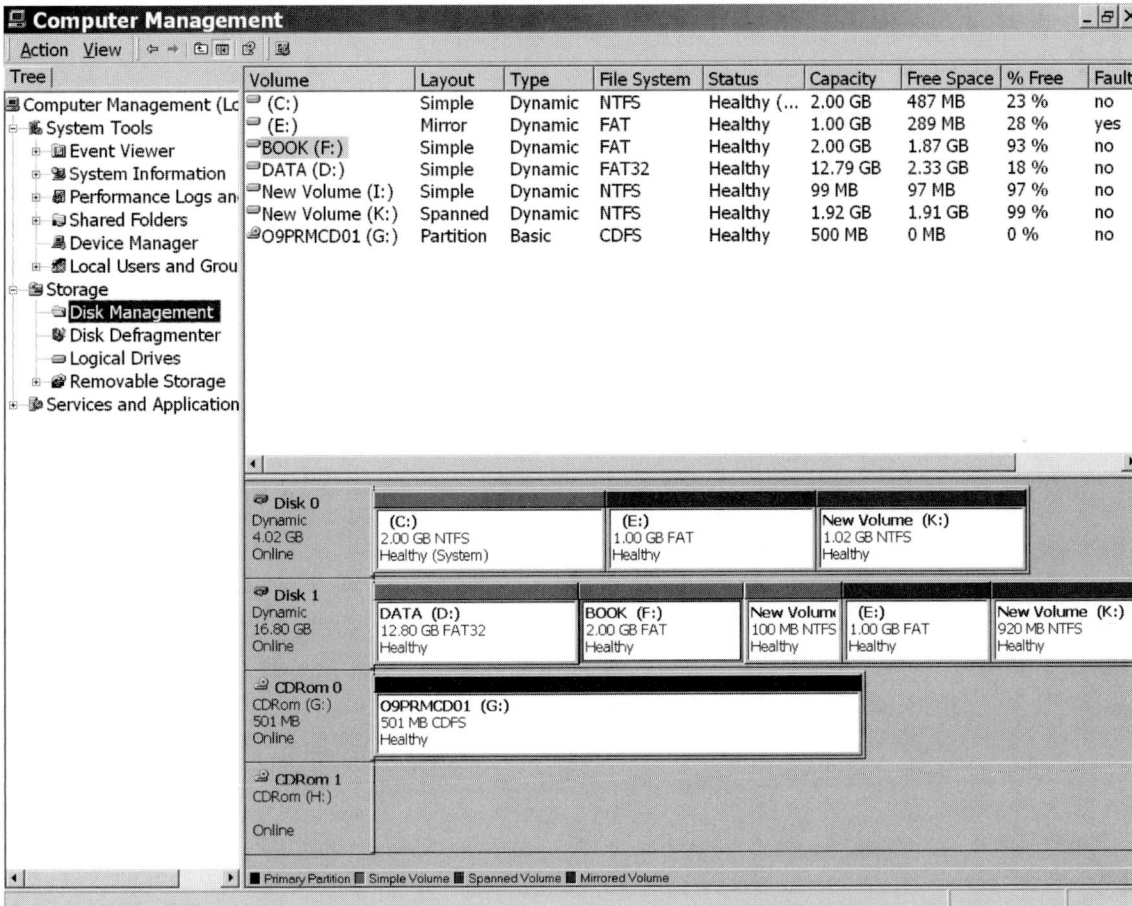

FIGURE 13.3 Disk Management console

that the folder or files are not at the designated location because drive letters have been assigned differently.

Windows also allows the static assignment of drive letters, during which a drive letter can be permanently assigned to a specific hard disk and volume. Then, when a hard disk is added to the existing computer system or partitions and volumes are created, it does not affect statically assigned drive letters.

Windows 2000 allows you to change drive letters; however, since many DOS and Windows programs refer to specific drive letters, be careful when making drive-letter assignments. You cannot change a drive letter that is assigned to a boot volume or system volume. To assign a drive letter to a volume that currently has no drive letter, do the following:

1. Open the Disk Manager.
2. Right-click the volume that you want to assign a drive letter to.
3. Select Change Drive Letter and Path option and click on the Add button.
4. Select a drive letter and click the OK button.

To remove a drive letter, do the following:

1. Open the Disk Manager.
2. Right-click the volume that you want to remove the drive letter from and click the Remove button.
3. Click Yes to confirm your actions.

To change a drive letter, do the following:

1. Open the Disk Manager.
2. Right-click the volume that you want to change the assigned drive letter to.

3. Select Change Drive Letter and Path option and click on the Edit button.
4. Select a new drive letter and click the OK button.
5. Click Yes to confirm your actions.

13.3.2 NTFS Mounted Drives

A NTFS mounted drive is attached to an empty folder on an NTFS volume, which is assigned a path, not a drive letter. Volumes that are created after the twenty-sixth drive letter have to be accessed as a mounted drive. When you open the drive or directory that holds the mounted drive and then open the mounted drive, you will be automatically redirected to the mounted volume. To the user, the mounted drive acts like a folder or directory at the assigned location.

To create a mounted drive, do the following:

1. Create a folder where you want the drive mounted.
2. Open Disk Manager.
3. Right-click the volume that you want to mount and select the Change Drive Letter and Path option.
4. Click on the Add button if it is a volume without a drive letter or is not already mounted, or click on the Edit button if the volume already has a drive letter or is already mounted.
5. Select the Mount in this NTFS Folder option and type or browse the path of the new folder. Click on the OK button. See Figure 13.4.

13.3.3 Formatting Volumes

Before you can store files and directories on a volume or partition, perform a high-level format on it. A high-level format creates the file allocation table and root directory. It can be performed when creating a volume or partitions with Disk Manager or it can be formatted at a later time.

To perform a high-level format, do the following:

1. Open Disk Manager.
2. Right-click on the partition or volume that you want to format, type the volume label if desired, select the file system, and select the allocation unit size. Click on the OK button.
3. If you get a warning that all information will be lost during a format, click on the OK button.

Instead of a normal format, you can perform a quick format. Although a quick format removes the files from the disk, it does not scan the disk for bad sectors. Therefore, its use is not recommended.

13.3.4 Deleting Volumes

Sometimes you may need to delete a volume so that you can reallocate the disk space. You cannot delete the boot volume, however, although you can delete the system volume. To delete the volume, do the following:

FIGURE 13.4 Mounting a drive

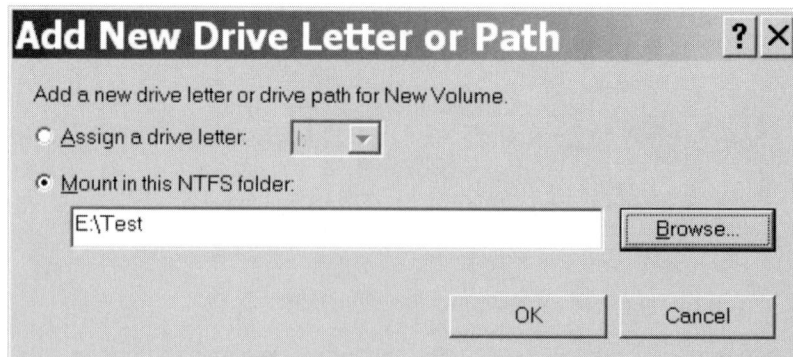

Add New Drive Letter or Path

Add a new drive letter or drive path for New Volume.

○ Assign a drive letter: [I:] ▼

● Mount in this NTFS folder:

E:\Test   [Browse...]

[OK] [Cancel]

1. Open the Disk Management console.
2. Right-click the volume that you want to delete and select the Delete Volume option.
3. At the warning that all data will be lost, click on the Yes button.

13.3.5 Converting FAT or FAT32 to NTFS

One way to convert a partition or volume to NTFS is to reformat it, but then, all information on the volume is lost. Another way to convert a partition or volume is to open a Command prompt window and use the following command:

```
CONVERT x: /FS:NTFS
```

where *x:* is the drive letter that you want to convert. If the CONVERT command cannot lock the drive, it will offer to convert it the next time the computer restarts, such as the system volume. You cannot, however, convert a NTFS partition or volume to FAT or FAT32.

13.4 WINDOWS 2000 DISK STORAGE TYPES

Windows 2000 has two disk storage types, basic disks and dynamic disks. A **basic disk** is a physical disk that contains primary partitions, extended partitions, or logical drives that can be accessed directly using DOS, Windows 95, or Windows 98. Basic disks may contain volume sets, mirrored sets, striped sets, and striped sets with parity if they were created using Windows NT 4.0 or earlier and upgraded to Windows 2000. Windows 2000 offers limited support of striped sets and striped sets with parity. You can repair, regenerate, and delete striped sets and striped sets with parity, but you cannot create them.

When you install Windows 2000, your hard disks are automatically initialized as basic. Using Disk Management console in Windows 2000, you can create, delete, and format partitions without having to restart your computer to make the changes effective.

To create a partition on a basic disk, do the following:

1. Open Disk Management console.
2. Right-click unallocated space on the basic disk where you want to create the partition, select the Create Volume option, and select the Next button.
3. Select the partition type and click on the Next button. The default size is the amount of unallocated space.
4. Specify the size for the selected disk and click on the Next button.
5. Assign the drive letter or path and click on the Next button.
6. Select the "Format this partition with the following settings"; specify the file system, allocation unit sizes, and volume name; and click on the Next button.
7. Click on the Finish button.

A **dynamic disk** is converted from basic to dynamic storage which contains simple volumes, spanned volumes, mirrored volumes, striped volumes, and RAID-5 volumes. Unlike basic disks, dynamic disks cannot contain partitions or logical drives or be accessed by DOS. The advantage of dynamic disks is they contain an unlimited number of volumes, and volumes can be extended to include noncontiguous space on available disks.

On a dynamic disk, the disk configuration information is stored on the disk and not in the registry. Therefore, to upgrade a drive successfully from basic to dynamic, you must have at least 1 MB of free space at the end of the disk. To upgrade a basic disk to a dynamic disk, right-click the disk using Computer Manager and select the Upgrade to Dynamic Disk option. When using the Computer Management program to create partitions or volumes on a disk, it will reserve the free space automatically. In addition, if you upgrade a boot disk or if a volume or partition is in use on the disk, then the computer must be restarted for the upgrade to succeed. Remember that removable media devices can only contain primary partitions, so you cannot upgrade them. Dynamic disks also are not supported on portable computers.

After you upgrade a basic disk to a dynamic disk, you cannot change the dynamic volumes back to partitions. Instead, you must delete all dynamic volumes on the disk and then use the Revert to Basic Disk command.

13.5 RAID

RAID is an acronym for redundant array of inexpensive disks. It is a category of disk drives that employs two or more drives in combination for fault tolerance and performance. RAID disk drives are used frequently on servers but are not generally necessary for personal computers. Ideally, you use a RAID system to ensure that no data are lost and to recover data from failed disk drivers without shutting the system down.

RAID was originally defined as a memory architecture that uses a subsystem of two or more hard disk drives treated as a single, larger logical drive. The purpose of this proposed architecture was twofold: to take advantage of data redundancy inherent in the multiple-drive design and to capitalize on the lower costs of smaller drives. When RAID was first proposed, it was considerably cheaper to buy five 200-MB hard drives than one 1-GB drive. Today, however, this is not true, so the current focus for RAID is data integrity and reliability instead of cost saving.

Originally, RAID had six levels (RAID 0 through RAID 5). See Table 13.7. A few more levels have been added to combine the features of other levels. Although RAID 6 follows the general numbering process, most new levels break with the number sequence, usually for marketing reasons. Not all RAID levels are commercially available, and some are supported by only a few products. Only RAID 0, 1, and 5 are supported by Windows NT 4.0 and Windows 2000 without additional hardware and software from RAID vendors. Those three levels, though, can be supported by a variety of hard disk and controller combinations.

RAID 0 is the base of RAID technology. RAID 0 stripes data across all drives. With striping, all available hard drives are combined into a single, large, virtual file system, with its blocks spread evenly across all the drives. For example, if you have three 500-MB hard drives, RAID 0 provides for a 1.5-GB virtual hard drive (sometimes referred to as a volume). When you store files, they are written across all three drives. When a large file, such as a 100-MB multimedia presentation, is saved to the virtual drive, part of it may be written to the first drive, the next chunk to the second, more to the third, and perhaps more wrapping back to the first drive to start the sequence again. The exact manner in which the

TABLE 13.7 RAID levels

RAID Type	Description
RAID 0—Disk Striping	Data striping is the spreading out of blocks of each file across multiple disks. It offers no fault tolerance, but it increases performance. Level 0 is the fastest and most efficient form of RAID.
RAID 1—Disk Mirroring/ Duplexing	Disk mirror duplicates a partition onto two hard drives. Information is written to both hard drives simultaneously. It increases performance and provides fault tolerance. Disk duplexing is a form of disk mirroring. Disk mirroring uses two hard drives connected to the same card; disk duplexing uses two controller cards, two cables, and two hard drives.
RAID 2—Disk Striping with ECC	Level 2 uses data striping plus ECC to detect errors. It is rarely used today since ECC is embedded in almost all modern disk drives.
RAID 3—ECC Stored as Parity	Level 3 dedicates one disk to error correction data. It provides good performance and some level of fault tolerance.
RAID 4—Disk Striping with Large Blocks	Level 4 offers no advantages over RAID-5 level and does not support multiple, simultaneously written operations.
RAID 5—Disk Striping with Parity	Raid-5 level uses disk striping and includes byte correction on one of the disks. If one disk goes bad, the system will continue to function. After the faulty disk is replaced, the information on the replaced disk can be rebuilt. This system requires at least three drives. It offers excellent performance and good fault tolerance.

chunks of data move from one physical drive to another depends on the way the virtual drive has been set up, which includes considering drive capacity and the way in which blocks are allocated on each drive. No parity control is used with RAID 0; therefore, it is a true form of RAID. RAID 0 has several advantages, however. Most important is that striping provides some increase in performance through load balancing. Windows NT 4.0 and Windows 2000 supports RAID 0 with two or more drives.

RAID 1 is known as disk mirroring or disk duplexing. With RAID 1, each hard drive on the system has a duplicate drive that contains an exact copy of the first drive's contents. Since every bit that is written to the file system is duplicated, data redundancy exists with RAID 1. If one drive in the RAID-1 array fails or develops a problem of any kind (such as a bad sector), the mirror drive can take over and maintain all normal file-system operations while the faulty drive is diagnosed and fixed.

Many RAID-1 disk controllers have software routines that will automatically take a faulty drive offline, run diagnostics on it, and if possible reformat the drive and copy all data back from the mirror image—all while the file system proceeds as if nothing has happened. Users are frequently unaware of faults with RAID-1 controllers. Alert messages can be triggered when a fault occurs.

One big disadvantage of RAID 1 is its use of disks. If you have two 2-GB drives, you have a total file system of only 2 GB (the other 2 GB is mirrored). In this situation, then, you are getting only half of the disk space that you are paying for, but you also have fully redundant drives. In case of catastrophic failure of a drive, controller, or motherboard, you can remove a mirror drive and boot on another controller or server.

RAID 1 offers an increase in read performance in most implementations, as the controller card allows both drives (primary and mirror) to be read at the same time, resulting in a faster read operation. Write operations are not faster, though, because data must be written to two drives. In many RAID-1 systems that do not use separate drive controllers for the primary and mirror drives, the speed of writing can even decrease, because the system must perform two complete write operations in sequence.

Implementations of RAID 1 usually require two drives of similar size. If you use a 1.5-GB and a 2-GB drive, for example, the extra 0.5 GB on the second drive is wasted. Some controllers let you combine drives of different sizes, with the extra space used for nonmirrored partitions. Windows NT 4.0 and Windows 2000 support RAID 1 with two drives.

RAID 5 is very similar to RAID 0, but one hard drive is used for parity (error correction) to provide fault tolerance. To increase performance, spread the error-correction drive across all hard drives in the array to avoid having the one drive do all the work in calculating the parity bits. RAID 5 is supported by NT 4.0 and most RAID vendors, because it is a good compromise between data integrity, speed, and cost. RAID 5 has better performance than RAID 1 (mirroring).

RAID 5 usually requires at least three drives, with more drives preferable. The overhead that RAID 5 imposes on RAM also can be significant, so Microsoft recommends at least an additional 16 MB RAM when RAID 5 is used. As with RAID 1, however, drives of disparate capacities may result in much unused disk space, because most RAID-5 systems use the smallest drive capacity in the array for all RAID-5 drives. Extra disk space can be used for unstriped partitions, but these are not protected by the RAID system.

The Microsoft Windows 2000 HCL contains numerous hardware RAID configurations for clusters. Many hardware RAID solutions provide power redundancy, bus redundancy, and cable redundancy within a single cabinet and can track the state of each component in the hardware RAID firmware. The significance of these capabilities is they provide data availability with multiple redundancies to protect against multiple points of failure. Hardware RAID solutions also can use an onboard processor and cache.

Because SCSI drives have command queuing and typically a higher throughput, they are the best choice for RAID systems. According to RAID vendors, SCSI subsystems represent more than 95% of the RAID market. Some RAID systems support hot swappable drives where a drive can be removed without powering down the system. SCSI controller cards that support RAID through hardware will allow for better performance, because they will do most of the calculations that would have been done through the processor as instructed by the software. Lastly, if performance is critical on some systems, use a system with a RAID controller card that has a relatively large amount of RAM to be used as cache for the controller card.

13.6 VOLUME DISKS

Windows 2000 supports several types of volumes. They include:

- Simple volume
- Spanned volume
- Mirrored volume
- Striped volume
- RAID-5 volume

13.6.1 Simple and Spanned Volumes

A simple volume is made up of disk space from a single physical disk (single region or multiple regions linked together). To create a simple volume, do the following:

1. Open Disk Management console.
2. Right-click unallocated space on the dynamic disk where you want to create the simple volume, select the Create Volume option, and select the Next button.
3. Select Simple Volume and click on the Next button.
4. Select the disk and the size for the selected disk and click on the Next button.
5. Assign the drive letter or path and click on the Next button.
6. Select the file system to use, select the allocation unit size, and enter the volume label. If you wish, you can also select Quick Format. Click on the Next button.
7. Click on the Finish button.

A spanned volume encompasses multiple disks. It can free drive letters for other uses and allows the creation of extremely large volume for file system use. A spanned volume can be extended up to a maximum of 32 disks and can be used on FAT, FAT32, or NTFS. To create a spanned volume, do the following:

1. Open Disk Manager.
2. Right-click unallocated space on the dynamic disk where you want to create the spanned volume, click the Create Volume option, and click on the Next button.
3. Select the Spanned Volume and click on the Next button.
4. Select the disks and the size for the selected disks and click on the Next button. (The selected disk size for the different drives can be different.)
5. Assign the drive letter or path and click on the Next button.
6. Select the file system to use, select the allocation unit size, and enter the volume label. Click on the Next button.
7. Click on the Finish button.

To extend or expand an existing volume (simple or spanned), right-click the volume, select the Extend Volume option, and use the wizard to specify which unallocated space to use and the amount of it. Only NTFS volumes can be extended—you cannot extend a system or boot volume. System and boot volumes are system and boot partitions that were contained in basic disks and upgraded to dynamic disks.

To delete a simple or spanned volume, right-click the volume and select the Delete option. When it asks if you are sure, click Yes and the volume is deleted.

13.6.2 Mirrored Volume

A mirrored volume (RAID 1) is two identical copies of a simple volume on separate hard disks. When data are written to one disk, they are also written to the other. If one hard drives fail, the other hard drive will continue to function without loss of data or downtime. Mirrored volumes also have better overall read performance than simple, spanned, and RAID-5 volumes.

When a disk fails that contains a mirrored volume, you must break the mirrored volume to separate the working volume with its own drive letters. Note that breaking the mirrored volume does not delete the information. You can then re-create a new mirrored volume with unused free space of the same size or greater on another disk. If the second volume is larger, then the remaining space becomes free space. Any existing volume, including the system and boot volumes, can be mirrored onto another volume of equal or greater size on another disk. When creating mirrored volumes, it is best to use disks that are the same size and model, and from the same manufacturer.

To create a mirror set, do the following:

1. Open Disk Management console.
2. Right-click unallocated space on the dynamic disk where you want to create the spanned volume, click the Create Volume option, and click on the Next button.
3. Select the Mirrored Volume and click on the Next button.
4. Select the disks and their sizes and click on the Next button.
5. Assign the drive letter or path and click on the Next button.
6. Select the file system to use, select the allocation unit size, and enter the volume label. Click on the Next button.
7. Click on the Finish button.

If a basic disk that contains part of a mirror set is disconnected or fails, the status of the mirror set becomes Failed Redundancy and the status of the disk remains Online. To repair a basic mirrored set, open Disk Management, right-click the mirrored volume you want to repair, select the Repair Volume option, and follow the instructions on the screen. The mirrored volume's status should change to Regenerating, then to Healthy. If the volume does not return to the Healthy status, right-click the volume and then click Resynchronize Mirror. If this does not work, right-click the volume and select the Remove Mirror option to delete one of the mirrors. Then rebuild the mirror by right-clicking the volume and selecting the Add Mirror option.

To delete a mirror set, right-click the mirror set and select the Delete the Volume option. To create two independent partitions or logical drives, right-click the mirror set and select the Break Mirror. If you want to make the system and boot volume fault tolerant, mirror the volume. As mentioned in Chapter 7, create two different boot disks that point to both the boot partitions by using the appropriate advanced RISC computing (ARC) name. First test each boot disk to make sure that they start Windows 2000 by booting the system with the boot disk.

13.6.3   Striped Volume

A striped set is similar to a volume set in that it is the combining of areas of free space on 2 to 32 disks onto one logical volume. The difference is that a striped volume has the data allocated alternately and evenly in 64-KB stripes through the various drives. Therefore, when data are read or written, it writes to all drives simultaneously. Striped volumes offer the best performance of all volumes available in Windows 2000, since the hard disks are working simultaneously. Like a volume set, if one disk in the striped volume fails, then the entire volume is lost. Striped volumes cannot be mirrored or extended.

To allocate striped volumes, do the following:

1. Open Disk Management.
2. Right-click unallocated space on the dynamic disk where you want to create the spanned volume, click the Create Volume option, and click on the Next button.
3. Select Striped Volume and click on the Next button.
4. Select the disks and the size for the selected disks and click on the Next button. (The selected disk size for the different drives can be different.)
5. Assign the drive letter or path and click on the Next button.
6. Select the file system to use, select the allocation unit size, and enter the volume label. Click on the Next button.
7. Click on the Finish button.

13.6.4   RAID-5 Volume

RAID-5 volumes are fault-tolerant striped volumes, except they use an extra drive to store parity information. In Windows NT, RAID-5 volume is very similar to stripe set with parity. See Table 13.8. A Raid-5 volume can use between 3 and 32 drives. If a drives fail, Windows 2000 uses the parity information in the stripe to reconstruct data, thus allowing the system to function. In addition, it still benefits from the performance offered by striping.

When a hard disk fails in a RAID-5 volume, repair or replace the failed hard drive, right-click the unallocated disk space that is the same size or larger than the other members of the RAID-5 volume, and select the Regenerate the Data option. Since the volume must be locked to regenerate, all network connections to the volume are lost and it may require a reboot.

TABLE 13.8 Comparison between RAID-5 and mirrored volumes

Mirrored Volume	RAID-5 Volume
Supports FAT and NTFS.	Supports FAT and NTFS.
Can mirror system or boot volumes.	Cannot stripe system or boot volumes.
Requires 2 hard disks.	Supports 3 to 32 hard disks.
Has higher cost per megabyte (50% utilization).	Has lower cost per megabyte.
Has good read and write performance.	Has moderate write performance and excellent read performance.
Uses less system memory.	Requires more system memory.

13.7 DISK UTILITIES

Several utilities or built-in functions can be used to help maintain and optimize disks and their data. These include scanning for disk errors, defragging a volume, and backing up the data. These functions can all be accessed by right-clicking the drive, selecting the Properties option, and clicking on the Tools tab. See Figure 13.5. Try to run the disk error-checking tool and the defragmentation tool during the night or during some other period of low usage, using the Task Scheduler.

13.7.1 Disk Error-Checking Tool

The disk error-checking tool will check the selected physical drive for damage. This includes invalid entries in the file allocation table, invalid directory entries, lost clusters, cross-linked files, files left in an open status, and bad sectors.

During this time, the disk is not available for use. To run the Check Disk tool, all files must be closed. If a volume is currently in use, a message will ask if you want to reschedule the disk checking for the next time you restart your computer. The disk-checking process can take a long time for large drives or drives with a large number of files. See Figure 13.6.

FIGURE 13.5 Disk tools

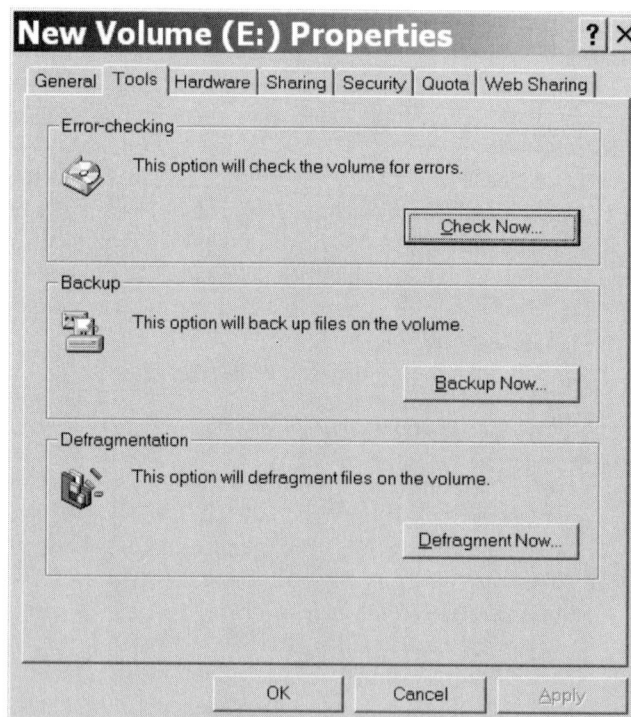

New Volume (E:) Properties

General | Tools | Hardware | Sharing | Security | Quota | Web Sharing

Error-checking
This option will check the volume for errors.
Check Now...

Backup
This option will back up files on the volume.
Backup Now...

Defragmentation
This option will defragment files on the volume.
Defragment Now...

OK   Cancel   Apply

FIGURE 13.6   Check Disk utility

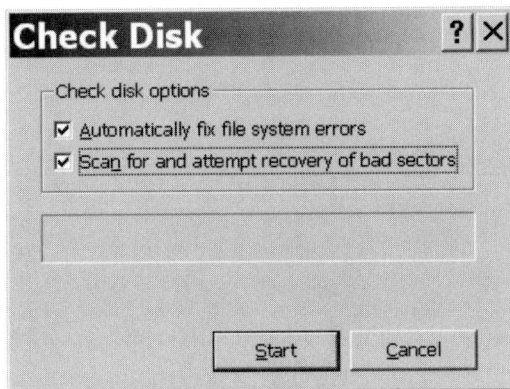

Check Disk   ? ✕

Check disk options
☑ Automatically fix file system errors
☑ Scan for and attempt recovery of bad sectors

Start   Cancel

FIGURE 13.6   Check Disk utility

13.7.2   Disk Defragmenter

When a file is created, it is assigned a certain number of clusters which hold the data. After the file is saved to the disk, other information is usually saved to the clusters following the saved file. Therefore, if you change the original file, add more information to it, and save it back to the disk, then the bigger file will not fit within the allocated clusters. Therefore, part of the file will be saved in the original clusters and the remaining amount will be placed elsewhere on the disk.

Over time, files become fragmented as they are spread across the disk. The fragmented files are still complete when opened, but the computer takes longer to read them. Opening them also causes more wear and tear on the hard disk. Defragging the drive fixes this fragmentation problem by reorganizing the drive, thus gathering all parts of a file and placing them in continuous sectors.

The Disk Defragmenter can be started using the Tools tab or by using the MMC with the Disk Defragmenter snap-in. If the drive is badly fragmented or is extremely large, then the program may take several hours to finish. In addition, when the defragmenting occurs, it will affect server performance, particularly when accessing the drive with the volume that you are defragging. Therefore, defrag only during periods of low usage. Lastly, defrag busy file servers more often than those on workstations and after deleting large numbers of files. See Figure 13.7.

FIGURE 13.7   Disk Defragmenter

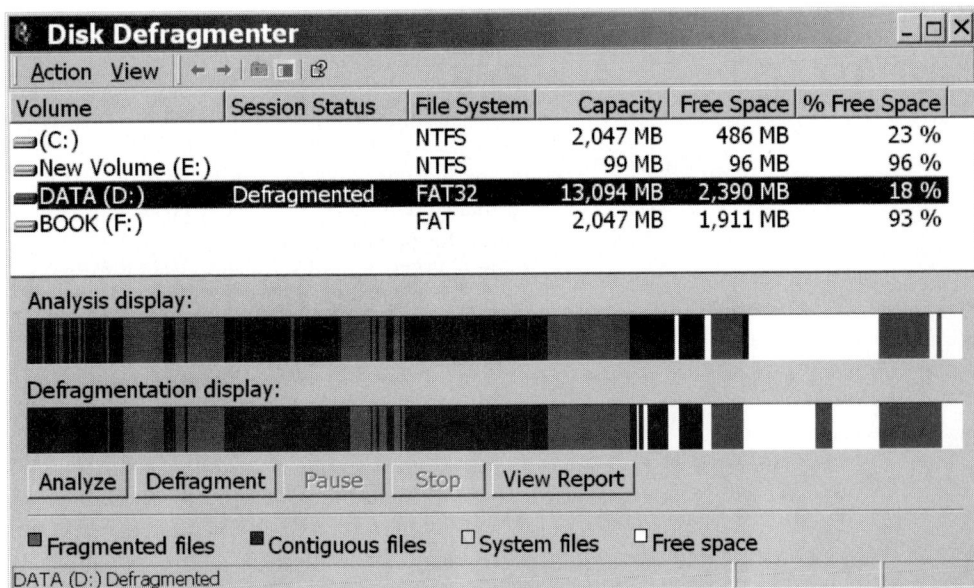

Disk Defragmenter   _ ☐ ✕

Action View   ← →

Volume	Session Status	File System	Capacity	Free Space	% Free Space
(C:)		NTFS	2,047 MB	486 MB	23 %
New Volume (E:)		NTFS	99 MB	96 MB	96 %
DATA (D:)	Defragmented	FAT32	13,094 MB	2,390 MB	18 %
BOOK (F:)		FAT	2,047 MB	1,911 MB	93 %

Analysis display:

Defragmentation display:

Analyze   Defragment   Pause   Stop   View Report

■ Fragmented files   ■ Contiguous files   ☐ System files   ☐ Free space

DATA (D:) Defragmented

1. The disk structure describes how it stores files on the disk.
2. Disks are divided into a circular grid consisting of tracks and sectors.
3. A volume is a logical unit of disk space that functions as though it was a physical disk.
4. For operating systems, the most basic storage unit is not a sector, but a cluster. Another name for cluster is allocation unit.
5. The file allocation table (FAT) is an index used to remember which file is located in which cluster.
6. Every volume has a starting point called the root directory, which is located at the top of the tree structure. The root directory will hold files and/or directories.
7. The directories under the root directory can also be referred to as subdirectories. Newer operating systems may refer to a directory as a folder. Each directory can also hold files and/or more subdirectories.
8. A file is a collection of related information that is referenced by name. Each file or directory on a volume can be uniquely identified by using the file name (including the file extension, if any) and the path (or location on the tree).
9. The most common include FAT, FAT32, and NTFS.
10. FAT is a simple and reliable file system, which uses minimal memory. It supports file names of 11 characters, which include the 8 characters for the file name and 3 characters for the file extension.
11. VFAT is an enhanced version of the FAT structure, which allows Windows 95, Windows 98, Windows NT, and Windows 2000 to support long file names (LFNs) up to 255 characters.
12. FAT32, which uses 32-bit FAT entries, was introduced in the second major release of Windows 95 (OSR2/Windows 95B) and is an enhancement of the FAT/VFAT file system.
13. NTFS is a file system for Windows NT and Windows 2000, designed for both the server and workstation. It provides a combination of performance, reliability, and compatibility.
14. In Windows 2000, to configure and manage disks, use the Disk Management console or the Computer Management console.
15. Windows 2000 allows as many volumes as you wish. Unfortunately, you cannot assign more than 26 drive letters (A–Z).
16. A NTFS mounted drive is attached to an empty folder on an NTFS volume, which is assigned a path, not a drive letter.
17. Before you can store files and directories on a volume or partition, perform a high-level format on it.
18. When converting a partition or volume, open a Command prompt window and use the CONVERT x: /FS:NTFS command.
19. Windows 2000 has two disk storage types, basic disks and dynamic disks.
20. A basic disk is a physical disk that contains primary partitions, extended partitions, or logical drives.
21. A dynamic disk is converted from basic to dynamic storage that contains simple volumes, spanned volumes, mirrored volumes, striped volumes, and RAID-5 volumes.
22. RAID is an acronym for redundant array of inexpensive disks. It is a category of disk drives that employs two or more drives in combination for fault tolerance and performance. RAID disk drives are used frequently on servers but are not generally necessary for personal computers.
23. Only RAID 0, 1, and 5 are supported by Windows NT 4.0 and Windows 2000, without additional hardware and software from RAID vendors.
24. A simple volume is made up of disk space from a single physical disk (single region or multiple regions linked together).
25. A mirror volume (RAID 1) is two identical copies of a simple volume on separate hard disks.
26. A striped set is similar to a volume set in that it is the combining of areas of free space on 2 to 32 disks onto one logical volume.
27. RAID-5 volumes are fault-tolerant striped volumes, except they use an extra drive to store parity information.
28. Several utilities or built-in functions can be used to help maintain and optimize disks and their data. These include the disk error-checking tool and the defragmentation tool.

QUESTIONS

1. Pat wants to convert one of the NTFS volumes on his computer to FAT so it can be accessed by workstations running Windows 98. The NTFS partition does not contain system or boot files. How can Pat convert the NTFS partition to a FAT partition using the least amount of administrative effort?
   a. Use the CONVERT.EXE program at the command prompt.
   b. Select the NTFS partition and choose Convert on the Partition menu in Disk Administrator.
   c. Select the NTFS partition and choose Reformat on the Partition menu in Disk Administrator.
   d. Right-click on the volume in Disk Administrator and select the Format option to change the format of the partition to FAT and restore all data from tape backup.

2. You want to convert one of the FAT partitions on your computer to NTFS while preserving all of the data currently contained on the partition. What is the best way to accomplish this?
   a. Use the CONVERT.EXE program at the command prompt.
   b. Select the FAT partition and choose Convert on the Tools menu in Disk Administrator.
   c. Select the FAT partition and choose Format on the Partition menu in Disk Administrator.
   d. Select the FAT partition and choose Convert on the Partition menu in Disk Administrator.

3. What does RAID stand for?
   a. Redundant And Inexpensive Disk
   b. Readable and Interruptible Disk
   c. Redundant Array of Inexpensive Disks
   d. Readable Array of Inexpensive Disks

4. Which RAID level has no redundant information and therefore provides no fault-tolerance?
   a. RAID 0
   b. RAID 1
   c. RAID 2
   d. RAID 5

5. Which RAID levels does Windows 2000 support?
   a. 0, 1, 2 and 5
   b. 0, 1, 2, 3 and 5
   c. 1, 2, 3 and 5
   d. 0, 1 and 5
   e. 0, 1, 2, 3, 4 and 5

6. You want to implement fault tolerance on your Windows 2000 server and you are most concerned with read performance. There are five hard drives on your Windows 2000 server. Which of the following will best satisfy your requirements?
   a. a volume set
   b. RAID-5 volume
   c. disk mirroring
   d. disk duplexing

7. You want to combine the free disk space from five hard disks into one logical drive. You are not concerned with providing fault tolerance, but you want to achieve the fastest possible read/write performance. Which of the following disk management strategies would be the best choice?
   a. a volume set
   b. a mirror set
   c. a striped volume
   d. a RAID-5 volume

8. Your file server contains a database that is accessed by all users in your domain and all trusted domains. Currently, the server only has one physical disk. You want to implement fault tolerance on your file server. Which of the following actions would accomplish this task?
   a. Install one additional physical disk and implement a RAID-5 volume
   b. Install one additional physical disk and implement a striped volume
   c. Install three additional physical disks and implement a volume set
   d. Install three additional physical disks and implement a RAID-5 volume

9. You want to combine the free disk space from six hard drives into a single logical drive. You are not concerned with providing fault tolerance, and you want to achieve the fastest possible read performance. Which of the following Windows 2000 disk management strategies would you implement?
   a. a volume set
   b. a mirror set
   c. a striped set
   d. a RAID-5 Volume

10. Suppose the following situation exists:

Your Windows NT Workstation computer has five 1-GB IDE hard drives. Each drive is formatted with a single FAT partition. The first hard drive contains the system and boot partition. It also contains the paging file. You want to optimize the performance of your computer.

Required result:
- You must speed access to files on your local hard drives.

Optional desired results:
- You want to optimize the paging file.
- You want to optimize network binding orders.

Proposed solution:

Create a second paging file on the drive that contains the system and boot partition. Combine the spaces from the remaining four disks into a volume set. Store all data and program files on the volume set. Move the most frequently used protocols to the top of the network binding order.

Which result does the proposed solution produce?

a. The proposed solution produces the required result and produces both of the optional desired results.

b. The proposed solution produces the required result and produces only one of the optional desired results.

c. The proposed solution produces the required result but does not produce either of the optional desired results.

d. The proposed solution does not produce the required result.

11. Jeremy's new computer has a 2-GB hard disk drive, and it will be configured with a single partition. The computer will be used mainly to store payroll records and other sensitive information. Which of the following are reasons for using the NTFS file system instead of FAT? Choose all that apply.

a. Only NTFS supports long file names.

b. Only NTFS allows permissions to be assigned to individual files and folders.

c. NTFS generally provides better performance for volumes greater than 400 MB.

d. NTFS partitions can be used by operating systems other than Windows NT, such as Windows 95 and MS-DOS.

12. The master boot record holds the _____ .
    a. partition table   c. volume boot sector
    b. file allocation table   d. root directory

13. Which utility identifies and corrects files that are not contiguous?
    a. SCANDISK   c. disk error-checking tool
    b. defrag   d. backup

14. Which best describes a fragmented hard drive?
    a. The platters are bad or cracked.
    b. The platters are slipping on the spindle.
    c. Data files are corrupted.
    d. Files are not stored in consecutive clusters.

15. The disk error-checking tool can be used to _____ .
    a. locate lost clusters and provide options to save or delete them
    b. look for viruses
    c. reconfigure and optimize programs in RMA
    d. locate a specific program on a disk or drive

16. A hard drive has slowed considerably over time. What should be done to improve the performance of the hard drive?
    a. use DriveSpace or DoubleSpace   d. use CHKDSK
    b. use Fast Disk   e. increase the maximum amount
    c. use Disk Defragmenter   of virtual memory

17. Pat must install a busy server with 32 volumes. While Pat is installing all of the drives and is assigning drive letters to all of the volumes, he realizes that he used up all of the drive letters, yet he still has several volumes that require drive letters. What can be done to make the other volumes available to the users?
    a. Map these drives to another server.
    b. Use NTFS mounted volume.
    c. Start using double-letter drive letters, starting with AA.
    d. Start using numbers with the drive letters, starting with A2.
    e. Do none of the above.

18. You have a hard drive with three partitions. The first partition is a FAT32 and the other two partitions are extended partitions formatted as logical drives with NTFS. You install a second hard drive with a NTFS partition. Which of the following statements are true? Assume the three partitions are assigned C, D, and E drive letters. Choose the one answer that applies.
    a. The C drive remains the same, the new partition becomes the D drive, and the other partitions are assigned E and F.
    b. The C, D, and E drive letters are still assigned to the three partitions and the drive letter can be assigned to any other drive letter.
    c. You need to use Disk Manager to change the D and E drive to E and F and the new partition will automatically be recognized as the D drive.
    d. You must use a NTFS mounted volume to use the partition.

EXERCISES

EXERCISE 1: CREATE A VOLUME

1. Start Computer Management.
2. In the left pane, click on Disk Management.
3. Right-click on Disk 0 and select the Properties option. Do not click on the partitions or volumes.
4. Click on the OK button to close the Disk 0 Properties dialog box.
5. Right-click the C drive and select the Properties option.
6. Click OK to close the Local Disk (C) dialog box.
7. You should have free disk space at the end of the first disk.
8. Right-click the Unallocated Disk Space and select Create Partition.
9. Select the Primary Partition and click on the Next button.
10. Use the entire disk space for the partition and click on the Next button.
11. Assign the M drive to the partition. Click on the Next button.
12. Select to use the FAT32 file system and click on the Next button. Do not select the Perform a Quick Format option. Instead, for a network environment, you would rather have the system perform a more thorough format to find problems with the drive.
13. Click on the Finish button.
14. Right-click the M drive and select to format the drive as a NTFS partition.
15. Right-click the M drive and select the Delete option.

EXERCISE 2: CONVERT A FAT OR FAT32 PARTITION TO A NTFS PARTITION

Note: It is always recommended to perform a backup of all important data before doing a conversion to NTFS.

1. Right-click the unallocated disk space and create a FAT32 partition. Assign the drive letter M to the drive.
2. Start a Command prompt.
3. At the Command prompt, execute the CONVERT M: /FS:NTFS.
4. At the Command prompt, execute the CONVERT C: /FS:NTFS.
5. After the CONVERT command displays a message saying that it has to reboot the computer to finish the conversion, complete a proper shutdown of Windows 2000 and reboot the computer.

EXERCISE 3: USING NTFS MOUNTED VOLUME

1. Right-click the C drive and select Open.
2. In the C drive, create a folder called TestMount.
3. Right-click the M drive and select the Change Drive Letter and Path option. Click on the M drive and click on the Remove button. When it asks if you are sure, click Yes.
4. Right-click the New Volume and select the Change Drive Letter and Path option. Click on the Add button and select the Mount in this NTFS Folder.
5. Either specify C:\TESTMOUNT or use the Browse button to point to the C:\TESTMOUNT folder. Click on the OK button.
6. Right-click the C drive and select the Properties button. Notice the amount of free disk space. Click on the OK button to close the C drive Properties dialog box.
7. Right-click the C drive and select the Open option.
8. Right-click the TESTMOUNT folder and select the Properties option. Click on the Properties button. Notice the amount of free disk space.

EXERCISE 4: CONVERT TO DYNAMIC DISKS

Note: It is always recommended to perform a backup of all important data before doing a conversion to Dynamic Disks.

1. In Disk Management, right-click Disk 0 and select Upgrade to Dynamic Disk.
2. Click OK to accept Disk 0. Click on the Upgrade button. Click the Yes button when it asks if you are sure. On the warning that disks will be forced to dismount, click on the Yes mount.
3. Right-click the last volume and delete it.
4. Right-click the unallocated disk space and select the Create a Volume option. Click on the Next button. With the simple volume selected, click on the Next button.
5. Since Disk 0 is already selected, specify the size of 500 MB. Click on the Next button.
6. Assign the M drive and click on the Next button.
7. Since the Format this Volume is selected, click on the Next button.
8. Click on the Finish button.
9. Right-click the M drive and select the Extend Volume option. Click on the Next button.
10. Since Drive 0 is already selected, specify 100 MB and click on the Next button.
11. Click the Finish button.
12. Right-click the unallocated disk space and create a 100-MB FAT volume. Assign the next available drive letter.
13. Right-click the M drive and select the Extend option. Use the rest of the disk space to extend the drive.

EXERCISE 5: USING ERROR-CHECKING AND DEFRAGMENTER UTILITIES

1. Right-click the C drive and select the Properties option.
2. Click on the Tools tab.
3. Click on the Check Now button.
4. Select the Automatically Fix System Errors option. If you have time in class, enable the "Scan for and attempt recovery of bad sectors." Click on the Start button.
5. From the Computer Management console, click on the C drive and click on the Defragment button.

14

Directory and File Management

INTRODUCTION

This chapter is the first to focus on enabling and managing a network resource, specifically the directories and files that are located on the hard drive. Because the files on the server need to be kept secure, this chapter will discuss NTFS and Share permissions, the Distribution File System, and encryption. In addition, it will also discuss file compression and quotas.

OBJECTIVES

1. Monitor, manage, and troubleshoot access to files and folders.
2. Configure, manage, and troubleshoot file compression.
3. Control access to files and folders by using permissions.
4. Optimize access to files and folders.
5. Manage and troubleshoot access to shared folders.
6. Create and remove shared folders.
7. Control access to shared folders by using permissions.
8. Manage and troubleshoot the use and synchronization of offline files.
9. Connect to shared resources on a Microsoft network.
10. Encrypt data on a hard disk by using the encrypting file system (EFS).

14.1 NTFS PERMISSIONS

A primary advantage of NTFS over FAT and FAT32 is that NTFS volumes have the ability to apply NTFS permissions to secure folders and files. By setting the permissions, you specify the level of access for groups and users for accessing files or directories. For example, to one user or group of users, you can specify that they can only read the file; another user or group of users can read and write to the file; and others have no access. No matter if you are logged on locally at the computer or accessing a computer through the network, NTFS permission always applies.

All folder and file NTFS permissions (also known as special permissions) are listed in Tables 14.1 and 14.2. To simplify the task of administration, the permissions have been logically grouped into the standard folder and file NTFS permissions as shown in Tables 14.3 and 14.4. The standard folder permissions include Full Control, Modify, Read & Execute, List Folder Contents, Read, and Write. The standard file permissions include Full Control, Modify, Read & Execute, Read, and Write. The folder permissions are shown in Tables 14.1 and 14.3 while the file permissions are shown in Tables 14.2 and 14.4.

The NTFS permissions that are granted are stored in an Access Control List (ACL) with every file and folder on an NTFS volume. The ACL contains an access control entry (ACE) for each user account and group that have been granted access for the file or folder as well as the permissions granted to each user and group.

To assign NTFS permissions, right-click a drive, folder, and file (using My Computer or Windows Explorer), select the Properties option, and click on the Security tab. To assign the special permissions, click on the Advanced button within the Security tab and click on the View/Edit button. See Figure 14.1.

14.1.1 Explicit Permission

Permissions are given to a folder or file as either explicit permissions or inherited permissions. In this section we discuss explicit permissions and in Section 14.1.2 we discuss inherited types. **Explicit permissions** are those granted directly to the folder or file. Some of these permissions are granted automatically, such as when a file or folder is created, while others have to be assigned manually.

TABLE 14.1 NTFS folder permissions

Special Permissions	Full Control	Modify	Read & Execute	List Fold Contents	Read	Write
Traverse Folder/Execute File	X	x	x	x		
List Folder/Read Data	X	x	x	x	x	
Read Attributes	X	x	x	x	x	
Read Extended Attributes	X	x	x	x	x	
Create Files/Write Data	X	x				x
Create Folders/Append Data	X	x				x
Write Attributes	X	x				x
Write Extended Attributes	X	x				x
Delete Subfolders and Files	X					
Delete	X	x				
Read Permissions	X	x	x	x	x	x
Change Permissions	X					
Take Ownership	X					
Synchronize	X	x	x	x	x	x

Note: Although List Folder contents and Read & Execute appear to have the same permissions, they are inherited differently. List Folder contents is inherited by folders but not files, and it should only appear when you view folder permissions. Read & Execute is inherited by both files and folders and is always present when you view file or folder permissions.

TABLE 14.2 NTFS file permissions

Special Permissions	Full Control	Modify	Read & Execute	Read	Write
Traverse Folder/Execute File	x	x	x		
List Folder/Read Data	x	x	x	x	
Read Attributes	x	x	x	x	
Read Extended Attributes	x	x	x	x	
Create Files/Write Data	x	x			x
Create Folders/Append Data	x	x			x
Write Attributes	x	x			x
Write Extended Attributes	x	x			x
Delete Subfolders and Files	x				
Delete	x	x			
Read Permissions	x	x	x	x	x
Change Permissions	x				
Take Ownership	x				
Synchronize	x	x	x	x	x

TABLE 14.3 Standard NTFS folder permissions

NTFS Folder Permissions	Allows the user to:
Read	Display the file's data, attributes, owner, and permissions.
Write	Create new files and subfolders within the folder, write to the file, append to the file, read and change folder attributes, and view ownership.
List Folder Contents	View the names of folders and subfolders in the folder.
Read & Execute	Display the folder's contents and the data, attributes, owner, and permissions for files within the folder and execute files within the folder. In addition, it allows the user to navigate a folder to reach other files and folders, even if the user does not have permissions for these folders.
Modify	Read the files, execute files, write and modify files, create folders and subfolders, delete subfolders and files, and change attributes of subfolders and files.
Full Control	Read the files, execute files, write and modify files, create folders and subfolders, delete subfolders and files, change attributes of subfolders and files, change permissions, and take ownership of the file.

Note: Groups or users who are granted full control for a folder can delete files and subfolders within that folder regardless of the permissions that are protecting the files and subfolders.

To explicitly grant a permission to a folder or file, select the permission by checking the respective box. To remove an explicit permission, deselect the permission to remove the check in the respective box. To remove a user or group from being assigned explicit permissions, click to highlight the user, and click on the Remove button.

Because users can be members of several groups, it is possible for them to have several sets of explicit permissions to a folder or file. When this occurs, the permissions are combined to form the **effective permissions,** which are the actual permissions when logging in and accessing a file or folder. They consist of explicit permissions plus any inherited permissions.

TABLE 14.4   Standard NTFS file permissions

NTFS File Permissions	Allows the user to:
Read	Display the file's data, attributes, owner, and permissions.
Write	Write to the file, append to the file, overwrite the file, change file attributes, and view file ownership and permissions.
Read & Execute	Display the data, attributes, owner, and permissions for file and execute the file.
Modify	Read the file, execute the file, write, modify and delete the files, and change attributes of file.
Full Control	Read the file, execute the file, write, modify and delete the files, change attributes of file, change permissions, and take ownership of the file.

Note: Groups or users who are granted full control for a folder can delete files and subfolders within that folder regardless of the permissions that are protecting the files and subfolders.

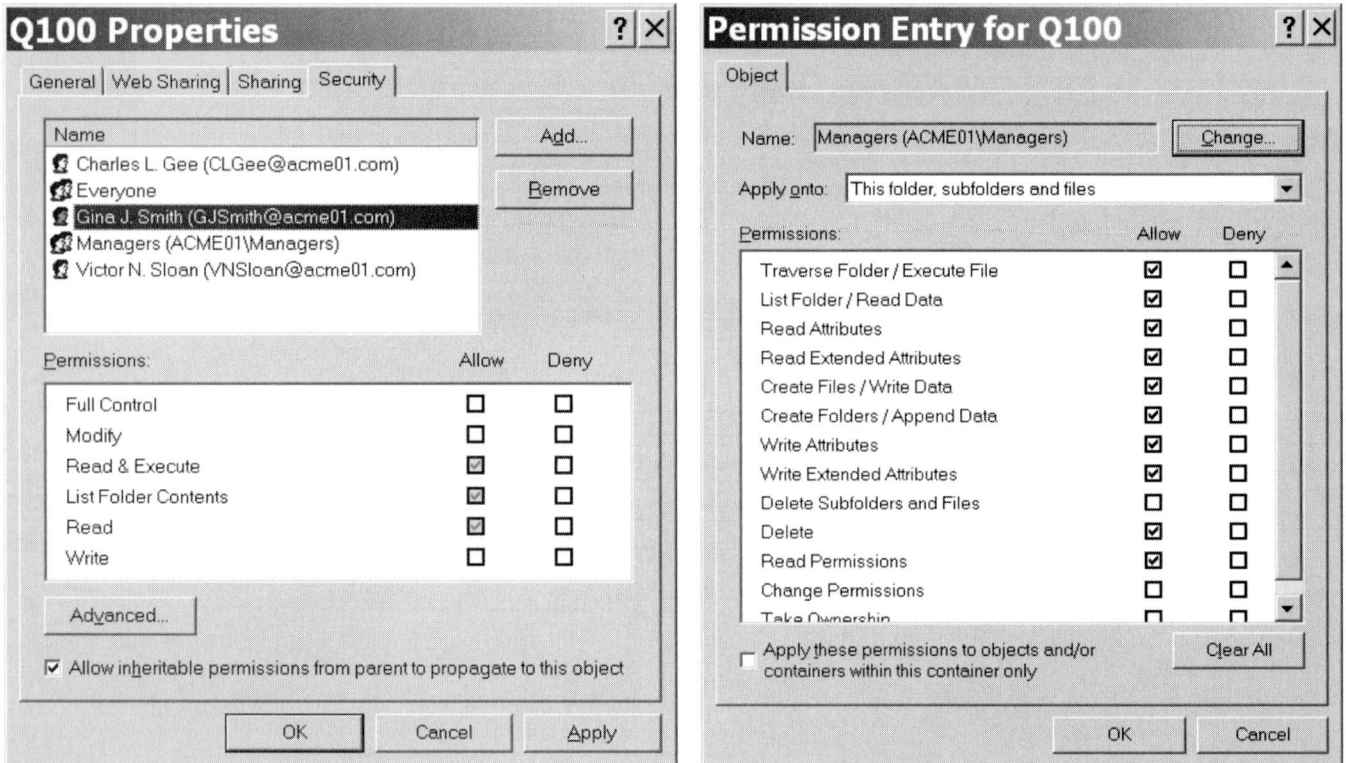

FIGURE 14.1   NTFS permissions

EXAMPLE 1   USER1 is a member of the local group called MANAGERS.

The MANAGERS group is assigned the Read (R) permission to DOC1.TXT and USER1 is assigned the Write (W) permission to DOC1.TXT. What are the effective rights to the DOC1.TXT file for USER1?

When USER1 accesses the DOC1.TXT file, the effective rights would be both the Read and Write permissions. Therefore, the user would be able to display the file's data, write to the file, append to the file, overwrite the file, view and change the file's attributes, and view the owner and permissions of the file.

DOC1.TXT		User1	Managers	Combined
	Explicit	W	R	
	Effective	W	R	RW

Windows 2000 offers the ability to deny individual permissions (standard or special). The Deny permission always overrides the permissions that have been granted, including when a user or group has been granted full control. For example, if the group has been granted read and write, yet a person has been denied the Write permission, the user's effective rights would be the Read permission.

EXAMPLE 2 USER1 is a member of the MANAGERS group.

The MANAGERS group is assigned the Modify (M) permission to DOC1.TXT and USER1 is assigned the Deny Write (W) permission to DOC1.TXT. What are the effective rights to the DOC1.TXT file for USER1?

In this case, the Deny Write permission will overwrite those rights granted by the Modify permission. If you use Table 14.2, take the special NTFS permissions that are granted for the Modify permissions and remove those permissions that are represented by the Write permission. Therefore, the only permissions that would be granted would be to read and change the attributes and delete the file. The execute file would also be granted, but the DOC1.TXT file is not an executable file.

DOC1.TXT		User1	Managers	Combined
	Explicit	~~W~~	M	
	Effective	~~W~~	M	See Below

Special Permissions	Modify	Write	Effective
Traverse Folder/Execute File	x		X
List Folder/Read Data	x		X
Read Attributes	x		X
Read Extended Attributes	x		X
Create Files/Write Data	x	x	—
Create Folders/Append Data	x	x	—
Write Attributes	x	x	—
Write Extended Attributes	x	x	—
Delete	x		X
Read Permissions	x	x	—
Synchronize	x	x	—

EXAMPLE 3 USER1 is a member of the MANAGERS group.

The MANAGERS group is assigned the Full Control (FC) permission to DOC1.TXT and USER1 is assigned the Deny R (R) permission to DOC1.TXT. What are the effective rights to the DOC1.TXT file for USER1?

If USER1 is denied access to read the file, it greatly reduces access to the file. Yet, USER1 can still write to the file, append to the file so long as the user does not have to read the file (such as adding information to the end of a log file), change attributes, delete the file, or take ownership of the file.

DOC1.TXT		User1	Managers	Combined
	Explicit	R	FC	
	Effective	R	FC	See Below

Special Permissions	Full Control	Read	Effective
Traverse Folder/Execute File	x		X
List Folder/Read Data	x	x	
Read Attributes	x	x	
Read Extended Attributes	x	x	
Create Files/Write Data	x		X
Create Folders/Append Data	x		X
Write Attributes	x		X
Write Extended Attributes	x		X
Delete Subfolders and Files	x		X
Delete	x		X
Read Permissions	x	x	
Change Permissions	x		X
Take Ownership	x		X
Synchronize	x	x	

NTFS file permissions override folder permissions. Therefore, if a user has access to a file, but not the folder that contains the file, the user will still be able to gain access to a file. The user who has no access to the folder cannot navigate or browse through the folder to get to the file. Therefore, the user would have to utilize the universal naming convention (UNC) or local path to open the file.

14.1.2 Inherited Permissions

When you set permissions to a folder (explicit permissions), the files and subfolders created in the folder inherit these permissions (called **inherited permissions**). In other words, the permissions flow down from the folder into the subfolders and files, indirectly giving permissions to a user or group. Inherited permissions ease the task of managing permissions and ensure consistency of permissions among the subfolders and files within the folder.

When viewing the permissions, the checkbox will be checked, cleared (unchecked), or shaded. If the box is checked, the permission was explicitly assigned to the folder or file. If the box is clear, the user or group does not have that permission explicitly granted to the folder or file. A user may still obtain permission through a group permission or a group may still obtain permission through another group. If the box is shaded, the permission is granted through inheritance from a parent folder. Review Figure 14.1.

EXAMPLE 4 On the E drive, you have a DATA folder and under the DATA folder, you have the FINANCE folder. Under the FINANCE folder, you have the DOC1.TXT file.

For USER1, you assign the Modify (M) permission for the DATA folder. What are the effective rights to the DOC1.TXT file for USER1?

The Modify permission was given to the DATA folder and flows down through the FINANCE folder to the DOC1.TXT file. Therefore, the user will have the modify permission for the DOC1.TXT.

		User1
DATA	Explicit	M
	Effective	M
FINANCE	Inherited	M
	Explicit	—
	Effective	M
DOC1.TXT	Inherited	M
	Explicit	—
	Effective	M

When assigning permissions to a folder, by default, the permissions apply to the folder being assigned and the subfolders and files of the folder. If you show the permission entries, you can specify how the permissions are applied to the folder, subfolder, and files. See Figure 14.2.

To stop permissions from being inherited, clear the "Allow inheritable permissions from parent to propagate to this object" checkbox. When the checkbox is clear, Windows 2000 will respond with a Security dialog box as shown in Figure 14.3. When you click on the Copy button, the explicit permissions will be copied from the parent folder to the subfolder or file. You can then change the subfolder's or file explicit permissions. If you click on the Remove button, it will remove the inherited permissions altogether. Another way to stop a permission from being inherited is to change the parent folder so that the file or subfolder will inherit different permissions or to select the opposite (Allow or Deny) to override the inherited permission.

EXAMPLE 5   On the E drive, you have a DATA folder and under the DATA folder, you have the FINANCE folder. Under the FINANCE folder, you have the DOC1.TXT file.

FIGURE 14.2   Selecting how the permissions are applied to a folder and its subfolders and files

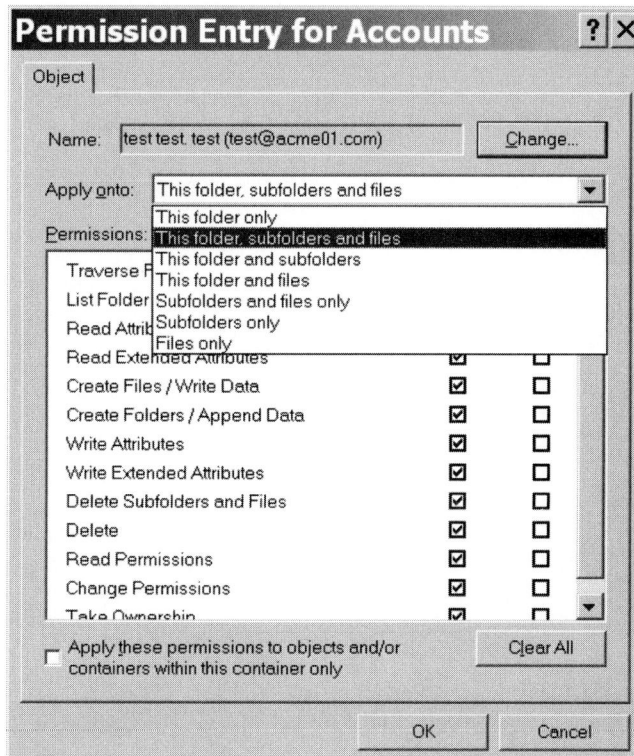

427

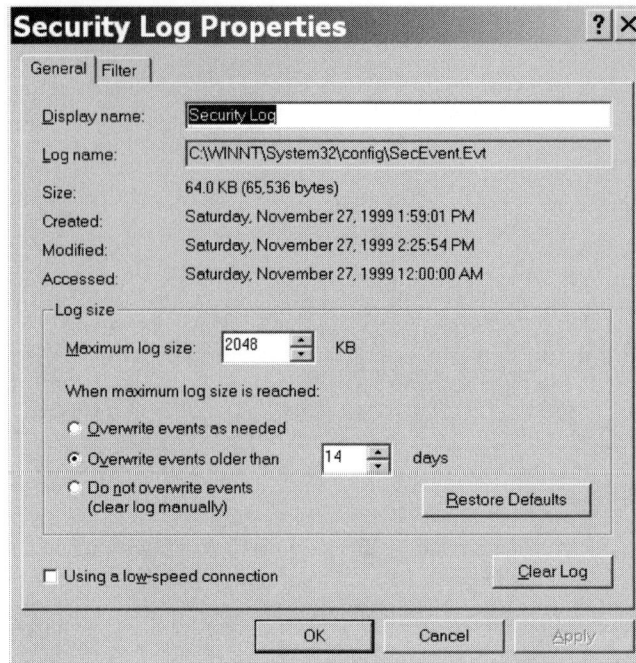

For USER1, you assign the Read (R) permission for the DATA folder. In addition, you assign the Modify (M) permission to the FINANCE folder. What are the effective rights to the DOC1.TXT file for USER1?

The Read permission was given to the DATA folder and flows down through the FINANCE folder. Assuming that the administrator is selected to overwrite the permissions and not copy, the Read permission is overwritten by the Modify permission, which then flows down to the DOC1.TXT file. Therefore, the user will have the Modify permission for the DOC1.TXT.

		User1
DATA	Explicit	R
	Effective	R
FINANCE	Inherited	R
	Explicit	M
	Effective	M
DOC1.TXT	Inherited	M
	Explicit	—
	Effective	M

EXAMPLE 6 On the E drive, you have a DATA folder and under the DATA folder, you have the FINANCE folder. Under the FINANCE folder, you have the DOC1.TXT file. USER1 is a member of the local group called MANAGERS.

For the MANAGER group, you assign the Read (R) permission to the DATA folder. For USER1, you assign the Write (W) permission to the FINANCE folder. What are the effective rights to the DOC1.TXT file for USER1?

For USER1, the Write permission flows down from the DATA folder to the FINANCE folder, down to the DOC1.TXT file. For the MANAGER group, the Read permission flows down from the DATA folder to the FINANCE folder, down to the DOC1.TXT file. The trick is to keep the permissions granted to USER1 and MANAGERS separate except when combining the writes to determine the effective rights for the object. Therefore, the effective rights for the DOC1.TXT would be to combine the permissions for the user and group; thus, when USER1 logs in, it will have both Read and Write permissions.

		User1	Managers	Combined
DATA	Explicit	—	R	
	Effective	—	R	R
FINANCE	Inherited	—	R	
	Explicit	W	—	
	Effective	W	R	RW
DOC1.TXT	Inherited	W	R	
	Explicit	—	—	
	Effective	W	R	RW

14.1.3 Folder and File Owners

Every folder and file has an **owner,** a person who controls how permissions are set on a folder or file and who grants permissions to others. When a folder or file is created, either automatically becomes the owner.

To be able to take ownership of a folder or file, the user has to be granted Take Ownership permission or be the administrator. After logging in, the user can take ownership by doing the following:

1. Right-click the folder or file and select the Properties option.
2. Click on the Security tab and then on the Advanced button.
3. Click on the Owner tab.
4. Click on the user that will take ownership.
5. If you want to replacc the owner for all subfolders and files, mark the checkbox.
6. Click on the OK button.

The administrator automatically has the Take Ownership of any folder or directory, which is under his or her administrative control. Therefore, if someone leaves the company or gets fired, the administrator can take control of that person's file. The administrator cannot transfer ownership to others, but can permit the person to take ownership. See Figure 14.4.

14.1.4 Copying and Moving Files

When you copy and move files and folders from one location to another, it is important to understand how the NTFS folder and file permissions are affected. If you copy a file or folder, the new folder and file will automatically acquire the same permissions as the original drive or folder.

If the folder or file is moved within the same volume, either will retain the originally assigned permissions. When the folder or file is moved from one volume to another, either will automatically acquire the permissions of the original drive or folder and file. An easy way to remember the difference is when you move a folder or file from within the same volume, the folder or file is not physically moved but the Master File Table is adjusted to indicate a different folder. When you move a folder or file from one volume to another, it

FIGURE 14.4 Taking ownership of a NTFS folder

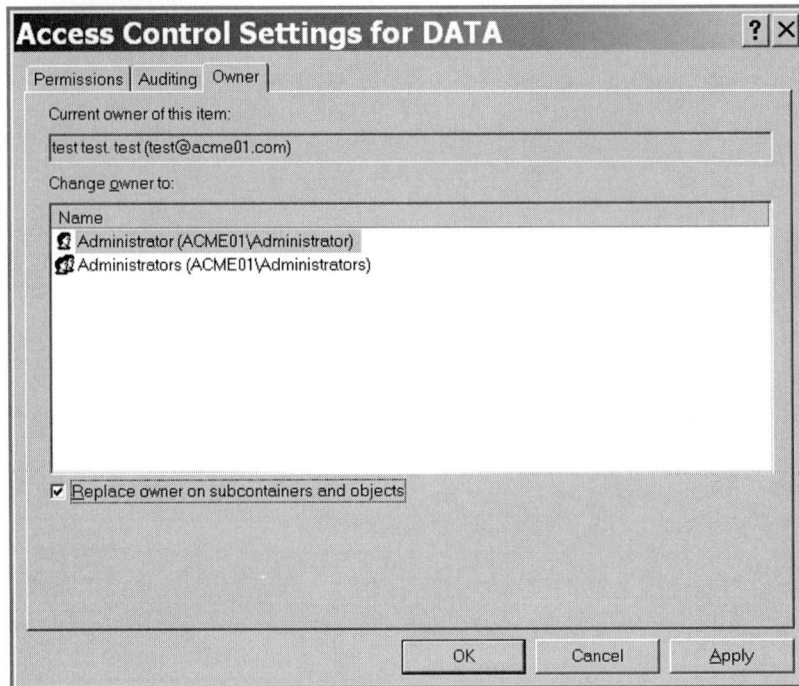

Access Control Settings for DATA

copies the folder or file to the new location and then deletes the old location. Therefore, the moved folder and files are new to the volume and acquire the new permissions.

14.2 SHARED FOLDERS

A **shared folder** on a computer makes the folder available for others to use on the network. A shared drive on a computer makes the entire drive available for others to use on the network. Shared drives and folders can be used on FAT, FAT32, and NTFS volumes. If used on an NTFS partition, the user will still need NTFS permissions before accessing the share. When the drive or directory is shared, it is indicated with a hand underneath the drive or folder.

To share a folder or drive, do the following:

1. Open Windows Explorer and locate the folder or drive that you want to share.
2. Right-click the folder or drive and select the Sharing option.
3. Click on the Sharing tab and select the Share This Folder.
4. The name of the folder becomes the default share name. If you wish to change the share name, type the name in the Shared Name text box. MS-DOS and Windows for Workgroups 3.11 machines are restricted to 8.3 characters.
5. To set the shared folder permissions on the shared folder or drive, click the Permissions button (not the Permissions tab).

If the folder is already shared, click New Share and then type the new share name. To share a folder and drive, you must be logged on as a member of the Administrators, Server Operators, or Power Users group. See Figure 14.5. If the Sharing tab is not visible, try using the Services snap-in to start the Server service.

If a Windows 2000 computer has limited bandwidth, you may want to use the User Limit option. In Windows 2000 Professional, the maximum user limit is 10, regardless of the number you type in the Allow box.

14.2.1 Connecting to a Shared Folder or Drive

The shared resource can be accessed using Network Neighborhood in Windows 95, Windows 98, and Windows NT; My Network Places in Windows 2000; or the universal nam-

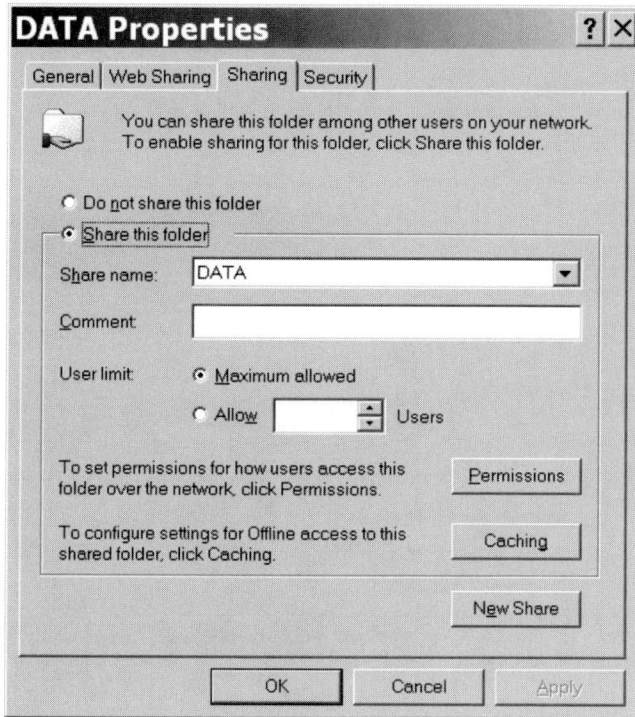

FIGURE 14.5   Sharing a folder

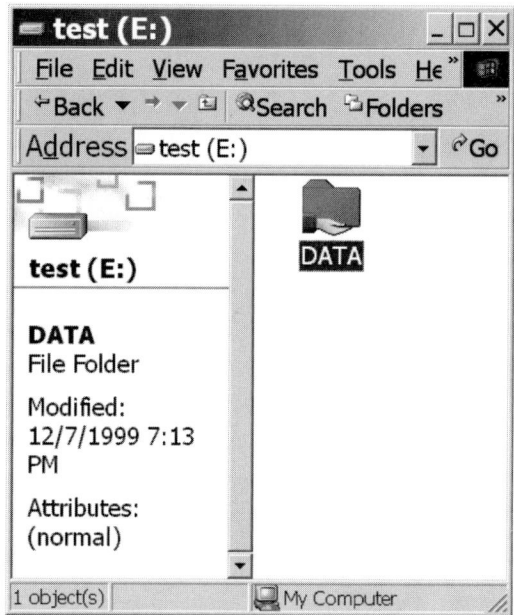

FIGURE 14.6   Access a shared folder using Windows Explorer

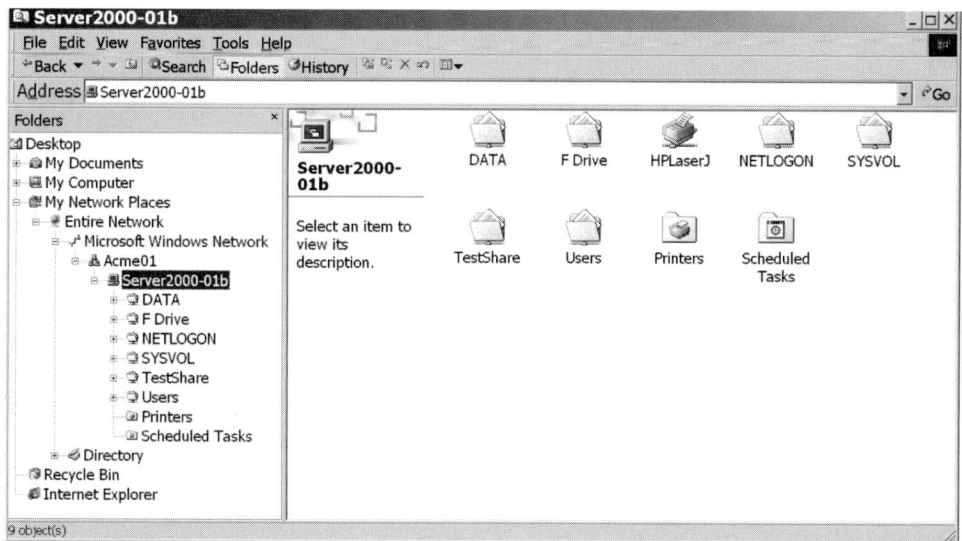

ing convention (\\servername\sharename). See Figure 14.6. Another way to access a shared drive or folder is to map a network drive. This can be accessed by starting Windows Explorer, opening the Tools menu, and selecting Map Network Drive. Then select the drive letter that will be used to point to the network shared drive or folder and enter the UNC name in the Folder text box. See Figure 14.7. When you copy a shared folder, only the original shared folder is still shared. When you move a shared folder, it is no longer shared.

EXAMPLE 7   You have a computer with Windows 2000 Server called SERVER1. On the E drive, you have a DATA folder and under the DATA folder, you have the FINANCE folder. Under the FINANCE folder, you have a GROUP1 folder.

If you share the DATA folder and you want to access the GROUP1 folder, then access the \\SERVER1\DATA\FINANCE\GROUP1 folder. Another way is to access the

FIGURE 14.7 Mapping
network drives

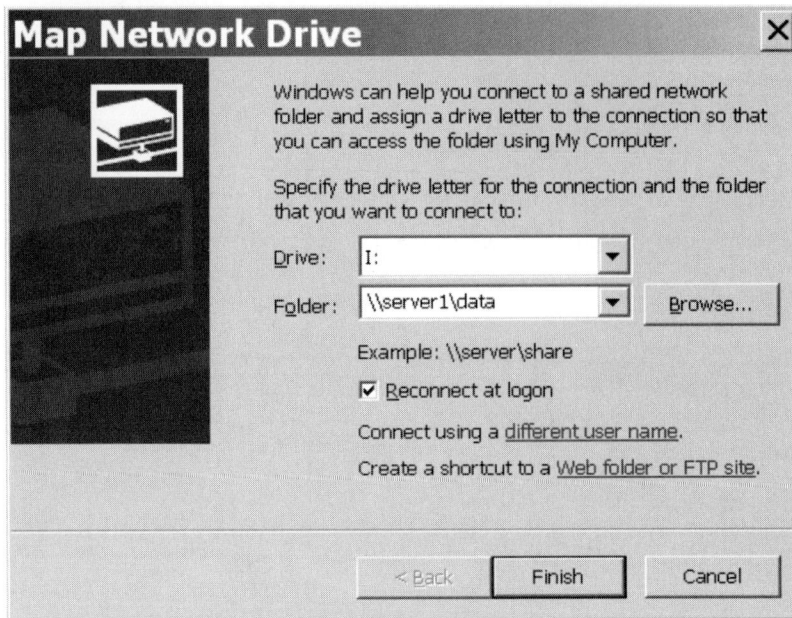

Map Network Drive

Windows can help you connect to a shared network folder and assign a drive letter to the connection so that you can access the folder using My Computer.

Specify the drive letter for the connection and the folder that you want to connect to:

Drive: I:

Folder: \\server1\data   Browse...

Example: \\server\share

☑ Reconnect at logon

Connect using a different user name.

Create a shortcut to a Web folder or FTP site.

< Back   Finish   Cancel

FIGURE 14.8 Accessing a
shared folder remotely and
locally on a first-level shared
folder

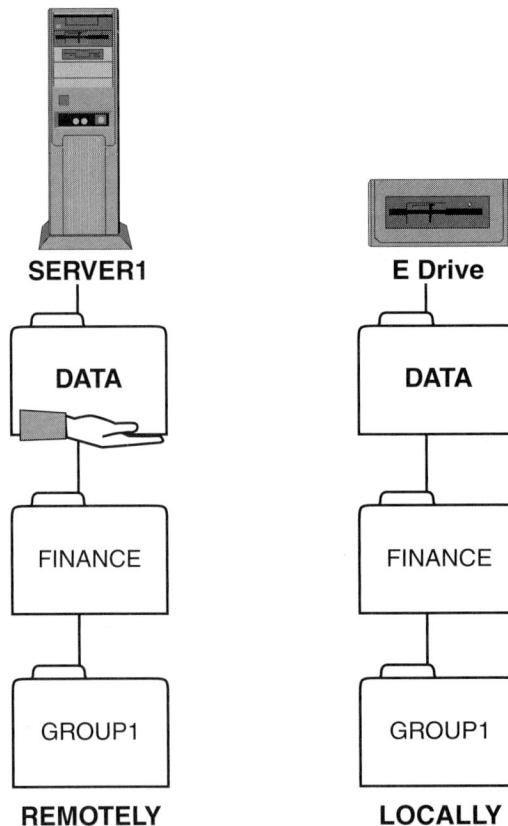

SERVER1   E Drive

DATA   DATA

FINANCE   FINANCE

GROUP1   GROUP1

REMOTELY   LOCALLY

\\SERVER1\DATA share. From there, open the FINANCE folder, which would then allow you to access the GROUP1 folder. See Figure 14.8.

EXAMPLE 8 You have a computer with a Windows 2000 Server called SERVER1. On the E drive, you have a DATA folder and under the DATA folder, you have the FINANCE folder. Under the FINANCE folder, you have a GROUP1 folder.

If you share the FINANCE folder, you would not have access to the DATA folder and FINANCE remotely. You can still access the folder locally by opening the E drive. To ac-

FIGURE 14.9 Accessing a
shared folder remotely and
locally on a third-level shared
folder

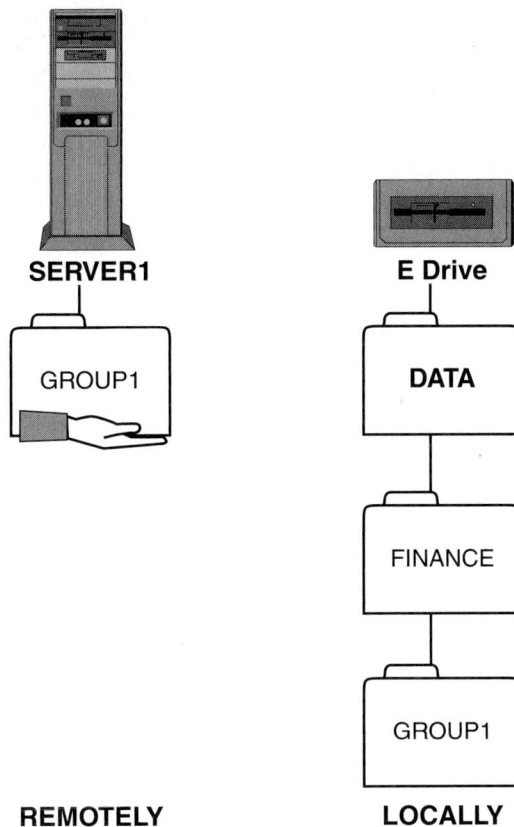

SERVER1

E Drive

GROUP1

DATA

FINANCE

GROUP1

REMOTELY

LOCALLY

**TABLE 14.5 Share
permissions**

Share Permissions	Description
Read	Allows the user to view folder names and file names, open and view subfolder files and their attributes, and navigate the tree structures.
Change	Allows all permissions granted by the Read permission and it also allows the user to create folders, add files to folders, change data in files, append data to files, change file attributes, and delete folders and files.
Full Control	Allows all permissions granted by the Change permission and it also allows the user to change file permissions and take ownership of files.

cess the GROUP1 share, use \\SERVER1\GROUP1, even though the GROUP1 folder is three levels deep. See Figure 14.9.

14.2.2 Shared Folder Permissions

To control how users gain access to a shared folder, assign Shared Folder permissions. These permissions are neither required nor utilized if the user is accessing the directory locally (logged on to the computer that has the shared drive or directory). The Shared Folder/Drive permissions are shown in Table 14.5.

To grant or change share permissions to a shared folder, right-click the shared folder or drive, select the Sharing option, and click on the Permissions button. To add a user or group, click on the Add button. To remove a user or group, click to highlight the user or group and click on the Remove button. To specify which permissions to grant, click to highlight the user or group and select or deselect the Allow and Deny options.

Different from NTFS permissions, a shared folder has only one level of permissions assigned to it. For example, on the SERVER1 server, you share a folder called FOLDER1 and assign the Full Control permission. In the FOLDER1 folder, you share a folder called FOLDER2 and assign the Modify permission. Therefore, if you access the \\SERVER1 \FOLDER1 folder, you use the Full Control share permission, even when accessing the \\SERVER1\FOLDER1\FOLDER2 folder. If you access the \\SERVER folder, you use the Change Share permission. The permissions assigned to FOLDER1 are not an issue.

Since a user can be a member of several groups, it is possible for the user to have several sets of explicit permissions to a shared drive or folder. The effective permissions are the combination of all user and group permissions. For example, if the user has a Write permission to the user and a Read permission to the group, of which the user is a member, the effective permission would be the Write permission. See Figure 14.10. Like NTFS permissions, Deny permissions override the granted permission.

14.2.3 Special Shares

Windows 2000 has several special shared folders that are automatically created by Windows 2000 for administrative and system use. See Table 14.6. Different from regular shares, these shares do not show when a user browses the computer resources using Network Neighborhood, My Network Places, or similar software. In most cases, special shared folders should not be deleted or modified. For the Windows 2000 Professional computer, only members of the Administrators or Backup Operators group can connect to these shares. For Windows 2000 Servers, members of Administrators, Backup Operators, and Server Operators groups are the only members who can connect to these shares.

An **administrative share** is a shared folder, typically used for administrative purposes. To make a shared folder or drive into an administrative share, the share name must end with a $. Since the share folder or drive cannot be seen during browsing, you would have to use a UNC name, which includes the share name (including the $). Instead, it would have to be accessed using the Start button, selecting the Run option, and typing the UNC name and clicking the OK button. By default, all volumes with drive letters automatically have administrative shares (C$, D$, E$, and so on). Other administrative shares can be created as needed for individual folders.

FIGURE 14.10   Share permissions

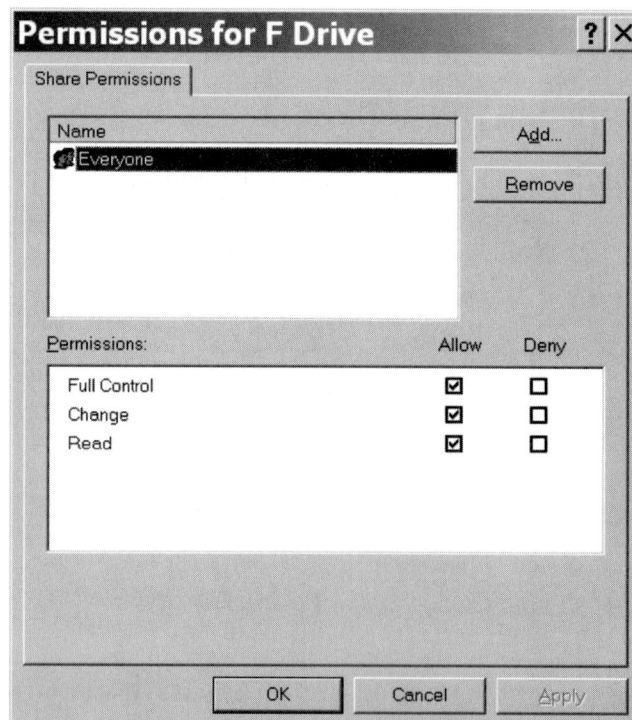

TABLE 14.6   Special shares

Special Share	Description
Drive letter$	A shared folder that allows administrative personnel to connect to the root directory of a drive, also known as an administrative share. It is shown as A$, B$, C$, D$, and so on. For example, C$ is a shared folder name by which drive C might be accessed by an administrator over the network.
ADMIN$	A resource used by the system during remote administration of a computer. The path of this resource is always the path to the Windows 2000 system root (the directory in which Windows 2000 is installed, for example, C:\Winnt).
IPC$	A resource sharing the named pipes that are essential for communication between programs. It is used during remote administration of a computer and when viewing a computer's shared resources.
PRINT$	A resource used during remote administration of printers.
NETLOGON	A resource used by the Net Logon service of a Windows 2000 Server computer while processing domain logon requests. This resource is provided only for Windows 2000 Server computers. It is not provided for Windows 2000 Professional computers.
FAX$	A shared folder on a server used by fax clients in the process of sending a fax. The shared folder is used to temporarily cache files and access cover pages stored on the server.

FIGURE 14.11   Windows 2000 will ask you to log in to the IPC$ if you are not connected to the IPC$ share or to the remote system using an account and password that exist in the remote system's domain or local user accounts, while the guest account is disabled

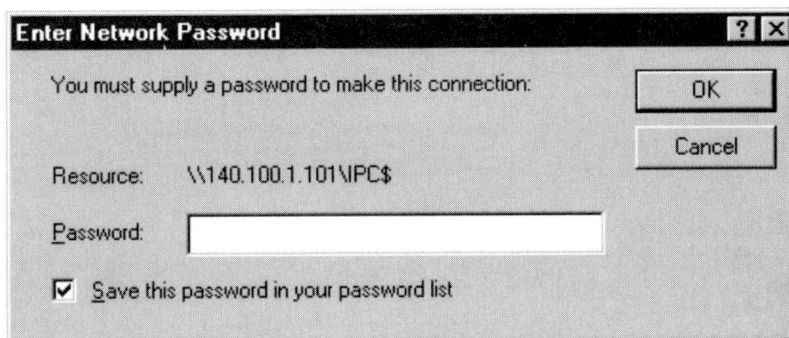

The IPC$ is used for remote administration of a computer and when viewing a computer's shared resources. Windows 2000 uses it to exchange information directly between applications on two different computers, without having to write to or read from the file system. If you do not have the guest account enabled, you have not connected to the IPC$ share. You are trying to access a shared folder or shared printer on a remote system of which you are not a member, the remote computer's domain or local user account Windows 2000 will ask you for the password for the IPC$ share, as shown in Figure 14.11.

Following are two ways to overcome this problem:

- Enable the guest account. This is typically not recommended for security reasons. To access any network resource, you must be logged in with a user name and password (the password may be blank). It does not matter if you are logging into the same domain, another domain, or logging in locally.
- Connect to the IPC$ share or the remote system using an account and password that exist in the remote system's domain or local user accounts. To connect to the remote IPC$ share, use the following command at the Command prompt:

```
NET USE \\computername\IPC$ /USER:accountname password
```

or by running (executing) the following command in the Run dialog box:

```
\\computername\IPC$ /USER:accountname password
```

Console Window Help

Action View Favorites

Tree	Favorites		Shared Fol...	Shared Path	Type	# Client Redirections	Comment
Console Root		ADMIN$	C:\WINNT	Windows	0	Remote Admin
Shared Folders (Local)		C$	C:\	Windows	0	Default share
Shares		D$	D:\	Windows	0	Default share
Sessions		DATA	E:\DATA	Windows	1	
Open Files		E$	E:\	Windows	0	
		F Drive	F:\	Windows	0	
		F$	F:\	Windows	0	Default share
		IPC$		Windows	1	Remote IPC
		NETLOGON	C:\WINNT\SYSVOL\sysvol\acm...	Windows	0	Logon server share
		print$	C:\WINNT\System32\spool\driv...	Windows	0	Printer Drivers
		SYSVOL	C:\WINNT\SYSVOL\sysvol	Windows	0	Logon server share
		TestShare	C:\TestShare	Windows	0	
		Users	C:\Users	Windows	0	

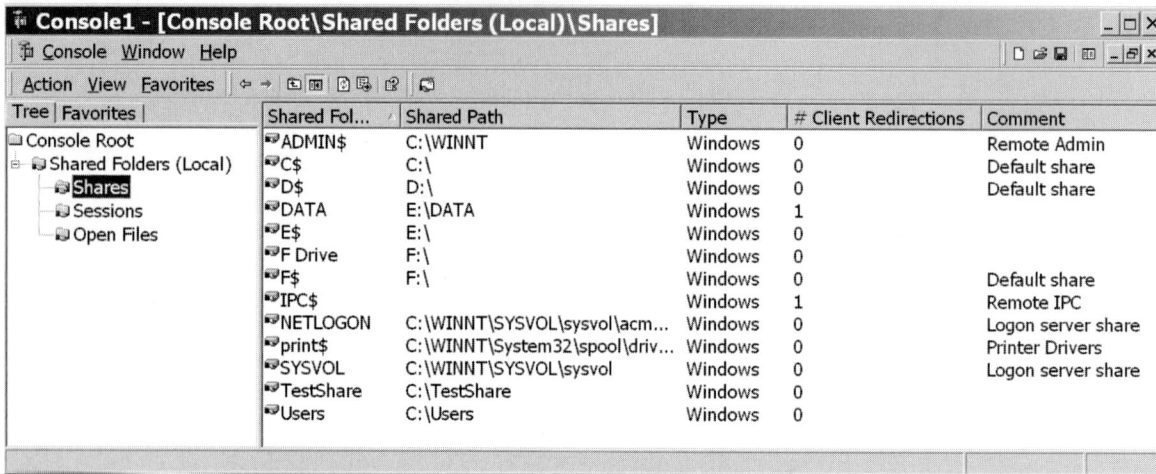

FIGURE 14.12   Shared folders console

14.2.4   Managing Shares

By using the Shared Folders snap-in (included in the Computer Management console), you can manage the server's shared folders. See Figure 14.12. With the Shared Folder snap-in, you can do the following:

- Create, view, and set permissions for shares, including shares on Windows 2000 computers.
- View a list of all users who are connected to the computer over a network and disconnect one or all of them.
- View a list of files opened by remote users and close one or all of the open files.

14.3   COMBINING SHARED FOLDER PERMISSIONS AND NTFS PERMISSIONS

As a review, NTFS permissions apply if you are logged on locally or if you connect to the computer using a network connection. Share permissions are the only protection for FAT and FAT32 drives. Unfortunately, share permissions only protect the network resource when you connect to the computer using a network connection. If you log on locally to the computer, share permissions do not matter. Remember that you help protect the local files by only allowing a small group of people to log on locally.

When accessing a shared folder on an NTFS volume, the effective permissions that a person can do in the share folder are calculated by combining the shared folder permissions with the NTFS permissions. When combining the two, first determine the cumulative NTFS permissions and the cumulative shared permissions and apply the more restrictive permission—the one that gives the fewest permissions.

EXAMPLE 9   On the E drive (NTFS partition), you have a DATA folder and under the DATA folder you have a FINANCE folder. In the FINANCE folder, you have the DOC1.TXT file. The DATA folder is shared.

For USER1, you assign the Read (R) NTFS permission to the DATA folder and Full Control (FC) Share permission to the DATA folder. What are the effective rights to the DOC1.TXT file for USER1?

To determine the effective rights when accessing the server remotely, take the more restrictive permission, or the more limiting permission. Since the Read permission is more limiting than the FC permission, the effective rights when USER1 is accessing the DATA folder over the network are those granted by the Read permission. As the rights flow down to the FINANCE folder, again the R is more limiting than the FC. Therefore, the effective rights when USER1 is accessing the DATA folder over the network are those granted by the Read permission. Lastly, the rights flow into the DOC1.TXT file and again the effective permission is Read.

Remote Access Effective Permissions

		NTFS	Shared	Combined
DATA (folder)	Explicit	R	FC	
	Effective	R	FC	R
FINANCE (folder)	Inherited	R	FC	
	Explicit	—	—	
	Effective	R	FC	R
DOC1.TXT (file)	Inherited	R	FC	
	Explicit	—	—	
	Effective	R	FC	R

If USER1 logs on locally to the server, the share permissions do not matter and the only permission worth looking at is the Read permission.

Local Access Effective Permssions

		NTFS	Shared	Combined
DATA (folder)	Explicit	R	FC	
	Effective	R	NA	R
FINANCE (folder)	Inherited	R	NA	
	Explicit	—	—	
	Effective	R	NA	R
DOC1.TXT (file)	Inherited	R	NA	
	Explicit	—	—	
	Effective	R	NA	R

EXAMPLE 10 On the E drive (NTFS partition), you have a DATA folder and under the DATA folder you have a FINANCE folder. In the FINANCE folder, you have the DOC1.TXT file. The DATA folder is shared. USER1 is a member of the local group called MANAGERS.

For USER1, you assign the Modify (M) NTFS permission to the DATA folder and Read and Execute (R&E) permission to the FINANCE folder. For the MANAGERS group, you assign the Write permission to the FINANCE folder. In addition, for the Managers group, you assign Full Control (FC) Share permission to the DATA folder. What are the effective permissions to the DOC1.TXT file for USER1 when USER1 logs on remotely?

To determine the effective rights when accessing the server remotely, take the more restrictive permission, or the more limiting permission. For the USER1 NTFS permission, the R&E permission will stop the FINANCE folder from inheriting the Modify permission from the DATA folder. The DOC1.TXT file will inherit the R&E permission. For MANAGERS, DOC1.TXT file will inherit the Write permission from the FINANCE folder. When you combine all NTFS permissions for the DOC1.TXT file, USER1 will have Read & Execute and Write permissions.

For the shared permissions, the MANAGERS were assigned Full Control to the shared Data folder. Therefore, the effective shared permission is the Full Control permission.

When you combine the two at the DOC1.TXT level, you must take the permissions that are the most restrictive. Since the R&E and Write permissions are more restrictive than Full Control, USER1 will have those permissions when accessing the DOC1.TXT file.

		USER1 NTFS	MANAGERS NTFS	Combined NTFS	USER1 Shared	MANAGERS Shared	Combined Shared	Combined Overall
DATA	Explicit	M	—		—	FC		
	Effective	M	—	M	—	FC	FC	M
FINANCE	Inherited	R&E	—		—	FC		
	Explicit	—	W		—	—		
	Effective	R&E	W	R&E & W	—	FC	FC	R&E & W
DOC1.TXT	Inherited	R&E	W		—	FC		
	Explicit	—	—		—	—		
	Effective	R&E	W	R&E & W	—	FC	FC	R&E & W

14.4 DISTRIBUTED FILE SYSTEM

The **distributed file system (DFS)** groups and organizes shared folders that reside on the same or different computers into a single hierarchical tree structure. By using DFS, a user can easily gain access to network resources without knowing the actual location of the underlying resources. To create a DFS share, you must first create a DFS root and then create child nodes that represent the shared folders, which can be physically located on different servers. See Figure 14.13. Note that any Windows 2000 Server can host one DFS root that is located in a FAT or NTFS volume; however, NTFS would be a better choice because of its extra security features.

The two types of DFS are as follows:

DFS Type	Description
Stand-alone DFS	Stores the DFS topology on a single computer. It does not provide any fault tolerance if the computer that stores the DFS topology or any of the shared folders fails.
Domain DFS (Fault Tolerance)	Provides fault tolerance by having the child nodes point to multiple, identical shared folders. Different than the stand-alone DFS, the domain DFS stores its topology in the Active Directory. By using Active Directory, it supports DNS. Each child node can have up to 32 replicas.

To create the stand-alone DFS root, do the following:

1. On the Action menu, click the New DFS Root and click on the Next button.
2. Select Create a Stand-alone DFS Root and click on the Next button.
3. Type in the name of the host server for the DNS root and click on the Next button.
4. Select an existing share or select the Create a New Share and specify the path to share and the share name and click on the Next button.
5. Specify a unique DFS share name and click on the Next button.
6. Click on the Finish button.

To create the domain (fault-tolerant) DFS root, do the following:

1. On the Action menu, click the New DFS Root and click on the Next button.
2. Select Create a Domain DFS Root and click on the Next button.
3. Type in the name of the host domain for the DFS root and click on the Next button.
4. Type in the name of the host server for the DNS root and click on the Next button.
5. Select an existing share or select the Create a New Share and specify the path to share and the share name and click on the Next button.

FIGURE 14.13 Distributed File
System

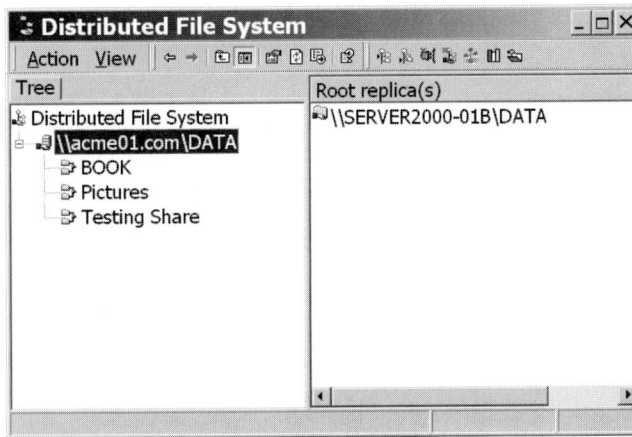

FIGURE 14.14 Setting up
replicas in DFS

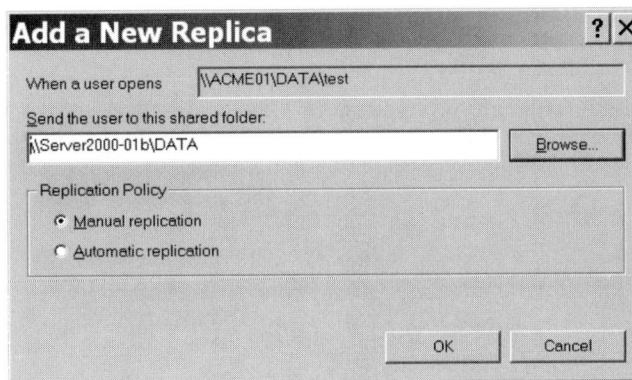

6. Specify a unique DFS share name and click on the Next button.
7. Click on the Finish button.

Replicating the DFS root to another server in the domain ensures that if the host server becomes unavailable for any reason, the distributed file system that is associated with that DFS root is still available to domain users. To set up a replication of the root, right-click the root, select the New Root Replica, and use the wizard.

By replicating a DFS shared folder, it stores a duplicate copy of the contents of the original shared folder into another shared folder. To replicate a DFS shared folder, right-click the DFS Link and select the New Replica option. Then specify an alternate shared folder that will act as a replication partner to the original shared folder, select the type of replication policy, and click on the OK button. If Automatic Replication is selected, it automatically will replicate the files from the one share to the other. Automatic Replication can only be used, however, for files stored on NTFS volumes on Windows 2000 servers. When manual replication is selected, you must manually copy the files. Also, when using stand-alone DFS, you can only select manual replication. See Figure 14.14.

To remove a link, right-click the link and select the Remove DFS Link option and click on the Yes button. To delete a DFS root share, right-click the DFS root share, select the Delete DFS Root option, and click on the Yes button.

To access DFS resources, the operating system must include DFS client software. This software is included in Windows 98, Windows NT 4.0, and Windows 2000. If you have a Windows 95 computer, you must download and install a DFS client. If your system has the DFS client, then access it like you would a shared folder or directory.

14.5 DEFAULT PERMISSIONS AND GUIDELINES

By now, it should be obvious that NTFS is a much better choice than FAT or FAT32. The only reason to use FAT or FAT32 is if you need to dual boot between two different operating systems, such as Windows 2000 and Windows 98.

TABLE 14.7   Recommended folder rights

Directory Purpose	Description	Recommended Rights
Application folder	Contains the programs that the user needs to run.	• If the user needs to read the files and run the applications, assign NTFS Read & Execute permission to the users/groups and the Read permission to the share. • If the user needs to upgrade and troubleshoot the applications, assign NTFS Read & Execute and write permissions to the users/groups and the Modify permission to the share. • If the user needs to administer the application folder, assign NTFS Full Control permission to the users/groups and Full Control permission to the share.
Public data folder	Used to store data files that belong to a group of people. For example, you may have a folder for an administrative group or a project.	• If the user needs to make changes to the file or create new files, assign NTFS Read & Execute and Write permissions to the users/groups that need it and the Modify permission to the share. • If the user needs to read the files without making changes or creating new files, assign NTFS Read & Execute to the users/groups that need it and the Read permission to the share. • If the user needs to manage the public data folder, assign NTFS Full Control permission to the users/group and Full Control permission to the share.
Home folder	Used to hold the individuals' data directories.	• Assign the Full control permission for the individual to the individual's home folder. • Refer to Section 12.1.4 on setting up user home directories.

When you format a volume with NTFS, or create a share, the Full Control permission is assigned to the Everyone group. Therefore, to make the server a secure system, either remove the Everyone group and assign groups and individuals to the drive and directories with the appropriate permissions or change the Full Control permission to something more appropriate.

When planning a directory structure, attempt to group files into application, public data folders, and home folders. See Table 14.7. Then you can assign the appropriate rights to the folder instead of the different files and the task of performing backups is simpler. Next, try to create groups according to their access, and assign only the level of access that the users need by first assigning rights to the group followed by assigning rights to the individual users who do not fit into the groups.

14.6 TROUBLESHOOTING FILE ACCESS PROBLEMS

When a user cannot gain access to a folder or file, the problem is usually caused by insufficient permissions (either NTFS or shared). Make sure that the user or group is assigned the proper permission and that the user is assigned to the proper group. In addition, when checking the permissions of the user and the group, of which the user is a member, check for denied permissions. Lastly, if you just added permissions to a user or group or just added a user to a group and the user still cannot access the folder or file, have the user log off and log on so that the person will get a new token that allows access to the folder or file.

14.7 FILE ENCRYPTION

Encryption is the process of converting data into a format that cannot be read by another user. Once a user has encrypted a file, it automatically remains encrypted when the file is stored on disk. **Decryption** is the process of converting data from encrypted format back to its original format. Once a user has decrypted a file, the file remains decrypted when stored on disk.

Windows 2000 includes the **encrypting file system (EFS),** which allows a user to encrypt and decrypt files that are stored on a NTFS volume. By using EFS, folders and files are still kept secure against those intruders who might gain unauthorized physical access to the drive such as by stealing a notebook computer or a removable drive. EFS is not intended to support the accessing of an encrypted file by multiple users nor does it decrypt or encrypt files that are transmitted over the network. Instead, another protocol would have to be used to secure the data sent over the network.

When EFS is enabled, the process of encrypting and decrypting a file is transparent to the user. As a file is encrypted, each file has a unique file encryption key, which is later used to decrypt the file's data. The file encryption key is also encrypted by the user's public key. When decrypting files, the file encryption key must first be decrypted. The file encryption is decrypted when the user has a private key that matches the public key. When a user who does not have the right private key tries to open, copy, move, or rename an encrypted file, the user will get an access denied message.

Be aware of the following guidelines when using EFS:

- You cannot encrypt files or folders that are compressed.
- Encrypted files can become decrypted if you copy or move the file to a volume that is not an NTFS volume.
- Use cut and paste to move files into an encrypted folder. If you use a drag-and-drop operation to move the files, they will not automatically be encrypted in the new folder.
- System files cannot be encrypted.
- Encrypting a folder or file does not protect against deletion. Anyone with delete permission can delete encrypted folders or files.
- Temporary files, which are created by some programs when documents are edited, are also encrypted as long as all the files are on an NTFS volume and in an encrypted folder. It is recommended that you encrypt the Temp folder on your hard disk for this reason.
- Encrypt the My Documents folder, if this is where you save most of your documents, to ensure that your personal documents are encrypted by default.

14.7.1 Encrypt Attribute

To encrypt a folder or file, do the following:

1. Right-click the folder or file and select the Properties option.
2. Click on the Advanced button. See Figure 14.15.
3. Select the Encrypt Contents to Secure Data option.

FIGURE 14.15 Encryption and compression attributes

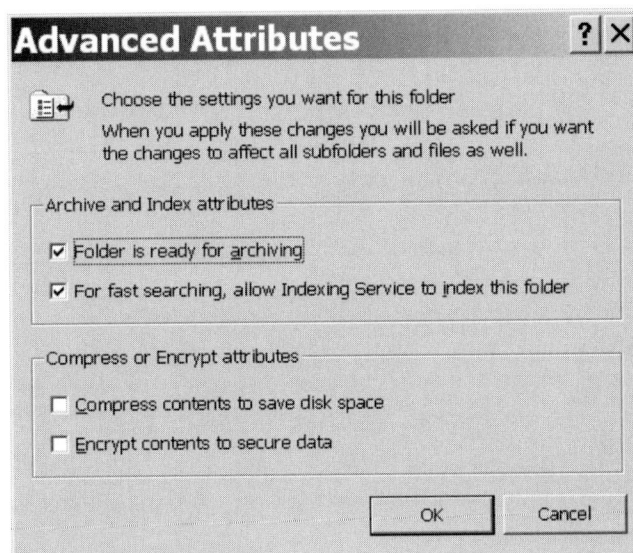

441

Note that when a folder is encrypted, the folder itself is not encrypted, but all files in the folder are encrypted. To decrypt the folder or file, unselect the Encrypt Contents to Secure Data option. You also can encrypt or decrypt a file or folder by using the CIPHER command. For more information on the CIPHER command, type CIPHER /? at the Command prompt. See Figure 14.16.

14.7.2 Recovering Encrypted Files

If a person leaves the company and his or her data are encrypted, the data will have to be decrypted before someone else can use the information. A recovery agent is an administrator who is authorized to decrypt data that was encrypted by another user. Before you can add a recovery agent for a domain, ensure that each recovery agent has been issued an X.509 Version 3 certificate.

The recovery agent has a special certificate and associated private key that allow data recovery for the scope of influence of the recovery policy. If you are the recovery agent, be sure to use the **Export** command from Certificates in Microsoft Management Console (MMC) to back up the recovery certificate and associated private key to a secure location. After backing them up, use Certificates in MMC to delete the recovery certificate. Then, when you need to perform a recovery operation for a user, first restore the recovery certificate and associated private key using the **Import** command from Certificates in MMC. After recovering the data, you should again delete the recovery certificate. You do not have to repeat the export process. See Figure 14.17.

FIGURE 14.16   Cipher /? command

```
C:\>cipher /?
Displays or alters the encryption of directories [files] on NTFS partitions.

  CIPHER [/E | /D] [/S:dir] [/A] [/I] [/F] [/Q] [/H] [/K] [pathname [...]]

   /E          Encrypts the specified directories. Directories will be marked so that
               files added afterward will be encrypted.
   /D          Decrypts the specified directories. Directories will be marked so that
               files added afterward will not be encrypted.
   /S          Performs the specified operation on directories in the given directory
               and all subdirectories.
   /A          Performs the operation for files and directories. The encrypted file
               could become decrypted when it is modified if the parent directory is
               not encrypted. It is recommended that you encrypt the file and the parent
               directory.
   /I          Continues performing the specified operation even after errors have
               occurred. By default, CIPHER stops when an error is encountered.
   /F          Forces the encryption operation on all specified objects, even those that
               are already encrypted. Already-encrypted objects are skipped by default.
   /Q          Reports only the most essential information.
   /H          Displays files with the hidden or system attributes. These files are
               omitted by default.
   /K          Creates a new file encryption key for the user who is running CIPHER. If
               this option is chosen, all other options will be ignored.
  pathname     Specifies a pattern, file, or directory.

Used without parameters, CIPHER displays the encryption state of the current directory and
any files it contains. You may use multiple directory names and wildcards. You must put
spaces between multiple parameters.
```

FIGURE 14.17 File Recovery certificates

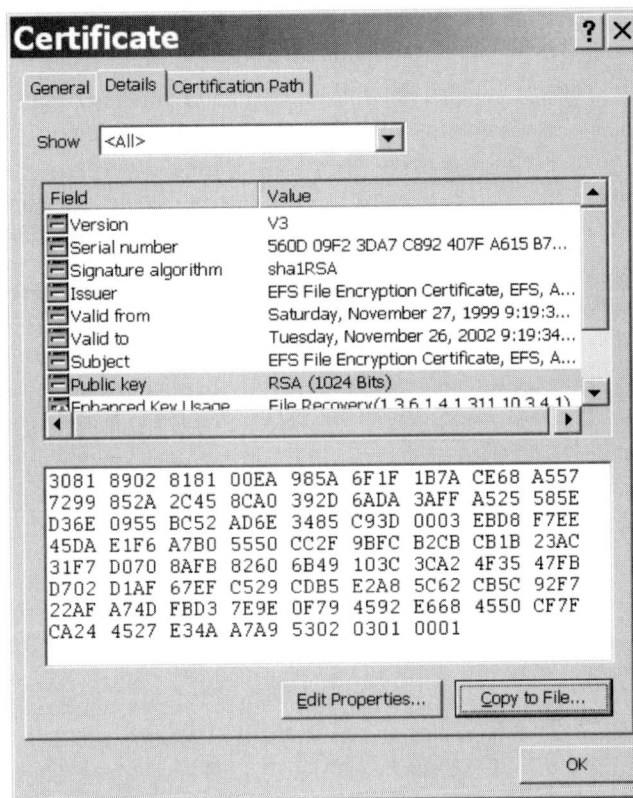

Certificate

General | Details | Certification Path

Show <All>

Field	Value
Version	V3
Serial number	560D 09F2 3DA7 C892 407F A615 B7...
Signature algorithm	sha1RSA
Issuer	EFS File Encryption Certificate, EFS, A...
Valid from	Saturday, November 27, 1999 9:19:3...
Valid to	Tuesday, November 26, 2002 9:19:34...
Subject	EFS File Encryption Certificate, EFS, A...
Public key	RSA (1024 Bits)
Enhanced Key Usage	File Recovery(1.3.6.1.4.1.311.10.3.4.1)

```
3081 8902 8181 00EA 985A 6F1F 1B7A CE68 A557
7299 852A 2C45 8CA0 392D 6ADA 3AFF A525 585E
D36E 0955 BC52 AD6E 3485 C93D 0003 EBD8 F7EE
45DA E1F6 A7B0 5550 CC2F 9BFC B2CB CB1B 23AC
31F7 D070 8AFB 8260 6B49 103C 3CA2 4F35 47FB
D702 D1AF 67EF C529 CDB5 E2A8 5C62 CB5C 92F7
22AF A74D FBD3 7E9E 0F79 4592 E668 4550 CF7F
CA24 4527 E34A A7A9 5302 0301 0001
```

Edit Properties...   Copy to File...

OK

14.8 COMPRESSION

NTFS compression is the ability to selectively compress the contents of individual files, entire directories, or entire drives on an NTFS volume. NTFS compression uses file compression that works by substitution. It starts by locating repetitive patterns and replaces the repetitive data with another pattern, which is shorter. Windows 2000 tracks which files and folders are compressed via a file attribute. As far as the user is concerned, the compressed drive, folder, or file is simply another drive folder or file that works like any other. Although you expand the amount of space for a volume, the performance of the PC will be slower because it has to process the compression and decompression of files. Therefore, do not use compression unless you are compressing files that are rarely used or when disk space is critical. If disk space is critical, use this as a temporary solution until you can delete or move files from the drive or can extend the volume.

To compress a file or folder on a NTFS drive, do the following:

1. Open Windows Explorer.
2. Right-click the file or folder that you want to compress and select the Properties option.
3. Select the Advanced button.
4. Select the "Compress contents to save disk space" checkbox.
5. Click on the OK or Apply button.
6. If you select to compress a folder, select "Apply changes to this folder only" or "Apply changes to the folder, subfolder and files" and click on the OK button.

To compress an NTFS drive, do the following:

1. Open My Computer.
2. Right-click the drive that you want to compress.
3. Select the "Compress drive to save disk space" checkbox.

To uncompress a drive, folder, or file, unselect the option.

To display compressed files and folders in a different color, do the following:

1. Open the Control Panel.
2. Double-click the Folder Options applet.
3. Click the View tab.

4. Select the "Display compressed files and folders with alternate color" checkbox.
5. Click on the OK or Apply button.

When you copy and move files and folders from one location to another, it is important to understand how the NTFS folder and file compressions are affected. If you copy a file or folder, the new folder and file will automatically acquire the compression attribute of the original drive or folder. If the folder or file is moved within the same volume, either will retain the originally assigned compression attribute. Thus, if it was compressed, it will remain compressed at the new location; if it was uncompressed, it will remain uncompressed at the new location. When the folder or file is moved from one volume to another, either will automatically acquire the compression attribute of the original drive or folder and file. An easy way to remember the difference is when you move a folder or file from within the same volume, the folder or file is not physically moved but the Master File Table is adjusted to indicate a different folder. When you move a folder or file from one volume to another, it copies the folder or file to the new location and then deletes the old location. Therefore, the moved folder and files are new to the volume and acquire the new compression attribute.

14.9   QUOTAS

When working as a network administrator, you will find that typical users will think of their network drives as their own personal drives. Therefore, as they receive files from a friend or find something they like on the Internet, they will save the files in the network drive. Unfortunately, after a period of time, a volume will become full.

To manage this type of situation, Windows 2000 has introduced **disk quotas,** which track and control disk space usage for a NTFS volume. When you enable disk quotas, you can set both the disk quota limit and the warning level. The limit specifies the user's allowable amount of disk space. The warning level specifies when it will log a system event when the warning level is met. Disk quotas are based on file ownership, not on where the file is located on the volume. In addition, file compression does not affect the quota statistics.

As soon as disk quota is activated on your volume, it will track all disk usage by the user. When users exceed their disk quota, you can deny them the ability to save any more files to the volume. Enabling quotas, not limited disk space use, is useful when you do not want to deny users access to a volume, but want to track disk space use on a per-user basis. See Figure 14.18.

FIGURE 14.18   Disk quotas

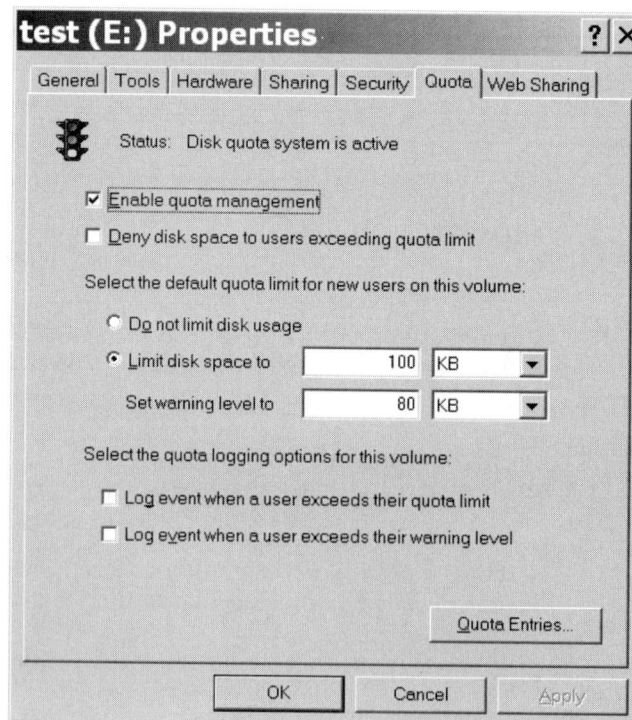

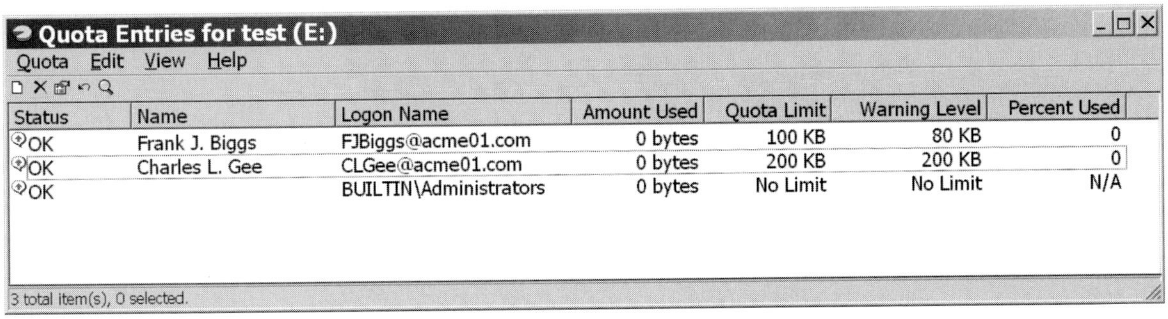

Quota Entries for test (E:)

Status	Name	Logon Name	Amount Used	Quota Limit	Warning Level	Percent Used
OK	Frank J. Biggs	FJBiggs@acme01.com	0 bytes	100 KB	80 KB	0
OK	Charles L. Gee	CLGee@acme01.com	0 bytes	200 KB	200 KB	0
OK		BUILTIN\Administrators	0 bytes	No Limit	No Limit	N/A

3 total item(s), 0 selected.

FIGURE 14.19   Managing disk quotas

To administer quotas on a NTFS volume, do the following:

1. Open My Computer.
2. Right-click the NTFS volume on which you want to enable disk quotas. See Figure 14.18.
3. Select Enable Quota Management checkbox.
4. If you want users to be denied access when they have exceeded their disk quota, select the "Deny disk spaces to users exceeding quota limit" checkbox.
5. If you wish, under the "Select the default quota limit for new users on this volume" option, specify the default limit disk space and the warning level for the new users. Note that the default value is in kilobytes.
6. Select the appropriate logging options.
7. Click on the OK or Apply button.

To administer quotas on a volume, you must be a member of the Administrators group on the computer where the drive resides. If you are not a member of the Administrators group or the volume is not a NTFS volume, the Quota tab will not be displayed. Lastly, new members are those who write to a disk for the first time after the disk quotas have been enabled.

To manage the quota, do the following:

1. Open My Computer.
2. Right-click the NTFS volume on which you want to enable disk quotas. See Figure 14.19.
3. Select the Quota Entries button.

To change a quota limit or warning level, double-click on the user. To create a quota entry, open the Quota menu and select the New Quota entry. To delete an entry, right-click the entry and select the Delete option.

SUMMARY

1. One primary advantage of an NTFS volume is its ability to apply NTFS permissions to secure folders and files.
2. The standard folder and file permissions consist of special permissions that are grouped logically.
3. The NTFS permissions that are granted are stored in an Access Control List (ACL) with every file and folder on an NTFS volume.
4. Explicit permissions are those granted directly to the folder or file.
5. The effective permissions are the actual permissions when logging in and accessing a file or folder.
6. NTFS file permissions override folder permissions.
7. The deny permission always overrides the permissions that have been granted, including when a user or group has been granted full control.
8. Every folder and file has an owner, the person who controls how permissions are set on a folder or file and who can grant permissions to others. When a folder or file is created, the folder or file automatically becomes the owner.
9. If you copy a file or folder, the new folder and file will automatically acquire the permissions of the original drive or folder.

10. If the folder or file is moved within the same volume, the folder or file will retain the same permissions that were already assigned. When the folder or file is moved from one volume to another, the folder or file will automatically acquire the permissions of the original drive or folder.

11. A shared folder on a computer makes the folder available to others on the network. A shared drive on a computer makes the entire drive available to others on the network.

12. To control how users gain access to a shared folder, assign Shared Folder permissions.

13. An administrative share is a shared folder typically used for administrative purposes. It does not show when a user browses the computer resources using Network Neighborhood, My Network Places, or similar software.

14. NTFS permissions apply if you log on locally or if you connect to the computer using a network connection.

15. The distributed file system (DFS) groups and organizes shared folders that reside on the same or different computers into a single hierarchical tree structure.

16. A stand-alone DFS stores the DFS topology on a single computer.

17. Domain DFS provides fault tolerance by having the child nodes point to multiple, identical shared folders.

18. Different than the stand-alone DFS, the domain DFS stores its topology in the Active Directory.

19. Once a user has encrypted a file, it automatically remains encrypted when the file is stored on disk.

20. When EFS is enabled, the process of encrypting and decrypting a file is transparent to the user.

21. You also can encrypt or decrypt a file or folder by using the CIPHER command.

22. NTFS compression is the ability to selectively compress the contents of individual files, entire directories, or entire drives on a NTFS volume.

23. If the folder or file is moved within the same volume, either will retain the same compression attribute that was already assigned.

24. When the folder or file is moved from one volume to another, either will automatically acquire the compression attribute of the original drive or folder and file.

25. Windows 2000 has introduced disk quotas, which track and control disk space usage for a NTFS volume.

QUESTIONS

1. You have created an uncompressed file called REPORT.DOC in the Documents folder on an NTFS partition. The Documents folder is not compressed. Now you move the file REPORT.DOC to a compressed folder called SCHOOLWORK on a different NTFS partition. Which of the following will occur?
   a. The file will stay uncompressed.
   b. The file will inherit the compression attribute from the target folder.
   c. The file will inherit the compression attribute from the target partition.
   d. A dialog box will appear, asking you whether you want to compress the file in the target folder.

2. What will occur if you copy a file from one folder to another folder within the same NTFS partition?
   a. The file will retain its permissions.
   b. The file will inherit the permissions of the source folder.
   c. The file will inherit the permissions of the target folder.
   d. All file permissions will be lost.

3. You want to control the permissions of files and directories on an NTFS drive on the network. Which application must you use?
   a. Windows Explorer
   b. Active Directory Users and Computers console
   c. Computer Management console
   d. Disk Administrator console

4. What will occur to the permissions and the long file name for a file when it is copied from an NTFS partition to a FAT partition on a Windows 2000 Server computer?
   a. Both the permissions and the long file name for the file will be lost.
   b. The permissions for the file will be lost, but the long file name will be retained.
   c. The permissions for the file will be retained, but the long file name will be lost.
   d. Both the permissions and the long file name for the file will be retained.

5. You have created an uncompressed file called MyPaper.doc in the Documents folder on an NTFS partition. The Documents folder is not compressed. Now you move the file MyPaper.doc to a compressed folder called TermPaper on a different NTFS partition. Which of the following will occur?
   a. The file will stay uncompressed.
   b. The file will inherit the compression attribute from the target folder.
   c. The file will inherit the compression attribute from the target partition.
   d. A dialog box will appear, asking you whether you want to compress the file in the target folder.

6. Pat was a member of the Research group. Now he has been reassigned to the Design team. As a result, he needs access to program source files in the \DATA directory on the file server. The Design Administrators group has Full Control permissions to the \DATA directory. The Everyone group has Read permission and the Research group has been denied access to the DATA. After logging on as a member of the Design group, Pat notices that he still cannot access the files in the \DATA directory on the file server. What is the best way for Pat's domain administrator to grant him the required access to the \DATA directory?
   a. Remove Pat's user account from the Research group.
   b. Remove Pat's user account from the Research group and the Everyone group.
   c. Grant all permissions for the \DATA directory to Pat's user account.
   d. Add Pat's user account to the Administrators group.

7. What will occur if you copy a file from one folder to another within the same NTFS partition?
   a. The file will retain its permissions.
   b. The file will inherit the permissions of the source folder.
   c. The file will inherit the permissions of the target folder.
   d. All file permissions will be lost.

8. What will occur if you move a file from one folder to another within the same NTFS partition?
   a. The file will retain its permissions.
   b. The file will inherit the permissions of the source folder.
   c. The file will inherit the permissions of the target folder.
   d. All file permissions will be lost.

9. Which of the following are features of the NTFS file system on Windows 2000? (Choose all that apply)
   a. support for on-disk compression
   b. support for file and folder-level security
   c. support for disk quotas
   d. support for on-disk file encryption

10. You have just modified the permissions on an NTFS folder. By default, which permissions will be changed?
    a. permissions on the folder and files in the folder, but not on subfolders
    b. permissions on the folder only
    c. permissions on the folder, files, and subfolders
    d. permissions on the folder and subfolders, but not on any files

11. You have created a DFS tree on your Windows 2000 servers. Which of the following network clients can utilize this DFS tree by default?
    a. Windows 98
    b. Windows 3.11
    c. Windows NT 4.0
    d. Windows 95

12. There are two domains in your company—Sales and Development—on a mixed domain forest. Five members of the Sales team are now assisting the Development team with verifying beta testing results. For support with this task, these individuals need access to the shared NTFS folder \RESULTS on a server named BETA, residing in the Development domain. What is the recommended approach to grant the five members of the Sales team Full Control permission to the \RESULTS folder?

a. Create a global group in the Sales domain called SALES_TESTERS. Assign each of the five individuals to this global group. Create a domain local group on BETA in the Development domain called BETA_RESULTS. Assign Full Control permission for the \RESULTS folder to this domain local group. Include the SALES_TESTERS global group in the BETA_RESULTS local group.

b. Create a global group in the Sales domain called SALES_TESTERS. Assign each of the five individuals to this global group. Create a global group in the Development domain called BETA_RESULTS. Assign Full Control permission for the \RESULTS folder to the BETA_RESULTS global group. Include the SALES_TESTERS global group in the BETA_RESULTS global group.

c. Remove the existing trust relationship. Establish a new trust relationship where the Sales domain trusts the Development domain. Create a global group in the Sales domain called SALES_TESTERS. Assign each of the five individuals to this global group. Create a local group on BETA in the Development domain called BETA_RESULTS. Assign Full Control permission for the \RESULTS folder to this local group. Include the SALES_TESTERS global group in the BETA_RESULTS local group.

d. Remove the existing trust relationship. Establish a new trust relationship where the Sales domain trusts the Development domain. Create a global group in the Sales domain called SALES_TESTERS. Assign each of the five individuals to this global group. Create a global group in the Development domain called BETA_RESULTS. Assign Full Control permission for the \RESULTS folder to the BETA_RESULTS global group. Include the SALES_TESTERS global group in the BETA_RESULTS global group.

13. Two of your users have connected to their home folders on the same file-and-print server. When the users query the amount of free disk space on the server, they receive different answers. What is the reason for this?

a. One of the users has a compressed home folder.

b. The users have different local hard disk sizes. The local hard disk size determines the disk cluster size, which, in turn, affects the calculation of the free disk space on the server.

c. One user has no access to some of the folders. These folders will not be included in the free disk space calculation.

d. The users have different amounts of unused disk quota on the server.

14. You have several shared folders on your network containing vital company information. Your users are complaining that having to navigate between these servers is confusing and time consuming. You would like to make it easier for your users to navigate around these shared folders. Which feature of Windows 2000 allows this?

a. DNS

b. RIS

c. DFS

d. IntelliMirror

15. Pat is a user in the Corp domain. He needs to access the \Management folder on a file server that is located in another domain called Sales. Both the Corp and Sales domains are part of the same forest. Pat belongs to the Managers global group in the Corp domain. You assign Read permission for the NTFS folder, \Management, on the file server in the Sales domain to the Managers global group. However, Pat is still not able to access the folder. What is the most likely cause of the problem?

a. The trust relationship is not set up correctly.

b. Permissions to resources located in another domain cannot be assigned to global groups.

c. There are overriding share permissions on the folder.

d. NTFS permissions do not apply when a user tries to access the resource remotely.

16. Pat is about to access a Word document from a shared folder in another domain that belongs to the same forest. The folder's share name is FINANCE. The Domain Accountants group of the trusted domain has Change permission to the folder and its Word document files. The folder is stored on an NTFS partition. Pat's user account only has Read permission for the FINANCE share, and the Accountants group has Full Control permission for the FINANCE share. Assuming Pat is not logging on locally to the server that contains the Word document files, what will be his level of access?

    a. No Access   c. Change

    b. Read   d. Full Control

17. A user moves a folder from an NTFS partition on his Windows 2000 Server to a shared NTFS folder on a remote Windows 2000 Server. What will be the permissions of the moved folder?

    a. No Access permission

    b. Full Access permission

    c. same permissions as its original, residing on the user's workstation

    d. same permissions as the folder to which it was copied

18. A Windows 2000 Server contains a shared folder on an NTFS partition. Which one of the following statements concerning access to the folder is correct?

    a. A user who is accessing the folder remotely has the same or more restrictive access permissions than if he accesses the folder locally.

    b. A user who is accessing the folder remotely has less restrictive access permissions than if he accesses the folder locally.

    c. A user who is accessing the folder remotely has the same access permissions as when accessing the folder locally.

    d. A user who is accessing the folder remotely has more restrictive access permissions than if he accesses the folder locally.

19. Pat is a member of the Manager group. There is a shared folder called DATA on an NTFS partition on a remote server. Pat is given the Write NTFS permission, the Manager group is given the Read & Execute NTFS permissions, and the Everyone group has the Read NTFS permission to the DATA folder. In addition, Pat, Manager, and Everyone are assigned the shared Change permission to the DATA folder. When Pat logs on his client computer and accesses the DATA folder, what would be Pat's permissions? (Choose all that apply)

    a. Read the files in the folder.   d. Delete the files in the folder.

    b. Write to the files in the folder.   e. Have no access to the files in the folder.

    c. Execute the files in the folder.

20. Pat is a member of the Manager group. There is a shared folder called DATA on an NTFS partition. Pat is given the Write NTFS permission, the Manager group is given the Read & Execute NTFS permission, and the Everyone group has the Read NTFS permission to the DATA directory. In addition, Pat, Manager, and Everyone are assigned the shared Change permission to the DATA folder. When Pat logs on at the server, what would be Pat's permissions? (Choose all that apply)

    a. Read the files in the folder.   d. Delete the files in the folder.

    b. Write to the files in the folder.   e. Have no access to the files in the folder.

    c. Execute the files in the folder.

21. Pat is a member of the Manager group. There is a shared folder called DATA on an NTFS partition on a remote server. Pat is given the Write NTFS permission, the Manager group is given the Deny All NTFS permissions, and the Everyone group has the Read NTFS permission to the DATA folder. In addition, Pat and Everyone are assigned the shared Change permission to the DATA folder and the Manager group is assigned the Full Control shared permission. When Pat logs on his client computer and accesses the DATA folder, what would be Pat's permissions? (Choose all that apply)

    a. Read the files in the folder.   d. Delete the files in the folder.

    b. Write to the files in the folder.   e. Have no access to the files in the folder.

    c. Execute the files in the folder.

22. You are sharing a directory called DOWNLOAD. You want the share to be hidden from the Network browse list. What should be the name of your share?
    a. DOWNLOAD
    b. %DOWNLOAD
    c. DOWNLOAD$
    d. %DOWNLOAD%

EXERCISES

EXERCISE 1: UNDERSTANDING NTFS RIGHTS

1. Delete the M drive and create a new FAT32 volume. Assign the M drive.
2. On your C drive (which should be NTFS volume), create a folder called DIRN1.
3. In the DIRN1 folder, create a DIRN2 folder.
4. In the DIRN2 folder, create a DIRN3 folder.
5. In the DIRN3 folder, create a text file called FILE1.TXT. To create a text file, open the DIRN3 folder and right click the empty space and select New Text file. In the FILE1.TXT file, type your first name.
6. Right-click the DIRN1 folder and select the Properties option. Click on the Security tab.
7. Click on the Add button. Pick Charlie Brown and click on the Add button. Click on the OK button.
8. Log out as Administrator and log in as Charlie Brown.
9. Try to open the FILE1.TXT file.
10. Try to add your last name to the FILE1.TXT file, save the changes, and exit the program.
11. Try to delete the FILE1.TXT file and try to delete the DIRN2 folder.
12. Try to create a new text file called FILE2.TXT in the DIRN1 folder.
13. Log out as Charlie Brown and log in as the Administrator.
14. Right-click the DIRN1 folder and click the Properties options.
15. From the Security tab, click to highlight Charlie Brown. Give Charlie Brown the Write right.
16. Try to open the FILE1.TXT file.
17. Try to add your last name to the FILE1.TXT file, save the changes, and exit the program.
18. Try to delete the FILE1.TXT file and the DIRN2 folder.
19. Try to create a new text file called FILE2.TXT in the DIRN1 folder.
20. Log out as Charlie Brown and log in as Administrator.
21. For the DIRN1 folder, assign the Modify right also to Charlie Brown.
22. Try to open the FILE1.TXT file.
23. Try to add your last name to the FILE1.TXT file, save the changes, and exit the program.
24. Try to delete the FILE1.TXT file and try to delete the DIRN2 folder.
25. Try to create a new text file called FILE2.TXT in the DIRN1 folder.
26. Log out as Charlie Brown and log in as Administrator.
27. In the DIRN1 folder, re-create the DIRN2 folder. In the DIRN2 folder, re-create the DIRN3 folder. In the DIRN3 folder, re-create the FILE1.TXT file.
28. Right-click the DIRN2 and select the Properties option. From the Security tab, specify Charlie Brown to have the Read permission.
29. Log out as the Administrator and log in as Charlie Brown.
30. Open the FILE1.TXT file and try to add your last name to it. Again try to save it.

EXERCISE 2: SHARE RIGHTS

1. Log in as Administrator.
2. In the M drive, create a folder called DIRS1.
3. In the DIRS1 folder, create a folder called DIRS2.
4. In the DIRS2 folder, create a text file called FILE1.TXT, with your first name listed in the file.
5. Right-click the DIRS1 folder and select the Sharing option.
6. Click the Share This Folder option. Keep the default share name and click on the Permissions button to set the share rights.
7. With the Everyone account highlighted, click on the Remove button.

8. Click on the Add button and add Charlie Brown.
9. For Charlie Brown share rights, assign only the Read permission. Click on the OK button.
10. From your partner's computer, log in as Charlie Brown and access the DIRS1 share through the My Network Places.
11. Close the DIRS1.
12. Click on the Start button, select the Run option, and execute \\SERVER2000-XXx\DIRS1.
13. Try to open the FILE1.TXT file.
14. Try to add your last name to the FILE1.TXT file, save the changes, and exit the program.
15. Try to delete the FILE1.TXT file.
16. Try to create a new text file called FILE2.TXT in the DIR1 folder.
17. As the Administrator back at your own computer, change the share rights to include Read and Change permissions for Charlie Brown.
18. From your partner's computer, try to open the FILE1.TXT file.
19. Try to add your last name to the FILE1.TXT file, save the changes, and exit the program.
20. Try to delete the FILE1.TXT file.
21. Try to create a new text file called FILE2.TXT in the DIR1 folder.
22. As the Administrator back at your own computer, change the rights to "Deny all rights for Charlie Brown."
23. From your partner's computer, try to access the shared folder.
24. Back at your own computer, log out as the Administrator and log in as Charlie Brown.
25. Try to delete the FILE2.TXT file. Try to figure out why Charlie Brown was able to delete the folder.

EXERCISE 3: GROUPS AND PERMISSIONS

1. Log in as the Administrator. Make sure that Charlie Brown is a member of the Sales group.
2. In the C drive, create a folder called GROUP1.
3. In the GROUP1 folder, create a GROUP2 folder.
4. In the GROUP2 folder, create a new text file with your name on it.
5. Assign the NTFS Read & Execute and List Folder Contents to the Sales group for the GROUP1 folder.
6. Assign the NTFS Modify to Charlie Brown for the GROUP1 folder.
7. Assign the NTFS List Folder Contents to Everyone for the GROUP1 folder.
8. Right-click the text file and select the Properties option. Click on the Permissions tab. Note the rights assigned to Charlie Brown. Of those rights, note those that are grayed out.
9. Assign Full Control to Charlie Brown for the GROUP1 folder.
10. Assign Full Control to Sales for the GROUP1 folder.
11. Deny all rights to Everyone for the GROUP1 folder.
12. Share the GROUP1 folder and assign Full Control share permissions to Charlie Brown, Sales group, and Everyone.
13. From your partner's computer, log in as Charlie Brown and try to access the text file.
14. Back on your own computer, log out as Administrator and log in as Charlie Brown. Try to access the text file in the GROUP2 folder.
15. Log out as Charlie Brown and log in as Administrator.
16. Remove Everyone from the NTFS permissions for the GROUP1 folder.
17. Log out as Administrator and log in as Charlie Brown. Try to access the text file.
18. Back on your partner's computer, log in as Charlie Brown, try to access the text file.

EXERCISE 4: TAKING OWNERSHIP OF A DIRECTORY

1. Log in as Charlie Brown.
2. Create a folder called OWNER in the C drive.
3. Inside the OWNER folder, create a text file.
4. Right-click the OWNER folder and assign Charlie Brown all NTFS permissions.

5. Note that the checkmarks for the NTFS rights for the Everyone group are grayed out. Click on the Everyone group and click on the Remove button.
6. The error appeared because the rights are inherited from above the folder. Click on the OK button.
7. Disable the "Allow inheritable permissions from parent to propagate to this object" option. Select the Remove button so that the inherited rights cannot flow into the OWNER folder.
8. Click on the OK button.
9. Log out as Charlie Brown and log in as Administrator.
10. Try to access the text file.
11. Right-click the OWNER folder and select the Properties option.
12. Click the Security tab. Read the message and click on the OK button.
13. Click on the Advanced button.
14. Click on the Owner tab. Click on the Administrators and enable the "Replace owner on subcontainers and objects." Click on the OK button. Click Yes to replace all permissions.
15. Click on the OK button.
16. Right-click the OWNER folder and select the Properties option. Click on the Security tab to view the NTFS permissions.
17. Click the OK button to close the OWNER Properties dialog box.
18. Open the OWNER folder.
19. Try to access the text document.
20. Right-click the text document and select the Properties option. Try to figure out why the administrator cannot access the text document.
21. In the Security tab, assign all NTFS permissions to the Administrators group. Click the OK button.
22. Try to access the text document.

EXERCISE 5: DISTRIBUTED FILE SYSTEM

Creating a Stand-alone DFS Root

1. In Administrative Tools, start the distributed file system (DFS) console.
2. On the Action menu, click the New DFS Root and then the Next button.
3. Select "Create a stand-alone DFS root" and click on the Next button.
4. Type in the name of your server for the DNS root and click on the Next button.
5. Select the GROUP1 share and click on the Next button.
6. Specify a unique DFS share name (if you can) and click on the Next button.
7. Click on the Finish button.
8. Share the DIRN1a directory that you created in an earlier exercise.
9. In the DFS console, right-click the \\SERVER2000-XXX\GROUP1 share and select New DFS Link.
10. In the Link Name text box, type in DIRN1a. Use the Browse button to find your server and specify your DIRN1 share. Click on the OK button.
11. In the DFS console, right-click the \\SERVER2000-XXX\GROUP1 share and select New DFS Link.
12. In the Link Name text box, type in DIRS1a. Use the Browse button to find your server and specify your DIRS1 share. Click on the OK button.
13. In the DFS console, right-click the \\SERVER2000-XXX\GROUP1 share and select New DFS Link.
14. In the Link Name text box, type in DIRN1b. Use the Browse button to find your partner's server and specify his or her DIRN1 share. Click on the OK button.
15. In the DFS console, right-click the \\SERVER2000-XXX\GROUP1 share and select New DFS Link.
16. In the Link Name text box, type in DIRS1b. Use the Browse button to find your partner's server and specify his or her DIRS1 share. Click on the OK button.
17. From either computer, use the Run option under the Start button and execute the \\SERVER2000-XXX\GROUP1.
18. Note that you can browse the shares without being concerned where the shares are located.

Creating a Domain DFS

19. Right-click the DFS Group1 share and delete the DFS root. Click on the Yes button.
20. On the Action menu, click the New DFS Root and click on the Next button.
21. Select "Create a domain DFS root" and click on the Next button.
22. Select your domain for the DFS root and click on the Next button.
23. Type in the name of your server for the DNS root and click on the Next button. Here you can use the Browse button.
24. Select the Create a New Share option. Specify C:\DFSROOT in the Path to Share text box and DFSROOT in the Share Name text box. Click on the Next button. Click on the Yes box to confirm creating the shared directory.
25. Keep the default DFS root name and click on the Next button.
26. Click on the Finish button.
27. Right-click the DFSROOT share in the DFS console and select the Add New Replica option.
28. In the DFS console, right-click the DFSROOT share and select the New Replica option.
29. In the Add a New Replica dialog box, in the Send the User to This Shared Folder box, type \\SERVER2000-XXx\DFSROOT.
30. Select the Automatic Replication option and click the OK button.
31. In the DFS console tree, right-click DFSROOT and select the "Check Status to ensure both replicas" option.
32. Create a text file in one of the shares. Check to see if the text file gets replicated to the other shared directory.

EXERCISE 6: COMPRESSED FOLDERS

1. Your C and D drives must be NTFS volumes to perform this exercise. If they are not NTFS partitions, use the CONVERT command to convert them to NTFS volumes.
2. In the D drive, create two folders called Uncompressed and Compressed.
3. Open the Compressed folder.
4. Open the Tools menu and select the Folder Options option. Make sure that the "Display compressed files and folders option with alternate color" is enabled. Click on the OK button.
5. In the Compressed folder, create a text file. Note the color of the text underneath the text file icon.
6. Close the Compressed folder.
7. Right-click the Compressed folder and select the Properties option.
8. Click the Advanced button. Select the "Compress contents to save disk space" option and click on the OK button.
9. Open the Compressed folder and note the color of the text underneath the text file icon.
10. Right-click the text file and select the Copy option.
11. Close the Compressed folder.
12. Open the Uncompressed folder.
13. Open the Tools menu and select the Folder Options option. Make sure that the "Display compressed files and folders option with alternate color" is enabled. Click on the OK button.
14. Right-click the background of the Uncompressed folder and select the Paste option. Note whether the file is compressed.
15. Delete the file in the Uncompressed folder.
16. Open the Compressed folder. Right-click the text file and select the Cut option.
17. In the Uncompressed folder, paste the text file. Note whether the file is compressed.
18. In the D drive, create a folder called Uncompress.
19. Open the Tools menu and select the Folder Options option. Make sure that the "Display compressed files and folders option with alternate color" is enabled. Click on the OK button.
20. Paste the text file into the Uncompress folder. Note whether it is compressed.

EXERCISE 7: FILE ENCRYPTION

1. Log in as Charlie Brown.
2. In the C drive, create a folder called ENCRYPTED.
3. In the ENCRYPTED folder, create a text file with your first and last name in it.
4. Right-click the ENCRYPTED folder in the C drive and select the Properties option.
5. Click on the Advanced button. Select the "Encrypt contents to secure data" option and click on the OK button.
6. Click on the OK button to close the Uncompressed properties dialog box. Select the "Apply Changes to this folder, subfolders and files." Click on the OK button.
7. Log out as Charlie Brown and log in as Administrator.
8. Try to access the text file in the ENCRYPTED folder.

EXERCISE 8: QUOTAS

1. Open My Computer. Right-click the C drive and select the Properties option. Note the amount of free disk space.
2. Click the Quota tab.
3. Enable the Enable Quota Management option.
4. Set the default quota limit to 100 MB and the warning to 90 MB. Click the Apply button. Click on the OK button.
5. Click the Quota Entries button.
6. Open the Quota menu and select the New Quota option.
7. Right-click the Administrator user and select the Properties option. Select the "Do not limit disk usage" option. Click on the OK button.
8. Do the same for the Administrators group and Domain Admins group.
9. Close the Quota Entries window.
10. Click the OK button to close the C Drive Properties dialog box.
11. Log out as Administrator and log in as Charlie Brown.
12. Right-click the C drive and select the Properties option. Note the amount of free disk space.

15

Printing in Windows 2000

INTRODUCTION

One basic network service that is available in Windows 2000 is network printing. Yet, for the users and sometimes the administrators, printing can easily become frustrating, especially if it is poorly planned, executed, and managed. This chapter discusses the basics of using Windows 2000 as a print server and allowing the various clients to send print jobs to printers that are connected to a Windows 2000 print server.

OBJECTIVES

1. Manage printers and print jobs.
2. Control access to printers by using permissions.
3. Connect to an Internet printer.
4. Connect to a local print device.
5. Connect to network print devices.

15.1 INTRODUCTION TO NETWORK PRINTING

When printing with a network printer, the printer communicates with a print server. The printer can be connected directly to the print server or to the network using a network interface. In Windows NT and Windows 2000, the printer is known as the **print device.**

When a client prints to the network printer, the print job is sent to the print server. If the print job comes from a Windows application, then the application calls the graphical device interface (GDI), which calls the printer driver that is associated with the target printer. Using the driver, the GDI transforms the data being sent to the printer into a language that is understood by the printer and passes it to the client print spooler. A **print spooler** is software that accepts a document, which is sent to a printer by the user, and then stores it on disk or in memory until the printer is ready for it. The word *spooler* is an acronym for "simultaneous print operations on line." For Windows 95 and 98 machines, the client spooler is the Print Manager. If the client is using a non-Windows operating system, a non-Windows application replaces the GDI to perform a similar task.

The client computer sends the print job to the print server. A **print server** can be any computer on a network that manages all printers on that network. For Windows NT 4.0 or Windows 2000 clients, the client spooler makes a remote procedure call (RPC) to the server spooler to send the data to the print server, and is stored on disk. The disk location that stores the print jobs is sometimes referred to as the print spooler or print queue.

On the print server, print jobs from Windows NT or Windows 2000 clients are of the **enhanced metafiles (EMF)** data type. Most non-Windows applications use the RAM (ready to print) data format.

The print server monitors the printers and when the printer is ready for another print job, the print server sends it to the printer. The printer receives the print job, converts each page into bitmap format, and then prints it. During the entire process, the status of the print job can be reported back to the client so that the user can monitor it.

While the print device is the actual hardware device that produces the printed document, the software interface between the operating system and the print device is the **printer.** The user prints to the printer (software interface or logical printer), which then sends the print job to the print server, which eventually sends the print job to the print device. The printer can be connected to a single print device or multiple print devices (known as a printer pool).

To set up printing on a Windows 2000 network, you must have at least one computer to operate the print server (Windows 2000 workstation or server). If you have more than 10 concurrent connections, use the Windows 2000 Server. In addition, if you have a large number of printers or users, then dedicate the server to printing. The Windows 2000 computer would need extra RAM beyond the minimum amount required for Windows 2000 and sufficient disk space to store all print jobs until they can be printed.

15.2 SHARING A LOCAL PRINTER

To share a local printer so that it can be used on the network, first connect the printer to the computer with a serial port, parallel port, or USB. Windows 2000 can sometimes automatically detect the printer and load the appropriate drivers. The steps to sharing a local printer are as follows:

1. Log on as the Administrator.
2. Click on the Start button, select the Settings option, and select the Printers option.
3. Double-click the Add Printer to start the Add Printer wizard.
4. Click on the Next button.
5. Select the Local Printer option and click on the Next button.
6. If you had the "Automatically detect and install my Plug and Play printer" option from the previous window, it will automatically detect the printer. If not, click on the Next button.
7. Specify the port that you want the printer to use and click on the Next button. See Figure 15.1.
8. Select the printer manufacturer and printer and click the Next button. If the printer is not listed, use either the Windows Update button or the Have Disk button. See Figure 15.2.
9. You would then input the name of the printer—a meaningful name that would easily identify the printer.

FIGURE 15.1   Selecting the printer port using the Add Printer wizard

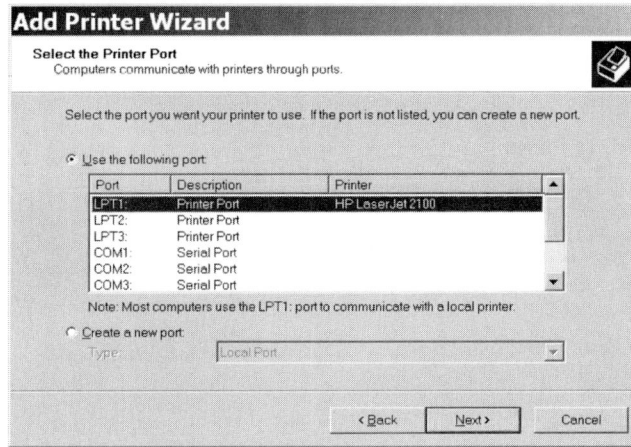

FIGURE 15.2   Selecting the printer using the Add Printer wizard

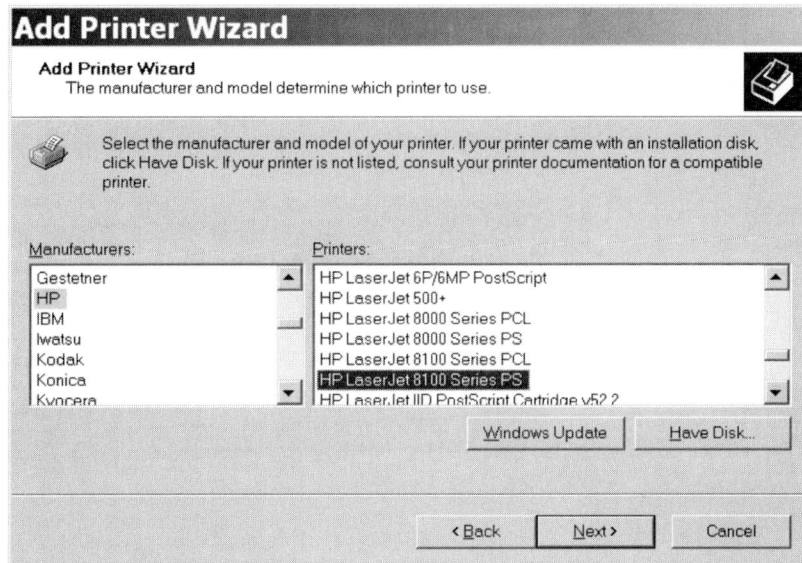

10. If you want your Windows-based program to use this printer as the default printer, select Yes. In either case, click on the Next button.

11. If you want to share the printer so that it can be available to other users, select the Share As option and provide a meaningful name for the printer. Click on the Next button.

12. Fill in the Location (optional) and Comments (optional) for the printer and click on the Next button.

13. If you want to print a test page, select the Yes option. In either case, click on the Next button.

14. Click on the Finish button.

If you have a printer already installed that is not shared, right-click the printer and select the Sharing option. Then select the Shared As option and give it a meaningful share name. If you want the name to be listed in the Active Directory, click the List in the Directory option.

15.3   ADDING AND SHARING A PRINTER FOR A NETWORK-INTERFACE PRINT DEVICE

In larger companies, most print devices will not be connected locally to the print server, especially if the print server provides other network services, which is typically locked in a secure area. Instead, they would be connected using a network interface such as built-in or add-in network port. Another reason to connect a printer directly to the network is that the parallel port or serial port is often the bottleneck for printing.

FIGURE 15.3 Creating a network printer using the printer wizard

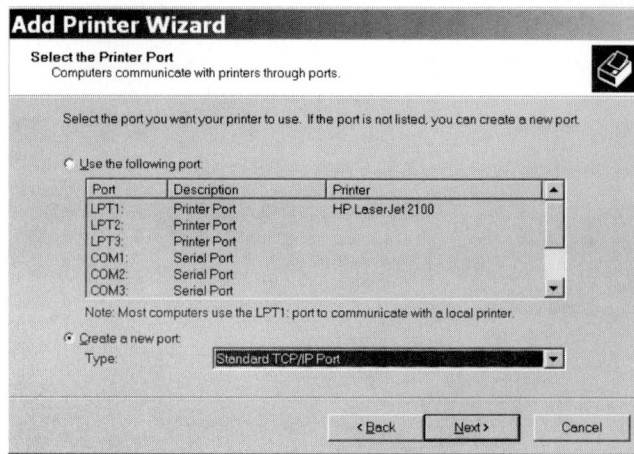

Add Printer Wizard

Select the Printer Port
Computers communicate with printers through ports.

Select the port you want your printer to use. If the port is not listed, you can create a new port.

○ Use the following port:

Port	Description	Printer
LPT1:	Printer Port	HP LaserJet 2100
LPT2:	Printer Port	
LPT3:	Printer Port	
COM1:	Serial Port	
COM2:	Serial Port	
COM3:	Serial Port	

Note: Most computers use the LPT1: port to communicate with a local printer.

◉ Create a new port:
Type: Standard TCP/IP Port

< Back   Next >   Cancel

FIGURE 15.4 Specifying the IP address of network printer

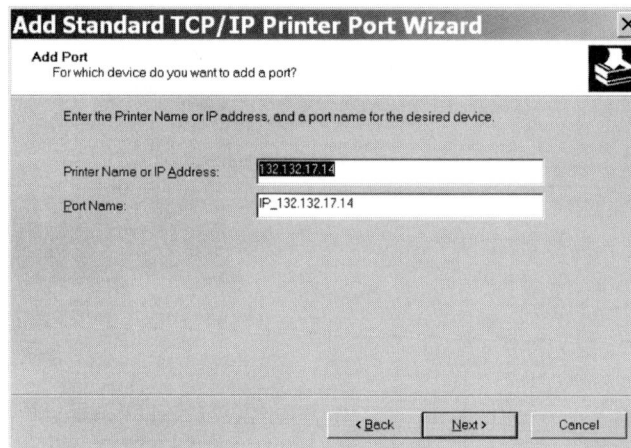

Add Standard TCP/IP Printer Port Wizard

Add Port
For which device do you want to add a port?

Enter the Printer Name or IP address, and a port name for the desired device.

Printer Name or IP Address: 132.132.17.14

Port Name: IP_132.132.17.14

< Back   Next >   Cancel

To connect to a network interface print device, do the following:

1. Start the Add Printer wizard and click on the Next button.
2. Select Local Printer and click on the Next button.
3. Select Create a New Port and select the Standard TCP/IP port. When you click on the Next button, it will start the Add Standard TCP/IP Printer Port wizard. See Figure 15.3.
4. In the TCP/IP Printer Port wizard, click on the Next button and enter the printer name or IP address and port name. Click on the Next button. See Figure 15.4.
5. From the device type, select the appropriate network interface used by the printer and click on the Next button.
6. Click on the Finish button to close the Add Standard TCP/IP Printer Port wizard.
7. Finish the Add Printer wizard.

15.4 PRINTER PRIORITY AND PRINT SCHEDULING

Documents can be sent to a printer with different priorities. Users with a higher priority can bypass other print jobs and thus be printed first. Priorities are assigned to the logical printers, which are associated with the same print device. Windows 2000 routes documents with the highest priority level to the printer first.

To set the print priority, do the following:

1. Open the Printer folder.
2. Install the printer and the first print driver.
3. Right-click the printer on which you want to set the priority (1–99) and select the Properties option.
4. Click the Advanced tab.

FIGURE 15.5 Setting printer priorities

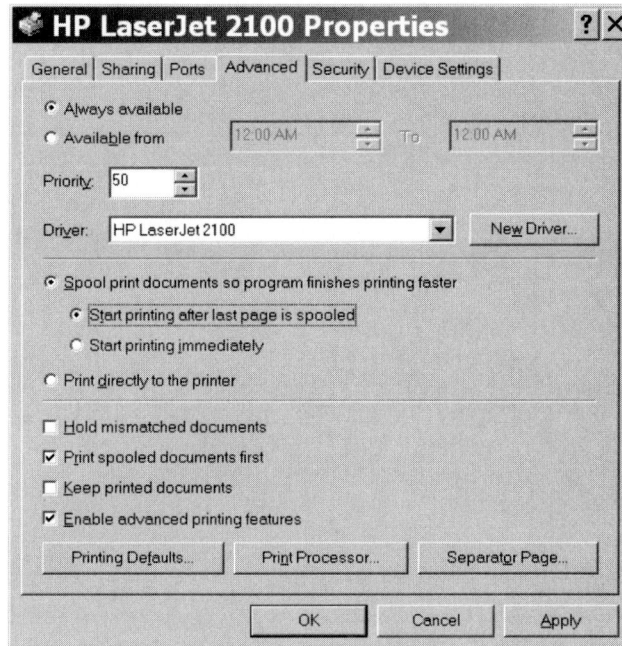

5. In the Priority section, set the priority.
6. Click on the OK button.
7. Install the second print driver that was pointing to the same printer.
8. Right-click the printer on which you want to set the priority and select the Properties option.
9. Click the Advanced tab.
10. In the Priority section, set a higher priority. See Figure 15.5. When loading the driver at the client stations, the normal users are assigned to the first logical printer name and the people with the higher priority use the second logical printer name.
11. Click on the OK button.

Another use for creating two logical printers that point to the same printer is to schedule print jobs. For example, say that when you print letters and small documents, you print to one logical printer that will print immediately to the print device. When you have a large report that will take some time to print, send the data to a second logical printer that is scheduled to print at night.

To schedule printing times, do the following:

1. Open the Printer folder.
2. Right-click the printer you want to set and click the Properties option.
3. Click the Advanced tab and click the Available From option.
4. Set the time period that the printer will be available.
5. Click on the OK button.

After you have two logical printers set up, you can load the drivers at the client computers that point to both logical printers. With large print jobs, then, you print the logical printer that is set to print at night. Be sure to check for sufficient disk space to hold the print jobs so that they can be printed later.

15.5 SETTING CLIENT COMPUTERS TO PRINT TO A NETWORK PRINTER

When a network printer becomes available, set up the clients so that they can access it. For Windows 95, Windows 98, Windows NT 4.0, and Windows 2000, use the computer's Add Printer wizard. Instead of selecting Local Printer, select Network Printer. If directory service is enabled, search for the printer, right-click the printer, and select Connect. In other cases, you will need to provide a UNC name (\\printservername\printersharename).

Additional Drivers   **?** ☒

You may install additional drivers so that users on the following
systems can download them automatically when they connect.

Environment	Version	Installed	
☐ Alpha	Windows 2000	No	
☐ Alpha	Windows NT 3.1	No	
☐ Alpha	Windows NT 3.5 or 3.51	No	
☐ Alpha	Windows NT 4.0 or 2000	No	
☑ Intel	Windows 2000	Yes	
☑ Intel	Windows 95 or 98	No	
☐ Intel	Windows NT 3.1	No	
☐ Intel	Windows NT 3.5 or 3.51	No	
☑ Intel	Windows NT 4.0 or 2000	No	
☐ MIPS	Windows NT 3.1	No	
☐ MIPS	Windows NT 3.5 or 3.51	No	
☐ MIPS	Windows NT 4.0	No	
☐ PowerPC	Windows NT 3.51	No	
☐ PowerPC	Windows NT 4.0	No	

OK   Cancel

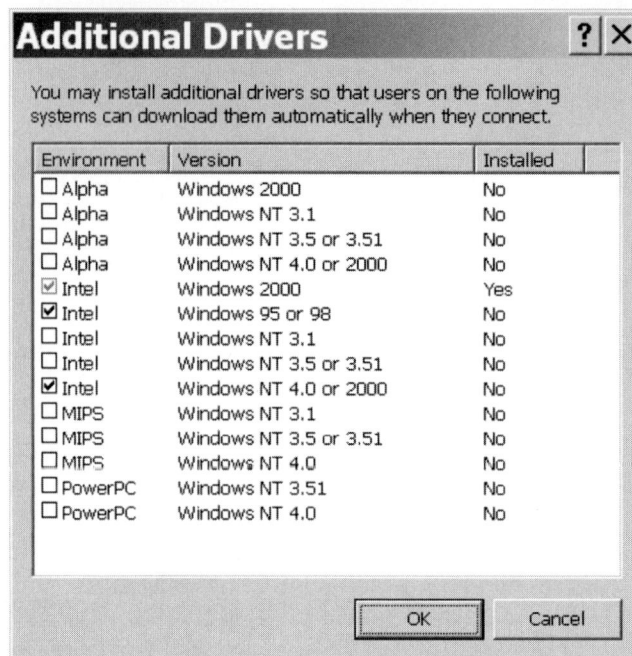

Windows 2000 has the capability to store drivers for the various operating systems. To add these to the Windows 2000 print server, right-click the printer in the Printer folder and select the Sharing option. Then click the Additional Drivers button and select the operating systems that you will be using. See Figure 15.6. You may need the operating system installation CD or disks for the driver for the specific operating system that you are trying to load.

If you have loaded the various operating system drivers on the print server, Windows 2000 and Windows NT 4.0 will automatically check the print drivers and configurations each time they connect to the print server. Therefore, if the print driver is newer, it will automatically update the client's driver. The print driver for Windows 95 and Windows 98 is not automatically kept current, but is loaded when you use the Add Printer wizard to load the driver. If you update the Windows 95 and Windows 98 drivers on the server, you will have to manually install the drivers on the Windows 95 and Windows 98 clients. Of course, it will be easier if they can install the driver from the print server instead of using floppy disks or CD-ROMs.

When using DOS or Windows for Workgroups, use the NET command to map a printer to a port. In so doing, you are setting up a redirector that takes any print jobs that are sent to the port and redirects them to the network printer. In addition, the Microsoft Network redirector can send jobs using NetBEUI, NWLink, or TCP/IP protocols.

15.6   PRINTER POOL

A **printer pool** is two or more print devices that use the same driver, are connected to one print server, and act as a single printer. When you print a document to the printer pool, the print job is sent to the first available printer in the pool. The user does not have to determine which printer is free. This is useful in a network with a high volume of printing, because it decreases the time that users wait for their documents. If a print device jams when in the middle of a print job, the job remains in the printer until the jam is fixed. It is not rerouted to the other printers.

The print server checks for an available port and sends documents to ports in the order that they are added. Adding a port to the fastest printer ensures that documents are first sent here before they are routed to slower printers in the printing pool. Lastly, since the user does not know which printer in the pool prints a given document, the print devices should be located in the same location.

FIGURE 15.7   Setting up a printer pool

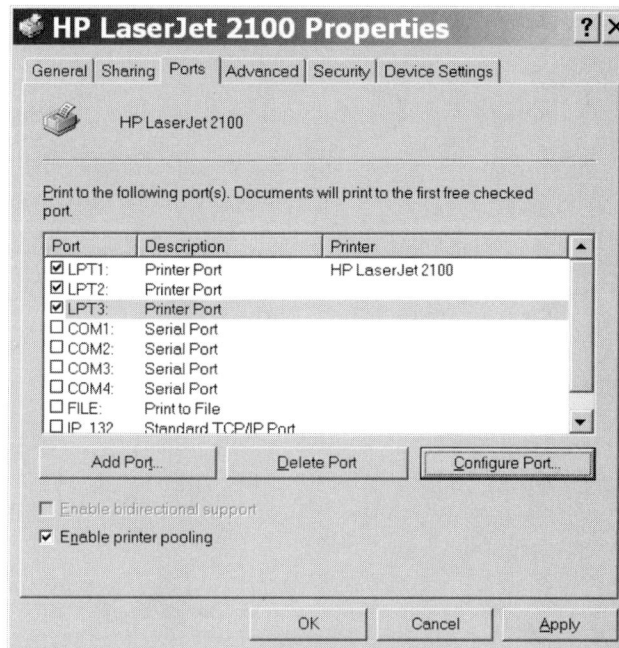

HP LaserJet 2100 Properties

General | Sharing | Ports | Advanced | Security | Device Settings |

HP LaserJet 2100

Print to the following port(s). Documents will print to the first free checked port.

Port	Description	Printer
☑ LPT1:	Printer Port	HP LaserJet 2100
☑ LPT2:	Printer Port	
☑ LPT3:	Printer Port	
☐ COM1:	Serial Port	
☐ COM2:	Serial Port	
☐ COM3:	Serial Port	
☐ COM4:	Serial Port	
☐ FILE:	Print to File	
☐ IP_132	Standard TCP/IP Port	

Add Port...   Delete Port   Configure Port...

☐ Enable bidirectional support
☑ Enable printer pooling

OK   Cancel   Apply

FIGURE 15.8   Displaying the shared printers using the HTML interface

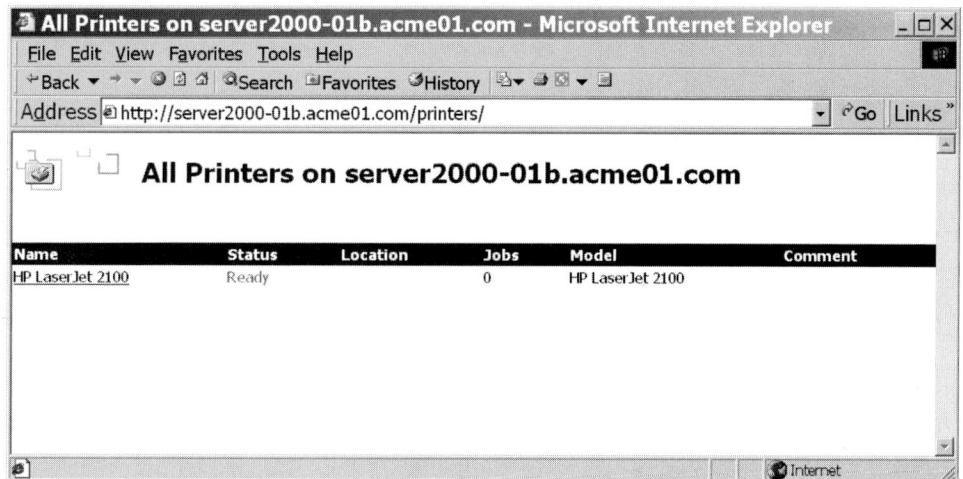

All Printers on server2000-01b.acme01.com - Microsoft Internet Explorer

File   Edit   View   Favorites   Tools   Help

Back ▼ → ▼ ⊘ ⚙ ⚙ ⚙Search ⚙Favorites ⚙History ⚙▼ ⊕ ⊠ ▼ ⊒

Address http://server2000-01b.acme01.com/printers/   Go Links

All Printers on server2000-01b.acme01.com

Name	Status	Location	Jobs	Model	Comment
HP LaserJet 2100	Ready		0	HP LaserJet 2100	

Internet

To set up printing to multiple print devices, perform the following:

1. Open the Printer folder.
2. Install a printer and print driver to be used on the network.
3. Right-click the printer and select the Properties option.
4. Click on the Ports tab and select the Enable Printer Pooling checkbox. See Figure 15.7.
5. Click each port where the printers you want to pool are connected.

15.7   PRINTING VIA WEB BROWSERS AND THE INTERNET

A new feature of Windows 2000, Internet printing, allows you to submit print jobs to a printer across the Internet by using the printer's URL as the name of the printer—such as when you use the Add Printer wizard. See Figures 15.8 and 15.9. For the Windows 2000 Server to process print jobs that contain URLs, it must be running Microsoft Internet Information Services (IIS). For print servers that are implemented on Windows 2000 Professional, use Microsoft Peer Web Service (PWS).

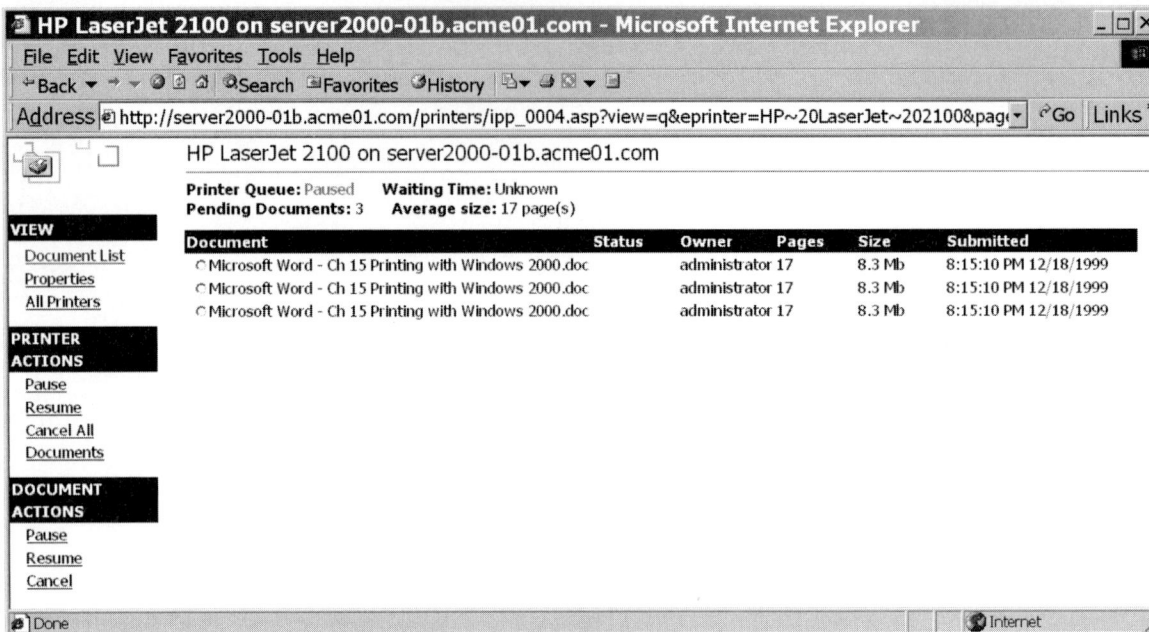

FIGURE 15.9   Managing print jobs using the HTML interface

To connect to a printer with a browser, do the following:

1. Start Internet Explorer 4.0 or later.
2. In the Address bar, type http://*PrinterServerName*/printers to receive a page listing of all printers and click the printer you want; or, if you know the printer name, type its URL using http://*PrintServerName*/*PrinterName*.
3. After you have connected to the printer, it will automatically copy the correct print driver to the client computer.

A web page designer can customize this page by displaying a floor plan which shows the location of the print devices.

15.8   PRINTER ADMINISTRATION

One high-problem area on a network is the printer, which can lead to headaches and frustration for the user and the administrator. After the printer is set up, you still need to manage the printer permissions, the printers, and the documents and to troubleshoot any print problems that arise.

15.8.1   Printer Permissions

Printer permissions control who can use the printer and which print tasks a user can perform. For example, if you have an expensive color printer, you might want only a handful of people to have access to it. For all printers, you will need someone to delete print jobs, change the printer properties, or change the printer permissions. Therefore, after you have added and shared a printer, assign the appropriate rights to the users.

There are three levels of printer permissions: Print, Manage Printer, and Manage Documents. By default, all users have the Print permission as members of the Everyone group. Administrators, print operators, and server operators on a domain controller have Manage Printer permission. Lastly, the owner of the document has Manage Documents permission for that document only. See Table 15.1.

To set or remove permissions for a printer, do the following:

1. Open the Printer folder.

TABLE 15.1   Print permissions

Print Permission Capabilities	Print	Manage Printer	Manage Documents
Print documents	X	X	X
Pause, resume, restart, and cancel the user's own document	X	X	X
Connect to a printer	X	X	X
Control job settings for all documents		X	X
Pause, restart, and delete all documents		X	X
Share a printer		X	
Change printer properties		X	
Delete printers		X	
Change printer permissions		X	

FIGURE 15.10   Printer permissions

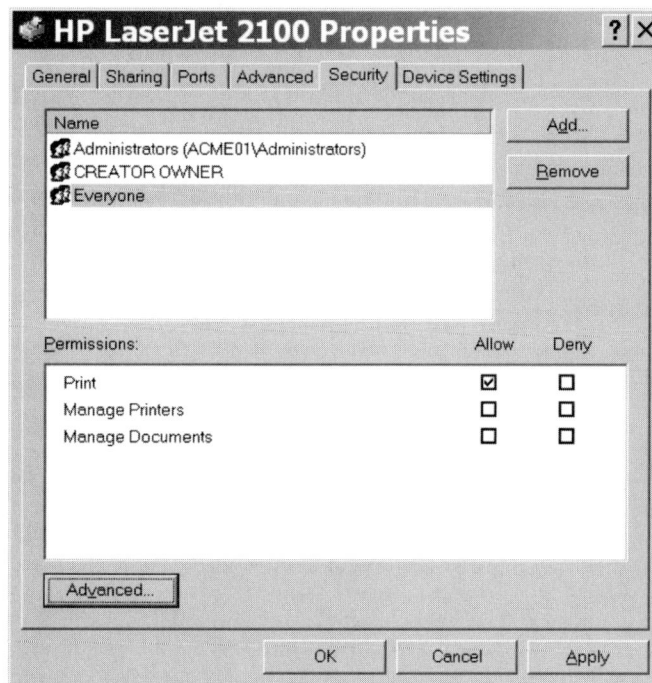

2. Right-click the shared printer for which you want to set permissions and click the Properties option.
3. Select the Security tab.
4. To change permissions from an existing user or group, click the name of the user or group and select and deselect the appropriate rights. To add a user or group, click on the Add button, select the users or group, and click the OK button. To remove an existing user or group, click its name and click the Remove button.
5. In Permissions, click Allow or Deny for each permission that you want to either allow or deny. See Figure 15.10.

15.8.2   Printer Device Settings

The processes of managing printers include assigning forms to paper trays, page orientation (portrait or landscape), selecting which tray to print from, configuring the amount of printer memory, and enabling printer duplexing (printing on both sides). The specific type of printer also determines some of the available options.

FIGURE 15.11 Device Settings
tab found in the Printer
Properties

HP LaserJet 2100 Properties

General | Sharing | Ports | Advanced | Security | Device Settings

HP LaserJet 2100 Device Settings
- Form To Tray Assignment
  - First Available Tray: Letter
  - Tray 1: Letter
  - Manual Paper Feed: Letter
  - Tray 2: Letter
  - Tray 3: Not Available
- Font Substitution Table
- Installed Font Cartridge(s)
  - Slot #1: Not Available
- Page Protection: Off
- External Fonts...
- Installable Options
  - Tray 3: Not Installed
  - Printer Memory: 20MB

OK   Cancel   Apply

Many laser printers have multiple trays that hold different types of forms. For example, one tray can be assigned to hold letterhead paper (typically used for the front page of a letter); the second tray, normal white paper; and a third tray, envelopes. Users can then select the form from within the program that is open. When users print a document, the print job is routed to the correct paper tray. Note that the default form setting for a paper tray is letter size ($8\frac{1}{2}'' \times 11''$ paper).

To match a form to a printer tray, do the following:

1. Open the Printer folder.
2. Right-click the printer you are using and click the Properties option. See Figure 15.11.
3. Click on the Device Settings tab.
4. Under the Form To Tray Assignment, click a printer tray.
5. Click one of the available paper sizes for the selected tray.

Laser printers are known as page printers, because they store an entire page in the memory of the printer before printing. Depending on the resolution of the printer, the size of the document, and the complexity of the printer, the printer may require extra memory. After the memory is added to the printer, tell the driver how much memory the printer has. To set the printer memory, do the following:

1. Open the Printer folder.
2. Right-click the printer that you want to configure and select the Properties option.
3. Click on the Device Settings tab.
4. Under the Installable Options, click Printer Memory and specify the amount of memory installed in your printer.

If you specify more or less memory than actually exists in your printer, then you may receive a memory error message or partial printed pages. To check the amount of memory, use the self-test feature of the printer.

15.8.3   Managing the Printer Spooler

Print jobs stored in the spooler are written to a spool (SPL) file, whereas administrative information such as user name, document name, and data type are stored in a shadow (SHD) file. This way, if a power failure or other disaster occurs before all jobs in the queue are printed, the print jobs are protected.

By default, the spooler files are written to the *%SystemRoot%*\System32\Spool\Printers folder. If the hard disk partition containing Windows 2000 does not have enough space for these files, you can change the location of the folder. To change the location of the default spool folder for all printers on a server, do the following:

1. In the Printers dialog box, on the File menu, click Server Properties.
2. In the Print Server Properties dialog box, click the Advanced tab.
3. Type the path and name for the new default spool folder, and then click OK.
4. The change is effective immediately. You do not need to restart your system.

Do not attempt to spool to a root (such as C:\). This causes the files to revert to the old default.
   To change the location of the spool folder for a specific printer, use the registry, as follows:

1. Create a new spool folder.
2. Start a registry editor (Regedt32.exe or Regedit.exe).
3. Add the following entry to the registry:
   Entry name: SpoolDirectory
   Path: HKEY_LOCAL_MACHINE\SOFTWARE\Microsoft\Windows NT\
   CurrentVersion\Print\Printers\<Printer-name>
   Data type: REG_SZ
   Value: <path to the new spool folder>
4. Restart the computer to make the change effective.

15.8.4   Printer Preferences and Defaults

To configure the page orientation, page order, pages per sheet, printing on both sides, and the correct tray, use Printing Preferences or Printing Defaults. Default settings apply to all users, but are overwritten by Printing Preferences. To set up personal printing preferences, do the following:

1. Open the Printers folder.
2. Right-click the printer you are using and click Printing Preferences.
3. To select the page orientation (portrait versus landscape), page order, pages per sheet, and printing on both sides (if the option is available to the printer), click on the Layout tab. To select the paper source, click on the Paper/Quality tab.
4. Click on the OK button to close the Printing Preferences dialog box.
5. Click on the OK button to close the Printing Properties dialog box.

   To set the printer defaults, do the following:

1. Open the Printers folder.
2. Right-click the printer on which you want to set the default printer settings and click the Properties option. See Figure 15.12.
3. Click on the Advanced tab.
4. Click the Printing Defaults button.
5. To select the page orientation (portrait versus landscape), page order, pages per sheet, and printing on both sides (if the option is available to the printer), click on the Layout tab. To select the paper source, click on the Paper/Quality tab.
6. Click on the OK button to close the Printing Defaults dialog box.
7. Click on the OK button to close the Printing Properties dialog box.

15.8.5   Separator Pages

A **separator page,** sometimes known as a banner page, separates print jobs and typically states who sent the document to the printer and the date and time of printing. You can use one of the standard separator pages or create a custom page.
   Windows 2000 provides three separator page files, located in the *systemroot*\ SYSTEM32 folder. Table 15.2 shows the names of the separator page files, the type of printer with which it is compatible, and the purpose of each page. You can edit or customize any of these separator pages. The separator page files provided by Windows 2000 may not work with some printers, however.

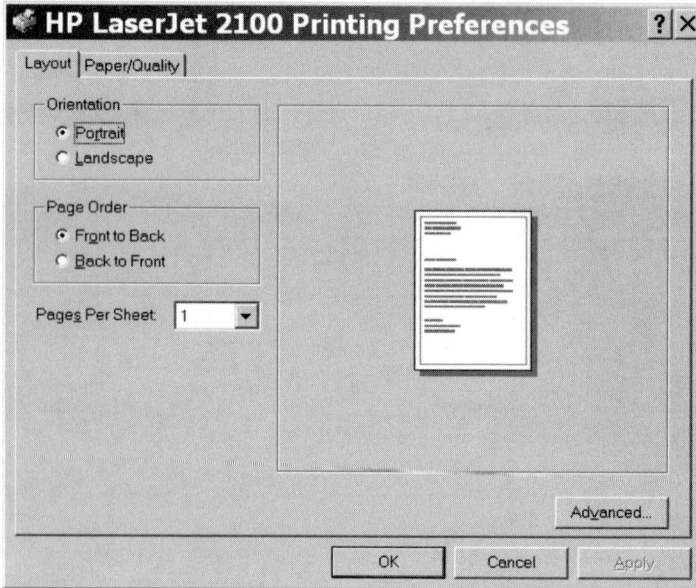

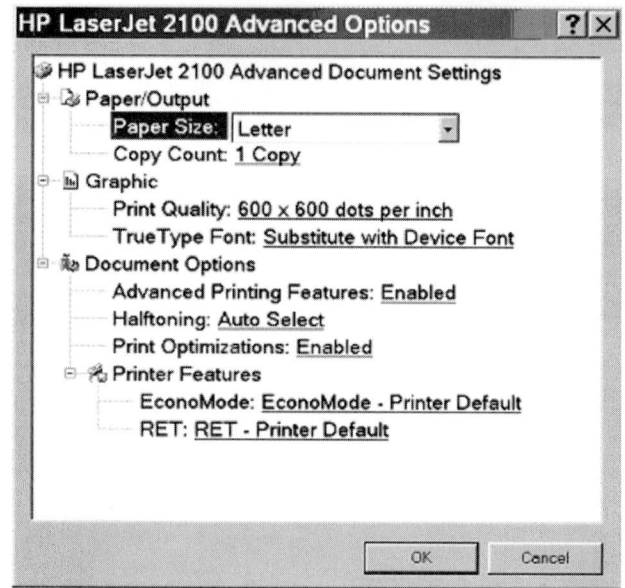

FIGURE 15.12   Printing Preferences

TABLE 15.2   Printer separator pages

File Name	Printer Language	Purpose
pcl.sep	Printer Control Language (PCL)	Switches the printer to PCL (Hewlett-Packard printing language standard) printing and prints a separator page before each document. Note: Pcl.sep is compatible with PCL printers, but may not work unless the printer also supports Printer Job Language (PJL).
pscript.sep	PostScript	Switches the printer to PostScript printing, but does not print a separator page before each document. Pscript.sep is compatible with PostScript printers that also support PJL.
sysprint.sep	PostScript	Switches the printer to PostScript printing and prints a separator page before each document. Sysprint.sep is compatible with PostScript printers.

Note: Some printers can support both PCL and PostScript printing. In these cases, the separator page would be used to switch the printer back and forth between the two printer languages.

To create your own separator file, copy and rename one of the supplied separator files. Table 15.3 shows the escape codes that you can include in a separator file. The first character of the separator page file must always be the escape character. This character is used throughout the separator page file in escape codes. The separator file interpreter replaces these escape codes with appropriate data, to be sent directly to the printer.

To choose a separator page, do the following:

1. Open the Printer folder.
2. Right-click the printer you are using and select the Properties option.
3. Click on the Advanced tab.

TABLE 15.3 Escape codes used in separator pages

Escape Code	Function
@N	Prints the user name of the person who submitted the job.
@I	Prints the job number.
@D	Prints the date the job was printed. The representation of the date is the same as the Date Format in the International section in Control Panel.
@T	Prints the time the job was printed. The representation of the time is the same as the Time Format in the International section in Control Panel.
@L*xxxx*	Prints all the characters (*xxxx*) following it until another escape code is encountered.
@F*pathname*	Prints the contents of the file specified by path name, starting on an empty line. The contents of this file are copied directly to the printer without any processing.
@H*nn*	Sets a printer-specific control sequence, where *nn* is a hexadecimal ASCII code sent directly to the printer. To determine the specific numbers, see your printer manual.
@W*nn*	Sets the width of the separator page. The default width is 80; the maximum width is 256. Any printable characters beyond this width will be truncated.
@B@S	Prints text in single-width block characters until \U is encountered.
@E	Ejects a page from the printer. Use this code to start a new separator page or to end the separator page file. If you get an extra blank separator page when you print, remove this code from your separator page file.
@*n*	Skips *n* number of lines (from 0 through 9). Skipping zero lines simply moves printing to the next line.
@B@M	Prints text in double-width block characters until \U is encountered.
@U	Turns off block character printing.

4. Click the Separator Page button.
5. Click Browse and select the separator page located in the *systemroot*\SYSTEM32 folder.

15.8.6 Pausing and Resuming a Printer

Normally you pause a printer if you have a printer problem or need to perform scheduled maintenance. To pause and then resume the printer, you need to have the Manage Printer option. To pause a printer and the printing of all its documents, do the following:

1. Open the Printers folder.
2. Right-click the printer on which you want to pause or resume printing, and select the Pause Printing option.

Note the checkmark that appears next to Pause Printing when the printer is paused. Any documents that are currently printing will finish, but other print jobs in the print queue will be put on hold.

To resume printing on the printer, do the following:

1. Open the Printers folder.
2. Right-click the printer on which you want to pause or resume printing, and select the Pause Printing option to remove the checkmark.

If you want to cancel all print jobs in the print queue, proceed as follows:

1. Open the Printers folder.
2. Right-click the printer on which you want to pause or resume printing, and select the Cancel All Documents option.

```
┌────────────────────────────────────────────────────────────────────────────────────────┐
│ ◆ HP LaserJet 2100 - Paused                                              _ □ ×          │
├────────────────────────────────────────────────────────────────────────────────────────┤
│ Printer   Document   View   Help                                                         │
├──────────────────────────────────────┬──────────┬────────────┬───────┬─────────┬────────┤
│ Document Name                         │ Status   │ Owner      │ Pages │ Size    │ Submitted         │
├──────────────────────────────────────┼──────────┼────────────┼───────┼─────────┼───────────────────┤
│ ▣ Microsoft Word - Ch 15 Printing with Windows 2000.doc │ Deleting -... │ administrator │ 8/24 │ 5.25 M... │ 4:29:30 PM  12/19/1999 │
│ ▣ Microsoft Word - Document2         │          │ administrator │ 1    │ 3.94 KB │ 4:31:26 PM  12/19/1999 │
│ ▣ Microsoft Word - document1.doc     │          │ administrator │ 1    │ 3.94 KB │ 4:31:35 PM  12/19/1999 │
│ ▣ Microsoft Word - Ch 15 Printing with Windows 2000.doc │ Spooling │ administrator │ 13 │ 11.2 MB │ 4:32:03 PM  12/19/1999 │
├──────────────────────────────────────┴──────────┴────────────┴───────┴─────────┴────────┤
│ ◀ │                                                                                  ▶ │
├──────────────────────────────────────────────────────────────────────────────────────────┤
│ 4 document(s) in queue                                                                    │
└────────────────────────────────────────────────────────────────────────────────────────┘
```

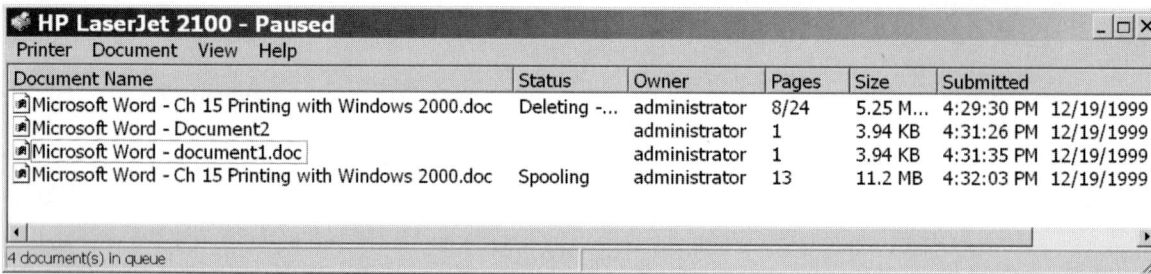

FIGURE 15.13   Viewing the print queue

15.9   MANAGING DOCUMENTS

As a user or an administrator, at times you may need to manage the individual print jobs or documents. To view documents waiting to print, do the following:

1. Open the Printers folder.
2. Double-click the printer on which you want to view the documents waiting to print.

The print queue shows information about a document such as print status, owner, and number of pages to be printed. See Figure 15.13. From the print queue, you can cancel or pause the printing of a document that you have sent to the printer. You also can open the print queue for the printer on which you are printing, by double-clicking the small printer icon in the status area on the taskbar.

To pause a document, do the following:

1. Open the Printer folder.
2. Double-click the printer you are using to open the print queue.
3. Right-click on the document you want to pause and select the Pause option.

To resume printing of a paused document, do the following:

1. Open the Printer folder.
2. Double-click the printer you are using to open the print queue.
3. Right-click on the document you want to pause and select the Resume option.

By default, all users can pause, resume, restart, and cancel printing of their own documents. To manage documents that are printed by other users, however, you must have the Manage Documents permission.

In general, once a document has started printing, it will finish printing even if you pause it. If something happens to the printing of the document such as the wrong type of paper was in the printer, you can restart the print job if it is still in the print queue, as follows:

1. Open the Printers folder.
2. Double-click the printer you are using to open the ping queue.
3. Right-click on the document that you want to restart printing and select the Restart option.

If there is a higher priority document, it will print first.

To cancel a print job, do the following:

1. Open the Printers folder.
2. Double-click the printer you are using to open the print queue.
3. Right-click on the document that you want to stop printing and select the Cancel option.

You can cancel the printing of more than one document by holding down the CTRL key and clicking on each document that you want to cancel.

To change the printing priority of documents waiting to print, proceed as follows:

1. Open the Printers folder.
2. Double-click the printer you are using to open the print queue.
3. Right-click on the document that you want to move in the print order, and click the Properties option. See Figure 15.14.
4. Click the General tab.
5. Drag the Priority slider to raise or lower the document priority.

FIGURE 15.14   Viewing the
properties of a print job

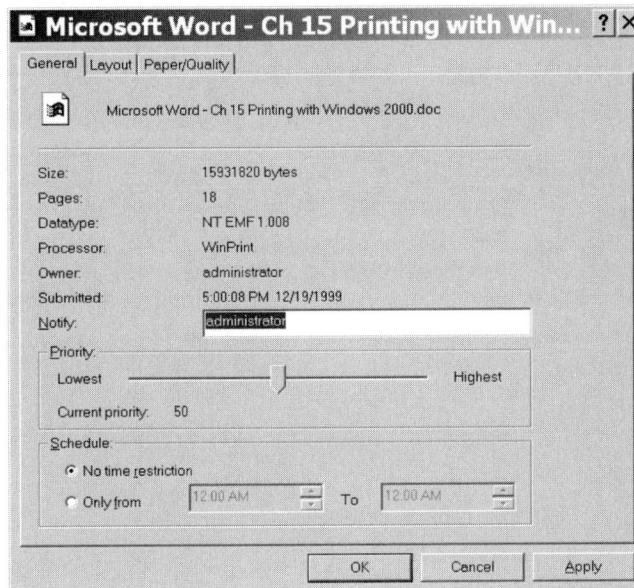

Once a document has started printing, any printing priority changes you make will not affect the document.

If a printer fails before printing and cannot be repaired quickly, then you may want to transfer the document to another printer, as follows:

1. Open the Printers folder.
2. Double-click the printer that holds the document(s) you want to redirect.
3. Open the Printer menu and select the Properties option.
4. To send documents to another printer on the same print server, click the port to which the other printer is assigned and click the OK button. To send documents to a printer on a different print server, click the Add Port button, select the Local Port option, and click on the New Port button. Type the name of the other print server and the share name using the UNC name. If you are adding to a network port, you can only add an existing port.

To redirect documents to another printer, you must have the Manage Printers permission on both printers. In addition, since the program has already converted the document into a language that is specific to a particular printer, you must redirect the document(s) that uses the same driver.

15.10   PRINTING USING LPR AND LPD

To print in a pure TCP/IP network, load the **line printer daemon (LPD)** on a computer designated as the print server. A daemon is a background process or service. The client computer sends print jobs to the **print spooler service** on another computer by using the **line printer remote (LPR)** service. LPR was originally developed as a standard for transmitting print jobs between computers running Berkeley UNIX as published in the RFC 1179. LPR clients are often on UNIX systems, but LPR software exists for most operating systems including Windows NT and Windows 2000. To view the print queue, then, use the **line printer query (LPQ).**

For Windows 2000 servers to accept LPR client print jobs, load **Print Services for UNIX,** as follows:

1. Open the Control Panel.
2. Double-click on the Add/Remove Programs applet.
3. Click on the Add/Remove Windows Components.
4. Click to highlight Other Network File and Print Services and click on the Details button.
5. Select the Print Services for UNIX option.
6. Click on the OK button.

469

7. Click on the Next button.
8. Click on the Finish button.
9. Click on the Close button to close the Add/Remove Programs dialog box.

To create a LPR port, do the following:

1. Open the Printers folder.
2. Double-click Add Printer and click on the Next button.
3. Click Local Printer, clear the "Automatically detect my printer" checkbox, and click on the Next button.
4. Click Create a New Port option, select the LPR port in the Type: box and click on the Next button.
5. Enter the DNS name and the IP address of the server in the "Name or address of server providing LPD."
6. Enter the name of printer or print queue on the server just listed.
7. Select the printer manufacturer and model and click on the Next button.
8. Enter a printer name. If you want it to be the default printer, click on the Yes option. Click on the Next button.
9. Select the Share As option, provide the share name of the printer, and click on the Next button.
10. Provide an optional location and comment and click on the Next button.
11. Select the Yes option if you want to print a test page. Click on the Next button.
12. Click on the Finish button.

If a printer is already shared, add LPR ports by using the Print Server Properties dialog box.

LPR clients traditionally print with the LPR command. The syntax for the LPR command is as follows:

LPR –S *Server* **–P** *Printer* [**–C** *Class*] [**–J** *Jobname*] [**–O** *option*] *filename*

–S *Server*
    Specifies the name or IP address of the computer that has the printer attached to it.
–P *Printer*
    Specifies the name of the printer for the desired queue.
–C *Class*
    Specifies the content of the banner page for the class.
–J *Jobname*
    Specifies the name of this job.
–O *option*
    Indicates the type of file. The default is a text file. Use **–Ol** (lowercase "**el**") for a binary file (e.g., PostScript).
    filename
    Is the name of the file to be printed.

The syntax for the LPQ is as follows:

LPQ –S *Server* **–P** *Printer* [**–l**(el)]

–S *Server*
    Specifies the name of the computer that has the printer attached to it.
–P *Printer*
    Specifies the name of the printer for the desired queue.
–l
    Specifies that a detailed status should be given.

15.11 NETWORK PRINTER HARDWARE

Although Windows 2000 handles printing quite well, there are advantages to using printer network cards, such as Hewlett Packard's **JetDirect cards,** and print server boxes, such as **Intel's Netport.**

The printer network card is installed into an empty slot located inside the printer, and others are included with the printer as a built-in port. Because these cards/ports connect directly to the network using either UTP cable or coaxial cable, they must have MAC and IP addresses. The MAC address is burned into a ROM chip. The IP address, including the subnet mask and gateway address, is set using the printer controls or by running software on a computer on the same subnet and connecting to the printer by its MAC address. The primary advantage of using a printer network card is to conceal the print server easily in a server room. Others are the cost is cheaper, because one print server can handle many printers throughout the network, and the network connection is faster than the parallel port.

The print server boxes provide connections to the network, multiple ports to connect printers, and memory to store the print jobs. Because they are similar to a dedicated print server, they also need a MAC and an IP address. Like the printer network card, the MAC address is burned into a ROM chip within the box. The IP address (including the subnet mask and gateway) is configured using a special software program that attaches using the MAC address or by connecting a cable from a computer directly to the box. After the box has an IP address, the print server can be configured using the Telnet program or web pages. The advantage of using a print server box is that the box is cheaper than an entire PC and often has better performance.

15.12 TROUBLESHOOTING PRINTERS

Printing problems occur because of trouble with the following:

- Printing device
- Connection between the printer and the network
- Other printing components in Windows 2000
- Network, protocols, and other communication components

A quick way to solve most problems associated with printing is to follow these steps:

- Verify that the physical printer is operational. If other users can print, it is probably not a problem with the printer, or with the print server.
- Check that the physical printer is in the ready state (online) and if the printer can print a test page.
- Verify that the printer on the print server is using the correct print driver. If print clients are using other operating systems, then install all necessary drivers for the other platforms. If the correct print driver is not selected, it will often print garbage.
- Verify that the print server is operational, that there is enough disk space for spooling, and that the print spool service is running.

SUMMARY

1. When a client prints to the network printer, the print job is sent to the print server.
2. A print spooler is software that accepts a document which is sent to a printer by the user and then stores it on disk or in memory until the printer is ready for it.
3. Documents can be sent to a printer with different priorities. Users with a higher priority can bypass other print jobs and thus be printed first.
4. Another use for creating two logical printers both pointing to the same printer is to schedule print jobs.
5. A printer pool is two or more print devices that use the same driver, are connected to one print server, and act as a single printer.
6. Internet printing allows you to submit print jobs to a printer across the Internet by using the printer's URL as the name of the printer.
7. The print server checks for an available port and sends documents to ports in the order that they are added.
8. Printer permissions control who can use the printer and which print tasks a user can do.
9. There are three levels of printer permissions: Print, Manage Printer, and Manage Documents.
10. A separator page, sometimes known as a banner page, separates print jobs and typically states who sent the document to the printer and the date and time of printing.

11. Pause a printer when you have a printer problem or need to perform scheduled maintenance. To pause and then resume the printer, you need to have the Manage Printer option.

12. The print queue shows information about a document such as print status, owner, and number of pages to be printed.

13. To print in a pure TCP/IP network, load the line printer daemon (LPD) on a computer designated as the print server.

14. A daemon is a background process or service.

15. In a pure TCP/IP network, the client computer sends print jobs to the print spooler service on another computer by using the line printer remote (LPR) service.

16. To view the print queue in a pure TCP/IP network, use the line printer query (LPQ).

17. Although Windows 2000 handles printing quite well, there are advantages to using printer network cards, such as Hewlett Packard's JetDirect cards, and print server boxes, such as Intel's Netport.

QUESTIONS

1. There are 50 Windows 2000 Professional computers on your network that print to an HP LaserJet IV SI printer attached to a Windows 2000 print server. You have just obtained an updated print driver. What is the best way to update the driver for all computers that print to this print server?
   a. Update the driver on the print server and do nothing more.
   b. Update the driver on all client computers; there is no need to update the server.
   c. Create a separate printer with the updated driver on the print server, and instruct all client computers to print to the new printer.
   d. Update the driver on the print server and instruct all client computers to download the updated driver from the server.

2. You receive several calls from users who are trying to print to a print server. They complain that they have sent several jobs to the print server, but the jobs have not printed and cannot be deleted. What must you do to resolve the problem?
   a. Delete all files from the spool folder on the print server.
   b. Select Services in the Services console on the print server, stop the spooler service, and then restart it.
   c. Be sure that the Pause Printing option is not checked on the print server.
   d. Delete the stalled printer from the print server and create a new printer.

3. Your boss has asked you to configure network printing so that all documents from the executives will print before other users' documents. What is the best way to accomplish this task?
   a. Create a separate printer for the executives, and set the printer priority to 99.
   b. Create a separate printer for the executives, and set the printer priority to 1.
   c. Create a separate printer for the executives, and configure the printer to start printing after the last page is spooled.
   d. Create a separate printer for the executives, and configure the printer to start printing immediately.
   e. Create a separate printer for the executives, and configure the printer to print directly to the print device.

4. Pat often sends large print jobs to his local printer. His system is running with 128 MB of RAM and a 4-GB hard disk. Pat's hard drive is approaching capacity and he has recently added another 10-GB hard disk. Pat has been experiencing problems when he tries to print large graphics files, because his default spool folder is located on his old hard drive. How can Pat change the location of the spool folder to the new hard drive?
   a. Edit the registry.
   b. Edit the user profile.
   c. Choose the Server option in the Services console.
   d. Choose the System option in the Control Panel.
   e. Choose Printers from My Computer.

5. One print device on your network has failed. There are several jobs waiting to be printed on that print device. How do you redirect the documents to a different print device so that the users do not need to resubmit their print jobs?
   a. Add a local port to the faulty printer and type the UNC name of another printer in the Port Name dialog box.
   b. Delete the faulty printer, create a new printer attached to another print device, and assign to the new printer the same name as the old printer.
   c. Pause the faulty printer first, then drag and drop all print jobs to another printer.
   d. Pause the faulty printer first, and choose Redirect Print Jobs from the Printer menu.

6. You have just installed a new printer on your print server. You send a print job to the printer, but it comes out as pages of nonsense words. What is the most likely cause of the problem?
   a. The print spooler is corrupt.
   b. The correct protocol is not installed.
   c. An incorrect printer driver is installed.
   d. There is not enough hard disk space for spooling.

7. You are the print operator on your network. You would like to switch the printing mode to PCL for an HP LaserJet print device. What is the easiest way to accomplish this task?
   a. Assign a different paper tray for printing in PCL mode.
   b. Assign a different printer port for printing in PCL mode.
   c. Create a separate printer with PCL drivers.
   d. Use a separator page.

8. You are creating a printer pool with four print devices. Which condition must be satisfied for you to create a printer pool? (Select all that apply)
   a. All print devices must be connected to the same print server.
   b. They must use the same print driver.
   c. All print devices must be located in the same room.
   d. All print devices must be able to use the same protocol.
   e. All print devices must use the same printer port.

9. You have already created a printer pool consisting of two print devices for the Sales department. What must you do on the print server to add a new print device to the existing printer pool?
   a. Enter the print driver and port for the new print device.
   b. Enter the name and port for the new print device.
   c. Enter the name for the new print device only.
   d. Enter the port for the new print device only.

10. Users on your network complained that some large print jobs monopolize the printer, causing unacceptable delays for their small print jobs. The large print jobs are usually not needed on the day they are sent to the printer. How would you improve the situation?
    a. Create a separate printer, configure the printer to print directly to the printer port, and direct users to print large documents to the newly created printer.
    b. Create a separate printer, configure the printer to start printing immediately, and direct users to print large documents to the newly created printer.
    c. Create a separate printer, configure the printer to be available only after hours, and direct users to print large documents to the newly created printer.
    d. Create a separate printer, set it to the highest priority, and direct users to print large documents to the newly created printer.

11. You have set up a printer to be available only during off-hours. What will happen to the print jobs sent to the printer during the day?
    a. They will be stored in the print spooler.
    b. They will be stored in a temporary directory.
    c. They will be stored in the print processor.
    d. They will be redirected to a functioning printer.

12. You have two printers connected in a printer pool. The first printer is printing a job and the second printer is idle. A paper jam occurs on the first printer. What will happen to the rest of the job that was being printed?
    a. The job will be completed on the second printer.
    b. The job will be completed on the second printer if the second printer has a higher priority level.
    c. The job is held for completion by the first printer until the device is fixed.
    d. The job is canceled.

13. You are trying to print a payroll report and the print device jams. The print device is attached to a Windows 2000 print server. Your current application does not allow you to resubmit print jobs. After clearing the paper jam, what would you do to reprint the document?
    a. Select the printer from the Printers folder and choose Resume from the Printer menu.
    b. Select the printer from the Printers folder and choose Restart from the Printer menu.
    c. Select the printer from the Printers folder and choose Resume from the Document menu.
    d. Select the printer from the Printers folder and choose Restart from the Document menu.

14. A local group called LaserJet_Users consists of user accounts that are currently accessing a printer in your domain. You have already removed the Everyone special group from the LaserJet_Users group so that only specified users have access to the printer. You would like to tighten security even further by allowing individuals who are using the printer to manage only their own print jobs. However, you want individuals to be able to view all jobs sent to the printer. The LaserJet_Users group currently has Full Control permission for the printer. What is the best way to enact this new security measure?
    a. Assign Read permission to the LaserJet5_Users group. Add the user accounts from the LaserJet5_Users group to the Print Operators group on the PDC of the printer's domain.
    b. Assign Read permission to the LaserJet5_Users group. Assign Manage Documents permission to the Creator Owner special group on the printer.
    c. Assign Read permission to the LaserJet5_Users group. Assign Print permission to the Creator Owner special group on the printer.
    d. Assign Print permission to the LaserJet5_Users group.

15. Users from the Accounting domain and the Sales domain share one print device. Users from the Accounting domain often print very large documents which can cause unacceptable delays for small print jobs sent by users in the Sales domain. The large accounting jobs are usually not needed on the day they are sent to the printer. How would you improve the situation?
    a. Create a separate printer, configure the printer to print directly to the printer port, and direct users from the Accounting domain to print large documents to the newly created printer.
    b. Create a separate printer, configure the printer to start printing immediately, and direct users from the Accounting domain to print large documents to the newly created printer.
    c. Create a separate printer, configure the printer to be available only after hours, and direct users from the Accounting domain to print large documents to the newly created printer.
    d. Create a separate printer, set the printer to the highest priority, and direct users from the Accounting domain to print large documents to the newly created printer.

16. Which one of the following statements best describes how Print permission and Manage Documents permission differ for a printer?
    a. Manage Documents permission allows users to submit jobs to a printer and change the status of their own print jobs. Print permission simply allows users to submit jobs to a printer.

b. Manage Documents permission allows users to change the status of any print job submitted by any user to a printer, but it does not allow users to submit their own jobs to a printer. Print permission allows users to submit their own jobs to a printer.

c. Manage Documents permission allows users to submit jobs to a printer and change the status of any print job submitted by any user. Print permission allows users to submit jobs to a printer and change the status of their own print jobs.

d. Manage Documents permission allows users to submit jobs to a printer and change the status of any print job submitted by any user. Print permission simply allows users to submit jobs to a printer.

17. Pat wants to use a Windows 2000 Server computer as the print server for a TCP/IP network print device. Users will send their documents to the server, which will then forward the documents to the print device. What is the best way for Pat to proceed?

a. Install the TCP/IP printing service on the network print device.

b. Install the TCP/IP printing service on the Windows 2000 Server computer.

c. Use the LPQ utility to redirect all print jobs to a printer port that is mapped to the print device's IP address.

d. Use the LPR utility to redirect all print jobs to a logical printer port that is mapped to the print device's IP address.

18. You configure a UNIX computer as an LPD server. Users on a Windows 2000 Professional computer want to send documents to a print device connected to the UNIX LPD server. How must they send their documents to this print device?

a. Install the TCP/IP printing service on each client computer.

b. Map a logical printer port to the UNC name for the printer.

c. Use the LPR utility.

d. Use the LPQ utility.

EXERCISES

EXERCISE 1: INSTALLING A PRINTER

Note: Most of these exercises can be done even if you do not have a printer connected to your computer.

1. Log in as the Administrator.
2. Click on the Start button, select the Settings option, and select the Printers option.
3. In the Printers window, double-click the Add Printer icon.
4. On the Welcome to the Add Printer Wizard page, click the Next button.
5. On the Local or Network Printer page, verify that Local Printer is selected and click the Next button. If the printer was not automatically detected because it is not a plug-and-play printer, then specify the port that the printer is going to use and select the manufacturer and the printer model.
6. On the Name Your Printer page, type in PRINTER1-XXx, where *XX* is your assigned partner number and *x* is either a or b in the Printer Name box, and click the Next button.
7. On the Printer Sharing page, type the PRINTER1-XXx in the Share As box and click on the Next button.
8. If you add spaces, a warning message will state that the share name may not be accessible from some MS-DOS workstations. In these situations, click the Yes button.
9. On the Location and Comment page, type the room number in the Location box, the row number of your classroom seat, and the actual manufacturer and type of printer.
10. To be sure that the printer is connected properly and functioning, print a test page. On the Print a Test Page page, leave the Yes option enabled and click the Next button. If Windows 2000 asks if the printer page prints out OK, click on the Yes button.
11. On the Completing the Add Printer Wizard page, confirm the summary of your installation choices and click the Finish button.
12. Right-click the printer you have just created and select the Pause Printing option.

EXERCISE 2: MANAGING THE PRINTER DOCUMENTS

The following should be done on your computer.

1. Make sure that the printer is paused.
2. Click on the Start button, select the Programs option, select the Accessories options, and select the WordPad program.
3. Type in your first and last name. Change the font to Arial, the style to Bold, and the font size to 48 points.
4. Print to your printer.
5. Click on the Start button, select the Settings option, and select the Printers option.
6. Double-click your printer icon. Note that your print job is there waiting to be printed.

The following should be done on your partner's computer.

7. Click on the Start button, select the Settings option, and select the Printers option.
8. Double-click the Add Printer icon.
9. When the Add Printer Wizard starts, click on the Next button.
10. Select the Network Printer option and select the Next option.
11. Select Type the Printer Name option, type in the name of your shared printer in the Name text box, and click on the Next button.
12. Select the Yes option to make the printer the default printer. Click on the Next button.
13. Click on the Finish button.
14. Click on the Start button, select the Programs option, select the Accessories option, and select the WordPad program.
15. Type in your first and last name. Change the font to Arial, the style to Bold, and the font size to 48 points.
16. Print to your printer.
17. Click on the Start button, select the Settings option, and select the Printers option.
18. Double-click your printer icon. Note that your print job is there waiting to be printed.
19. Right-click the second print job and select the Cancel option to cancel the print job.
20. Log out as the Administrator and log in as Charlie Brown.
21. In the Printer folder, double-click your printer. As Charlie Brown, try to delete the first print job.

The following should be done at your computer.

22. Right-click your printer, select the Properties option, and select the Security tab.
23. Click the Add button and add Charlie Brown to the list. Give Charlie Brown the Manage Documents and Print permissions. Click on the OK button.

The following should be done at your partner's computer.

24. Logged in as Charlie Brown, try to delete the first print job.

EXERCISE 3: SCHEDULING PRINT JOBS

1. Click on the Start button, select the Settings option, and select the Printers option.
2. In the Printers window, double-click the Add Printer icon.
3. On the Welcome to the Add Printer Wizard page, click the Next button.
4. On the Local or Network Printer page, verify that Local Printer is selected and click the Next button. If the printer was not automatically detected because it is not a plug-and-play printer, then specify the port that the printer is going to use and select the manufacturer and the printer model.
5. The Use Existing Driver page displays because the printer you are installing can use the drivers from a printer that has already been installed on this computer. Therefore, enable the "Keep existing driver" option (recommended) and then click the Next button.
6. On the Name Your Printer page, type PRINTER2-XXx, where *XX* is your assigned partner number and *x* is either a or b in the Printer Name box, and click the Next button.

7. The first printer installed on the computer automatically became the default printer. For printers installed subsequently, change the default printer. Therefore, to keep the first printer as the default, select the No option and click the Next button.

8. On the Printer Sharing page, type PRINTER2-XXx in the Share As box and click on the Next button.

9. If you add spaces, a warning message will state that the share name may not be accessible from some MS-DOS workstations. In these situations, click the Yes button.

10. On the Location and Comment page, type the room number in the Location box, the row number of your classroom seat, and the actual manufacturer and type of printer.

11. To be sure that the printer is connected properly and functioning, print a test page. On the Print a Test Page page, leave the Yes option enabled and click the Next button. If Windows 2000 asks if the printer page prints out OK, click on the Yes button.

12. On the Completing the Add Printer Wizard page, confirm the summary of your installation choices and click the Finish button.

13. Right-click the printer and select the Properties option.

14. In the Advanced tab, select the Available From option and specify "12:00 am to 5 am." Click on the OK button.

15. Create the WordPad document again with your name.

16. Open the File menu and select the Print option.

17. With the General tab, select the second printer.

18. Double-click on the printer icon to view the print queue. Although the printer is not paused, the document will remain until 12:00 am before it starts printing.

19. Cancel the print jobs.

EXERCISE 4: CONNECTING A PRINTER THAT IS CONNECTED DIRECTLY TO THE NETWORK

1. Click on the Start button, select the Settings option, and then the Printers option.

2. In the Printers window, double-click the Add Printer icon.

3. On the Welcome to the Add Printer Wizard page, click the Next button.

4. On the Local or Network Printer page, verify that Local Printer is selected and disable the "Automatically detect and install my Plug and Play printer" option.

5. On the Select the Printer Port page, select the "Create a new port" option and select the Standard TCP/IP Port in the Type box. Click the Next button.

6. When the Add Standard TCP/IP Printer Port Wizard starts, click on the Next button.

7. On the Add Port page, type in 140.100.2XX in the Printer Name or IP Address box, where *XX* is your computer number. Click on the Next button to accept the default port name.

8. Verify the configuration options and click the Finish button.

9. On the Add Printer Wizard page, select the manufacturer for the printer under Manufacturer and select the model of the printer under printers. Click the Next button.

10. On the Name Your Printer page, type in PRINTER3-XXx, where *XX* is your assigned partner number and *x* is either a or b in the Printer Name box, and click the Next button.

11. The first printer installed on the computer automatically became the default printer. For printers installed subsequently, you can change the default printer. Therefore, to keep the first printer as the default, select the No option and click the Next button.

12. On the Printer Sharing page, type the PRINTER3-XXx in the Share As box and click on the Next button.

13. If you add spaces, a warning message will state that the share name may not be accessible from some MS-DOS workstations. In these situations, click the Yes button.

14. Right-click the printer you have just created and select the Pause Printing option.

15. On the Location and Comment page, type the room number in the Location box, the row number of your classroom seat, and the actual manufacturer and type of printer.

16. To be sure that the printer is connected properly and functioning, print a test page. On the Print a Test Page page, leave the Yes option enabled and click the Next button. If Windows 2000 asks if the printer page prints out OK, click on the Yes button.

17. On the Completing the Add Printer Wizard page, confirm the summary of your installation choices and click the Finish button.

16 Disaster Recovery

TOPICS COVERED IN THIS CHAPTER

INTRODUCTION

For some companies, every minute that the network is down means a loss of thousands of dollars in business and productivity. The loss of data on the network also may be irreplaceable. Although network administrators never hope for disasters to occur, they should be ready to deal with such possibilities. Therefore, this chapter's focus is twofold: first, on how to avoid disaster, and second, on how to deal with certain kinds of disasters, if they occur.

OBJECTIVES

1. Choose a disaster recovery plan for various situations.
2. Select the appropriate hardware and software tools to monitor trends in the network.
3. Resolve broadcast storms.
4. Identify and resolve network performance problems.
5. Monitor and configure removable media, such as tape devices.
6. Optimize and troubleshoot performance of the Windows 2000, including memory performance, processor utilization, disk performance, network performance, and application performance.
7. Recover systems and user data by using Windows Backup.
8. Troubleshoot system restoration by using safe mode.
9. Recover systems and user data by using the Recovery console.
10. Implement, configure, manage, and troubleshoot auditing.
11. Recover from disk failures.
12. Perform an authoritative restore of Active Directory.
13. Recover from a system failure of Active Directory.

16.1 PLANNING DISASTER RECOVERY

When a computer has a problem, the user's first response is "Oh no, not now!" Any computer user will agree that there is no *good* time for a computer to break down. A network failure can affect many people and can literally cost a company thousands of dollars of business or productivity for every hour the network is down. One primary job of the administrator is to deal with such disasters and to plan ahead as to minimize the frequency and degree of failures. This task includes:

1. Establishing a backup plan and then performing the backup
2. Documenting the network so that all team members can find information quickly about the network
3. Maintaining a log that lists all problems and their solutions, to be used to determine trends, plan network and personnel resources, and provide a quick reference of solutions for when the same problem recurs.

16.2 TAPE DRIVES

Tape drives read and write to a long magnetic tape. They are relatively inexpensive and offer large storage capacities, which makes them ideal for backing up hard drives on a regular basis. To back up a hard drive, insert a tape into the drive, start a backup software package, and select the drive/files you want to back up and it will be done. If a drive or file is lost, the backup software can be used to restore the data from the tape to the hard drive. If the correct tape drive and backup software are chosen, the drive could automatically back up the hard drive at night, when it is used the least. You must remember, however, to replace the tape each day. Also note that tapes do fail and oftentimes users fail to properly back up a drive or file. When disaster occurs, users are surprised by their blank tapes. Therefore, it is important to occasionally test the tapes by choosing an unimportant file and restoring it back to the hard drive.

Tapes come in different sizes and shapes and offer different speeds and capacities. As a result, several standards have been developed including the quarter-inch cartridge (QIC) and the digital audio tape (DAT).

In 1972, 3M Company introduced the first **quarter-inch cartridge (QIC)** designed for data storage. The tape cartridge measured $6'' \times 4'' \times \frac{5}{8}''$. Although the cartridge became the standard, each tape drive manufacturer used different encoding methods, varied the number of tracks, and varied the data density on the tape, thus causing all kinds of compatibility problems.

As a result, in 1982, a group of manufacturers formed the QIC Committee to standardize tape drive construction and applications. The full-size quarter-inch cartridge standardized by the QIC Committee is also referred to as the DC (data cartridge) 6000.

The first tape that the QIC Committee approved was the QIC-24, which used serpentine recording. It had nine tracks and a density of 8000 bits per inch, giving a total storage capacity of 60 MB. It achieved 90 inches per second and offered 720 kilobits per second with the QIC-02 interface.

Throughout the years, data density was increased and more tracks were added. The last QIC-1000 DC packed 30 tracks across the tape at 36,000 bits per inch, allowing up to 1.2 GB per tape cartridge. Its speed was also increased 2.8 megabits per second. Yet, the QIC-1000 DC drives could read previous QIC tapes.

By 1989, the QIC Committee revamped the original QIC standard by using 1,7 RLL encoding and higher coercivity. This allowed higher bit density, a high number of tracks, and faster transfer rates. By 1995, the QIC introduced the QIC-5210 DC, which had 144 tracks and 76,200 bits per inch, allowing for 25 GB of storage space.

Because the full-size QIC is too large to fit into a drive bay, the committee created the **minicartridge (MC),** which was 3.25″ × 2.5″ × 1.59″. The minicartridges are also called DC 2000 cartridges. The QIC-40 MC was the first standard to be adopted. It fit into the 5.25-inch drive bay and it connected to the computer via the floppy drive controller. As floppy disk drives use MFM encoding, so did the QIC-40 MC. Different from previous tapes, QIC specified the format of the data on the tape, including how sectors were assigned to files and FAT—to list bad sectors. This feature required formatted tapes, which is a time-consuming process. Consequently, you could buy tapes as formatted or unformatted. Another advantage was the tapes could be accessed randomly. Although it had to move the tape to the proper sector, it did not have to read each file sequentially.

The newest type of tape is the **digital audio tape (DAT),** which uses the same technology as VCR tapes (helical scan). The 8-mm DAT allows capacities up to 35 GB. The DAT standard has primarily been developed and marketed by Hewlett-Packard. This company chairs the Digital Data Storage (DDS) Manufacturers Group, which led the development of the DDS standards.

Data are not recorded on the tape in the MFM or RLL formats, but rather, bits of data received by the tape drive are assigned numerical values, or digits. These digits are then translated into a stream of electronic pulses that are placed on the tape. Later, when information is being restored to a computer system from the tape, the DAT drive translates these digits back into binary bits that can be stored on the computer.

The **digital data storage (DDS)** tape is currently the newest standard of the digital audio tape. The DDS-3 can hold 24 GB (or equivalent) of over 40 CD-ROMs and supports data transfer rates of 2 Mbps, all in a space that is slightly larger than a credit card. In a DDS drive, the tape barely creeps along, requiring about 3 seconds to move an inch. The head drum spins rapidly at 2000 revolutions per minute, putting down 1869 tracks across a linear inch of tape which allow 61 kilobits per inch. The main advantage of the DAT is its access speed and capacity.

One of the newest tapes is the **digital linear tape (DLT),** designed for high capacity, high speed, and highly reliable backup. A DLT is ½″ wide, has capacities of 35 GB to 70 GB compressed, and a data transfer rate of 5 Mbps to 10 Mbps or more. Unfortunately, the drives are quite expensive and are used primarily for network server backup.

16.3 BACKUP

Data are the raw facts, numbers, letters, or symbols that the computer processes into meaningful information. Examples of data include a letter to a company or a client, a report for your boss, a budget proposal of a large project, or an address book of your friends and business associates. Regardless of the data type, it can be saved (or written to disk) so that it can be retrieved at any time. Data can be printed on paper or e-mailed to someone else.

Data stored on a computer or stored on the network are vital to the users and probably the company. Data represent hours of work, which are sometimes irreplaceable. Data loss can be caused by many things including hardware failure, viruses, user error, and malicious users. When disaster occurs, the best method to recover data is backup, backup, backup, because without backup of important files, it is often too late to recover them.

A **backup** of a system is a copy of all data and/or programs. As a technician, consultant, or support person, you need to emphasize at every moment to back up on servers and client systems. In addition, clients should save their data files to a server so that you have a single, central location to store backup. This process may even entail selecting and installing the equipment, doing the backup, or training other people in doing the backup. Having proper equipment and methods will help ensure complete and regular backups; and having good personnel to perform backups is essential in a data protection system.

When developing a backup, follow these three steps:

1. Develop a backup plan
2. Stick to the backup plan
3. Test the backup plan

When developing a backup plan, consider the following:

1. What equipment will be used?
2. How much data need to be backed up?
3. How long will the backup take?
4. How often must the data be backed up?
5. When will the backup take place?
6. Who will do the backup?

Whatever equipment, person, or method is chosen, ensure that the backup is done. If you choose the best equipment, the best software, and the brightest person, but the backup is not done for whatever reason, then you have wasted your resources and put your data at risk.

Backups can be done with floppy disks, extra hard drives (including network drives), compact disk drives, tape drives, and other forms of removable media. The best method of file backup is probably tape drives, which can store 25+ GB.

Question: How often should the backup be done?

Answer: Frequency of backup depends on the importance of the data. If you have many customers loaded into a database that is constantly changed or your files represent the livelihood of your business, then you should back them up every day. If you send only a few letters throughout the week with nothing vitally important, then you can back up once a week.

16.3.1 Types of Backups

All types of backups can be arranged into four categories. See Table 16.1.

TABLE 16.1 Types of backup

Normal/Full	The full backup will back up all files selected and shut off the archive file attribute indicating the file has been backed up.
Incremental	An incremental backup will back up the files selected if the archive file attribute is on (files since the last full or incremental backup). After the file has been backed up, it will shut off the file attribute indicating that the file has been backed up.
Differential	A differential backup will back up the files selected if the archive file attribute is on (files since the last full backup). Different from the incremental backup, it does not shut off the archive attribute.
Copy	A copy backup will back up the files selected regardless of the archive attribute and will not shut off the archive attribute after the file has been copied. The copy backup is not typically used as part of a backup procedure. Instead, it would be used as an interim backup in case something goes wrong during major changes or upgrades.
Daily	A daily backup copies all selected files that have been modified on the day that the daily backup is performed. During the backup, files are not marked as having been backed up (archive attribute is not cleared).

EXAMPLE 1 You decide to back up the entire hard drive once a week on Friday, using the full backup method. Therefore, you perform a full backup every Friday. If the hard drive goes bad, you use the last backup to restore the hard drive.

EXAMPLE 2 You decide to back up the entire hard drive once a week on Friday, using the incremental method. Therefore, you perform a full backup on week 1. This will shut off all archive attributes, indicating that all files have been backed up. On week 2, week 3, and week 4, you perform incremental backups using a different tape or disk. Since the incremental backup turns the archive attribute, it backs up only new files and changed files; thus, the four backups comprise the entire backup. It is much quicker to back up a drive using an incremental backup than a full backup. If the hard drive fails, however, you must restore backup 1, backup 2, backup 3, and backup 4 to restore the entire hard drive.

EXAMPLE 3 You decide to back up the entire hard drive once a week on Friday, using the differential method. Therefore, you perform a full backup on week 1. This will shut off all archive attributes, indicating that all files have been backed up. On week 2, week 3, and week 4, you perform differential backups using a different tape or disk. Since the differential backup does not turn the archive attribute, it backs up the new files and the changed files since the last full backup; thus, the full backup and the last differential backup comprise the entire backup. It is much quicker to back up a drive using a differential backup than a full backup, but slower than an incremental backup. If the hard drive fails, you only restore backup 1 and the last differential in order to restore the entire hard drive.

After the backups are complete, check if they are working. For example, pick a nonessential file and restore it to the hard drive to check if the backups are empty or a backup/restore device is faulty.

As a manager, keep more than one backup. (Tapes and disks do fail.) One technique is to rotate three sets of backups. If you perform a full backup once a week, then use three sets of backup tapes or disks. For example, during week 1, use tape/disk 1; during week 2, use tape/disk 2; and during week 3, use tape/disk 3. On week 4, begin again with tape/disk 1. If you have to restore a hard drive and the tape or disk fails, you can always go to the tape or disk from the previous week. In addition, perform monthly backups and store them elsewhere. You would be surprised how many times a person loses a file but may not be aware of it for several weeks. If the data are especially important, consider keeping a backup set in a fireproof safe offsite. Lastly, when Windows 2000 is initially installed and you make any major changes to the system's configuration, always make two backups before proceeding. This way, if anything goes wrong, you have the ability to restore everything back to the way it was before the changes. The reason for the two backups is that tapes have been known to go bad on occasion.

After completing a backup, properly label the tape or disk before removing it and then store it in a secure, safe place. Also keep a log of what backups have been done, especially if you need to rebuild the server, so you know what was backed up and when. Keeping a log will also let you know if someone is forgetting to do the backup.

16.3.2 Performing a Backup

To start the backup program, click on the Start button, select the Programs option, select the Accessories option, select the System Tools option, and select Backup. The Backup program can be used to back up and restore files and to create an emergency repair disk. The Welcome tab has buttons to start Backup, Restore, and Emergency Repair Disk wizards. See Figure 16.1.

To back up or restore files, you must be an administrator or a member of the Backup Operator group. To back up files to a file or tape, do the following:

1. Open the Backup program and click on the Backup tab. See Figure 16.2.
2. Open the Job menu and click New.
3. Select the files and folders that you want to back up by clicking the box to the left of a file or folder. You can back up files on other computers by using the My Network Places located at the bottom of the tree.

FIGURE 16.1   Windows 2000
Backup and Recovery Tools
welcome screen

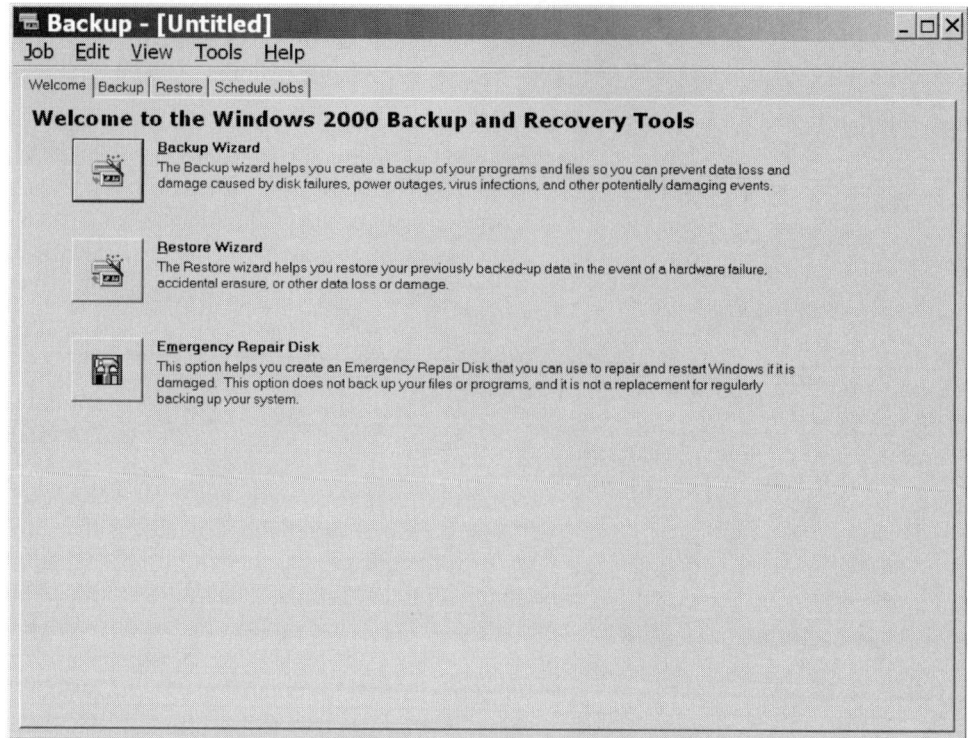

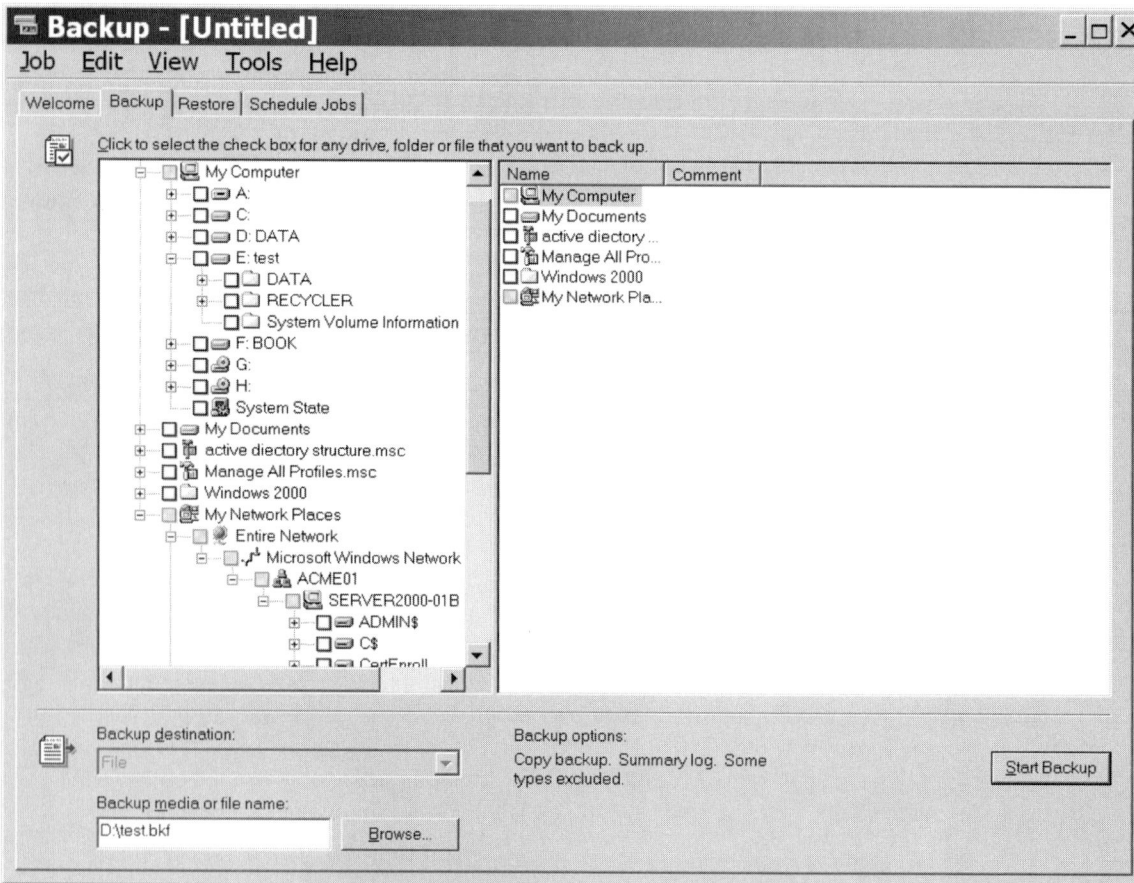

FIGURE 16.2   Using the Windows 2000 Backup program

4. In Backup Destination, select Tape Device if you want to back up to a tape drive or File if you want to back up to a file. If you do not have a tape device installed on your computer, File is selected by default.

5. In Backup Media or File Name, select the tape that you want to use or the path and file name. Backup files have a BKF file name extension.

6. Open the Tools menu and select the Options menu to select the backup type and log file type.

7. Click on the Start Backup button.

8. After making any changes to the Backup Job Information dialog box, click on the Start Backup button.

9. If you want to set data verification or hardware compression, or change the backup type, click on the Advance button. Click on the OK button to close the Advanced Backup options.

10. Click on the Start Backup button.

If you are attempting a backup and the Backup program indicates that there is no unused media available, you may have to use the Removable Storage console to add your tape to the Backup Media pool. You can access the Removable Storage console by loading the Computer Management console that is located under Administrative Tools.

16.3.3   Scheduling Backup

Because backup is performed on a regular basis and the best time to do a backup is when the network traffic is slow, you can schedule the backup to occur automatically. To schedule an automatic backup, do the following:

1. Start the Backup program.

2. Using the Backup tab, open the Job menu and select the New option.

3. Select the files and folders that you want to back up by clicking the box to the left of a file or folder, under "Click to select the check box for any drive, folder or file that you want to back up."

4. In Backup Destination, select File or a tape device in Backup Destination, and then save the file and folder selections by clicking the Job menu and then clicking Save Selections.

5. In Backup Media or File Name, type a path and file name for the backup file or select Tape.

6. If you want to select any Backup options, open the Tools menu and select the Options menu. Click OK to close the Options dialog box.

7. Click Start Backup and make any changes you want to the Backup Job Information dialog box.

8. If you want to set Advanced Backup options such as data verification or hardware compression, click the Advanced button. To close the Advanced Backup options, click the OK button.

9. Click Schedule in the Backup Job Information dialog box.

10. In the Set Account Information dialog box, enter the user name and password that you want the schedule backup to run under.

11. In the Scheduled Job Options dialog box, in Job Name, type a name for the scheduled backup job and then click Properties to set the data, time, and frequency parameters for the scheduled backup. When finished, click twice on the OK button.

If you are scheduling a tape backup, you may have to use the removable storage option to make sure that your tape is available in the Backup Media pool. In addition, you must have the Task Scheduler service running before you can schedule a backup. You can change the settings of a scheduled backup job or delete the job after you schedule it by clicking on the Schedule Jobs tab and then clicking on the job. See Figure 16.3.

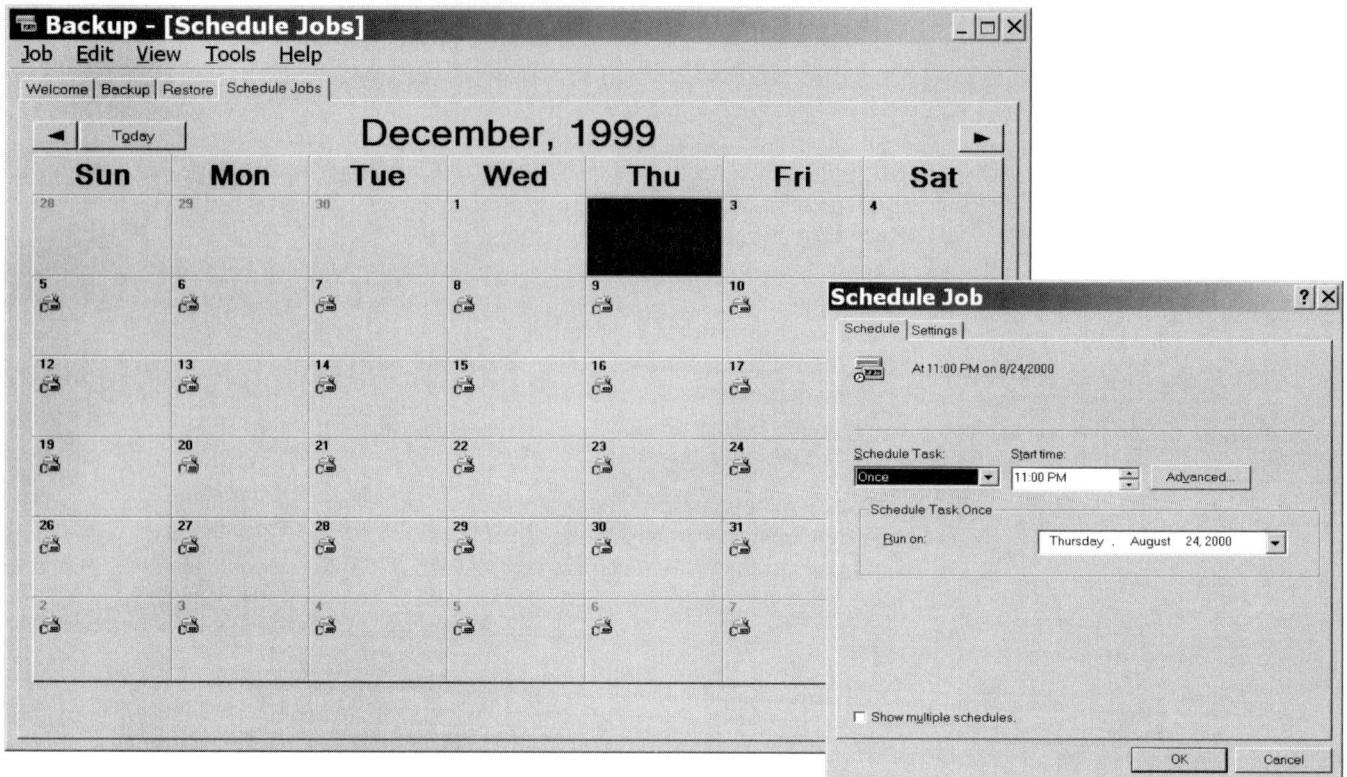

FIGURE 16.3   Scheduling backup jobs

16.3.4   System State

The **System State** data are a collection of system-specific data that can be backed up and restored. For Windows 2000, it includes the following:

- Registry
- COM+ Class Registration database
- System boot files
- Certificate Services database on servers that operate as a certificate server
- Active Directory directory services database on domain controller
- SYSVOL directory on domain controllers

Since the System State components depend on each other, when you choose to back up or restore the System State data, you must back up all of the System State data. To do so, put a check in the System State checkbox, located after all local drive letters. See Figure 16.4.

When you restore the individual components of the System State components, you can restore the registry files, SYSVOL directory files, cluster database information, and system boot files to an alternate location. When you back up and restore the System State data, you can only back up and restore on a local computer, not on remote computers.

16.3.5   Catalogs

When performing backups, users typically make a catalog. The **catalog** is a summary of the files and folders that have been saved in a backup set. The on-disk catalog is stored on the local disk drive, whereas the on-media catalog is stored on the backup storage media such as the backup tapes.

If you have all the tapes in the backup set and the tapes are not damaged or corrupted, you can speed the cataloging processor if you open the Tools menu, select the Options option, click the General tab, and note if the "Use the catalogs on the media to speed up build-

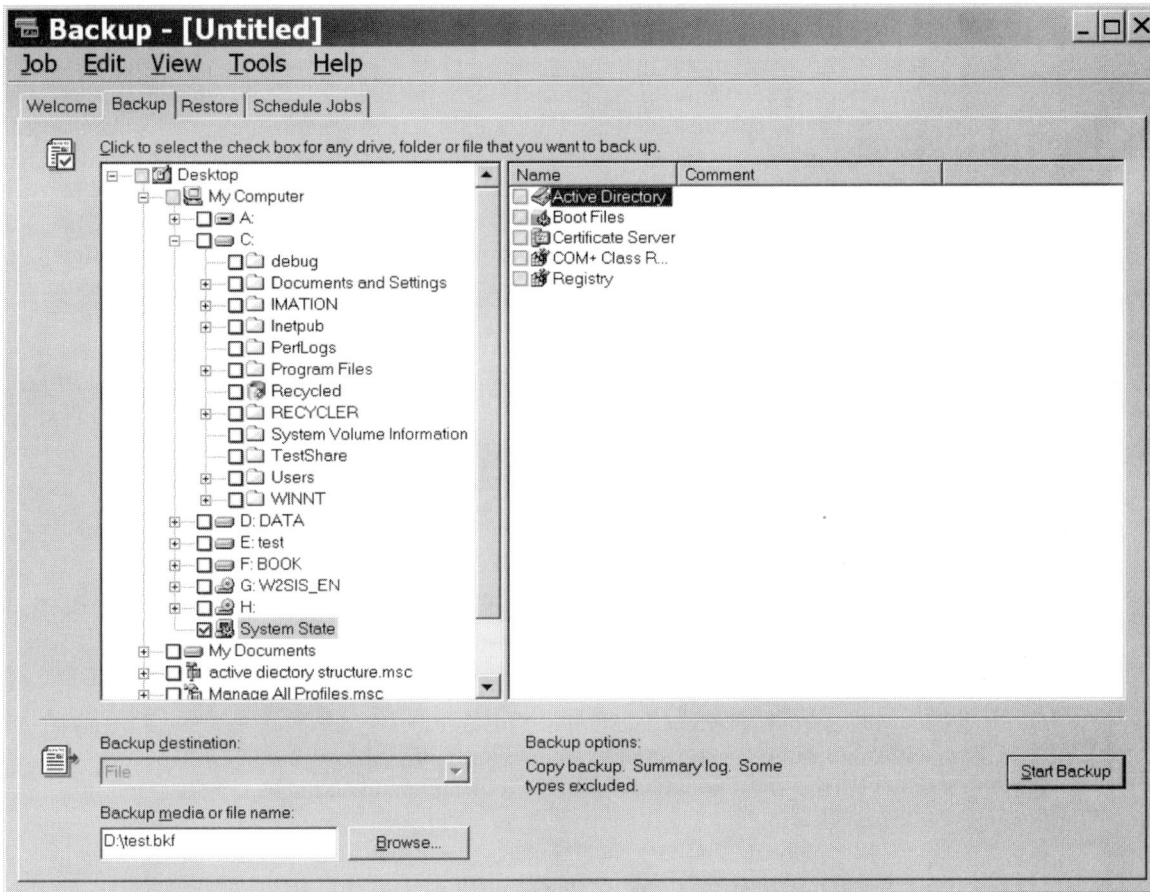

FIGURE 16.4   Backing up the System State

FIGURE 16.5   Backup's general options

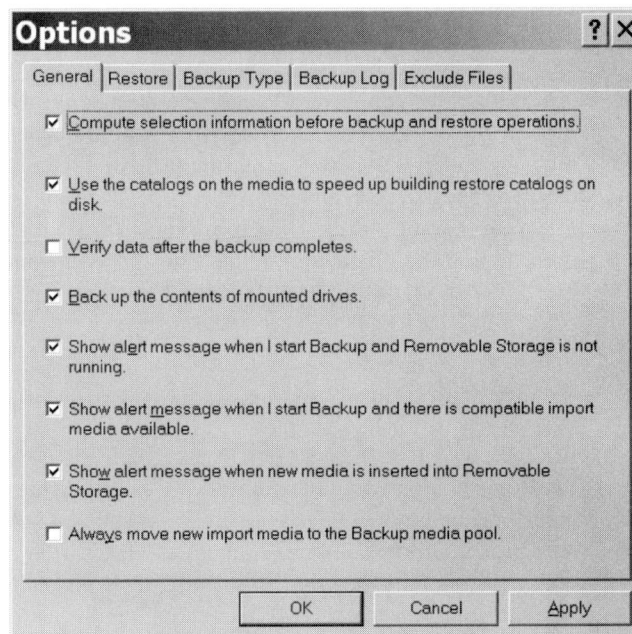

ing restore catalogs on disk" checkbox is marked. See Figure 16.5. However, if you want to restore data from several tapes and the tape with the on-media catalog is missing or you want to restore data from media that are damaged, do not select this option. Instead, the Backup program will scan the backup set and build an on-line disk catalog. Unfortunately, if your backup set is quite large, then this process could take several hours.

16.3.6 Performing a Recovery

When a Windows 2000 computer fails and you lose all information, you must first correct any hardware problems and then reinstall Windows 2000. Much like the backup, you can restore the computer using the Restore wizard or the Restore tab, both of which are located in the Backup program. The restore process will reload the Windows 2000 operating system, the Active Directory including database and registry settings, and any other services that were loaded. If you have more than one domain controller, the active directory will be replicated throughout the domain controllers. Therefore, if a domain controller goes down, you bring the domain controller up and let the active directory replicate to the domain controller.

An advantage of using today's backup programs is you can restore individual files or directories. For example, if someone accidentally deletes a file or a file becomes corrupted, you can then restore that file without overwriting the other files.

To restore files, do the following:

1. Open the Backup program.
2. Click on the Restore tab.
3. Click to select the drive, folder, or file that you want to restore.
4. In the Restore Files To: text box, select the location that the files will be restored:
   Original location
   Alternate location
   Single folder
   If you selected Alternate location or Single folder, type a path for the folder under Alternate location, or click the Browse button to find the folder.
5. Click on the Tools menu, click the Options option, click the Restore tab, and select one of the following:
   Do not replace the file on my computer
   Replace the file on disk only if the file on disk is older
   Always replace the file on my computer
6. Click on the Start Restore button.
7. If you want to change any of the advanced restore options, such as restoring security settings, click on the Advanced button. When done, click on the OK button.
8. Click OK to start the restore.

Let us say that you have a Windows 2000 Active Directory object, which has been corrupted. Of course, you can choose to restore the System State data. If you have several domain controllers, they would automatically overwrite the newly restored Active Directory database. To overcome this problem, perform an authorative restore using NTDSUTIL utility so that the Active Directory objects will be replicated to the other domain servers even though the objects are older than those currently in the Active Directory database. The NTDSUTIL utility can be found on the Windows 2000 installation CD under the \SUPPORT\RESKIT\NETMGMT folder.

16.3.7 Backup and Restore at the Command Line

Another way to do a backup is to use batch files that utilize the NTBACKUP command. The advantage of using batch files is you can run the batch file quickly if you need to do a quick backup before making any major changes. With the NTBACKUP command. You can only back up folders, not individual files. In addition, you cannot use wildcards like you can with the Command prompt commands, such as the COPY command.

The syntax for the NTBACKUP command is as follows:

ntbackup backup [systemstate] *"bks file name"* **/J** {*"job name"*} [**/P** {*"pool name"*}] [**/G** {*"guid name"*}] [**/T** { *"tape name"*}] [**/N** {*"media name"*}] [**/F** {*"file name"*}] [**/D** {*"set description"*}] [**/DS** {*"server name"*}] [**/IS** {*"server name"*}] [**/A**] [**/V:**{**yes|no**}] [**/R:**{**yes|no**}] [**/L:**{**f|s|n**}] [**/M** {*backup type*}] [**/RS:**{**yes|no**}] [**/HC:**{**on|off**}]

NTBACKUP Parameters

systemstate   Specifies that you want to back up the System State data. The backup type will be forced to Normal or Copy.

bks file name   Specifies the name of the backup selection file (.bks file) to be used for this backup operation. A backup selection file contains information on the files and folders you have selected for backup. Create the file using the graphical user interface (GUI) version of Backup.

/J {*"job name"*}   Specifies the job name to be used in the log file.

/P {*"pool name"*}   Specifies the media pool from which you want to use media. This selection may not be used with the following switches: **/A /G /F /T**.

/G {*"guid name"*}   Overwrites or appends to this tape. Do not use this switch in conjunction with **/P**.

/T {*"tape name"*}   Overwrites or appends to this tape. Do not use this switch in conjunction with **/P**.

/N {*"media name"*}   Specifies the new tape name. Do not use **/A** with this switch.

/F {*"file name"*}   Logical disk path and file name. You must not use the following switches with this switch: **/P /G /T**.

/D {*"set description"*}   Specifies a label for each backup set.

/DS {*"server name"*}   Backs up the directory service file for the specified Microsoft Exchange Server.

/IS {*"server name"*}   Backs up the Information Store file for the specified Microsoft Exchange Server.

/A   Performs an append operation. Either **/G** or **/T** must be used in conjunction with this switch. Do not use in conjunction with **/P**.

/V:{yes|no}   Verifies the data after the backup is complete.

/R:{yes|no}   Restricts access to this tape to the owner or members of the Administrators group.

/L:{f|s|n}   Specifies the type of log file: **f**=full, **s**=summary, **n**=none (no log file is created).

/M {*backup type*}   Specifies the backup type. It must be one of the following: Normal, Copy, Differential, Incremental, or Daily.

/RS:{yes|no}   Backs up the Removable Storage database.

/HC:{on|off}   Uses hardware compression, if available, on the tape drive.

EXAMPLE 4   NTBACKUP BACKUP C:\ /J "JOB1" /A /T "Tape 1" /M COPY

This example will perform a copy backup type called JOB1 of the entire C drive (logical drive). The back up files and folders will be appended (/A) to the tape called TAPE1.

16.4 EMERGENCY REPAIR

If Windows 2000 cannot load or start, accessing certain tools may help you solve the problem. They are the Advanced Boot menu, Windows 2000 Recovery console, and emergency repair disk.

16.4.1 Advanced Boot Menu

As discussed in Chapter 7, if you press F8 while the Windows 2000 Boot menu is displayed, it will display the Advanced Boot menu. Safe mode lets you start your system with a minimal set of device drivers and services (PS/2 mouse, monitor, keyboard, mass storage, base video, default system services) with no network connection. Typically, you would use safe mode if you installed a device driver or software that prevents the computer from starting, in the hopes that safe mode would remove either from your system.

Using the Enable VGA mode is another option if the wrong video driver is loaded and you need to force the video system into 640x480 with 16 colors, and the Last Known Good Configuration to start Windows 2000 using the last saved configuration which is stored in the registry.

The Directory Service Restore mode is a special version of safe mode that loads all drivers and services, and performs a CHKDSK on all volumes. After logging in, it allows you to restore the Active Directory of the domain controller from Backup Media. After getting into this mode, it will automatically set the computer to have the Active Directory check all indices next time you boot the domain controller normally.

Lastly, the Debugging mode has Windows 2000 send debug information through a serial cable to another computer. This allows you to monitor the process of a server's boot from another server.

16.4.2 Windows 2000 Recovery Console

If your computer fails to boot even in safe mode, the best tool to use is the **Recovery console.** The Recovery console provides a command line interface that will let you repair system problems using a limited set of command line commands, including enabling or disabling services, repairing a corrupted master boot record, and reading and writing data on a local drive (FAT, FAT32, or NTFS). Because Recovery console is a powerful tool, it can only be used by advanced users from the Administrators group who have a thorough knowledge of Windows 2000.

There are two ways to start the Recovery console. (1) You can run it from your Windows 2000 Setup disks or from the Windows 2000 Professional CD; or (2) you can install it on your computer so it is available in case you are unable to restart Windows 2000. To start the computer and use the Recovery console, do the following:

1. Insert the Windows 2000 Setup CD, or the first floppy disk you created from the CD, in the appropriate drive. If you cannot boot from the CD drive, you must use a floppy disk, and restart the computer. If you are using a floppy disk, you will be asked to switch disks.
2. When the text-based part of Setup begins, follow the prompts and choose the Repair or Recover option by pressing R. When prompted, choose the Recovery console by pressing C. Again, if you are using a floppy disk, you will be asked to switch disks.
3. If you have a dual-boot or multiple-boot system, choose the Windows 2000 installation that you need to access from the Recovery console.
4. When prompted, type the Administrator password.
5. At the System prompt, type the Recovery console commands.
6. To exit the Recovery console and restart the computer, type Exit.

To install the Recovery console as a Startup Menu option, do the following:

1. Insert the Windows 2000 Setup CD into your CD-ROM drive.
2. Click No when prompted to upgrade to Windows 2000.
3. At the Command prompt, switch to your CD-ROM drive, and then type the following: \I386\WINNT32.EXE /CMDCONS.
4. When it asks if you want to install the Recovery console, click on the Yes button.
5. Click OK when Setup is complete.
6. To run the Recovery console, restart your computer and select the Recovery Console option from the Boot menu.

To get a list of the commands that are available in the Recovery console, type in Help while in the console. See Table 16.2. To get help about a specific command, type in Help *commandname.* While the console resembles DOS, these commands are very limited. By default, you can copy from removable media to hard disk but not vice versa. In addition, wild characters in the copy command don't work, and you cannot read or list files on any partition except the system partition.

16.4.3 Emergency Repair Disk

The **emergency repair disk (ERD)** contains information about your current Windows system settings. You can use this disk to repair your computer if it will not start or your system files are damaged or erased. To create an emergency repair disk, do the following:

1. Open the Backup program.
2. On the Tools menu, click Create an Emergency Repair Disk. Or, you can use the Emergency Repair Disk wizard.
3. Insert a blank, formatted floppy disk into the drive. Mark the checkbox if you want a backup copy of the registry copied to the REPAIR folder (highly recommended) and click on the OK button.

TABLE 16.2   List of commands used in the Recovery console

Command	Description
ATTRIB	Changes or displays attributes of files or directories
CD	Changes directory
CHKDSK	Executes a consistency check of the specified disk
CLS	Clears the screen
COPY	Copies a file
DEL	Deletes a file
DIR	Lists Directory contents
DISKPART	Adds and deletes partitions
ENABLE	Starts or enables a system service or a device driver
DISABLE	Stops or disables a system service or device driver
EXTRACT	Extracts a file from a compressed file
FIXBOOT	Writes a new partition boot sector onto the system partition
FIXMRB	Repairs the master boot record of the partition boot sector
FORMAT	Formats a disk
LISTSVC	Lists the services and device drivers available on the computer
LOGON	Logs on to a Windows 2000 computer
MAP	Displays the drive letter mappings
MD	Creates a directory
TYPE	Displays a text file
RMDIR	Deletes a directory
REPAIR	Updates an installation with files using the Windows 2000 installation CD
REN	Renames a file
SYSTEMROOT	Changes to the installation system root directory (typically \WINNT)

4. Click on the OK button when the disk has been created.
5. Label the disk as Emergency Repair Disk and store in a safe place.

When using an emergency repair disk, the disk relies on information that has been saved to the *systemroot*\REPAIR folder. Therefore, do not change or delete this folder. You also should re-create the ERD after each service pack, system change, or updated driver.

To use an emergency repair disk for system repairs, do the following:

1. Insert the Windows 2000 Setup CD, or the first floppy disk you created from the CD, in the appropriate drive.
2. When the text-based part of Setup begins, follow the prompts; choose the Repair or Recover option by pressing R.
3. When prompted, insert the Windows 2000 Setup CD in the appropriate drive.
4. When prompted, choose the Emergency Repair Process by pressing R.
5. When prompted, choose between the following:
   - Manual Repair (press M): This should be used only by advanced users or administrators. Use this option to choose whether you want to repair system files, partition-boot sector problems, or start up environment problems.
   - Fast Repair (press F): This is the easiest option, and does not require input. This option will attempt to repair problems that are related to system files, the partition boot sector on your system disk, and your startup environment (if you have a dual-boot or multiple-boot system).
6. Follow the instructions on the screen and, when prompted, insert the emergency repair disk in the appropriate drive.

7. During the repair process, missing or corrupted files are replaced with files from the Windows 2000 CD or from the *systemroot*\Repair folder on the system partition. Be sure to follow the instructions on the screen; you may want to write the names of files that are detected as faulty or incorrect, to help you diagnose how the system was damaged.

8. If the repair was successful, allow the process to complete; it will restart the computer. The restarting of the computer indicates that replacement files were successfully copied to the hard disk.

If you have performed the emergency repair process and the computer still does not operate normally, then use the Windows 2000 Setup CD to perform an **in-place upgrade** over the existing installation. This is a last resort before reinstalling the operating system. An in-place upgrade takes the same amount of time as a reinstallation of the operating system. To perform an in-place upgrade of Windows 2000, do the following:

1. Install the Windows 2000 Setup CD.
2. Press Enter to install a copy of Windows 2000.
3. When prompted to repair the existing Windows 2000 installation, press R.

16.5 WINDOWS 2000 ERROR MESSAGES

Much like Windows 3.XX, Windows 95, Windows 98, and Windows NT, Windows 2000 may encounter a program error or a stop error. These errors are similar to Windows 95 or Windows 98 general protection errors.

16.5.1 Invalid Page Faults and Exception Errors

A program error or stop error signifies that something unexpected has happened within the Windows environment. This situation usually involves a program that tried to access a memory area which belongs to another program (**invalid page fault**) or an application that tried to pass an invalid parameter to another program (**exception error**). **Dr. Watson** is a program that (1) starts automatically when a program error occurs, (2) detects information about the system and the program failure, and (3) records the information into a log file. Dr. Watson will not prevent errors, but its log file can be used by technical support personnel to diagnose the problem. To open Dr. Watson, click the Start button, select the Run option, and execute the DRWTSN32 program. See Figure 16.6. For more information about the Dr. Watson log file, see Dr. Watson Log File Overview using Windows 2000 Help program.

FIGURE 16.6 Dr. Watson utility

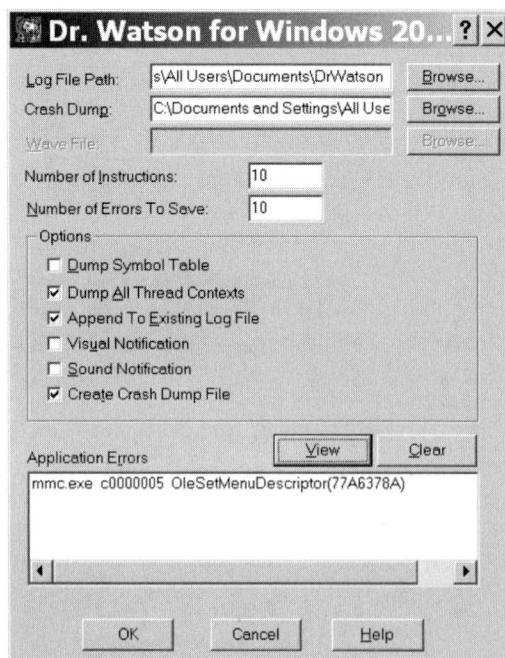

16.5.2 Stop Errors

Applications run in the user mode layer (Intel 386 protection model—ring 3). When an application causes an error, Windows 2000 halts the process and generates an illegal operation error. Because every Win32 application has its own virtual protected memory space, this error condition does not affect any other Win32 program that is running. If the application tries to access the hardware without going through the correct methods, Windows 2000 takes note and generates an exception error. In these cases, if the application faults, then either Windows 2000 or the user can close the offending program and resume work without affecting the other programs.

The Windows 2000 Kernel runs in the kernel mode layer (Intel 386 protection model—ring 0). When the kernel encounters a fatal error—such as a hardware problem, inconsistencies within data that are necessary for its operation, or a similar error—Windows can display a **stop error** (sometimes known as a "blue screen of death"). During this time, the system is stopped to prevent data corruption. These errors can be caused by hardware errors, corrupted files, a corrupted file system, or a software glitch.

Even though a stop screen may look intimidating, only a small amount of data on the screen is important in determining the cause of the error. At the top of the screen you will find the error code and parameters. In the middle of the screen is the list of modules that have been successfully loaded and initialized. At the bottom of the screen is a list of modules that are currently on the stack. See Figure 16.7.

You can configure Windows 2000 to write a memory dump file each time it generates a kernel STOP error. To make a memory dump file, you need sufficient space on a hard disk partition for the resulting file, which will be as large as your RAM memory. For example, if your system has 256 MB of RAM, you need 256 MB of free disk space.

To configure the Windows 2000 to save STOP information to a memory dump file, double-click on the System applet in the Control Panel, click on the Advanced tab, and click on the Startup and Recovery button. Within the Startup/Shutdown tab, mark the Write Debugging Information checkbox in the Recovery section. You can then specify the file location and name. If the "Overwrite any existing file" option is not checked and there is a file with that name, Windows 2000 will not overwrite the file.

Since most of the data in the dump file are useless to most support technicians who examine the file to troubleshoot the cause of the crash (instead it is aimed at programmers), Windows 2000 introduces a new option. This new alternative saves only the Windows 2000 Kernel information, which contains the state of the system at the time of a crash, active applications, loaded device drivers, and the executing code. A kernel-data-only crash dump is much smaller in size compared with the full dump. To choose the type of kernel memory dump, right-click My Computer and select Properties or double-click the System applet in the Control Panel. From the Advanced tab, click the Startup and Recovery button. See Figure 16.8.

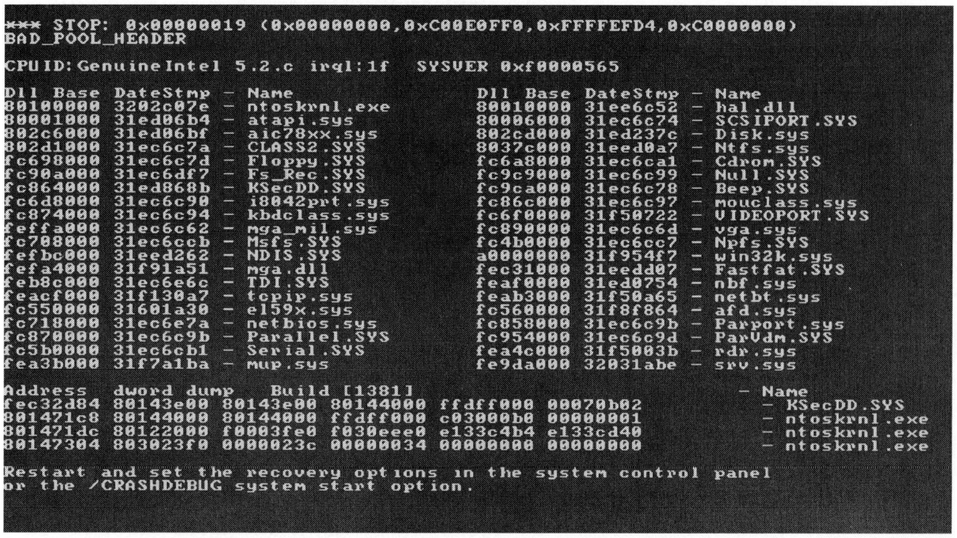

FIGURE 16.7 A typical stop error

FIGURE 16.8 Configure the Memory Dump that occurs during a stop error

Startup and Recovery

System startup

Default operating system:

"Microsoft Windows 2000 Server" /fastdetect

☑ Display list of operating systems for 30 seconds.

System Failure

☑ Write an event to the system log

☑ Send an administrative alert

☑ Automatically reboot

Write Debugging Information

Kernel Memory Dump

Dump File:

%SystemRoot%\MEMORY.DMP

☑ Overwrite any existing file

OK   Cancel

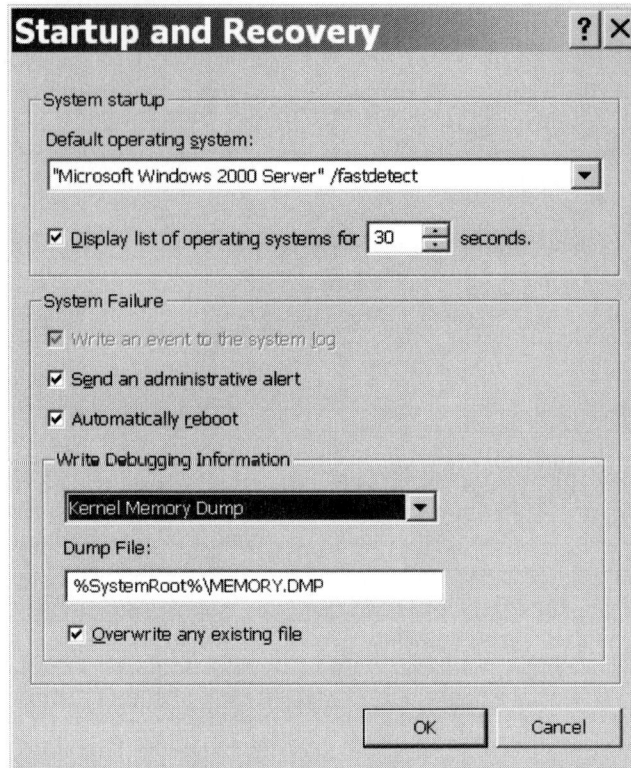

Many of the common stop errors are already listed and explained in Windows 2000 Server Help files. Other sources of information on stop errors include Microsoft's website and/or Microsoft Technet for possible solutions. In any case, make sure that the system has the most updated system BIOS and the newest Windows 2000 Service Pack.

16.6 VIRUSES

A **virus** is a program designed to replicate and spread, generally without the knowledge or permission of the user. Computer viruses spread by attaching themselves to other programs or to the boot sector of a disk. When an infected file is executed or accessed or the computer is started with an infected disk, the virus spreads onto the computer. Virus forms differ in that some are cute, some are annoying, and others are disastrous. Some of the disastrous symptoms of a virus include the following:

1. Computer fails to boot.
2. Disks have been formatted.
3. The partitions are deleted or the partition table is corrupt.
4. Cannot read a disk.
5. Data or entire files are corrupt or are disappearing.
6. Programs no longer run.
7. Files become larger.
8. System is slower than normal.
9. System has less available memory than it should.
10. Intercept information is being sent to and from a device.

If a virus gets on a server, it has the potential to infect every computer on the network. See Table 16.3.

Question: How does a virus spread?

Answer: Because viruses are small programs that are made to replicate themselves, they spread quite easily. For example, you are handed an infected disk or you download a file from the Internet or a bulletin board. When the disk or file is accessed,

TABLE 16.3   Virus facts

Viruses cannot infect a write-protected disk.	Viruses can infect read-only, hidden, and system files.
Viruses do not typically infect a document (except macro viruses).	Viruses typically infect boot sectors and executable files.
Viruses do not infect compressed files.	A file within a compressed file could have been infected before being compressed.
Viruses do not infect computer hardware such as monitors or chips.	Viruses can change your CMOS values causing your computer not to boot.
You cannot get a virus simply by being on the Internet or a bulletin board. Unfortunately, this fact is beginning to change, because ActiveX and Java controls are being used by today's web pages.	You can download an infected file.

TABLE 16.4   General information about viruses

Computer Virus Information and Virus Description Database	http://www.datafellows.com/vir-info/
Hoax Warnings on the Run	http://www.datafellows.com/news/hoax/
Symantec Antivirus Research Center	http://www.symantec.com/avcenter

the virus replicates itself to RAM. When you access any files on your hard drive, the virus again replicates itself to your hard drive. If you shut off your computer, the virus in the RAM will disappear. Unfortunately, since your hard drive is infected, the RAM becomes infected every time you boot from the hard drive. When you insert and access another disk, the disk also becomes infected. You then hand the disk or send an infected file to someone else and the cycle repeats itself.

For more information on viruses, how they work, how they affect your computer, and for descriptions of particular viruses, check out the websites given in Table 16.4.

Antivirus software is a package that will detect and remove viruses and help protect your computer against viruses. Whichever software package you choose, it should include a scanner/disinfector and an interceptor/resident monitor. The scanner/disinfector software will look for known virus patterns in the RAM, the boot sector, and the disk files. If a virus is detected, the software will typically attempt to remove the virus. The interceptor/resident monitor software is loaded and remains in the RAM. Every time a disk is accessed or a file read, the interceptor/resident monitor software will check the disk or file for the same virus patterns as the scanner/disinfector software. In addition, some interceptor/resident monitor software will detect files as you download them from the Internet or a bulletin board.

Unfortunately, scanner/disinfector software has three disadvantages. First, the software can only detect viruses that it knows about. Therefore, you must constantly update your antivirus software package, the easiest way being through the Internet. Second, the virus cannot always be removed. Therefore, you may need to delete the file or perform a low-level format on the hard drive. Lastly, if the virus is removed, the file or boot sector may still be damaged. Therefore, you need to delete/replace the infected file, re-create the boot sector, or perform a low-level format on the disk.

If you think you have a virus, even though you have interceptor software installed, then boot from a clean write-protected disk to ensure that the RAM is virus-free. Without changing to or accessing the hard drive, run an updated virus scanner/disinfector from the floppy disk. If a virus is detected and removed, it is then best to reboot the computer when the scanner/disinfector is done checking the hard drive. You also should boot from a bootable floppy and check the hard drive before installing any antivirus software.

To avoid viruses, do the following:

1. Avoid using pirated software.
2. Treat files that are downloaded from the Internet and bulletin boards with suspicion.
3. Never boot from or access a floppy disk of unknown origin.
4. Educate your fellow users.
5. Use an updated antivirus software package that constantly detects viruses.
6. Back up files on a regular basis.
7. Do not use the Administrative accounts for general use.
8. Give no more rights than what are needed.

If you suspect a virus, immediately check your hard drive and disk with a current antivirus software package. If you think that your hard drive is infected, obtain a noninfected, write-protected bootable floppy disk that contains the antivirus software; then boot the computer with the noninfected disk without accessing the hard drive. Next, run the software to check if the virus is gone from the hard drive. If you suspect a virus on the floppy disk, boot your computer using the hard drive. If you have been using the possibly infected disk on your computer, then first check the hard drive for viruses. Lastly, execute the antivirus program to check the floppy drive.

16.7 AUDITING

Auditing is a feature of Windows 2000 that monitors various security-related events so that you can detect intruders and attempts to compromise data on the system. Some events that you can monitor are access to an object such as a folder or file, management of user and group accounts, and logging on and off a system. The security events are provided in the Event Viewer. See Figure 16.9. Therefore, auditing is one way to find security holes in your network and to ensure accountability for people's actions—including the administrators.

Events are not audited by default. If you have Administrator permissions, you can specify what types of system events to audit using group policies (Computer Configuration\

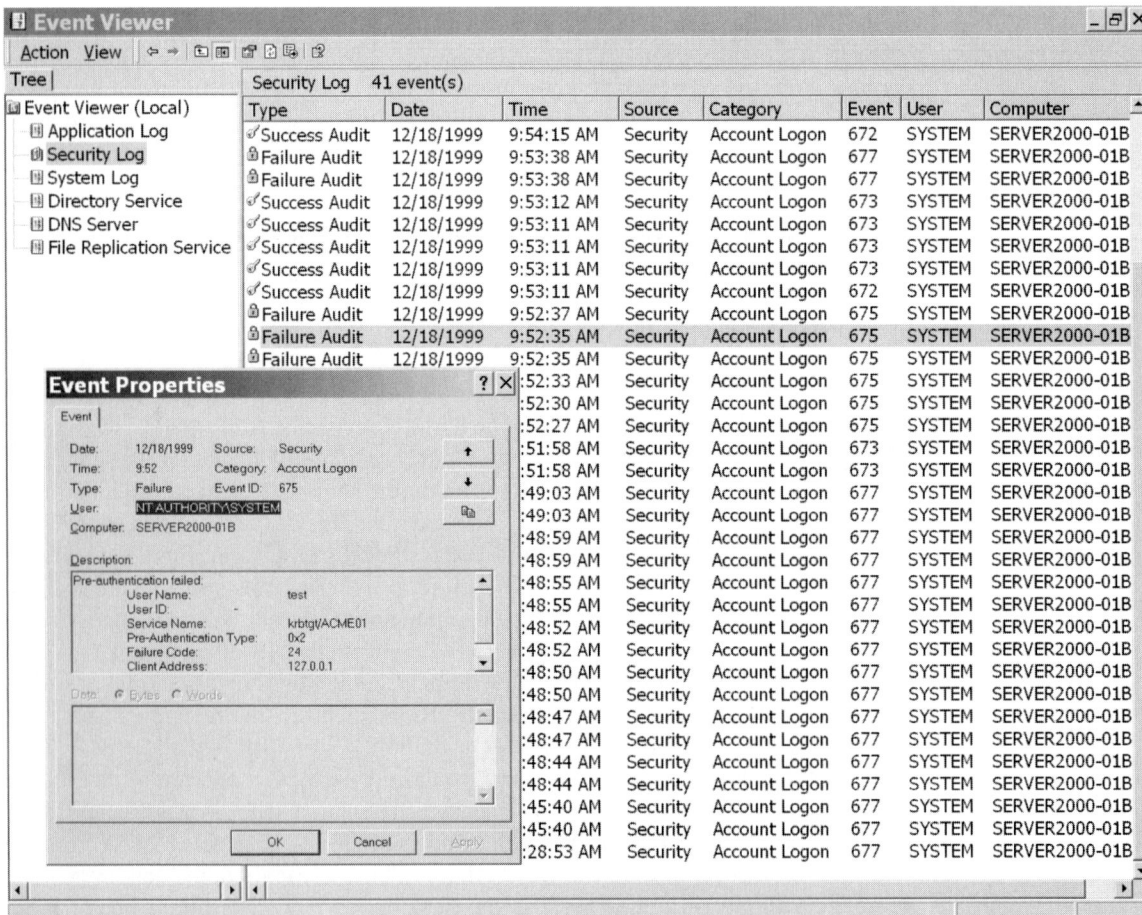

FIGURE 16.9   Viewing the Security Log in the Event Viewer

496

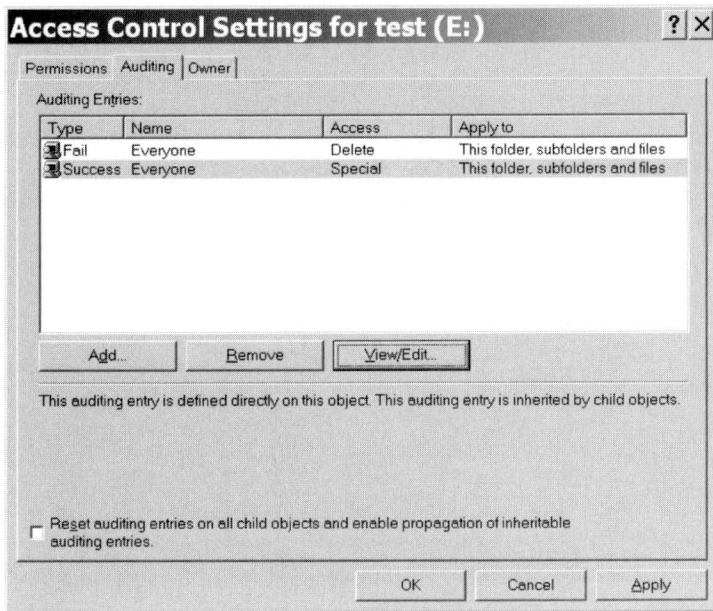

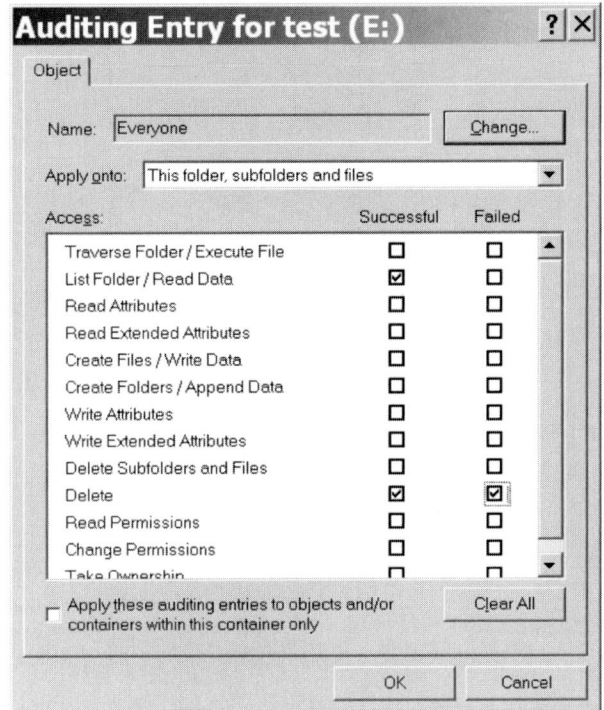

FIGURE 16.10   Configuring the Audit Policy

Windows Settings\Security Settings\Local Policies\Audit Policy). See Chapter 12 for more information on group policies and also Figure 16.12.

For files and folders, you can only audit those volumes that are formatted with NTFS. To set, view, or change auditing a file or folder, do the following:

1. Open Windows Explorer and locate the file or folder that you want to audit.
2. Right-click the file or folder and select the Properties option. See Figure 16.10.
3. Click the Security tab, click on the Advanced button, and click on the Auditing tab.
   - To set up auditing for a new group or user, click Add and specify the name of the user you want, and click the OK button to open the Auditing Entry dialog box.
   - To view or change auditing for an existing group or user, click the name and then the View/Edit button.
   - To remove auditing for an existing group or user, click the name and then the Remove button.

FIGURE 16.11   Configuring Security Log Properties

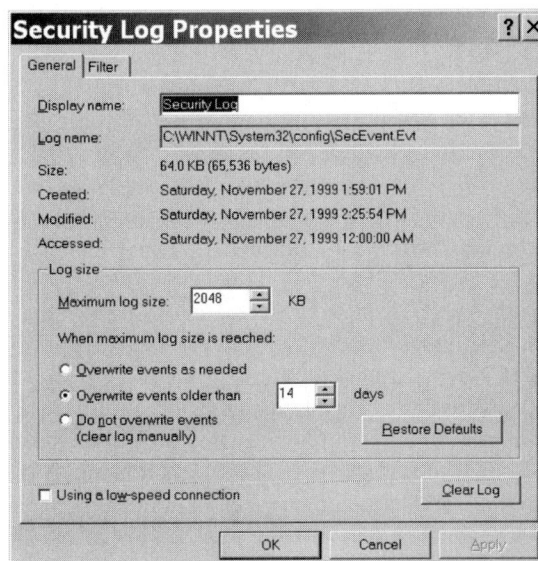

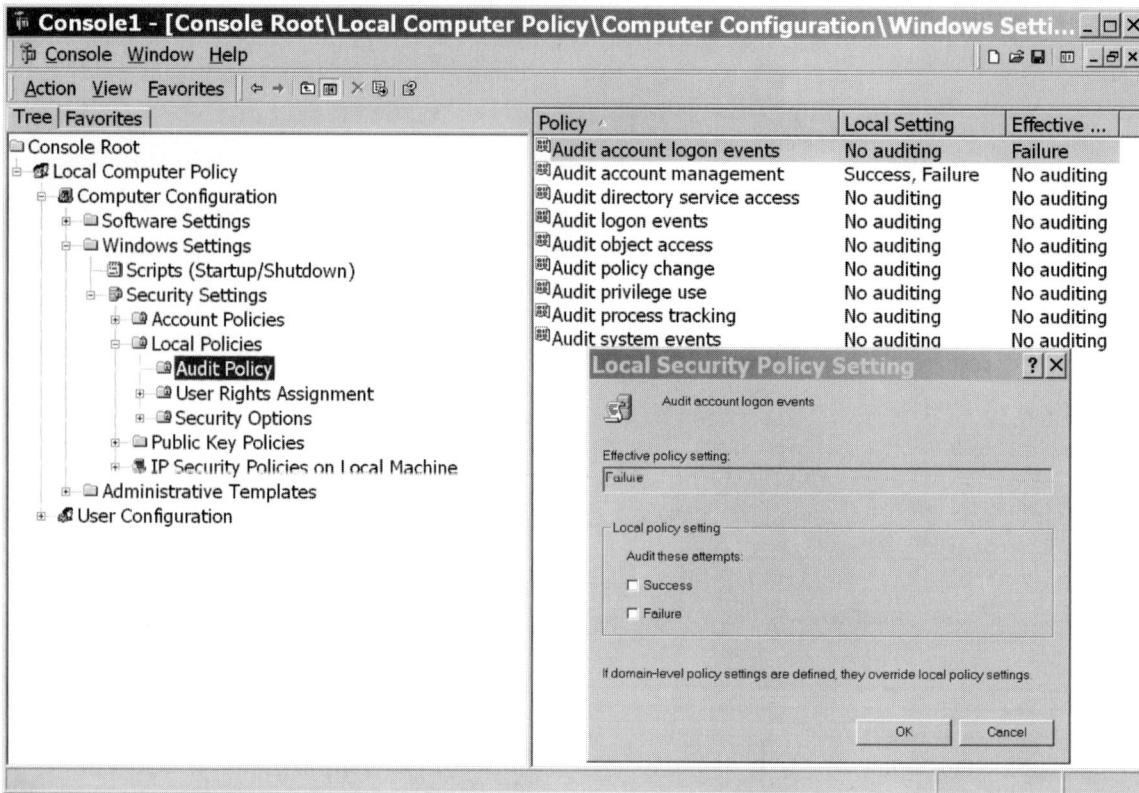

FIGURE 16.12   Configuring Security Log Properties

To perform auditing, you must be logged on as a member of the Administrator group or have been granted the Manage auditing and security log right in Group Policy.

Because the security log is limited in size, select only those objects that you need to audit and also consider the amount of disk space that the security log will need. The maximum size of the security log is defined in Event Viewer by right-clicking Security Log and selecting the Properties option. See Figure 16.11.

Taking certain auditing steps helps to minimize the risk of security threats. Table 16.5 lists various events that should be audited, as well as the specific security threat that the audit event monitors.

16.8   WINDOWS 2000 PERFORMANCE

Performance is the overall effectiveness of how data move through the system. To improve performance, determine the part of the system that is slowing the throughput—it could be the speed of the processor, the amount of RAM on the machine, the speed of the disk system, the speed of your network adapter card, or some other factor. This limiting factor is referred to as the bottleneck of the system. With Performance Monitor, you can measure the performance of your own computer or other computers on the network.

When you first start Performance Monitor, you see only a blank screen. You must select the objects, instances, and counters that you want to monitor. An object is any Windows 2000 System component that possesses a set of measurable property. It can be a physical part of the system such as the processor, RAM, disk subsystem, and network interface; a logical component such as a disk volume; or a software element such as a process or a thread. An instance shows how many occurrences of an object are available in the system. A counter represents one measurable characteristic of an object. See Figure 16.13.

TABLE 16.5   Audit best practices

Potential Threat	Audit Event
To watch for users who are trying random passwords to bypass security	Failure audit for logon/logoff
To watch for stolen password usage	Success audit for logon/logoff
To watch for misuse of privileges including those of administrators	Success audit for user rights, user and group management, security change policies, restart, shutdown, and system events
To watch the use of sensitive files	Success and failure audit for file-access and object-access events. File Manager success and failure audit of read/write access by suspect users or groups for the sensitive files.
To watch for the improper use of printers	Success and failure audit for file-access printers and object-access events. Print Manager success and failure audit of print access by suspect users or groups for the printers.
To watch for a virus outbreak	Success and failure write access auditing for program files (.EXE and .DLL extensions). Success and failure auditing for process tracking. Run suspect programs; examine security log for unexpected attempts to modify program files, or create unexpected processes. Perform this when you are actively monitoring the system log.

FIGURE 16.13   Adding Counters to the Performance Monitor

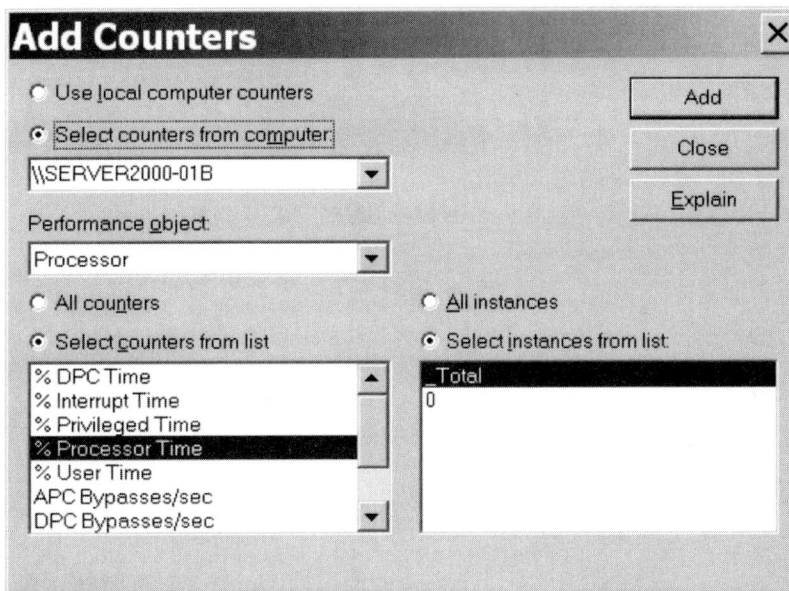

16.8.1   Performance Counters

As mentioned in Chapter 2, the computer is centered around the processor. Therefore, the performance of the computer is greatly affected by the performance of the processor. See Table 16.6.

Also mentioned in Chapter 2, RAM is an important factor in PC performance. You can typically increase PC performance by adding more RAM. The seven best RAM counters are listed in Table 16.7.

One task of a server is to provide file access, thus making a hard drive system an important factor in server performance. Important hard drive counters are listed in Table 16.8.

By default, the system is set to collect physical drive data. Logical drive data are not collected by default—you must enable them specifically. To enable the disk counters, execute the DISKPERF –Y command at the Command prompt and restart the computer.

The network interface counters measure the performance of the network interface cards. The more important counters are listed in Table 16.9.

TABLE 16.6 Specifics of a processor

Processor Counter	Description
Processor: % Processor Time	% Processor Time measures how busy the processor is. Although the processor may jump to 100% processor usage, overall average is still important. If the processor is at 80% all the time, you should upgrade the processor (creating a faster processor or adding a number of processors) or move some of the services to other systems. If you want to see what percentage of the processor utilization each process is using, utilize the Task Manager. Do not use 3-D screen savers, because they can consume 90% of the processor utilization. Instead, use a blank screen saver.
Processor: Interrupts/Second	Interrupts/Second measures how many hardware interrupts per second are occurring. A server that is running 100 interrupts/second can be normal. If the counter is increasing without a corresponding increase in the server load, then the problem may be due to hardware. You should always view this counter when you load a new device driver to make sure that it functions properly. A poorly written device driver can cause huge increases in interrupt activity.
System: Processor Queue Length	The number of threads indicated by the processor queue length is a significant indicator of system performance, because each thread requires a certain number of processor cycles. If demand exceeds supply, long processor queues develop and the system response suffers. Therefore, a sustained processor queue length greater than 2 on a single processor generally indicates that the processor is a bottleneck. This counter is always 0 unless you are monitoring a thread counter as well.

TABLE 16.7 Best RAM counters

Memory Counter	Description
Memory: Available Bytes	Available Bytes measures the amount of available virtual memory. It is calculated by summing space on the zeroed, free, and standby memory lists. Free memory is ready for use. Zeroed memory involves pages of memory filled with zeros to prevent later processes from seeing data used by a previous process. Standby memory is removed from a process's working set (physical memory) on route to disk, but is still available to be recalled. If it is less than 4 MB, consider adding more RAM.
Memory: Page Faults/Sec	A page fault occurs when a process attempts to access a virtual memory page that is not available in its working set in RAM. Hard page faults must be retrieved from disk, which greatly slows performance. Soft page faults are those that can be retrieved from the standby list, and therefore do not require disk I/O. If this value is high (20) or increasing, it is a sign that you need to add physical RAM to your server.
Memory: Cache Bytes	Cache Bytes monitors the size of the file system cache. When memory is scarce, the system trims the cache; when memory is ample, the system enlarges the cache. Compare this with general memory availability. In addition, track how small the cache gets, and how often that occurs, because this information is useful when associating the size of the cache with its performance.
Memory: Pages/sec	Pages/sec indicates the number of requested pages not immediately available in RAM but read from the disk or written to the disk to make room in RAM for other pages. If your system experiences a high rate of hard page faults, the value for Pages/sec can be high.
Memory: Committed Bytes	Committed Bytes is the amount of committed virtual memory, in bytes. Committed memory is physical memory for which space has been reserved on the disk paging file in case it needs to be written back to disk. This counter displays the last observed value only; it is not an average.
Paging File: % Usage	% Usage of the Paging file is the percentage of space allocated to the page file (virtual memory) that is actually in use. To calculate how much more RAM to add to a server to minimize paging, multiply the percentage by the size of the page file.
Paging File: % Usage Peak	% Usage Peak is the highest percentage usage of the page file. If this value frequently exceeds 90%, allocate more space to the page file and possibly more RAM. A 100% usage peak indicates that the server has, at least momentarily, run out of both physical and virtual memory.

TABLE 16.8 Hard drive counters

Disk Counter	Description
PhysicalDisk: % Disk Time	% Disk Time measures the percentage of elapsed time that a disk drive is actually occupied in reading data from and writing data to the disk. A value greater than 90% indicates that the disk is the bottleneck.
PhysicalDisk: % Avg. Disk Queue Length	% Avg. Disk Queue Length is the average number of read requests and writes queued for the disk in question. A sustained average higher than 2 queue length indicates that the disk is being over utilized.
LogicalDisk: % Free Disk Space	Free Disk Space reports the percentage of unallocated disk space to the total useable space on the logical volume.
LogicalDisk: Avg. Disk Bytes/Transfer and PhysicalDisk: Avg. Disk Bytes/Transfer	The Avg. Disk Bytes/Transfer measures the size of I/O operations. The disk is efficient if it transfers large amounts of data relatively quickly.
LogicalDisk: Avg. Disk sec/Transfer and PhysicalDisk: Avg. Disk sec/Transfer	Avg. Disk sec/Transfer indicates how fast data are being moved (in seconds). It measures the average time of each data transfer, regardless of the number of bytes read or written. A high value for this counter might mean that the system is retrying requests due to lengthy queuing or, less commonly, disk failures.
LogicalDisk: Disk Bytes/sec and PhysicalDisk: Disk Bytes/sec	Disk Bytes/sec indicates the rate at which bytes are transferred and is the primary measure of disk throughput.
LogicalDisk: Disk Transfers/sec and PhysicalDisk: Disk Transfers/sec	Disk Transfers/sec measures disk utilization by indicating the number of read and writes completed per second, regardless of how much data they involve. If the value exceeds 50 transfers/sec (per physical disk in the case of a striped set), then a bottleneck might be developing.

TABLE 16.9 Network interface counters

Network Interface Counter	Description
Network Interface: Output Queue Length	Output Queue Length indicates the length of the output packet queue. The value should be low. Queues of one or two items constitute satisfactory performance; longer queues indicate the adapter is waiting for the network and cannot keep pace with server requests.
Network Interface: Bytes Total/Sec	Bytes Total/Sec measures the rate that bytes, including data and framing characters, are sent and received on the network interface. If the value is close to or matching the network capacity, the network may be saturated.
Network Interface: Packets Outbound Errors	Packets Outbound Errors is the number of outbound packets that could not be sent because of errors. If the amount of errors is increasing, then more sophisticated troubleshooting tools are needed to determine the exact problem.
Network Interface: Packet Received Errors	Packet Received Errors is the number of inbound packets that had to be discarded because they contained errors that prevented them from being delivered to the proper destination protocol stack.

16.8.2 Performance Monitor

There are several views in Performance Monitor including chart and report. A real-time activity chart displays the value of the counter over time in a graph. Charts help to investigate why a computer or application is slow or inefficient, to continuously monitor the system to find intermittent performance problems, and to discover why the capacity of a subsystem needs to be increased. See Figure 16.14. A report view allows you to display constantly changing counter and instance values for selected objects. See Figure 16.15.

To add counters, right-click the chart or report pane (right pane) and select the Add Counters option. After you have identified the counters you want to monitor, save the information to reuse later. To save chart settings, right-click the chart or report pane and select the Save As option.

FIGURE 16.14 Chart view using the Performance Monitor

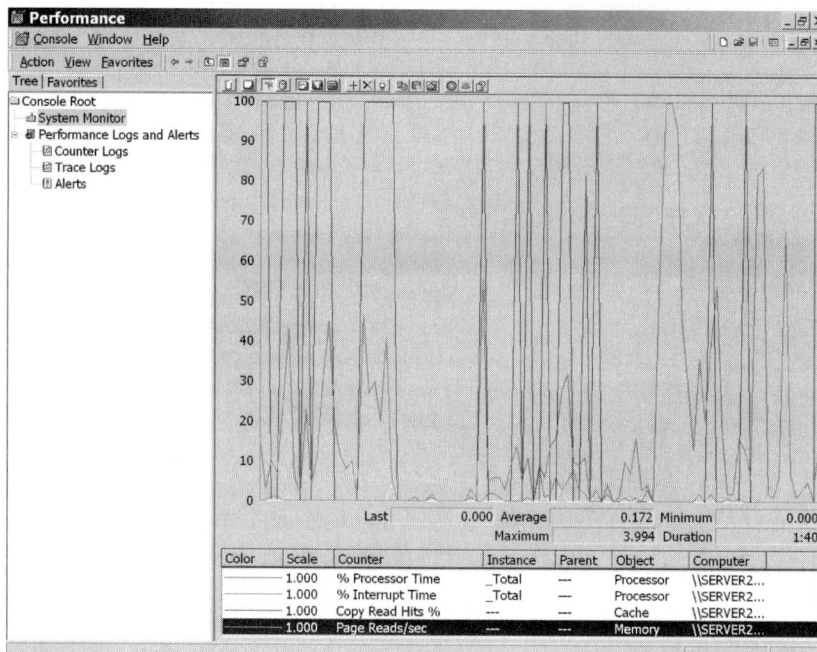

FIGURE 16.15 Report view using the Performance Monitor

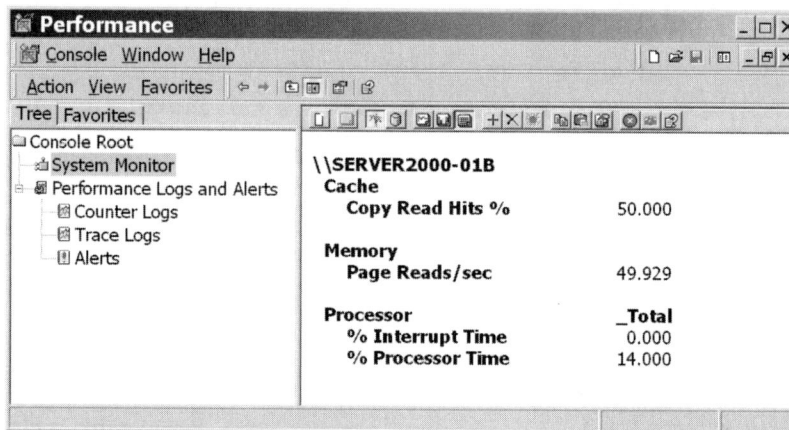

16.8.3 Logging Performance Data

The Performance Monitor supports three types of logs: counter logs, trace logs , and alert logs. **Counter logs** record data from local or remote computers about hardware usage and system service activity. **Trace logs** are event-driven, recording monitored data such as disk I/O or page faults. When a traced event occurs, it is recorded in the log. **Alert logs** take trace logs one step further in that they monitor counters, wait for them to exceed user-defined tolerances, and then log the event. You can even set up the alert log to send a message or run an application when a particular value is exceeded.

To create a log, click on the Performance Logs and Alert in the left pane of Performance Monitor, and open the folder for the type of log that you want. Right-click the empty space in the Details window and choose an option from Creating a New Log from the pop-up menu that appears. If you save the file as a binary file (*.BLG), you can use System Monitor to open the log and view it later. If you save the file as a comma-delimited file (*.CSV) or a tab-delimited file (*.TSV), you can open it with Excel to perform your own data analysis.

16.9 NETWORK PERFORMANCE

Network monitoring software enables you to detect and troubleshoot problems on LANs, identify network traffic patterns and network problems such as computers that make many more requests than other computers, and identify unauthorized users on your network. In addition, you can capture frames (packets) directly from the network or display, filter, save, and print the captured frames so that you can analyze them later. In Windows NT and Windows 2000, the network monitoring software is Network Monitor.

Network Monitor reads all data that are transferred over a network at any given time through a network card and passes the data to a temporary capture file. See Figure 16.16. Each frame or packet read contains the source address of the computer that sent the message, the destination address of the computer that receives the frame, the header information of the computer that receives the frame, and the data being sent to the destination computer. See Figure 16.17. To make it easier to analyze the data, isolate certain information by using filters depending on the basis of source and destination addresses, protocols, protocol properties, and pattern offset. You also can have Network Monitor respond to specific conditions as soon as it detects them and perform specified actions such as starting an executable file.

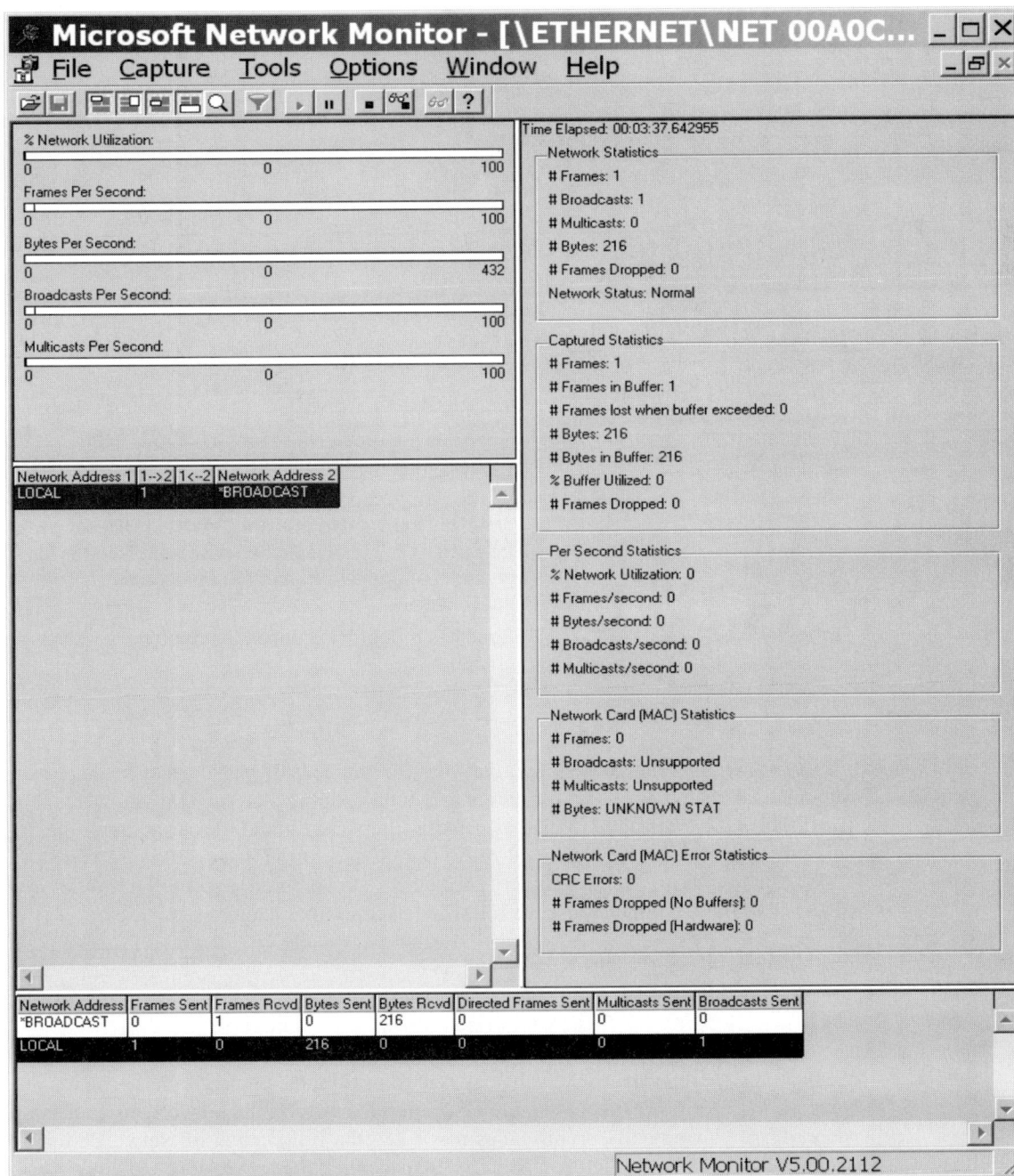

FIGURE 16.16 Network Monitor

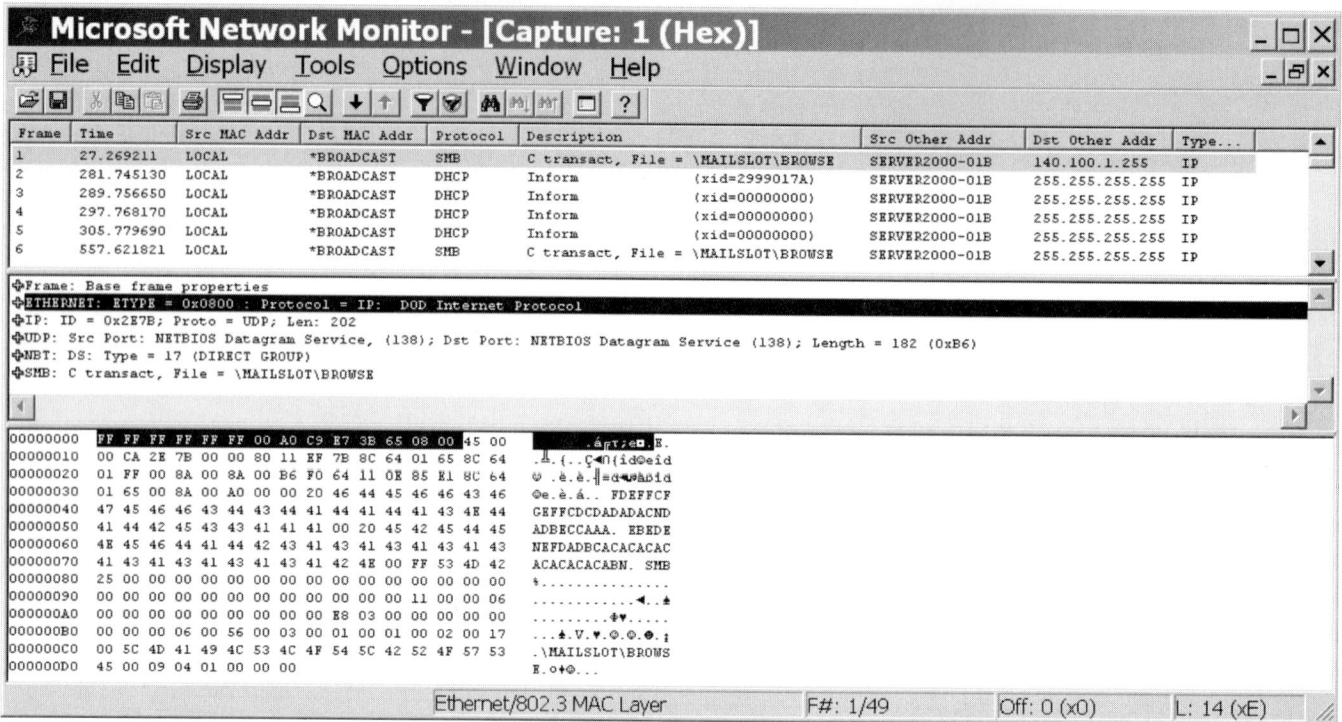

FIGURE 16.17 Viewing network packets using Network Monitor

TABLE 16.10 Network Monitor Filters

Filter Item	Description
Protocol	Specifies the protocols or protocol properties.
Address Filter (default is ANY <− −> ANY)	Specifies the computer addresses on which you want to capture data. Arrows specify the traffic direction you want to monitor. The INCLUDE or EXCLUDE keyword indicates how Network Monitor should respond to a frame that meets a filter's specifications.
Property	Specifies property instances that match your display criterion.

When analyzing the packets on a network, the number of packets can be quite overwhelming. Therefore, you can use filters to isolate only the material that you want. Table 16.10 shows the filter items. Whereas capture filters are limited to four address filter expressions, display filters can also utilize AND, OR, and NOT logic. See Figure 16.18.

To capture all traffic from PREGAN-DESK's computer, except the traffic from CBROWN-DESK computer, use the following capture filter address section:

include PREGAN-DESK <- - - -> Any

exclude PREGAN-DESK <- - - -> CBROWN-DESK

The Network Monitor program that comes with Windows 2000 is a limited software package that only captures the packets that are sent to or from the computer that is running the Network Monitor program. To capture all frames detected on the network, use a network adapter card that supports promiscuous mode and the full version of Network Monitor, which is available with Microsoft Systems Management Server version 2.0. Promiscuous mode is a state in which a network adapter card copies all frames that pass over the network, regardless of the destination address to a local buffer.

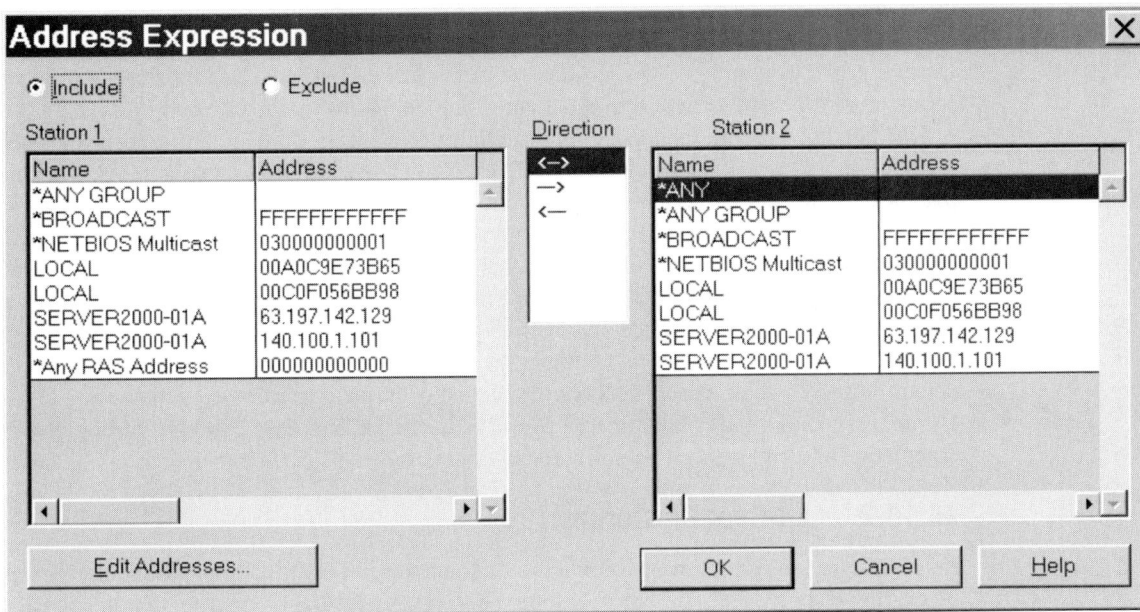

FIGURE 16.18   Configuring filtering for Network Monitor

16.10   CLUSTERING

Windows 2000 has many features to assist in overcoming failure points, as shown in Table 16.11. Redundant network cards were discussed in Chapter 4; UPS was mentioned in Chapter 2 and in Chapter 8. The WAN links and dial-up connections will be discussed in future chapters. This section will discuss server clustering.

Clustering is connecting two or more computers, known as nodes, in such a way that they behave like a single computer. Clustering is used for parallel processing, load balancing, and fault tolerance. The computers that form the cluster are physically connected by cable and are logically connected by cluster software. As far as the user is concerned, the cluster appears as a single system to end users.

16.10.1   Network Load Balancing

Network load balancing clusters distribute client connections over multiple servers. Internet clients access the cluster using a single IP address (or a set of addresses for a multihomed host). The clients are unable to distinguish the cluster from a single server. Server

TABLE 16.11   Common points of failures in a server

Failure Point	Failure Solution
Network hub and network card	Redundant network cards and hubs
Power problems	Uninterruptible power supply (UPS) and putting cluster nodes on separate electrical circuits
Disk	Hardware RAID
Other server hardware, such as CPU or memory	Fail-over clustering
Server software, such as the operating system or specific applications	Fail-over clustering
WAN links, such as routers and dedicated lines	Redundant links over the WAN, to provide secondary access to remote connections
Dial-up connection	Multiple modems

programs do not identify that they are running in a cluster; however, a network load balancing cluster differs significantly from a single host running a single server program, because it provides uninterrupted service even if a cluster host fails. The cluster also can respond more quickly to client requests than a single host (for load-balanced ports).

The heart of network load balancing is the driver WLBS.SYS, which is loaded into each member server, or host, in the cluster. WLBS.SYS includes the statistical mapping algorithm that the cluster hosts use collectively to determine which host handles each incoming request. To coordinate their actions, the hosts periodically exchange multicast or broadcast messages within the cluster to thus monitor the status of the cluster. When the state of the cluster changes (such as when hosts fail, leave, or join the cluster), network load balancing invokes a process known as convergence, in which the hosts exchange messages to determine a new, consistent state of the cluster and to elect the host with the highest host priority as the new default host. When all cluster hosts have reached consensus on the correct new state of the cluster, they record the completion of convergence in the Windows 2000 Event Log. At the completion of convergence, the traffic for a failed host is redistributed to the remaining hosts.

16.10.2   Fail-Over Clustering

In a **fail-over** configuration, two or more computers serve as functional backups for each other. If one should fail, the other automatically takes over the processing that is normally performed by the failed system, thus eliminating downtime. If the server is to share common data, the cluster servers are connected to at least one shared SCSI bus with a storage device connected to both servers, and at least one storage device that is not shared. Fail-over clusters are highly desirable for supporting mission-critical applications.

Storage Area Network (SAN) is a high-speed subnetwork of shared storage devices. A SAN's architecture works in a way that makes all storage devices available to all servers on a LAN. If an individual application in a server cluster fails (but the node does not), the Cluster service will typically try to restart the application on the same node. If that fails, it moves the application's resources and restarts them on another node of the server cluster. This process is called fail-over. The Cluster Administrator can use a graphical console to set various recovery policies, such as dependencies between applications, whether to restart an application on the same server, and whether to automatically rebalance, or fail back, workloads when a failed server comes back online.

16.10.3   Enabling Cluster Services

To install Cluster service after installing Windows 2000, do the following:

1. On one cluster node only, allow Windows 2000 Advanced Server to start. Open the Add/Remove Programs in the Control Panel.
2. Use the Windows Components wizard to select and install the Cluster service.
3. Ensure that the Cluster service is running successfully on the first node before starting the operating system on another node.

It is important to carry out the installation one node at a time. Make sure that Windows 2000 Advanced Server and the Cluster service are installed and running on one node before starting an operating system on another node. If the operating system is started on multiple nodes before the Cluster service is running on one node, the cluster disks could be corrupted.

SUMMARY

1. One primary job of the administrator is to deal with disasters and to plan ahead as to minimize the frequency and degree of failures.
2. Data stored on a computer or stored on the network are vital to the users and probably the company.
3. A backup of a system is a copy of all data and/or programs.
4. The best method for data protection and recovery is backup, backup, backup.
5. The System State data are a collection of system-specific data that can be backed up and restored, and include the registry, COM+ class registration database, system boot

files, Certificate Services database, Active Directory directory services database (domain controller), and SYSVOL directory (domain controllers).

6. The catalog is a summary of the files and folders that have been saved in a backup set.

7. If you have other domain controllers, they would have replicas of the Active Directory, and if you had distributed directories (file replication), the Active Directory data and replicated directories would automatically be updated by the other servers.

8. If Windows 2000 cannot load or start, accessing certain tools may help you solve the problem. They are the Advanced Boot menu, Windows 2000 Recovery console, and emergency repair disk.

9. If you press F8 while the Windows 2000 Boot menu is displayed, it will display the Advanced Boot menu.

10. Using the Enable VGA mode is another option if the wrong video driver is loaded and you need to force the video system into 640x480 with 16 colors, and the Last Known Good Configuration to start Windows 2000 using the last saved configuration is stored in the registry.

11. The Directory Service Restore mode is a special version of safe mode that loads all drivers and services, and allows you to restore the Active Directory of the domain controller from backup media.

12. To monitor the process of a server's boot from another server using a serial cable, use the Debugging mode.

13. If your computer fails to boot even in safe mode, the best tool to use is the Recovery console.

14. The Recovery console provides a command line interface that will let you repair system problems using a limited set of command line commands, including enabling or disabling services, repairing a corrupted master boot record, and reading and writing data on a local drive (FAT, FAT32, or NTFS).

15. The emergency repair disk (ERD) contains information about your current Windows system settings. You can use this disk to repair your computer if it will not start or your system files are damaged or erased.

16. If you have performed the emergency repair process and the computer still does not operate normally, then use the Windows 2000 Setup CD to perform an in-place upgrade over the existing installation.

17. A program error (user mode) or stop error (kernel mode) signifies that something unexpected has happened within the Windows environment.

18. Dr. Watson is a program that starts automatically when a program error occurs, detects information about the system and the program failure, and records the information into a log file.

19. A virus is a program designed to replicate and spread, generally without the knowledge or permission of the user.

20. Auditing is a feature of Windows 2000 that monitors various security-related events so that you can detect intruders and attempts to compromise data on the system.

21. Performance is the overall effectiveness of how data move through the system. To improve performance, you must determine the part of the system that is slowing the throughput.

22. Network monitoring software enables you to detect and troubleshoot problems on LANs.

QUESTIONS

1. The most important part of the computer is the _____ .
   a. microprocessor
   b. hard drive
   c. RAM
   d. data

2. The best method for protecting data is _____ .
   a. RAID
   b. backup, backup, backup
   c. a surge protector and UPS
   d. antivirus software

3. Which is ideal for backing up an entire hard drive?
   a. zip drive
   b. second Hard drive
   c. RAID
   d. tape Drive

4. After backing up a drive, you should occasionally _____ .
   a. restore a nonessential file to the hard drive
   b. reformat the hard drive
   c. shut down the system
   d. reformat the tape

5. The archive attribute indicates whether a file is backed up. Which one of the following backups does not shut off the archive attribute?
   a. full
   b. incremental
   c. differential
   d. none of the above

6. Which backup method requires you to provide the full backup and a tape for each day you want to go back and restore?
   a. incremental
   b. differential
   c. full
   d. daily

7. How do you save the local registry using Windows NT Backup?
   a. Type "Ntbackup /r" at the Command prompt.
   b. Select the drive and directory containing the registry and run Windows NT Backup.
   c. Select the drive containing the registry, check the Backup Local Registry box, and run Windows 2000 Backup.
   d. Select the System State using the Windows 2000 Backup program and run the Backup program.

8. You have modified some value entries in the registry. Now when you reboot the computer, Windows 2000 will not start. What is the best way to correct the problem?
   a. Boot from the Last Known Good Configuration.
   b. Start the computer from a boot disk and copy Ntldr, Ntdetect.com, Ntoskrnl.exe, and Bootsect.dos from a coworker's machine.
   c. Start the computer from a boot disk and restore the registry from tape backup.
   d. Start the computer from the emergency repair disk and let Windows NT automatically attempt to correct the problem.

9. Which of the following does the System State include? (choose all that apply)
   a. registry
   b. system boot files
   c. Active Directory files
   d. WINNT folder
   e. SYSVOL directory

10. Your Windows 2000 Server has encountered a series of stop errors. How do you configure Windows 2000 to save stop error information to a memory dump file?
    a. Specify the recovery option in Dr. Watson.
    b. Specify the recovery option in Performance Monitor.
    c. Specify the proper recovery option on the Startup/Shutdown tab in System Properties.
    d. Specify the recovery option in Server Manager.

11. Pat wants to use Performance Monitor to view logical disk object counters on his computer from a remote machine. What operations must Pat perform on his machine to allow him to monitor these counters remotely?
    a. Pat must install Network Monitor Agent on his computer.
    b. Pat must run the DiskPerf utility on his computer with the –Y switch.
    c. Pat must install Network Monitor Agent on his computer and run the DiskPerf utility on his computer with the –Y switch.
    d. Pat is not required to do anything more. Simply installing Windows NT Server on his machine is sufficient to allow other machines to monitor his logical disk counters remotely.

12. Which one of the following applications is best suited for producing a baseline of all network activities?
    a. Network Monitor
    b. ARP command
    c. Performance Monitor
    d. TRACERT command

13. Which application is best suited for identifying the computers on a network that are contributing most to network traffic?
    a. Network Monitor
    b. Packet Monitor
    c. Performance Monitor
    d. Server Manager

14. Your Windows 2000 Server is suffering from poor performance due to excessive paging. What is the best way to alleviate the excessive paging on your server?
    a. Add RAM to your server.
    b. Implement a disk management strategy on your server and create multiple paging files.
    c. Upgrade the disk drive containing the paging file to one with a faster data access speed.
    d. Upgrade the CPU on your server.

15. You want to use Network Monitor to capture all frames being sent from a computer named SERVER1 on your network. How must you specify the address inclusion line when designing your capture filter?
    a. INCLUDE NetBIOS= =SERVER1
    b. INCLUDE SMB= =SERVER1
    c. INCLUDE ANY <--> SERVER1
    d. INCLUDE SERVER1 <--> ANY
    e. INCLUDE SERVER1--> ANY

16. You suspect that a group of hackers is trying to gain access to your network. What is the best way to confirm your suspicions?
    a. Start auditing failures of directory and file access in the Directory Auditing window for key system files and directories.
    b. Start auditing failed logon/logoff actions in the Audit Policy window.
    c. Enable SLIP/PPP monitoring on all RAS servers.
    d. Alter your network's account policy by increasing the minimum required password length.

17. In planning for the expansion of your company's network, you want to capture and decode TCP/IP packets on your Windows 2000 Server computer. What must you use?
    a. ICMP
    b. Network Monitor
    c. Performance Monitor
    d. UPD

18. Working at his Windows 2000 Server, Pat wants to accumulate and view the TCP/IP protocol statistics on the server. What should he use?
    a. ARP
    b. NBTSTAT
    c. Network Monitor
    d. SNMP

19. Which Windows 2000 packages support clustering? (choose all that apply)
    a. Windows 2000 Professional
    b. Windows 2000 Server
    c. Windows 2000 Advanced Server
    d. Windows 2000 DataCenter

20. What are the two types of clustering offered by Windows 2000 Advanced Server and Windows 2000 DataCenter?
    a. fail-over
    b. load balancing
    c. startup synchronizing cluster
    d. RAID cluster

EXERCISES

EXERCISE 1: PERFORMING A BACKUP AND RESTORE

For this lab, you will need a blank, formatted floppy disk.

Using the Backup Wizard

1. Log in as the Administrator.
2. On your D drive, create a Data directory.
3. From the C:\WINNT\WEB\WALLPAPER folder, copy all files to the D:\DATA folder.

4. To start the Microsoft Backup program, click on the Start button, select the Programs option, select the Accessories option, select the System Tools option, and select the Backup option.
5. At the Welcome to Windows 2000 Backup and Recovery Tools page, click on the Backup wizard button. Click on the Next button.
6. Select the "Backup selected files, drives, network data" option and click on the Next button.
7. In the Items Backup page, find the DATA folder in the C drive. Click on the Data Drive to show its contents in the box to the right. To select the DATA folder to be backed up, put a checkmark in the box next to the DATA folder. Click on the Next button.
8. In a real-world network, you would typically back up to a tape drive. Because many schools do not have tape drives for every student nor do students have them at home, we will back up to the floppy drive. Therefore, for the Backup Media or File Name text box, keep the A:\BACKUP.BKT and click on the Next button.
9. Before clicking on the Finish button, click on the Advanced button.
10. Open the "Select the type of backup operation to perform" text box and note the options. When done, choose the Normal option and click on the Next button.
11. Select the Verify Data after Backup option and click on the Next button.
12. If the archive media already contains a backup, we will replace the data on media with this backup. Therefore, select this option and click on the Next button.
13. Accept the default labels and click on the Next button.
14. Keep the default of doing the backup now and click the Next button.
15. Insert the blank floppy disk in drive A.
16. Click on the Finish button.
17. When the backup is complete, click on the Close button.
18. When a backup is done, you would typically remove the tape (or disk) and label the name of the backup. It also is recommended to log the backup in a table or notebook or on the disk label. Therefore, remove the disk and write Backup—Data Folder and today's date on the label of the disk.

Using the Restore Wizard

19. Delete the Date folder on the C drive.
20. Back at the Welcome to the Windows 2000 Backup and Recovery Tools page, click on the Restore wizard button. Click on the Next button.
21. Note that because of the catalogs, Windows 2000 remembers the backups that were done. Therefore, click on the + sign next to File, click on the next plus sign, and put a checkmark for the C drive. Click on the Next button.
22. Before clicking on the Finish button, click on the Advanced button. This is where you can specify the files to be restored to a different location, if desired. Click on the Next button.
23. In this case, since we no longer have a Data folder, it does not matter which option we choose. Therefore, keep the default option and click on the Next button.
24. Since we have no system or security information to restore, click on the Next button.
25. Click on the Finish button.
26. Click on the OK button to restore from the Default file.
27. Click on the Close button.

Using the Backup Program without the Wizards

28. Click on the Backup tab.
29. To back up the entire C drive, click on the checkbox next to the C drive.
30. Note that while the System State (listed after the drives) is on the C drive, it is not considered normal data. Therefore, it was not included, so select the System State to be backed up.
31. Open the Tools menu and select the Options option. Click the OK button.
32. Click the Backup program.
33. At the bottom of the screen you will find the Backup Media or File Name text box. Type in M:\Backup.bkf in the text box and click on the Start Backup button.

34. Leave the defaults and click the Start Backup button.
35. After the backup is done, click on the Close button.
36. Click on the Restore tab. Note the backups have already been completed as listed by the catalogs on the system.
37. Double-click the first backup listed in the right pane. Double-click the C drive listed in the right pane. Double-click the Data folder listed in the right pane.
38. On the left pane, deselect the Data folder. On the right pane, select the first file listed.
39. Open the Tools menu and select the Options option. On the Restore tab, select "Always replace the file on my computer." Click the OK button.
40. Click on the Start Restore button.
41. Click on the OK button to confirm.
42. Insert the floppy disk from the original backup. Since we did the backup on the A drive, use the Browse button to find the backup file there.
43. After the backup file is selected, click on the OK button to close the Enter Backup File Name dialog box.
44. When the restore is complete, click on the Close button.

Re-creating a Catalog

45. In the Microsoft Backup program, click on the Restore tab.
46. Right-click each of the catalogs listed in the right pane and select the Delete Catalog option.
47. To re-create the catalog, open the Tools menu and select the Catalog a Backup File option.
48. Insert the original backup disk in drive A.
49. Use the Browse button to select the backup file in the A drive. Click the OK button to close the Backup File Name dialog box. Remember, if this is a large tape, the process can take some time.

EXERCISE 2: SCHEDULING BACKUP JOBS

1. Start the Microsoft Backup program and click the Scheduled Jobs tab.
2. To start the Backup wizard, double-click on today's date in the calendar. Click on the Next button.
3. Select the "Back up selected files, drives, or network data" option and click on the Next button.
4. Select the C:\DATA DIRECTORY. Click on the Next button.
5. Keep the default Backup Media or File Name of the A:\Backup.bkt, and click on the Next button.
6. Keep the default type of backup operation to perform and click on the Next button.
7. On the How to Back Up page, select the Verify Data after Backup option and click on the Next button.
8. Select the "Replace the data on the media with this backup" option and click on the Next button.
9. On the Backup Label page, click on the Next button.
10. If it asks for the Set Account Information dialog box, provide a user account and password and click on the OK button. You can use the Administrator account.
11. On the When to Back Up page, select the Later option, provide the job name of Test, and select the Set Schedule button.
12. On the Scheduled Task option, select the Daily option. Keep the Default Start Time at 12:00 AM and the Scheduled Task Daily at every 1 day.
13. Click on the Advanced button. This is where you would specify the start and end dates and how often you want to repeat the task. Click on the Cancel button.
14. Click on the OK button.
15. At the When to Back Up page, click on the Next button.
16. Click on the Finish button.
17. Change the time to 11:58 PM.
18. Wait 3 minutes and check if the backup is performed.

19. Close the Microsoft Backup program.
20. Reset the clock to its correct time.

EXERCISE 3: CREATING AND USING ERD DISKS

Create an Emergency Repair Disk

1. Open the Backup program.
2. On the Tools menu, click the Create an Emergency Repair Disk button. Or, use the Emergency Repair Disk wizard.
3. Insert a blank, formatted floppy disk into the drive. Mark the checkbox to back up the copy of the registry and click on the OK button.
4. Click on the OK button when the disk has been created.
5. Label as Emergency Repair Disk and store in a safe place.

Using the Emergency Repair Disk

6. If you have a computer with a bootable CD-ROM drive, insert the Windows 2000 Setup CD. If not, boot with the first installation floppy disk.
7. When the text-based part of Setup begins, follow the prompts and choose the Repair or Recover option by pressing R.
8. When prompted, insert the Windows 2000 Setup CD in the appropriate drive.
9. When prompted, choose the Emergency Repair Process by pressing R.
10. When prompted, choose the Fast Repair option by pressing F.
11. Follow the instructions on the screen. When prompted, insert the emergency repair disk in the appropriate drive.
12. During the repair process, missing or corrupted files are replaced with files from the Windows 2000 CD or from the *systemroot*\Repair folder on the system partition. Be sure to follow the instructions on the screen; you may want to write the names of files that are detected as faulty or incorrect to help you diagnose how the system was damaged.
13. If the repair was successful, allow the process to complete; it will restart the computer. The restarting of the computer indicates that replacement files were successfully copied to the hard disk.

EXERCISE 4: WINDOWS 2000 RECOVERY CONSOLE

1. Insert the Windows 2000 Setup CD into your CD-ROM drive. Click No when prompted to upgrade to Windows 2000.
2. At the Command prompt, switch to your CD-ROM drive, and then type the following:

   \I386\WINNT32.EXE /CMDCONS

3. When it asks if you want to install the Recovery console, click on the Yes button.
4. Click OK when the setup is complete.
5. Restart your computer and select the Recovery Console option from the Boot menu.
6. When it asks which Windows 2000 installation would you like to enter, select the number listed next to the C:\WINNT directory and pass the Enter key. When it asks you for the Administrator password, type in the password and press the Enter key.
7. At the C:\WINNT> Command prompt, execute the Help command. Press the space bar to go back to the prompt.
8. Type in the DIR command. Keep pressing the space bar until you return to the Command prompt.
9. At the Command prompt, execute the FIXBOOT command. When it asks if you are sure, type in a Y and press the Enter key.
10. At the Command prompt, execute the LISTSVC command. Keep pressing the space bar until you return to the Command prompt.
11. At the Command prompt, execute the EXIT command to reboot the computer.

EXERCISE 5: AUDITING

Enable Auditing

1. From the Administrative Tools, start the Domain Controller Security Policy console.
2. In the left pane, open Security Policies, open Local Policies, and click the Audit Policy options.
3. Double-click the "Audit directory service access" option in the right pane and enable the Success and Failure options.
4. Double-click the Audit Logon Events option and enable the Failure options.
5. Close the Domain Console Security Policy console.

Auditing Failed Login Attempts

6. If the Account Lockout option has not been set, use the Domain Security Policy console to enable the Account Lockout Threshold to four invalid logon attempts. If it asks, set the account logout duration to 30 minutes and the reset account lockout counter after 30 minutes.
7. Log out as the Administrator and incorrectly log in as Charlie Brown three times.
8. Log in as the Administrator.

Audit the Access of a File and Invalid Logons

9. Right-click the C:\DATA folder and select the Properties option.
10. On the Auditing tab, click the Add button.
11. Select Everyone and click on the OK button.
12. Select the following options for successful and failed attempts:
    - Create Files/Write Data
    - Create Folders/Append Data
    - Write Attributes
    - Write Extended Attributes
    - Delete Subfolders and Files
    - Delete
13. Select the "Apply these auditing entries to objects and/or containers within this container only" option and click on the OK button.
14. Click on the OK button again.
15. Delete the Windows 2000 .jpg file from the C:\DATA folder.

Using Event Viewer to View Security Log

16. Open the Event Viewer from the Administrative Tools.
17. Select the Security Log option in the left pane.

EXERCISE 6: PERFORMANCE MONITOR

1. Start the Performance console from the Administrative Tools.
2. Right-click the right pane and select the Add Counters option.
3. The % Performance Time option is already selected, so click on the Explain button.
4. Click on the Add button to add % Performance Time Counter to Performance Monitor.
5. Select the Interrupts/sec and click on the Add button.
6. Select the Cache for the Performance Object. Select the Copy Read Hits % from the Counters list and click on the Add button.
7. Select the Paging File for the Performance Object. Select the % Usage from the Counters list and click on the Add button.
8. Select the PhysicalDisk for the Performance Object. Select the % Disk Time from the Counters list and click on the Add button.
9. Click on the Close button.
10. Start the Pinball game and play for 1 minute. Observe the graph.

11. Right-click on the graph and select the Histogram option. Click on the OK button.
12. Play the Pinball game for 1 minute. Observe the histograph.
13. Right-click the graph and select the Report option. Click on the OK button.
14. Play the Pinball game for 1 minute. Observe the report.
15. Close the Pinball game.
16. In the left pane, right-click the Counter Logs. In the New Log Setting dialog box, type in Log1 and click the OK button.
17. Click the Add button and add the % Processor Time Counter.
18. Click the Close button. Click the OK button.
19. Start the Pinball game and play for 1 minute.
20. Right-click the Log1 option in the right pane, and select the Stop option.
21. Click the System Monitor option in the left pane.
22. Right-click the chart and select the Properties option.
23. In the Source tab, click the Browse button and open the C:\PERFLOGS\ LOG1_000001.blg file. Click the OK button.
24. Add the % Processor Time Counter.
25. Right-click the Log1 from the left pane of the Performance console and select the Properties option.
26. In the Log Files tab, change the log file type to Text file—CSV. Click on the OK button.
27. Right-click the Log1 from the left pane of the Performance console and select the Start option.
28. Start the Pinball game and play for 1 minute.
29. Right-click the Log1 from the left pane of the Performance console and select the Stop option.
30. If you have access to Microsoft Excel, view the LOG1_000001.CSV file that is located in the C:\PERFLOGS folder.

EXERCISE 7: NETWORK MONITOR

Installing Network Monitor

1. Start the Add/Remove Programs applet in the Control Panel.
2. Click the Add/Remove Windows Components button. Click on the Management and Monitoring Tools option and then click the Details button. Select the Network Monitor Tools and click on the OK button. Click on the Next button and install the Network Monitor software. Click on the Finish button.
3. Close the Add/Remove Programs and Control Panel.

Using Network Monitor

4. In Network Monitor, open the Capture menu and select the Start option.
5. Click the Start button and select the Run option. In the Open text box, type in \\Server2000-XXx to access your partner's computer.
6. In the Network Monitor program, open the Capture menu and select the Stop option.
7. Open the Capture menu and select the Display Captured Data option.
8. Scroll the list of captured frames and view the addresses that are listed in the Src MAC Addr and Dst MAC Addr columns.
9. View the protocols that are listed in the Protocol column.
10. Select a frame and double-click on it.

Unit 4: Advanced Network Topics

17

Understanding WAN Technology

INTRODUCTION

In Chapters 3 and 4 we briefly described a WAN and the concept of a router. In this chapter we look at the WAN technology that is used to link individual computers or entire networks that cover a relatively broad geographic area. These connections are usually through telephone companies or some other common carrier. Whereas LANs typically are multiple point-connections, WANs are typically made of multiple point-to-point connections.

OBJECTIVES

1. List the characteristics, requirements, and appropriate situations for WAN connection services, including X.25, ISDN, frame relay, and ATM.
2. Explain the differences between circuit switching and packet switching.
3. Given a scenario, recommend the type of technology that should be used.
4. Compare and contrast permanent point-to-point lines and leased lines.
5. Compare and contrast cell relay with packet switching.
6. Explain how SONET and SDH relate to ATM.
7. List and define the three implementations of ATM.
8. Compare and contrast xDSL and T1 lines.
9. Compare and contrast wireless technology and wire technology.

17.1 DATA SWITCHING TECHNIQUES

Large internetworks can have multiple paths that link source with destination devices, switching information as it travels through the various communication channels. The data switching techniques can be divided into circuit switching and packet switching.

17.1.1 Circuit Switching

The **circuit switching** technique connects the sender with the receiver via a single path for the duration of a conversation. Once a connection is established, a dedicated path exists (that is, always consumes network capacity) between both ends, even when there is no active transmission taking place, such as when a caller is put on hold. Once the connection has been made, the destination device acknowledges that it is ready to carry on a transfer. When the conversation is complete, the connection is terminated. See Figure 17.1. Therefore, circuit switching networks are sometimes called connection-oriented networks. Examples of circuit switching include phone systems and data that are transmitted in live video and sound.

17.1.2 Packet Switching

The **packet switching** technique involves messages that are broken into small parts, or packets. Each packet is tagged with source, destination, and intermediary node addresses, as is appropriate. Packets can have a defined maximum length and be stored in RAM instead of on hard disk. Packets can take a variety of possible paths through the network in an attempt to keep the network connections filled at all times. Because the message is broken into multiple parts, however, sending packets via different paths adds to the possibility that packet order could get scrambled. Therefore, each path receives a sequencing number.

Packets are sent over the most appropriate path. Each device chooses the best path at that time for every packet. If one path is too busy, it can send the packet over another path. Be-

516

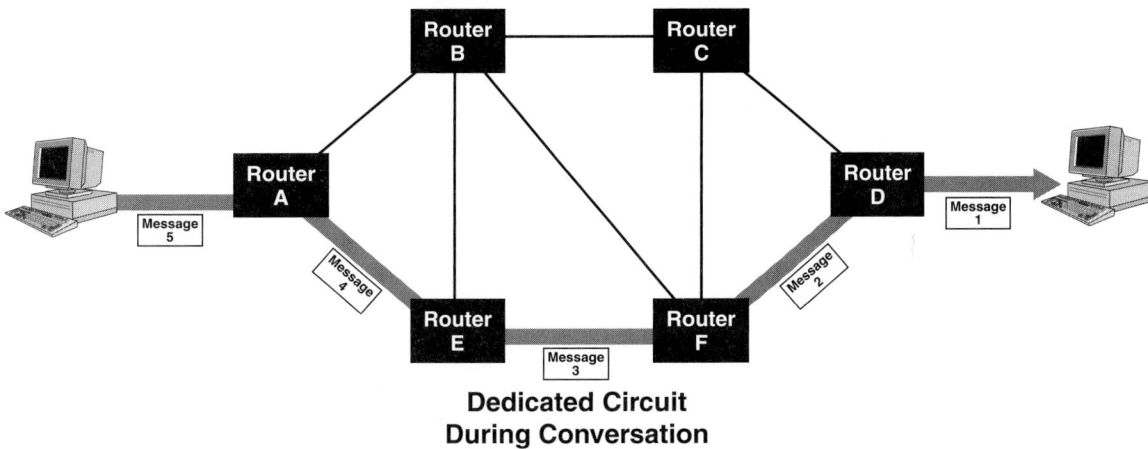

FIGURE 17.1   Circuit switching

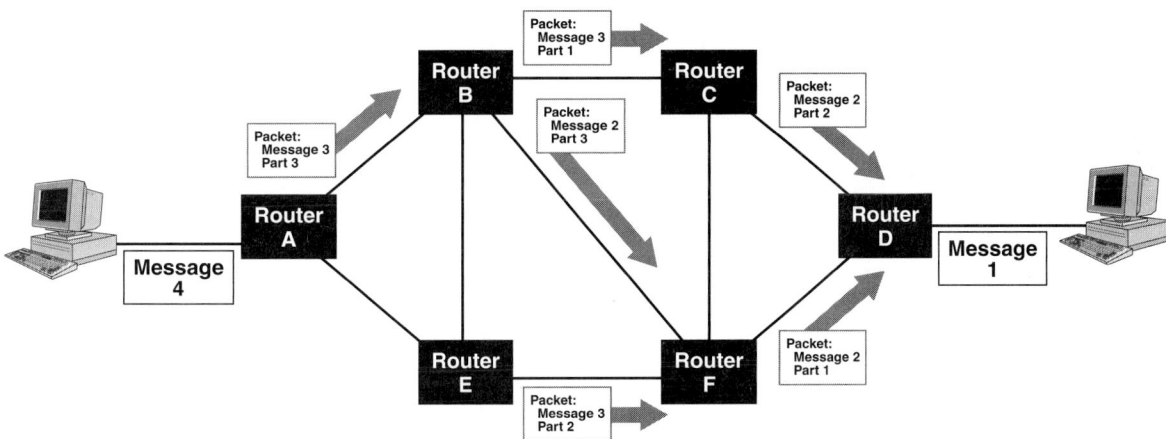

FIGURE 17.2   Packet switching

cause some packets may be delayed, which causes the packets to arrive out of order, the device will reorder them by sequence number to reconstruct the original message. See Figure 17.2. The Internet is based on a packet switching protocol. Other examples include asynchronous transfer mode (ATM), frame relay, switched multimegabit data service (SMDS), and X.25. Message switching is typically used to support services such as e-mail, web pages, calendaring, or workflow information.

17.1.3   Virtual Circuits

A **virtual circuit** is a logical circuit that is created to ensure reliable communications between two network devices. It acts as a direct connection even though it may not be directly connected. It is used most frequently to describe connections between two hosts in a packet switching network. In this case, the two hosts can communicate as though they have a dedicated connection even though the packets may actually travel very different routes before arriving at their destination.

Virtual circuits provide a bidirectional communications path from one device to another and are uniquely identified by some type of identifier. Several virtual circuits can be multiplexed into a single physical circuit for transmission across the network. This capability often can reduce the equipment and network complexity that is required for multiple device connections. A virtual circuit can pass through any number of intermediate devices or switches that are located within the virtual circuit.

Note: A network switch is a device that filters and forwards packets between LAN segments, which operate at the data link layer of the OSI reference model. In this chapter, a switch will be referred to as a routing switch, which performs routing operations much like a router, although the routing switch also performs many of the functions found in the network layer of the OSI model. The routing switch has its routing capabilities implemented with hardware, not software, so it is faster but not as powerful as a full-fledged router.

Virtual circuits can be either permanent or temporary (or switched). A **permanent virtual circuit (PVC)** is a permanently established circuit that consists of one mode: data transfer. PVCs are used in situations when data transfer between devices is constant. PVCs decrease the bandwidth use associated with the establishment and termination of virtual circuits, but increase costs due to constant virtual circuit availability. PVCs are more efficient than temporary circuits for connections between hosts that communicate frequently, and they play a central role in frame relay and X.25 networks.

A **temporary (switched) virtual circuit (SVC)** is dynamically established on demand and terminated when transmission is complete. Communication over an SVC consists of three phases: circuit establishment, data transfer, and circuit termination. The establishment phase involves creating the virtual circuit between the source and destination devices. Data transfer involves transmitting data between the devices over the virtual circuit, and the circuit termination phase involves tearing down the virtual circuit between the source and destination devices. SVCs are used in situations in which data transmission between devices in sporadic, largely because SVCs increase bandwidth that is used due to the circuit establishment and termination phases, but decease the cost associated with constant virtual circuit availability.

When leasing a PVC or SVC end-to-end circuit, there is a monthly fee regardless if you send data. If you choose not to lease the line but sign with a network provider, the provider then bills you only for the amount of data packets you send and the distance the packets travel. The cost also will be based on the technology used, and there is often a minimum charge per call.

17.1.4 WAN Devices

Typically, when talking about the various WAN connection devices, they can be divided into the following two categories:

- data terminal equipment (DTE); or
- data circuit-terminating equipment (DCE).

Data terminal equipment (DTE) devices are end systems that communicate across the WAN. They are usually terminals, PCs, or network hosts and are located on the premises of individual subscribers. **Data circuit-terminating equipment (DCE)** devices are special communication devices that provide the interface between the DTE and the network. Examples include modems and adapters. The purpose of the DCE is to provide clocking and switching services in a network and data transmission through the WAN. Therefore, the DCE controls data flowing to or from a computer.

Another term often used when discussing WAN connections is the **channel service unit/data service unit (CSU/DSU).** The DSU is a device that performs protective and diagnostic functions for a telecommunications line; the CSU connects a terminal to a digital line. Typically, the two devices are packaged as a single unit, basically as a high-powered, expensive modem. Please see Table 17.1 for a summary of the information in this chapter.

17.2 PUBLIC SWITCHED TELEPHONE NETWORK

The **Public Switched Telephone Network (PSTN)** is the international telephone system that is based on copper wires (UTP cabling) carrying analog voice data. The PSTN, also known as the **plain old telephone service (POTS),** is the standard telephone service used in most homes. The PSTN is a huge network with multiple paths that link source with des-

TABLE 17.1   Carrier technology

Carrier Technology	Speed	Physical Medium	Connection Type	Comment
Plain Old Telephone Service (POTS)	Up to 56 Kbps	Twisted pair	Circuit switch	Used by home and small business
Asymmetrical Digital Subscriber Line Lite (ADSL Lite)	Up to 1 Mbps downstream Up to 512 Kbps upstream	2 twisted pair	Circuit switch	Used by home and small business
Asymmetrical Digital Subscriber Line (ADSL)	1.5–8 Mbps downstream Up to 1.544 Mbps upstream	2 twisted pair	Circuit switch	Used by small to medium business
High Bit-Rate Digital SubscriberLine (HDSL)	1.544 Mbps full duplex (T1) 2.048 Mbps full duplex (E1)	2 pairs of twisted pair	Circuit switch	Used by small to medium business
DS0 Leased Line	64 Kbps	1 or 2 pairs of twisted pair	Dedicated point-to-point	The base signal on a channel in the set of digital signal levels
Switched 56	56 Kbps	1 or 2 pairs of UTP	Circuit switch	Used by home and small business
Switched 64	64 Kbps	1 or 2 pairs of UTP	Circuit switch	Used by home and small business
Fractional T1 Leased Line	64 Kbps to 1.536 Mbps in 64-Kbps increments	1 or 2 pairs of twisted pair	Dedicated point-to-point	Used by small to medium business
T1 Leased Line (DS-1)	1.544 Mbps (24–64 Kbps channels)	2 pairs of UTP or coaxial or optical fiber	Dedicated point-to-point	Used by medium to large business, ISP to connect to the Internet
T3 Leased Line (DS-3)	44.736 Mbps	2 pairs of UTP or optical fiber	Dedicated point-to-point	Used by large business, large ISP to connect to the Internet or as the backbone of the Internet
E1	2.048 Mbps	Twisted pair, coaxial cable, or optical fiber		32-channel European equivalent of T1
ISDN—BRI	64 to 128 Kbps	1 or 2 pairs of UTP	Circuit switch	Used by home and small business
ISDN—PRI	23–64 Kbps channels plus control channel up to 1.544 Mbps (T1) or 2.048 (E1)	2 pairs of UTP	Circuit switch	Used by medium to large business
X.25	Up to 64 Kbps	1 or 2 pairs of twisted pair	Packet switch	Older technology still used in areas where newer technology is still not available
Frame Relay	56 Kbps to 1.536 Mbps (using T1) or 2.048 Mbps using E-1.	1 or 2 pairs of twisted pair	Packet switch	Popular technology used mostly to connect LANs
FDDI/FDDI-2	100 Mbps/200 Mbps	Optical fiber	Packet switch	Large, wide-range LAN usually in a large company or a larger ISP
SDH/Sonet	51.84 Mbps and up	Optical fiber	Dedicated point-to-point	Used by ATM
SMDS	1.544 to 34 Mbps using T-1 and T3 lines	1 or 2 pairs of twisted pair	Cell relay	Popular growing technology used mostly to connect LANs; is the connectionless component of ATM
ATM	Up to 622 Mbps	T1, T3, E1, E3, SDH, and SONET	Cell relay	The faster network connection up to date
Cable Modems	500 Kbps to 1.5 Mbps or more	Coaxial cable	Leased point-to-point	Used by home and small business

FIGURE 17.3 Public switched telephone network

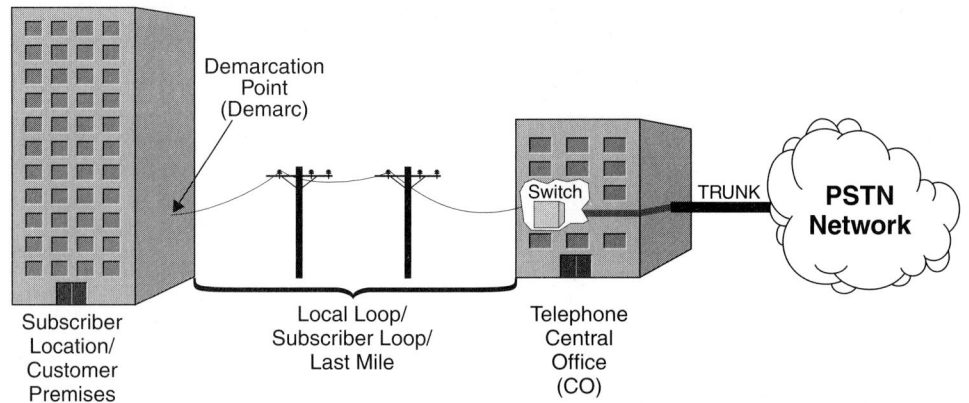

tination devices. Because PSTN uses circuit switching when a call is made, the data are switched to a dedicated path throughout the conversation.

The original concept of the Bell System was a series of PSTN trunks connecting the major U.S. cities. The PSTN originally began with human operators who sat at a switchboard and manually routed calls. Today, the PSTN systems still use analog signals from the end node (phone) to the first switch. The switch then converts the analog signal to a digital signal and routes the call on to its destination. Since the digital signal travels on fiber-optic cabling, the signals are switched at high speeds. Once the call is received on the other end, the last switch in the loop converts the signal back to analog, and the call is initiated. The connection will stay active until the call is terminated (user hangs up). The active circuit enables one caller to hear the other caller almost instantaneously.

The **subscriber loop,** or **local loop,** is the telephone line that runs from a home or office to the telephone company's central office (CO) or neighborhood switching station (often a small building with no windows). Although its cable length can be as long as 20 miles, it is referred to as the **last mile,** not because of its length, but because it is the slow link in the telecommunications infrastructure as it carries analog signals on a twisted-pair cable. The point where the local loop ends at the customer's premise is called the **demarcation point (demarc).** Unless you have an agreement with them, phone companies are only responsible for the loop from the central office to the demarc. See Figure 17.3.

The standard home phone communicates over the local loop using analog signals. Therefore, when a PC needs to communicate over the local loop, it must use a modem to convert the PC's digital data to analog signals. Unfortunately, the analog lines can only reach a maximum speed of 56 Kbps due to FCC regulations that restrict the power output, during which time the speed is not guaranteed and is often not reached.

17.3 LEASED DIGITAL LINES

The **T-carrier system** was introduced by the Bell System in the United States in the 1960s as the first successful system to convert the analog voice signal to a digital bit stream. Although the T-carrier system was originally designed to carry voice calls between telephone company central offices, today it is used to transfer voice, data, and video signals between different sites and to connect to the Internet.

The human voice can produce sounds in the range between 50 Hz and 1500 Hz, while a person can hear in the range between 300 Hz and 3400 Hz. To carry a voice signal on a digital carrier, the analog voice signal needs to be converted to a digital signal. The **Nyquist theorem** states that to ensure accuracy, a signal should be sampled at least twice the rate of its frequency. Therefore, the voice signal is sampled (measured) 8000 times per second. Each measurement is then quantized or converted into an 8-bit number (256 different combinations). As a result,

$$8000 \text{ samples/second} \times 8 \text{ bits/sample} = 64 \text{ Kbps}$$

to represent a digital voice signal.

17.3.1 T-Carrier and E-Carrier Systems

The T-carrier and E-carrier systems are entire digital systems that consist of permanent, dedicated point-to-point connections. These digital systems are based on 64-Kbps channels (DS-0 channel, where DS stands for digital signal), and each voice transmission is assigned a channel.

In North America and Japan, you would typically find a T1 line that has 24-Kbps to 64-Kbps channels for a bandwidth of 1.544 Mbps, and a T3 line that has 672-Kbps to 64-Kbps channels for a bandwidth of 44.736 Mbps. In Europe, you will find E1 lines with 32-Kbps to 64-Kbps channels for a bandwidth of 2.048 Mbps, and an E3 line with 512-Kpbs to 64-Kbps channels for a bandwidth of 34.368 Mbps. T1 lines are a popular leased line option for businesses that are connecting to the Internet and for Internet service providers (ISPs) that are connecting to the Internet backbone. T3 connections comprise the Internet backbone and are used by larger ISPs to connect to the backbone. See Table 17.2.

If your company is not ready for a full T1 line, then it can lease a fractional T1 line and thus use only a portion of the 24 channels. Since the hardware already exists to use a full T1 line, you simply call the carrier to increase the number of channels. Therefore, a fractional T1 line leaves room for growth.

The T-carrier system is a bipolar, framed format, time-division digital communication system. It will normally use one of two encoding methods, alternate mark inversion (AMI) or bipolar 8-zero substitution (B8ZS).

AMI bipolar encoding uses no voltage to indicate a digital 0 and uses alternating positive and negative voltages to represent digital 1s. The reason that unipolar encoding (a positive voltage indicates a digital 1 and no voltage indicates a digital 0) was not used is there is no "return to zero," which can lead to lost synchronization of the signal as the system loses track of the many sequential digital 1s. Therefore, the bipolar signal alternates between positive voltage and negative voltage and thus adds redundancy to the timing of the circuit. See Figure 17.4.

TABLE 17.2   Digital carrier systems

Digital Signal Designator	T-Carrier	E-Carrier	Data Rate	DS0 Multiple
DS0	—	—	64 Kbps	1
DS1	T1	—	1.544 Mbps	24
—	—	E1	2.048 Mbps	32
DS1C	—	—	3.152 Mbps	48
DS2	T2	—	6.312 Mbps	96
—	—	E2	8.448 Mbps	128
—	—	E3	34.368 Mbps	512
DS3	T3	—	44.736 Mbps	672
—	—	E4	139.264 Mbps	2048
DS4/NA	—	—	139.264 Mbps	2176
DS4	—	—	274.176 Mbps	4032
—	—	E5	565.148 Mbps	8192

FIGURE 17.4   AMI bipolar encoding

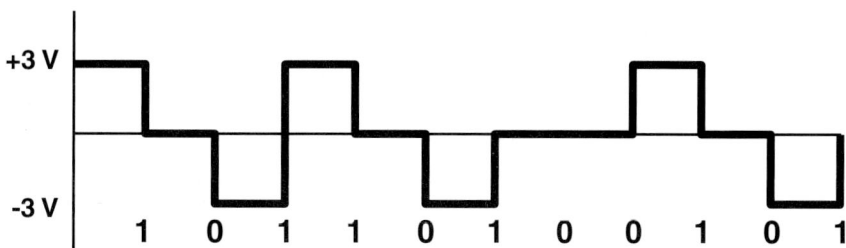

AMI coding suffers the drawback that a long run of zeros produces no transitions in the data stream, which lead to a loss of synchronization. Therefore, successful transmission relies on the user not sending long runs of zeros, but unfortunately, this type of encoding is not transparent to the sequence of bits being sent.

American T1 lines use bipolar 8-zero substitution. B8ZS is like AMI, but to avoid the loss of synchronization with a large number of sequential zeros, B8ZS replaces eight consecutive zeros with a fictional word. To differentiate between a real word and a fictional one, the hardware creates a bipolar violation where it will send two consecutive positive pulses or two consecutive negative pulses. See Figure 17.5. A T3 system uses a bipolar 3-zero substitution (B3ZS), which works the same as B8ZS but modifies any group of three zeros instead of eight. E-carrier systems use high-density bipolar 3 (HDB3), which replaces a four-zero bit pattern with two zeros followed by a bipolar violation.

A T1 line uses a D4 framing format, in which a frame bit prefixes 24 bytes of data, each from one of the 24 channels. The framing bit follows a special 12-bit pattern, called the frame alignment signal. Every 12 frames, the signal repeats, allowing the hardware on either side of the connection to signal changes in line status. This group of 12 D4 frames is called a superframe. The T3 uses a framing called M13, which uses a 4760-bit frame, compared with a 193-bit D4 frame. Of this, 56 bits are used for frame alignment, error correction, and network monitoring.

Traditionally, the T1 system uses two unshielded twisted-pair copper wires to provide full-duplex capability (one pair to receive and one pair to send). Today, the T-carrier system can also use coaxial cable, optical fiber, digital microwave, and other media. See Table 17.3.

While analog signals can travel up to 18,000 feet from the subscriber to the central office or switching station, T1 lines require cleaner lines for the faster speeds. Therefore, repeaters are placed approximately every 6000 feet.

17.3.2 T-Carrier Hardware

To complete the T-carrier connection, you also must use a CSU/DSU, and possibly a multiplexer and router. The CSU/DSU physically and electrically terminates the connection to the telephone company, controls timing and synchronization of the signal, and handles line coding and framing. By law, all T1s require that a CSU be connected between the DTE and the T1 line to act as a surge protector and to monitor the line.

FIGURE 17.5 B8ZS encoding used by T1 lines

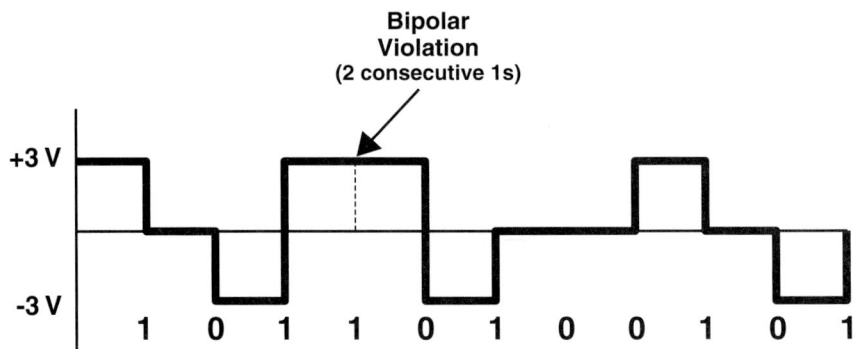

TABLE 17.3 Different media used in T-carrier systems

Media	T-Carrier Capacity
Twisted pair	Up to 1 T1 circuit
Coaxial cable	Up to 4 T1 circuits
Microwave	Up to 8 T3 circuits
Fiber optics	Up to 24 T3 circuits

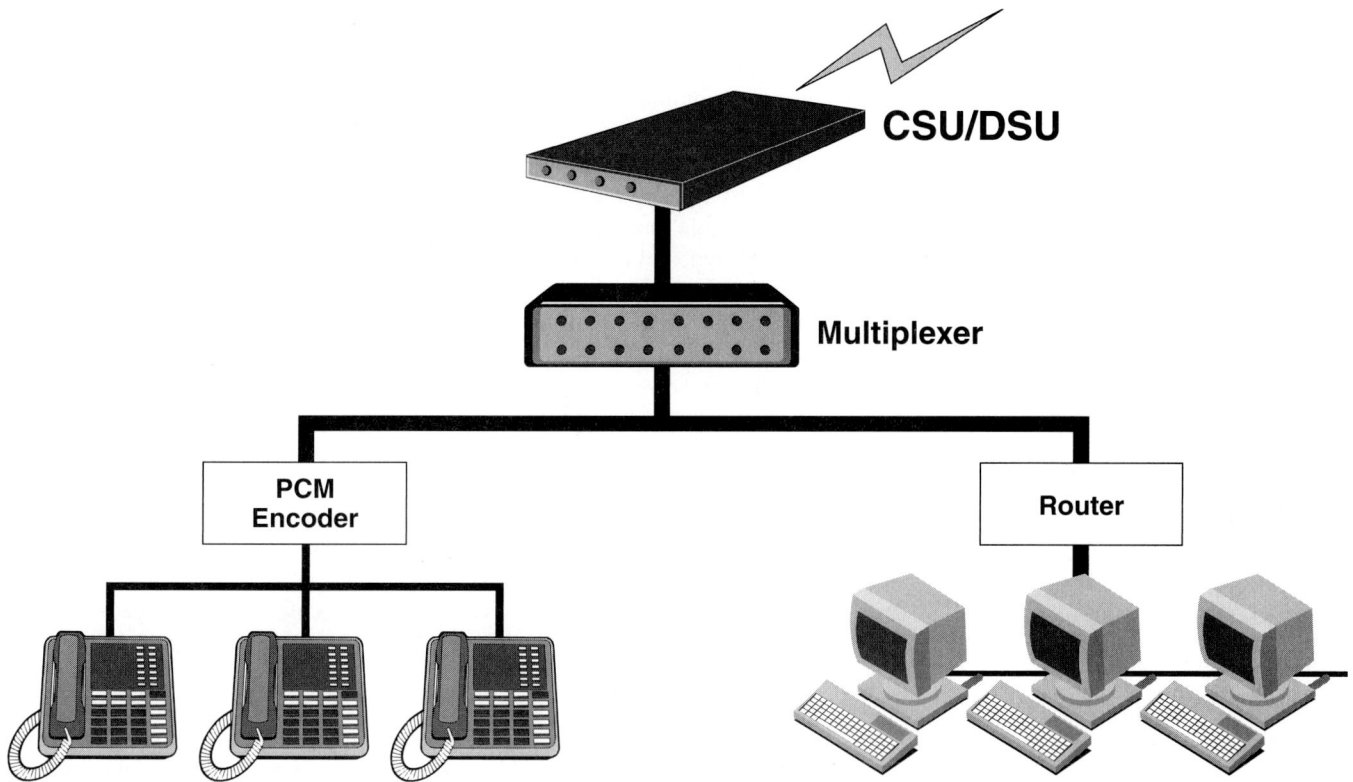

FIGURE 17.6   A multiplexer on a T1 line

A multiplexer is used to split the circuit into the various channels and some can also provide PCM encoding for analog devices such as an analog phone. Remember that most business systems use digital phone systems. The router can connect an Ethernet LAN to a WAN. A router would generally provide one or more RJ-45 connectors for the Ethernet LAN(s) and will include a serial connection for the CSU/DSU, or include a built-in CSU/DSU. See Figure 17.6.

17.4   CIRCUIT SWITCHED LINES

As you see from the previous information, digital lines are much faster than analog lines. The disadvantages of the leased digital lines are their higher cost and their limitation as a point-to-point connection. If you have multiple sites, then put several lines connecting one site to all other sites.

The advantage of the POTS system is its circuit switching ability, in which it has multiple paths linking many devices together. When one device needs to communicate with another, a circuit is established that consists of a dedicated path. When the conversation is finished, the connection is terminated. This circuit is established on demand. Circuit switch technology can be referred to as **dial-up technology.**

Combining the best features of both systems provides a basic switched digital technology. Examples of such technology are switched 56, switched 64, and ISDN.

17.4.1   Switched 56 and 64 Technology

T-carrier system lines are permanent leased lines. **Switched 56** and **switched 64** lines are digitally switched or "dial-up lines" that provide a single digital channel for dependable data connectivity. While it supports sporadic high-speed applications, it is typically cheaper than having a dedicated DS-0 line and the cost of the termination equipment also is cheaper.

The 56-Kbps value set with the switched 56 technology is based on the DS-0 64-Kbps technology. To avoid a loss of synchronization with many consecutive zeros, it uses pulse stuffing where every eighth bit is taken out of the signal and forced to be a pulse.

17.4.2 Integrated Services Digital Network

The **Integrated Services Digital Network (ISDN)** is the planned replacement for POTS to provide voice and data communications worldwide using circuit switching as well as the same wiring that is currently being used in homes and businesses. Because ISDN is a digital signal from end to end, it is faster, more dependable, and has no line noise. ISDN has the ability to deliver multiple, simultaneous connections—in any combination of data, voice, video, or fax—over a single line and allows for multiple devices to be attached to the line.

The ISDN uses two types of channels, a B channel and a D channel. The **bearer channels (B channels)** transfer data at a bandwidth of 64 Kbps for each channel. The **data channels (D channels)** use a communications language called DSS1 for administrative signaling, such as to instruct the carrier to set up or terminate a B-channel call, to ensure that a B channel is available to receive a call, or to provide signaling information for such features as caller identification. Since the D channel is always connected to the ISDN, the call setup time is greatly reduced to 1 to 2 seconds (versus 10 to 40 seconds using an analog modem) as it establishes a circuit. The bandwidths listed here are the uncompressed speeds of the ISDN connections. The compressed bandwidth has a current maximum transmission speed of four times the uncompressed speed. You will probably see a much lower speed, however, as much of the data flowing across the network are already compressed.

Today the two well-defined standards used are the **basic rate interface (BRI)** and the **primary rate interface (PRI).** The BRI defines a digital communications line consisting of three independent channels: two bearer (or B) channels, each carrying 64 KB/s, and one data (or D) channel at 16 Kbps. For this reason, the ISDN basic rate interface is often referred to as **2B+D.** BRIs were designed to enable customers to use their existing wiring in their homes or businesses. This provides a low-cost solution for customers and explains why it is the most basic type of service today for small business or home use.

H channels are used to specify a number of B channels. The following list shows the implementations:

- **H0**—384 Kbps (6 B channels)
- **H10**—1472 Kbps (23 B channels)
- **H11**—1536 Kbps (24 B channels)
- **H12**—1920 Kbps (30 B channels)—Europe

In BRI, ISDN devices use two binary/one quaternary (2B1Q) encoding and time-division multiplexing (TDM) to create the 2B+D basic ISDN channels. The 2B1Q means that the input voltage level can be one of four distinct levels, or quaternaries. See Figure 17.7. Each quaternary represents two data bits—since there are four possible ways to represent two bits—and two bits are carried per baud. See Table 17.4.

The PRI is a higher-level network interface defined at the rate of 1.544 Mbps for North America and Japan. The primary rate, used on T1 digital lines, consists of 23 B channels, each at 64 Kbps, and one 64-Kbps D channel for signaling. These B channels can interconnect with the BRI, or carry services to any POTS line. Note that some switches limit B channels to 56 Kbps. European countries support a different kind of ISDN standard for PRI that consists of 30 B channels and one 64-Kbps D channel for a total of 1984 Kbps. A new

FIGURE 17.7   Two binary/one quaternary used by BRI ISDN

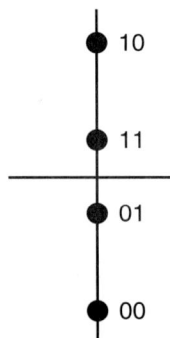

technology, nonfacility associated signaling (NFAS), also is available to enable users to support multiple PRI lines with one 64-Kbps D channel.

17.4.3 ISDN Equipment

To use BRI services, subscribe to ISDN services through a local telephone company or provider. By default, you must be within 18,000 feet (about 3.4 miles) of the telephone company's central office. Repeater devices are available for ISDN service to extend this distance, but these devices can be quite expensive. The various ISDN equipment and interfaces are shown in Table 17.5 and Figure 17.8.

TABLE 17.4 Quaternaries

Bits	Quaternary Symbol	Voltage Level
1 1	+1	+0.833
1 0	+3	+2.5
0 1	−1	−0.833
0 0	−3	−2.5

TABLE 17.5 ISDN equipment and interfaces

Equipment and Interface Types	Descriptions
U INTERFACE	The U interface is the two-wire interface between the customer premise and the CO. In North America, the customer is responsible for supplying all the equipment from the U interface forward.
NT1	At the customer premise, the ISDN line is terminated by an NT1 at the demarcation (boundary between the customer premise and the phone company's network). It converts the physical wiring interface delivered by the telephone company to the wiring interface needed by your ISDN equipment and provides a testing point for troubleshooting. The NT1 interface combines the B channels and the D channel into a single bit stream at the physical level and is also capable of supporting more than one device attached to an ISDN line, sometimes referred to as a multidrop configuration.
S/T INTERFACE	The S/T interface is between the NT1 and the ISDN networking equipment, which can support up to seven devices. To allow full-duplex interface, the S/T interface uses two pairs of wires, one pair to receive data and the other to transmit data.
TE1	Terminal equipment type 1 (TE1) devices are manufactured from the outset to be completely ISDN compatible. Examples are ISDN phones and integrated video devices.
R Interface	The R reference point provides a non-ISDN interface between equipment that is not ISDN compatible and the rest of the ISDN network.
TA	The terminal adapter allows nonnative ISDN devices (TE2), such as PCs, to connect to the S interface so that it may communicate over the ISDN network.
TE2	All terminal equipment type 2 devices (TE2) are non-ISDN compatible. Therefore, they require a terminal adapter (TA) to provide ISDN functionality to the rest of the network. Examples are personal computers and 3270 terminals.

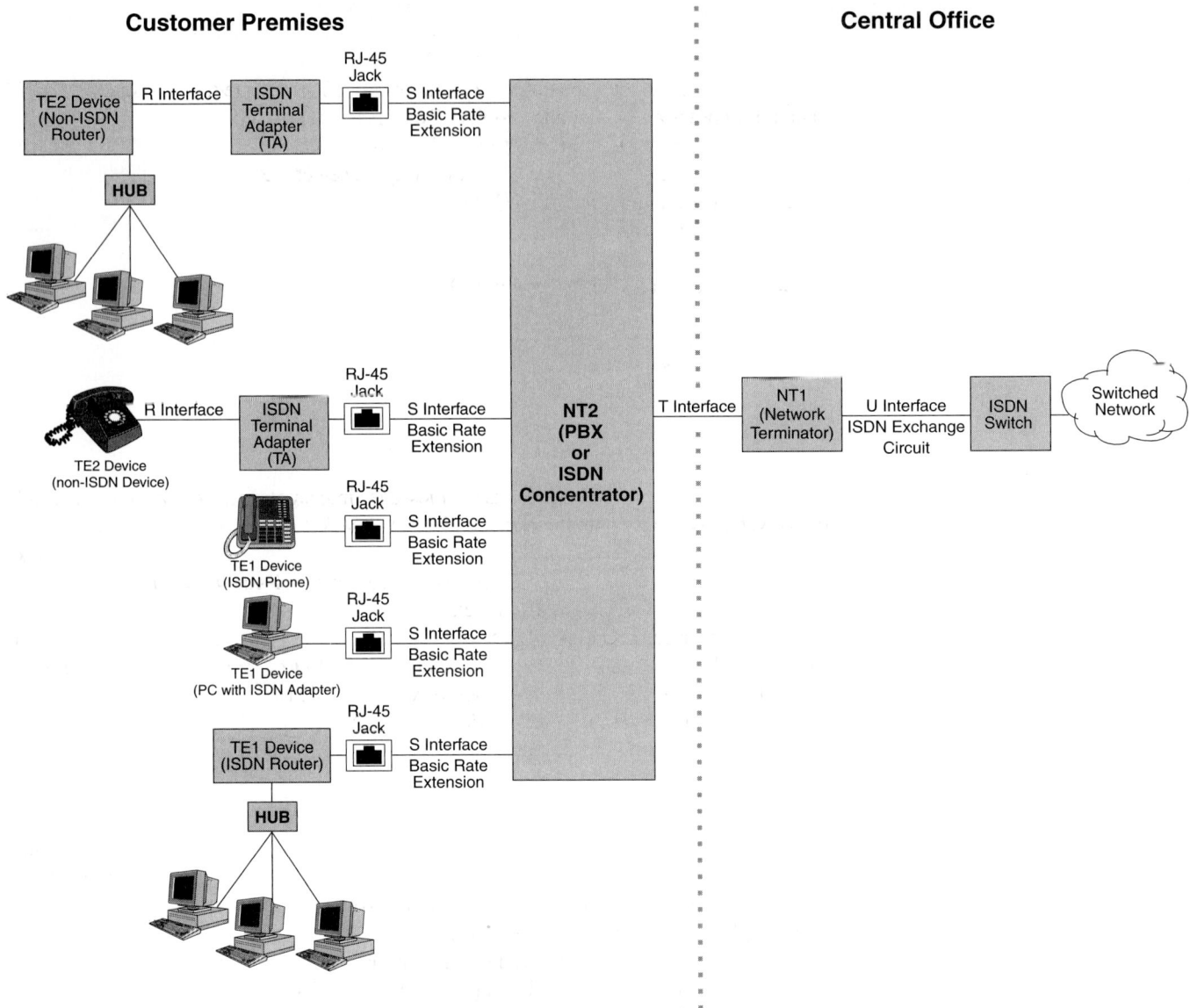

FIGURE 17.8   Integrated Services Digital Network

The **directory number (DN)** is the 10-digit phone number that the telephone company assigns to any analog line. A **service profile identifier (SPID)** includes the DN and an additional identifier that is used to identify the ISDN device to the telephone network. However, depending on the type of service switch and the utilization of the ISDN service, you may not need a SPID or you may need a SPID for each B channel or each device. Unlike an analog line, a single DN can be used for multiple channels or devices, or up to eight DNs can be assigned to one device. Therefore, a BRI can support up to 64 SPIDs. Most standard BRI installations, however, include only two directory numbers, one for each B channel.

17.4.4   ISDN and Windows 2000

To connect your Windows 2000 computer to an ISDN line, physically install your ISDN adapter and start the computer. After you load the appropriate drivers for the adapter, then use the Device Manager to specify the type of switch to which the adapter is connected: ATT (AT&T), NI-1 (National ISDN-1), and NTI (Northern Telecom). If your ISDN adapter is internal to your computer, then it appears in Network adapters; if the ISDN adapter is external to your computer, it appears in Modems.

To connect to the Internet by using an ISDN device, do the following:

1. Install the ISDN hardware in your computer.
2. On your desktop, right-click My Computer, click Properties, and on the Hardware tab, click Hardware Wizard.
3. After the ISDN adapter is added and Windows 2000 is restarted, verify that the ISDN adapter appears as a port under Ports in Routing and Remote Access.
4. To configure the ISDN port, right-click the port, and then click Properties.
5. In the Port Properties dialog box, click Configure, select the "Demand-dial routing connections (inbound and outbound)" check box, and then click OK.
6. To create a demand-dial interface and configure it to use the ISDN device to dial in to the ISP, right-click Routing Interfaces, and then click New Demand Dial Interface.
7. A default route has a destination of 0.0.0.0, a network mask of 0.0.0.0, and a metric of 1. Add a default route that uses the newly created demand-dial interface. Note, however, because the dial-up connection to the ISP is a point-to-point link, the gateway IP address is not configurable.

17.5 PACKET SWITCHED LINES

As mentioned, packet switching breaks the messages into small parts, or packets, and sends each packet individually onto the network. Since the packets have their own source and destination addresses, each can travel a different path than the other packets. When the packets reach their destination, they must be assembled. Two well-known examples of packet switched networks are X.25 and frame relay.

When a subscriber accesses a WAN network such as X.25, frame relay, and ATM, the leased network is sometimes referred to as a cloud. A **cloud** represents a logical network with multiple pathways used with a black box approach. Subscribers who connect to the cloud do not worry about the details inside the cloud, because subscribers know they connect at one edge of the cloud and the data are received at the other edge.

17.5.1 X.25

X.25 network is a packet switched network that allows remote devices to communicate with each other across digital links without the expense of individual leased lines. It typically is used in the packet switched network of common carriers, such as the telephone companies that are implemented at speeds below 64 Kbps.

The user end of the network is the data terminal equipment (DTE), and the carrier's equipment is the data circuit-terminating equipment (DCE). Connections occur on logical channels using either switched or permanent virtual circuits.

You can directly access the X.25 data network by using an X.25 smart card or a modem, which would dial into a packet assembler/dissembler (PAD). A PAD is used when a DTE device such as a character mode terminal is too simple to implement the full X.25 functionality. The PAD is located between a DTE device and a DCE device, and it performs three primary functions: buffering, packet assembly, and packet disassembly. See Figure 17.9.

Remote access clients who run Windows 2000 Professional or Windows 2000 Server can use either an X.25 card or dial in to an X.25 PAD to create connections. To accept incoming connections on a computer that is using X.25 and running Windows 2000 Professional or Windows 2000 Server, use an X.25 smart card.

Switched virtual circuits (SVCs) are temporary connections used for sporadic data transfers. They require that two DTE devices establish, maintain, and terminate a session each time the devices need to communicate. To establish a connection on a SVC, the calling DTE sends a call request packet, which includes the address of the remote DTE to be contacted. The destination for each packet is identified by means of the logical channel identifier (LCI) or logical channel number (LCN). This allows the network switches to route each packet to its intended DTE.

The calling DCE device examines the packet headers to determine which virtual circuit to use and then sends the packets to the closest switch in the path of that virtual circuit. The X.25 switches pass the traffic to the next intermediate node in the path, which may be another switch or the remote DCE device.

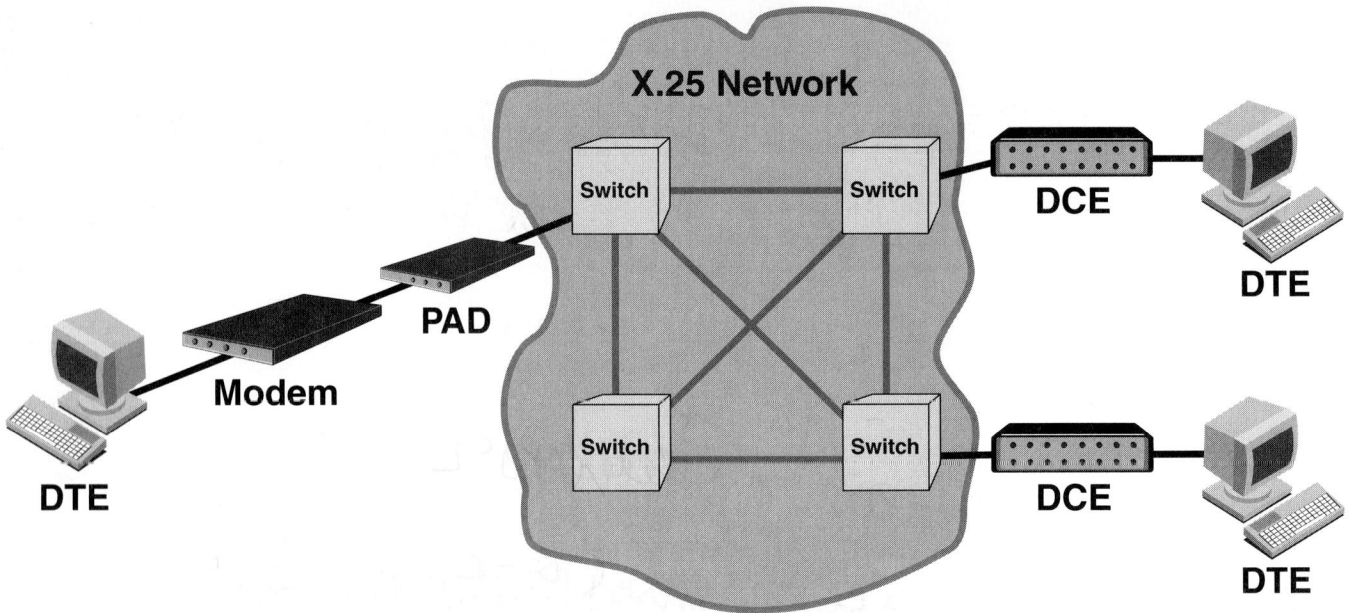

FIGURE 17.9 X.25 Network

The destination DTE decides whether to accept the call—the call request packet includes the sender's DTE address and other information that the called DTE can use to decide. A call is accepted by issuing a call accepted packet, or cleared by issuing a clear request packet. Once the originating DTE receives the call accepted packet, the virtual circuit is established and data transfer may take place. When either of the DTEs wish to terminate the call, a clear request packet is sent to the remote DTE, which responds with a clear confirmation packet.

Permanent virtual circuits (PVCs) are permanently established connections used for frequent and consistent data transfers. PVCs do not require that sessions be established and terminated. Therefore, DTEs can begin transferring data when necessary, because the session is always active.

X.21bis is a physical layer protocol used in X.25 that defines the electrical and mechanical procedures for using the physical medium. X.21bis handles the activation and deactivation of the physical medium connecting DTE and DCE devices. It supports point-to-point connections, speeds up to 19.2 Kbps, and synchronous, full-duplex transmission over four-wire media.

Link access procedure, balanced (LAPB) is a data link layer protocol that manages communication and packet framing between DTE and DCE devices. LAPB is a bit-oriented protocol that ensures frames are correctly ordered and error free. See Figure 17.10.

There are three types of LAPB frames: information, supervisory, and unnumbered. The information frame (I-frame) carries upper-layer information and some control information. I-frame functions include sequencing, flow control, and error detection and recovery. I-frames carry send-and-receive sequence numbers. The supervisory frame (S-frame) carries control information. S-frame functions include requesting and suspending transmissions, reporting on status, and acknowledging the receipt of I-frames. S-frames carry only receive sequence numbers. The unnumbered frame (U-frame) carries control information. U-frame functions include link setup and disconnection, as well as error reporting. U-frames carry no sequence numbers.

Packet layer protocol (PLP) is the X.25 network layer protocol. PLP manages packet exchanges between DTE devices across virtual circuits. PLPs also can run over logical link control 2 (LLC2) implementations on LANs and over the ISDN interfaces that run link access procedure on the D channel (LAPD). The PLP operates in five distinct modes: call setup, data transfer, idle, call clearing, and restarting.

To make a dial-up connection by using X.25 in Windows 2000, do the following:

1. Click on Start, select the Settings option, and then select the Network and Dial-up Connections option.
2. Double-click the Make New Connection icon, and then click the Next button.

FIGURE 17.10   X.25 Protocol suite

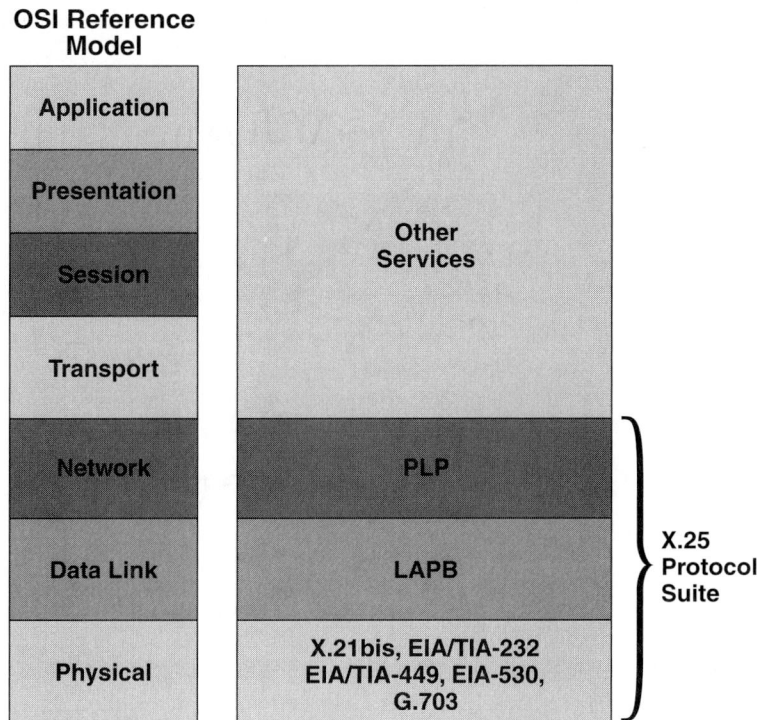

3. Select the Dial-up to Private Network option, click on the Next button, and then follow the instructions in the Network Connection wizard.
4. When the wizard is finished, right-click the new connection, and then click the Properties option.
5. Click on the Options tab and select the Click X.25 button.
6. In Network, click your X.25 network provider.
7. At the X.121 address, type the X.121 address (the X.25 equivalent of a phone number) for the server you want to call.
8. If required by the X.25 host computer, in User Data, type additional connection information.
9. In Facilities, type any additional facility options you want to request from your X.25 provider (e.g., some providers support the /r option to specify reverse charging).

17.5.2   Always On/Dynamic ISDN

Always on/dynamic ISDN (AO/DI) is a networking application that uses the ISDN D-channel X.25 packet service of 16 Kbps to maintain an "always on" connection between an ISDN end user and an information service provider. Low bandwidth requirements, such as sending and receiving simple e-mail, news feeds, automated data collection, and credit card verification, can be met using this constant virtual connection, which accommodates speeds up to 9.6 Kbps. The rest of the 16-Kbps D channel continues to be used for call setup and teardown signaling.

When additional bandwidth is necessary (for example, to download a graphics intensive file from the Internet), AO/DI automatically adds circuit switched B channels of 64 Kbps each, for a speed total of 128 Kbps when both B channels are in use. When the additional bandwidth is no longer required, one or both B channels are dropped, leaving the D-channel connection in place. Because the D-channel is an always-available connectionless packet-oriented link between the subscriber and the central office, it is possible to offer an always-available service that is based on it.

17.5.3   Defining Frame Relay

Frame relay is a direct descendant of X.25 and is defined by the International Telecommunications Union (ITU-T) Q.922 and Q.933 standards. **Frame relay** is a packet switching

FIGURE 17.11 Dedicated
point-to-point versus frame
relay

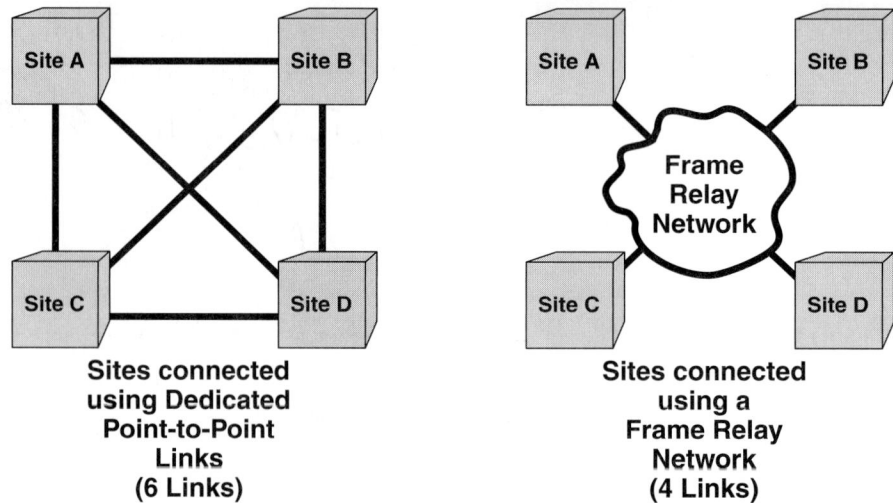

Sites connected
using Dedicated
Point-to-Point
Links
(6 Links)

Sites connected
using a
Frame Relay
Network
(4 Links)

protocol designed to use high-speed digital backbone links to support modern protocols for error handling and flow control for connecting devices on a WAN. It only defines the physical and data link layers of the OSI model, which allows for greater flexibility of upper-layer protocols to be run across frame relay. Frame relay networks are typically found using a switched 56 line, a fractual T1 line, or a full T1 line.

17.5.4 Accessing a Frame Relay Network

The most common use of a frame relay network is to connect individual LANs together. Data terminal equipment (DTE) is user terminal equipment that creates information for transmissions such as PCs and routers. Using a frame relay network, the cost is typically less than using dedicated point-to-point circuits, since customers can use multiple virtual circuits on a single physical circuit. For example, Figure 17.11 shows how it would take six dedicated point-to-point links to connect four sites. With a frame relay network, it would only take four links to connect the same sites. If you have five sites, it would take ten links using dedicated point-to-point links versus five links in a frame relay network.

To connect the DTEs to the frame relay network, use a **frame relay access device (FRAD),** sometimes referred to as a frame relay assembler/dissembler. It multiplexes and formats traffic for entering a frame relay network. A FRAD is a device that allows non–frame relay devices to connect to a frame relay network. FRADs can be included in routers, bridges, or be stand-alone. You also may need a CSU/DSU to connect to common links such as a T1 line which may be included with the FRAD.

Frame relay routers are the most versatile of the three, as they can handle traffic from other WAN protocols, reroute traffic if a connection goes down, and provide flow and congestion control. Frame relay bridges, which are used to connect a branch office to a hub, are basically low-cost unintelligent routers. Stand-alone FRADs are designed to aggregate (gather) and convert data, but have no routing capabilities. They are typically used on sites that already have bridges and routers or for sending mainframe traffic. Distinctions between these devices are blurring as vendors combine their functions into a single device.

Initially, frame relay gained acceptance as a means to provide end users with a solution for LAN-to-LAN connections and other data connectivity requirements. Due to advances in areas such as digital signal processing and faster backbone links within the frame relay network, end users are beginning to see viable methods being developed that incorporate nondata traffic such as voice or video over the frame relay network.

If you have a common private frame relay network and you are using both frame relay and non–frame relay interfaces (such as phone system), you will need a special multiplexer. The multiplexer will forward the frame relay traffic to the frame relay interface and on to the data network while non–frame relay traffic is forwarded to the appropriate application or service [such as **private branch exchange (PBX)** for telephone service or a video-teleconferencing application]. See Figure 17.12.

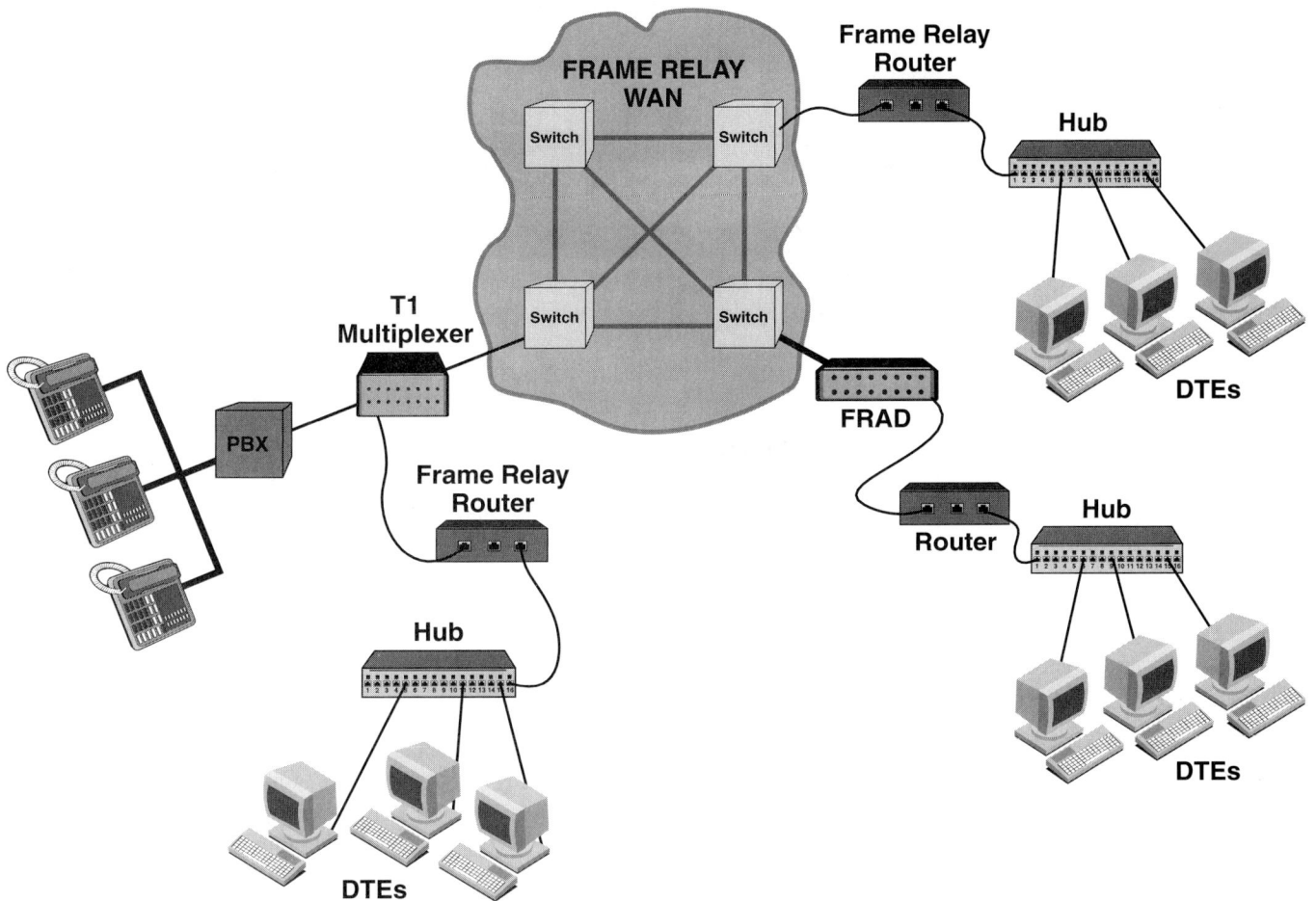

FIGURE 17.12   Frame relay network

To get to a frame relay network, first obtain a frame relay bearer service (FRBS), which is offered by common carriers such as your telephone company. If you use switched virtual circuit (SVC) as provided by a public carrier, the frame relay switching equipment is located in the central offices of a telecommunications carrier. The administering and maintaining of the frame relay network equipment and service is provided by the carrier, which relieves the customer of such duties. The majority of today's frame relay networks are public carrier–provided networks.

Today, many organizations worldwide are using PVC to deploy private frame relay networks. In such cases, the administration and maintenance of the network is the responsibility of the customer. The customer owns all the equipment, including the switching equipment.

When a packet is being sent on the frame relay network, it reaches the first switch. The packet is then switched or forwarded to the next switch toward its final destination, using the PVCs or SVCs. The network is connection oriented, because the originating and terminating ends are programmed into the network switches as identified by data link connection identifiers (DLCIs). DLCI values typically are assigned by the frame relay service provider (e.g., the telephone company). Frame relay DLCIs have local significance, which means that the values themselves are not unique in the frame relay WAN. Two DTE devices that are connected by a virtual circuit, for example, may use a different DLCI value to refer to the same connection. See Figure 17.13.

17.5.5   Characteristics of Frame Relay

At the physical layer, all signals in frame relay networks are binary and so are represented by a one or a zero. The FRAD is responsible for receiving these binary signals from the telephone carrier. To transmit binary signals on electronic media, voltage is raised or lowered. Most encoding methods are bipolar.

Virtual Circuits

DLC   DLC

15 ◄ · · · · · · · · · · · · · · · · ► 27

FRAME
RELAY
NETWORK

50 ◄ · · · · · · · · · · · · · · · · ► 80

72 ◄ · · · · · · · · · · · · · · · · ► 50

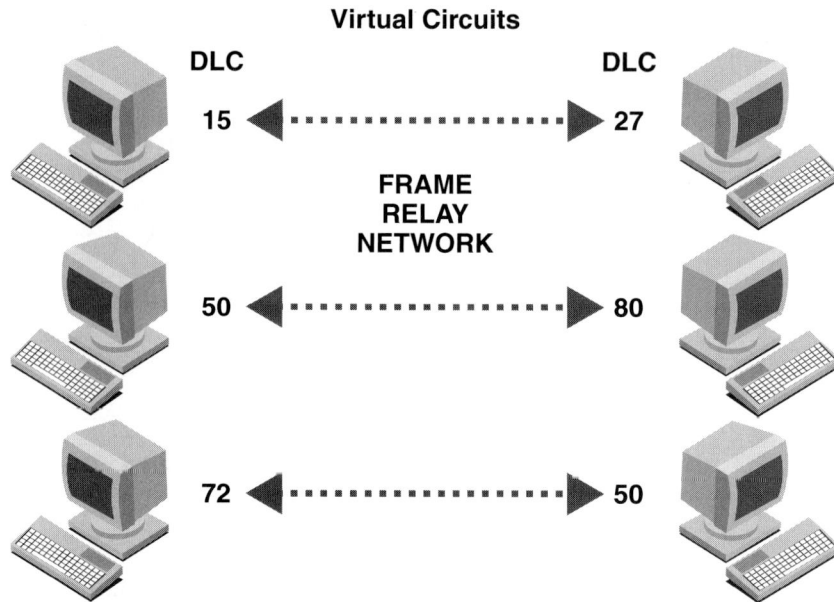

Frame relay networks are statistically time-division multiplexed, in which they assign time slots dynamically to make them more efficient by not using empty time slots. As a result, statistical time-division multiplexing uses the bandwidth more efficiently, particularly for bursty data traffic among clients and servers and for inter-LAN links. These networks also use variable-length packets to make efficient and flexible transfers.

Frame relay networks do not provide any error handling or control. When an error occurs in a frame, typically caused by noise on the line, it is detected upon receipt of the frame using the frame check sequence (FCS), a form of the cyclic redundancy check (CRC), and the frame is simply thrown away. Instead, the error handling is left to the destination devices such as the router and the upper protocols (e.g., TCP/IP or IPX). Because the frame relay is built on top of an all-digital network, the network discards very few frames, particularly when the networks are operating at well below design capacity. Because the frame relay network does not have to perform any error handling and the network automatically preserves the order of data, there are no frame retransmissions, sequence numbers, or acknowledgments. Therefore, the frame relay network has low overhead resulting in a faster network.

Much like network cards, frame relay cards have a buffer or temporary memory that holds the incoming frames until they are processed or holds the outgoing frames until they are sent. Like any other network connection, a frame relay network may experience network congestion, such as when a network node or switch receives more frames than it can process or when a network node or switch needs to send more frames across a given line than the speed of the line permits. As a result, the buffer fills up and the node is forced to discard frames. Since LAN traffic is extremely bursty, the probability of congestion occurring occasionally is high. Therefore, it is important that the frame relay network have excellent congestion management features both to minimize the occurrence and severity of congestion and to minimize the effect of the discards when they are required.

When you subscribe to a 56-Kbps, fractual T1 or full T1 line, you will be asked to specify a **committed information rate (CIR)** for each DLCI. This value specifies the maximum average data rate that the network undertakes to deliver under "normal conditions." If you send faster than the CIR on a given DLCI, the network will flag some frames with a **discard eligibility (DE)** bit. The network will do its best to deliver all packets but will discard any DE packets first if there is congestion. Many inexpensive frame relay services are based on a CIR of zero, which means that every frame is a DE frame, and the network will throw away any frame when necessary.

Help with frame relay to provide indications that the network is becoming congested comes from the **forward explicit congestion notification (FECN)** bit and the **backward explicit congestion notification (BECN)** bit, which are embedded within data frames.

These are used to tell the application to slow, hopefully before packets start to be discarded. For these bits to be used, the router or bridge would have to recognize and process them.

Frame relay offers bursting, a unique advantage over leased lines. Since the leased line bandwidth is fixed, if a device attempts to send data at a rate higher than the line bandwidth, the packets will not get through and performance will be degraded. Frame relay, however, allows a device to transmit data at a higher rate than the CIR for a few seconds at a time.

Devices using the extra free bandwidth run a risk in that any data beyond the CIR is eligible for discard, depending on network congestion. The greater the network congestion, the greater the risk that frames transmitted above the CIR will be lost. While the risk is typically very low up to the CIR, if a frame is discarded it will have to be resent. Data can even be transmitted at rates higher than the CIR, but doing this has the greatest risk of lost packets. If you find that you consistently need long bursts, then consider purchasing a higher CIR.

17.5.6 Frame Relay and Windows 2000

To connect to the Internet by using frame relay on Windows 2000, do the following:

1. Physically install the frame relay adapter in your computer.
2. On your desktop, right-click My Computer, select the Properties option, click on the Hardware tab, then click the Hardware Wizard. Follow the prompts to configure the frame relay adapter.
3. After the frame relay adapter is added and Windows 2000 is restarted, verify that the frame relay adapter appears as a routing interface under Routing Interfaces in Routing and Remote Access.
4. The default network route has a destination of 0.0.0.0, Network mask of 0.0.0.0, and metric of 1. Add a default network route that uses the frame relay adapter. Because the frame relay connection to the ISP is a point-to-point link, the gateway IP address is not configurable.

17.6 SYNCHRONOUS DIGITAL HIERARCHY AND SYNCHRONOUS OPTICAL NETWORK

The **Synchronous Digital Hierarchy (SDH)** and the **Synchronous Optical Network (SONET)** are transmission technology standards of synchronous data transmission over fiber-optic cables. They provide a high-speed transfer of data, video, and types of information across great distances without regard to the specific services and applications they support. Service providers who must aggregate (combine) multiple T1s to provide connections across the country or around the world typically use both. SDH is an international standard. The International Telecommunications Union (ITU), formerly known as CCITT, coordinates the development of the SDH standard. **SONET** is the North American equivalent of SDH, which is published by the American National Standards Institute (ANSI).

When referring to SONET or SDH, neither apply directly to switches. Instead, they specify the interfaces between switches that are linked by fiber-optic cable. A SONET or SDH switch does not actually exist, but instead, an ATM switch or a similar switch (such as FDDI, ISDN, or SMDS) is used. ATM, which we will discuss later in the chapter, can then support a variety of other services including frame relay and voice. SDH and SONET map the physical layer of the OSI model.

17.6.1 Building Blocks of SDH and SONET

The basic foundation of SONET consists of groups of DS-0 signals (64-Kbps) that are multiplexed to create a 51.84-Mbps signal, also known as a synchronous transport signal (STS-1). STS-1 is an electrical signal rate that corresponds to the optical carrier line rate of OC-1, SONET's building block. OC-1 has enough bandwidth to support 28 T1 links or one T3 link. Higher rates of transmission are a multiple of this basic rate. So, for example, STS-3 is three times the basic rate, or 155.52 Mbps. OC-12 is 12 times the basic rate, giving a rate of 622 Mbps, and OC-192, which is 9.95 Gbps. The SDH standard is based on the STM-1, which is equivalent to OC-3. Therefore, SDH uses multipliers of 155.52 Mbps. See Table 17.6.

TABLE 17.6 SONET/SDH digital hierarchy

Optical Level	SDH Equivalent	Electrical Level	Line Rate (Mbps)	Payload Rate (Mbps)	Overhead Rate (Mbps)
OC-1	—	STS-1	51.840	50.112	1.728
OC-3	STM-1	STS-3	155.520	150.336	5.184
OC-9	STM-3	STS-9	466.560	451.008	15.552
OC-12	STM-4	STS-12	622.080	601.344	20.736
OC-18	STM-6	STS-18	933.120	902.016	31.104
OC-24	STM-8	STS-24	1,244.160	1,202.688	41.472
OC-36	STM-13	STS-36	1,866.240	1,804.032	62.208
OC-48	STM-16	STS-48	2,488.320	2,405.376	82.944
OC-96	STM-32	STS-96	4,976.640	4,810.752	165.888
OC-192	STM-64	STS-192	9,953.280	9,621.504	331.776
OC-768	STM-256	STS-768	39,813.12	38,186.01	1,327.10

In essence, SONET and SDH are the same technology. There are minor differences in header information, payload size, and framing, but at 155 Mbps and above, the two are completely interoperable.

The basic level of SONET starts at approximately 51 Mbps, but the lower bit rate asynchronous signals are not ignored. The basic STS-1 frame contains 810 DS-0s, 783 of which are used for sending data (including slower asynchronous signals) and 27 of which are overhead. The overhead in this case is information concerning framing, errors, operations, and format identification.

Signals with speeds below STS-1, such as DS-1 and the European E1 (2.048 Mbps) can be accommodated by dividing the STS-1 payload into smaller segments known as virtual tributaries (VTs). The lower data rate signals are combined with overhead information, which leads to the creation of synchronous payload envelopes (SPEs). SPEs allow these signals to be transported at high speeds without compromising integrity. Each VT on an STS-1 signal includes its own overhead information and exists as a distinct segment within the signal.

17.6.2 SDH and SONET Lines and Rings

Local SDH and SONET services are sold in two forms: point-to-point dedicated lines or dual fiber rings. Both deliver high speed, but only dual fiber rings guarantee automatic rerouting around outages. Long-distance SDH and SONET connections employ multiple rings within the public network. Some are even using dual ring pairs (four redundant rings on one circuit). See Figure 17.14.

When using dual rings, if one circuit is broken, the traffic reverses and flows in the opposite direction on the same ring, thus avoiding the cut altogether. If the ring is broken in two places, the traffic is automatically rerouted onto the second circuit. This takes a few milliseconds longer than merely reversing direction, but it still happens nearly instantaneously.

Long-distance point-to-point SDH and SONET services are similar to standard leased lines. A company buys a connection between two points, but its traffic is sent over multiple rings within the public network. Although the customer is not buying rings, he or she gets all the benefits associated with rings within a long-distance portion of the carrier network. Customers who need high reliability end to end should use local access rings.

17.7 CELL RELAY

Data transmission technology is based on transmitting data in relatively small, fixed-size packets, or **cells.** Each cell contains only basic path information that allows switching devices to route the cell quickly. Cell relay systems can reliably carry live video and audio, because cells of fixed size arrive in a more predictable way than systems with packets or frames of varying size. Examples of cell relay are SMDS and ATM.

FIGURE 17.14 Carrier SDH/SONET networks use dual rings to provide redundant pathways.

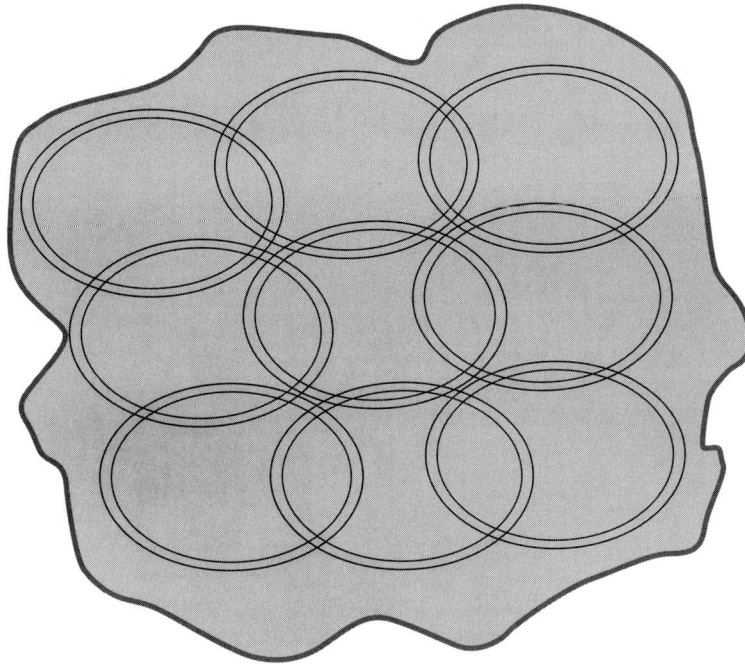

FIGURE 17.14 Carrier SDH/SONET networks use dual rings to provide redundant pathways.

17.7.1 Switched Multimegabit Data Service

Switched multimegabit data service (SMDS) is a high-speed, cell relay WAN service designed for LAN interconnection through the public telephone network. SMDS can use fiber- or copper-based media. While it mostly supports speeds between 1.544 Mbps and 34 Mbps, it has been extended to support lower and higher bandwidth to broaden its target market. Although SMDS and frame relay are both desirable ways of gaining Internet access for your business, frame relay is popular for speeds of T1 and below and SMDS is popular for speeds above T1 up to T3. An SMDS circuit is committed at its specified speed. Bursting of the circuit to full bandwidth is never required.

Different from frame relay and ATM, SMDS is a connectionless service, in that there is no predefined path or virtual circuit setup between devices. Instead, SMDS simply sends the traffic into the network for delivery to any destination on the network. With no need for a predefined path between devices, data can travel over the least congested routes in an SMDS network. As a result, it provides faster transmission for the networks bursty data transmissions and greater flexibility to add or drop network sites. SMDS currently supports only data, and is being positioned as the connectionless part of ATM services.

An SMDS digital service unit (DSU) or channel service unit (CSU) takes frames that can be up to 7168 bytes long (large enough to encapsulate the entire IEEE 802.3, IEEE 802.5, and FDDI frames) from a router and breaks them up into 53-byte cells. The cell is then passed to a carrier switch. Each cell has 44 bytes of payload or data and 9 bytes for addressing, error correction, reassembly of cells, and other control features. The switch then reads addresses and forwards cells one by one over any available path to the desired endpoint. SMDS addresses ensure that the cells arrive in the right order.

Each destination is assigned up to 16 unique SMDS addresses (10-digit numbers similar to a telephone number) as needed. This feature gives the subscribers of SMDS to indicate individual users within a business. This will allow you to create the illusion of dedicated access for specific customers when in fact you are using a single SMDS connection.

Another feature is group addressing, in which files or information can be sent to multiple users at one time. This capability allows one member to make last-minute changes and then send all members of the group the updated copy in a matter of minutes.

SMDS implements two security features: source address validation and address screening. Source address validation ensures that the source address is legitimately assigned to the device, which it originated. This helps prevent address spoofing, in which illegal traffic assumes the source address of a legitimate device. Address screening allows a subscriber to establish a private virtual network that excludes unwanted traffic. If an address is disallowed, the data unit is not delivered.

To connect to an SMDS network, communicate with a common carrier to provide T1 connection. To connect to SMDS, use a CSU/DSU (or some other connection device) and router. Often the CSU/DSU and router can be found packaged together.

17.7.2　Introduction to ATM

Asynchronous transfer mode (ATM) is both a LAN and a WAN technology, which is generally implemented as a backbone technology. It is a cell switching and multiplexing technology that combines the benefits of circuit switching and packet switching. The small, constant cell size allows ATM equipment to transmit video, audio, and computer data. Current implantations of ATM support data transfer rates of 25 Mbps to 622 Mbps. The ATM standards allow it to operate in virtually every transmission medium including T1, T3, E1, E3, SDH, and SONET.

Because of its asynchronous nature, ATM is more efficient than synchronous technologies, such as time-division multiplexing (TDM). With TDM, users are assigned to time slots, and no other station can be sent in that time slot. If a station has numerous data to send, it can send only when its time slot comes up, even if all other time slots are empty. If, however, a station has nothing to transmit when its time slot comes up, then the time slot transmits as empty and is wasted. Because ATM is asynchronous, time slots are available on demand.

17.7.3　ATM Overlay Model

The protocol reference model used for ATM is taken from a model that was developed by the ITU for the **Broadband-Integrated Services Digital Network (B-ISDN).** Because ATM is the transport mode used for B-ISDN, this model applies directly to ATM and is often used as the protocol to describe it.

The ATM overlay model is three-dimensional and consists of three planes and four layers. The three planes that span all layers are as follows:

- **Control**—Responsible for generating and managing signaling requests to perform connection administration, such as call setup and call teardown.
- **User**—Responsible for managing the transfer of data across the network.
- **Management**—Maintains the network and carries out operational functions. This plane is further subdivided into layer management and plane management to manage the different layers and planes.
- **Layer Management**—Manages layer-specific functions, such as the detection of failures and protocol problems
- **Plane management**—Manages and coordinates functions related to the complete system

The ATM reference mode is composed of the following ATM layers:

- **Physical layer**—Analogous to the physical layer of the OSI reference model, the ATM physical layer manages the medium-dependent transmission.
- **ATM layer**—Combined with the ATM adaptation layer, the ATM layer is roughly analogous to the data-link layer of the OSI reference model. The ATM layer is responsible for establishing connection and passing cells through the ATM network. To do this, it uses information in the header of each ATM cell.
- **ATM Adaptation Layer (AAL)**—Combined with the ATM layer, the AAL is roughly analogous to the data-link layer of the OSI mode. The AAL is responsible for isolating higher-layer protocols from the details of the ATM processes.

The ATM physical layer has four functions: bits are converted into cells, the transmission and receipt of bits on the physical medium are controlled; ATM cell boundaries are tracked; and cells are packaged into the appropriate type of frame for the physical medium. The ATM physical layer is divided into two parts, the physical medium-dependent (PMD) sublayer and the transmission-convergence (TC) sublayer. The PMD sublayer synchronizes transmission and reception by sending and receiving a continuous flow of bits with associated timing information and it specifies the physical details for the physical medium used,

FIGURE 17.15 ATM Model

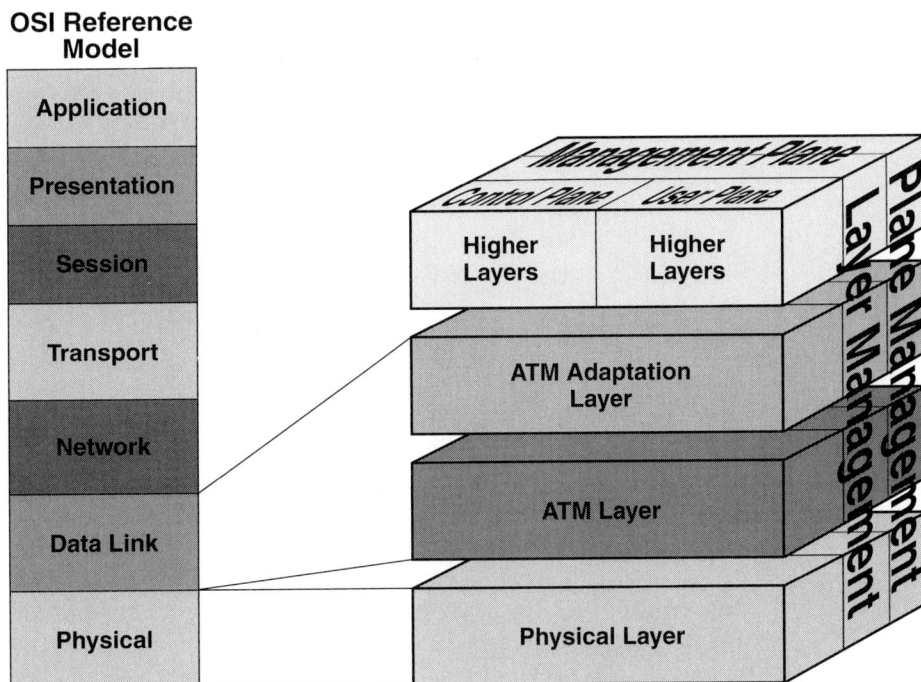

TABLE 17.7 ATM Model layers

Layer	Description
AAL1	Supports circuit emulation using ATM cells for traditional voice, T1, and T3 carrier services. AAL1 uses two methods: structured data transfer (SDT) and synchronous residual time stamp (SRTS). Requirements include a constant bit rate and connection-oriented, isochronous (time-dependent) service.
AAL2	Supports packet-based video and other time-dependent applications that can use variable bit rate. AAL2 is not currently defined by the ATM standards.
AAL3/4	Supports multiplexing of data streams, including connection-oriented and connectionless services. AAL3/4 was used in early ATM implementations but is no longer favored because of additional overhead cost resulting from the segmentation and reassembly (SAR) sublayer. The SAR header and SAR trailer are each 2 bytes long, effectively reducing available cell payload for data to 44 bytes.
AAL5	Designed to support efficient transport of LAN traffic. Widely implemented today in most ATM products. AAL5 is the adaptation layer used most often, offering the best performance when compared to other AALs for LAN traffice.

such as cable and connector types. The TC sublayer performs cell delineation, header error-control (HEC) sequence generation and verification, cell-rate decoupling and transmission-frame adaptation. See Figure 17.15.

The ATM layer describes how cells are transported through the network and how quality of service is enforced so that connections provide the necessary bandwidth for the specified service. The ATM layer provides creation of cells, multiplexing and demultiplexing of cells, managing of cell flow and sequencing, handling of dropped cells, switched-based routing using virtual paths and virtual circuits. See Table 17.7.

The ATM adaptation layer (AAL) is where user information is created and received as 48-byte payloads. The adaptation layer resolves any disparity between services provided by the cell-based technology of the ATM layer to the bit-stream technology of digital services (such as telephones and video cameras) and the packet-stream technology of traditional data networks (such as Frame relay and X.25 used in WANs and LAN protocols such as Ethernet and TCP/IP).

The International Telecommunication Union (ITU) first determined the need to provide several standard AALs (classes of service) to satisfy the requirements of encapsulating different information types into ATM layer cells.

The upper layers of the ATM protocol reference model include optional protocol layers that are used to further encapsulate ATM service for use with TCP/IP and other protocols. Some examples of upper-layer protocols include those specified by RFC 1483 (multiprotocol encapsulation over AAL5) and RFC 1577 (classical IP over ATM).

17.7.4 ATM Cell

The ATM cell has a size of 53 bytes. This cell contains 48 bytes of payload or data and 5 bytes to hold the control and routing information. A fixed cell was chosen for two reasons. First, the switching devices that are necessary to build the ATM network require very fast silicon chips to perform the cell switching at such high speeds. These chips can run much faster if they do not vary, but remain a fixed, known value. By using a payload length of 48 bytes for data, ATM offers a compromise between a larger cell size (such as 64 bytes) optimized for data, and a smaller cell size (such as 32 bytes) optimized for voice. See Figure 17.16. Secondly, ATM is designed to carry multimedia communications, composed of data, voice, and video. Unlike data, however, voice and video require a highly predictable cell arrival time for sound and video to appear natural. These cells must arrive one after the other, in a steady stream with no late cells. If the cells varied in size, a very long cell could block the timely arrival of a shorter cell. Therefore, a constant cell size was chosen.

The ISDN data communications discussed earlier in this chapter is known as narrowband ISDN, which was designed to operate over the current communications infrastructure based on the twisted-pair copper cable. With broadband-ISDN (B-ISDN), the end user gets true bandwidth on demand, only paying for the bandwidth used. Bandwidth of B-ISDN is faster than a T1 line (1.544 Mbps), which generally is found on fiber-optic cables that are running ATM or SMDS.

17.7.5 ATM Network Operation

An ATM network consists of ATM switches (sometimes referred to as network node interfaces or NNI) and ATM endpoints (sometimes referred to as user network interfaces or UNI), consisting of PCs with ATM adapter cards, routers, bridges, CSU/DSUs, and video

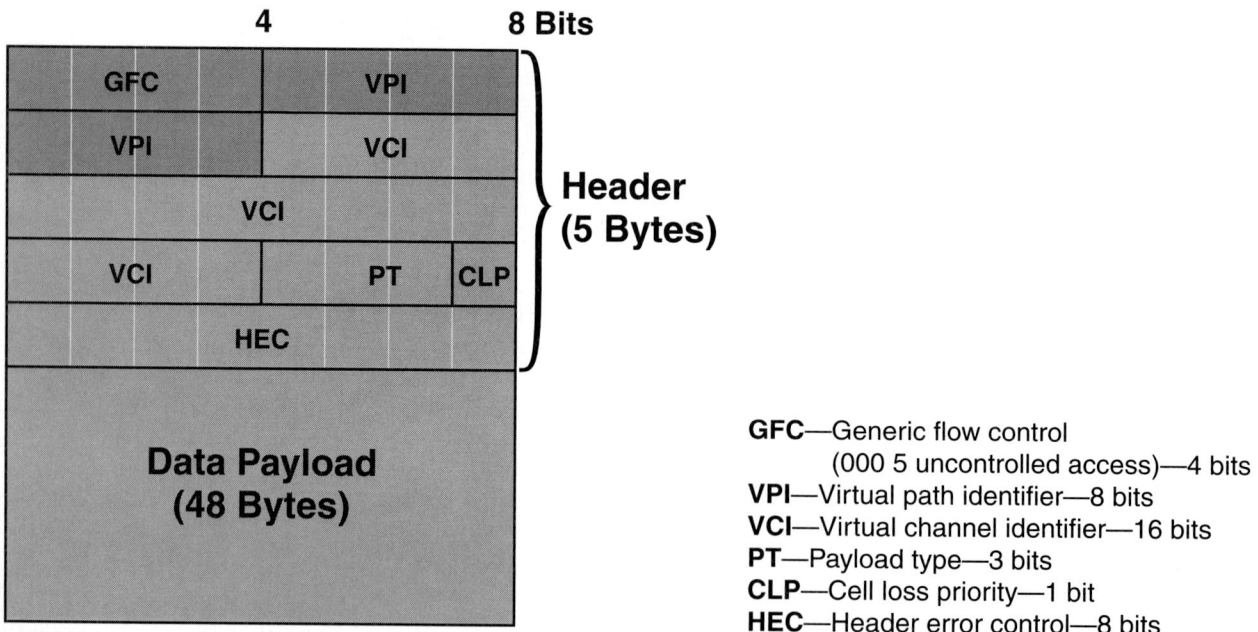

FIGURE 17.16 ATM cell content

GFC—Generic flow control
(000 5 uncontrolled access)—4 bits
VPI—Virtual path identifier—8 bits
VCI—Virtual channel identifier—16 bits
PT—Payload type—3 bits
CLP—Cell loss priority—1 bit
HEC—Header error control—8 bits

538

coder/decoders. When a packet initially enters the ATM network, it is segmented into ATM cells, a process known as cell segmentation.

An ATM switch is responsible for moving cells through an ATM network. It accepts the incoming cell from an ATM endpoint or another ATM switch. It then reads and updates the cell header information and quickly switches the cell to an output interface toward its destination. An ATM endpoint (or end system) contains an ATM network interface adapter. Examples of ATM endpoints are workstations, routers, data service units, LAN switches, and video coder/decoders. ATM cells used by UNI and NNI have a slightly different cell format. NNI cells have no generic flow control (GFC) field. Instead, the first 4 bits of the cell are used by an expanded 12-bit VPI space.

On an ATM network, each ATM-attached device (workstations, servers, routers, or bridges) has immediate, exclusive access to the switch. Because each device has access to its own switch port, devices can send cells to the switch simultaneously. Latency becomes an issue only when multiple streams of traffic reach the switch at the same time. To reduce latency at the switch, the cell size must be small enough so the time it takes to transmit a cell has little effect on the cells waiting to be transmitted. Since each ATM device has immediate, exclusive access to a switch port, device connection to an ATM switch does not require complete media access schemes to determine which device has access to the switch.

On an ATM network, each ATM-attached device (workstations, servers, routers, and bridges) is attached directly to a switch. When an ATM-attached device requests a connection with a destination device, the switches to which the two devices are attached set up the connection. While setting up the connection, the switches determine the best route to take. Therefore, the ATM switch is a routing switch.

Connection setup standards for the ATM layer define virtual circuits and virtual paths. An ATM virtual circuit is the connection between two ATM endstations for the duration of the connection. The virtual circuit is bidirectional, which means that once a connection is established, each endstation can send to or receive from the other endstation.

Whereas a virtual circuit is a connection that is established between two endstations for the duration of a connection, a virtual path lies between two switches and exists all the time, regardless of whether a connection is being made. In other words, a virtual path is a permanent path that all traffic from a single switch can take to reach another switch.

When a user requests a virtual circuit, the switches determine which virtual path to use in order to reach the endstations. Traffic for more than one virtual circuit may travel the same virtual path at the same time. For example, a virtual path with 120 Mbps of bandwidth can be divided into four simultaneous connections of 30 Mbps each.

After the connection is established, the switches between the endstations receive translation tables, which specify where to forward cells based on the port from which the cells enter and the special values in the cell headers that identify the virtual circuit identifiers (VCIs) and virtual path identifiers (VPIs). The translation table also specifies which VCIs and VPIs the switch should include in the cell headers before the switch sends the cells.

ATM can use three types of virtual circuits—permanent virtual circuits (PVCs), switched virtual circuits (SVCs), and smart permanent virtual circuits (SPVCs). The PVCs and the SVCs are the same as discussed earlier in this chapter. The permanent virtual circuits are manually established pathways between two endstations that do not require setup or tear down of the connection. Most early ATM devices supported only PVCs, whereas today's devices support all three.

The SVC is established dynamically when an endstation needs to send data. First, the endstation requests a connection with another endstation. An SVC is established between two endstations on an as-needed basis and expires after an arbitrary amount of time. In ATM, the process of determining whether to establish a connection is called connection admission control (CAC). SVCs are often the preferred mode of operation because they can be dynamically established, thus minimizing reconfiguration complexity. See Figure 17.17.

The **smart permanent virtual circuit (SPVC)** is a hybrid of a PVC and an SVC. Like a PVC, an SPVC is established manually when the network is configured. However, the ATM service provider or network administrator sets up only the endstations. For each transmission, the network determines through which switches the cells will pass.

When the virtual circuits are established, the application or service specifies the average and peak traffic rates, peak traffic duration, and burstiness of the traffic. By setting these

539

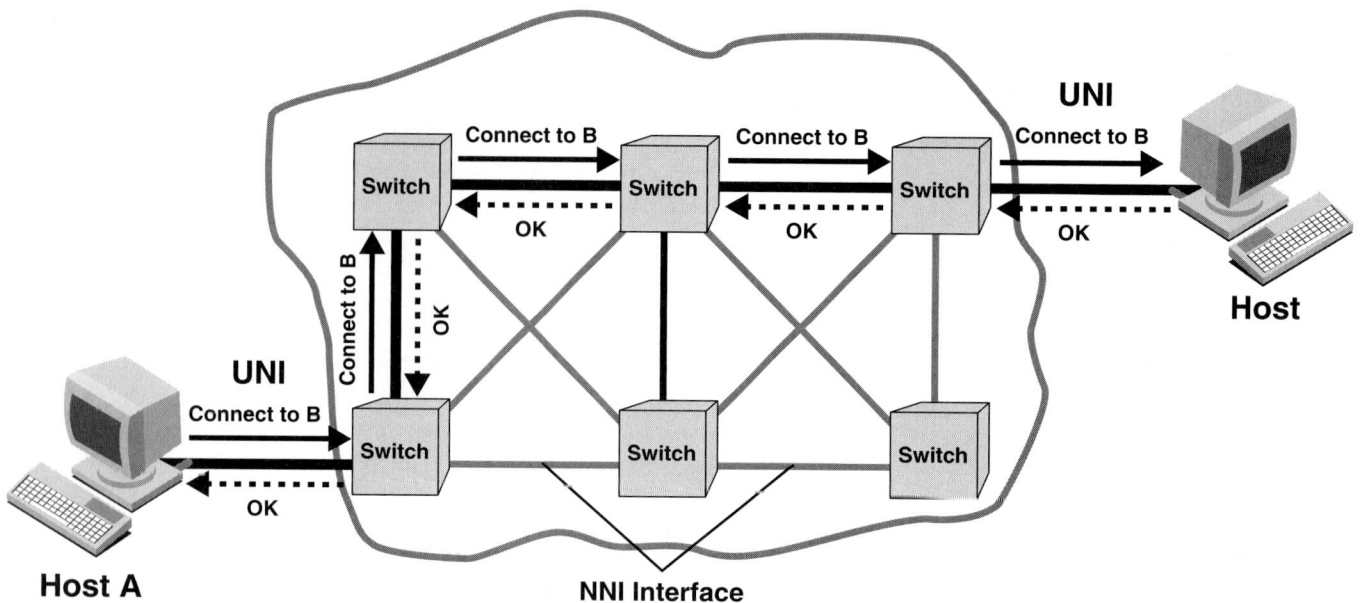

FIGURE 17.17   ATM virtual circuit

parameters, network designers can ensure that voice, video, and data traffic get the required quality of service (QoS). For example, the network can respond to traffic burst by automatically allocating additional bandwidth to a particular virtual circuit. Certain types of traffic or calls can be prioritized according to their importance or sensitivity to time delay. The ATM adaptation layer defines the following service categories:

- **Constant bit rate (CBR)** specifies a fixed bit rate so that data are sent in a steady stream for applications and services that require a small amount of bandwith at all times and are sensitive to delay and cell loss. Some typical CBR applications are circuit emulation, voice, video, and any type of data that are time-defined. CBR applications are not burst. One example is supporting voice traffic connected to a PBX.
- **Real-time variable bit rate (rt-VBR)** provides a specified throughput capacity when services and applications require high bandwidth but data are not sent evenly. This is a popular choice for voice and video conferencing data.
- **Non-real-time variable bit rate (nrt-VBR)** is similar to the nr-VBR except without the latency requirement. An application that may use this category is routers that connect LANs.
- **Unspecified bit rate (UBR)** does not guarantee any throughput levels. It is for non-real-time rates with no required service guarantees. UBR gives no guarantee that the cells will even be delivered. Therefore, ATM provides the carrier with the option of dropping certain cells if the network gets congested. This is used for applications such as file transfer and e-mail that can tolerate delays.
- **Available bit rate (ABR)** provides a guaranteed minimum capacity but allows data to be bursted at higher capacities when the network is free. This is used for applications such as file transfer and e-mail that can tolerate delays.

Although ATM is connection oriented, many applications such as mail services and other data services are characterized by small amounts of data, sent sporadically. To save time and expense, no connection is established (connectionless service). User information is sent in a message that contains all necessary addressing and routing information, such as those used in local area networks. SMDS provides the connectionless service for ATM networks.

17.7.6   Classical IP Over ATM

Classical IP (CIP) over ATM is a mature standard that enables user to route IP packets over an ATM network or cloud using ATM as either a backbone or a workgroup technology. Classical IP over ATM maps IP network layer addresses to ATM addresses and enables ATM-attached devices to send IP packets over an ATM network.

The IP over ATM client is an endstation (computer, router, or bridge) with an ATM adapter installed and connected to the ATM network. These endstations may be located in physical separate LANs, but because they are both connected to the ATM network, they are part of the same logical subnetwork, or virtual subnetwork. ATM networks are partitioned into logical IP subnets (LISs) that communicate with each other via routers. Because there is no native broadcast capability in ATM, the traditional broadcast address resolution protocol (ARP) is replaced by a client-server-based ATMARP protocol.

The ATMARP server maintains a database of IP and ATM addresses of all clients on the network, resolving multicast and broadcast IP addresses to ATM addresses. ATMARP only supports unicast traffic, allowing the client to set up point-to-point connections with other endstations. The multicast address resolution server (MARS) maintains a database of broadcast and multicast addresses to all members of the network. The MARS can pass a list of addresses directly to a client so that the client can set up a point-to-point call. If a multicast server is available, the MARS can pass the address of the server to the client. In this case, the client contacts the MARS, and the MARS creates a point-to-multipoint connection and distributes the packets to the endstations. The ATM ARP/MARS server with an integrated multicast server is shipped with Windows 2000 Server.

An endstation can use either PVC or SVC connections. If an endstation is using a PVC, the ATM service provider or network administrator manually maps the IP addresses of reachable destination endstations to ATM virtual circuits. To establish a connection using classical IP over ATM, a sending station sends an ATMARP request to the ATMARP server on the logical IP subnet. The server reconciles the IP address of the receiving station with its ATM address and returns the ATM address to the sending client. The sending client then uses the address to establish an ATM link. When the receiving computer receives the first packet, it also sends an ATMARP request to the server to find the location of the sender. Once the server returns the appropriate address, the link is established and the two clients can communicate directly without further involvement by the server. After the PVC or SVC has been established, the sending endstation converts its packets into ATM adaptation layer (AAL) 5 cells and sends them over the virtual circuit to the destination endstation. The destination endstation then converts the AAL 5 cells into IP packets.

A station can have more than one active virtual circuit at a time. A file server or e-mail server may have hundreds of connections within a short period of time, depending on how many client systems it serves. Connections that go unused for a specified amount of time (the default is 15 to 20 minutes) are automatically cleared to recover adapter and ATM network resources. To use classical IP over ATM as a backbone to connect physically separate LANs, route IP packets over ATM by using routers as endstations.

Classical IP over ATM has several disadvantages. Because the ATMARP server can reach only one IP subnetwork, IP hosts can communicate directly with only destination IP hosts in that subnetwork. To send packets to a destination IP host in another virtual subnetwork, a sending IP host must send the packets through a router. The sending IP host uses a virtual circuit to connect to the router, and the router uses another virtual circuit to connect to the destination IP host. Routers create a bottleneck because they are slower than switches.

Another disadvantage is that classical IP over ATM can route only IP. If you want to route other protocols such as IPX, you cannot use classical IP over ATM. Furthermore, it does not solve delay or congestion problems because it cannot take advantage of ATM's qualities of service.

17.7.7 ATM LAN Emulation

LAN emulation (LANE) as defined by the ATM Forum, is an ATM standard that emulates an Ethernet or token ring LAN on top of an ATM network. In other words, the LANE protocol makes an ATM network look and behave like an Ethernet or token ring LAN. Since the LANE emulates the media access control (MAC) sublayer of the OSI model, LANE enables any network layer protocol that works with the OSI model (including IPX, IP, and NetBIOS) to travel over an ATM network without modification. Users can then run applications over the ATM network, as they would over an Ethernet or token ring LAN; the ATM network is invisible to LAN users.

An **emulated LAN (ELAN)** consists of a set of LAN emulation clients that reside as ATM endpoints. The ELAN is equivalent to a virtual LAN (VLAN). The end systems (ATM network interface cards and internetworking and LAN switching equipment) have LAN emulation service that is implemented as device drivers, which will resolve MAC addresses to ATM addresses, set up direct connections between the clients, and forward the data. The LANE protocol operates transparently over and through the ATM switches and does not directly impact ATM switches. End stations on the same physical segment can be part of different emulated LANs, and endstations can be members of more than one emulated LAN.

Each ELAN supports connectivity only between an ATM network and a single traditional (Ethernet or token ring) LAN environment. LANE cannot serve as a gateway between Ethernet and token ring LANs. An ELAN consists of the following components:

- **LAN emulation client (LEC)**—The LEC is implemented in ATM end system computers so that they can forward data between the ATM and LAN environments, process LAN broadcasts, and register MAC addresses with the LAN emulation server (LES). The LEC also provides a standard LAN interface to a high-level protocol on legacy LANs. The LEC is jointly identified by two addresses: its ATM address when communicating to the ATM network and a MAC address when communicating with the Ethernet or token ring LAN. Note that ELANs will have one LEC per ELAN.
- **LAN emulation server (LES)**—The LES provides a central control point for LECs to include registration of MAC addresses for new LECs, resolution of MAC addresses to ATM addresses for existing LECs, and maintenance of a mapped pairing of ATM addresses-to-media access control addresses for all ELAN member devices.
- **LAN emulation configuration server (LECS)**—The LECS is responsible for assigning LAN emulation clients to specific emulated LANs. When a client requests configuration information for joining an emulated LAN, the LECS supplies the ATM address of the LES for the emulated LAN. Clients are directed to an appropriate LES, based on either their physical location (as interpreted by analyzing the client ATM address) or the identity of a particular destination LAN. The configuration server can, as an option, allow a client to configure and directly assign itself to a specific LES and emulated LAN. Each internetwork has only one LECS that serves all LAN emulation clients within it.
- **Broadcast and unknown server (BUS)**—The BUS is a multicast server that is used to flood unknown destination address traffic and to forward multicast and broadcast traffic to clients within a particular ELAN. Each client is associated with a single BUS, identified by a unique ATM address, and located within a particular ELAN.

Suppose that you were working on an emulated LAN and wanted to access a file stored on a server that was located on a physically separate LAN. First, you would send the file request, and your LANE client would determine if it knew the ATM address of its LANE server. If your LANE client did not know this address, the client would query the LANE configuration server, asking for the ATM address of the LANE server. Second, after your LANE client received the correct address, your client would query the LANE server for the ATM address of the LANE server on which the file was stored. If the LANE server knew this address, the LANE server would send the address to your LANE client. If the LANE server did not know this address, the server would query the LANE BUS. The LANE BUS, in turn, would ask all LANE clients on the emulated LAN for their ATM addresses. The LANE BUS would then return the correct address to the LANE server, which would return the address to your LANE client. Finally, your LANE client would establish a virtual circuit to the server on which the file was stored. Then the LANE client would convert its Ethernet or token ring frames into cells and send these cells over the virtual circuit to the server.

With LANE, you can also send cells to other LANE clients without establishing a virtual circuit. Every LANE client that joins the emulated LAN establishes a permanent SVC with the LANE BUS. As a result, all LANE clients are connected to the LANE BUS via SVCs. If you want to send cells to another LANE client, your LANE client would forward cells to the BUS via an existing SVC, and the BUS would forward the cells to the destination LANE client. In this way, LANE enables the connection-oriented ATM network to mimic a connectionless network. LANE also allows you to broadcast cells to the entire emulated LAN. Your LANE client sends the cells to the LANE BUS, and the BUS forwards these cells to all other LANE clients.

Since existing LANs mainly support connectionless services, the LAN emulation service uses available bit rate (ABR) service. As discussed, ABR service is important not only for LAN emulation, but also for applications involving bursty data traffic in which the bandwidth requirements cannot be predicted.

Because the LANE 1.0 standard does not specify a standard way for servers to communicate, each emulated LAN can have only one LANE server. Therefore, you cannot have redundant LANE servers, and the LANE server may become a bottleneck in large emulated LANs. In addition, because the LANE server must use virtual circuits to route traffic and because ATM switches can support only a limited number of virtual circuits (depending on your switches' capacities), using only one LANE server can overburden the switches. The LANE 2.0 standard will solve this problem by defining a way for the components of the LANE services to be distributed and will define the interface between these components, thus allowing up to 20 LANE BUS servers.

LANE has its limitations, however. As with IP hosts on classical IP over ATM subnetworks, an endstation in one emulated LAN must use a router to communicate with an endstation in another emulated LAN. The sending endstation must establish a virtual LAN with a router, which establishes a virtual circuit with the destination endstation. Routers can create bottlenecks because they are slower than switches.

Because LANE is a bridging technology, it limits you to a maximum of 2000 endstations per emulated LANE. Of course, the more endstations you have, the worse your network will perform. More endstations mean more broadcasts and more virtual circuits, which could be a problem since an ATM network can support only a certain number of virtual circuits at one time.

Because LANE is an OSI data link layer technology and is transparent to the higher layers of the OSI model, it cannot take advantage of ATM's quality of service. In addition, the LANE 1.0 standard supports only unspecified bit rate (UBR). LANE 2.0 enables you to specify the type of traffic your emulated LAN will carry: the constant bit rate (CBR), the variable bit rate (VBR), or the available bit rate (ABR). Unfortunately, the LANE 2.0 standard requires all virtual circuits on the emulated LAN to use the specified type of traffic.

17.7.8 Multiprotocol over ATM

Multiprotocol over ATM (MPOA) was developed by the ATM Forum to route protocols such as IP, IPX, and NetBIOS from traditional LANs over a switched ATM backbone. MPOA incorporates LANE to provide bridging capabilities while allowing the MPOA to route between virtual subnetworks without using traditional routers. Using routers in an ATM network significantly slows packet throughput, because each router has to reassemble cells of the OSI network layer packets for routing and then segment the packets into cells again for forwarding.

MPOA consists of route servers and edge devices. Route servers, also known as MPOA servers, maintain routing tables and calculate routes on behalf of edge devices. Route servers also communicate with traditional routers and with outer route servers. Route servers can be one piece of hardware or be built into routers and switches. Edge devices, also known as MPOA clients, can be intelligent switches that forward packets and cells between LANs and ATM networks or network interface boards that forward packets and cells between ATM-attached devices and ATM networks. Together, route servers and edge devices act as distributed routes, in which route servers determine where to send packets and edge devices forward these packets.

If an endstation on a LAN wanted to communicate with an ATM-attached device, the endstation would send a packet to the edge device. This device would check the destination MAC address or the network layer address of the packet. The device would then check its cache to see if it knew the corresponding ATM address. If the edge device did not know the ATM address, this device would query the route server. If the route server knew the ATM address, this server would simply respond with the address. If the route server did not know the ATM address, this server could use one of several routing protocols to communicate with other routers (both traditional routers and other route servers) to determine this address.

When the edge device knew the ATM address, it would establish a virtual circuit with the appropriate destination endstation, convert its LAN packets into ATM cells, and send these cells to the destination endstation. The edge device can establish a virtual circuit even if the destination endstation is on a different subnetwork; the edge device bypasses the route server when sending cells, sending them directly to the destination endstation. This process is called cut-through routing or one-hop routing.

For short transmissions, however, cut-through routing may not be the best approach because connection setup takes a long time relative to the size of the transmission. Using a process called hop-by-hop routing, MPOA can eliminate connection setup. With hop-by-hop routing, edge devices can forward packets to the route server just as LANE clients can forward packets to the LANE BUS. Edge devices can also perform flow detection: They can forward packets to the route server, but if they detect a flow (a long transmission), they can set up a virtual circuit to the destination endstation.

MPOA's biggest disadvantage is its relative newness. ATM Forum has not yet finalized the standard. Depending on the implementation, MPOA can also add complexity to your network; however, MPOA provides many capabilities that neither classical IP over ATM nor LANE provides. Because MPOA is an OSI network layer technology, it has access to important network layer information such as traffic characteristics and ATM's qualities of service. When establishing a connection, the edge device can use this network layer information to chart the best path to a destination endstation based on the Q of S that the sending endstation requests.

MPOA also provides routing capabilities that no other ATM interoperability architecture provides. With MPOA, you can route between traditional LANs that are connected by a high-speed ATM backbone, thus creating a high-speed internetwork without the bottleneck of a traditional router. You also can use cut-through and hop-by-hop routing to optimize both short and long transmissions.

17.7.9 ATM and Windows 2000

Windows 2000 can support a maximum of four ATM network adapters per computer. Most of these are plug-and-play cards. Therefore, after physically installing the card, it is automatically detected during boot-up and the correct driver is automatically installed.

After your ATM adapter is recognized and installed, Windows ATM servers are installed and available. By default, the connection is set up as a LAN emulation client (LEC). To view or configure additional services, do the following:

1. Click the Start button, point to Settings, and select the Network and Dial-up Connections option.
2. Click the ATM connection that corresponds to the ATM network adapter installed on this computer.
3. Open the File menu and select the Properties option.
4. In the list of network components used in this connection, select ATM LAN Emulation, and click Properties.
5. If needed, configure the list of emulated LANs available for use with this ATM connection.

To enable a simple TCP/IP-over-ATM connection, do the following:

1. Click the Start button, point to Settings, and select the Network and Dial-up Connections option.
2. Select the ATM connection that corresponds to the ATM network adapter installed on this computer.
3. Open the File menu and select the Properties option.
4. From the list of network components, select Internet Protocol (TCP/IP), and then click OK.

17.8 DIGITAL SUBSCRIBER LINE

The **digital subscriber line (DSL)** is a special communications line that uses sophisticated modulation technology to maximize the amount of data that can be sent over plain twisted-pair copper wiring, which is already carrying phone service to subscribers' homes. DSL was originally developed to transmit video signals to compete against the cable companies, but soon found use as a high-speed data connection with the explosion of the Internet. DSL is sometimes expressed as xDSL, because of the various kinds of digital subscriber line technologies including ADSL, R-DSL, HDSL, SDSL, and VDSL. See Table 17.8.

TABLE 17.8 Various xDSL technologies

Technology	Speed	Distance Limitation (24-gauge wire)	Applications
Asymmetrical digital subscriber line lite (ADSL lite)	Up to 1 Mbps downstream Up to 512 Kbps upstream	18,000 feet	Internet/intranet access, Web browsing, IP telephony, video telephony
Asymmetrical digital subscriber line (ADSL)	1.5–8 Mbps downstream Up to 1.544 Mbps upstream	18,000 feet (12,000 feet for fastest speeds)	Internet/intranet access, video-on-demand, remote LAN access, VPNs, VoIP
Rate adaptive digital subscriber line (R-ADSL)	1.5–8 Mbps downstream Up to 1.544 Mbps upstream	18,000 feet (12,000 feet for fastest speeds)	Internet/intranet access, video-on-demand, remote LAN access, VPNs, VoIP
ISDN digital subscriber line (IDSL)	Up to 144 Kbps full duplex	18,000 feet (additional equipment can extend the distance)	Internet/intranet access, Web browsing, IP telephony, video telephony
High bit-rate digital subscriber line (HDSL)	1.544 Mbps full duplex (T1) 2.048 Mbps full duplex (E1) (uses 2–3 wire pairs)	12,000–15,000 feet	Local, repeated T1/E1 trunk replacement, PBX interconnection, frame relay traffic aggregator, LAN interconnect
Single-line digital subscriber line (SDSL)	1.544 Mbps full duplex (T1) 2.048 Mbps full duplex (E1) (uses 1 wire pair)	10,000 feet	Local, repeated T1/E1 trunk replacement, collaborative computing, LAN interconnect
Very high bit-rate digital subscriber line (VDSL)	13–52 Mbps downstream 1.5–2.3 Mbps upstream (up to 34 Mbps if symmetric)	1000–4500 feet (depending on speed)	Multimedia Internet access, high-definition television program delivery

17.8.1   DSL Technology

The best quality of DSL technologies is their ability to transport large amounts of information across existing copper telephone lines. This capability is possible because of the DSL modem's leverage signal processing techniques that insert and extract more digital data onto analog lines. The key is modulation, a process in which one signal modifies the property of another.

In the case of digital subscriber lines, the modulating message signal from a sending modem alters the high-frequency carrier signal so that a composite wave, called a modulated wave, is formed. Because this high-frequency carrier signal can be modified, a large digital data payload can be carried in the modulated wave over greater distances than on ordinary copper pairs. When the transmission reaches its destination, the modulating message signal is recovered, or demodulated, by the receiving modem.

There are many ways to alter the high-frequency carrier signal that results in a modulated wave. For ADSL, there are two competing modulation schemes: carrierless amplitude phase (CAP) modulation and discrete multitone (DMT) modulation. CAP and DMT use the same fundamental modulation technique—quadrature amplitude modulation (QAM)—but apply it in different ways.

QAM, a bandwidth conservation process routinely used in modems, enables two digital carrier signals to occupy the same transmission bandwidth. With QAM, two independent message signals are used to modulate two carrier signals that have identical frequencies, but differ in amplitude and phase. QAM receivers are able to discern whether to use lower or higher numbers of amplitude and phase states to overcome noise and interference on the wire pair.

CAP modulation is a proprietary modulation of AT&T Paradyne. Income data modulates a single carrier channel that is sent down a telephone line. The carrier is suppressed before transmission and reconstructed at the receiving end. Generating a modulated wave that carries amplitude and phase state changes is not easy. To overcome this challenge, the CAP version of QAM stores parts of a modulated message signal in memory and then reassembles the parts in the modulated wave. The carrier signal is suppressed before transmission, because it contains no information and is reassembled at the receiving modem (hence the word *carrierless* in CAP). At start-up, CAP also tests the quality of the access line and implements the most efficient version of QAM to ensure satisfactory performance for individual signal transmissions. CAP is normally FDM based.

DMT is actually a form of frequency-division multiplexing (FDM). Because high-frequency signals on copper lines suffer more loss in the presence of noise, DMT discretely divides the available frequencies into 256 subchannels, or tones. As with CAP, a test occurs at start-up to determine the carrying capacity of each subchannel. Incoming data are then broken down into a variety of bits and distributed to a specific combination of subchannels based on their abilities to carry the transmission. To rise above noise, more data reside in the lower frequencies and less in the upper frequencies.

To create upstream and downstream channels, ADSL modems divide the phone line's available bandwidth using one of two methods: FDM or echo cancellation. FDM assigns one band or frequency for upstream data and another for downstream data. The downstream path is further divided by time-division multiplexing (TDM) into one or more high-speed channels for data and one or more low-speed channels, one of which is for voice. The upstream path is multiplexed into several low-speed channels.

Echo cancellation, the same technique used by V.32 and V.34 modems, means that the upstream and the downstream signals are sent on the wire at the same frequencies. The advantage of echo cancellation is that the signals are both kept at the lowest possible frequencies (since cable loss and crosstalk noise both increase with frequency) and therefore achieve greater cable distance for a given data rate. An ADSL receiver will see an incoming signal that is both the incoming signal from the far end and the outgoing signal from the local transmitter. These are mixed over the same frequency range. In other words, the received signal consists of not only the signal to be recovered from the far end but also a local echo due to the local transmitter. The local echo must be accurately modeled by DSP circuitry, and then this replica echo is electronically subtracted from the composite incoming signal. If done properly, all that remains are the incoming data from the far-end ADSL system. Although echo cancellation uses bandwidth more efficiently, the process of modeling the echo is quite complicated and is more costly. Therefore, only a few vendors have implemented the process.

17.8.2 ADSL

Currently, existing POTS data are transmitted over a frequency spectrum that ranges from 0 kHz to 4 kHz. Copper phone lines can actually support frequency ranges much greater than these, and ADSL takes advantage of these ranges by transmitting data in the ranges between 4 kHz and 2.2 MHz. Therefore, ADSL relies on advanced digital signal processing (DSP) and complex algorithms to compress all information into the phone line. In addition, ADSL modems correct errors that are caused by line conditions and attenuation. With this technology, any computer or network can easily become connected to the Internet at speeds comparable with T1 access for a fraction of the cost and it can serve as a suitable medium for video streaming and conferencing.

As its name implies, the asymmetrical digital subscriber line (ADSL) transmits an asymmetric data stream, with much more going downstream to the subscriber than is coming back. The reason for this has less to do with transmission technology than with the cable plant itself. Twisted-pair telephone wires are bundled together in large cables. Fifty pair to a cable is a typical configuration toward the subscriber, but cables coming out of a central office (CO) may have hundreds or even thousands of pairs bundled together. An individual line from a CO to a subscriber is spliced together from many cable sections as they fan out from the CO. (Bellcore claims that the average U.S. subscriber line has 22 splices.) Alexander Bell invented twisted-pair wiring to minimize crosstalk. A small amount of crosstalk does occur, and its amount increases as the frequencies and the length of line increases. Therefore, if you place symmetric signals in many pairs within the same cable, the crosstalk significantly limits the data rate and length of line.

Most people download information as they view web pages and download files, and the amount of information downloaded is far greater than the amount of information that a user uploads or transfers to other computers. This asymmetry, then, combined with always-on access (which eliminates call setup), makes ADSL ideal for Internet/intranet surfing, video-on-demand, and remote LAN access.

ADSL modems usually include a POTS splitter, which enables simultaneous access to voice telephony and high-speed data access. Some vendors provide active POTS splitters, which enable simultaneous telephone and data access. However, if the power fails or the modem fails with an active POTS splitter, then the telephone fails. A passive POTS splitter, on the other hand, maintains lifeline telephone access even if the modem fails (due to a power outage, for example), since the telephone is not powered by external electricity. Telephone access in the case of a passive POTS splitter is a regular analog voice channel, the same as customers currently receive to their homes.

Downstream, ADSL supports speeds between 1.5 Mbps and 8 Mbps; whereas upstream, the rate is between 640 Kbps and 1.544 Mbps. ADSL can provide 1.544-Mbps transmission rates at distances of up to 18,000 feet over one wire pair. Optimal speeds of 6 Mbps to 8 Mbps can be achieved at distances of 10,000 feet to 12,000 feet using standard 24-gauge wire.

Currently, the ADSL lite specification, also known as g.lite, is a low-cost, easy-to-install version of ADSL that is specifically designed for the consumer marketplace. ADSL lite is a lower-speed version of ADSL that will eliminate the need for the telephone company to install and maintain a premises-based POTS splitter. ADSL lite should also work over longer distances than full-rate ADSL, making it more widely available to mass market consumers. It will support both data and voice and provide an evolution path to full-rate ADSL.

17.8.3 HDSL

The **high-bit-rate digital subscriber line (HDSL)** derives its name from the high bandwidth that is transmitted in both directions over two copper loops. HDSL has proven to be a reliable and cost-effective means for providing repeaterless T1 and E1 services over two twisted-pair loops. HDSL transceivers can reliably transmit a 1.544- and 2.048-Mbps data signal over two nonloaded, 24-gauge (0.5 mm), unconditioned twisted-pair wire loops at a distance of up to 13,100 feet (4.2 km) without the need for repeaters. Different from ADSL, HDSL does not provide standard voice telephone service on the same wire pair.

Eliminating the need for repeater equipment and removal of bridged taps greatly simplifies the labor and engineering effort to provision the service. This attribute eliminates the need to identify, modify, and verify a controlled environment, with power, secured access, and other factors needed to support repeater equipment. It also reduces the time, cost, and effort of isolating faults and taking corrective action when a failure does occur. Studies by some service providers have indicated that troubleshooting and replacing defective repeater equipment often costs significantly more than the cost of the equipment itself. These attributes translate into increased network up-time and reduced engineering time, thus making possible T1 provisioning in a matter of days as opposed to weeks. Faster service provisioning and greater up-time leads to increased customer satisfaction and increased service revenues. To provision a 12,000- foot (3.6-km) local loop with traditional T1 transmission equipment requires two transceivers and two repeaters. To provision the same loop with HDSL requires only two HDSL transceivers, one at each end of a line.

HDSL2 is the next generation of HDSL, designed to accomplish three primary goals: (1) full T1 rates (1.544 Mbps) on a single pair of copper wires with the same spectral compatibility as traditional HDSL; (2) vendor interoperability, meaning that service providers are no longer tied to proprietary solutions; and (3) a full extension of CSA reach to 12,000 feet. HDSL2 can be thought of as offering everything that traditional HDSL offers, but on a single pair of copper wires. HDSL2 has become the DSL of choice for corporations.

17.8.4 Connecting to an ADSL Line

To connect to an ADSL line, either install an ADSL modem via an expansion card or a USB port, or external stand-alone device, connected to a network card. If you are using the expansion card or USB device, assign and configure the TCP/IP. If you are using the external stand-alone device, assign and configure the TCP/IP for the network card.

17.9 CABLE MODEMS

Cable systems were originally designed to deliver broadcast television signals efficiently to subscribers' homes. To ensure that consumers could obtain cable service with the same TV sets that they use to receive over-the-air broadcast TV signals, cable operators re-create a portion of the over-the-air radio frequency (RF) spectrum within a sealed coaxial cable line. Traditional coaxial cable systems typically operate with either 330 MHz or 450 MHz of capacity, whereas modern hybrid fiber/coax (HFC) systems are expanded to 750 MHz or more. Logically, downstream video programming signals begin at approximately 50 MHz, the equivalent of channel 2 for over-the-air TV signals. Each standard TV channel occupies 6 MHz of the RF spectrum. Thus, a traditional cable system with 400 MHz of downstream bandwidth can carry the equivalent of 60 analog TV channels, and a modern HFC system with 700 MHz of downstream bandwidth has the capacity for about 110 channels.

Regular cable is analog, whereas digital cable uses digital signals. The digital signals can be compressed much more than analog signals. Digital cable can provide 200 to 300 channels of the same bandwidth as analog cable. To deliver data services over a cable network, one television channel in the 50-MHz to 750-MHz range is typically allocated for downstream traffic to homes and another channel in the 5-MHz to 42-MHz band is used to carry upstream signals. A cable modem termination system (CMTS) communicates through these channels with cable modems located in subscriber homes to create a virtual LAN connection. Most cable modems are external devices that connect to a PC through a standard 10Base-T Ethernet card and twisted-pair wiring, through external universal serial bus (USB) modems and internal PCI modem cards.

A single downstream 6-MHz television channel may support up to 27 Mbps of downstream data throughput from the cable using 64 QAM transmission technology. Speeds can be boosted to 36 Mbps using 256 QAM. Upstream channels may deliver 500 Kbps to 10 Mbps from homes using 16 QAM or QPSK (quadrature phase shift key) modulation techniques, depending on the amount of spectrum allocated for service. This upstream and downstream bandwidth is shared by the active data subscribers who are connected to a given cable network segment, typically 500 to 2000 homes on a modern HFC network.

An individual cable modem subscriber may experience access speeds from 500 Kbps to 1.5 Mbps or more, depending on the network architecture and traffic load—blazing performance compared with dial-up alternatives. When surfing the Web, however, performance can be affected by Internet backbone congestion. In addition to speed, cable modems offer another key benefit: constant connectivity. Because cable modems use connectionless technology, much like in an office LAN, a subscriber's PC is always online with the network, so there is no need to dial in to begin a session. Users, therefore do not have to worry about receiving busy signals or tying up their telephone lines when going online.

17.10 WIRELESS LAN

A **wireless LAN (WLAN)** is a flexible data communications system that is implemented as an extension to or an alternative for a wired LAN without a building or campus. Using electromagnetic waves, WLANs transmit and receive data over the air, thus minimizing the need for wired connections. A mobility-wireless LAN system can provide LAN users with access to real-time information anywhere in their organizations. Installing a wireless LAN system can be fast and easy and can eliminate the need to pull cable through walls and ceilings. In addition, wireless technology allows the network to go where wire cannot go. Thus, WLANs combine data connectivity with user mobility and, through simplified configuration, enable movable LANs.

Although the initial investment required for wireless LAN hardware can be higher than the cost of wired LAN hardware, overall installation expenses and life-cycle costs can be significantly lower. In addition, long-term cost benefits are greatest in dynamic environments requiring frequent moves, adds, and changes. Manufacturers of wireless LANs have a range of technologies from which to choose when designing a wireless LAN solution. Each technology comes with its own set of advantages and limitations.

17.10.1 Cellular Topology

Most wireless technology uses a **cellular topology,** which divides an area into cells. A broadcast device located at the center broadcasts in all directions to form an invisible cir-

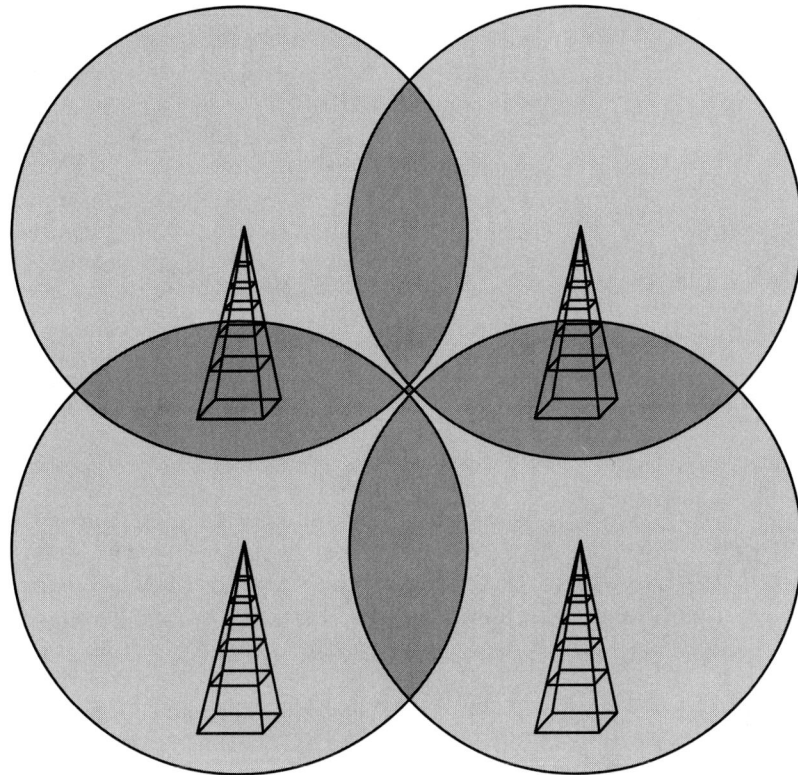

FIGURE 17.18   Cellular topology

cle (cell). All network devices located within the cell communicate with the network through the central station or hub, which is interconnected with the rest of the network infrastructure. If the cells are overlapped, then devices may roam from cell to cell while maintaining connection to the network. See Figure 17.18. Today the best-known example of cellular topology is cellular phones.

17.10.2   Radio Systems

Radio frequency (RF) resides between 10 KHz and 1 GHz on the electromagnetic spectrum. It includes shortwave radio, very-high-frequency (VHF) waves, and ultrahigh-frequency (UHF) waves. Radio frequencies have been divided between regulated and unregulated bandwidths. Users of the regulated frequencies require a license from the regulatory bodies that have jurisdiction over the desired operating area (the FCC in the United States and the CDC in Canada). Although the licensing process can be difficult, licensed frequencies typically guarantee clear transmission within a specific area.

In unregulated bands, users must operate at regulated power levels (less than 1 watt in the United States) to minimize interference with other signals. If a device broadcasts using less power, then the effective area will be smaller. The Federal Communications Commission (FCC) allocated the following bands for unregulated broadcast: 902–928 MHz, 2400–2483.5 MHz, and 5752.5–5850 MHz. Users are not required to obtain a license from the FCC to use these frequencies, but they must meet the FCC regulations, which include power limits, to minimize interference.

A **band** is a contiguous group of frequencies that are used for a single purpose. Commercial radio stations often refer to the band of frequencies they are using as a single frequency; however, typical radio transmissions actually cover a range of frequencies and wavelengths. Because most tuning equipment is designed to address the entire bandwidth at the kilohertz or megahertz level, the distinction between one frequency and a band is often overlooked.

A **narrowband radio system** transmits and receives user information on a specific radio frequency. Narrowband radio keeps the radio signal frequency as narrow as possible just to pass the information. Undesirable crosstalk between communications channels is avoided by carefully coordinating different users on different channel frequencies. In a radio system, privacy and noninterference are accomplished by using separate radio frequencies. The radio receiver filters out all radio signals except the ones on its designated frequency. Depending on the power and frequency of the radio signal, the range could be across a room, throughout an entire building, or over long distances.

Spread-spectrum signals are distributed over a wide range of frequencies and then collected onto their original frequencies at the receiver. Different from narrowband signals, spread-spectrum signals use wider bands, which transmit at a much lower spectral power density (measured in watts per hertz). Unless the receiver is not tuned to the right frequency or frequencies, a spread-spectrum signal resembles noise, thus making the signals harder to detect and jam. As an additional bonus, spread-spectrum and narrowband signals can occupy the same band, with little or no interference. There are two types of spread-spectrum signals: direct sequence and frequency hopping.

A **direct-sequence spread-spectrum (DSSS) signal** generate a redundant bit pattern for each bit to be transmitted. This bit pattern is called a chip (or chipping code). The intended receiver knows which specific frequencies are valid and deciphers the signal by collecting valid signals and ignoring the spurious signals. The valid signals are then used to reassemble the data. Because multiple subsets can be used within any frequency range, direct sequence signals can coexist with other signals. Although direct sequence signals can be intercepted almost as easily as other RF signals, eavesdropping is ineffective because it is quite difficult to determine which specific frequencies comprise the bit pattern, retrieve the bit pattern, and interpret the signal. Because of modern error-detection and correction methods, the longer the chip, the greater the probability that the original data can be recovered, even if one or more bits in the chip are damaged during transmission.

Frequency hopping is switching quickly between predetermined frequencies, often many times each second. Both the transmitter and receiver must follow the same pattern and maintain complex timing intervals to be able to receive and interpret the data being sent. Similar to direct sequence spread spectrum, intercepting the data being sent is extremely difficult unless the device reading the signal know the signals to monitor and the timing pattern. In addition, dummy signals can be added to increase security and confuse eavesdroppers. The length of time that the transmitter remains on a given frequency is known as the dwell time.

17.10.3   Infrared Systems

Another form of wireless technology is the **infrared (IR) system,** which is based on infrared light (light that is just below the visible light in the electromagnetic spectrum). Similar to your TV or VCR remote controls, infrared links use light-emitting diodes (LEDs) or injection laser diodes (ILDs) to transmit signals, and photodiodes to receive signals. Since IR is essentially light, it cannot penetrate opaque objects. Infrared devices work as using either directed or diffused technology. Directed IR uses line-of-sight or point-to-point technology. Diffused (or reflective) technology spreads the light over an area to create cells, limited to individual rooms. Because infrared signals are not capable of penetrating walls or other opaque objects and are diluted by strong light sources, they are most useful in small or open indoor environments.

SUMMARY

1. The circuit switching technique connects the sender with the receiver via a single path for the duration of a conversation.
2. The packet switching technique involves messages that are broken into small parts, or packets.
3. A virtual circuit is a logical circuit created to ensure reliable communications between two network devices.
4. A virtual circuit can be either a permanent virtual circuit (PVC) or a switched virtual circuit (SVC).
5. A PVC is a permanently established virtual circuit that consists of one mode: data transfer.
6. An SVC is a virtual circuit that is dynamically established on demand and terminated when transmission is complete.

7. Data terminal equipment (DTE) devices are end systems that communicate across the WAN.
8. Data circuit-terminating equipment (DCE) devices are special communication devices that provide the interface between the DTE and the network.
9. The DSU is a device that performs protective and diagnostic functions for a telecommunications line. The CSU is a device that connects a terminal to a digital line. Typically, the two devices are packaged as a single unit (CSU/DSU).
10. The Public Switched Telephone Network (PSTN) is the international telephone system that is based on copper wires (UTP cabling) carrying analog voice data. The PSTN, also known as the plain old telephone service (POTS), is the standard telephone service used in most homes.
11. The subscriber loop, or local loop, is the telephone line that runs from a home or office to the telephone company's central office (CO) or neighborhood switching station (often a small building with no windows).
12. The T-carrier system was introduced by the Bell System in the United States in the 1960s as the first successful system that converted the analog voice signal to a digital bit stream.
13. The disadvantages of the leased digital lines are their higher cost and their limitation as a point-to-point connection.
14. Circuit switch technology can be referred to as dial-up technology.
15. T-carrier system lines are permanent leased lines. Switched 56 and switched 64 lines are digitally switched or "dial-up lines" that provide a single digital channel for dependable data connectivity.
16. The Integrated Services Digital Network (ISDN) is the planned replacement for POTS so that it can provide voice and data communications worldwide using circuit switching as well as the same wiring that is currently being used in homes and businesses.
17. The ISDN uses two types of channels, a B channel and a D channel. The bearer channels (B channels) transfer data at a bandwidth of 64 Kbps for each channel. The data channels (D channels) use a communications language called DSS1 for administrative signaling.
18. Today, the two well-defined standards used are the basic rate interface (BRI) and the primary rate interface (PRI).
19. X.25 network is a packet switched network that allows remote devices to communicate with each other across digital links without the expense of individual leased lines.
20. Frame relay is a packet switching protocol designed to use high-speed digital backbone links to support modern protocols for error handling and flow control for connecting devices on a WAN.
21. SDH and SONET are transmission technology standards of synchronous data transmission over fiber-optic cables. They provide a high-speed transfer of data, video, and types of information across great distances without regard to the specific services and applications they support.
22. A data transmission technology is based on transmitting data in relatively small, fixed-size packets, or cells.
23. Switched multimegabit data service (SMDS) is a high-speed, cell relay, WAN service designed for LAN interconnection through the public telephone network.
24. Asynchronous transfer mode (ATM) is both a LAN and a WAN technology, which is generally implemented as a backbone technology.
25. The digital subscriber line (DSL) is a special communication line that uses sophisticated modulation technology to maximize the amount of data that can be sent over plain twisted-pair copper wiring, which is already carrying phone service to subscriber's homes.
26. Asymmetrical DSL (ADSL) transmits an asymmetric data stream, with much more going downstream to the subscriber than is coming back.
27. Cable modems communicate through the cable system to create a virtual LAN connection.
28. A wireless LAN (WLAN) is a flexible data communication system that is implemented as an extension to or an alternative for a wired LAN without a building or campus.
29. Most wireless technology uses a cellular topology, which divides an area into cells.

1. PSTN is an acronym for _____ .
   a. Partial Switched Telephone Network
   b. Public Switched Transmission Network
   c. Partial Switched Transmission Network
   d. Public Switched Telephone Network

2. Which one of the following technologies is intended to replace analog phone lines?
   a. PSTN/POTS
   b. ATM
   c. frame relay
   d. ISDN

3. T1 is a widely used type of digital communication line. What does the T1 technology offer?
   a. transmission speed of up to 1.544 Mbps
   b. transmission speed of up to 45 Mbps
   c. point-to-point, full-duplex transmission
   d. two 64 Kbps B channels and one 16 Kbps D channel per line

4. T1 lines utilize technology that combines signals from different sources onto one cable for transmission. Which one of the following devices can be used to combine multiple data signals onto a single transmission line?
   a. data assembler
   b. transceiver
   c. multiplexer
   d. redirector

5. How many separate devices can be connected to a BRI?
   a. 1
   b. 8
   c. 16
   d. 64
   e. no physical limitation

6. What is the maximum number of individual phone numbers that could be assigned to an ISDN BRI?
   a. 1
   b. 8
   c. 16
   d. 64

7. What is another representation for a standard basic rate interface (BRI) ISDN line?
   a. 64B2+16D
   b. B2+D
   c. 2B+D
   d. 2B64+D16

8. You decide to implement PPP multilink over multiple ISDN BRI lines. How many BRIs without compression will you need to achieve your minimum-required rate of 384 Kbps?
   a. 1
   b. 2
   c. 3
   d. 4
   e. You cannot use PPP multilink with ISDN.

9. What is the compressed theoretical maximum transmission speed you could achieve with compression if you implemented three BRIs to achieve a minimum uncompressed transmission speed of 384 Kbps?
   a. 512 Kbps
   b. 768 Kbps
   c. 1536 Kbps
   d. 2048 Kbps

10. One of your offices is 2 miles away from a local telephone central office. The other is approximately 5 miles away from the same office. What will you need in order to use an ISDN BRI?
    a. one repeater between the telephone central office and your first office and two repeaters between the telephone central office and your second office
    b. one repeater between the telephone central office and your second office
    c. two repeaters between the telephone central office and your second office
    d. no repeaters

11. Suppose the following situation exists:
    Your company is based in Sacramento, and it has branch offices in Los Angeles and New York City. Each of the three offices has a 10Base-T network. Users must access resources in all three offices.

Required result:

- You must implement a networking solution, which will offer WAN communications between the three sites.

Optional desired results:

- The WAN connection must support approximately 256 Kbps of data and several analog telephone conversations between sites.
- The WAN connection must continue operations even if one of the WAN links should fail.

Proposed solution:

- Use three T1 connections—one between Sacramento and Los Angeles, one between Los Angeles and New York, and one between New York and Sacramento.

Which result does the proposed solution produce?

a. The proposed solution produces the required result and both of the optional desired results.

b. The proposed solution produces the required result but only one of the optional desired results.

c. The proposed solution produces the required result but does not produce either of the optional desired results.

d. The proposed solution does not produce the required result.

12. Suppose the following situation exists:
    Your company is based in Atlanta, and it has branch offices in Los Angeles and New York City. Each of the three offices has a 10Base-T network. Users must access resources in all three offices.

    Required result:

    - You must implement a networking solution which would offer WAN communications between sites.

    Optional desired results:

    - The WAN connection must support approximately 1.5 Mbps of data.
    - The WAN connection must continue operations at 1.5 Mbps even if one of the WAN links should fail.

    Proposed solution:

    - Use two T1 connection lines—one between Atlanta and Los Angeles, and one between Los Angeles and New York.

    Which result does the proposed solution produce?

    a. The proposed solution produces the required result and both of the optional desired results.

    b. The proposed solution produces the required result but only one of the optional desired results.

    c. The proposed solution produces the required result but does not produce either of the optional desired results.

    d. The proposed solution does not produce the required result.

13. A wide area network requires complex and expensive packet switching equipment. Which one of the following WAN technologies is actually a protocol suite that uses packet assemblers and disassemblers?
    a. X.25
    b. ATM
    c. ISDN
    d. frame relay

14. Frame relay is a form of packet switching technology that evolved from X.25. Which one of the following statements best describes the frame relay technology?
    a. It transmits fixed-length packets at the physical layer through the most cost-effective path.
    b. It transmits variable-length packets at the physical layer through the most cost-effective path.

    c. It transmits fixed-length frames at the data link layer through the most cost-effective path.

    d. It transmits variable-length frames at the data link layer through the most cost-effective path.

15. Why is CRC needed in frame relay?
    a. It helps detect connection failures.
    b. It helps prevent congestion.
    c. It ensures transmission speed.
    d. It ensures frame accuracy.

16. Which one of the following WAN technologies can provide subscribers with bandwidth as needed?
    a. frame relay   c. ISDN
    b. X.25   d. T1

17. SONET systems are _____ technology.
    a. twisted-pair, copper-based   c. fiber-optic
    b. thinnet cabling   d. wireless

18. SONET's base signal (STS-1) operates at a bit rate of _____ .
    a. 64 Kbps   c. 51.840 Mbps
    b. 1.544 Mbps   d. 155.520 Mbps

19. _____ is the standard for North America, whereas _____ is the standard for the rest of the world.
    a. SONET; SDH   c. ATM; SONET
    b. SDH; SONET   d. ATM; SDH

20. In ATM networks, all information is formatted into fixed-length cells consisting of _____ bytes.
    a. 32   c. 64
    b. 48   d. 128

21. Packet switching networks divide data into packets and send them over a common transmission line using virtual circuits. Which one of the following is an implementation of a packet switching technology?
    a. T1   c. switched 56
    b. ISDN   d. ATM

22. The basic connection unit in an ATM network is known as the _____ .
    a. virtual channel connection   c. DSL connection
    b. virtual path connection   d. DS-0 connection

23. CBR is used primarily for _____ .
    a. multimedia e-mail   c. data transport
    b. videoconferencing   d. simple e-mail

24. Which service category is most likely to suffer cell loss due to bandwidth constraints?
    a. CBR   c. UBR
    b. ABR   d. VBRnrt

25. ADSL increases existing twisted-pair access capacity by _____ .
    a. twofold   c. thirtyfold
    b. threefold   d. fiftyfold

26. A modem translates _____ .
    a. analog signals into digital signals   c. both of the above
    b. digital signals into analog signals   d. none of the above

27. A digital subscriber line (DSL) refers to a _____ .
    a. specific gauge of wire used in modem communications
    b. modem enabling high-speed communications

   c. connection created by a modem pair enabling high-speed communications

   d. specific length of wire

28. Circle True or False—T1/E1 and HDSL are essentially equivalent technologies.

29. The practical upper limit of length of ADSL is _____ feet.

   a. 6000   c. 18,000

   b. 12,000   d. 36,000

30. What is the major road block to providing full digital service over POTS?

   a. The "last mile" of telephone cable is still copper.

   b. Competition is not strong enough to warrant full digital service.

   c. Fiber-optic lines can be brought out to rural areas.

   d. It is too expensive for the telephone companies to implement.

31. Name two advantages of xDSL technology.

   a. It runs on fiber-optic cables and increased bandwidth.

   b. It is inexpensive and uses commonly available modems.

   c. It provides high-speed digital access and is inexpensive.

   d. It provides fast analog service and always-on Internet access.

   e. It is widely available in rural areas for a modest cost.

18 Remote Access

INTRODUCTION

Networks are very common in today's workplace. Most people who want to extend their networks away from their offices or to connect to their networks from home or on the road can now access applications and data that are stored on the network. For years, Windows NT has offered remote access and has marketed that fact as a major feature. Windows 2000 continues to offer remote access with additional features and functionality.

OBJECTIVES

1. Connect to computers by using a virtual private network (VPN) connection.
2. Create a dial-up connection to connect to a remote access server.
3. Connect to the Internet by using dial-up networking.
4. Configure inbound connections.
5. Create a remote access policy.
6. Configure a remote access profile.
7. Configure authentication protocols.
8. Configure encryption protocols.
9. Configure and troubleshoot IPSec.
10. Install, configure, monitor, and troubleshoot Terminal Services.

18.1 REMOTE ACCESS SERVICE

Today, modern networks offer **remote access service (RAS),** which allows users to connect remotely using various protocols and connection types. Current projections call for the number of remote users—telecommuters, road warriors, and other mobile users—to grow from 30 million in 1999 to more than 100 million by the year 2002 as remote LAN and Internet access continues to grow.

Typically, remote access costs anywhere from 63% to 157% more to support than its office-bound counterparts. About one-third of these costs are attributed to incremental charges such as equipment costs and access fees. The remaining costs are devoted to ongoing administration and technical support for the act of keeping remote users connected to the corporate network. These drawbacks are offset by reduced real estate costs and increased productivity. To make remote access a truly sound investment, organizations must find a way to simplify remote desktop management and support while providing the same level of performance and reliability that is currently available with local solutions. There are two types of remote access, dial-up networking and virtual private networking.

Dial-up networking is used when a remote access client makes a nonpermanent, dial-up connection to a physical port on a remote access server by using the service of a telecommunications provider such as an analog phone, ISDN, or X.25. The best example of dial-up networking is that of a dial-up networking client who dials the phone number of one port/modem of a remote access server. A **remote access server** is the computer and associated software that is set up to handle users who are seeking access to the network remotely. Sometimes called a **communication server,** a remote access server usually includes or is associated with a firewall server to ensure security and a router that can forward the remote access request to another part of the corporate network.

Virtual private networking is the creation of secured, point-to-point connections across a private network or a public network such as the Internet. A virtual private networking client uses special TCP/IP-based protocols, called tunneling protocols, to make a virtual call to a virtual port on a virtual private networking server. The best example of virtual private networking is that of a virtual private networking client who makes a virtual private network connection to a remote access server that is connected to the Internet. The remote access server answers the virtual call, authenticates the caller, and transfers data between the virtual private networking client and the corporate network.

Dial-up networking over an analog phone or ISDN is a direct physical connection between the dial-up networking client and the dial-up networking server. You can encrypt data sent over the connection, but it is not required. In contrast to dial-up networking, virtual private networking is always a logical, indirect connection between the virtual private networking client and the virtual private networking server. To ensure privacy, you must encrypt data that are sent over the connection.

18.2 TRADITIONAL REMOTE ACCESS

The typical dial-in session actually consists of six distinct steps or stages, each of which must be completed successfully before the user gains access to centralized resources. Each step is subject to its own unique set of problems or failures that individually could jeopardize the entire connection. These steps are as follows:

1. Modem connection
2. Dial-out
3. Handshake
4. Authentication
5. IP address negotiation
6. Access to resources

During the modem connection, the modem checks the phone line for a dial tone. If the modem fails or cannot detect a dial tone, the connection process fails immediately. A "no dial tone" error can be caused by problems with the phone cable, the phone jack, the modem, the dialer, or any combination of conditions related to these items or devices.

If a dial tone is detected, the remote PC proceeds with the connection process by dialing the remote access number. Any number of events, such as no answer, a busy signal, or no carrier, could cause the connection process to fail. If the call gets through, the remote (dialing) modem will begin negotiating a common data rate and other transmission parameters with the answering modem. This step is referred to as the **handshaking** stage, because both modems must agree on the same parameters before continuing. Technology advancements have produced a variety of ways to encode, compress, modulate, and transfer bits over the phone line, adding complexity—and processing time—to the process. These factors make it critical for IT managers to monitor and measure both the final outcome of the negotiation process and the length of time it takes for the negotiation to be completed. Connection failures and delays are typically the result of the local and remote modems disagreeing on one or more transmission parameters.

Once authentication is complete, the remote PC is assigned a dynamic IP address for identification purposes. The remote PC must agree to use the specified IP address or the local system will refuse to establish the link. The remote PC is also assigned a DNS server at this time; if the assignment is not recorded or is unsuccessful for some reason, attempts to connect will appear as application problems.

Once the previous five stages of the remote connection process have been successfully completed, the dial-in client PC is granted access to the corporate network. At this stage, the user can then check e-mail, send messages, copy or transfer files, run application programs, and so forth, but the connection is still not entirely safe. Connectivity problems can occur if the call is abnormally terminated, if the modem speed is too slow, or if the dial-up network has an excessive amount of line noise.

18.3 REMOTE AUTHENTICATION DIAL-IN USER SERVICE

Adding a remote access point to the centralized corporate network increases the chance of a break-in. Therefore, a secure authentication scheme is required to provide security and protect against remote client impersonation. During the authentication stage, the remote access server collects authentication data from the dial-in client and checks it against its own user database or against a central authentication database server, such as the **remote authentication dial-in user service (RADIUS).**

RADIUS is the industry standard client-server protocol and software that enables remote access servers to communicate with a central server to authenticate dial-in users and authorize their access to the requested system or service for authenticating remote users. It is defined in RFCs 2138 and 2139, "Remote Authentication Dial-in User Service (RADIUS)" and "RADIUS Accounting." RADIUS allows a company to maintain user profiles in a central database that all remote servers can share. It provides better security, allowing a company to set up a policy that can be applied at a single administered network point. Because it has a central server, RADIUS also makes it easier to perform accounting of network usage for billing and for keeping network statistics.

When a user dials in to a remote access device/server, this server then communicates with the central RADIUS server to determine if the user is authorized to connect to the LAN. The RADIUS server performs the authentication and responds with an "accept" or a "reject." If the user is accepted, the RAS routes the user on to the network; if not, the RAS will terminate the user's connection. If the user's login information, user name, and/or password does not match the entry in the RAS, or if the RAS is unable to contact the RADIUS server, then the connection will be denied. Besides being deployed in remote access servers, RADIUS can also be deployed on routers and firewalls.

18.4 WINDOWS 2000 DIAL-UP NETWORKING

Windows 2000 Dial-up Networking includes the following components:

1. Dial-up networking server
2. Dial-up networking clients
3. LAN and remote access protocols

18.4.1 Windows 2000 RAS Server

Windows 2000 Server Routing and Remote Access is a mature, full-featured, third-generation service of Windows-based server operating systems. It provides a rich complement of authentication services and protocols that simplify connectivity for clients running Windows CE, Windows 95, Windows 98, Windows NT Workstation, and Windows 2000 Professional, as well as Novell-, Apple-, and UNIX-based clients. However, only client computers running Windows 2000 Professional give remote workers the full spectrum of networking and communication services, technologies, and features.

A RAS server running Windows 2000 must have a modem or multiport adapter and analog telephones (or other WAN connections). A port is a channel of a device (such as modem) that can support a single point-to-point connection. For single port devices, the device and the port are indistinguishable. For multiport devices, the port is the subdivision of the device over which a separate point-to-point communication is possible. For example, BRI ISDN adapters support two separate B channels. The ISDN adapter is a device. Each B channel is a port, because a separate point-to-point connection occurs over each B channel. If the server provides access to the local network, a separate network adapter card must be installed and connected to the local network. Note that the processor is heavily used by RAS, especially when encryption and routing are implemented.

RAS software is installed automatically when you install Windows 2000; however, you must enable and configure it for your environment. To enable the remote access server, do the following:

1. Click on the Start button, select the Programs option, select Administrative Tools, and select Routing and Remote Access.
2. Right-click the server name for which you want to enable remote access, and then click Properties. See Figure 18.1.
3. On the General tab, select the Remote Access Server checkbox.

To initially configure the RAS server, complete the following steps:

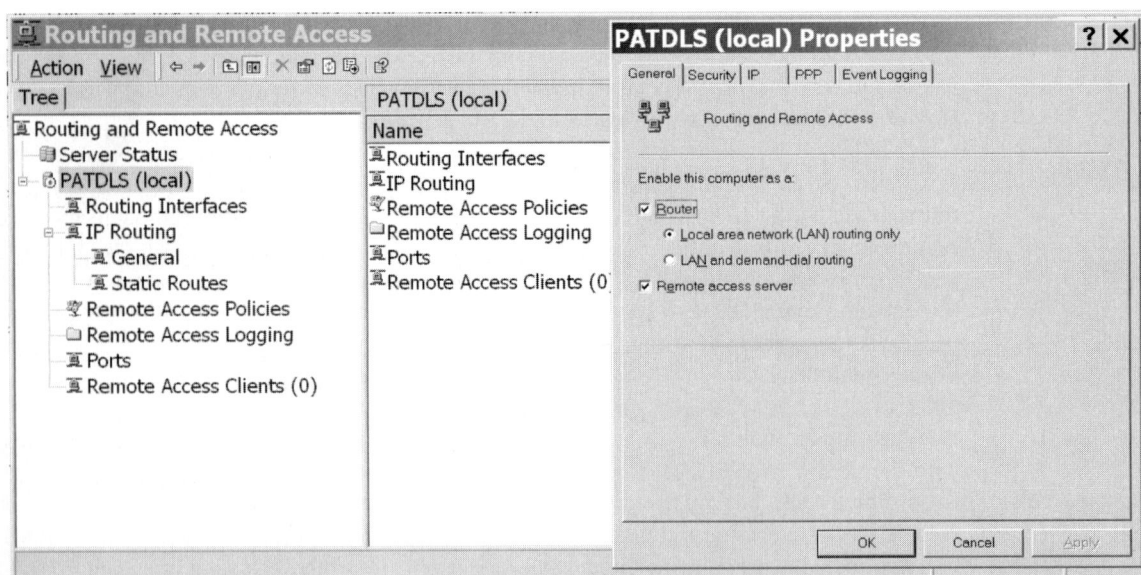

FIGURE 18.1 Configuring the Routing and Remote Access console

1. Click the Start button, select the Programs option, select the Administrative Tools, and click Routing and Remote Access.
2. In the console tree, right-click on your server and select the Disable Routing and Remote Access option.
3. When asked to continue, click on the Yes button.
4. In the console tree, right-click on your server, and select the Configure and Enable Routing and Remote Access option.
5. When the Routing and Remote Access Server Setup wizard starts, click on the Next button.
6. Select the Remote Access Server option and then select the Next button. See Figure 18.2.
7. Select the appropriate protocols and click on the Next button. Note that to select a protocol, it must already be installed on the system.
8. Remote clients must be assigned to one network for addressing, dial-up purposes, and other purposes. Select the appropriate network connection from the list and click on the Next button.
9. Select the method for which IP addresses will be assigned to the remote hosts. If you select the Automatically, then select the Next button.
10. If you want to use a RADIUS server, select the Yes option. In either case, click on the Next button.
11. If you select No to the RADIUS server, click on the Finish button.
12. If you select to use a DHCP server for remote access, the wizard will display a message saying that you have to configure the DHCP relay agent with the IP address of your server. Click on the OK button.

18.4.2 SLIP and PPP Protocols

The first protocol used for carrying IP packets over dial-up lines was the **serial line interface protocol (SLIP).** SLIP is simple protocol, in which you send packets down a serial link delimited with special END characters. SLIP, however, fails in certain functions that data link protocols can perform. First, it only works with TCP/IP, so it cannot be used with other protocols such as IPX. It does not perform error checking at the OSI data layer or authenticate users who are dialing in to an access router. When using SLIP, you must know the IP address assigned to you by your service provider, and the IP address of the remote system you will be dialing into. If IP addresses are dynamically assigned (which depends on your service provider), your SLIP software must be able to pick up the IP assignments automatically or you will have to set them up manually. Lastly, you will have to configure

FIGURE 18.2 Configuring the Remote Access Server

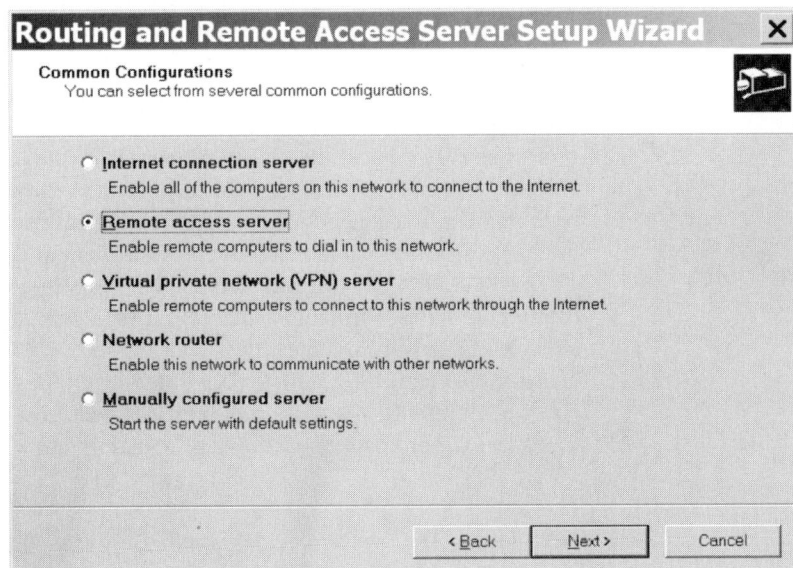

Routing and Remote Access Server Setup Wizard

Common Configurations
You can select from several common configurations.

- Internet connection server
  Enable all of the computers on this network to connect to the Internet.

- Remote access server
  Enable remote computers to dial in to this network.

- Virtual private network (VPN) server
  Enable remote computers to connect to this network through the Internet.

- Network router
  Enable this network to communicate with other networks.

- Manually configured server
  Start the server with default settings.

< Back   Next >   Cancel

certain parameters of the device such as the maximum transmission unit (MTU) and the maximum receive unit (MRU) and the use of compression.

The **point-to-point protocol (PPP)** has become the predominant protocol for modem-based access to the Internet. It provides full-duplex, bidirectional operations between hosts and can encapsulate multiple network layer LAN protocols to connect to private networks. Furthermore, a multilink version of PPP is also used to access ISDN lines and inverse multiplexing analog phone lines and high-speed optical lines. PPP is one of the many variants of an early, internationally standardized data link protocol known as the high-level data link control (HDLC) protocol. To enable PPP to transmit data over a serial point-to-point link, the following three components are used.

- **High-level data link control (HDLC) protocol**—Encapsulates its data during transmission.
- **Link control protocol (LCP)**—Establishes, configures, maintains, terminates point-to-point links [including multilink PPP (MP) sessions], and optionally tests link quality prior to data transmission. In addition, user authentication is generally performed by LCP as soon as the link is established.
- **Network control protocols (NCPs)**—Used to configure the different communications protocols, including TCP/IP and IPX, which are allowed to be used simultaneously.

One typical IP-specific function of the NCP is the provision of dynamic IP addresses for remote dial-in users. ISPs commonly employ this PPP feature to assign temporary IP addresses to their modem-based customers. The PPP mechanism is not as complete as DHCP, however, which can set up a lease period and provide such things as subnet masks, default router addresses, DNS server addresses, domain names, and much more to a newly connected computer. An ISP or other provider of remote connectivity services can use DHCP as the source of IP addresses that NCP hands out.

The negotiation of a PPP connection involves four distinct phases. Each of these phases must be successfully completed before the PPP connection is ready to transfer user data. The four phases of a PPP connection are as follows:

1. PPP configuration
2. Authentication
3. Callback (optional)
4. Protocol configuration

PPP configures its protocol parameters using the LCP. During the initial LCP phase, each device on both ends of a connection negotiates communication options that are used to send data and include PPP parameter addresses and control field compression, protocol ID compression, authentication protocols as used to authenticate the remote access client, and multilink options. Note that an authentication protocol is selected but not implemented until the authentication phase begins. After LCP is complete, the authentication protocol agreed upon by the remote access server and the remote access client is implemented. The nature of this traffic is specific to the PPP authentication protocol.

The Microsoft implementation of PPP includes an optional callback phase using the callback control protocol (CBCP) immediately after authentication. For a remote access client user to be called back, the dial-in properties of the user account must be enabled for callback and either the remote access client or the remote access server must specify the callback number. If a connection is implementing callback, both PPP peers hang up and the remote access server calls the remote access client at the negotiated number.

Once the PPP is configured and callback is complete (optional), network layer protocols can be configured. With remote access on Windows 32-bit operating systems, the remote access server sends the remote access client configuration-request messages for all LAN protocols that are enabled for remote access on the remote access server. The remote access client either continues the negotiation of the LAN protocols enabled at the remote access client or sends an LCP protocol-reject message containing the configuration-request message.

18.5 AUTHENTICATION PROTOCOLS

A Windows 2000 Server decides whether to let a Windows 2000 remote access client connect after evaluating the client's credentials, accepting a connection only after authenticating and authorizing the client attempt. Authentication is the process of verifying client's credentials. A client uses an authentication protocol to send either encrypted or unencrypted credentials to the server, depending on the protocol.

In Windows 2000, when a client dials in to a Windows 2000 RAS server, the **Internet authentication service (IAS)** verifies the client's credentials for authentication. IAS is the server's security subsystem/software service that provides security and authentication for dial-in users. Windows 2000 uses the client's user account properties and remote access policies to authorize the connection. If authentication and authorization succeed, Windows 2000 allows a connection.

When a remote access server receives a connection attempt, it negotiates with the user different authentication types enabled at the server. If the client accepts one of them, it sends the appropriate credentials for the accepted authentication type. If the user refuses authentication, the routing and remote access service checks its properties to verify if unauthenticated access is enabled and, if enabled, forwards the access-request packet to IAS. This access-request packet does not contain a user name attribute or any other credentials.

Several of the PPP authentication protocols are supported by the RADIUS protocol. Each protocol has advantages and disadvantages in terms of security, usability, and breadth of support. See Table 18.1 and Figure 18.3.

TABLE 18.1 Various security protocols

Protocols	Security	Use When . . .
PAP	Low	The client and server cannot negotiate by using a more secure form of validation.
SPAP	Medium	Connecting to a Shiva LanRover, or when a Shiva client connects to a Windows 2000–based remote access server.
CHAP	High	You have clients that are not running Microsoft operating systems.
MS-CHAP	High (most secure)	You have clients running Windows 2000, Windows NT version 4.0, or Microsoft Windows 95 or later. MS-CHAP is the most secure form of authentication.

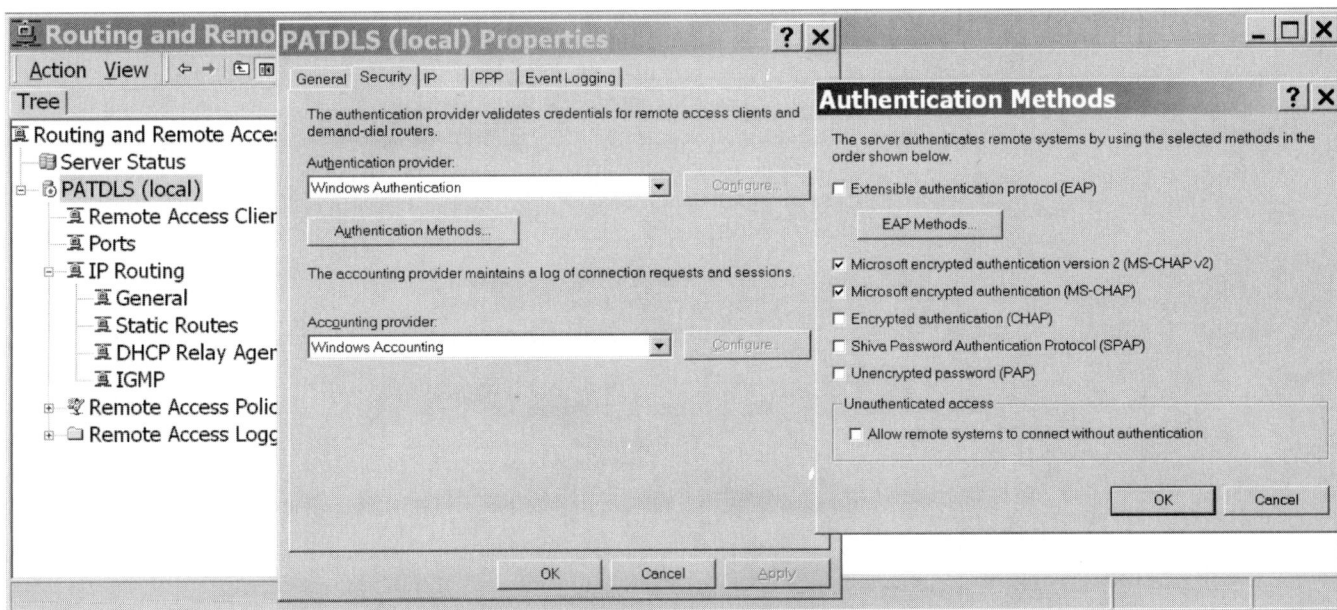

FIGURE 18.3 Enabling the PPP authentication protocols

18.5.1 Password Authentication Protocol

The **password authentication protocol (PAP)** is the least secure authentication protocol because it uses **clear text** (plaintext) passwords. The steps when using PAP are as follows:

1. The remote access client sends a PAP authenticate-request message to the remote access server containing the remote access client's user name and clear text password. Clear text, also referred to as plaintext, is textual data in ASCII format.
2. The remote access server checks the user name and password and sends back either a PAP authenticate-acknowledgment message when the user's credentials are correct, or a PAP authenticate-no acknowledgment message when the user's credentials are not correct.

Therefore, the password can easily be read with a protocol analyzer. In addition, PAP offers no protection against replay attacks, remote client impersonation, or remote server impersonation. So, to make your remote access server more secure, ensure that PAP is disabled. Another disadvantage of using PAP is if your password expires, PAP does not have the ability to change your password during authentication. In Windows 2000, PAP is disabled by default.

If you are wondering when to use PAP, generally, it is used only when attempting to connect to a non-Windows OS that does not support encrypted passwords (e.g., when dialing in to an ISP that cannot negotiate a higher level of authentication). You can also use PAP to dial in to SLIP servers, which do not support encrypted passwords.

18.5.2 Shiva Password Authentication Protocol

The **Shiva password authentication protocol (SPAP),** which is Shiva's proprietary version of PAP, offers a little more security than PAP's plaintext password with its reversible encryption mechanism. SPAP is more secure than PAP but less secure than CHAP or MS-CHAP. Someone capturing authentication packets will be unable to read the SPAP password, but this authentication protocol is susceptible to playback attacks (i.e., an intruder records the packets and resends them to gain fraudulent access). Playback attacks are possible because SPAP always uses the same reversible encryption method to send the passwords over the wire. Like PAP, SPAP does not have the ability to change the password during the authentication process. Typically, you will use SPAP for connecting a Shiva client to a Windows 2000 Server (Win2K Server) running RAS, or to connect a Windows 2000 Professional (Win2K Pro) client to a Shiva LAN Rover.

18.5.3 Challenge Handshake Authentication Protocol

Historically, **challenge handshake authentication protocol (CHAP)** is the most common dial-up authentication protocol used, and it uses an industry message digest 5 (MD5) hashing scheme to encrypt authentication. A **hashing scheme** scrambles information in such a way that makes it unique and prevents it from being reversed to the original format.

CHAP does not send the actual password over the wire, but rather, uses a three-way challenge response mechanism with one-way MD5 hashing to provide encrypted authentication without sending the password over the link. The three-way mechanism works as follows:

1. The remote access server sends a CHAP challenge message that contains a session ID and an arbitrary challenge string.
2. The remote access client returns a CHAP response message containing the user name in clear text and a hash of the challenge string, session ID, and the client's password using the MD5 one-way hashing algorithm.
3. The remote access server duplicates the hash and compares it to the hash in the CHAP response. If the hashes are the same, the remote access server sends back a CHAP success message. If the hashes are different, a CHAP failure message is sent.

Because standard CHAP clients use the plaintext version of the password to create the CHAP challenge response, passwords must be stored on the server to calculate an equivalent response. Therefore, Windows 2000 stores these passwords using reversibly encrypted passwords.

Since CHAP uses an arbitrary challenge string per authentication attempt, it protects against replay attacks. However, CHAP does not protect against remote server impersonation. In addition, because the algorithm for calculating CHAP responses is well known, it is very important that passwords be carefully chosen and sufficiently long. CHAP passwords that are common words or names are vulnerable to dictionary attacks if they can be discovered by comparing responses to the CHAP challenge with every entry in a dictionary. Passwords that are not sufficiently long can be discovered with brute force by comparing the CHAP response to sequential trials until a match to the user's response is found.

A Windows 2000 Server that is configured for CHAP can negotiate plaintext authentication with another RAS server or a client that does not support CHAP. However, keep in mind that a client you configure to require encrypted authentication will not succeed in connecting to a server that accepts only plaintext passwords.

18.5.4 Microsoft Challenge Handshake Authentication Protocol

Microsoft challenge handshake authentication protocol (MS-CHAP) is Microsoft's proprietary version of CHAP. Windows 2000 supports both MS-CHAP 1 and MS-CHAP 2, and both versions are enabled by default. Unlike PAP and SPAP, MS-CHAP lets you encrypt data that are sent using the point-to-point protocol (PPP) or PPTP connections using Microsoft point-to-point encryption (MPPE). The challenge response is calculated with an MD4 hashed version of the password and the NAS challenge. The two flavors of MS-CHAP (versions 1 and 2) allow for error codes including a "password expired" code and password changes. Therefore, these are the only authentication protocols in Windows 2000 that support password changes during authentication. The process is as follows:

1. The remote access server sends an MS-CHAP challenge message containing a session ID and an arbitrary challenge string.
2. The remote client must return the user name and an MD4 hash of the challenge string, the session ID, and the MD4-hashed password.
3. The remote access server duplicates the hash and compares it with the hash in the MS-CHAP response. If the hashes are the same, the remote access server sends back a CHAP success message. If the hashes are different, a CHAP failure message is sent.

Because MS-CHAP 1 supports only one-way authentication, it does not provide protection against remote server impersonation, which means that a client cannot determine the authenticity of a RAS server to which it connects. MS-CHAP 2 provides stronger security for remote access connections and allows for mutual authentication when the client authenticates the server. The process is as follows:

1. The remote access server sends an MS-CHAP 2 challenge message to the remote access client that consists of a session identifier and an arbitrary challenge string.
2. The remote access client sends an MS-CHAP 2 response that contains the user name, an arbitrary peer challenge string, an MD4 hash of the received challenge string, the peer challenge string, the session identifier, and the MD4 hashed versions of the user's password.
3. The remote access server checks the MS-CHAP 2 response message from the client and sends back an MS-CHAP 2 response message containing an indication of the success or failure of the connection attempt. An authentication response is based on the sent challenge string, the peer challenge string, the client's encrypted response, and the user's password.
4. The remote access client verifies the authentication response and if it is correct, uses the connection. If the authentication response is not correct, the remote access client terminates the connection.
5. If a user authenticates by using MS-CHAP 2 and attempts to use an expired password, MS-CHAP prompts the user to change the password while connecting to the server. Other authentication protocols do not support this feature, effectively locking out the user who used the expired password.

If you configure your connection to use only MS-CHAP 2 and the server you are dialing in to does not support MS-CHAP 2, then the connection will fail. This behavior is different from Windows NT, where the RAS servers negotiate a lower-level authentication if possible. In addition, MS-CHAP passwords are stored more securely at the server but have the same vulnerabilities to dictionary and brute force attacks as CHAP. When using MS-CHAP, it is important to ensure that passwords are well chosen and long enough that they cannot be calculated readily. Many large customers require passwords to be at least six characters long with upper- and lowercase characters and at least one digit.

By default, MS-CHAP 1 for Windows 2000 supports LAN Manager authentication. If you want to prohibit the use of LAN Manager authentication with MS-CHAP 1 for older Microsoft operating systems such as Windows NT 3.5x and Windows 95, then set Allow LM Authentication (HKEY_LOCAL_MACHINE\SYSTEM\CurrentControlSet\Services \RemoteAccess\Policy) to zero on the authenticating server. Windows 95 and Windows 98 support MS-CHAP 2 only for virtual private network (VPN) connections. Windows 95 and Windows 98 do not support MS-CHAP 2 for dial-up connections.

18.5.5 Extensible Authentication Protocol

Security and authentication is a constantly changing field; thus, embedding authentication schemes into an operating system is impractical at times. To solve this problem, Microsoft has included support for the **extensible authentication protocol (EAP),** which allows new authentication schemes to be plugged in as needed. EAP allows third-party vendors to develop custom authentication schemes such as retina scans, voice recognition, fingerprint identification, smart cards, Kerberos, and digital certificates. In addition, EAP offers mutual authentication. To add, click EAP authentication methods on the Security tab of the remote access server's Properties dialog box.

The **extensible authentication protocol message digest 5 challenge handshake authentication protocol (EAP-MD5 CHAP)** is an EAP type that uses the same challenge handshake protocol as PPP-based CHAP, but the challenges and responses are sent as EAP messages. A typical use for EAP MD5 CHAP is to authenticate the credentials of remote access clients by using user name and password security systems. See Figure 18.4.

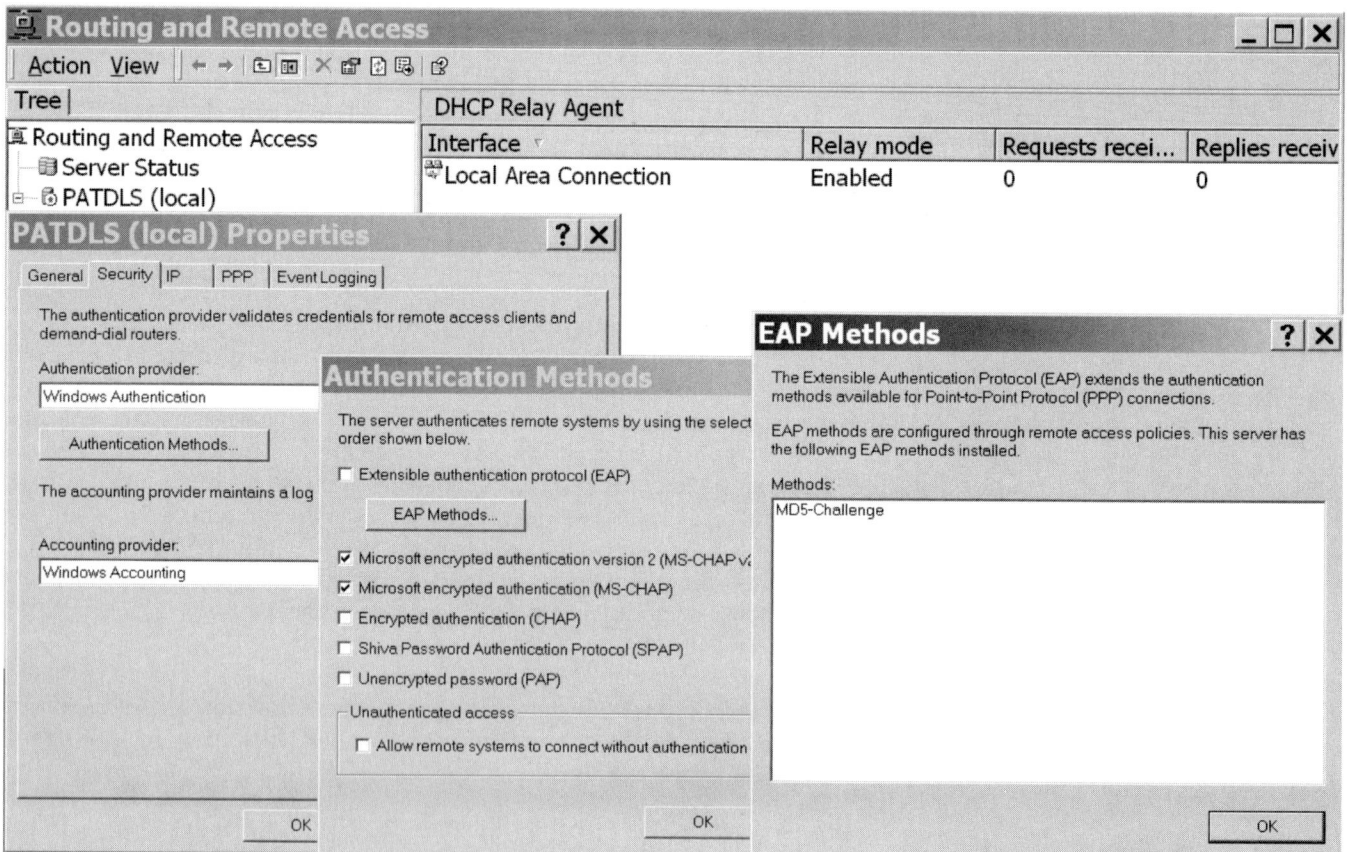

FIGURE 18.4   Select the EAP protocol and choose its method of authentication

EAP-Transport level security (EAP-TLS) is an EAP type that is used in certificate-based security environments. If you are using smart cards for remote access authentication, then use the EAP-TLS authentication method. The EAP-TLS exchange of messages provides mutual authentication, negotiation of the encryption method, and secured private key exchange between the remote access client and the authenticating server. EAP-TLS provides the strongest authentication and key exchange method, and is supported only on a remote access server that is running Windows 2000 and is a member of a Windows 2000 mixed or native domain.

EAP-RADIUS is not an EAP type, but the passing of EAP messages of any EAP type by a remote access server to a RADIUS server for authentication. The EAP messages sent between the remote access client and remote access server are encapsulated and formatted as RADIUS messages between the remote access server and the RADIUS server. EAP-RADIUS is used in environments where RADIUS is used as the authentication provider. An advantage of using EAP-RADIUS is that EAP types do not need to be installed at each remote access server, only at the RADIUS server. In a typical use of EAP-RADIUS, a remote access server is configured to use EAP and to use RADIUS as its authentication provider. When a connection is made, the remote access client negotiates the use of EAP with the remote access server. When the client sends an EAP message to the remote access server, the remote access server encapsulates the EAP message as a RADIUS message and sends it to its configured RADIUS server. The RADIUS server processes the EAP message and sends a RADIUS-encapsulated EAP message back to the remote access server. The remote access server then forwards the EAP message to the remote access client. In this configuration, the remote access server is only a pass-through device. All processing of EAP messages occurs at the remote access client and the RADIUS server.

18.5.6 Unauthenticated Access Method

The unauthenticated access method allows remote access users to log on without checking their credentials. For example, IAS does not verify the user's name and password. The only user validation performed in the unauthenticated access method is authorization. Enabling unauthenticated access presents security risks that must be carefully considered when deciding whether to enable this authentication method.

18.5.7 Enabling Authentication Protocols in Windows 2000

To enable authentication protocols, do the following:

1. Click on the Start button, select the Programs option, select Administrative Tools, and select the Routing and Remote Access program.
2. Right-click the server name for which you want to enable authentication protocols, and click the Properties option.
3. On the Security tab, click the Authentication button.
4. In the Authentication Methods dialog box, select the appropriate checkboxes for the authentication protocols that the remote access server will use to authenticate remote clients and then click the OK button.

18.6 VIRTUAL PRIVATE NETWORKING

As mentioned earlier in the chapter, a virtual private network (VPN) is the creation of secured, point-to-point connections across a private network or a public network such as the Internet. In other words, a VPN connects the components of one network using another network. The basic technology that defines a VPN is tunneling. **Tunneling** is the method for transferring data packets over the Internet or other public network, providing security and features formerly available only on private networks. A tunneling protocol encapsulates the data packet in a header that provides routing information to enable the encapsulated payload to securely traverse the network. The entire process of encapsulation and transmission of packets is tunneling, and the logical connection through which the packets travel is the tunnel.

As a comparison, tunneling is similar to sending a letter from one company's building in Los Angeles to the same company's building in Chicago. The letter is initially sent through the corporation's mail services. When the letter gets to the mail room, it will then be sent to Chicago via the U.S. mail. The U.S. mail carrier then delivers it to the second building. The letter is then sent through the corporate mail service to the correct office.

18.6.1 Point-to-Point Tunneling Protocol

Remote users who can gain access to their networks via the Internet are probably using the **point-to-point tunneling protocol (PPTP).** PPTP, developed by Microsoft and Ascend Communications, uses the Internet as the connection between remote users and a local network, as well as between local networks. It is easy to see why the PPTP is the most popular implementation of the VPN, as it is an inexpensive way to create WANs with PSTN, ISDN, and X.25 connections. PPTP wraps various protocols inside an IP datagram, an IPX datagram, or a NetBEUI frame. This process lets the protocols travel through an IP network tunnel, without user intervention. In addition, PPTP saves companies the work of building proprietary and dedicated network connections for their remote users and instead lets them use the Internet as their conduit.

PPTP is based on the point-to-point protocol. The difference between PPP and PPTP is that PPTP allows Internet access as the connection medium, rather than requiring a direct connection between the user and the network. In other words, instead of having to dial up the corporate network directly, a remote user could log in to a local Internet service provider, and PPTP will make the connection from that provider to the corporate network's Internet connection. From there, it continues into the corporate network the same as if the user dialed in directly. The process is as follows:

1. The remote client makes a point-to-point connection to the front-end processor via a modem.
2. The front-end processor connects to the remote access server, establishing a secure "tunnel" connection over the Internet. This connection then functions as the network backbone.
3. The remote access server handles the account management and supports data encryption through IP, IPX, or NetBEUI protocols.

PPTP has certain drawbacks. PPTP has weak encryption technology, so it is not a good choice for highly secure transmission over the Net. Its authentication features (the same used by PPP) are also weak. Therefore, bit corporations tend to prefer IP security.

18.6.2 IP Security

While the Internet is inherently insecure, businesses still need to preserve the privacy of data as they travel over the network. To help, the Internet Engineering Task Force has developed a suite of protocols called **Internet Protocol Security (IPSec).** IPSec creates a standard platform to develop secure networks and electronic tunnels between two machines. It also encapsulates each data packet in a new packet that contains the information necessary to set up, maintain, and tear down the tunnel when it is no longer needed.

Networks that use IPSec to secure data traffic can automatically authenticate devices by using digital certificates, which verify the identities of the two users who are sending information back and forth. IPSec can be an ideal way to secure data in large networks that require secure connections among many devices.

Users who are deploying IPSec can secure their network infrastructure without affecting the applications on individual computers. The protocol suite is available as a software-only upgrade to the network infrastructure, which allows security to be implemented without costly changes to each computer. Most important, IPSec allows interoperability among different network devices, PCs, and other computing systems. The process is as follows:

1. Send message—Data within an application are sent to a message authentication and integrity function with optional digital signature provision.
2. Data encryption—Data are encrypted with algorithms using a public key.
3. Secure packets—Data are sent over the Internet as encrypted packets in a secure tunnel.
4. Session key—Data are encrypted again in the tunnel using a session key.
5. Data decryption—At the end of the tunnel, data are decrypted using a private key.

6. Verification—When the data hit the receiver's PC, they are checked and the sender is verified using a message integrity function and an authentication provision.
7. Message received—The receiver can now read the text.

Rather than using applications or the operating system, use IPSec policies to configure IPSec security services. The policies provide variable levels of protection for most traffic types, in most existing networks. Your network security administrator can configure IPSec policies to meet the security requirements of a user, group, application, domain, site, or global enterprise. Windows 2000 provides an administrative interface, IP Security Policy Management, to create and manage IPSec policies (centrally at the group policy level for domain members, or locally on a nondomain computer). IP Security Policy Management is a snap-in that you can add to any Microsoft Management Console (MMC).

To configure TCP/IP to use an IPSec, do the following:

1. Click on the Start button, select the Settings option, and select the Network and Dial-up Connections.
2. Right-click the network connection that you want to configure, and then click Properties.
3. On the General tab (for a local area connection) or the Networking tab (for all other connections), click Internet Protocol (TCP/IP), and then click Properties.
4. Click on the Advanced button.
5. On the Options tab, click IP Security, and then click on the Properties button.
6. To enable IP security, click Use This IP Security Policy, and then click the name of a policy.
7. To disable IP security, click Do Not Use IPSec.

The Intel Corporation has developed network adapter cards, which include an encryption coprocessor that offloads IPSec encryption and authentication from the processor in Windows 2000 and Windows NT 4.0 machines. As a result, machine performance and connection speed will be better than when doing IPSec encryption and authentication with the processor doing all the work.

18.6.3 Layer Two Tunneling Protocol

To compete against the PPTP protocol, Cisco developed the layer 2 forwarding (L2F) protocol, which allows a server to frame dial-up traffic using PPP and transmit it over WAN links to an L2F server or router, which then unwraps the packets before releasing them to the network.

The **layer 2 tunneling protocol (L2TP)** is a combination of the L2F and PPTP, and enables remote users to access networks in a secure fashion. Therefore, L2TP represents the next generation of tunneling protocols. L2TP provides tunneling and supports header compression, but it does not provide encryption. Instead, it provides a secure tunnel by cooperating with other encryption technologies such as IPSec. IPSec does not require L2TP, but its encryption functions complement L2TP to create a secure VPN solution. Since the L2TP-based VPN connections are a combination of L2TP and IPSec, both L2TP and IPSec must be supported by both routers. L2TP is installed with the routing and remote access service and, by default, is configured for five L2TP ports.

18.6.4 Multilink and Bandwidth Allocation Protocol

The **PPP multilink protocol (MP)** is an extension that is used to aggregate or combine multiple physical links into a single logical link. A good example is the aggregation of both B channels of an ISDN BRI connection or having two modems connect to a RAS server that has two modems. The MP fragments, sequences, and reorders alternating packets sent across multiple physical connections so that the end result is a single logical link with the combined bandwidth of all aggregated physical links. MP must be supported on both sides of the connection, however.

In Windows 2000, **bandwidth allocation protocol (BAP)** and **bandwidth allocation control protocol (BACP)** enhance multilink by dynamically adding or dropping links on demand. BAP is especially valuable to operations that have carrier charges based on bandwidth

FIGURE 18.5 Enabling a
multilink connection

PATDLS (local) Properties   **? X**

General | Security | IP | PPP | Event Logging |

This server can use the following Point-to-Point Protocol (PPP) options. Remote
access policies determine which settings are used for an individual connection.

☑ Multilink connections
    ☑ Dynamic bandwidth control using BAP or BACP
☑ Link control protocol (LCP) extensions
☑ Software compression

OK | Cancel | Apply

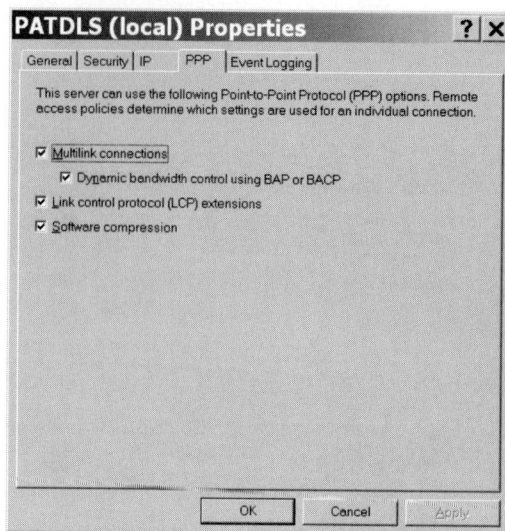

utilization. BAP and BACP are sometimes used interchangeably to refer to bandwidth-on-demand functionality. Both protocols are PPP control protocols and work together to provide bandwidth on demand. BAP provides an efficient mechanism for controlling connection costs while dynamically providing optimum bandwidth. For example, a multilink and BAP-enabled remote access client and remote access server create a multilink connection that consists of a single physical link. As the utilization of the single link rises to a configured level, the remote access client uses a BAP request message to request an additional link. The BAP request message specifies the type of link desired, such as analog phone, ISDN, or X.25. The remote access server then sends a BAP response message that contains the phone number of an available port on the remote access server of the same type as specified by the remote access client in the BAP request. When multilink and BAP are used in combination with the callback feature set to always call back to the same number, a concentrator must exist on the caller side that can distribute incoming calls to the same number on various ports. See Figure 18.5.

You can enable multilink and BAP protocols on a server-wide basis from the PPP tab of each remote access server's Properties dialog box. Configure BAP settings through remote access policies. Using these policies, specify that an extra line be dropped if link utilization drops below 75% for one group and below 25% for another group. Remote access policies are described later in this module.

18.7 CONFIGURING OUTBOUND CONNECTIONS

You can configure all outbound connections in Windows 2000 with the Network Connection wizard. Much of the work of configuring protocols and services is automated when you use this process. There are three basic types of outbound connections:

- Dial-up connections, which include connections to a private network, private server, stand-alone computer in someone's home, modem pool in a corporate intranet, or connection to an Internet service provider (ISP)
- Connections to a VPN
- Direct connections to another computer through a cable

18.7.1 Creating a Dial-up Connection to a Private Network

To create and configure an outbound dial-up connection to a private network using a phone, ISDN, or X.25 connection, use the Network Connection wizard. To connect to a private network, perform the following steps:

1. Click on the Start button, select the Settings option, select the Network and Dial-up Connections option, and select the Make New Connection option.

2. The Network Connection wizard will appear. Click on the Next button. Select the Dial-up to Private Network option and click on the Next button.

3. Enter the phone number of the computer to which you are connecting and click on the Next button. This may be an ISP for an Internet connection or the modems for your private network. If you need to use an area code or select the country/region code, enable Use Dialing Rules.

4. If you want this connection to be made available to all users of this computer, click For All Users, and then click the Next button. If you want to reserve the connection for yourself, click Only for Myself, and then click on the Next button.

5. If you select the For All Users option, and you want to let other computers gain access to resources through this dial-up connection, then select the "Enable shared access for this connection" checkbox. In either case, click on the Next button.

6. If you enable Internet connection sharing, the LAN adapter will use an IP address of 192.168.0.1. Click on the Yes button. Select the local network connection that will access resources through this connection. Click on the Next button.

7. Type a name for the connection and click the Finish button.

18.7.2 Creating a Dial-up Connection to an ISP

To create and configure an outbound dial-up connection to an Internet service provider (ISP) or other private network using a phone, ISDN, or X.25 connection, use the Network Connection wizard. To connect to a private network, perform the following steps:

1. Click on the Start button, select the Settings option, select the Network and Dial-up Connections option, and select the Make New Connection option.

2. The Network Connection wizard will appear. Click on the Next button. Select the Dial-up to the Internet, and click on the Next button.

3. The Internet Connection wizard will start. If you have an account, choose either "I want to set up my Internet connection manually," or "I want to connect through a local area network (LAN)" option. Click on the Next button.

4. If you are connecting using the phone, select the "I want to connect through a phone and a modem" option. Click on the Next button.

5. Enter the phone number of the ISP and click on the Next button. If you need to use an area code or select the country/region code, enable Use Dialing Rules. In addition, the default dial-up protocol is PPP. If you specify the SLIP or C-SLIP protocol or specify an IP address for your system, click on the Advanced button.

6. Type in the user name and password used to log on to the ISP.

7. Type a name for the connection and click the Next button.

8. If you have an e-mail account through the ISP, select the Yes option and click on the Next button.

9. Select the "Create a new Internet mail account" option and click the Next button.

10. Specify the name that will show in the e-mail when you send messages and click on the Next button.

11. Specify your e-mail addresses so that other people reply to you and click on the Next button.

12. Specify the type and name of the incoming mail server and click on the Next button.

13. Type the account name and password for your e-mail account. Click on the Next button.

14. Click on the Finish button.

18.7.3 Configuring Virtual Private Network Connections

To set up a virtual private network connection, do the following:

1. Click on the Start button, select the Settings option, select the Network and Dial-up Connections option, and select the Make New Connection option.

2. The Network Connection wizard will appear. Click on the Next button. Select the "Connect to a private network through the Internet" option and click on the Next button.

3. Specify the host name (such as Microsoft.com) or an IP address and click on the Next button.

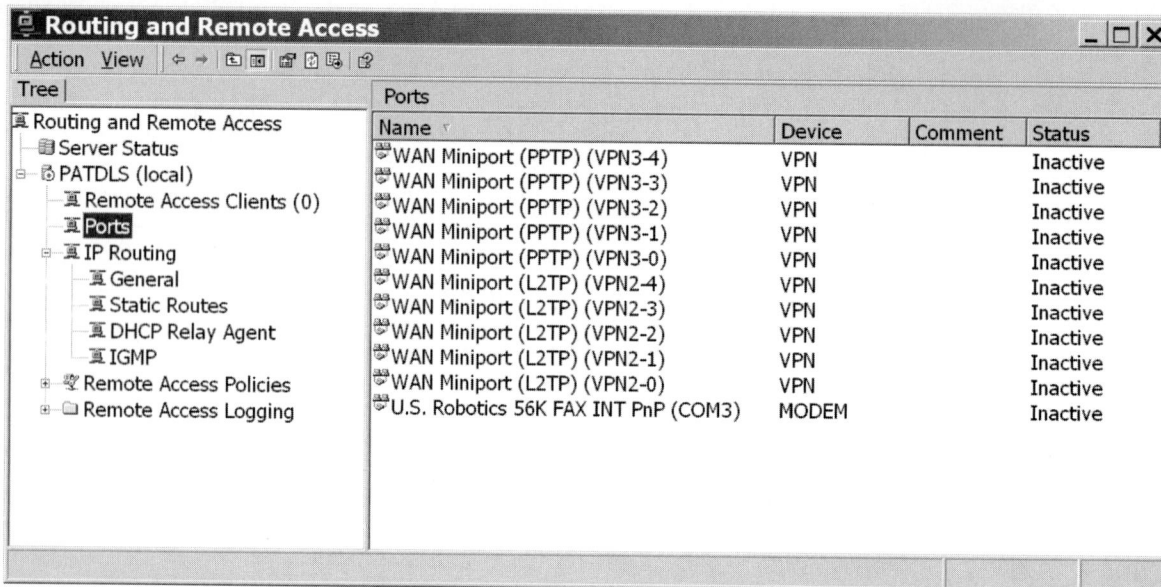

FIGURE 18.6 Configuring a VPN port

4. If you want this connection to be made available to all users of this computer, click For All Users, and then click the Next button. If you want to reserve the connection for yourself, click Only for Myself, and then click on the Next button.

5. If you select the For All Users option, and you want to let other computers gain access to resources through this dial-up connection, select the "Enable shared access for this connection" checkbox. In either case, click on the Next button.

6. If you enable Internet connection sharing, the LAN adapter will use an IP address of 192.168.0.1. Click on the Yes button. Select the local network connection that will access resources through this connection. Click on the Next button.

7. Type a name for the connection and click the Finish button.

8. The Connect Virtual Private Connection dialog box appears. Type a user name and password. To specify the virtual private connection parameters, click on the Properties button and click the Security tab. Click Advanced (custom) settings following by the Settings button to specify the authentication protocol. When complete, click on the Connect button.

In Windows 2000, five PPTP and five L2TP ports are created when the routing and remote access service is started for the first time. In addition, Windows 2000 automatically detects any modems that are installed and creates modem ports for them. Windows 2000 also creates ports for each parallel or serial cable that it detects.

To configure the VPN ports, do the following:

1. Click on the Start button, select the Program option, select the Administrative Tools option, and select the Routing and Remote Access program.

2. In the console tree, right-click Ports, then select the Properties option.

3. In the Ports Properties dialog box, select the device you want to configure and click on the Configure button.

4. In the Configure Ports dialog box, select the Remote Access (inbound) checkbox to enable inbound VPN connections. In addition, set the number of virtual ports available on the server. Click on the OK button to close the Configure Ports dialog box. See Figure 18.6.

5. Click OK to close the Ports Properties dialog boxes.

18.8 REMOTE ACCESS PERMISSIONS

In Windows NT, authorization was based on a simple grant dial-in permission to user option in User Manager or the Remote Access Admin utility. Callback options were also configured on a per-user basis. In Windows 2000, authorization can be granted to the individ-

FIGURE 18.7 Configuring a
user to access a remote access
server

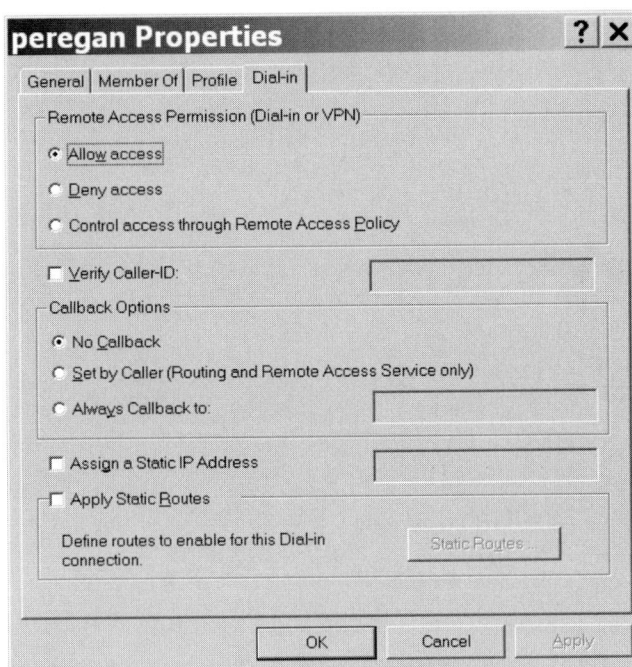

ual user account or through the configuration of specific remote access policies. **Remote access policies** are a set of conditions and connection settings that give network administrators more flexibility in authorizing connection attempts.

To grant remote access permission to a user, do the following:

If the remote access server is part of a Windows 2000 domain:

1. In Administrative Tools, open the Active Directory Users and Computers console.
2. In the console tree, open Domain and click Users.
3. In the Details pane, right-click a user name, and select the Properties option. See Figure 18.7.

If the remote access server is a stand-alone server, which is not a part of a Windows NT 4.0 or Windows 2000 domain or is a member server and you want to use a local user account:

1. In Administrative Tools, open the Computer Management console.
2. In the console tree, open System Tools and Local Users and Groups and click on Users.
3. In the Details pane, right-click a user name, and then click Properties.
4. On the Dial-in tab, under Remote Access Permission (Dial-in or VPN), click either Allow Access or Control Access through Remote Access Policy, and then click OK.

If the remote access server is a member of a Windows NT 4.0 domain or a Windows 2000 mixed-mode domain, use the Windows NT 4.0 User Manager for Domains administrative tool to grant or deny dial-in access for user accounts.

In the first part of the dialog box, specify if a user is to be allowed access, denied access, or if the access will be controlled through the remote access policy. The Verify Caller-ID option specifies whether this user is required to dial in from a particular number. If access is explicitly allowed, remote access policy conditions, user account properties, or profile properties can still deny the connection attempt. The Control Access through Remote Access Policy is only available on user accounts for stand-alone Windows 2000 remote access servers or members of native Windows 2000 domain.

If the Verify Caller-ID option is enabled, the server verifies the caller's phone number. If the caller's number does not match the configured phone number, the connection attempt is denied. For this to work, all parts of the connection must support caller ID. Caller ID support on the remote access server consists of caller ID answering equipment and the driver that passes caller ID information to routing and remote access. If you configure a caller ID

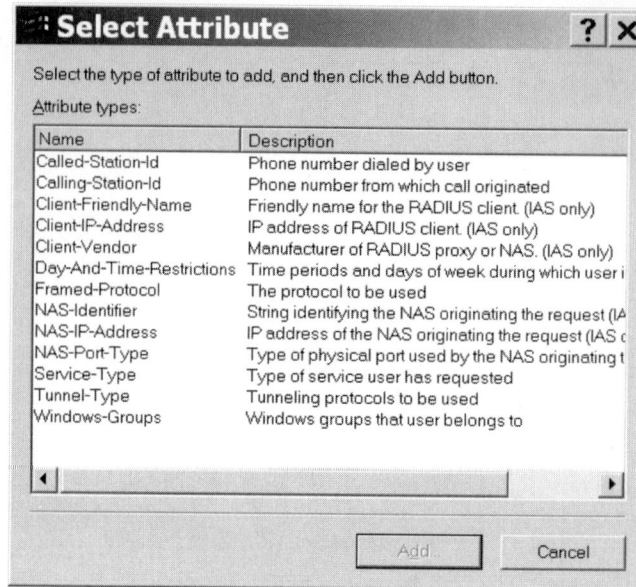

phone number for a user and you do not support for the passing of caller ID information from the caller to routing and remote access, then the connection attempt is denied.

In the Callback options, you may determine if the server calls back a specific phone number. If the No Callback option is selected, the remote access server establishes a connection as soon as the connection attempt has been accepted. The Set by Caller option is useful for clients who call from various locations and phone numbers. It also minimizes telephone charges for these users. When the Always Callback To option is selected, it provides additional security by calling back to the preset number. Set this option for stationary remote computers, such as those used by telecommuters that do not move around.

To configure a remote access policy to grant or deny access, follow these steps:

1. Do one of the following:
   - In the Administrative Tools, open the Routing and Remote Access console and, if necessary, double-click Routing and Remote Access and the server name.
   - Open Internet Authentication Service and, if necessary, double-click Internet Authentication Service.
2. In the console tree, click Remote Access Policies.
3. In the Details pane, right-click the policy you want to configure, and then click Properties.
4. Under "If a user matches the conditions,"
   - To grant dial-up permission to these users, click "Grant remote access permission."
   - To deny dial-up permission to these users, click "Deny remote access permission."

Remote access policy conditions are assigned attributes that are compared with the settings of a connection attempt. If there are multiple conditions in a policy, then all conditions must correspond to the settings of the connection attempt to result in a match.

You can create a remote access policy and an associated profile under the Remote Access Policies node of the Routing and Remote Access console tree. To add a remote access policy, perform the following steps:

1. Open Routing and Remote Access.
2. Right-click Remote Access Policies, and then select the New Remote Access Policy.
3. In the Policy Properties dialog box, type the name of the profile in the Policy Friendly Name text box.
4. To configure a new condition, click Add, and then do the following:
   - In the Select Attribute dialog box, click the attribute to add, and then click OK.
   - In the Attribute dialog box, enter the information required by the attribute, and then click OK. See Figure 18.8.
5. If a user matches the conditions section, then do the following:

- To grant access to these users, click "Grant remote access permission."
- To deny access to these users, click "Deny remote access permission."

To edit the policy's profile, perform the following steps:

1. In the Policy Properties dialog box, click Edit Profile.
2. In the Edit Dial-in Profile dialog box, configure the settings on any of the six tabs, then click OK.
3. Click OK to close the Policy Properties dialog box.

The profile specifies what kind of access the user will be given if the conditions match. This access will only be granted if the connection attempt does not conflict with the settings of the user account or the profile. See Figure 18.9. Six different tabs can be used to configure a profile as follows:

- **Dial-in Constraints** —These options include settings for idle-time disconnect, maximum session time, day and time, phone number, and media type (ISDN, VPN, etc.).
- **IP**—These settings configure client IP address assignments and TCP/IP packet filtering. Separate filters can be defined for inbound or outbound packets.
- **Multilink**—These settings configure multilink and BAP. A line can be dropped if bandwidth drops below a certain level for a given length of time. Multilink can also be set to require the use of BAP.
- **Authentication**—These settings define the authentication protocols that are allowed for connections using this policy. The protocol selected must also be enabled in the server's properties.
- **RAS Encryption**—These settings specify that IPSec or Microsoft Point-to-Point Encryption (MPPE) is prohibited, allowed, or required.
- **Advanced**—This tab allows for the configuration of additional network parameters that could be sent from non-Microsoft RADIUS servers. You can create a remote access profile from the Policy Properties dialog box.

A connection is authorized only if the settings of the connection attempt match at least one of the remote access policies (subject to the conditions of the dial-in properties of the user account and the profile properties of the remote access policy). If the settings of the connection

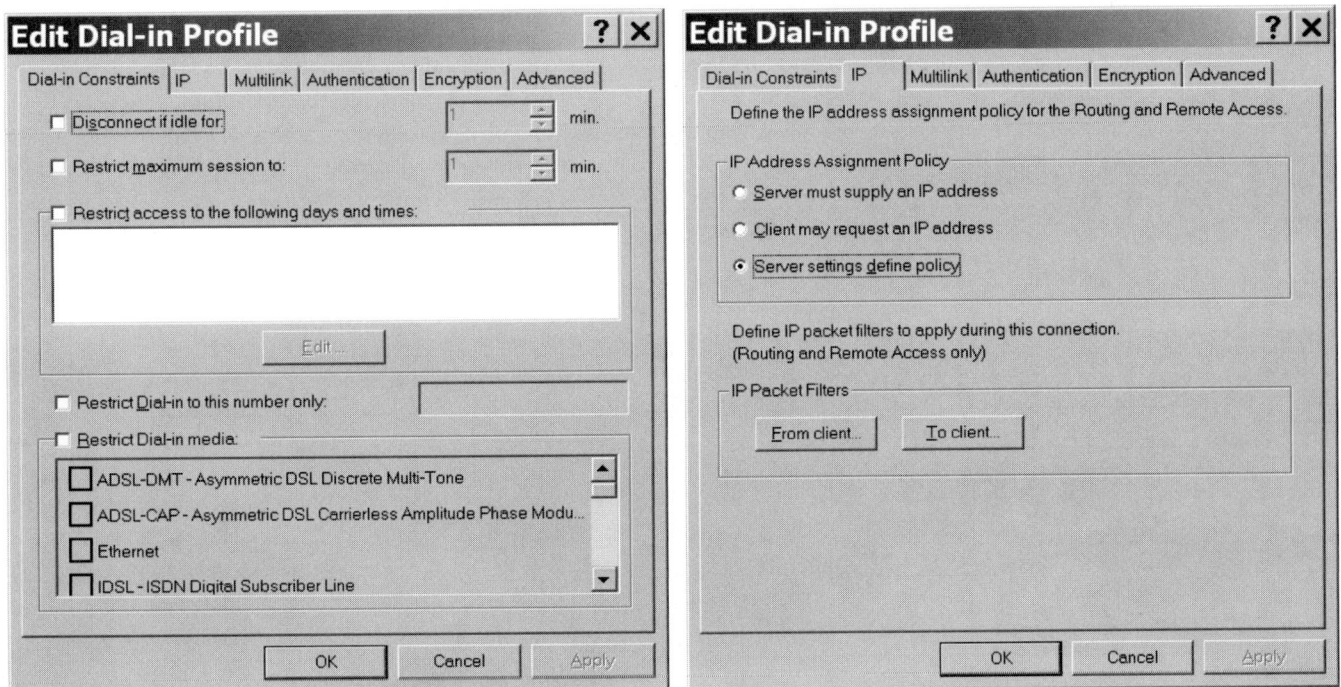

FIGURE 18.9   Configuring a Dial-in Profile

attempt do not match at least one of the remote access policies, then the connection attempt is denied regardless of the dial-in properties of the user account. The list of parameters checked are time of day, user groups, caller IDs, or IP addresses, matched to the parameters of the client connecting to the server.

If you set the dial-in permission on every user account to "Control access through Remote Access Policy," and if you do not change the default remote access policy, then all connection attempts will be rejected. However, if one user's dial-in permission is set to "Allow access," then that user's connection attempts will be accepted. If you change the permission setting on the default policy to "Grant remote access permission," then all connection attempts are accepted.

18.9 TERMINAL EMULATION

Telnet is a user command and an underlying TCP/IP protocol for accessing remote computers, as specified in RFC 854. **Terminal emulation (Telnet)** is used to access someone else's computer, assuming that person has given you permission. Such a computer is frequently called a host computer, whereas the client is called a dumb terminal. With Telnet, you log on as a regular user with whatever privileges you may have been granted to the specific applications and data on that computer. You can then enter commands through the Telnet program and they will be executed as if you were entering them directly on the server console; thus, you control the server and communicate with other servers on the network. Telnet is a common way to remotely control web servers.

RFC 854 states: "The Telnet Protocol is built upon three main ideas: first, the concept of a 'Network Virtual Terminal'; second, the principle of negotiated options; and third, a symmetric view of terminals and processes." The default port for Telnet is TCP port 23.

A popular terminal program/brand of terminal from DEC is Visual Terminal 100 (vt100). It was extremely popular with certain companies and universities that ran Berkeley UNIX on their VAXes (also from DEC). Most communication packets support vt100.

To start Telnet Server on a Windows 2000 server, do the following:

1. Open Computer Management
2. In the console tree, click Services located under Services and Applications
3. In the Details pane, right-click Telnet, and then select the Start option

To connect to a UNIX computer or Windows 2000 with the Telnet Server using a Telnet program, Windows 2000 computers provide a DOS command called TELNET.EXE. The TELNET command can be used with or without a computer name. If no computer name is used, Telnet provides Command mode and provides a prompt to the user. The activation of Telnet results in a TCP connection to the server and to the Telnet daemon, TLNTSVR. After a connection is established, Telnet enters input mode. Depending on the remote computer, typed text is sent from the client either one character at a time or one line at a time. Since Telnet is used for UNIX computers, it does not allow the executing of GUI-based programs. See Figure 18.10.

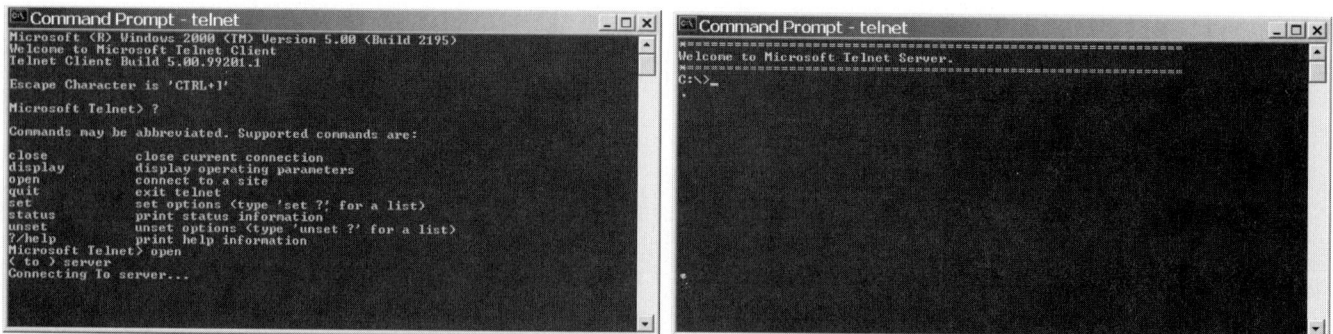

FIGURE 18.10   Using the Windows 2000 Telnet client software

18.10 THIN CLIENTS

A **thin client** is a computer that is between a dumb terminal and a PC, and is designed to be especially small so that the bulk of the data processing occurs on the server. Although the term *thin client* usually refers to software, it is increasingly used for computers, such as network computers and Net PCs that are designed to serve as the clients for client-server architectures.

A **fat client** performs the bulk of the data processing operations. The data are stored on the server. Although the term *fat client* also refers to software, it can apply to a network computer that has relatively strong processing abilities.

18.10.1 Introduction to Terminal Services

Windows 2000 **Terminal Services** is a fully integrated component of Windows 2000 Server, which enables the latest Windows-based applications to be accessible from any Microsoft-based client operating system. It gives you the opportunity to extend remote solutions to your customers that are more efficient and easier to manage.

When a client runs an application on a Terminal Services–enabled Windows 2000 Server, all client application execution, data processing, and data storage occur on the server. The Terminal Services component in Windows 2000 Server provides a bridging technology for customers who are moving to a Windows 2000 Professional desktop. It allows a homogeneous set of applications to run on a heterogeneous set of desktop hardware through terminal emulation. Since only screen (via the remote desktop protocol, RDP), keyboard, and mouse information is exchanged between the client and server, it is an ideal solution for the remote dial-up network or for using a shared application on a single server that many people need to access from distant locations (across the Internet or dial-up).

There are two main aspects, or modes, of Terminal Services that you can enable: the applications server mode and the remote administration mode. The application server mode allows end users to log on to Terminal Services from any connected device and access the applications that are installed on the server. The remote administration mode allows up to two concurrent administrators to remotely manage the server. By choosing one over the other, users are able to assign a priority to the way Terminal Services operates. When remote administration mode is enabled, memory and CPU utilization settings are unaffected, application compatibility settings are disabled, and there is minimal impact on the server.

Remote administration mode allows end users the ability to manage their server from any location. There are no licensing requirements when using this mode, as only administrator accounts can attach to the server. While in this mode, administrators can view event logs, run commands, check databases such as SQL or Exchange, or upgrade a domain controller. This capability can benefit customers who do not have the budget to upgrade their desktops in order to run Windows 2000 Professional, as they are still able to utilize advanced applications such as Microsoft Office 2000. Terminal Services will also benefit roaming users who do not work on the same computer all the time but need to access their desktop environments and information stored on the server. Companies that require remote management solutions for their networks are another type of potential customer.

18.10.2 Installing and Removing Terminal Services

To add Terminal Services to Windows 2000 Server, do the following:

1. Open the Add/Remove Programs applet in the Control Panel.
2. Choose Add/Remove Windows Components.
3. From the list available, check the box next to Terminal Service. Click on the Next button. You do not need to install Terminal Services Licensing to run terminal sessions, although you will need to install the licensing tool on some Windows 2000 servers in the network to keep track of terminal license use. If you click the Details button, you will see that you can install either client-created files or files that allow a user to connect to the terminal server and that are needed to enable Terminal Services.
4. As shown in Figure 18.11, in the Terminal Services Setup dialog box, you have two choices for setting up Terminal Services. Choose the desired option and click the Next button.

FIGURE 18.11 Selecting the
Terminal Services mode

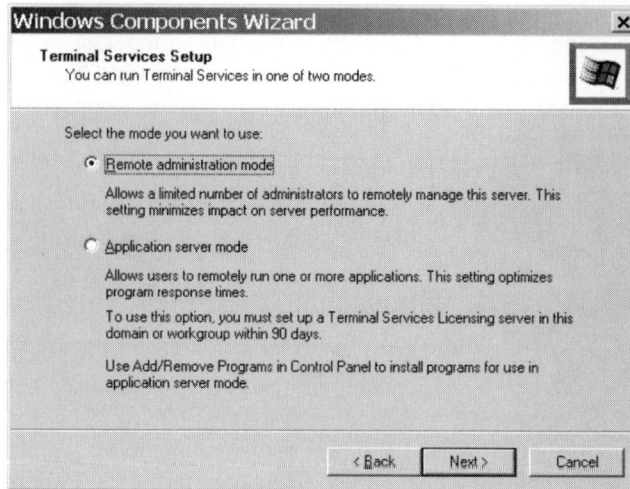

Windows 2000 will then copy the files from an installation source. When the copying is done, you will be prompted to restart the computer to effect the changes. Upgrading from Windows NT Server 4.0, Terminal Server Edition to Windows 2000 with Terminal Services will preserve user and applications configuration information. The system will be placed in application server mode and Terminal Server 4.0-based security will be used. Upgrading directly from Citrix WinFrame is not supported; the system must be upgraded to Terminal Server 4.0 as an intermediate step, and then upgraded to Windows 2000. A clean install is recommended but not required.

To remove Terminal Services from a Windows 2000 server, open the Add/Remove Programs applet in the Control Panel and choose Add/Remove Windows components.

18.10.3 Configuring and Using the Terminal Services Client

Windows 2000 Server includes new clients for Terminal Services, with enhancements to the remote desktop protocol (RDP) and new features. Clients included on the Windows 2000 Server CD are compatible with the following platforms: 16-bit client (runs on Windows 3.11) and 32-bit client (runs on Windows 95, Windows 98, Windows NT 3.51, Windows NT 4.0, and Windows 2000). Earlier Windows NT Server 4.0, Terminal Server Edition clients will also work with Windows 2000 Server with Terminal Services enabled; however, some of the new RDP features will not be available unless the new Windows 2000 Terminal Services clients are used.

To connect a PC-based client to Terminal Services, run a Terminal Services Client installation program. You can either create installation disks with the Terminal Services Client Creator or share the client files on the network and run the Setup program from the client. The Terminal Services Client Creator can be found in the Administrator Tools program.

As shown in Figure 18.12, the default client types available are Terminal Services for 16-bit Windows (which requires four disks) and Terminal Services for 32-bit x86 Windows (which require two disks).

FIGURE 18.12 Creating
Terminal Services Client
installation disks

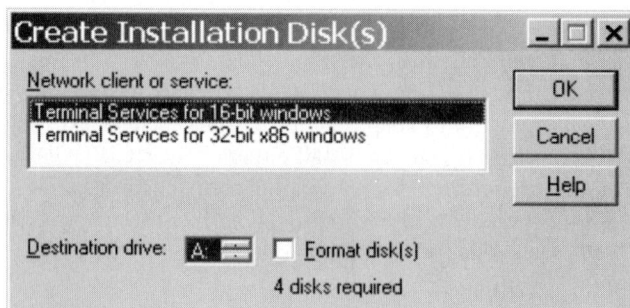

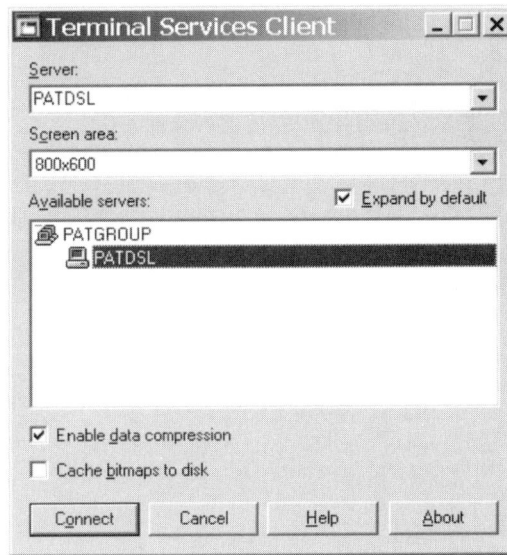

One way to connect to a terminal server is to run the Terminal Services Client found in the Terminal Services Client program group. See Figure 18.13. From the Terminal Services Client dialog box, select the screen area size in resolution and the server. Using a lower resolution will give you a smaller window on the desktop. If the client gets an error message saying that the terminal server is busy, try again later, check the client TCP/IP settings, and make sure the IP address and subnet are valid. After the connection is made, a window will appear that allows you to log in to the server to which you are connecting. After you log on, run programs at the server from your computer. See Figure 18.14.

To create a customized connection, you can use the Client Connection Manager. To define a new connection, open the File menu and select the New option. Then using the Client Connection Manager wizard, specify the server, the information to do an automatic login, and the session display settings. After the connection has been defined, then choose to create a shortcut in the Programs option using the Start button or by using the Client Connection Manager.

To manage the Terminal Services connection in trusted domains, administrators can use the Terminal Services Manager. The Terminal Services Manager is used to monitor users, sessions, and applications on each terminal server, and allows you to carry out assorted actions to manage the server.

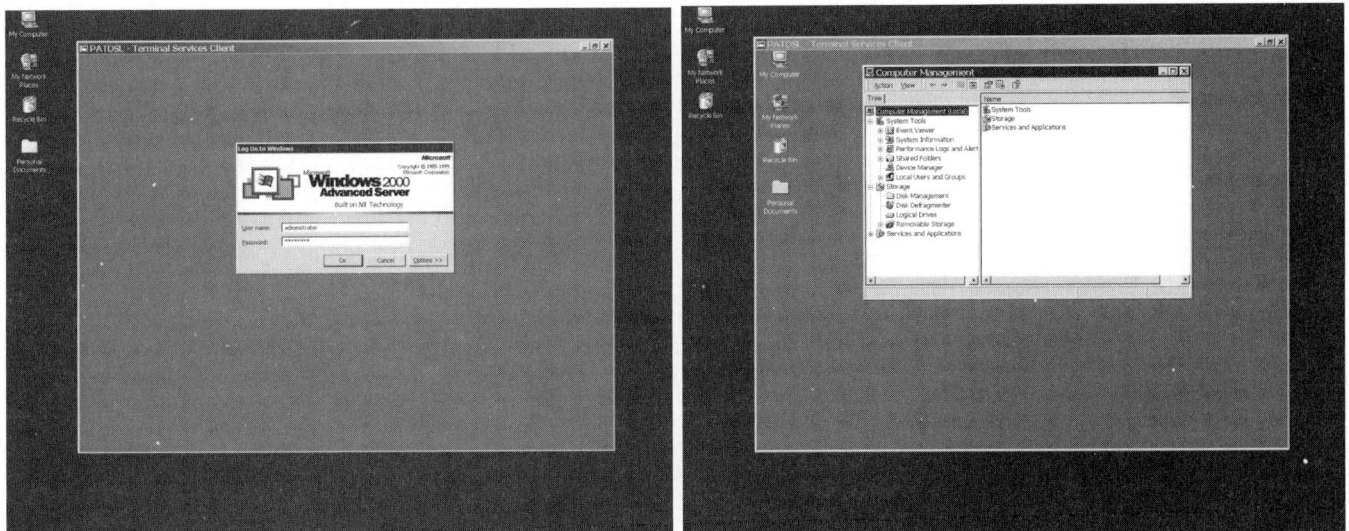

FIGURE 18.14   Connecting to a second computer using Windows 2000 Terminal Services

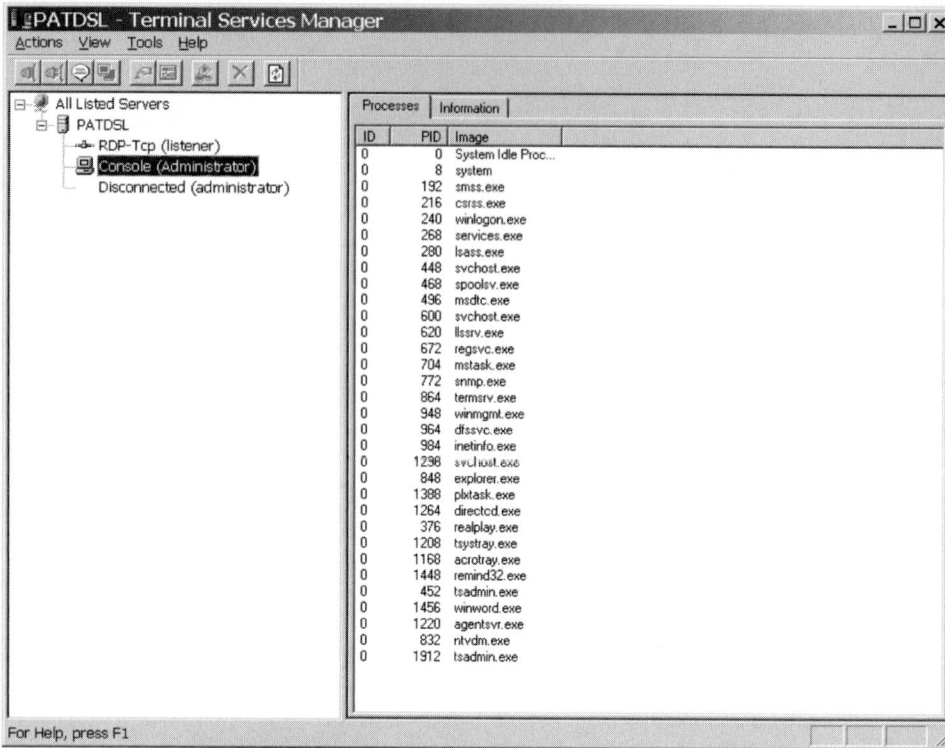

FIGURE 18.15   Terminal Services Manager

When a user creates a session by connecting to the terminal server from a client computer, the session appears in the Session list in Terminal Services Manager. In addition, the name of the user who logs on using the session appears in the Users list. Any applications run at the user's session can be monitored on the Processes list. Therefore, you can oversee all users, sessions, and processes on a terminal server from one location. See Figure 18.15.

18.10.4   Applications under Terminal Services

If you plan to run Windows 2000 Server in the application server mode (a traditional terminal server), then you must install applications correctly in multiuser mode in one of the following two ways.

1. Install the applications using the Add/Remove Programs tool in the Control Panel. Using this tool is the easiest option. It will properly install the applications for use in a Terminal Services environment.
2. Alternatively, put the server into Install mode manually by using the Change User/Install command line utility before installing applications on Windows 2000 with Terminal Services enabled. After installing the application, the server should be put back into execute mode by using the Change User/Execute command line option.

Note that some programs require an application compatibility script to be run in the Terminal Services environment. Microsoft supplies such a script for Office 2000 in the Office 2000 Resource Kit.

When running applications, you will need to have sufficient processor power and memory. Microsoft states that a quad processor Pentium Pro with 512 MB of RAM will concurrently support about 60 typical users. Of course, the server would typically be dedicated to this task.

18.10.5   Licensing for Terminal Services

The licensing for Terminal Services is a bit more complicated than normal licenses for the server. In a nutshell, each device that initiates a Windows 2000 Terminal Services session must be licensed with the following:

1. Windows 2000 Professional license or Windows 2000 Terminal Services Client Access license
2. Windows 2000 Server Client Access license or BackOffice Family Client Access license

For more information about the Terminal Services licenses, visit the Microsoft.com website.

SUMMARY

1. Modern networks offer remote access service (RAS), which allows users to connect remotely using various protocols and connection types.
2. A remote access server is the computer and associated software that is set up to handle users who are seeking access to the network remotely.
3. Virtual private networking is the creation of secured, point-to-point connections across a private network or a public network such as the Internet.
4. The typical dial-in session actually consists of six distinct steps or stages, each of which must be completed successfully before the user gains access to centralized resources.
5. If the call gets through, the remote (dialing) modem will begin negotiating a common data rate and other transmission parameters with the answering modem. This step is referred to as the handshaking stage, because both modems must agree on the same parameters before continuing.
6. Remote authentication dial-in user service (RADIUS) is the industry standard client-server protocol and software that enables remote access servers to communicate with a central server to authenticate dial-in users and authorize their access to the requested system or service for authenticating remote users.
7. Windows 2000 Server Routing and Remote Access is a mature, full-featured, third-generation service of Windows-based server operating systems.
8. The first protocol used for carrying IP packets over dial-up lines was the serial line interface protocol (SLIP).
9. The point-to-point protocol (PPP) has become the predominant protocol for modem-based access to the Internet.
10. A client uses an authentication protocol to send either encrypted or unencrypted credentials to the server, depending on the protocol.
11. The password authentication protocol (PAP) is the least secure authentication protocol because it uses clear text (plaintext) passwords.
12. Historically, challenge handshake authentication protocol (CHAP) is the most common dial-up authentication protocol used, and it uses an industry message digest 5 (MD5) hashing scheme to encrypt authentication.
13. Microsoft challenge handshake authentication protocol (MS-CHAP) is Microsoft's proprietary version of CHAP, which is the most secure within Windows 2000.
14. Microsoft has included support in Windows 2000 for the extensible authentication protocol (EAP), which allows new authentication schemes to be plugged in as needed.
15. Point-to-point tunneling protocol (PPTP) is the most popular implementation of the virtual private network (VPN) and is an inexpensive way to create wide area networks (WANs).
16. IPSec has stronger encryption and authentication than PPTP.
17. The PPP multilink protocol (MP) is an extension that is used to aggregate or combine multiple physical links into a single logical link.
18. In Windows 2000, bandwidth allocation protocol (BAP) and bandwidth allocation control protocol (BACP) enhance multilink by dynamically adding or dropping links on demand.
19. In Windows 2000, authorization is granted based on the dial-in properties of a user account and remote access policies.
20. Terminal emulation (Telnet) is a user command and an underlying TCP/IP protocol for accessing remote computers.
21. A thin client is designed to be especially small so that the bulk of the data processing occurs on the server.

22. Windows 2000 Terminal Services is a fully integrated component of Windows 2000 Server, which enables the latest Windows-based applications to be accessible from any Microsoft-based client operating system.

23. In Windows 2000 Terminal Services, remote administration mode allows end users the ability to manage their servers from any location.

QUESTIONS

1. Which of the following are dial-up communication protocols? (Choose all that apply)
   a. PSTN
   b. Telnet
   c. PPP
   d. SLIP
   e. TCP

2. SLIP is an acronym for _____ .
   a. serial line internet protocol
   b. service layer internal protocol
   c. system level information protocol
   d. serial line information protocol

3. PPP is an acronym for _____ .
   a. point to present protocol
   b. physical point protocol
   c. point-to-point protocol
   d. present point protocol

4. What is the most common network protocol that is run over a PPP dial-up connection?
   a. TCP/IP
   b. IPX
   c. NetBEUI
   d. PPTP

5. Which of the following functions does PPP LCP perform? (Choose all that apply)
   a. establish link connection
   b. terminate link connection
   c. configure link connection
   d. test link connection

6. Which of these are authentication protocols for PPP? (Choose all that apply)
   a. CHAP
   b. PAP
   c. LCP
   d. NCP

7. What is RADIUS used for?
   a. dial-up authentication
   b. WAN-based encryption
   c. remote access compression algorithm
   d. dynamic network configuration

8. Pat has set up a RAS server that connects to an Internet service provider (ISP) using a dial-up connection. Pat wants Windows 98 users on the local network to access the Internet through the RAS server. How should Pat set up the Internet connection?
   a. On the RAS server, leave the default gateway address blank.
   b. On the RAS server, use the subnet mask provided by the ISP.
   c. On the Windows 98 computers, select ISDN as the dial-up networking type.
   d. On the Windows 98 computers, set up the RAS server for multilink.
   e. On the Windows 98 computer, use the subnet mask provided by the ISP.

9. A number of employees at your company have Windows 2000 computers installed on their notebook computers. They need to dial up over the Internet to access client-server applications on your network's Windows 2000 Server computer. You need to provide adequately encrypted security for these connections and prevent unauthorized users from accessing the server. How can you best do this?
   a. Enable IP address filtering on the server.
   b. Implement the point-to-point tunneling protocol (PPTP).
   c. Set up FTP with user-level security.
   d. Set up RAS with SSL security.

10. Your network connects via a RAS server to an Internet service provider (ISP). Employees make a PPTP connection to your network through the ISP. A remote dial-up user calls to say she is having difficulty in connecting to your network. What should you do first?
    a. Have the user run TRACERT.
    b. Have the user try a standard PPP connection.

c. Reboot the RAS server.

d. RUN IPCONFIG /ALL on the RAS server to verify the ISP's DNS address.

11. Pat has just set up PPTP on his RAS server, which connects his local network to the Internet. With PPTP filtering enabled, how does the RAS server provide for network security?

a. The RAS server allows the administrator to set the RAS server port for Dial Out Only, Receive Calls Only, or Dial Out and Receive Calls connections.

b. The RAS server allows only PPTP packets to enter the local network.

c. The RAS server allows only PPTP packets to leave the local network.

d. The RAS server enforces callback security.

12. Over which of the following networks can the point-to-point tunneling protocol (PPTP) support a virtual private network (VPN) using a RAS server?

a. ISDN only

b. PSTN (Public Switched Telephone Network) only

c. ISDN and PSTN only

d. PSTN, ISDN, and X.25

13. Which one of the following protocols does point-to-point tunneling protocol (PPTP) support?

a. AppleTalk   c. IPX

b. IP   d. NetBEUI

14. You are deciding whether to use PPTP or L2TP to allow network connections to your LAN to be tunneled across the Internet. Which of the following are true of L2TP? (Choose all that apply)

a. L2TP provides data encryption.   c. L2TP supports tunnel authentication.

b. L2TP requires an IP-based network.   d. L2TP supports header compression.

15. You are concerned about hackers capturing network packets and reading their contents. Which feature of Windows 2000 allows you to secure your network traffic so that it cannot be easily read by anyone who may be capturing the packets?

a. IPSec   c. PAP

b. Dynamic DNS   d. BAP

16. Your remote access server has multiple modems as do your RAS clients. You would like to use the multilink capabilities of this configuration but are concerned that you may be paying for extra phone calls when multilink connections are unnecessary. Which feature of Windows 2000 allows you to use the multilink features dynamically, so that extra links are only made when the extra bandwidth is required?

a. BAP   c. L2TP

b. PPTP   d. EAP

17. Pat, a field technician, is required to travel from client site to client site. He does not always have the opportunity to convey his travel itinerary to the network support staff of his company. What is the best way to configure the RAS server at the home office of Pat's company to allow him remote access from various workstations at his client sites while still providing network callback security?

a. Reset the remote access permissions in the Routing and Remote Access console to the default setting.

b. Set the remote access permissions in the Routing and Remote Access console to Preset To and enter his callback number.

c. Set the remote access permissions in the Routing and Remote Access console to Set By Caller.

d. Set the remote access permissions in the Routing and Remote Access console to Preset To, enter a callback number, but invoke the MD5 RAS encryption protocol.

e. Install third-party software that captures Pat's caller ID and configures his Preset To callback number accordingly.

18. Which protocols can be used as dial-out protocols by a Windows 2000 Server using the remote access service?
    a. NetBEUI only
    b. TCP/IP, NetBEUI, or NWLink
    c. TCP/IP and NetBEUI, but not NWLink
    d. TCP/IP and NWLink, but not NetBEUI

19. Which dial-up protocol requires that devices on each end know the other's address?
    a. SLIP
    b. PPP
    c. TCP/IP
    d. POTS
    e. xDSL

20. Which dial-up protocol is capable of passing multiple LAN protocols across the link?
    a. SLIP
    b. PPP
    c. TCP/IP
    d. POTS
    e. xDSL

21. Which protocol provides an encrypted connection between devices using virtual private networking?
    a. PPTP
    b. PPP
    c. TCP/IP
    d. SLIP
    e. xDSL

22. Which dial-up protocol is considered self-configuring?
    a. SLIP
    b. PPP
    c. TCP/IP
    d. POTS
    e. xDSL

EXERCISES

EXERCISE 1: REMOTE ACCESS SERVER

Configure the Remote Access Server

Perform the following on your computer.

1. Log on as the Administrator for your domain.
2. In the Administrative Tools, start the Open Routing and Remote Access console.
3. In the left pane, right-click your server and select the Configure and Enable Routing and Remote Access option. Click on the Next button.
4. On the Common Configurations page, select the Remote Access Server option and select the Next button.
5. On the Remote Client Protocols page, TCP/IP is listed. Keep the default option of "Yes, all of the required protocols are on this list," and click the Next page.
6. On the Network Selection page, select your network connection and click on the Next button.
7. On the IP Address Assignment page, select the "From a specified range of addresses" option and click on the Next button.
8. On the Address Range Assignment page, click the New button.
9. In the Start IP address box, type 10.x.
10. Use 140.100.100.XX for the start address and 140.100.100.XX+1 for the end address, where *XX* is your computer number. Click on the OK button. Click on the Next button.
11. On the Managing Multiple Remote Access Servers page, keep the default setting of "No, I don't want to set up this server to use RADIUS now" and click the Next button. Click on the Finish button. Click OK to close the Routing and Remote Access message box.
12. Close the Routing and Remote Access console.

Granting Dial-in Permissions

Perform the following on your computer.

1. In the Administrative Tools, open the Active Directory Users and Computers console from the Administrative Tools menu.
2. In the console tree in the left pane, expand your domain.

3. Click on Users in the left pane and double-click Administrator in the right pane.
4. On the Dial-in tab, select the Allow Access option and click the OK button.
5. Close the Active Directory Users and Computers console.

Configure a VPN Connection

Perform the following on your partner's computer.

1. Right-click My Network Places and then click Properties.
2. In the Network and Dial-up Connections, double-click the Make New Connection icon. Click on the Next button.
3. On the Network Connection Type page, select the "Connect to a private network through the Internet" option and click the Next button.
4. On the Destination Address page, type in the address of your partner's computer and click the Next button.
5. On the Connection Availability page, click the "Only for myself" option. Click the Next button followed by the Finish button.
6. Click on the Cancel button.

Making a VPN Connection

Perform the following on your partner's computer.

1. Right-click My Network Places and select the Properties button.
2. Double-click the Virtual Private Connections icon.
3. Connect as the Administrator with the Connect Virtual Private Connection dialog box.
4. After a message appears indicating that you are connected, you will see an icon in the system tray. Click the OK button.
5. Close Network and Dial-up Connections.
6. At the Command prompt, execute IPCONFIG to verify the IP address for the connection.
7. Double-click the Connection icon in the system tray.
8. Click the Disconnect button.

EXERCISE 2: TERMINAL SERVICES

Installing Terminal Services

Perform the following on your server.

1. Open the Add/Remove Programs applet in the Control Panel.
2. Choose Add/Remove Windows Components.
3. From the list available, check the box next to Terminal Service. Click on the OK button.
4. In the Terminal Services Setup page, select Remote Administration Mode and click on the OK button.
5. Windows 2000 will then copy the files from an installation source. When the copying is done, click on the Finish button. When you are prompted, restart the computer.

Creating Terminal Services Client Disks

Perform the following on your server. You will need two floppy disks.

1. In the Administrative Tools, select the Terminal Services Client Creator.
2. Select the Terminal Services for 32-bit x86 Windows and click the OK button.
3. Insert the first floppy disk in drive and click on the OK button.
4. When it prompts you, insert the second floppy and click on the OK button.
5. Click the OK button when finished.
6. Click the Cancel button to close the Terminal Services Client Creator.

Installing Terminal Services Client Software

Perform the following on your partner's computer.

1. Insert the first Terminal Services client disks in drive A.
2. Execute the SETUP.EXE in the A drive. Click on the Continue button.
3. Enter your name and organization and click on the OK button. Click on the OK button a second time.
4. Click on the I Agree button to agree to the license agreement.
5. Click on the Large button to start the installation.
6. When it asks if you want all Terminal Services users of this computer to have all the same initial settings, click on the Yes button.
7. When prompted, insert the second disk and click on the OK button.
8. When the installation is complete, click on the OK button.

Note that if you have Terminal Services loaded on your partner's machine, the Terminal Service Client software can be executed from the \WINNT\SYSTEM32\CLIENTS\ TSCLIENT\DISKS\DISK1 folder.

Using Terminal Services to Connect to a Terminal Server

Perform the following on your partner's computer.

1. Click the Start button and select the Programs option, the Terminal Services Client option, and the Terminal Services Client option.
2. In Server, type the name or IP address of the terminal server to which you want to connect.
3. Select the Screen Area of 640x480.
4. Click the Connect button.
5. In the Terminal Services Client window, log in as the Administrator.
6. In the Terminal Services Client window, execute the Active Directory Users and Computers program located in the Administrative Tools. Note that if the taskbar is set to Autohide, move the mouse pointer to the bottom of the Terminal Services Client window to show the taskbar.
7. Close the Active Directory Users and Computers program.
8. Close the Terminal Services Client window. When it asks to disconnect, click on the OK button.

19

Introduction to Routers and Routing

**TOPICS COVERED
IN THIS CHAPTER**

INTRODUCTION

As mentioned in earlier chapters, the router is used to connect different LANs together. This chapter discusses routers and the routing protocols used to connect the various LANs. In addition, it discusses multitasking and gives an introduction to Cisco routers.

OBJECTIVES

1. Compare distant-vector routers with link-state routers.
2. Compare RIP and OSPF protocols.
3. Define an autonomous system and explain how it relates to EGP and IGP.
4. Update a Windows 2000–based routing table by means of static routes.
5. Manage and monitor border routing.
6. Manage and monitor internal routing.
7. Manage and monitor IP routing protocols.

19.1   ROUTER OVERVIEW

As local area networks become more and more popular and increasingly vital to the daily operation of an organization, the need to connect multiple LANs has become as crucial as it once was to link individual PCs into a workgroup. Although segmenting the network can solve all kinds of problems—including increasing available bandwidth on the local LANs, linking distant networks, and breaking a large network into smaller, more manageable subnets—you must have a mechanism for the different segments to communicate with each other.

A **router,** which works at the network ISO layer, is a device that connects two or more LANs. As multiple LANs are connected together, multiple routes are created to get from one LAN to another. The primary role of a router is to transmit similar types of data packets from one wide area communications link (such as a T1 line or fiber link) to another T1 line or fiber link. The second role of a router is to select the best path between the source and destination.

When you send a packet from one computer to another computer, the computer first determines if the packet is sent locally to another computer on the same LAN or if the packet is sent to a router so that it can be routed to the destination LAN. If the packet is meant to go to a computer on another LAN, it is sent to the router (or gateway). The router will then determine what is the best route to take and forward the packets to that route. The packet will then go to the next router and the entire process will repeat itself until it gets to the destination LAN. The destination router will then forward the packets to the destination computer.

Determining the best route requires complex routing algorithms, which consider a variety of factors including the number or fastest set of transmission media, the number of network segments, and the network segment that carries the least amount of traffic. Routers then share status and routing information to other routers so that they can provide better traffic management and bypass slow connections. In addition, routers provide additional functionality, such as the ability to filter messages and forward them to different places based on various criteria. Most routers are multiprotocol routers, because they can route data packets using many different protocols.

A **metric** is a standard of measurement, such as hop count, that is used by routing algorithms to determine the optimal path to a destination. A **hop** is the trip a data packet takes from one router to another router (or another intermediate point). On a large network, the number of hops that a packet has taken toward its destination is called the **hop count.** When a computer communicates with another computer via four routers, then the computer would have a hop count of 4. With no other factors considered, a metric of four would be assigned. If a router had to choose between a route with four metrics and a route with six metrics, it would choose the route with the least amount of metrics. If you want to force the router to choose the route with six metrics, then overwrite the metric for the route with four hops in the routing table to one with six hops.

To track the various routes in a network, the routers will create and maintain routing tables. The routers communicate with one another to maintain their routing tables through a routing update message. This message can consist of all or a portion of a routing table. By analyzing routing updates from all other routers, a router can build a detailed picture of network topology.

19.2
CHARACTERISTICS OF ROUTER PROTOCOLS

There are various types of routing algorithms, each one having a different impact on network and router resources. Routing algorithms use a variety of metrics that affect calculation of optimal routes.

Routing algorithms can be differentiated based on several key characteristics. First, the capability of the routing algorithm to optimally choose the best route is important. One routing algorithm may use the number of hops and the length of delays, but may weigh the length of delay more heavily in the calculation. To maintain consistency and predictability, routing protocols use strict metric calculations.

Routing algorithms are designed to be as simple as possible. In other words, the routing algorithm must offer efficiency with a minimum of software and utilization overhead. Aside from being efficient, the routing algorithm must be robust so that it can quickly change routes when a route goes down because of hardware failure or a route has a high load (high amount of traffic). During these types of situations, the routing algorithm must be stable.

Routing algorithms also must converge rapidly. **Convergence** is the process of agreement by all routers to which routes are the optimal routes. When a route goes down, routers distribute the new routes by sending routing update messages. The time that it takes for all routers to get new routes to which they all agree upon should be quick. If not, routing loops or network outages can occur. A routing loop occurs when a packet is forwarded back and forth between several routers without ever getting to its final destination.

A **count to infinity,** which typically happens when a network has slow convergence, is a loop that occurs when a link in a network goes down and routers on the network update their routing tables with incorrect hop counts. For example, a loop can occur if the link to router C goes down. Router B then advertises that the link is down and that it has no route to C. Because router A has a route to C with a metric of 2, it responds to router B and sends its link to C. Router B then updates its table to include a link with metric 3, and the routers continue to announce between the two routers and update their routing entries to C until they reach metric 16, a count to infinity.

To handle the count to infinity problem, networks use either split horizon or poison reverse. **Split horizon** is a route-advertising algorithm that prevents the advertising of routes in the same direction in which they were learned. **Poison reverse** is a process that, used with split horizon, improves RIP convergence over simple split horizon by advertising all network IDs. However, the network IDs learned in a given direction are advertised with a hop count of 16, indicating that the network is unavailable.

19.2.1   Static versus Dynamic Routes

Static routing algorithms are hardly algorithms at all, but are table mappings established by the network administrator prior to the beginning of routing. These mappings do not change unless the network administrator alters them. Algorithms that use static routes are simple to design and work well in environments where network traffic is relatively predictable and where network design is relatively simple.

Because static routing systems cannot react to network changes, they generally are considered unsuitable for today's large, changing networks. Most dominant routing algorithms in the 1990s are dynamic routing algorithms, which adjust to changing network circumstances by analyzing the incoming routing update messages. If the message indicates that a network change has occurred, the routing software recalculates routes and sends out new routing update messages. These messages flow through to permeate the network, stimulating routers to rerun their algorithms and change their routing tables accordingly.

Dynamic routing algorithms can be supplemented with static routes where appropriate. A router of last resort (a router to which all unroutable packets are sent), for example, can be designated to act as a repository for all unroutable packets, ensuring that all messages are at least handled in some way.

19.2.2   Single-Path versus Multipath Algorithms

Some sophisticated routing protocols support multiple paths to the same destination. Unlike single-path algorithms, these multipath algorithms permit traffic multiplexing over multiple lines where the traffic can be split among the different routes. In addition, it provides redundancy in case one of the routes fails. Networks connected with redundant routes can also be referred to as a spanning-tree algorithm.

19.2.3 Flat versus Hierarchical Algorithms

In a flat routing system, the routers are peers of all others. In a hierarchical routing system, some routers form what amounts to a routing backbone. Packets from nonbackbone routers travel to the backbone routers, where they are sent through the backbone until they reach the general area of the destination. At this point, they travel from the last backbone router through one or more nonbackbone routers to the final destination.

Routing systems often designate logical groups of nodes—called domains, autonomous systems, or areas. In hierarchical systems, some routers in a domain can communicate with routers in other domains, while others can communicate only with routers within their own domain. In expansive networks, additional hierarchical levels may exist, with routers at the highest hierarchical level forming the routing backbone.

The primary advantage of hierarchical routing is that it mimics the organization of most companies and therefore supports their traffic patterns well. Most network communication occurs within small company groups (domains). Because intradomain routers need to know only about other routers within their domain, their routing algorithms can be simplified, and, depending on the routing algorithm being used, routing update traffic can be reduced accordingly.

19.2.4 Distance-Vector versus Link-State Algorithms

Routers use **distance-vector-based routing protocols** to periodically advertise or broadcast the routes in their routing tables, but they send it only to their neighboring routers. Routing information exchanged between typical distance-vector-based routers is unsynchronized and unacknowledged. Distance-vector-based routing protocols are simple, easy to understand, and easy to configure. The disadvantages are as follows: (1) Multiple routes to a given network can reflect multiple entries in the routing table, thus increasing the routing table. (2) If you have a large routing table, network traffic increases as it periodically advertises the routing table to the other routers, even after the network has converged. (3) Distance vector protocol convergence of large internetworks can take several minutes.

Link-state algorithms are also known as shortest path first (SPF) algorithms. Instead of using broadcast, link-state routers send updates directly (or by using multicast traffic) to all routers within the network. Each router, however, sends only the portion of the routing table that describes the state of its own links. In essence, link-state algorithms send small updates everywhere. Because they converge more quickly, link-state algorithms are somewhat less prone to routing loops than distance-vector algorithms. In addition, link-state algorithms do not exchange any routing information when the internetwork has converged. They have small routing tables since they store a single optimal route for each network ID. Conversely, link-state algorithms require more CPU power and memory than distance-vector algorithms. Link-state algorithms, therefore, can be more expensive to implement and support and are considered harder to understand.

19.2.5 Metrics

The various routing protocols use different metrics, including path length, hop counts, routing delay, bandwidth, load, reliability, and cost. Path length is the most common routing metric. Some routing protocols allow network administrators to assign arbitrary costs to each network link. In this case, path length is the sum of the costs associated with each link traversed. Other routing protocols define hop counts, as described earlier in this chapter.

Routing delay refers to the length of time required to move a packet from source to destination through the internetwork. Delay depends on many factors, including the bandwidth of intermediate network links, the port queues at each router along the way, network congestion on all intermediate network links, and the physical distance to be traveled. Because delay is a conglomeration of several important variables, it is a common and useful metric.

Bandwidth refers to the available traffic capacity of a link. All other things being equal, a 10-Mbps Ethernet link would be preferable to a 64-kbps leased line. Although bandwidth is a rating of the maximum attainable throughput on a link, routes through links with greater bandwidth do not necessarily provide better routes than routes through slower links. If, for example, a faster link is busier, the actual time required to send a packet to the destination could be greater. Load refers to the degree to which a network resource, such as a router, is busy. Load can be calculated in a variety of ways, including CPU utilization and packets processed per second. Monitoring these parameters on a continual basis can also be resource intensive.

Reliability, in the context of routing algorithms, refers to the dependability (usually described in terms of the bit-error rate) of each network link: Some network links may go down more often than others. After a network fails, certain network links might be repaired more easily or more quickly than other links. Any reliability factors can be taken into account in the assignment of the reliability ratings, which are arbitrary numeric values usually assigned to network links by network administrators.

Communication cost is another important metric, especially because some companies may not care about performance as much as they care about operating expenditures. Even though line delay may be longer, they will send packets over their own lines rather than through the public lines that cost money for usage time.

19.3 INTERNET ROUTING PROTOCOL

As discussed earlier in the book, the Internet protocol (IP) is a connectionless, unreliable protocol that is responsible for addressing and routing between hosts. The term *connectionless* means that a session is not established before exchanging data. Each packet that travels through the Internet is treated as an independent unit of data, which is not affected by other data packets. The term *unreliable* means that the delivery is not guaranteed. IP always makes a best-effort attempt to deliver a packet. An IP packet might be lost, delivered out of sequence, duplicated, or delayed. IP does not attempt to recover from these types of errors. Instead, the acknowledgment of packets delivered and the recovery of lost packets is the responsibility of a higher-layer protocol, such as TCP.

19.3.1 IP Packet

The IP specifies the format of the packets (also called datagrams). An IP packet, or IP datagram, consists of an IP header and an IP payload. The IP header consists of the following packets (see Figure 19.1):

- The IP version indicates the version of IP currently being used.
- The IP header length (IHL) indicates the datagram header length in 32-bit words.
- The type of services specifies how an upper-layer protocol would like a current datagram, and assigns to datagrams various levels of importance.
- The total length specifies the length, in bytes, of the entire IP packet including the data and header.

FIGURE 19.1   IP packet

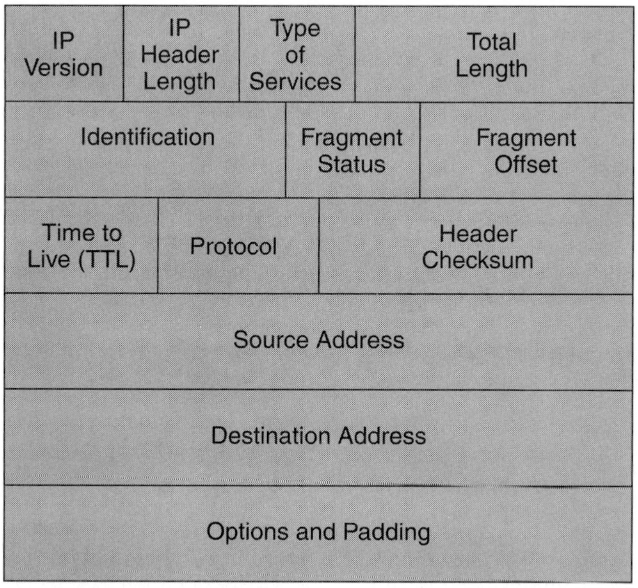

591

- The identification section contains an integer that identifies the current datagram. This field is used to help piece together datagram fragments.
- In fragment status, the flags consist of a 3-bit field of which the two least-significant bits control fragmentation. The low-order bit specifies whether the packet can be fragmented. The middle bit specifies whether the packet is the last fragment in a series of fragmented packets. The third bit is not used.
- Fragment offset indicates the position of the fragment's data relative to the beginning of the data in the original datagram. This allows the destination IP process to properly reconstruct the original datagram.
- The time to live (TTL) designates the number of network segments on which the datagram is allowed to travel before being discarded by a router. The TTL is set by the sending host and is used to prevent packets from endlessly circulating on an IP internetwork. When forwarding an IP packet, routers are required to decrease the TTL by at least one.
- The protocol section indicates which upper-layer protocol receives incoming packets after IP processing is complete.
- The header checksum helps ensure IP header integrity.
- The source address is the IP address of the original source of the IP datagram.
- The destination address is the IP address of the final destination of the IP datagram.
- The options section allows IP to support various options, such as security, whereas padding is to help create the datagram in the correct format and maintain a minimum size.

19.3.2 IP Addresses and Routing

TCP/IP hosts use a routing table to maintain knowledge about other IP networks and hosts. As discussed in Chapter 5, networks and hosts are identified with an IP address and a corresponding subnet mask to specify which bits of the address comprise the network address and which bits comprise the host address.

When a computer prepares to send an IP datagram, it inserts its own source IP address and the destination IP address of the recipient into the IP header. The computer then examines the destination IP address, compares it with a locally maintained IP routing table, and takes appropriate action based on what it finds. The computer does one of three things:

- It passes the datagram up to a protocol layer above IP on the local host.
- It forwards the datagram through one of its attached network interfaces.
- It discards the datagram.

IP searches the routing table for the route that is the closest match to the destination IP address (either a route that matches the IP address of the destination host or a route that matches the IP address of the destination network). If a matching route is not found, IP discards the datagram.

To determine if a packet is to be sent to a host on the local network or to determine if the packet needs to be a remote network via the router, the TCP/IP protocol will complete a couple of calculations. When TCP/IP is initialized on a host, the host's IP address is ANDed (logical bit-wise AND) with its subnet mask at a bit-by-bit level. Before a packet is sent, the destination IP address is ANDed with the same subnet mask. If both results match, IP knows that the packet belongs to a host on the local network. If the results do not match, then the packet is sent to the IP address of an IP router so that it can be forwarded to the host.

When ANDing an IP address and subnet mask, each bit in the IP address is compared with the IP address of the corresponding bit in the subnet mask. If both bits are 1s, the result bit is 1. If there is any other combination, the resulting bit is 0. Be aware that: by ANDing the address and subnet mask, you are essentially isolating the network addresses.

EXAMPLE 1   You have a host 145.17.202.56 with a subnet mask of 255.255.224.0 with data that must be delivered to 145.17.198.75 with a subnet mask of 255.255.224.0. Are the two hosts located on one local network or are they on two remote networks?

To solve the problem, first convert the addresses and subnet mask to binary.

145.17.202.56	10010001.00010001.11001010.00111000
145.17.198.75	10010001.00010001.11000110.01001011
255.255.224.0	11111111.11111111.11100000.00000000

592

TABLE 19.1 ICMP messages

ICMP Message	Description
Echo request	Determines whether an IP node (a host or a router) is available on the network.
Echo reply	Replies to an ICMP echo request.
Destination unreachable	Informs the host that a datagram cannot be delivered.
Source quench	Informs the host to lower the rate at which it sends datagrams because of congestion.
Redirect	Informs the host of a preferred route.
Time exceeded	Indicates that the time-to-live (TTL) section of an IP datagram has expired.

Next AND the source address and the subnet mask.

145.17.202.56	10010001.00010001.11001010.00111000
255.255.224.0	<u>11111111.11111111.11100000.00000000</u>
	10010001.00010001.11000000.00000000

Then AND the destination address and the subnet mask.

145.17.198.75	10010001.00010001.11000110.01001011
255.255.224.0	<u>11111111.11111111.11100000.00000000</u>
	10010001.00010001.11000000.00000000

If you compare the two results, you will find that they are the same. Therefore, the packet will be sent directly to the host and not the router. If they were different, it would then send the packet to the router.

19.3.3 Internet Control-Message Protocol

As mentioned, IP does not provide reliable communication, or either end-to-end or hop-by-hop acknowledgments. There is no error control for data, other than a header checksum, or error or flow control. There are no retransmissions, as these are left to the higher protocols such as TCP. To report errors and other information regarding IP packet processing, the **Internet control-message protocol (ICMP)** sends notices back to the source. ICMP messages are usually sent automatically in one of the following situations:

- An IP datagram cannot reach its destination.
- An IP router (gateway) cannot forward datagrams at the current rate of transmission.
- An IP router redirects the sending host to use a better route to the destination.

Review: ICMP documentation in RFC 792. See Table 19.1.

When you use the PING command, an ICMP echo request message is sent and the receipt of ICMP echo replay messages is recorded. This allows you to detect network or host communication failures and troubleshoot common TCP/IP connectivity problems.

By using its messages, the ICMP can be used by local routers to provide a way for hosts to sense routers that are down and to perform router solicitation and advertisement. Router solicitations are sent by hosts to discover routers on their networks. Router advertisements are sent by routers in response to a router solicitation and periodically to notify hosts on the network that the router is still available. ICMP Router Discovery is documented in RFC 1256, "ICMP Router Discovery Messages."

19.4 INTERIOR GATEWAY PROTOCOLS

As IP provides addressing and connectionless services for packet forwarding, it relies on other protocols for address resolution, dynamic route discovery, and prioritization. Since IP networks can be quite large and complex internetworks, TCP/IP allows the IP internetwork to be divided into logical groups called autonomous systems. The routing protocols that manage routing information within an autonomous system are called **interior gateway protocols (IGPs).** Examples of an IGP are RIP, OSPF, and IGRP (to be discussed in the chapter). Autonomous systems are interconnected using an exterior gateway protocol (EGP), such as **border gateway protocol (BGP).** RIP and OSPF protocols are standard TCP/IP routing protocols, which can run as a stand-alone autonomous system or within an autonomous system that belongs to a larger network. See Figure 19.2.

19.4.1   Router Information Protocol

The **routing information protocol (RIP)** is designed for exchanging routing information within a small to medium-size network. The biggest advantage of RIP is that it is extremely simple to configure and deploy.

RIP uses a single routing metric of hop counts to measure the distance between the source and a destination network. Each hop in a path from source to destination is assigned a hop-count value, which is typically 1. When a router receives a routing update that contains a new or changed destination-network entry, the router adds one to the metric value indicated in the update and enters the network in the routing table. The IP address of the sender is used as the next hop.

RIP prevents routing loops from continuing indefinitely by implementing a limit on the number of hops allowed in a path from the source to a destination. The maximum number of hops in a path is 15 hops. If a router receives a routing update that contains a new or changed entry, and if increasing the metric value by 1 causes the metric to be infinity (in this case, 16), the network destination is considered unreachable. Thus, RIP is unable to scale to large or very large internetworks.

Initially, the routing table for each router includes only the networks that are physically connected to it. A RIP router periodically sends announcements that contain its routing table entries so that the other routers can update their routing tables. RIP version 1 uses IP

FIGURE 19.2   Connecting autonomous systems

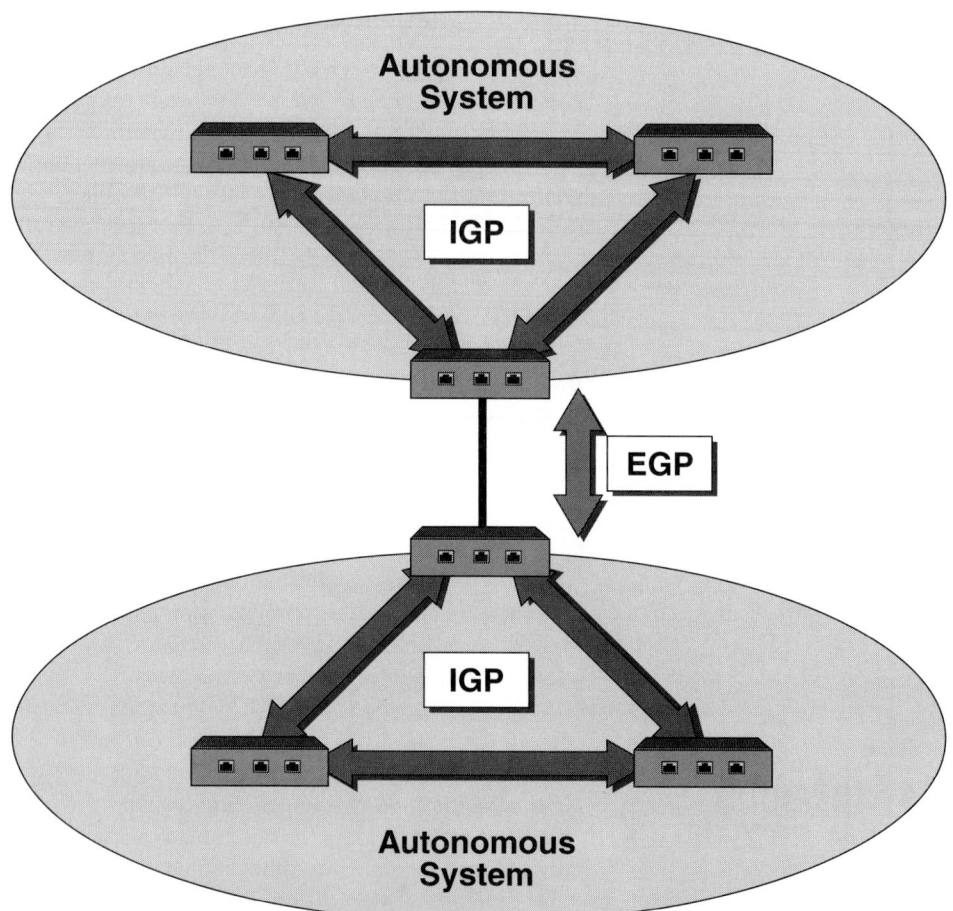

broadcast packets for its announcements. RIP version 2 uses multicast or broadcast packets for its announcements. All RIP messages are sent over UDP port 520.

RIP routers also can communicate routing information through triggered updates, which are set off when the network topology changes. Different from the scheduled announcements, the triggered updates are sent immediately rather than held until the next periodic announcement. For example, when a router detects a link or router failure, it updates its own routing table and sends the updated routes. Each router that receives the triggered update modifies its own routing table and propagates the change to the other routers.

Since the RIP is a distance-vector protocol, as internetworks grow larger in size, the periodic announcements by each RIP router can cause excessive traffic. Another disadvantage of RIP is its high convergence time. When the network topology changes, it may take several minutes before the RIP routers reconfigure themselves to the new network topology. While the network reconfigures itself, routing loops may form that result in lost or undeliverable data. To help prevent routing loops, RIP implements split horizon.

To overcome some of the RIP shortcomings, **RIP version 2 (RIP II)** was introduced to provide the following features:

- You can use a password for authentication by specifying a key that is used to authenticate routing information to the router.
- RIP II includes the subnet mask in the routing information and supports variable-length subnets. Variable-length subnet masks can be associated with each destination, allowing an increase in the number of hosts or subnets that are possible on your network.
- The routing table can contain information about the IP address of the router that should be used to reach each destination. This helps prevent packets from being forwarded through extra routers on the system.
- Multicast packets only speak to RIP II routers and are used to reduce the load on hosts that are not listening to RIP II packets. The IP multicast address for RIP II packets is 224.0.0.9.

RIP II is supported on most routers and end nodes. Make sure that the mode you configure is compatible with all implementations of RIP on your network.

19.4.2 Autonomous Systems Topology

For small or medium networks, distributing data throughout the network and maintaining a route table at each router is not a problem. When the network grows to a size that includes hundreds of routers, the routing table can be quite large (several megabytes) and require significant time to recalculate routes as router interfaces go up or down.

Some protocols, such as OSPF, allow areas (groupings of contiguous networks) to be gathered into an **autonomous system (AS).** See Figure 19.3. Areas that comprise the system usually correspond to an administrative domain such as a department, a building, or a geographic site. An AS can be a single network or a group of networks, owned and administered by a common network administrator or group of administrators.

The **backbone** (given an address of 0.0.0.0) of the AS is a high-bandwidth logical area where all areas are connected. The routers that attach an area to the backbone are called **area border routers (ABRs).** Because the ABR is supposed to be a high-bandwidth pathway connecting the areas and know the autonomous system's overall topology, it requires additional memory and processing power compared with regular routers located within an area. Therefore, do not use a low-end router as an ABR.

Areas that are not backbone OSPF areas can be classified as either a stub area or a transit area. A **stub area** has only one ABR. All routes to the destination outside the area must pass through the single router. Since the ABR is the only router that routes data in and out of the area, the ABR advertises itself as the default router to all external destinations, which reduces traffic and the size of the routing database. A **transit area,** on the other hand, contains more than one ABR.

OSPF routers within an area are not required to maintain a database of routers and networks located inside of the other areas. Since the topology of an area is invisible from outside of the area and routers internal to a given area know nothing of the detailed topology outside of the area, the isolation reduces the network traffic since not all routing information has to be sent to the routers in other areas.

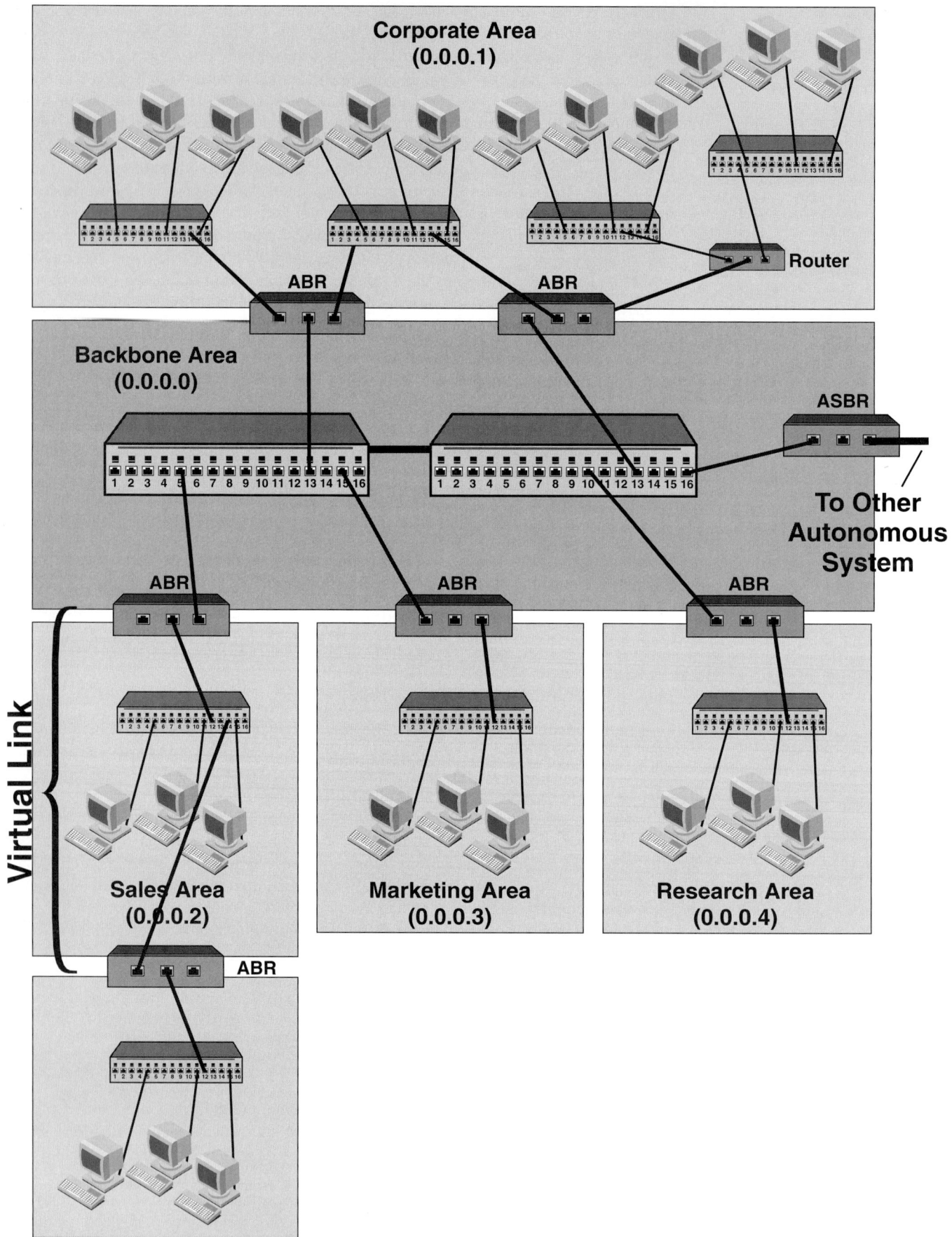

FIGURE 19.3 Autonomous system

Routing in the AS takes place on two levels, depending on whether the source and destination of a packet reside in the same area (intra-area routing) or different areas (inter-area routing). In intra-area routing, the packet is routed solely on information obtained within the area; no routing information obtained from outside the area needs to be used. This protects intra-area routing from the injection of bad routing information and helps simplify the routing tables.

When routing a packet between two areas, the packet will travel on the backbone. The path on which the packet will travel can be subdivided into three contiguous pieces: an intra-area path from the source to an area border router, a backbone path between the source and destination areas, and then another intra-area path to the destination.

A **virtual link** is a logical link between a backbone area border router, an unspecified number of routers, and a second border router that is not connected to the backbone. Virtual links are not recommended, because they can cause routing and other problems, and can be difficult to configure. Instead, always make an effort to connect new areas in your AS directly to the backbone by planning ahead before your AS is implemented.

To connect autonomous systems to other autonomous systems, use an **autonomous system boundary router (ASBR).** An ASBR exchanges routing information with routers belonging to other autonomous systems. The ASBR accesses its routing information from the Internet using EGP and accesses routing information about the corporate network (autonomous system) from OSPF. For further information regarding OSPF routers, see Table 19.2. EGP will be explained later in this chapter.

19.4.3 Open Shortest Path First Protocol

The **open shortest path first (OSPF)** protocol is a link-state routing protocol used in medium and large networks and calculates routing table entries by constructing a shortest-path tree. OSPF is considered superior to RIP for the following reasons:

- It is a more efficient protocol than RIP and does not have the restrictive 16-hop-count problem, which causes data to be dropped after the sixteenth hop. An OSPF network can have an accumulated path cost of 65,535, which enables you to construct huge networks (within the maximum TTL value of 255) and assign a wide range of costs. An OSPF network is best suited for a large infrastructure with more than 50 networks.
- OSPF networks can detect changes to the network and calculate new routes quickly (rapid convergence). The period of convergence is brief and involves minimal overhead. The count-to-infinity problem does not occur in OSPF internetworks.

TABLE 19.2 OSPF routers

Router Type	Description
Internal router	Internal routers send packets only within a single area. The internal router can also be a backbone router if is has no interfaces to other areas.
Area border router (ABR)	Area border routers have interfaces in multiple areas, and route packets between these areas. Area border routers condense topological information before passing it to the backbone, which reduces the amount of routing information that is passed across the backbone.
Backbone router	A backbone router has an interface on the backbone area. This includes all routers that interface to more than one area (i.e., area border routers). However, backbone routers do not have to be area border routers. Routers with all interfaces connected to the backbone are considered to be internal routers.
AS boundary router (ASBR)	AS routers exchange routing information with other autonomous systems. A router boundary exchanges routing information with routers belonging to other autonomous systems. Such a router has AS external routes that are advertised throughout the autonomous system. The path to each AS border router is known by every router in the AS. This classification is completely independent of the previous classifications: AS border routers may be internal or area border routers, and perhaps participate in the backbone.

- Less network traffic is generated by OSPF than by RIP, which requires each router to broadcast its entire database every 30 seconds. OSPF routers only broadcast link-state information when it changes.
- Link-state advertisements (LSAs) include subnet mask information about networks. Assign a different subnet mask for each segment of the network (variable-length subnetting) to increase the number of subnets and hosts that are possible for a single network address.

The shortest path first (SPF) routing algorithm is the basis for OSPF operations. When an SPF router is started, it initializes its routing-protocol data structures and then waits for indications from lower-layer protocols that its interfaces are functional.

After a router is assured that its interfaces are functioning, it uses the OSPF hello protocol to acquire neighbors, which are routers with interfaces to a common network. The router sends hello packets to its neighbors and receives their hello packets. In addition to helping acquire neighbors, hello packets also act as keep-alives to let routers know that other routers are still functional.

Adjacency is a relationship formed between selected neighboring routers for the purpose of exchanging routing information. When the link-state databases of two neighboring routers are synchronized, the routers are said to be adjacent. Not every pair of neighboring routers becomes adjacent. From the topological database generated from these hello packets, each router calculates a shortest-path tree, with itself as root. The shortest-path tree, in turn, yields a routing table. To maintain the table, the established adjacencies are compared with the hello messages sent out so that the router can quickly detect failed routers and the network's topology can be altered appropriately.

19.4.4   Interior Gateway Routing Protocol

The **interior gateway routing protocol (IGRP)** is an extension of the open RIP standard, which was developed in the mid-1980s by Cisco Systems, Inc. Cisco's principal goal in creating IGRP was to provide a robust protocol for routing within an AS having arbitrarily complex topology and consisting of media with diverse bandwidth and delay characteristics. Cisco developed enhanced IGRP in the early 1990s to improve the operating efficiency of IGRP.

IGRP is a distance-vector protocol that allows gateways to build up their routing table by exchanging information with other gateways. The metric used by IGRP includes the topological delay time, the bandwidth of the narrowest bandwidth segment of the path, the channel load of the path, and the reliability of the path. To get the most out of the network, it will split traffic between two or more paths that are proportional to its bandwidth. Therefore, if you have a 10-Mbps pathway and a 20-Mbps pathway, the 20-Mbps path will get roughly twice as much traffic as the 10-Mbps pathway.

Periodically, each gateway broadcasts its entire routing table (with some censoring because of the split-horizon rule) to all adjacent gateways. When a gateway gets this broadcast from another gateway, it compares the table with its existing table. Any new destinations and paths are added to the gateway's routing table. Paths in the broadcast are compared with existing paths. If a new path is better, it may replace the existing one. Information in the broadcast is also used to update channel occupancy and other information about existing paths.

In 1994, Cisco augmented IGRP with a product called **enhanced interior gateway routing protocol (EIGRP).** EIGRP combines the advantages of link-state protocols with those of distance-vector protocols. The key features include the diffusing update algorithm (DUAL) finite-state machine, variable-length subnet masks, partial routing table updates rather than full updates, and bounded updates. DUAL allows routers to synchronize route changes and does not involve routers unaffected by the change. Different from IGRP is that DUAL stores all of the neighbors' routing tables so that they can more intelligently recalculate alternate paths in order to speed convergence. Bounded updates is a method whereby routing update messages are sent only to those routers affected by the topology change, which saves overhead and helps speed convergence.

19.5   GATEWAY PROTOCOLS

As mentioned earlier, autonomous systems are interconnected using an exterior gateway protocol. Two common exterior gateway protocols are the border gateway protocol (BGP) and the exterior gateway protocol (EGP). BGP is an enhancement of EGP and is one of several new interdomain routing protocols designed to give administrators more control. These newer protocols also scale better in large internetworks. Both EGP and BGP are discussed further in this section.

19.5.1   Exterior Gateway Protocols

Most companies and organizations group all their routers into one AS, but some large companies will use more than one AS. The local routing information of an autonomous area is gathered using an internal gateway protocol such as RIP or OSPF. In these autonomous areas, one or more routers are chosen to use an **exterior gateway protocol (EGP)** to talk to other autonomous areas by providing a way for two neighboring routers that are located at the edges of their respective autonomous area to exchange routing information. EGP routers only forward routing table information to routers on the edge of their own autonomous area.

The first step that EGP takes to establish communication between exterior routers is to perform neighbor acquisition. During neighbor acquisition, one exterior router makes a request to another exterior router to agree that they share router table information. After communications have been established, the router then continually tests if the EGP neighbors are responding. Periodically, routing table information is exchanged at intervals between 120 and 480 seconds between EGP neighbors using routing update messages. The neighbor responds by sending its complete routing table.

EGP restricts exterior gateways by allowing them to advertise only those destination networks reachable entirely within that gateway's autonomous system. Thus, an exterior gateway using EGP passes along information to its EGP neighbors, but does not advertise itself to routers outside the autonomous system.

EGP has one major limitation: The distance indicated for a particular destination does not specify the cost to the destination. EGP only reports whether a destination is reachable. Because of this limitation, EGP can only be used in a tree-type network. All routing protocol domains must connect to the same central network; thus, EGP cannot support a looped topology. EGP can advertise only one route to a given network. There can be no load sharing for traffic between any given pair of machines, and packets can take nonoptimal paths when certain traffic conditions cause congestion in the selected route. As a result, it is difficult for EGP to switch to an alternate route if the primary route fails. The newest version of EGP is EGP-2.

19.5.2   Border Gateway Protocol

Because of its limitations, EGP has been made obsolete by the **border gateway protocol (BGP).** Border gateway protocol version 4 (BGP-4), documented in RFC 1771, is the current exterior routing protocol used on the Internet. Other BGP-related documents are RFC 1772 (BGP Application), RFC 1773 (BGP Experience), RFC 1774 (BGP Protocol Analysis), and RFC 1657 (BGP MIB). BGP uses TCP as its transport protocol, on port 179.

BGP is a distance-vector protocol, but unlike traditional distance-vector protocols such as RIP in which there is a single metric, BGP determines a preference order by applying a function that maps each path to a preference value and selects the path with the highest value. Where there are multiple viable paths to a destination, BGP maintains all of them but only advertises the one with the highest preference value. This approach allows a quick change to an alternate path, should the primary path fail. BGP-4 supports CIDR, IP prefixes, and path aggregation.

19.6   IP MULTICAST

IP multicast allows a host to communicate with several hosts simultaneously by transmitting information via one data stream. Multicast can greatly reduce the network traffic that bandwidth-hungry applications such as videoconferencing, software distribution, and Webcast create. Network vendors widely support IP multicast in their routers. Internet service providers (ISPs) now offer multicast service in their backbones.

The purpose of IP multicast is to deliver a data stream from a source to a group of receivers. Receivers interested in a particular multicast application join the application's multicast group. Receivers can join and leave the group at will, and the group disappears when no receivers remain. Group members listen to and receive data that the source delivers to the group's IP multicast address. Individual class D IP addresses in the range from 224.0.0.0 to 239.255.255.255 represent all multicast groups. The multicast's source does not need to join its group, nor does the source know who and where its receivers are. The source simply transmits multicast streams to the IP multicast address of its multicast group, then lets the network handle multicast data delivery. An IP multicast-enabled network can efficiently forward and route multicast data to receivers.

19.6.1 Multicast Scoping

Multicast scoping determines how far a multicast stream can travel from its source. Limiting the range of a multicast can prevent business data from reaching outside a network, thereby providing security. In multicast group management, multicast-enabled routers track multicast group membership through subnets to which the routers directly attach. The multicast-enabled routers forward multicast data only to the subnets that have group members, thereby saving network bandwidth. Multicast distribution trees define data delivery from a source to a multicast group, then build an optimized distribution tree that contains a set of routers and links to let group members receive data from the source.

Traditionally, IP multicast uses a time-to-live (TTL) parameter in an IP multicast application and multicast routers to control the multicast distribution. When you define the TTL value in an IP multicast application, contents do not transmit beyond the TTL value. For example, if you set Site Server's Active Channel Multicaster TTL value to 16, you ensure that Site Server's Web contents do not multicast beyond 16 router hops. Each multicast packet carries a TTL value in its IP header. Just as in unicast, every time a multicast router forwards a multicast packet, the router decreases the packet's TTL by 1. By default, a router will not forward packets with a value of TTL = 1. You can modify the default TTL threshold to another value on each interface in a multicast router. For example, if you set the TTL threshold to 10 on a router interface, only packets with a TTL value greater than 10 can pass that router interface.

TTL-based multicast scoping has two shortcomings. First, defining a proper TTL value in a multicast application can be difficult. If the value is too large, your multicast data may leave your network. If the TTL value is too small, your multicast data may fail to reach interested receivers beyond the multicast scope. Second, if someone sets custom TTL thresholds in certain router interfaces, your multicast range can be unpredictable. For example, you would not be able to divide your network into multicast regions to limit multicast applications to those regions, because if one router has an interface TTL threshold that is lower than a packet's TTL value, the router will forward the packet.

To overcome these limitations, the Internet Engineering Task Force (IETF) proposed Administratively Scoped IP Multicast as an Internet standard in its Request for Comments (RFC) 2365 in July 1998. Administrative scoping lets you scope a multicast to a certain network boundary (e.g., within your organization) by using an administratively scoped address. IETF has designated IP multicast addresses between 239.0.0.0 and 239.255.255.255 as administratively scoped addresses for local use in intranets. You can configure routers that support administratively scoped addressing on the border of your network to confine your private multicast region. You also can define multiple isolated multicast regions in your network so that sensitive multicast data will travel only within a designated area.

A multicast network forwards multicast data only to network subnets that have receivers in the corresponding multicast group in a scoped multicast region. To forward multicast data to receivers in a scoped multicast region, routers need information about group membership on their local subnets. One router on each subnet periodically multicasts membership query messages to all computers on the local subnet. Computers on the local subnet who are group members respond to the router's query message with a membership report about the group to which they belong. The router keeps this membership information in its group database. The local subnet computers also multicast membership reports to their own groups. When other group members receive a member computer's report, the group members postpone their membership reports and wait for a variable period of time. This waiting period reduces membership report traffic and router processing time. As long as a router knows one group member on the local subnet, the router forwards multicast data to that subnet, and other group members will receive the multicast data. When a new member joins a group, the member does not have to wait for the next membership query from the router. Instead, the new member immediately sends a membership report as if in response to a membership query. When the router receives this report, it immediately forwards multicast data to the subnet on which the new member resides, if the new member is the first member of that subnet's multicast group.

19.6.2 Internet Group-Membership Protocol

Routers and computers use the **Internet group-management protocol (IGMP)** to exchange membership information. IGMP is an integral part of IP. The two standard versions of IGMP are IGMPv1 (RFC 1112) and IGMPv2 (RFC 2236). IETF released IGMPv2 as an enhanced version of IGMPv1 in November 1997. Today, many routers and OSs support IGMPv2 including Windows 2000, Windows NT with Service Pack 4, Windows 98, and Windows 95 with WinSock 2.

IGMPv2's biggest enhancement is a group notification feature. In IGMPv1, a receiver that leaves a multicast group does not automatically notify the router. Rather, the router assumes no group member is on the local subnet if the router does not receive a membership report after several queries and waiting intervals. Several minutes or more can then pass before the router stops forwarding data to that subnet. In IGMPv2, receivers leaving a group directly inform the router. The router then queries the subnet to see whether any other group members remain. If the router does not receive a response, it assumes that no other group members exist on that subnet and stops multicast forwarding to that subnet.

IETF is working on IGMPv3 to further improve IGMPv2. IGMPv3 will include several new features. One such feature will let a computer specify from which sources in a specific group the computer will receive data.

19.6.3 Multicast Routing Protocols

Five multicast routing protocols currently exist, and you can classify each into either the source-based tree (dense mode) or shared-tree (sparse mode) protocol. The three dense-mode protocols are the distance-vector multicast routing protocol (DVMRP), multicast open shortest path first (MOSPF), and protocol independent multicast–dense mode (PIM-DM).

19.7 ROUTING IN WINDOWS 2000

For two hosts to exchange IP datagrams, they must both have a route to each other, or they must use a default gateway that knows a route between the two. Normally, routers exchange information using a protocol such as RIP or OSPF. RIP listening service is available for Microsoft Windows 2000 Professional, and full routing protocols are supported by Windows 2000 Server in the routing and remote access service.

19.7.1 ROUTE Command

The ROUTE command is used to view and modify the network routing tables of an IP network. The ROUTE PRINT command will display a list of current routes that the host knows. See Figure 19.4.

Routes added to a routing table are not made persistent unless the –p switch is specified. Nonpersistent routes last only until the computer is restarted or until the interface is deactivated. The interface can be deactivated when the plug-and-play interface is unplugged (such as for laptops and hot swap PCs), when the wire is removed from the media card (if the adapter supports media fault sensing), or when the interface is manually disconnected from the adapter in the Network and Dial-up Connections folder.

The usage for the ROUTE command is

ROUTE [–f] [–p] [command [destination]] [MASK netmask] [gateway] [METRIC metric]

The commands usable in the syntax above are Print, Add, Delete, and Change.

–f Clears the routing tables of all gateway entries. If this is used in conjunction with one of the commands, the tables are cleared prior to running the command.

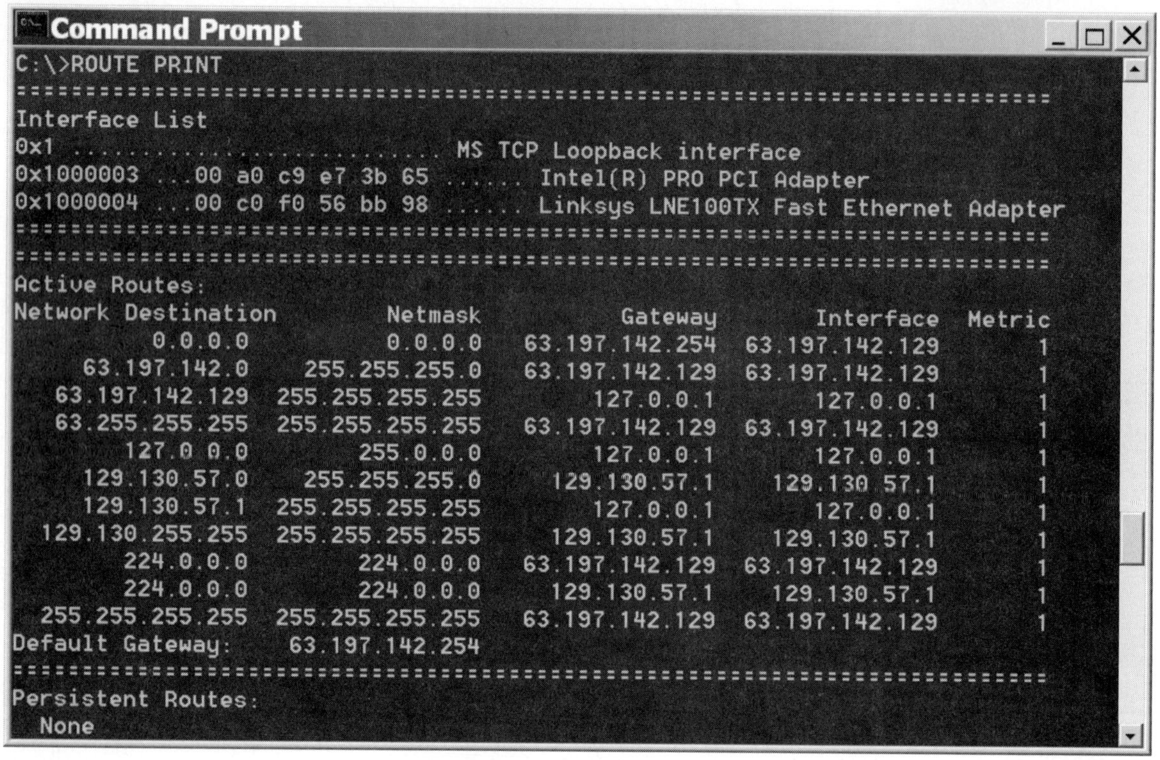

```
Command Prompt                                           _ □ X
C:\>ROUTE PRINT
===================================================================
Interface List
0x1 ........................... MS TCP Loopback interface
0x1000003 ...00 a0 c9 e7 3b 65 ...... Intel(R) PRO PCI Adapter
0x1000004 ...00 c0 f0 56 bb 98 ...... Linksys LNE100TX Fast Ethernet Adapter
===================================================================
Active Routes:
Network Destination        Netmask          Gateway        Interface  Metric
          0.0.0.0          0.0.0.0    63.197.142.254  63.197.142.129      1
     63.197.142.0    255.255.255.0    63.197.142.129  63.197.142.129      1
   63.197.142.129  255.255.255.255        127.0.0.1       127.0.0.1      1
  63.255.255.255  255.255.255.255    63.197.142.129  63.197.142.129      1
        127.0.0.0        255.0.0.0        127.0.0.1       127.0.0.1      1
    129.130.57.0    255.255.255.0     129.130.57.1    129.130.57.1      1
    129.130.57.1  255.255.255.255        127.0.0.1       127.0.0.1      1
  129.130.255.255  255.255.255.255     129.130.57.1    129.130.57.1      1
        224.0.0.0        224.0.0.0    63.197.142.129  63.197.142.129      1
        224.0.0.0        224.0.0.0     129.130.57.1    129.130.57.1      1
  255.255.255.255  255.255.255.255    63.197.142.129  63.197.142.129      1
Default Gateway:      63.197.142.254
===================================================================
Persistent Routes:
  None
```

FIGURE 19.4   Using the ROUTE command to display the current routing tables on a Windows 2000 multihomed computer

–p   When used with the Add command, this makes a route persistent or permanent, even if you reboot the system. By default, routes are not preserved when the system is restarted. When used with the Print command, it displays the list of registered persistent routes. It is ignored for all other commands, which always affects the appropriate persistent routes.

Commands	
PRINT	View or display a route
ADD	Adds a route
DELETE	Deletes a route
CHANGE	Modifies an existing route
Destination	Specifies the computer to send a command.
Mask subnetmask	Specifies a subnet mask to be associated with this route entry. If a netmask value is not specified, it defaults to 255.255.255.255.
Gateway	Specifies gateway or router.
METRIC metric	Assigns an integer cost metric (ranging from 1 to 9999) to be used in calculating the fastest, most reliable, and/or least expensive routes.

All symbolic names used for destination or gateway are referenced in both the network database file called Networks, and the computer name database file called Hosts. If the command is Print or Delete, wildcards may be used for the destination and gateway; or the gateway argument may be omitted.

For example, to create a static route, type in the following:

route ADD 132.133.200.0 MASK 255.255.255.0 63.197.142.1 METRIC 2

After this command is executed, any packet that is sent to the 132.133.200.0 network or host with an IP address ranging between 132.133.200.1 and 132.133.200.254 will be forwarded to the router with a local host address of 63.197.142.1. If there are multiple entries that specify these destination addresses, this route has a metric of 2 hops.

19.7.2 Routing and Remote Access Service

The routing and remote access service for Windows 2000 Server continues the evolution of multiprotocol routing and remote access services for the Microsoft Windows platform. New features of the routing and remote access service for Windows 2000 include the following:

- Internet group management protocol (IGMP) and support for multicast boundaries
- Network address translation with addressing and name resolution components that simplify the connection of a small office/home office (SOHO) network to the Internet
- Integrated AppleTalk routing
- Layer 2 tunneling protocol (L2TP) over IP Security (IPSec) support for router-to-router VPN connections
- The graphical user interface program is the routing and remote access administrative utility, a Microsoft Management Console (MMC) snap-in. The command line utility is Netsh

The combined features of the Windows 2000 Routing and Remote Access service make a Windows 2000 Server–based computer function as shown in Table 19.3.

An advantage of the routing and remote access service is its integration with the Windows 2000 Server operating system. The service works with a wide variety of hardware platforms and hundreds of network adapters; the result is a lower-cost solution than many mid-range dedicated router or remote access server products. The routing and remote access service is extensible with application programming interfaces (APIs) that third-party developers can use to create custom networking solutions and that new vendors can use to participate in the growing business of open internetworking. See Figure 19.5.

TABLE 19.3   Windows 2000 Routing and Remote Access service features

Feature	Description
Multiprotocol router	A routing and remote access service computer can route IP, IPX, and AppleTalk simultaneously. All routable protocols and routing protocols are configured from the same administrative utility.
Demand-dial router	A routing and remote access service computer can route IP and IPX over on-demand or persistent WAN links, such as analog phone lines or ISDN, or over VPN connections using either PPTP or L2TP over IPSec.
Remote access server	A routing and remote access service computer can act as a remote access server providing remote access connectivity to dial-up or VPN remote access clients using IP, IPX, AppleTalk, or NetBEUI.

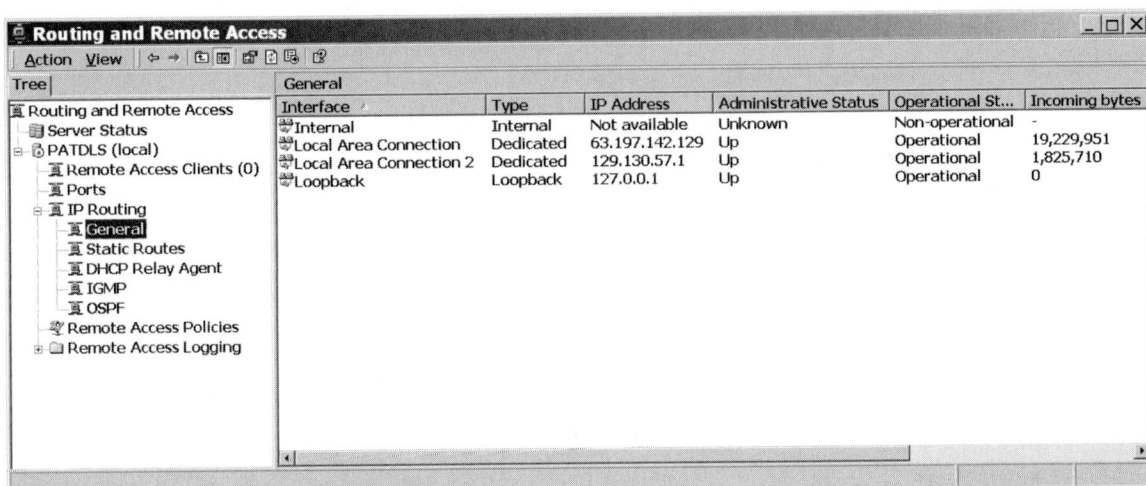

FIGURE 19.5   Using the Routing and Remote Access Console to configure a multihomed router

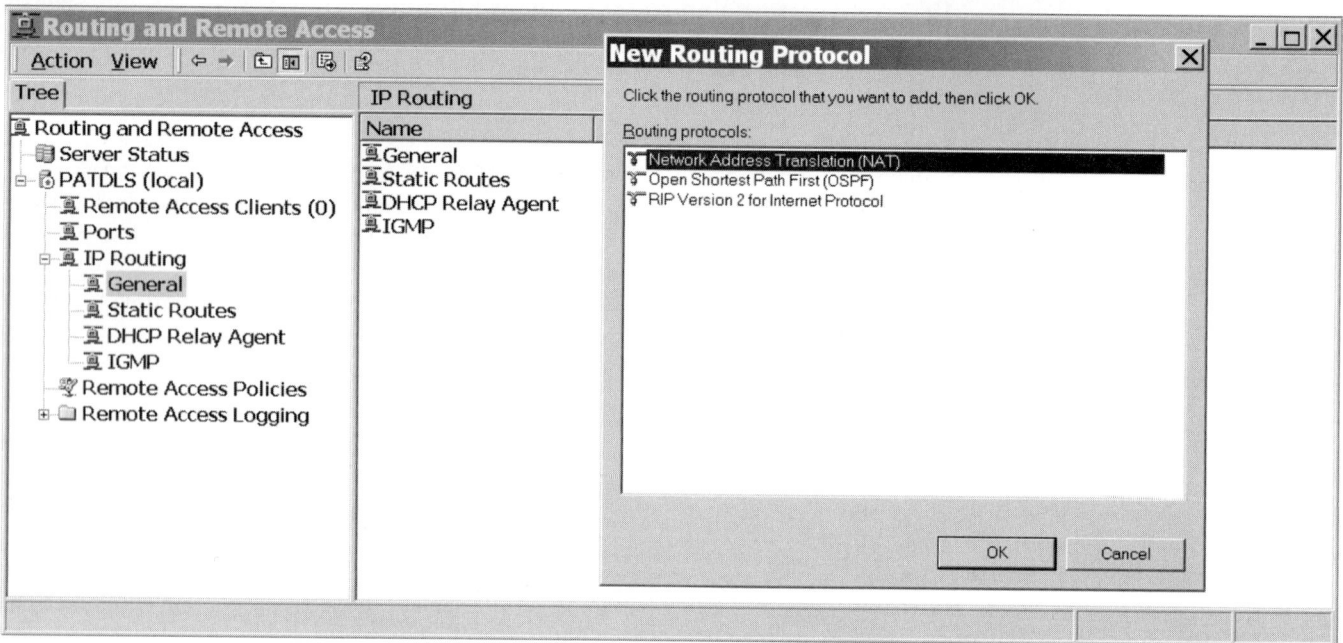

FIGURE 19.6   Adding a new network routing protocol using the Routing and Remote Access console on a multihomed computer

To enable a routing protocol such as RIP or OSPF to an interface, do the following:

1. Open Routing and Remote Access (see Figure 19.6).
2. In the Console tree, click IP Routing which is located under the server.
3. Right-click General in the Detail pane and then select the New Routing Protocol option.
4. Click to select the protocol that you want to enable and click on the OK button.

19.7.3   Creating Static Routes

To add a static route to a Windows 2000 multihomed computer, use the Routing Access and Remote Access program located under Administrative Tools or use the appropriate MMC snap-in. Next, right-click Static Routes, and select "New Static route for IP networks" or "New route for IPX routes." See Figure 19.7.

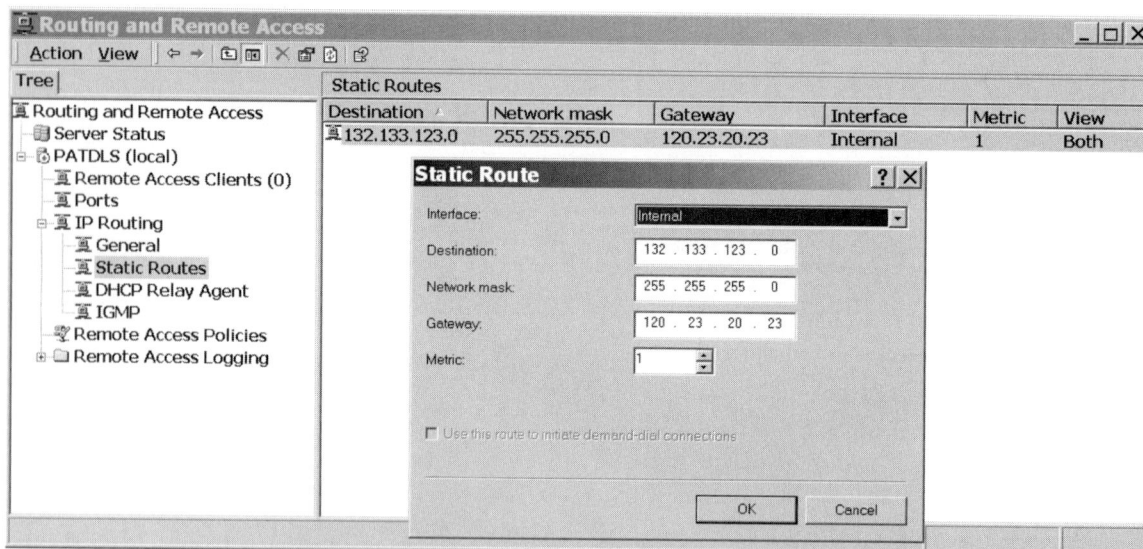

FIGURE 19.7   Configuring a static route using the Routing and Remote Access Console on a Windows 2000 multihomed computer

For a static IP route in Interface, Destination, Network Mask, Gateway, and Metric, enter the corresponding information into each route. If this is a demand-dial interface, Gateway is unavailable. You can also select the "Use this route to initiate demand-dial connections" checkbox to initiate a demand-dial connection for traffic that matches the route. For a static IPX route in Network Number (hex), Next Hop MAC Address (hex), Tick Count, Hop Count, and Interface, enter the corresponding information. To delete the static route, right-click the appropriate route, and then select the Delete option. To make changes to a static route entry, double-click the correct static route in the Detail pane.

For IP static addresses, the destination provides a space to type a destination for the route. The destination can be a host address, subnet address, network address, or the destination for the default route (0.0.0.0). The subnet mask provides a space to type the network mask for the static route. The network mask number is used in conjunction with the destination to determine when the route is used. The mask of 255.255.255.255 means that only an exact match of the destination number can use this route. The mask of 0.0.0.0 means that any destination can use this route. The gateway provides a space to type the forwarding IP address for this route. For LAN interfaces, the gateway address must be configured and must be a directly reachable IP address for the network segment of the selected interface. Again, for demand-dial interfaces, the gateway address is not configured or used. The metric provides a space to type the cost associated with this route to reach the destination. The metric is commonly used to indicate the number of routers (hops) to the destination. When deciding between multiple routes to the same destination, the route with the lowest metric is selected as the best route.

19.7.4 Planning a Static Route

If appropriate for your IP internetwork, perform the following steps to deploy static routing.

1. Draw a map of the topology of your IP internetwork to show the separate networks and the placement of routers and hosts.
2. For each IP network, assign a unique IP network ID (also known as an IP network address).
3. Assign IP addresses to each router interface. It is a common industry practice to assign the first IP addresses of a given IP network to router interfaces. For example, for an IP network ID of 192.168.100.0 with a subnet mask of 255.255.255.0, the router interface is assigned the IP address of 192.168.100.1.
4. For peripheral routers, configure a default route on the interface that has a neighboring router. The use of default routes on peripheral routers is optional.
5. For each nonperipheral router, compile a list of routes that need to be added as static routes to the routing table for that router. Each route consists of a destination network ID, a subnet mask, a gateway (or forwarding) IP address, a metric (number of router hops to reach the network), and the interface to be used to reach the network.
6. For nonperipheral routers, add the static routes compiled in step 5 to each router. You can add static routes by using Routing and Remote Access or the ROUTE command. If you use the ROUTE command, use the –p option to make the static routes persistent.
7. When your configuration is complete, use the PING and TRACERT commands to test connectivity between host computers so that all routing paths are checked.

19.7.5 RIP

Windows 2000 RIP for IP is RFC 1058 and 1723 compliant and has the following features:

- Split horizon, poison reverse, and triggered updates convergence algorithms
- Ability to modify the announcement interval (default is 30 seconds)
- Ability to modify the routing table entry timeout value (default is 3 minutes)
- Ability to be a silent RIP host
- Peer filtering: Ability to accept or discard updates of announcements from specific routers identified by IP address
- Route filtering: Ability to accept or discard updates of specific network IDs or from specific routers
- RIP neighbors: Ability to unicast RIP announcements to specific routers to support non-broadcast technologies such as frame relay (A RIP neighbor is a RIP router that receives unicasted RIP announcements.)
- Ability to announce or accept default routes or host routes

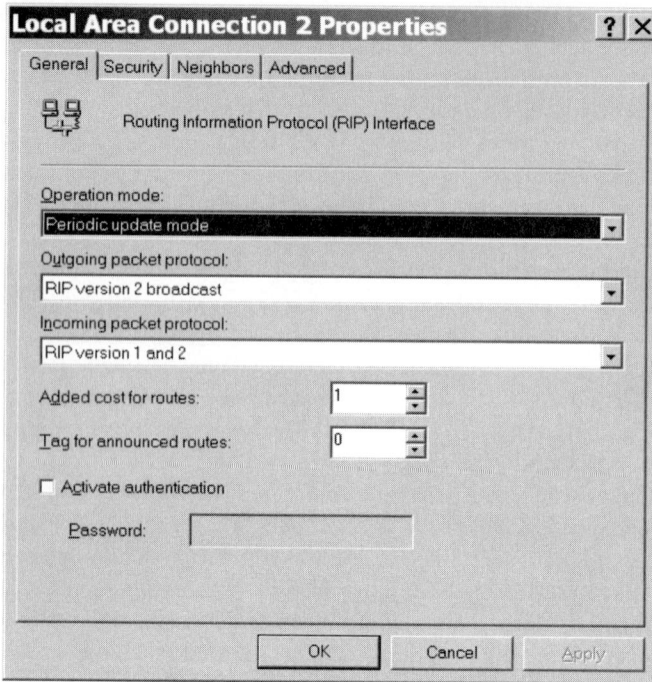

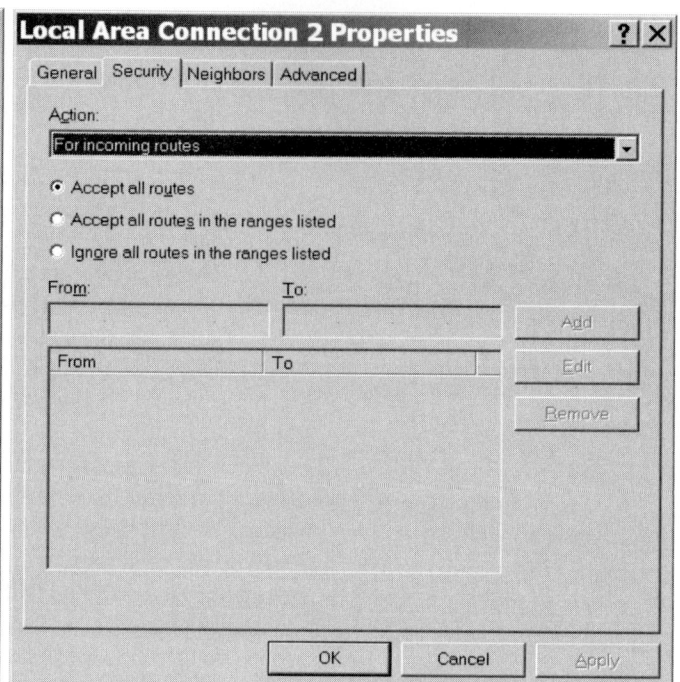

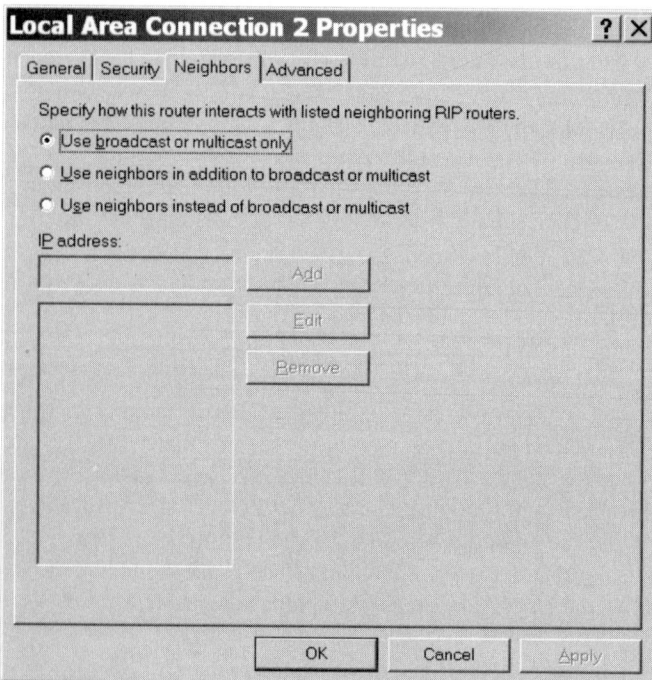

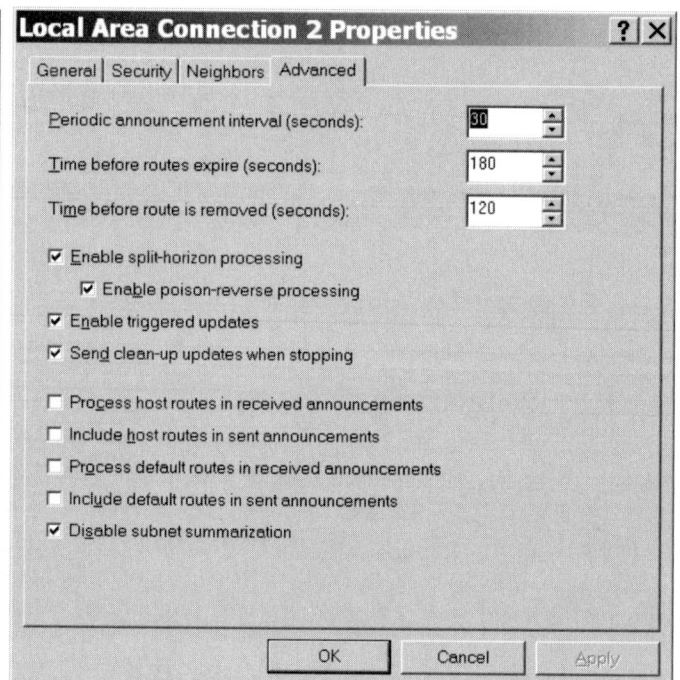

FIGURE 19.8   Configuring the RIP protocol on a Windows 2000 multihomed computer

When a Windows 2000 Router advertises a non-RIP learned route, it advertises it with a hop count of two. Non-RIP learned routes include static routes (even for directly attached networks), OSPF routes, and SNMP routes.

To configure RIP version 2, do the following:

1. Open Routing and Remote Access.
2. In the Console tree, click RIP.
3. In the Details pane, right-click the interface you want to configure for RIP version 2, and then click Properties. See Figure 19.8.

19.7.6 OSPF

To create an OSPF area, do the following:

1. Start the Routing and Remote Access console.
2. In the Console tree, click OSPF.
3. Right-click OSPF, and then click Properties.
4. On the Areas tab, click Add.
5. On the General tab, in Area ID, type the dotted decimal number that identifies the area.
6. To use a plaintext password, verify that the "Enable plaintext password" checkbox is selected.
7. To mark the area as a stub, select the Stub Area checkbox. You cannot configure the backbone area (area ID 0.0.0.0) as a stub area or virtual links through stub areas.
8. In Stub Metric, click the arrows to set the stub metric.
9. To import routes of other areas into the stub area, select the "Import summary advertisements" checkbox. See Figure 19.9.

To configure ranges for an OSPF area, do the following:

1. Start the Routing and Remote Access console.
2. In the Console, click OSPF.
3. Right-click OSPF, and then click Properties.
4. On the Areas tab, click Add.
5. On the Ranges tab, in Destination, type the IP network ID for the range.

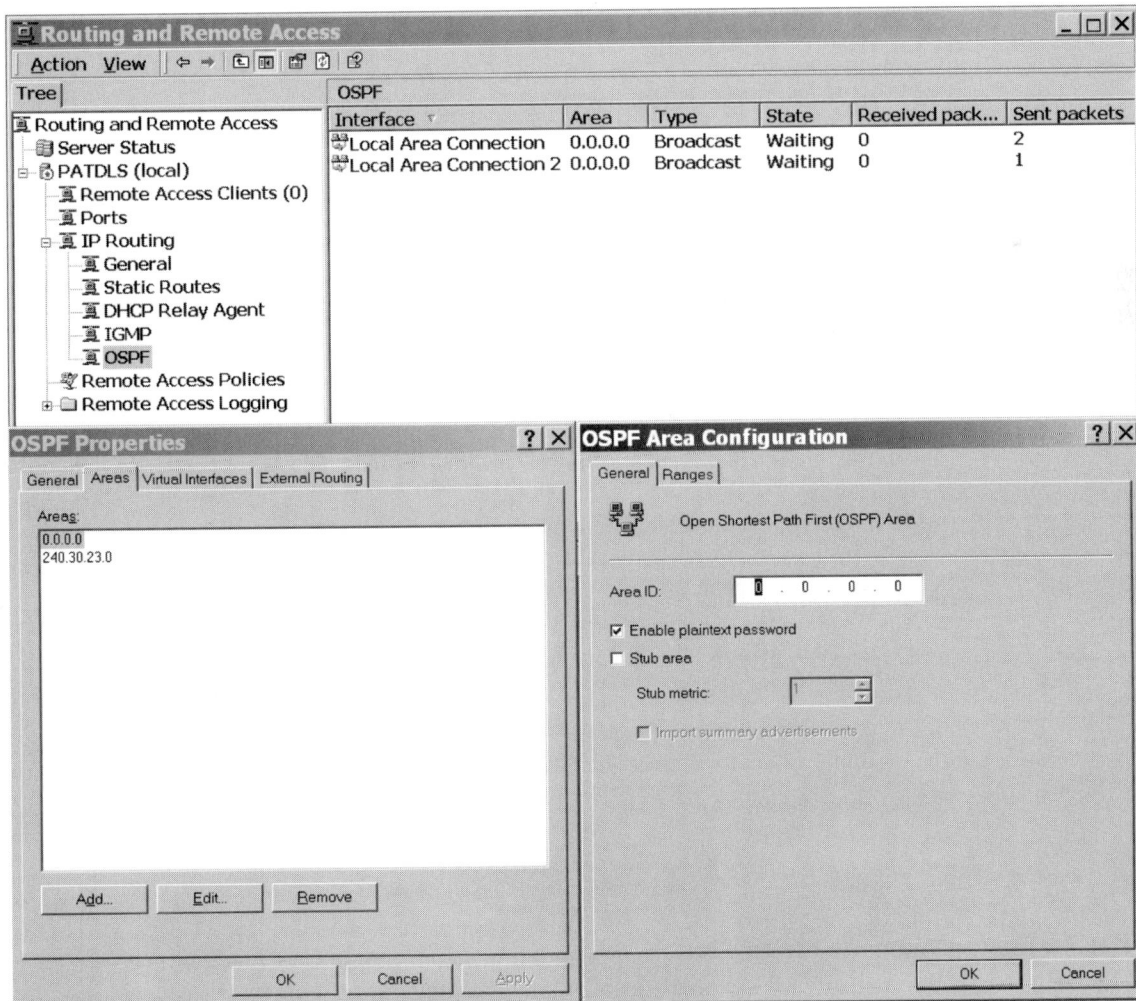

FIGURE 19.9   Configuring the OSPF protocol on a Windows 2000 multihomed computer

FIGURE 19.10 Configuring OSPF area on a multihomed computer

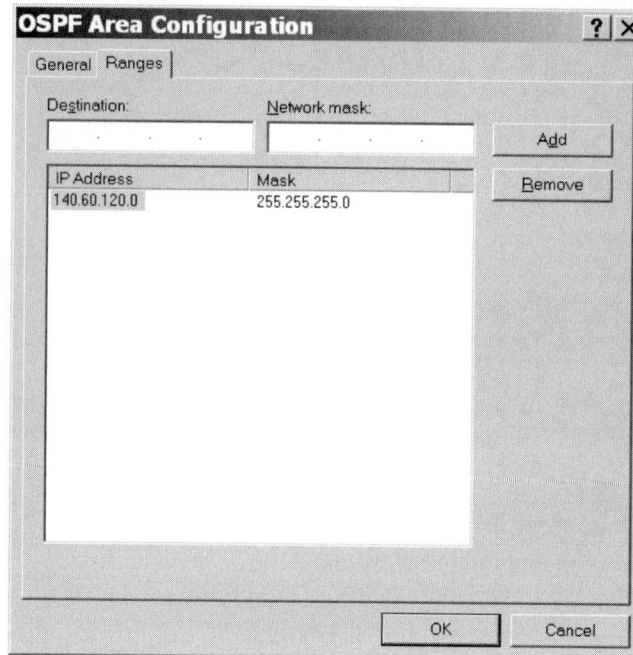

OSPF Area Configuration   ? X

General Ranges

Destination: Network mask:
[   ] [   ]   Add

IP Address	Mask
140.60.120.0	255.255.255.0

Remove

OK   Cancel

6. In Network Mask, type the associated mask for the range, and then click Add.
7. To delete a range, click the range you want to delete, and then click Remove. See Figure 19.10.

To configure an ASBR, do the following:

1. Open the Routing and Remote Access console.
2. In the Console tree, click OSPF.
3. Right-click OSPF, and then click Properties.
4. On the General tab, click "Enable autonomous system boundary router."
5. To configure external route sources, do the following:
   - On the External Routing tab, click either "Accept routes from all route sources except those selected" or "Ignore routes from all route sources except those selected."
   - Select or clear the appropriate checkboxes next to the route sources.
6. To configure external route filters, do the following:
   - On the External Routing tab, click Route Filters.
   - In Destination and in Network mask, type the route you want to filter, and then click Add.
   - Repeat the preceding step for all the routes you want to filter.
   - Click either "Ignore listed routes" or "Accept listed routes" depending on the appropriate filtering action, and then click OK.

19.7.7 EGP and BGP

Windows 2000 Router does not provide EGP and BGP, but Windows 2000 does have an application programming interface (API) support that provides a platform to support additional routing protocols such as the BGP for IP. Therefore, an EGP and BGP can be written by third-party independent software vendors.

19.8 INTRODUCTION TO CISCO ROUTERS

Cisco Systems is a leading manufacturer of network equipment including routers, bridges, and switches. Cisco routers are able to build routing tables, execute commands, and route packets across network interfaces using routing protocols; thus, the router must have processing power, storage capacity, and RAM. Appropriate software such as an operating system that can be used to configure routed and routing protocols is also necessary.

19.8.1 Cisco Hardware Components

Depending on the Cisco router model, the router comes with different processors. For example, the Cisco 2505 router contains a 20-MHz Motorola 68EC030 processor, and a higher-end router such as the Cisco 7010 router contains a 25-MHz Motorola MC68040 CPU.

Cisco routers need a place to store configuration information, a place to boot the router operating system (IOS), and memory that can be used to hold dynamic information as the router does its job of moving packets on the internetwork. Cisco routers actually contain different types of memory components that provide the storage and dynamic caching required. The different memory components found in a Cisco router are as follows:

- **ROM**—Contains the power-on self-test (POST) and the bootstrap program for the router. The ROM chip also contains either a subset or the complete router IOS.
- **NVRAM** (nonvolatile RAM)—Stores the start-up configuration file for the router. NVRAM can be erased and can copy running configuration on the router to NVRAM. The great thing about NVRAM is that it retains information even if the router is powered down.
- **Flash RAM**—Flash is a special ROM that can actually erase and reprogram. Flash is used to store the Cisco IOS that runs on your router. You can also store alternative versions of the Cisco IOS on the Flash such as an upgrade of your current IOS, which makes it easy for you to upgrade the router. Flash RAM actually comes in the form of SIMMS and depending on which router you have, additional Flash RAM may be installed.
- **RAM**—Similar to the dynamic memory on your PC, RAM provides the temporary storage of information. Packets are held in RAM when their addressing information is examined by the router. In addition, RAM also holds the current routing table and current-running router configuration.

19.8.2 Router Interfaces

Cisco routers can support LAN and WAN interfaces. The most common LAN router interfaces are Ethernet, Fast Ethernet, IBM token ring, and FDDI. WAN interfaces, sometimes called serial router interfaces, are connected via leased lines and other third-party connectivity technologies. Some of the commonly used WAN data link layer technologies are HDLC, X.25, frame relay, ISDN, and PPP.

Cisco routers such as those in the 2500 Series family basically are off-the-shelf routers that come with a predetermined number of LAN, WAN, and serial ports. Higher-end routers such as the Cisco 4500 are modular and actually contain empty slots that can be filled with several different interface cards. Built-in ports are designed by their connection type, followed by a number. For example, the first Ethernet port on a router is E0; the second Ethernet port on a router would be E1, and so on. Serial ports are designated the same way with the first serial port being S0. Modular routers designate their ports by connection type, followed by slot number, and then port number. For example, the first Ethernet port on an Ethernet card that is placed in the router's first slot would be designated as Ethernet 1/0. The slot is designated first, followed by the port number.

19.8.3 Router Boot Sequence

When you power the router on, the ROM chip runs a POST that checks the router's hardware such as the processor, interfaces, and memory. The next step in the router boot-up sequence is the execution of a bootstrap program that is stored in the router's ROM. This bootstrap program searches for the CISCO IOS. The IOS can be loaded from the ROM itself, the router's Flash RAM, or from a TFTP server on the network. The IOS is typically stored in the router's Flash RAM.

After the router's IOS is loaded, the router searches for the configuration file, which is normally held in NVRAM. As with the IOS, however, the configuration file can be loaded from a TFTP server. After the router loads the configuration file, the information in the file enables the interfaces and provides parameters related to the routing protocols.

19.8.4 Communicating with the Router

Every Cisco router has a console port on its back, which is there to provide a way to hook up a terminal to the router to work on it. The console port, which is sometimes referred to as a management port, is used by administrators to log in to a router directly without using a network connection. A stand-alone CRT, PC, or workstation can be used as a console. Console terminals must run a character-based user interface. They cannot directly use a graphical user interface (GUI). Instead, you must use a PC or workstation as a console, and terminal emulator software such as the HyperTerminal.

Most Cisco routers have an auxiliary port. The auxiliary port is sometimes referred to as the AUX port. Like the console port, the AUX port makes possible a direct, non-network connection to the router.

Once a router is installed on a network, access to it is almost always via Telnet sessions, not via the console or AUX ports. Telnet is a way to log in to a router as a virtual terminal. Telnet connections are instead made through the network. In the most basic terms, a virtual terminal session is composed of IP packets being routed over a network, pretending to be bits streaming over a serial line.

When using Telnet to access a router, do so over a virtual line provided by the Cisco IOS software. These are called VTY lines. IOS supports up to five virtual terminal sessions running simultaneously on a router. It is rare that you will have more than one virtual terminal session running on a router at the same time.

A more recent router access method is the HTTP server. Do not be misled by the name; no computer server is involved in using HTTP server. The term *server* in HTTP server refers to a small software application running inside the Cisco IOS software. HTTP server first became available with IOS Released 10.3. HTTP server makes it possible to interact with the router through a Web browser.

19.8.5 Internetworking Operating System

Cisco's **internetworking operating system (IOS)** is software that provides the router hardware with the capability to route packets on an internetwork. The IOS, like any operating system, provides the command sets and software functionality that you use to monitor and configure the router, and it also provides the functionality for the various protocols that make the internetwork a reality.

After the router contains a basic configuration, you can then begin to examine the different router modes available. The router supplies you with three basic levels of access: user mode, privileged mode, and configuration mode, as defined in the following list.

- **User mode**—This modes provides limited access to the router. You are provided with a set of nondestructive commands that allow examination of certain router configuration parameters. You cannot, however, make any changes to the router configuration.
- **Privileged mode**—This mode (also known as the enabled mode) allows greater examination of the router and provides a more robust command set than the user mode. After you enter the privileged mode using the secret or enable password, you have access to the configuration commands supplied in the configuration mode, meaning that you can edit the configuration for the router.
- **Configuration mode**—This mode (also known as the global configuration mode) is entered from the privileged mode and supplies the complete command set for configuring the router.

Configuring the router means that you enable the various interfaces and protocols on the router. You must use commands that bring your various hardware interfaces such as Ethernet or serial interfaces to life. You must provide configuration information for the protocols that are routed, such as IP or IPX; and you must configure routing protocols such as RIP, OSPF, and IGRP. After the router is configured, then mange your configuration files.

Cisco provides a **command line interface (CLI)** that you can use to configure and maintain your router. You can access the CLI using a router console or by Telnetting to a router using a virtual terminal. The IOS has hundreds of commands, some of which can be used anywhere in IOS, whereas others can be used within a specific area only. A few major root commands handle most tasks that are associated with configuring routers, as follows:

show	Examine router status.
configure	Make changes to config file parameters.
no	Negate a parameter setting.
copy	Put configure file changes into effect.

SUMMARY

1. A router, which works at the network ISO layer, is a device that connects two or more LANs.
2. The primary role of a router is to transmit similar types of data packets from one wide area communications link (such as a T1 line or fiberlinks) to another T1 line or fiber link. The second role of a router is to select the best path between the source and destination.
3. To determine the best route, the routes use complex routing algorithms, which take into account a variety of factors including the number of the fastest set of transmission media, the number of network segments, and the network segment that carries the least amount of traffic.
4. A metric is a standard of measurement, such as hop count, that is used by routing algorithms to determine the optimal path to a destination. A hop is the trip a data packet takes from one router to another router or a router to another intermediate point to another in the network.
5. To track the various routes in a network, the routers will create and maintain routing tables.
6. Routing algorithms must converge rapidly. Convergence is the process of agreement by all routers to which routes are the optimal routes.
7. Static routing algorithms are hardly algorithms at all, but are table mappings established by the network administrator prior to the beginning of routing. These mappings do not change unless the network administrator alters them.
8. Because static routing systems cannot react to network changes, they generally are considered unsuitable for today's large, changing networks.
9. Most dominant routing algorithms in the 1990s are dynamic routing algorithms, which adjust to changing network circumstances by analyzing incoming routing update messages.
10. Routers use distance-vector-based routing protocols to periodically advertise or broadcast the routes in their routing tables, but they only send it to their neighboring routers.
11. Link-state algorithms are also known as shortest path first algorithms. Link-state routers send updates directly (or by using multicast traffic) to all routers within the network. Each router, however, sends only the portion of the routing table that describes the state of its own links.
12. The various routing protocols use different metrics, including path length, hop counts, routing delay, bandwidth, load, reliability, and cost.
13. To report errors and other information regarding IP packet processing, the Internet control-message protocol (ICMP) sends notices back to the source.
14. The Routing Information Protocol (RIP) is designed for exchanging routing information within a small to medium-size network.
15. To overcome some of the RIP shortcomings, RIP Version 2 (RIP II) was introduced.
16. Some protocols, such as OSPF, allow areas (groupings of contiguous networks) to be gathered into an autonomous system (AS).
17. The backbone (given an address of 0.0.0.0) of the AS is a high-bandwidth logical area where all areas are connected.
18. Areas that are not backbone OSPF areas can be classified as either a stub area or a transit area.
19. A stub area has only one ABR. All routes to the destination outside the area must pass through the single router.
20. A transit area contains more than one ABR.
21. The open shortest path first (OSPF) protocol is a link-state routing protocol used in medium and large networks and calculates routing table entries by constructing a shortest-path tree.
22. The interior gateway routing protocol (IGRP) is a distance-vector protocol that allows gateways to build up their routing table by exchanging information with other gateways.

23. The local routing information of an autonomous area is gathered using an internal gateway protocol such as RIP or OSPF.

24. In these autonomous areas, one or more routers are chosen to use an exterior gateway protocol (EGP) to talk to other autonomous areas by providing a way for two neighboring routers that are located at the edges of their respective autonomous area to exchange routing information.

25. IP multicast allows a host to communicate with several hosts simultaneously by transmitting information via one data stream.

26. Routers and computers use the Internet group-membership protocol (GMP) to exchange membership information.

27. Cisco Systems is a leading manufacturer of network equipment including routers, bridges, and switches. Cisco routers are able to build routing tables, execute commands, and route packets across network interfaces using routing protocols.

QUESTIONS

1. Which one of the following communications devices would be used to transport packets between autonomous systems that are over 1000 km apart?
   a. concentrator   c. router
   b. hub   d. transparent bridge

2. Your company is based in Atlanta. It has one branch office in Los Angeles and one in New York City. Each office is networked. The WAN uses three T1 connections to connect the three sites. Which one of the following connectivity devices should be used to connect the LANs to the multiple paths in the WAN?
   a. routers   c. repeaters
   b. bridges   d. gateways

3. Which one of the following is the smartest network device?
   a. bridge   c. concentrator
   b. gateway   d. router

4. How does a router determine where a packet should be delivered?
   a. It examines the MAC address.
   b. It examines the source network address.
   c. It examines the telephone number.
   d. It examines the destination network address.

5. What are the two types of routers? (Select two answers)
   a. passive   c. static
   b. active   d. dynamic

6. Routes and routing tables are built by routing algorithms using _____ .
   a. cost of the router   c. metrics
   b. speed of the router   d. whichever router is available first

7. Pat has four multihomed Windows 2000 Servers that are functioning as routers on his network. He wants to configure the routing tables on these servers with the least amount of administrative effort possible. How should Pat proceed?
   a. Install DHCP relay agent.   c. Run NETSTAT.EXE.
   b. Install RIP for IP.   d. Run ROUTE.EXE.

8. What type of packets does TRACERT send out?
   a. ICMP echo packets   c. TCP packets
   b. UDP packets   d. IP packets

9. Pat has been asked to set up a small network with three subnets as shown in the following figure. He plans to connect the subnets with two multihomed gateways. The subnet mask on the network is 255.255.255.0. What command would be used to configure the two routers? (Choose two answers)
   a. On Router1: ROUTE ADD 132.4.25.0 MASK 255.255.255.0 132.4.24.2
   b. On Router2: ROUTE ADD 132.4.23.0 MASK 255.255.255.0 132.4.25.2

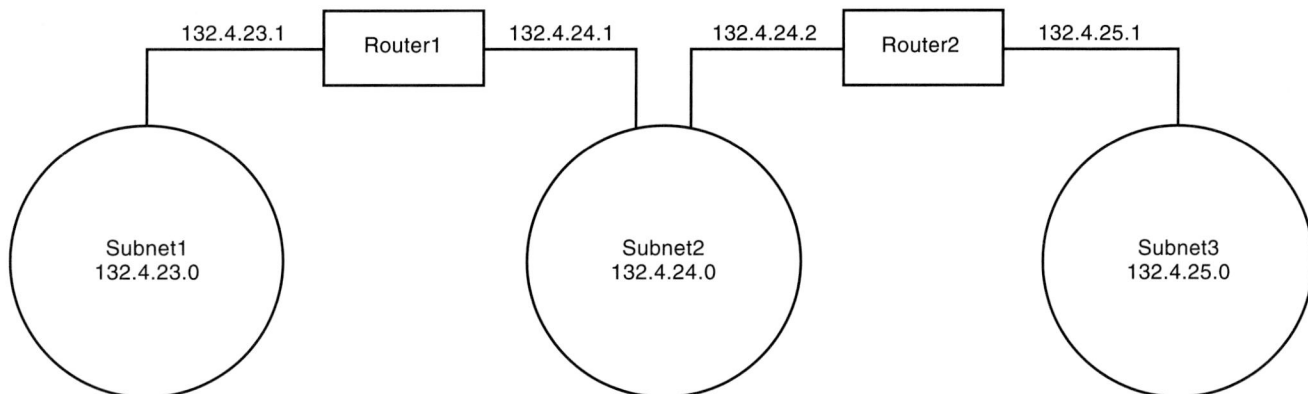

```
132.4.23.1    Router1    132.4.24.1    132.4.24.2    Router2    132.4.25.1
```

Subnet1
132.4.23.0

Subnet2
132.4.24.0

Subnet3
132.4.25.0

    c. On Router1: ROUTE ADD 132.4.25.0 MASK 255.255.255.0 132.4.24.1
    d. On Router2: ROUTE ADD 132.4.23.0 MASK 255.255.255.0 132.4.23.1

10. _____ is the process of agreement by all routers to which routes are the optimal routes.
    a. Coordinating   c. Convergence
    b. Synchronization   d. Ripping

11. _____ is a distant-vector routing protocol.
    a. RIP   c. ICMP
    b. OSPF   d. Static route

12. _____ is a link-state routing protocol.
    a. RIP   c. ICMP
    b. OSPF   d. Static route

13. You just added 10 routers to your network and suddenly the network performance has dramatically decreased. What is the problem?
    a. The network is probably using static routes.
    b. The network is probably using RIP.
    c. The network is probably using OSPF.
    d. One of the routing links is down.

14. What type of protocol connects autonomous systems together?
    a. EGP   c. RIP
    b. IGP   d. OSPF

15. _____ allows a host to communicate with several hosts simultaneously by transmitting information via one data stream.
    a. Multicasting   c. Multiplying
    b. Multibroadcasting   d. Converging

16. What is the address of the backbone of an OSPF network?
    a. 0.0.0.0   c. 240.0.0.0
    b. 127.0.0.1   d. 255.255.255.255

20 Introduction to Web Servers

INTRODUCTION

Because the Internet and Web servers are essential to many businesses, it would make sense that Windows 2000 would include a Web server. Microsoft's Web server is called Internet Information Services (IIS), which has been available for Windows NT. Since the Internet has grown more popular during the last few years, IIS is now included with Windows 2000 Server. This chapter will discuss how to configure IIS.

1. Monitor, configure, troubleshoot, and control access to files and folders via Web services.
2. Manage and troubleshoot Web server resources.
3. Monitor, configure, troubleshoot, and control access to Web sites.

20.1 INTRODUCTION TO WEB PAGES AND SERVERS

The terms *Internet* and *World Wide Web* are often used interchangeably, but they do have different meanings. As mentioned in earlier chapters, the **Internet** refers to the huge global WAN. Until the early 1990s, a person who wanted to use the computer to communicate with another computer via the Internet needed a good deal of knowledge and the ability to understand and use some fairly unfriendly commands. The **World Wide Web (WWW)** was created in 1992, and refers to the means of organizing, presenting, and accessing information over the Internet.

To access the Web, a user needs the following technologies:

1. HTML
2. Web server
3. Web browser
4. http and ftp

Web pages are written using the **hypertext markup language (HTML).** This language is fairly simple and is implemented as special ASCII tags or codes that you embed within your document to give the browser a general idea of how the information should be displayed. The browsers understand the standard HTML tags, although they may display the same document a little differently. If you want your documents to be accessible by people using different browsers, you should stick with the standard tags. The HTML standard is still actively evolving, so new tags are constantly becoming available to support new browser features.

A **Web server** is a computer that is equipped with the server software that uses Internet protocols such as hypertext transfer protocol (HTTP) and file transfer protocol (FTP) (both to be discussed) to respond to Web client requests on a TCP/IP network via Web browsers. One server can service a large number of clients. There are several free server programs available on the Internet. Most Web browsers are built to process two basic types of requests—file server and database server requests. New features are always being added to provide additional support for new technology. A Web server acting as a file server simply accepts a request for a document, validates the request, and sends the requested files back to the browser. In addition, the browser can act as a front-end tool or interface to collect data and feed it in a database or script. The database can be located either on the same server as the Web server or on a different server. When the database responds with the results, it will then send the results back to the browser.

The **Web browser** is the client program/software that you run on your local machine to gain access to a Web server. It receives the HTML commands, interprets the HTML, and displays the results. It is strictly a user-interface/document presentation tool. It knows nothing about the application to which it is attached and only knows how to take the information from the server and present it to the user. It is also able to capture data entry made into a form and get the information back to the server for processing. The most common browsers are Microsoft Internet Explorer and Netscape Communicator. Both of these tools are available for little or no charge on the Internet.

The application protocol that makes the Web work is **hypertext transfer protocol (HTTP).** Whereas the HTML is the language used to write Web pages, HTTP is the protocol that Web browsers and Web servers use to communicate with each other over the Internet. It is an application level protocol, because it sits on top of the TCP layer in the protocol stack and is used by specific applications to talk to one another. In this case, the applications are Web browsers and Web servers.

HTTP is a text-based protocol. Clients (Web browsers) send requests to Web servers for Web elements such as Web pages and images. After the request is serviced by a server, the connection between client and server across the Internet is disconnected. A new connection must be made for each request. Most protocols are connection oriented, which means that the two computers communicating with each other keep the connection open over the Internet—HTTP does not. Before an HTTP request can be made by a client, a new connection must be made to the server. Another common protocol is the **file transfer protocol (FTP),** which is used on the Internet to send files.

Currently, most Web browsers and servers support HTTP 1.1. One main feature of HTTP 1.1 is it supports persistent connections, which means that once a browser connects to a Web server, it can receive multiple files through the same connection. This protocol should improve performance by as much as 20%.

HTTP is called a stateless protocol because each command is executed independently, without any knowledge of the commands that came before it. This explains why it is difficult to implement Web sites that react intelligently to user input. This shortcoming of HTTP is being addressed in several new technologies, including ActiveX, Java, JavaScript, and cookies.

Typically when accessing a Web page, you will specify the Web page location by using a **uniform resource locator (URL).** The first part of the address indicates what protocol to use; the second part specifies the IP address or the domain name where the resource is located; and the last part indicates the folder and file name. See Figure 20.1 and the other URL examples listed here.

> ftp://www.acme.com/files/run.exe
> http://www.acme.com/index.html

When you type a URL into a Web browser, the following occurs:

1. If the URL contains a domain name, the browser first connects to a DNS server and retrieves the corresponding IP address for the Web server.
2. The Web browser connects to the Web server and sends an HTTP request (via the protocol stack) for the desired Web page.
3. The Web server receives the request and checks for the desired page. If the page exists, the Web server sends it. If the server cannot find the requested page, it will send an HTTP 404 error message (404 means "Page Not Found," as anyone who has surfed the Web probably knows).
4. The Web browser receives the page back and the connection is closed.
5. The browser then parses through the page and looks for other page elements it needs to complete the Web page, including images, applets, etc.
6. For each element needed, the browser makes additional connections and HTTP requests to the server for each element.
7. When the browser has finished loading all images, applets, etc., the page will be completely loaded in the Browser window.

A **gopher** was an early system that predates the World Wide Web for organizing and displaying files on Internet servers. A gopher server presents its contents as a hierarchically structured list of files. Gophers are basically obsolete, as they have primarily been replaced by search engines.

Since the Internet is so vast, consisting of countless Web pages, it is often difficult to know the URL that a person wants. Therefore, the World Wide Web has several search engines that can be used to locate desired documents. A **search engine** is a program that searches documents for specified keywords and returns a list of the documents where the keywords were found. Typically, a search engine works by sending out a spider to fetch as many documents as possible. Another program, called an indexer, then reads these documents and creates an index based on the words contained in each document. Each search engine uses a proprietary algorithm to create its indices such that, ideally, only meaningful results are returned for each query.

FIGURE 20.1   Typical URL

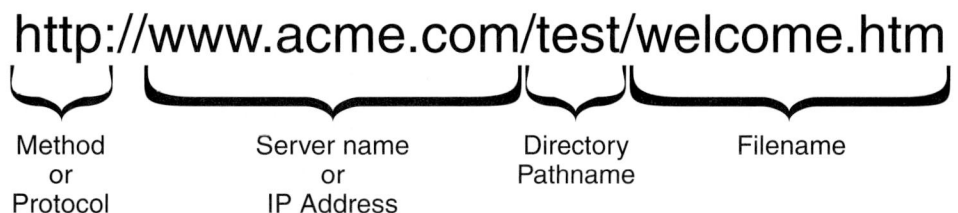

http://www.acme.com/test/welcome.htm

Method
or
Protocol

Server name
or
IP Address

Directory
Pathname

Filename

For more information on the HTTP protocol, go to W3C Architecture Domain Web site located at http://www.w3.org/Protocols/

20.2 INTRODUCTION TO HTML LANGUAGE

To create or modify a Web page, use a text editor such as DOS's EDIT or Windows Notepad. The text file will begin with the <HTML> tag and end with the </HTML> tag. Web browsers can identify an HTML file by the file extension, and many browsers will display pages that do not carry the <HTML> tag. When viewing your document with your own browser, you may be able to get away without the tag, but you cannot be sure that everyone who wants to view your page will be using the same browser that you use. It is best to stick with good programming habits and avoid the temptation to take shortcuts.

The HTML document is divided into two sections—the head and the body. The head begins with the <H> tag and ends with the </H> tag, and the body begins with <BODY> and ends with </BODY>. Although these tags are not case sensitive, try to type them in uppercase so that they are easy to identify. Most HTML tags are used in pairs, with an opening tag and an ending tag to delineate which text you want handled in a particular way. Ending tags are the same as opening tags, but with the addition of a forward slash (/).

20.2.1 The Head

The head will include the title, which is defined with the <TITLE> </TITLE> tags. The information enclosed within the title tags will be displayed by your browser at the top of the Web page in the title bar. Beside the title, the head can also include META tags, which are used for including information about the document. This can be valuable when listing the contents of the page in either a series of keywords or a description that can be targeted by search engines. The <META> tag is a singlet with no closing tag. See Figure 20.2.

```
<HTML>
<HEAD>
<TITLE>THIS IS THE TITLE</TITLE>
<META NAME="KEYWORD" CONTENTS="HOT,IMPORTANT">
<META NAME="DESCRIPTION" CONTENTS="This is a sample website">
</HEAD>

<BODY>
This is the body of the web page
<BODY>
</HTML>
```

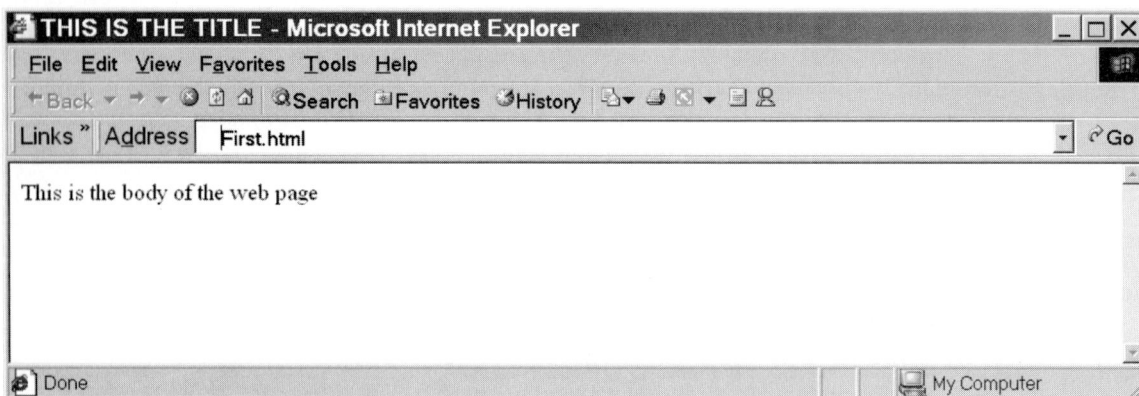

FIGURE 20.2   HTML Web page that defines the title and body

20.2.2 The Body

Everything that shows up on the Web page itself is found in the body of the HTML document. The HTML equivalent to hitting the Return key in a word-processing program is the use of the paragraph<P> </P> tags or the break tag <BR>. The paragraph tag <P> </P> is like hitting the Return key twice, creating a blank line and a break. The break tag <BR> is like hitting the Return key once.

To center the text on the screen, use the <CENTER> </CENTER> tag. One excellent way to organize materials on a Web page is through the use of a list. There are two types of lists: ordered lists and unordered lists. Ordered lists are usually numbered and often used when items are listed in some kind of sequence. Unordered lists are not numbered and are usually bulleted. The tags<OL> </OL> are used to create an ordered list, and <UL> </UL> are used to indent the text. To make the list into a bulleted list or a numbered list, add the list tag <LI> in front of each item on the list. A <BLOCKQUOTE> </BLOCK-QUOTE> is used when you want to set off a quotation or any paragraph by indenting on the left and the right.

Standard text in a browser will usually appear as 12-point text. The font that will appear depends on which font is the default font for the browser. To vary the size of the text on the Web page, use the heading tab <H> </H>. The heading tab always includes a number 1 through 6, which identifies the size of the text: <H1> </H1> is the largest, <H6> </H6> is the smallest.

The font tags <FONT> </FONT> can also be used to change the size of the text, specify the font, and change the color of the text. Unlike the <H> </H> tag, the font tag does not force a break after the text that it encloses. By itself, the font tag does nothing. It must be given attributes for size, color, or font. An **attribute** is an option that can be used within the tag, and is usually followed by a value, indicating a specific choice within the range of choices for that attribute. All font attributes can be nested within a single font tag.

To change the size of text, enclose it with the font tags. After the word FONT in the opening tag, leave a space and include SIZE=N where N is any number between 1 and 7. The number 1 is the smallest text size and 7 is the largest text size. The default text size is about 3 on the font scale.

To change the font, include a <FONT FACE="name of font"> inside the font tags. For example, to change the font to Arial, use <FONT> <FONT FACE="Arial">text </FONT>. To do this exercise, you must have the font Arial on your computer.

Changing the text color is a bit more complicated than changing the size and font of the text. Inside the font tags, the color is specified with the <FONT COLOR= "rrggbb"> where rr is two hexadecimal numbers that specify the amount of red; gg specifies the amount of green; and bb specifies the amount of blue. If you want the text to be red, then specify FF0000 where FF is the hcxadecimal equivalent to 255. If you want black text, then use 000000, and for white use FFFFFF. To change the background color of the page, use the <BODY BG COLOR="rrggbb"> inside the body tag. To change the default text color, use <BODY TEXT="rrggbb"></BODY>. See Figure 20.3.

Many colors can be indicated by entering the name of the color rather than its hexadecimal number value. Some samples of colors include black, navy, silver, blue, maroon, purple, red, fuchsia, green, teal, lime, aqua, olive, gray, yellow, and white.

20.2.3 Adding Hyperlinks

Links on a Web page are most often achieved through the use of hypertext. Hypertext is text that is connected or hyperlinked to other information on the WWW. When you click on Hypertext, it will take you to the information where the link is anchored. Hypertext is colored and underlined to set it off from the rest of the text. The default color for hypertext is blue, but it can be changed.

Essential to creating hypertext links is the anchor tag <A> </A>. The anchor tags enclose text that will become the hypertext link. For example, if you want to create a hypertext link to the Acme Toy Company, start by enclosing the text with the anchor tabs. By themselves, the anchor tags do nothing; they must contain information that indicates where

File Edit View Favorites Tools Help

Back → Search Favorites History

Links » Address First.html Go

TITLE

This is the body of the webpage

1. First
2. Second
3. Third

RED GREEN BLUE

END

Done   My Computer

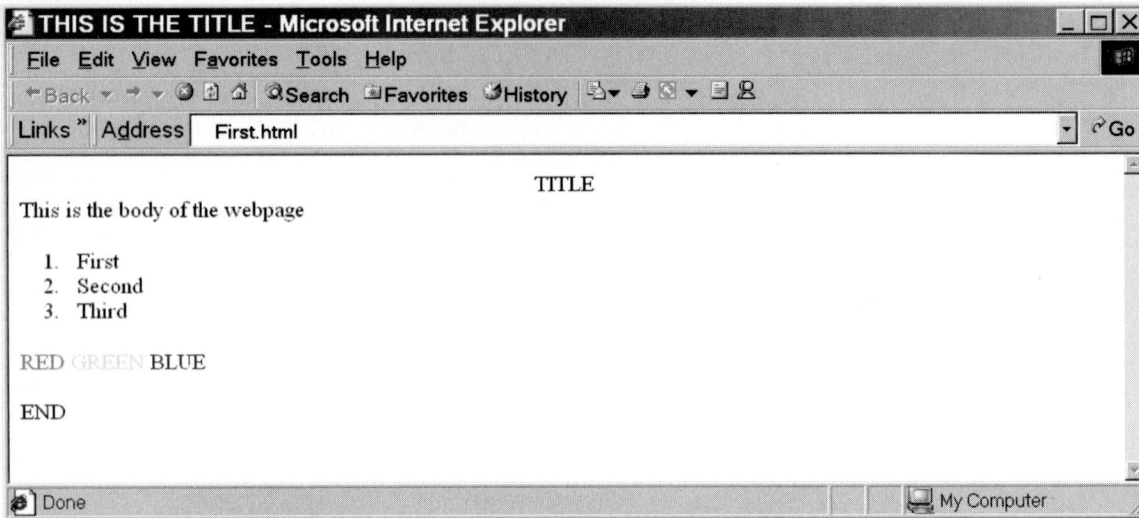

FIGURE 20.3   HTML Web page with a centered text, ordered list, and colors

```
<HTML>

<HEAD>
  <TITLE>THIS IS THE TITLE</TITLE>

  <META NAME="KEYWORD" CONTENTS="HOT,IMPORTANT">
  <META NAME="DESCRIPTION" CONTENTS="This is a sample website">
</HEAD>

<BODY>
  <CENTER>TITLE</CENTER>
  This is the body of the webpage  <OL>
    <LI>First
    <LI>Second
    <LI>Third
  </OL>

  <FONT COLOR="FF0000">RED
  <FONT COLOR="00FF00">GREEN
  <FONT COLOR="0000FF">BLUE

  <BR><BR>
  <FONT COLOR="000000">END
<BODY>
</HTML>
```

the Acme home page is located. We need to add an attribute to the tag that has the address. This attribute is called the hypertext reference or HREF. For example, to have the Acme Toy Company link to the Acme home page, use

<A HREF= "http://www.acme.com\list.htm">Acme Toy Company</A>

When links are made between files that recode on the same server and in the same directory, only a partial address or URL is required. The partial address is called a relative URL address. For example, if you have a file called LIST.HTM in the same directory, you can then use

```
<A HREF= "list.htm">Acme Toy Company</A>
```

20.2.4 Adding Pictures

Using pictures helps create an attractive Web page. To add a picture (either a GIF or JPG file), use the <IMG tab>, which is another singlet. Inside the <IMG> tag, preceded by a space, include the attribute SRC, which refers to the source or location of the image. The SRC attribute is followed by an equal sign (=) and the URL, in quotations, for the location of the graphic file. For example, to add a picture of the LOGO.JPG, use

```
<IMG SRC= "LOGO.JPG">
```

   or

```
<IMG SRC= "PICTURES\LOGO.JPG">
```

if the LOGO.JPG is located in the PICTURES directory, which is located underneath the directory of the current Web page that you are viewing or using.

Here we assume that the LOGO.JPG is located in the same directory as the current page. The Web page does not have the picture embedded into the Web page, but rather, it is a separate file shown by a link. Therefore, if you copy a Web page, you will also need to copy the individual pictures.

Hyperlinks are not limited to text; they can also be images. The following shows how to display the LOGO.JPG which is linked to the ACME.COM home page.

```
<A HREF="http://www.acme.com"> <IMGSRC="LOGO.JPG"></A>
```

20.2.5 Forms and Common Gateway Interface

The common gateway interface (CGI) is a standard for interfacing external applications with Web servers. A plain HTML Web page is a static document because it does not change. A CGI program, on the other hand, is executed in real time, so that it can output dynamic information. For example, say that you wanted to "hook up" your server database to the World Wide Web, to allow people from all over the world to query it. Basically, you need to create a CGI program that the PC with the Web page will execute to transmit information to the database engine, receive the results back, and display them to the client.

When a Web server executes a CGI program, it typically begins with displaying the HTML form on the screen. The form provides space for specific data items in text boxes, pulldown boxes, checkboxes, radio buttons, and so on. The form typically has a Submit button that a user can click to submit the entry, and some will have a Clear button to reset the data values to their default values. The FORM tag in the HTML Web page specifies a form to complete. More than one fill-out form can be in a single document, but forms cannot be nested (one inside of another).

To create a form, use the </FORM></FORM> tags. Inside the form, identify inputs using the <input> tag. The <input> tag has several inputs including the type, value, size, and maximum length. See Table 20.1.

TABLE 20.1   Input tag

Attribute	Description
Type	Specifies the type of input. Choices include Text, Checkboxes, and Radio (radio buttons). The default type is TEXT.
Value	Inserts labels on the form to indicate the input.
Size	Specifies the size of the input.
Maximum Length	Specifies the maximum length of the input.

To create a Submit Query button, use <INPUT TYPE="SUBMIT">, which will submit the data into the specified database. To create a Reset button, which will reset all values back to their default values, use <INPUT TYPE="RESET">. See Figure 20.4.

```
<html>

<FORM action="http://www.acme.com/script/"
method="GET">

  Name: <INPUT TYPE="TEXT" NAME="Name"
  Size="35"><BR>
  <BR>

  E-mail: <INPUT TYPE="TEXT" NAME="E-mail"
  SIZE="35"><BR>
  <BR><BR>

  Sex: <BR>
  Male<INPUT TYPE="RADIO" NAME="sex">
  Female<Input Type="RADIO": NAME="sex">
  <BR><BR>

  Check if you want to be added to our mailing list:
  <Input Type="checkbox" NAME="mailinglist">
  <BR><BR><BR>

  <INPUT TYPE="SUBMIT">
  <INPUT TYPE="RESET">
</FORM>

</html>
```

FIGURE 20.4   HTML Web page acting as a form

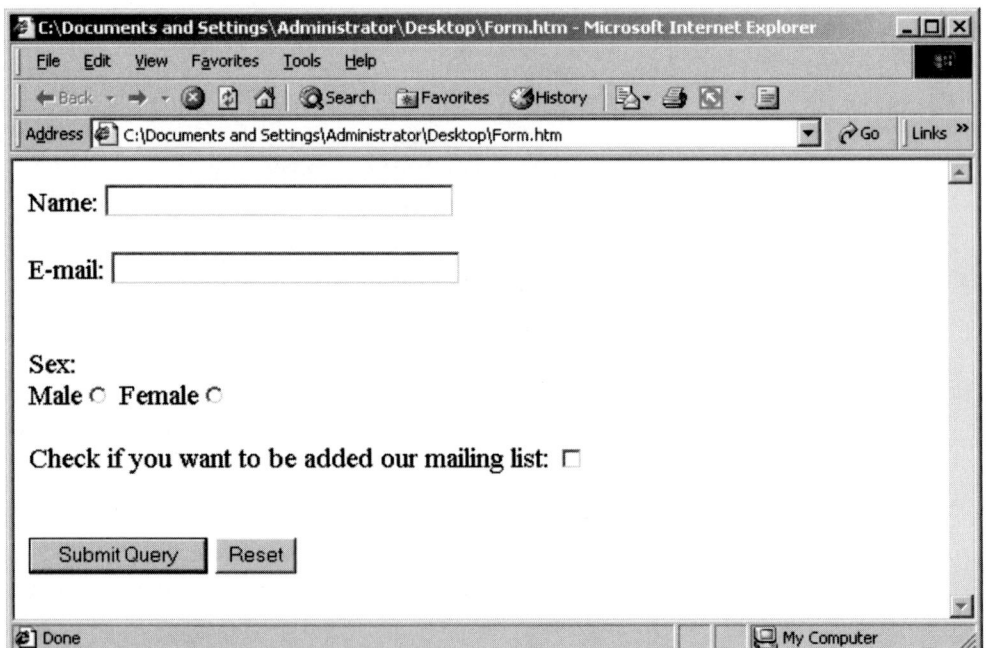

The METHOD attribute, which is entered into the <FORM> tag, specifies what to do with the data that are entered into the form. The <FORM METHOD=POST> will send the information to the database. For example:

```
<FORM ACTION="http://www.acme.com/script" METHOD=POST>
```

The practical extraction and report language (PERL), is a programming language especially designed for processing text. Because of its strong text-processing abilities, Perl has become a popular language for writing CGI scripts. Perl is an interpretive language and so makes it easy to build and test simple programs.

20.2.6   Dynamic HTML and Active Server Pages

For the most part, HTML creates static and unanimated documents. Once a Web page is loaded into a browser window, it does not change in content or form without being reloaded. This truly limits HTML's potential as an interactive multimedia format. To make the Web a more compelling medium, HTML must provide Web authors with the ability to dynamically update content; change the appearance of content; and hide, show, and animate content.

Active server page (ASP) is a specification for a dynamically created Web page with an .ASP extension that utilizes ActiveX scripting. When a browser requests an ASP, the Web server generates a page with HTML code and sends it back to the browser. Thus, ASPs are similar to CGI scripts, but they enable Visual Basic programmers to work with familiar tools.

Control over page layout is one of HTML's most obvious limitations. Even with today's third-generation HTML specifications, Web authors do not have the ability to lay out pages with the pixel-level accuracy that is available to the traditional desktop publishing author. A standard HTML document cannot specify that text and images be located at exact coordinates, on top of each other, or even that the text be displayed in a particular point size. Since the beginning of HTML's existence, Web authors have struggled to get browsers to display HTML content exactly the way they want it.

Dynamic HTML (DHTML) allows content to be displayed with more design flexibility and accuracy through the use of a cascading style sheet (CSS). Using CSS, a standard from the World Wide Web Consortium (W3C), Web authors can define fonts, margins, and line spacing for different parts of an HTML document. In addition to these stylistic improvements, CSS allows the absolute positioning of content by specifying x,y coordinates (and even a z-index, which allows different elements to overlap).

The Microsoft implementation of DHTML also adds built-in multimedia and data objects (treated as properties of CCS) that can be controlled through scripting languages, allowing for stereo sound, on-the-fly image manipulation, and even access to serverside databases.

20.2.7   ActiveX, VBScript, and JavaScript

An ActiveX control can be automatically downloaded and executed by a Web browser. ActiveX is not a programming language, but rather a set of rules for how applications should share information. Programmers can develop ActiveX controls in a variety of languages, including C, C++, Visual Basic, and Java.

An ActiveX control is similar to a Java applet. Unlike Java applets, however, ActiveX controls have full access to the Windows operating system, giving them more power than Java applets. With this power, however, comes a certain risk that the applet may damage software or data on your machine. To control this risk, Microsoft developed a registration system so that browsers can identify and authenticate an ActiveX control before downloading it. Another difference between Java applets and ActiveX controls is Java applets can be written to run on all platforms, whereas ActiveX controls are currently limited to Windows environments.

Related to ActiveX is a scripting language called VBScript that enables Web authors to embed interactive elements in HTML documents. Just as JavaScript is similar to Java, so VBScript is similar to Visual Basic. Currently, Microsoft Internet Explorer supports Java, JavaScript, and ActiveX, whereas Netscape Navigator browsers support only Java and JavaScript (though plug-ins can enable support of VBScript and ActiveX).

623

VBScript is an abbreviated version of Visual Basic Scripting Edition, a scripting language developed by Microsoft and supported by Microsoft's Internet Explorer Web browser. VBScript is based on the Visual Basic programming language, but is much simpler. It enables Web authors to include interactive controls, such as buttons and scroll bars, on their Web pages.

JavaScript is a programmable API that allows cross-platform scripting of events, objects, and actions. It allows the page designer to access events such as start-ups, exits, and users' mouse clicks. JavaScript extends the programmatic capabilities of Netscape Navigator (and to a slightly lesser extent, Microsoft's Internet Explorer) to a wide range of authors, and is relatively easy for anyone who can compose HTML.

20.2.8   Web Page Editors

Today, many Web pages are created using a Web page editing program such as Microsoft Front Page. Using these programs you can create a Web page similar to one in Microsoft Word or Microsoft PowerPoint. In fact, both of these programs also have the capability to save as an HTML format, but they do not have the full capability of creating a Web page like Web page editors.

20.3   SECURE COMMUNICATION OVER THE INTERNET

Most companies with connections to the Internet have implemented firewall solutions to protect the corporate network from unauthorized users coming in—and unauthorized Internet sessions going out—via the Internet. Although firewalls do guard against intrusion and unauthorized use, they cannot guarantee the security of the link between a workstation and server on opposite sides of a firewall nor the security of the actual message being conveyed.

To create this level security, two related Internet protocols, secure sockets layer (SSL) and secure hypertext transfer protocol (S-HTTP), ensure that sensitive information passing through an Internet link is safe from prying eyes. Whereas SSL is designed to establish a secure connection between two computers, S-HTTP is designed to send individual messages securely. SSL and S-HTTP have very different designs and goals, so it is possible to use the two protocols together.

Secure socket layer (SSL) operates at the transport layer, such as at the HTTP, FTP, NNTP, and SMTP levels. When both a client (usually in the form of a Web browser) and a server support SSL, any data transmitted between the two become encrypted. Microsoft Internet Explorer and Netscape Navigator both support SSL at the browser level.

When an SSL client wants to communicate with an SSL-compliant server, the client initiates a request to the server, which in turn sends a X.509 standard certificate back to the client. The certificate includes the server's public key and preferred cryptographic algorithms, or ciphers.

The client then creates a key to be used for that session, encrypts the key with the public key sent by the server, and sends the newly created session key to the server. After it receives this key, the server authenticates itself by sending a message encrypted with the key back to the client, proving that the message is coming from the proper server. After this handshake process, which results in the client and server agreeing on the security level, all data transfer between that client and that server for a particular session is encrypted using the session key. When a secure link has been created, the first part of the URL will change from http:// to https://. SSL supports several cryptographic algorithms to handle the authentication and encryption routines.

Secure hypertext transfer protocol (S-HTTP) is simply an extension of HTTP, and is created by SSL running under HTTP. The protocol was developed as an implementation of the RSA encryption standard. While SSL operates at the transport layer, S-HTTP supports secure end-to-end transactions by adding cryptography to messages at the application layer. Whereas SSL is application independent, S-HTTP is tied to the HTTP protocol.

An S-HTTP message consists of three parts:

- HTTP message
- Sender's cryptographic preferences
- Receiver's preferences

The sender integrates both preferences, which results in a list of cryptographic enhancements to be applied to the message.

S-HTTP, like SSL, can be used to provide electronic commerce without customers worrying about who might intercept their credit card numbers or other personal information. To decrypt an S-HTTP message, the recipient must look at the message headers, which designate which cryptographic methods were used to encrypt the message. Then, to decrypt the message, the recipient uses a combination of his or her previously stated and current cryptographic preferences and the sender's previously stated cryptographic preferences. S-HTTP does not require that the client possess a public key certificate, which means that secure transactions can take place at any time without needing individuals to provide a key (as in session encryption with SSL).

20.4 INTRODUCTION TO INTERNET INFORMATION SERVICES

Internet Information Services (IIS) 5.0 is a built-in Web server found in the Windows 2000 Server family. IIS makes it easy to share documents and information across a company intranet or the Internet. IIS 5.0, the fastest Web server for Windows 2000 Server, is completely integrated with Microsoft Active Directory directory service. This combination of the Web and operating system services makes it possible to deploy scalable and reliable Web-based applications.

The standard Internet services (the Web and FTP servers) reside in a process called Inetinfo. In addition to these Internet services, this process contains the shared thread pool, cache, and logging services of IIS 5.0. IIS includes the following components:

- WWW Server—Supports HTTP, allowing users to publish content to the Internet.
- File transfer protocol (FTP) Server—Supports the FTP protocol to transfer files between computers on a TCP/IP network.
- Simple mail transfer protocol (SMTP) Server—Allows IIS to be an e-mail client so that users can access their e-mail using Web-based applications.
- Network News Transfer Protocol (NNTP) Server—Allows a commercial-grade electronic discussion group or forum. The NNTP is beyond the scope of this book.

By sharing the same security model (user accounts) as Windows 2000 Server, IIS 5.0 eliminates the need for additional user-account administration. IIS 5.0 administration also borrows existing Windows 2000 Server tools such as System Monitor, Event Viewer, and MMC to conduct similar administrative procedures.

Microsoft World Wide Web (WWW) Service supports HTTP, allowing users to publish content to the Internet. The Web is the most graphical service on the Internet and has the most sophisticated linking capabilities. Whether your site is on the intranet or on the Internet, the principles of providing content are the same. You place your files in directories on your Web site so users can view files with a Web browser, such as Microsoft Internet Explorer or Netscape Communicator. The documents include HTML format which includes text, graphics, animation, or video.

20.4.1 Installing IIS

Internet Information Services is installed on Windows 2000 Server by default. You can remove IIS or select additional components by using the Add/Remove Programs application in Control Panel.

To install IIS, add components, or remove components, do the following:

1. Click the Start button, select the Settings option, select the Control Panel option.
2. Double-click the Add/Remove Programs application.
3. Select the Add/Remove Windows Components option
4. Enable the IIS option by marking the checkbox next to the IIS option and click on the Next button. If so desired, change the IIS options by clicking on the Details button.
5. Click on the Finish button.

If you upgrade to Windows 2000, IIS 5.0 will be installed by default only if IIS was installed on your previous version of Windows.

The following directories containing user content will remain on your system after you completely uninstall IIS:

\Inetpub

systemroot\Help\iisHelp

systemroot\system32\inetsrv

The online documentation is also available when performing remote administration tasks. To reach the documentation, start a browser and type http://*servername*/iishelp/iis/misc/ default.asp, where servername is the name of the computer running IIS. If you are at the IIS server, you can also use http://localhost/iishelp/iis/misc/default.asp.

20.4.2 Administering IIS

The IIS snap-in is an administration tool for IIS 5.0 that has been integrated with other administrative functions of Windows 2000. In previous releases, this tool was called the Internet Service Manager.

To launch the IIS console, do the following:

1. Click the Start button, select the Programs option, select the Administrative Tools option, and select the Computer Management option.
2. Under the Server Applications and Services node, expand Internet Information Services.

Another way to launch the IIS is to start the MMC console and use the Internet Information Services snap-in. See Figure 20.5.

With IIS, you can also manage the sites remotely. You can use either the IIS console or you can use a browser-based Internet Services Manager (HTML). Lastly, you can also use Microsoft Terminal Services over a network connection to remotely administer IIS. Although Internet Services Manager (HTML) offers many of the same features as the snap-in, property changes that require coordination with Windows utilities, such as certificate mapping, cannot be made with Internet Services Manager (HTML).

Internet Services Manager (HTML) uses a Web site listed as Administration Web Site to access IIS properties. When IIS is installed, a port number between 2000 and 9999 is ran-

FIGURE 20.5   MMC console with Internet Information Services snap-in

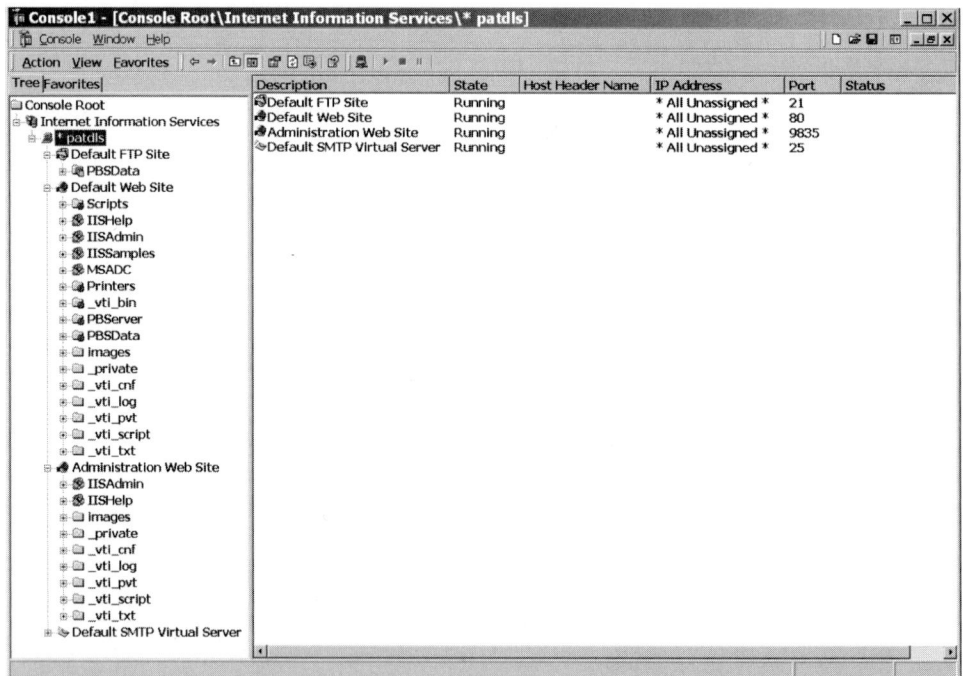

domly selected and assigned to this Web site. The site responds to Web browser requests for all domain names installed on the computer, provided the port number is appended to the address. If basic authentication is used, the administrator will be asked for a user name and password when the site is reached. Only members of the Windows Administrators group can use the site. Website Operators can also administer Web sites remotely.

To enable the browser-based Internet Services Manager (HTML), do the following:

1. In the IIS snap-in on the local computer, open the property sheets for the Administration Web Site and note the TCP port number on the Web Site property sheet.
2. In the Directory Security property sheet, click the Edit button under IP address and domain name restrictions to set permissions for computers that will be used to administer IIS remotely.

To start the browser-based Internet Services Manager (HTML), start a browser and type the domain name for the Web site followed by the assigned port number of the administration site. For example, http://www.acme.com:5453/. See Figure 20.6.

Website Operators are a special group of users who have limited administrative privileges on individual Web sites. Operators can administer properties that affect only their respective sites. They do not have access to properties that affect IIS, the Windows server computer hosting IIS, or the network. Therefore, each Operator can act as the site administrator and can change or reconfigure the Web site as necessary. For example, the Operator can set Web site access permissions, enable logging, change the default document or footer, set content expiration, and enable content ratings features. The Web site Operator is not permitted to change the identification of Web sites, configure the anonymous user name or password, throttle bandwidth, create virtual directories or change their paths, or change application isolation. Because Operators have more limited privileges than Web site Administrators, they are unable to remotely browse the file system and therefore cannot set properties on directories and files, unless a UNC path is used.

FIGURE 20.6   The browser-based Internet Services Manager (HTML)

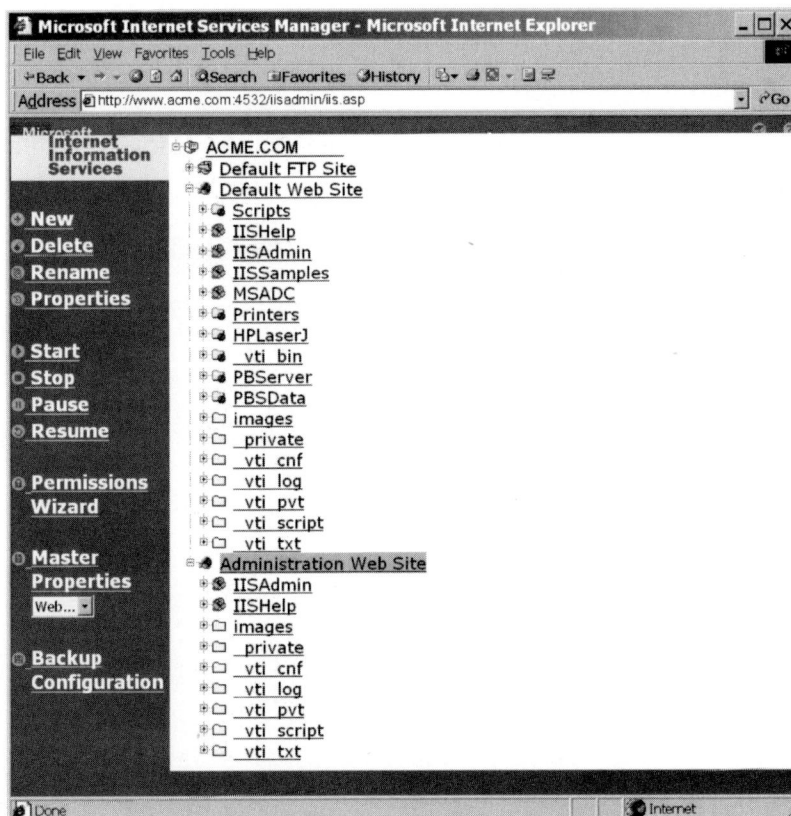

To add an Operator, do the following:

1. In the Internet Information Services snap-in, select the Web site and open its property sheets.
2. On the Operators property sheet, under Web Site Operator, click the Add button. This opens the Add Users and Groups window.
3. Either select a user or group from the Names list or select another name list from the List Names From box.
4. Select a member from a group of users by clicking the Members button and selecting the member from the window.
5. Search for a user or group on a network by clicking the Search button.

To remove an Operator, do the following:

1. In the Internet Information Services snap-in, select the Web site and click the Properties button to display its property sheets.
2. On the Operators property sheet, under Web Site Operator, select the user or groups and click the Remove button. See Figure 20.7.

Note that Web Site Operator accounts need not be members of a Windows Administrators group.

20.4.3 Home Directories

Each Web site or FTP site must have one home directory—a central location for published pages. It contains a home page or index file that welcomes customers and contains links to other pages in your site. The home directory is mapped to your site's domain name or to your server name. For example, if your site's Internet domain name is www.acme.com and your home directory is C:\INETPUB\WWWROOT, then browsers use the URL http://www.acme.com to access files in your home directory. If you have an intranet and the server name is ACME, then the browsers would use the http://acme URL to access files in your home directory.

When you install IIS and create a new Web site, a default home directory is created. The default directory for the World Wide Web server is C:\INETPUB\WWWROOT directory and the default directory for the FTP server is C:\INETPUB\FTPROOT directory. To change the home directory do the following:

1. In the Internet Information Services snap-in, right-click a Web site or FTP site and select the Properties option.

FIGURE 20.7 Specifying Web Site Operators

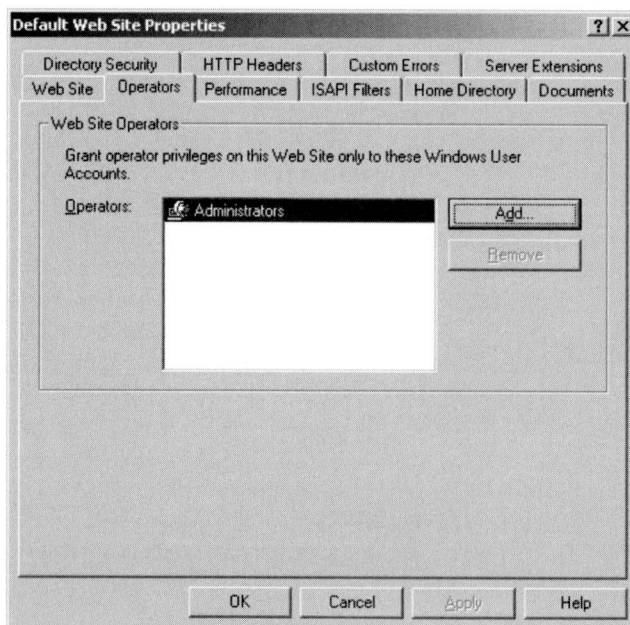

2. Click the Home Directory tab, and then specify where your home directory is located. You can select any of the following:
   - Directory located on a hard disk on your computer
   - Shared directory located on another computer
   - Redirection to a URL (Browsers requesting this URL are forwarded to a new URL. You cannot redirect an FTP directory.)
3. In the text box, specify the path name, share name, or URL of your directory.

If you select a directory on a network share, you may need to enter a user name and password to access the resource. We recommend that you use the *IUSR_computername* account. If you use an account that has administration permissions on the server, clients can gain access to server operations, which would seriously jeopardize the security of your network.

20.4.4 Virtual Servers

Normally, a domain name such as WWW.ACME.COM refers to a single computer host. You can, however, configure a single server to appear as different hosts. You might, for example, create virtual servers, one for each department in your organization, or be an ISP who is hosting several sites for various clients. Each of the sites would be recognized by a different IP address in your DNS server, but all would be hosted on the same server. They are called virtual servers because they are hosted on the same physical server.

To create a virtual server, you must have a unique IP address for the virtual server. If you are acting as your own ISP, simply choose an available IP address from your own subnet. If you connect through an outside ISP, then contact the ISP and request an IP address for each virtual server.

After identifying the IP address of each server, you need to plan how you will create the directory structure to contain the files for the virtual servers. Each virtual server needs its own root directory, which can be located on the same disk as all other virtual servers' directories or on a separate disk. For example, your primary server Web root for www.acme.com may be D:\INETPUB\WWWROOT, and the root directory for CLIENT2.COM may be E:\INETPUB\CLIENT2ROOT; or, if you are placing them on the same disk, you may use D:\INETPUB\WWWROOT for www.acme.com and D:\INETPUB\CLIENT2 for client2.com.

To add new sites to a computer by launching the site wizard, do the following:

1. In the Internet Information Services snap-in, select the computer or a site and click the Action button.
2. Click New and then Web Site or FTP Site to launch the site wizard. Click on the Next button.

FIGURE 20.8 Specifying the folder that will hold the Web pages

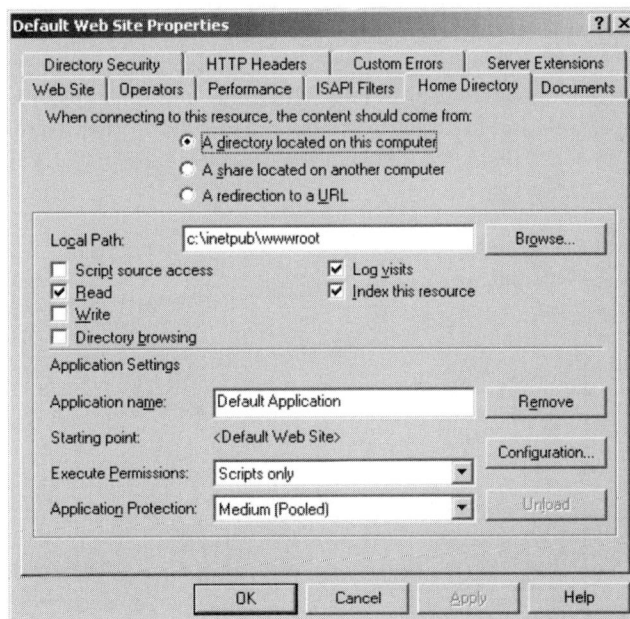

3. Type in the description or name of the Web or FTP site.
4. Enter the IP address and TCP port to be used by the Web or FTP site. Click on the Next button.
5. Specify the home directory to be used by the site. Click on the Next button.
6. Specify the Access Permissions and click on the Next button.
7. Click on the Finish button.

The default Web site uses all of the IP addresses that are not assigned to other sites. Only one site can be set to use unassigned IP addresses.

Multiple domain names can be hosted on a single computer that is running Microsoft IIS using virtual servers. The three ways of hosting virtual servers on IIS are as follows:

1. IP address
2. Port number
3. Host header name

To use multiple IP addresses, add the host name and its corresponding IP address to your DNS servers. Then clients need only type the text name in a browser to reach your Web site. If you use multiple IP addresses, you will need an additional network card for each IP address.

In TCP/IP networking, a **port** is a mechanism that allows a computer to simultaneously support multiple communication sessions with computers and programs on the network. A port directs the request to a particular service that can be found at that IP address. The destination of a packet can be further defined by using a unique port number. The port number is determined when the connection is established. See Table 20.2.

The Internet Assigned Numbers Authority (IANA) defines the unique parameters and protocol values necessary for operation of the Internet and its future development. In the past, these numbers were documented through the RFC document series. Since that time, the assignments have been listed on the IANA Web site (http://www.isi.edu/in-notes/iana/assignments/port-numbers), which is constantly updated and revised when new information is available and new assignments are made. By definition, dynamic ports are randomly assigned and therefore cannot be known until they are assigned. Private ports are not registered with the IANA, but are used by software applications.

By using appended port numbers, your site would only need one IP address to host many sites. For clients to reach your site, then, they would append a port number at the end of the static IP address (except for the default Web site, which uses port 80).

If a visitor attempts to contact your site with an older browser (before Microsoft Internet 3.0 or Netscape Navigator 2.0) that does not support host headers, that visitor is directed to the default Web site assigned to that IP address (if a default site is enabled), which may not necessarily be the site requested. Also, if a request from any browser is received for a site that is currently stopped, the visitor receives the default Web site instead. For this reason, carefully consider what the default Web site displays. Typically, ISPs display their own home page as the default, and not one of their customers' Web sites. This prevents requests for a stopped site from reaching the wrong site. Additionally, the default site can include a script that supports the use of host header names for older browsers.

TABLE 20.2   Default port setting

Service	Port
FTP	21
Telnet	23
SMTP	25
WWW	80
NNTP	119
SSL	443
NNTP with SSL	563

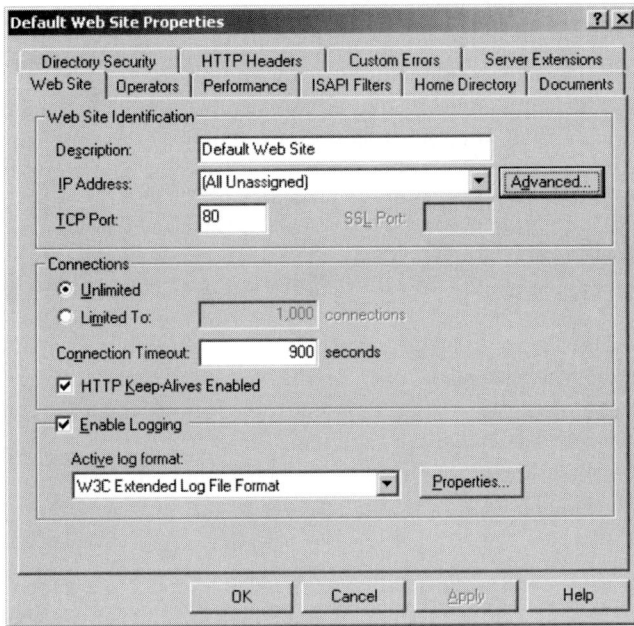

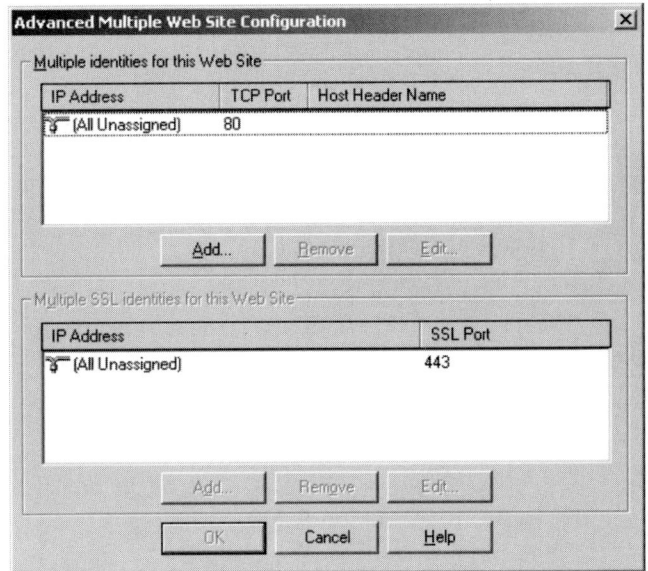

FIGURE 20.9   Web site TCP ports

You can also use host header names with a single static IP address to host multiple sites. Like the previous method, you would still add the host name to your DNS server. The difference is that once a request reaches the computer, IIS uses the host name passed in the HTTP header to determine which site the clients are requesting.

To change or assign an IP address, port, or host header name, right-click the site and select the Properties option. The IP address and TCP port can be set in the Web Site tab. See Figure 20.9. To specify multiple site configuration, click on the Advanced button of the host header name, then click on the Advanced button. When using SSL, you can assign only one host header name to an IP address because the domain name is specified in the server certificate; however, you can have multiple server certificates, multiple IP addresses, and multiple SSL ports per Web site.

20.4.5   Virtual Directories

Before installing IIS, you must do a little planning. First, plan the amount of disk space that you anticipate you will need by determining the total size of files to be included in each folder and add an estimated amount for growth. Next, decide where your Web and FTP files will be stored. By default, IIS creates a folder named INETPUB on drive C. Create a folder on your data disk that will serve as the root folder for all of your Internet resources. For example, if you are using drive D to store the data, create D:\INETPUB or any other name of your choosing. After you have created the root folder for the Internet resources, create root folders within that folder for the Web and FTP services. For example, create folders named \INETPUB\WWWROOT and \INETPUB\FTPROOT.

In addition to creating virtual servers, you can create virtual directories. A virtual directory exists as a physical directory on disk, but it appears under the root service directory even if it is located on another disk. In effect, virtual directories are like a mountable file system in which directories from disparate physical directories and disks can appear to be a homogenous directory structure. For example, assume that your Web root is D:\INETPUB\WWWROOT and contains the general resources for your Web site. You also want to host three different product catalogs, but for fault tolerance, you want to place each catalog on a separate physical disk. So, you might create E:\INETPUB\CAT1ROOT, F:\INETPUB\CAT2ROOT, and G:\INETPUB\CAT3ROOT. Then specify each of these three directories as virtual directories under D:\INETPUB\WWWROOT. When the client browses the site, it appears that the three catalog directories are subdirectories of WWWROOT.

An important point to understand about virtual directories, however, is that they do not appear in directory listings. For example, if users connect to your FTP site, they will see the root folder, its contents, and any physical subdirectories of that root directory. Virtual directories, even though they are considered logical subdirectories of that root, do not appear in the directory listing. To access virtual directories, users must enter the resource name. Typically, these virtual directories will be accessed by using links. Therefore, moving from one virtual directory to another would be transparent to the user.

To create a virtual directory, do the following:

1. In the Internet Information Services snap-in, select the Web site or FTP site to which you want to add a directory.
2. Click the Action button, and then point to New, and select Virtual Directory.
3. When the New Virtual Directory wizard starts, click on the Next button.
4. Provide the alias of the virtual directory and click on the Next button. The alias is the name by which the directory is known under the root home directory.
5. Specify the location of the virtual directory and click on the Next button.
6. Specify the access permissions and click on the Next button.
7. Click on the Finish button.

If you are using NTFS, you can also create a virtual directory by right-clicking a directory in Windows Explorer, clicking Sharing, and then selecting the Web Sharing property sheet.

To delete a virtual directory, do the following:

1. In the Internet Information Services snap-in, select the virtual directory you want to delete.
2. Click the Action button and select Delete.

Deleting a virtual directory does not delete the corresponding physical directory or files.

20.4.6 Default Documents

When a URL is entered into a browser that does not specify a file name, the default document is generated automatically by the server or is designated by the administrator. Default documents can be a directory's home page or an index page containing a site document directory listing. To add a new default document, do the following:

1. Right-click the Web site and select the Properties option
2. On the Documents tab, click the Add button, specify the default document, and click the OK button. See Figure 20.10.

FIGURE 20.10 Specifying the default Web page

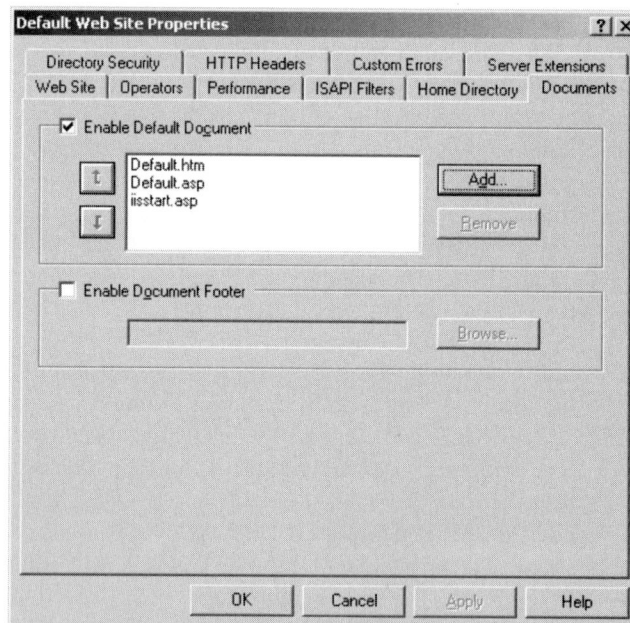

632

Since you can specify more than one default document, they are served in the order in which the names appear in the list. To change the search order, select a document and click on the arrow buttons. To remove a default document from the list, click Remove.

20.4.7 IIS Security

Microsoft IIS has multiple security options for keeping a server and its data from possible intruders and hackers. The four methods of security that you can apply to your IIS computer are as follows:

- IP Access
- User Authentication
- Web Permission
- File Security (NTFS)

You can configure your Web server to prevent specific computers, a group of computers, or an entire network from accessing your Web server content. See Figure 20.11. When a user initially tries to access your Web server content, the server checks the IP address of the user's computer against the server's IP address restriction settings.

To grant access to computers, groups of computers, or domains, do the following:

1. In the IIS snap-in, select a Web site, directory, or file, and open its property sheets.
2. Select the appropriate Directory Security or File Security property sheet. Under IP Address and Domain Name Restrictions, click Edit.
3. In the IP Address and Domain Name Restrictions dialog box, select the Denied Access option. When you select this option, you deny access to all computers and domains, except those to which you specifically grant access.
4. Click Add.
5. In the Grant Access On dialog box, select Single Computer, Group of Computers, or Domain Name options. For more information about these options, click Help.
6. Click the DNS Lookup button to search for computers or domains by name, rather than by IP address. Type in a name, then click OK to close both dialog boxes.

To deny access to computers, groups of computers, or domains, do the following:

1. In the IIS snap-in, select a Web site, directory, or file, and open its property sheets.
2. Select the appropriate Directory Security or File Security property sheet. Under IP Address and Domain Name Restrictions, click Edit.

FIGURE 20.11   IIS security options

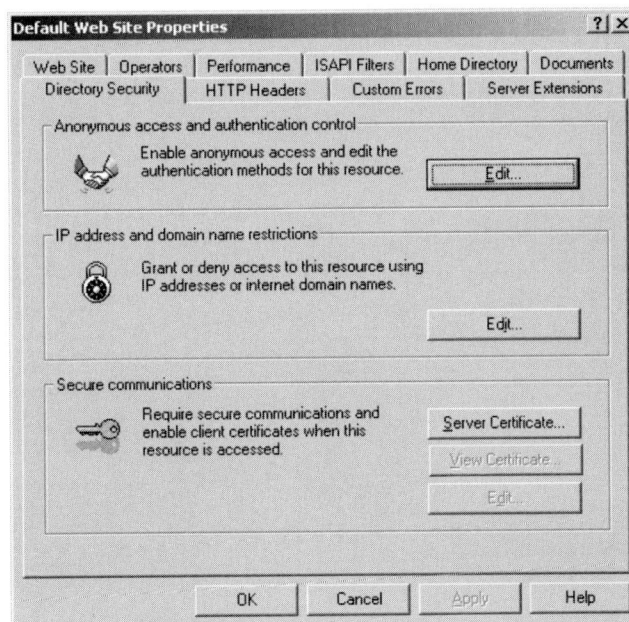

3. In the IP Address and Domain Name Restrictions dialog box, select the Granted Access option. When you select this option, you grant access to all computers and domains, except those to which you specifically deny access.
4. Click Add.
5. In the Deny Access On dialog box, select Single Computer, Group of Computers, or Domain Name options. For more information about these options, click Help.
6. Click the DNS Lookup button to search for computers or domains by name, rather than by IP address. IIS will search on the current domain for the computer, and if found, will enter its IP address in the IP address text box. Click OK to close both dialog boxes.

A group of computers can be either denied or granted access based on the group's network ID and a subnet mask. The network ID is the IP address of a host computer—usually a router for the subnet. The subnet mask determines which part of the IP address is a subnet ID, and which part is a host ID. All computers in a subnet have the same subnet ID, but their own host ID. By specifying a network ID and a subnet mask, you can select a group of computers. See Figure 20.12.

EXAMPLE 1 If the host computer has an IP address of 130.200.16.1 and a subnet mask of 255.255.0.0, all computers in that subnet would have IP addresses that begin with 130.200. Therefore using this IP address and subnet mask, specify all computers from 130.200.0.0 to 130.200.255.255.

EXAMPLE 2 If you have a network ID of 200.200.200.128 with a subnet mask of 255.255.255.192, you will be specifying the IP addresses of from 200.200.200.128 to 200.200.200.191.

EXAMPLE 3 If you have a network ID of 150.150.64.0 with a subnet mask of 255.255.224.0, you will be specifying the IP address of from 150.150.64.0 to 150.150.95.255.

EXAMPLE 4 If you have a network ID of 222.34.42.15 with a subnet mask of 255.255.255.240, you will be specifying the IP address of from 222.34.42.16 to 222.34.42.31.

20.4.8 Authentication

You can require users to authenticate themselves by providing a valid Windows user name and password before accessing any information on your server. For IIS, authentication can be set at the Web site, directory, or file level. IIS provides the following authentication methods for controlling access to content on your server.

WWW Methods
1. Anonymous authentication
2. Basic authentication
3. Digest authentication
4. Integrated Windows authentication
5. Certificate authentication

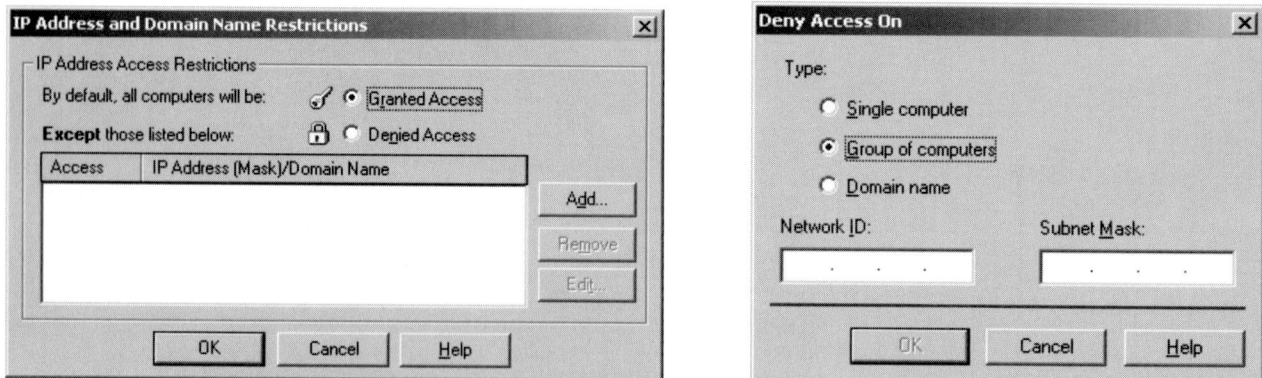

FIGURE 20.12   Specifying the IP address and domain name restrictions

FTP Methods

1. Anonymous FTP authentication
2. Basic FTP authentication

In most situations, all users who attempt to establish a connection with a Web server will log on as an anonymous user. **Anonymous authentication** gives users access to the public areas of your Web or FTP site without prompting them for a user name or password. When a user attempts to connect to your public Web or FTP site, your Web server assigns the user to the Windows user account called IUSR_*computername,* where *computername* is the name of the server on which IIS is running.

By default, the IUSR_computername account is included in the Windows user group called Guests. This group has security restrictions, imposed by NTFS permissions, that designate the level of access and the type of content available to public users.

If you have multiple sites on your server, or if you have areas of your site that require different access privileges, then you can create multiple anonymous accounts, one for each Web or FTP site, directory, or file. By giving these accounts differing access permissions, or by assigning these accounts to different Windows user groups, you can grant users anonymous access to different areas of your public Web and FTP content.

IIS uses the IUSR_*computername* account in the following way:

1. The IUSR_*computername* account is added to the Guests group on the computer.
2. When a request is received, IIS will impersonate the IUSR_*computername* account before executing any code or accessing any files. IIS is able to impersonate the IUSR_*computername* account because the user name and password for this account are known by IIS.
3. Before returning a page to the client, IIS checks NTFS file and directory permissions to see if the IUSR_*computername* account is allowed access to the file.
4. If access is allowed, authentication completes and the resources are available to the user.
5. If access is not allowed, IIS will attempt to use another authentication method. If none is selected, IIS returns an "HTTP 403 Access Denied" error message to the browser.

If anonymous authentication is enabled, IIS will always try to authenticate using it first, even if other methods are enabled. In some cases, the browser will prompt the user for a user name and password.

The anonymous account must have the user right to log on locally. If the account does not have the Log On Locally permission, IIS will be unable to service any anonymous requests. The IIS installation specifically grants the Log On Locally permission to the IUSR_computername account. The IUSR_computername accounts on domain controllers are not given to guest accounts by default and must be changed to Log On Locally to allow anonymous logons.

The **basic authentication** method is a widely used, industry standard method for collecting user name and password information. Basic authentication proceeds as follows:

1. The user's Web browser displays a dialog box where users can enter their previously assigned Windows 2000 account user names and passwords.
2. The Web browser then attempts to establish a connection using this information. (The password of Base64 is encoded before being sent over the network).
3. If the server rejects the information, the Web browser repeatedly displays the dialog box until the user either enters a valid user name and password or closes the dialog box.
4. When your Web server verifies that the user name and password correspond to a valid Windows user account, a connection is established.

The advantage of basic authentication is that it is part of the HTTP specification, and is supported by most browsers. The disadvantage is that Web browsers using basic authentication transmit passwords in an unencrypted form. By monitoring communications on your network, someone could easily intercept and decipher these passwords by using publicly available tools. Therefore, basic authentication is not recommended unless you are confident that the connection between the user and your Web server is secure, such as a direct cable connection or a dedicated line. Note that integrated Windows authentication takes precedence over basic authentication. The browser will choose integrated Windows authentication and will attempt to use the current Windows logon information before prompting the user for a user name and password. Currently, only Internet Explorer, version 2.0 and later, supports integrated Windows authentication.

New to IIS 5.0, **digest authentication** offers the same features as basic authentication but involves a different way of transmitting the authentication credentials. These pass through a one-way process, often referred to as hashing. The result of this process is called a hash, or message digest, and it is not feasible to decrypt it; that is, the original text cannot be deciphered from the hash.

Digest authentication proceeds as follows:

1. The server sends the browser certain information, which will be used in the authentication process.
2. The browser adds this information to its user name and password plus some other information and performs a hash on it. The additional information will help to prevent someone from copying the hash value and reusing it.
3. The resulting hash is sent over the network to the server along with the additional information in clear text.
4. The server then adds the additional information to a plaintext copy it has of the client's password and hashes all of the information.
5. The server then compares the hash value it received with the one it just made.
6. Access is granted only if the two numbers are absolutely identical.

The additional information is added to the password before hashing so that no one can capture the password hash and use it to impersonate the true client. Values are added that help to identify the client, the client's computer, the realm or domain to which the client belongs, and a time stamp. Because digest authentication is a new HTTP 1.1 feature, not all browsers support it. If a noncompliant browser makes a request on a server that requires digest authentication, the server will reject the request and send the client an error message. Digest authentication is supported only for domains with a Windows 2000 domain controller.

Integrated Windows authentication (formerly called NTLM or Windows NT Challenge/ Response authentication) is a secure form of authentication, because the user name and password are not sent across the network. When you enable integrated Windows authentication, the user's browser proves its knowledge of the password through a cryptographic exchange with your Web server, involving hashing.

Integrated Windows authentication can use both the Kerberos v5 authentication protocol and its own challenge/response authentication protocol. If Directory Services is installed on the server, and the browser is compatible with the Kerberos v5 authentication protocol, then both the Kerberos v5 protocol and the challenge/response protocol are used; otherwise only the challenge/response protocol is used.

The Kerberos v5 authentication protocol is a feature of the Windows 2000 Distributed Services architecture. For Kerberos v5 authentication to be successful, both the client and server must have a trusted connection to a key distribution center (KDC) and be Directory Services compatible.

Unlike basic authentication, Integrated Windows authentication does not initially prompt users for a user name and password. The current Windows user information on the client computer is used for the integrated Windows authentication. However, if the authentication exchange initially fails to identify the user, the browser will prompt the user for a Windows user account user name and password, which it will process by using integrated Windows authentication. Internet Explorer will continue to prompt the user until the user enters a valid user name and password, or closes the Prompt dialog box.

Although integrated Windows authentication is secure, it does have two limitations. Only Microsoft Internet Explorer, version 2.0 or later, supports this authentication method. Integrated Windows authentication does not work over HTTP Proxy connections. Therefore, integrated Windows authentication is best suited for an intranet environment, where both user and Web server computers are in the same domain, and where administrators can ensure that every user has Microsoft Internet Explorer, version 2.0 or later.

You can also use your Web server's secure sockets layer (SSL) security features for two types of authentication. You can use a server certificate to allow users to authenticate your Web site before they transmit personal information, such as a credit card number; or you can use client certificates to authenticate users who are requesting information on your Web site. SSL authenticates by checking the contents of an encrypted digital identification submitted by the user's Web browser during the logon process. Users obtain client certificates from a mutually trusted third-party organization. Server certificates usually contain information about your company and the organization that issued the certificate. Client certificates usually contain identifying information about the user and the organization that issued the certificate.

You can associate, or map, client certificates to Windows user accounts on your Web server. After you create and enable a certificate map, each time a user logs on with a client certificate, your Web server automatically associates that user to the appropriate Windows user account. This way, you can automatically authenticate users who log on with client certificates, without requiring the use of either basic, digest, or integrated Windows authentication. You can either map one client certificate to one Windows user account or many client certificates to one account. For example, if you have several different departments or businesses on your server, each with its own Web site, then you can use many-to-one mapping to map all client certificates of each department or company to its own Web site. In this way, each site would provide access only to its own clients.

You can configure your FTP server to allow anonymous access to FTP resources. If anonymous authentication is enabled, IIS will always try to use it first, even if basic authentication is enabled. If you select anonymous authentication for a resource, all requests for that resource will be taken without prompting the user for a user name or password. This is possible because IIS automatically creates a Windows user account called IUSR_*computername,* where *computername* is the name of the server on which IIS is running. This process is similar to Web-based anonymous authentication.

To establish an FTP connection with your Web server by using basic authentication, users must log on with a user name and password that corresponds to a valid Windows user account. If the FTP server cannot verify a user's identity, the server returns an error message. FTP authentication is not very secure, because the user transmits the password and user name across the network in an unencrypted form.

20.4.9 Web Server and NTFS Permissions

Web server permissions allow you to control how users access and interact with specific Web sites. You can use these permissions to control whether users visiting your Web site are allowed to view a particular page, upload information, or run scripts on the site. Different from NTFS permissions, Web server permissions apply to all users who access your Web site.

To change the Web server permissions, right-click the directory, select the Properties option, and select the Directory tab. If you select the Web site, the Web server permissions are under the Home Directory tab. See Tables 20.3 and 20.4.

For example, disabling the Web server's read permission for a particular file will prevent all users from viewing that file, regardless of the NTFS permissions applied to those user

TABLE 20.3 Web file/directory permissions

Permission	Description
Script source access	Allows users to access a source code if either read or write permissions are set. The source code includes scripts in ASP applications.
Read	Allows users to read or download files or directories and their associated properties.
Write	Allows users to upload files and their associated properties to the enabled directory on your server, or changes content in a write-enabled file. Write can only be done with a browser that supports the PUT feature of the HTTP 1.1 protocol standard.
Directory browsing	Allows the user to see a hypertext listing of the files and subdirectories in this virtual directory. Virtual directories will not appear in directory listings; users must know a virtual directory's alias.

TABLE 20.4 Web execute permissions

Permission	Description
None	Only static files, such as HTML or image files, can be accessed.
Scripts only	Only scripts, such as ASP scripts, can be run.
Scripts and Executables	All file types can be accessed or executed.

accounts. However, enabling read permission will allow all users to view the file, unless NTFS permissions that restrict access have also been applied. After setting NTFS permissions, your Web server must be configured with an authentication method to identify users prior to granting access to restricted files.

When you select the Write and Execute checkboxes, you enable users to upload and execute programs on your Web server. In this case, a user could upload and then run a potentially destructive program on your server. When possible, select the Script option rather than the Execute option, because the Script option limits users to executing programs associated with an installed script engine, not any executable application.

20.4.10 Throttling Bandwidth

If the network or Internet connection used by your Web server is also used by other services such as e-mail or news, you may want to limit the bandwidth used by your Web server so bandwidth is available for the other services. If you are running more than one Web site, you can individually throttle the bandwidth used by each site.

To determine how much bandwidth your server is using, use System Monitor to examine the bytes total/sec or current bandwidth counter in the Network Interface object. With a normal load, your server should not use more than 50% of its total available bandwidth of the network connection. The remaining bandwidth is used during peak periods. If your server is subject to large peaks or extreme spikes, it is wise to keep the normal load at even lower than 50%.

To throttle the bandwidth used by IIS, do the following:

1. In the Internet Information Services snap-in, select the computer that is running IIS.
2. On the Internet Information Services property sheet, select the Enable Bandwidth Throttling checkbox.
3. In the Maximum Network Use box, type the maximum number of kilobytes per second (KB/s) that you want to be used by IIS.

To throttle the bandwidth used by an individual Web site, do the following:

1. In the Internet Information Services snap-in, select the Web site and click the Properties button to display its property sheets.
2. On the Performance property sheet, select the Enable Bandwidth Throttling option.
3. In the Maximum Network Use box, type the maximum number of kilobytes per second you want to be used by the site.

20.4.11 Denial of Service

Denial of Service (DoS) is a type of attack on a network that is designed to bring the network to its knees by flooding it with useless traffic. Many DoS attacks, such as the ping of death and teardrop attacks, exploit limitations in the TCP/IP protocols. In the worst cases, for example, a Web site accessed by millions of people can occasionally be forced to temporarily cease operation. A DoS attack can also destroy programming and files in a computer system. Although usually intentional and malicious, a DoS attack can sometimes happen accidentally.

For all known DoS attacks, there are software fixes that system administrators can install to limit the damage caused by the attacks. See Table 20.5. Like viruses, however, new DoS attacks are constantly being invented by hackers.

TABLE 20.5 Types for DoS attacks

Form of Denial of Service Attacks	Description
Buffer Overflow Attacks	The most common kind of DoS attack is simply to send more traffic to a network address than the programmers who planned its data buffers anticipated someone might send. The attacker may be aware that the target system has a weakness that can be exploited or the attacker may simply try the attack in case it might work. A few of the better known attacks based on the buffer characteristics of a program or system include: • Sending e-mail messages that have attachments with 256-character file names to Netscape and Microsoft mail programs • Sending oversized Internet control message protocol (ICMP) packets (also known as the "ping of death") • Sending to a user of the Pine e-mail program a message with a "From" address larger than 256 characters
SYN Attack	When a session is initiated between the transport control program (TCP) client and a server in a network, a very small buffer space exists to handle the usually rapid handshaking exchange of messages that sets up the session. The session-establishing packets include a SYN field that identifies the sequence in the message exchange. An attacker can send a number of connection requests very rapidly and then fail to respond to the reply. This leaves the first packet in the buffer so that other, legitimate connection requests cannot be accommodated. Although the packet in the buffer is dropped after a certain period of time without a reply, the effect of many of these bogus connection requests is to make it difficult for legitimate requests for a session to get established. In general, this problem depends on the operating system providing correct settings or allowing the network administrator to tune the size of the buffer and the timeout period.
Teardrop Attack	This type of denial of service attack exploits the way that the Internet protocol (IP) requires that a packet which is too large for the next router to handle be divided into fragments. The fragment packet identifies an offset to the beginning of the first packet that enables the entire packet to be reassembled by the receiving system. In the teardrop attack, the attacker's IP puts a confusing offset value in the second or later fragment. If the receiving operating system does not have a plan for this situation, it can cause the system to crash.
Smurf Attack	In this attack, the perpetrator sends an IP ping (or "echo my message back to me") request to a receiving site. The ping packet specifies that it be broadcast to a number of hosts within the receiving site's local network. The packet also indicates that the request is from another site—the target site that is to receive the denial of service. (Sending a packet with someone else's return address in it is called spoofing the return address.) The result will be lots of ping replies flooding back to the innocent, spoofed host. If the flood is great enough, the spoofed host will no longer be able to receive or distinguish real traffic.
Viruses	Computer viruses, which replicate across a network in various ways, can be viewed as denial of service attacks where the victim is not usually specifically targeted but simply a host unlucky enough to get the virus. Depending on the particular virus, the denial of service can range from hardly being noticeable to being disastrous.
Physical Infrastructure Attacks	Here, someone may simply snip a fiber-optic cable. This kind of attack is usually mitigated by the fact that traffic can sometimes quickly be rerouted.

20.5 WEBDAV

Web-distributed authoring and versioning (WebDAV) extends the HTTP/1.1 protocol to allow clients to publish, lock, and manage resources on the Web. Since it is integrated into IIS, WebDAV allow clients to do the following:

• Manipulate resources in a WebDAV publishing directory on your server.
• Modify properties associated with certain resources.
• Search the content and properties of files in a WebDAV directory.

To set up a publishing directory, do the following:

1. On the Windows 2000 Desktop, click My Computer.
2. In the Inetpub directory, create a physical directory. You can actually put this directory anywhere you want, except under the Wwwroot directory. Wwwroot is an exception because its default discretionary ACLs are different from those on other directories.
3. In the IIS snap-in, create a virtual directory.
4. Type WebDAV as the alias for this virtual directory, and link it to the physical directory that you created earlier in the second step.
5. Grant read, write, and browsing access permissions for the virtual directory to the users who need to publish and manage those documents. Granting write access does not give clients the ability to modify active server pages (ASPs) or any other script-mapped files. To allow these files to be modified, grant write permission and script source access after creating the virtual directory.

You can access a WebDAV publishing directory through one of the following Microsoft products.

- Windows 2000 connects to a WebDAV server through the Add Network Place wizard and displays the contents of a WebDAV directory as if it were part of the same file system as on your local computer. Once connected, you can drag and drop files, retrieve and modify file properties, and do many other file system tasks.
- You can connect to a WebDAV directory through Internet Explorer 5 on the Windows 2000, Windows NT 4.0, Windows 95, or Windows 98 operating systems. From the File menu in Internet Explorer 5, click Open. In the Open box, type the URL for the desired WebDAV directory. Select the Open as Web Folder box, and click OK.
- Create a document in any Office 2000 application. From the File menu, click Save As. In the left column of the Save As dialog box, click My Network Places.

Because WebDAV is integrated with Windows 2000 and IIS 5.0, it borrows the security features offered by both. These features include the IIS permissions specified in the IIS snap-in and the discretionary access control lists (DACLs) in the NTFS file system.

20.6 HTTP ERROR CODES

HTTP status codes are returned by Web servers to indicate the status of a request. The status code is a three-digit code indicating the particular response. The first digit of this code identifies the class of the status code. The remaining two digits correspond to the specific condition within the response class. Table 20.6 outlines all status codes defined for the HTTP/1.1 draft specification outlined in IETF RFC 2068.

The 4xx errors indicate that the client has sent bad data or a malformed request to the server. Client errors are generally issued by the Web server when a client tries to gain access to a protected area using a bad user name and password. For example, the 403 indicates that the requested resource is forbidden, which generally means you have no privileges to access that page. A 404 indicates that a host server responded to your browser but it cannot find the Web page on the server, which usually means that the Web page was moved or deleted or that you typed in the wrong URL.

The 5xx errors indicate a server error, which indicates that the client's request could not be successfully processed due to some internal error in the Web server. These error codes may indicate something is seriously wrong with the Web server. See Table 20.7.

IIS features its own set of custom errors that provide more informative or "friendly" feedback than the default HTTP 1.1 errors which are returned to client browsers. For example, the HTTP 1.1 404 error message, which by default simply reads "Object Not Found" is expanded to read "The Web server cannot find the file/script you asked for. Please check the URL to ensure that the path is correct. Please contact the server's administrator if this problem persists." These friendly custom error messages are set by default in the default Web site in Internet Service Manager.

To configure friendly error messages in IIS, do the following:

1. In Internet Service Manager, select the Web site, virtual directory, directory, or file in which you would like to customize HTTP errors and click the Properties button.

TABLE 20.6 HTTP error codes (4xx)

Code Number	Code Description
400	Bad Request
401	Unauthorized
	401.1 Logon failed
	401.2 Logon failed due to server configuration
	401.3 Unauthorized due to ACL on resource
	401.4 Authorization failed by filter
	401.5 Authorization failed by ISAPI/CGI application
402	Payment Required
403	Forbidden
	403.1 Execute access forbidden
	403.2 Read access forbidden
	403.3 Write access forbidden
	403.4 SSL required
	403.5 SSL 128 required
	403.6 IP address rejected
	403.7 Client certificate required
	403.8 Site access denied
	403.9 Too many users
	403.10 Invalid configuration
	403.11 Password change
	403.12 Mapper denied access
	403.13 Client certificate revoked
	403.14 Directory listing denied
	403.15 Client access licenses exceeded
	403.16 Client certificate untrusted or invalid
	403.17 Client certificate has expired or is not yet valid
404	Not Found
405	Method Not Allowed
406	Not Acceptable
407	Proxy Authentication Required
408	Request Timeout
409	Conflict
410	Gone
411	Length Required
412	Precondition Failed
413	Request Entity Too Long
414	Request-URI Too Long
415	Unsupported Media Type

2. Select the Custom Errors property sheet. See Figure 20.13.
3. Select the default HTTP error that you would like to change.
4. Click the Edit Properties button.
5. Select URL from the Message Type box.
6. Type /iisHelp/common/<file name>, where <file name> is the HTML file name of the friendly error message.
7. Friendly error messages are installed by default to the following location: <drive letter>:\ WINNT\Help\common. The file names are numbers that correspond to the specific HTTP errors, for example, 400.htm, 401-1.htm, and so on.

20.7 FIREWALLS

A firewall is a system designed to prevent unauthorized access to or from a private network. Firewalls can be implemented in both hardware and software, or a combination of both. Firewalls are frequently used to prevent unauthorized Internet users from accessing private

TABLE 20.7 HTTP error codes (5xx)

Code Number	Code Description
500	Internal Server Error—An internal server error has caused the server to abort your request. This error condition may also indicate a misconfiguration with the Web server. However, the most common reason for 500 server errors is when you try to execute a script that has syntax errors.
501	Not Implemented—This code is generated by a Web server when the client requests a service that is not implemented on the server. Typically, non-implemented codes are returned when a client attempts to POST data to a non-CGI (i.e., the form action tag refers to a nonexecutable file).
503	Service Unavailable—The server is too busy to handle your request or there is some problem along the Internet. Try again later.
502	Bad Gateway—The server, when acting as a proxy, issues this response when it receives a bad response from an upstream or support server.
503	Service Unavailable—The Web server is too busy processing current requests to listen to a new client. This error represents a serious problem with the Web server (normally solved with a reboot).
504	Gateway Timeout—This code is normally issued by proxy servers when an upstream or support server fails to respond to a request in a timely fashion.
505	HTTP Version Not Supported—The server issues this status code when a client tries to talk using an HTTP protocol that the server either does not support or is configured to ignore.

FIGURE 20.13 Customizing the standard HTTP errors

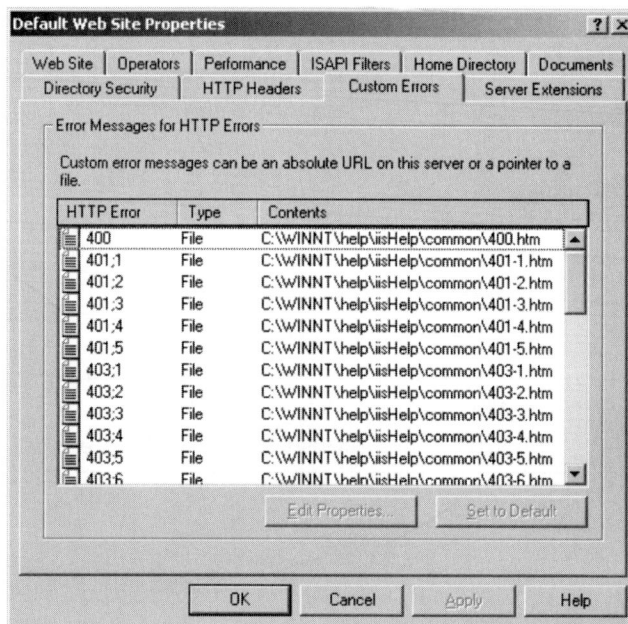

networks that are connected to the Internet, especially intranets. All messages entering or leaving the intranet pass through the firewall, which examines each message and blocks those that do not meet the specified security criteria.

The types of firewall techniques are as follows:

- Packet filter—Looks at each packet entering or leaving the network and accepts or rejects it based on user-defined rules. Packet filtering is fairly effective and transparent to users, but it is difficult to configure. In addition, it is susceptible to IP spoofing.
- Application gateway—Applies security mechanisms to specific applications, such as FTP and Telnet servers. This technique is effective, but can impose a performance degradation.

- Circuit-level gateway—Applies security mechanisms when a TCP or UDP connection is established. Once the connection has been made, packets can flow between the hosts without further checking.
- Proxy server—Intercepts all messages entering and leaving the network. The proxy server effectively hides the true network addresses.

In practice, many firewalls use two or more of these techniques in concert. A firewall is considered a first line of defense in protecting private information.

A proxy server has the authority to act on behalf of other computers on the network. The proxy server serves as proxy by providing access to the TCP/IP networks such as the Internet while keeping the workstation address anonymous. The client workstation makes a TCP/IP-based protocol request, such as to enter a URL into a Web browser to pull up a Web page. The client sends the request to the proxy server and waits for the reply. Then the proxy server receives the request and sends it to the destination address, substituting its server address for the client address. This substitution maintains the anonymity of the client address. Next, the destination processes the request and sends the results back to the proxy server. Finally, the proxy returns the results to the client.

Since the workstation is kept anonymous, the proxy server gives the internal network two major advantages. First, it makes intruder attacks on a machine within the internal network much more difficult since the intruder will not know the address of the machine. Second, you can assign real Internet addresses to the proxy server and assign any addresses to the internal network. Therefore, if you are assigned a class C network, which allows up to 254 computers, and you have a network with 1000 computers, then you only have to assign the class C addresses to the proxy server and to those computers connected directly to the Internet. The other computers can be assigned any network addressing scheme as long as they are configured to go through the proxy server to access the Internet.

A proxy server can dramatically improve performance for groups of users, because it saves the results of all requests for a certain amount of time. Since certain files are constantly being accessed, these files will already be copied to the proxy server. Therefore, when those files are being accessed, the access will be quick. In addition, since these files do not have to be downloaded through the Internet connection, the connection will be a little less congested. Real proxy servers support hundreds or thousands of users. The major online services such as CompuServe and America Online, for example, employ an array of proxy servers.

Proxy servers can also be used to filter requests. For example, a company might use a proxy server to prevent its employees from accessing a specific set of Web sites.

Microsoft offers Microsoft Proxy Server 2.0, which allows clients and servers to access the Internet while keeping your intranet free from intruders. Microsoft Proxy Server 2.0 does not ship with Windows 2000; instead, it can be found as a stand-alone package or on compact disk 3 of the Microsoft BackOffice 4.0 or 4.5. BackOffice is a set of Windows NT Server and related software. Much like IIS and the other administrative tools of Windows 2000, Microsoft Proxy Server is managed from the MMC. Before Windows 2000 Server can use the Microsoft Proxy Server 2.0, download a patch with an update, which can be found on the www.microsoft.com site, for it will function properly.

20.8 BROWSERS

The most common two browsers are Internet Explorer and Netscape Navigator. Not only will this book discuss how to use the browser, but it will also discuss how to configure certain features of the browser that are relevant to managing a network.

20.8.1 Plug-ins

A **plug-in** is a software module that adds a specific feature or service to the browser such as to display or play different types of audio or video messages. The two most common plug-ins are Shockwave and Real Media, but the number of plug-ins available is countless.

Shockwave is a technology developed by Macromedia, Inc. that enables Web pages to include multimedia objects. To create a shockwave object, use Macromedia's multimedia authoring tool called Macromedia Director, and then compress the object with a program called Afterburner. You then insert a reference to the "shocked" file in your Web page. To see a Shockwave object, you need the Shockwave plug-in, a program that integrates seamlessly with your Web browser. The plug-in is freely available from Macromedia's Web site. Shockwave supports audio, animation, and video and even processes user actions such as mouse clicks. It runs on all Windows platforms and on Macintosh.

Streaming technology is a technique for transferring data such that it can be processed as a steady and continuous stream. Streaming technologies are becoming increasingly important with the growth of the Internet, because most users do not have fast enough access to download large multimedia files quickly. With streaming, the client browser or plug-in can start displaying the data before the entire file has been transmitted.

For streaming to work, the client side receiving the data must be able to collect the data and send it as a steady stream to the application that is processing the data and converting it to sound or pictures. Thus, if the streaming client receives the data more quickly than required, the client needs to save the excess data in a buffer. If the data does not come quickly enough, however, the presentation of the data will not be smooth.

There are a number of competing streaming technologies emerging. For audio data on the Internet, the de facto standard is Progressive Network's RealAudio. RealAudio was developed by RealNetworks and supports FM-stereo-quality sound. To hear a Web page that includes a RealAudio sound file, you need a RealAudio player or plug-in. RealVideo is a streaming technology developed by RealNetworks for transmitting live video over the Internet. It uses a variety of data compression techniques and works with both normal IP connections and IP multicast connections.

20.8.2 Cookies

A **cookie** is a message given to a Web browser by a Web server, which is typically stored in a text file on the PC's hard drive. The message is then sent back to the server each time the browser requests a page from the server.

A command line in the HTML of a document tells the browser to set a cookie of a certain name or value. An example of using a cookie is:

```
Set-Cookie: NAME=VALUE; expires=DATE; path=PATH;
domain=DOMAIN_NAME;
```

The main purpose of cookies is to identify users and possibly prepare customized Web pages for them. When you enter a Web site using cookies, you may be asked to fill out a form providing such information as your name and interests. This information is packaged into a cookie and sent to your Web browser, which stores it for later use. The next time you go to the same Web site, your browser will send the cookie to the Web server. The server can use this information to present you with custom Web pages. So, for example, instead of seeing a generic welcome page, you might see a welcome page with your name on it. Some uses of cookies include keeping track of what a person buys using online ordering systems, personalizing a Web site, storing a person's profile, storing user IDs, and providing support to older Web browsers that do not support host header names.

A cookie cannot be used to get data from your hard drive, get your e-mail address, or steal sensitive information about your person. Early implementations of Java and JavaScript could allow people to do this, but for the most part these security leaks have been plugged. Although a cookie typically makes things easier for the user, the feature is not a necessity.

If you want to disallow cookies, do the following:

If you have Internet Explorer 4.0 or greater:
1. Go to the Tools menu and select the Internet Options menu.
2. Click on the Security tab and click on the Custom Level button.
3. In the "Allow cookies that are stored on your computer" section, select Disable.

If you have Netscape 4.0 or greater:
1. Go to the Edit menu and select the Preferences option.
2. Select the Advanced option.
3. In the Cookies section, select Disable Cookies.

20.8.3 Connection Configuration

For the browser to go through a proxy server, the browser must be configured. These configuration settings include automatic configuration, script configuration, or manual specification. The automatic and script configurations enable you to change settings after you deploy Internet Explorer. By providing a pointer to configuration files on a server, you can change the settings globally without having to change each user's computer. This can help reduce administrative overhead and potentially reduce help desk calls about browser settings.

To use automatic configuration, create an Internet Explorer administration kit (IEAK) profile by using the IEAK Profile Manager. The profile consists of an .ins file and any cabinet (.cab) files generated by the Profile Manager. The profile contains information that is used to configure users' browsers. After creating the profile, place it on a server.

To configure the proxy selection and proxy bypass settings in Internet Explorer, do the following:

1. Open the **Tools** menu, and then click on **Internet Options.**
2. Click the **Connections** tab, and then click **LAN Settings.**
3. In the proxy server area, select the **Use a Proxy Server** checkbox.
4. Click **Advanced,** and then fill in the proxy location and port number for each Internet protocol that is supported.

In the cases when only a single proxy server is used for all protocols, enter the proxy location and port number for the HTTP setting, and then select the "Use the same proxy server for all protocols" checkbox. If you want to manually set the addresses, then enable the "Use a proxy server" and specify the address and port number of the proxy server. If you need to specify different addresses and/or port numbers for the various Internet services, click on the Advanced button. See Figure 20.14.

For Netscape, the same basic options are available when you do the following:

1. Open the Edit menu and select the Preferences option.
2. Find and open the Advanced option.
3. Click on the Proxy option. See Figure 20.15.

20.8.4 Internet Explorer Administration Kit

The Internet Explorer administration kit (IEAK) is a software product that makes it easy to create a custom browser that can be deployed across a network or via Web download. See Table 20.8 for features and descriptions.

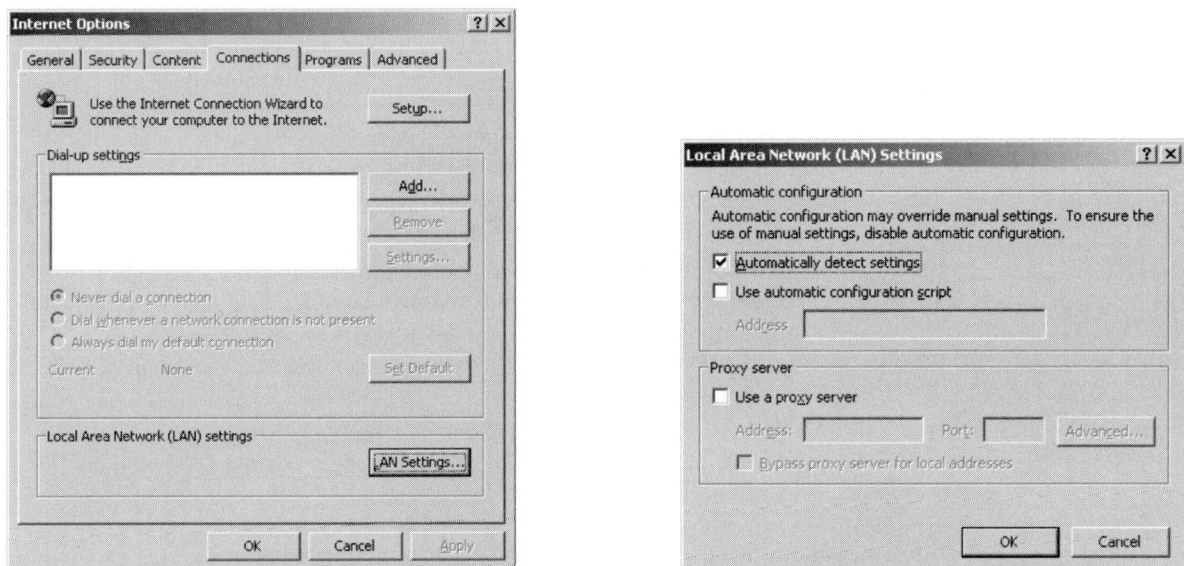

FIGURE 20.14   Configuring the proxy selection and proxy bypass settings in Internet Explorer   (continued)

FIGURE 20.14 Configuring the proxy selection and proxy bypass settings in Internet Explorer _(continued)_

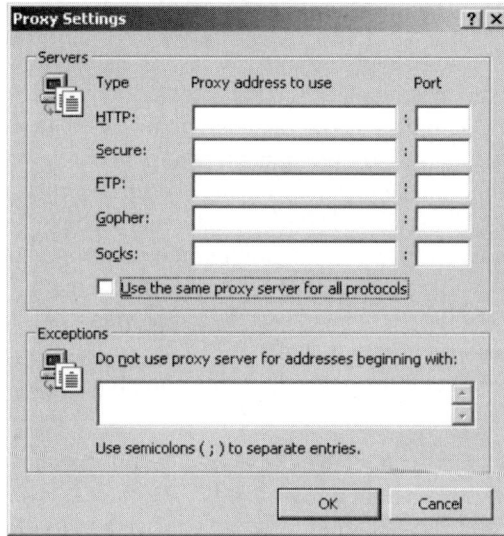

FIGURE 20.15 Configuring the proxy settings for Netscape

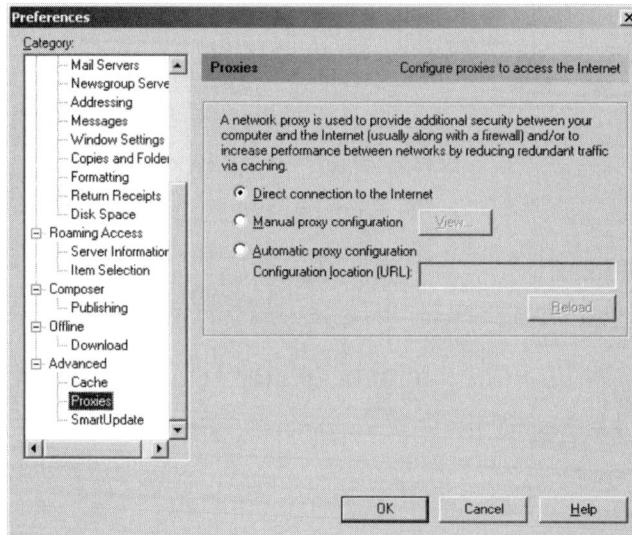

TABLE 20.8 Internet Explorers administration kit

Feature	Description
Customization Wizard	Provides a simple processor for administrators to generate customized versions of Internet Explorer.
Custom Browser Settings	To control the cost of managing the browser, customizes and locks down virtually every setting.
Enhanced Setup and Deployment Options	Provides greater flexibility and choice about how to distribute your custom browser to your subscribers.
On-Demand Install	Lets administrators include components in an IEAK distribution that will not be installed until the user requires them.
Easier Remote Connections	The connection manager administration kit (CMAK) allows you to provide a custom dialer and phone book to simplify deployment of preconfigured dial-up/VPN configurations.
Manage Browser Settings	The Internet Explorer Profile Manager makes it easy for administrators to manage browser installations.

SUMMARY

1. Web pages are written using the hypertext markup language (HTML).
2. A Web server is a computer that is equipped with the server software that uses Internet protocols such as hypertext transfer protocol (HTTP) and file transfer protocol (FTP) to respond to Web client requests on a TCP/IP network via Web browsers.
3. The Web browser is the client program/software that you run on your local machine to gain access to a Web server.
4. A gopher was an early system that predates the World Wide Web for organizing and displaying files on Internet servers.
5. The World Wide Web has several search engines that can be used to locate desired documents. A search engine is a program that searches documents for specified keywords and returns a list of the documents where the keywords were found.
6. To create or modify a Web page, use a text editor such as DOS's EDIT or Windows Notepad.
7. To create this level security, two related Internet protocols, secure sockets layer (SSL) and secure hypertext transfer protocol (S-HTTP), ensure that sensitive information passing through an Internet link is safe from prying eyes.
8. Internet Information Services (IIS) 5.0 is a built-in Web server found in the Windows 2000 Server family.
9. You can use either the IIS console or a browser-based Internet Services Manager (HTML).
10. Web site Operators are a special group of users who have limited administrative privileges on individual Web sites.
11. Each Web site or FTP site must have one home directory—a central location for published pages.
12. Normally, a domain name such as WWW.ACME.COM refers to a single computer host. You can, however, configure a single server to appear as different hosts.
13. When a URL is entered into a browser that does not specify a file name, the default document is generated automatically by the server or is designated by the administrator.
14. Microsoft IIS has multiple security options for keeping a server and its data from possible intruders and hackers including restricting IP address, user authentication, Web permission, and NTFS file security.
15. Web-distributed authoring and versioning (WebDAV), extends the HTTP/1.1 protocol to allow clients to publish, lock, and manage resources on the Web.
16. HTTP status codes are returned by Web servers to indicate the status of a request.
17. A firewall is a system designed to prevent unauthorized access to or from a private network. Firewalls can be implemented in both hardware and software, or a combination of both.
18. A plug-in is a software module that adds a specific feature or service to the browser such as to display or play different types of audio or video messages.
19. A cookie is a message given to a Web browser by a Web server, which is typically stored in a text file on the PC's hard drive.

QUESTIONS

1. You have created a Web site for your company on the Internet using TCP/IP with both DNS and WINS enabled. Users complain, however, that they can only access the corporate Web page by its IP address, 142.33.112.86, not by its domain name, www.acme.com. What is the most likely reason that users cannot connect using http://www.acme.com?
   a. NetBEUI must be installed.
   b. The IP address is not mapped to www.igminc.com in DNS.
   c. The virtual server for the IP address points to the wrong directory.
   d. DHCP is not enabled.

647

2. You want to create a Web site to which the general Internet public will be denied access; only company employees with specific IP addresses will be granted access to the site. You open the Advanced tab of the IIS WWW Service Properties page, choose Denied Access, then Add. In the Grant Access On dialog box, you select Group of Computers. You enter the network ID 200.200.200.64 and the subnet mask 255.255.255.224. Which IP addresses will be allowed to access your Web site?
   a. 200.200.200.32 through 200.200.200.39
   b. 200.200.200.32 through 200.200.200.95
   c. 200.200.200.33 through 200.200.200.64
   d. 200.200.200.64 through 200.200.200.95

3. Pat has no trouble logging on remotely to her company's network with her Windows 2000 user account. However, when she tries to browse the company's Web site she finds she can access several of the site's virtual directories but not the \WWWROOT directory. What is the best way to correct this problem?
   a. Disable Windows NT challenge/response authentication.
   b. Enable Directory Browsing Allowed on \WWWROOT.
   c. Grant read permission to \WWWROOT.
   d. Add Pat to the Administrators group since access to the \WWWROOT directory is restricted to the Administrators group.

4. You want to block a range of IP addresses from your company's IIS Web site. Opening the Advanced tab of the WWW Service Properties page, you choose Granted Access, then Add. In the Deny Access On dialog box, you select Group of Computers. Which combination of network ID and subnet mask would you enter to deny access to IP addresses 195.49.200.64 through 195.49.200.95?
   a. network ID: 195.49.200.95; subnet mask: 255.255.255.192
   b. network ID: 195.49.200.64; subnet mask: 255.255.255.224
   c. network ID: 195.49.200.95; subnet mask: 255.255.255.240
   d. network ID: 195.49.200.64; subnet mask: 255.255.255.248

5. One way that Internet browser users can access a virtual Web directory is to click on an HTML page link. What other method can Internet browser users utilize to access a virtual Web directory?
   a. Enter the appropriate virtual server path in the browser address field.
   b. Enter the appropriate virtual directory path in the browser address field.
   c. Enter the appropriate virtual directory's IP address in the browser address field.
   d. Enter the appropriate URL address in the browser address field.

6. Internet users complain that they cannot access your Web site. Of the following, which one is the most likely cause of the problem?
   a. They are not accessing port 21.   c. They are not accessing port 70.
   b. They are not accessing port 23.   d. They are not accessing port 80.

7. You would assign the same host name to multiple IP addresses in DNS to _____ .
   a. allow replication between IIS servers
   b. increase WINS name resolution performance
   c. perform load balancing
   d. maximize available network bandwidth

8. What permissions must you grant so that anonymous users can upload, but not download, files using FTP?
   a. execute   c. read only
   b. read and write   d. write only

9. You enable anonymous user access to your IIS Web site WEBWORLD. The IUSR_WEBWORLD account is given read permission on the \WWWROOT directory. The Windows NT default permissions are not changed. Who can access the \WWWROOT directory?
   a. everyone
   b. everyone except users with valid Windows NT user accounts

c. only anonymous users

d. only guests

10. Paul has an SSL-enabled client laptop that he takes with him on the road to access the Internet. Which one of the following directories can he access?

a. http:// directories and https:// directories

b. http:// directories, https:// directories, and httpssl:// directories

c. only https:// directories

d. only httpssl:// directories

11. All your company's employees use their Windows 2000 accounts to access files on the corporate FTP site. The FTP site is hosted on an IIS computer named Web on the intranet. What must you do to allow all Internet users to view directories and download files from your corporate FTP site? (Select all that apply)

a. In the FTP site's properties, assign the Directory Browsing Allowed permission.

b. Enable anonymous access for the FTP site.

c. Assign the NTFS read permission to IUSR_Web user account for the FTP site's content directories.

d. On the Directory Security tab of the FTP site's properties, grant access to the near-side IP address of the router on the Internet connection.

12. You are a Web site administrator for a company that publishes on the Internet. One user has been trying to bring your Web server down by performing DoS attacks. By monitoring network traffic on your Internet connection, you determine that the attacking user connects to the Internet through an Internet service provider (ISP). What is the easiest and most reasonable way to protect your Web site from that user's attack?

a. Disable anonymous access to the Web site.

b. Configure the Web site to use SSL encryption.

c. Configure the Web site to deny access to the IP address of the attacking user's computer.

d. Configure the Web site to deny access to the network to which the attacking user's computer belongs.

13. You want to administer Web and FTP sites on an IIS computer in a remote branch of your company. The branch's network is protected by a firewall that does not allow you to use Internet Service Manager. How can you connect to the administration Web site on the branch's IIS computer?

a. Specify the remote IIS computer's DNS name and the IISAdmin virtual directory's alias.

b. Specify the remote computer's DNS name and TCP port 8827.

c. Specify the remote IIS computer's DNS name and the TCP port that is assigned to the administration Web site.

d. Specify the host header name that is assigned to the default Web site on the remote IIS computer and the TCP port that is assigned to the administration Web site.

14. Which one of the following statements correctly describes one way that cookies can be used?

a. Cookies are special viruses that can be used by malicious individuals to cause damage to Internet users' systems through their Web browsers.

b. Cookies can be used to encrypt the contents of Web pages without using encryption key pairs.

c. Cookies are special antivirus programs that can be used by Web servers to ensure that client computers that connect to the server's Web sites do not contain viruses.

d. Cookies can be used to provide support to older Web browsers that do not support host header names.

15. Your Web site on an IIS computer is experiencing a high volume of user requests. What should you do to provide the best load balancing and fault tolerance?

a. Install an additional Web site.

b. Install an additional virtual server.

c. Install an additional virtual directory.

d. Install an additional IIS computer.

16. You are setting up an Internet Web site for a government agency that has multiple branches nationwide. The information that is published on the Web site is highly sensitive, and only authorized officials are allowed to access it. Each user must be granted access to only the information that is pertinent to their work duties. All the data transmitted over the Internet must be encrypted. How should you configure the Web site?

   a. Implement digital signature on the Web server to ensure data authenticity.

   b. Grant each user account the appropriate NTFS permissions for the Web site's files and user client certification to authenticate users; use SSL data encryption.

   c. Use cookies to authenticate users, and instruct all the authorized users to upgrade their Web browsers to use 128-bit encryption.

   d. Grant each user account the appropriate NTFS permission for the Web site's files and use Windows NT challenge/response authentication.

EXERCISES

EXERCISE 1: CREATING A WEB PAGE

1. Create a folder called ACME on the C drive.
2. Click the Start button, select Programs, select Accessories, and select Notepad.
3. In Notepad, type in the following:

```
<HTML>
<HEAD>
<TITLE>THIS IS THE TITLE</TITLE>
<META NAME="KEYWORD" CONTENTS="HOT, IMPORTANT">
<META NAME="DESCRIPTION" CONTENTS="This is a sample website">
</HEAD>

<BODY>This is the body of the web page<BODY>
</HTML>
```

4. Save the file on the ACME folder and name it DEFAULT.HTM. Ensure that the file is called DEFAULT.HTM. Leave Notepad open.
5. Double-click on the DEFAULT.HTM file. Leave Internet Explorer open.
6. Change back to Notepad and modify the file to the following:

```
<HTML>
<HEAD>
  <TITLE>THIS IS THE TITLE</TITLE>

  <META NAME="KEYWORD" CONTENTS="HOT,IMPORTANT">
  <META NAME="DESCRIPTION" CONTENTS="This is a sample website">
</HEAD>

<BODY>
  <CENTER>TITLE</CENTER>
  This is the body of the webpage
  <OL>
    <LI>First
    <LI>Second
    <LI>Third
  </OL>

  <FONT COLOR="FF0000">RED
  <FONT COLOR="00FF00">GREEN
  <FONT COLOR="0000FF">BLUE
```

```
   <BR><BR>
   <FONT COLOR="000000">END
<BODY>
</HTML>
```

7. Save the file in Notepad.
8. Change back to Internet Explorer and notice that the Web page has not changed.
9. For Internet Explorer to reread the file, click on the Refresh button.

EXERCISE 2: CREATING A WEB SITE

1. Open the Internet Services Manager from the Administrative Tools menu.
2. In Internet Information Services, expand the server in the Console tree.
3. Right-click Default Web Site, and then click Stop.
4. Right-click Server, select the New option, and click Web Site. Click the Next button.
5. On the Web Site Description page, type ACMEXX Web Site where XX represents your computer number. Click on the Next button.
6. On the IP address and Port Settings page, verify that the Web site uses all unassigned IP addresses and TCP port 80. Click the Next button.
7. On the Web Site Home Directory page, in the Path box, type C:\ACME. Verify that the "Allow anonymous access to this Web site" checkbox is selected, and then click the Next button.
8. On the Web Site Access Permissions page, verify that the read and run script permissions checkboxes are selected. Click the Next button and click the Finish button.
9. Right-click the ACMEXX Web Site and select Properties.
10. On the Operators tab, verify that only the Administrators group is listed as an operator.
11. On the Documents tab, verify that Default.htm is listed.
12. Click on the OK button.
13. Start Internet Explorer and type in http://localhost.
14. In Internet Explorer, type in http://localhost/iishelp/iis/misc/default.asp.
15. In the Internet Services Manager, right-click the Administration Web Site and select Properties. In the Web Site tab, record the TCP port number.
16. In Internet Explorer, type 3463

EXERCISE 3: CONFIGURE BASIC SECURITY

1. In the Internet Information Services, right-click the AcmeXX Web Site and select Properties.
2. On the Directory Security tab, under Anonymous Access and Authentication Control, click the Edit button.
3. In the Authentication Methods dialog box, clear the Anonymous Access and Integrated Windows Authentication checkboxes. Select the Basic Authentication check box. When the Internet Services Manager dialog box appears, read the warning and click the Yes button.
4. Click on the OK button to close the Authentication Methods dialog box and click the OK button to close the AcmeXX Web Site Properties dialog box.
5. In Internet Explorer, type in http://localhost.
6. In the Enter Network Password dialog box, type ACMEXX\CBROWN for user name and Cbrown's password. Click on the OK button.
7. Minimize Internet Explorer.
8. In Internet Services Manager, right-click the ACMEXX Web Site and select the Explore option.
9. In Windows Explorer, right-click Default.htm and select Properties.

10. On the Security tab, clear the "Allow inheritable permissions from parent to propagate to this object" checkbox. In the Security dialog box, click Remove.

11. Click the Add button. In the Select Users, Computers, or Groups dialog box, under Name, click Administrators, click Add, and then click OK.

12. Allow the Administrators group the full control permission, and then click OK.

13. Close Windows Explorer.

14. Restore Internet Explorer.

15. Click the Refresh button.

16. When Internet Explorer prompts you for your network credentials, try to log in as ACMEXX\CBROWN.

17. Log in as ACMEXX\ADMINISTRATOR with the appropriate password.

18. Close Internet Explorer.

19. Right-click ACME Web Site, and then select Properties.

20. On the Directory Security tab, under Anonymous Access and Authentication Control, click Edit.

21. Clear the Basic Authentication checkbox, and then click OK.

22. In the Authentication Methods dialog box, select the Integrated Windows Authentication checkbox. Click the OK button.

23. Close Internet Information Services.

24. Start Internet Explorer and connect to http://localhost. Note that this worked because you used your current login information to access the Web site.

25. Close all open windows.

21 Introduction to E-mail

INTRODUCTION

To make any corporate network complete, you must have an e-mail system. Today's modern e-mail systems are complex software applications and services. For Windows NT and Windows 2000, you would typically use Microsoft Exchange to host the e-mail system. Unfortunately, because Microsoft Exchange is so complex, this chapter is designed only as an overview of e-mail systems.

OBJECTIVES

1. Explain how the e-mail system uses the Internet to deliver messages.
2. Explain the function of MIME and how it relates to the Internet.

21.1 E-MAIL OVERVIEW

Electronic mail (e-mail) is the transmission of messages over communications networks. The messages can be notes entered from the keyboard or electronic files stored on disk. Most mainframes, minicomputers, and computer networks have an e-mail system. Some e-mail systems are confined to a single computer system or network, but others have gateways to several computer systems, enabling users to send electronic mail anywhere in the world. Companies that are fully computerized make extensive use of e-mail, because it is fast, flexible, and reliable. In recent years, the use of e-mail has exploded. By some estimates, there are now 25 million e-mail users sending 15 billion messages per year.

Most e-mail systems include a rudimentary text editor for composing messages, but many allow you to edit your own messages using any editor you want. You then send the message to the recipient by specifying the recipient's address. You also can send the same message to several users simultaneously.

All online services and Internet service providers (ISPs) offer e-mail, and most also support gateways so that you can exchange mail with users of other systems. Usually, it takes only a few seconds or minutes for mail to arrive at its destination. This is a particularly effective way to communicate with a group, because you can broadcast a message or document to everyone in the group at once.

21.1.1   E-mail and the Internet

No matter which e-mail package or service you use to host the e-mail, your e-mail travels the same road as all Internet-based information including web page downloads. That is, your e-mail traverses the Internet backbone. The sender creates an e-mail message on an application. The client system is the **user agent (UA).** When the user sends the message, it is transmitted to the user's Internet mail server.

Once the message reaches the Internet mail server, it enters the Internet's **message transfer system (MTS).** The MTS relies on other Internet mail servers to act as **message transfer agents (MTAs),** which relay the message toward the receiving UA. Once an MTA passes the message to the recipient's Internet mail server, the receiving UA can access the message.

RFC 822 defines the standard format for e-mail messages, dividing the message into two parts: an envelope and its contents. The **envelope** contains information that is needed to transmit and deliver an e-mail message to its destination. The **contents** is the message that the sender wants delivered to the recipient.

The envelope contains the e-mail address of the sender, the e-mail address of the receiver, and a delivery mode, which in our case states that the message is to be sent to a recipient's mailbox. We can divide the contents of the message into two parts, a header and a body. The header is a required part of the message format, and the sending UA automatically includes it at the top of the message; the user does not input this information. The receiving UA may reformat the header information or delete it entirely to make the message easier for the recipient to read.

The header contains detailed information about who sent the message, who received the message, and how the message got from the sending point to the receiving point. In addition, the header displays the date of the message, the times at which the different MTAs received the message, and the unique ID of the message.

The body of the message contains the actual text that the sender typed and is separated from the header by a "null" line. RFC 822 does not define the message body, because it can be anything the user enters, so long as it is ASCII text.

A standard e-mail address should be set up as follows:

```
<mailbox ID>@<domain name>
```

The mailbox ID is the name of an individual mailbox on a local machine. The domain name is the name of a valid domain registered in the domain name service (DNS). The DNS servers are key to Internet e-mail, because they allow MTAs to find the machine specified in the recipient's e-mail address.

21.1.2   Simple Mail Transfer Protocol

The transmission of an e-mail message through the Internet relies on the **simple mail transfer protocol (SMTP),** which is defined in RFC 821. SMTP specifies the way a UA establishes a connection with an MTA and the way it transmits its e-mail message. MTAs also use SMTP to relay the e-mail from MTA to MTA, until it reaches the appropriate MTA for delivery to the receiving UA.

The interactions that occur between two machines, whether a UA to an MTA or an MTA to another MTA, have similar processes and follow a basic call-and-response procedure. The main difference between a UA-to-MTA transaction and an MTA-to-MTA transaction is that with the latter, the sending MTA must locate a receiving MTA.

To do this, the sending MTA contacts the DNS to look up the domain name specified in the recipient e-mail address. The DNS may return the IP address of the domain name—in which case the sending MTA tries to establish a mail connection to the host at that domain—or the DNS may return a set of mail-relaying records that contain the domain names of intermediate MTAs that can act as relays to the recipient. In this case, the sending MTA tries to establish a mail connection to the first host listed in the mail-relaying record.

When an MTA is sending to an MTA, the sending MTA chooses a receiving MTA, which may be the final destination of the message or an intermediate MTA that will relay the message to another MTA. Next, the sending MTA requests a TCP connection to the receiving MTA. The receiving MTA responds with a server ID and a status report, which indicates whether it is available for the mail transaction. If not, the transaction is over; the sending MTA can try again later or attempt another route. If the receiving MTA is free to handle a session, it will accept the TCP connection.

The sending MTA then sends a Hello command followed by its domain name information to the receiving MTA, which responds with a greeting. Next, the sending MTA sends a Mail From command that identifies the e-mail address from which the message originated, as well as a list of the MTAs through which the message has passed. This information is also known as a return path. If the receiving MTA can accept mail from that address, it responds with an OK reply.

The sending MTA then sends a Rcpt To command, which identifies the e-mail address of the recipient. If the receiving MTA can accept mail for that recipient (it may perform a DNS lookup to verify this, specifically using the mail exchange (MX) records), then it responds with an OK reply. If not, it rejects that recipient. (An e-mail message may be addressed to more than one recipient, in which case this process is repeated for each recipient address.)

Once the receiving MTA identifies the recipient's address, the sending MTA sends the Data command. The receiving MTA accepts command by responding with OK. It then considers all succeeding lines of data to be the message text. Once the sending MTA gets an OK reply, it starts sending the message. The sending MTA signals the end of the message by transmitting a line that contains only a period (.).

When the receiving MTA receives the signal for the end of the message, it replies with an OK to signal its acceptance of the message. If for some reason the receiving MTA cannot process the message, it will signal the sending MTA with a failure code. After the message has been sent to the receiving MTA and the sending MTA gets an OK reply, the sending MTA can either start another message transfer or use the Quit command to end the session.

Once the receiving MTA accepts the message, it reverses its role and becomes a sending MTA, contacting the MTA next in line for the relay of the message. The process stops once the message reaches the Internet mail server that services the recipient specified in the Rcpt To e-mail address.

If at any point along the way an MTA cannot deliver the e-mail, it generates an error report, also known as an undeliverable mail notification. The MTA uses MTAs identified in the return path to relay the error report back to the original sender.

21.1.3   MIME

Initially, the Internet e-mail system was limited to simple text messages because SMTP, the protocol used to transport mail across the Internet, could carry only 7-bit ASCII text. In the United States, the main ASCII standard used for e-mail is US-ASCII. This version of ASCII offers only a basic set of characters—128 characters in all, each represented as a 7-bit binary number. This ASCII set was designed to cover the English alphabet, including both uppercase and lowercase letters, and the numbers 0 through 9, as well as some other characters.

If you use e-mail today, you know that it is not just simple text anymore. You can use rich text with italics, boldface, bullets, and other types of enriched formatting. You can even embed graphics in documents. But because SMTP was developed to handle only basic text messages in a 7-bit format—as laid out by RFC 822, which defines the standard for Internet text messaging—these more sophisticated data formats cannot be sent via e-mail. The problem with the 128-character set is it cannot accommodate rich text. In addition, you cannot use foreign language characters that are not represented in US-ASCII. Therefore, the 7-bit format required by SMPT prevents users from sending other types of data via e-mail, for example, the 8-bit binary data found in many executable files and in files created by applications such as Microsoft Word. For e-mail to handle diverse data such as text, word-processing documents, and images, users could share all sorts of data without having to ship disks or make any actual real-time network connections to copy or download files.

To overcome the limitation of SMPT's 7-bit ASCII format, today's e-mail utilizes the **multipurpose Internet mail extension (MIME).** MIME makes the data appear as standard e-mail messages to the Internet's SMTP servers regardless of the data it contains, even though it did not change the way it handles such data. In other words, the solution did not require that all Internet mail servers be upgraded to a new version of SMTP. In effect, the transport system remained untouched.

As defined in RFC 2045, MIME provides three main enhancements to standard e-mail. First, with MIME, e-mail can contain text that goes beyond basic US-ASCII, including various keystrokes such as different line and page breaks, foreign language characters, and enriched text. Second, users can attach different types of data to their e-mail, including such files as executables, spreadsheets, audio, and images. Third, users can create a single e-mail message that contains multiple parts, and each part can be in a different data format. For example, you can compose a single e-mail message that consists of a plaintext message, an image file, and a binary-based document such as a Word file.

RFC 2045 defines seven types of e-mail contents that MIME can package and pass across the Internet: text, image, audio, video, application, message, and multipart. Each of these data types comes in various formats, or subtypes. (MIME also augments types and subtypes with certain parameters, which are specified in RFC 2045; However, these parameters are too detailed to include in this overview of MIME.)

Obviously the text type of e-mail content supports messages that are carrying text; however, within the text type, MIME also supports the plain subtype, which is usually standard 7-bit ASCII. MIME also supports the rich text subtype, which allows for some simple formatting features such as page breaks.

The image type supports image files, and its subtypes include the graphics interchange format (GIF) and a compressed image format developed by the Joint Photographic Experts Group (JPEG). In the words of RFC 2045, the video type supports, "time-varying picture images," and for now, a compressed video format developed by the Motion Picture Experts Group (MPEG) is its only subtype. The audio type supports audio data, and its only subtype is basic.

According to the RFC, there is no one ideal audio format in use today, so the developers of MIME tried to define a subtype that would be the lowest common denominator. The basic subtype for audio signifies "single channel audio encoded using 8-bit ISDN mu-law" at a sampling rate of 8 KHz.

The application type supports two types of data: data that are meant to be processed by an application and data that do not fall into any of the other categories. For now, it supports the octet-stream subtype, which means the message can carry arbitrary binary data. It also supports the postscript subtype, meaning the message can be sent to print as a PostScript file. If a mail agent receives a message whose content subtype it does not recognize, then by default it will attempt to pass the message on as an application-type message with a subtype of octet-stream (or, application/octet-stream).

The remaining two content types allow for special handling of an e-mail message. For instance, the message type allows an e-mail to contain an encapsulated message. The external-body subtype allows an e-mail to indicate an external location where the intended body of the message resides. That way, the user can choose whether to retrieve the message body. The message type also allows MIME to send a large e-mail message as several small ones (the subtype for this is partial). The receiving MIME-enabled mail agent can then open the smaller e-mail messages and reassemble them into the original long version.

Finally, the multipart type allows an e-mail message to contain more than one body of data. The mixed subtype allows users to mix different data formats into one e-mail message. The alternative subtype allows a message to contain different versions of the same data, each version in a different format. MIME mail agents can then select the version that works best with the local computing environment. The digest subtype allows users to send a collection of messages in one e-mail, such as the kind used with Internet mailing lists sent in digest form. Finally, the parallel subtype allows mixed body parts, but the ordering of the body parts is not important.

To package the different data formats into the 7-bit ASCII format, MIME uses five different encoding schemes: 7 bit, 8 bit, binary, quoted printable, and base64. As you will soon see, however, only the quoted-printable and base64 schemes actually encode data.

The 7-bit scheme tells mail agents that the message contents are in plain ASCII. For this reason, no encoding is necessary as all mail systems should support ASCII. The 8-bit scheme indicates that the contents contain 8-bit characters. The mail agents, then, encode this information using their preferred means, if available. Because not all mail agents use the same encoding method for 8-bit characters, there is a good chance the 8-bit characters will not appear correctly when the e-mail is opened. For this reason, the 8-bit scheme currently is not a reliable encoding scheme. The binary scheme, because it is similar to the 8-bit scheme, shares the same problem.

The quoted-printable scheme is used for text that contains a mixture of 7-bit and 8-bit characters. Essentially, it allows 7-bit characters to go unencoded and converts each 8-bit character into a set of three 7-bit characters. As a result, mail servers and mail agents see an e-mail that contains only 7-bit characters.

The base64 scheme is used for data that are not text, such as data that constitute an executable file. Base64 works by breaking the data down into sets of three octets, each containing 24 bits. It then converts each set of 24 bits into a four-character sequence. (In other words, every 6 bits of data are represented by a character.) The characters used in the sequencing come from a set of 65 characters, all of which can be found in any version of ASCII. Because it is in ASCII, data encoded by base64 should be readable by any mail server or mail agent.

Using MIME is simple. If your mail system supports MIME, it automatically chooses the data type and encoding scheme. It then adds MIME headers to the body of the message. These headers tell receiving mail agents that they have received a MIME message and indicate how the mail agents should handle the message.

The main headers are MIME Version, Content Type, and Content Transfer Encoding. The last two headers refer to the data type found in the message and the scheme used to encode the data, respectively. Some other headers are Content Description, which lets you type in a description of the message (much like SMTP's subject header), and Content ID, which is similar to SMTP's Message ID.

21.1.4 Retrieving E-mail

When the e-mail arrives at the recipient's Internet mail server, the user agent can employ different methods to access, or retrieve, its e-mail from the MTA. For instance, the majority of companies rely on proprietary e-mail packages, such as Microsoft Exchange or cc:Mail, to handle e-mail operations on the local network. There are a few different models for this operation and a few different protocols that can handle the task, including proprietary protocols found in commercial e-mail packages and two Internet standards-based protocols, post office protocol 3 (POP-3) and the Internet message access protocol 4 (IMAP-4).

When an e-mail reaches the designated recipient's mail server, the message is placed in a message store, which is also called a post office. In its most basic form, the message store, located on a server, is usually some type of file system that holds delivered mail for access by users. You can think of the message store as one large directory with subdirectories dedicated to each user. These subdirectories are also known as mailboxes. A more advanced form of message store would allow users to create personal folders to store read and unread messages, and it would allow users to create archives for groups of messages. Other advanced features include the ability to perform keyword searches and to create a hierarchy of folders.

The message store component of an e-mail system is usually located on a different machine than the message transport component, which handles the delivery of incoming messages and the transmission of outgoing messages. The reason for their separation is that the transport component usually handles a high volume of operations. So, if both systems were on the same machine, the traffic of inbound and outbound mail would experience slower performance, as would users when they try to retrieve and manipulate mail in their mailboxes.

Once an e-mail message is delivered to the message store, it is ready for retrieval by the recipient. There are three basic models for message access: offline, online, and disconnected.

The offline model is the most basic of the three. In the offline model, a client connects to the mail server and downloads all messages designated for that particular recipient. Once downloaded, the messages are erased from the server. The user then processes, manipulates, and stores the messages locally at the client machine. The advantages of the offline model are (1) it requires a minimal amount of server connect time since the client accesses the server only periodically for mail downloads; (2) because the mail processing is performed at the client, the offline model will not devour server resources; and (3) it requires less server storage space because mail is deleted from the server once it has been downloaded by a client. The offline model also has certain drawbacks. (1) Because mail is downloaded to a particular machine, you must use that machine to access processed mail. Therefore, if you downloaded e-mail to your desktop computer and then went on the road with your laptop, you would not be able to read those same messages off the mail server from your laptop. (2) Because all processing and storage is performed at the client machine, the client must have enough resources to perform the task.

With the online model, all e-mail processing and manipulation is performed at the server. In fact, all the messages remain on the server even after users have read them. In some implementations, users can save e-mail to the local client. As you might expect, the online model requires a constant connection to the server whenever users need to access and work with their e-mail. Consequently, this model's high connect time could mean that more users are sharing bandwidth at any given time. Because the server is burdened with all message processing and storage, you will need a more powerful server to handle an online e-mail system than you would with an offline system. In addition, users cannot work with their e-mail unless they are online, meaning they must be at a client that can access the mail server even to do the simplest task such as reread an old message. Conversely, because all mail is stored and processed at the server, you can access your e-mail from any client that can connect to the server. For the same reason, this model requires fewer resources at the client machine. In addition, online-type systems usually offer enhanced features such as the ability to create a multitude of personal folders to organize e-mail and the ability to create archives for storing messages.

The disconnected model combines elements of both the offline and online models. Using this system, a user connects to a mail server to retrieve and download e-mail. Users can then process the mail on the local client. Once the user is finished with the messages, the user once again connects to the mail server and uploads any changes. With this model, the mail server acts as the main repository of the user's e-mail. The disconnected model's strengths are that users can access e-mail from a variety of clients that have access to the mail server. Users can also process mail offline, which translates to a shorter server connect time. The model does, however, require a sufficient amount of resources on both the server and the client.

The majority of e-mail packages—Microsoft Mail, Microsoft Exchange, and Lotus cc:Mail—follow the online model of e-mail access. These products use their own protocols to retrieve messages from Internet mail servers and store them on local servers. Users can then use the products' client programs to access and work with mail from message stores on the proprietary e-mail servers. The attractiveness of these proprietary packages lies in the added features they offer. For instance, they usually offer advanced mail manipulation such as folders, hierarchical folders, and archives. Some offer the ability to create bulletin board systems for general corporate communications. In addition, some offer message notification when a new e-mail arrives for a particular user. They also usually offer searching abilities, so users can find e-mails with particular headings, or look up the addresses of other users on the system.

One of the first e-mail access protocols developed was the **post office protocol (POP),** the latest version of which is POP-3. In essence, **POP-3** follows the offline model of message access and revolves around send-and-receive types of operation. There have been recent attempts, however, to remodel POP-3 so that it has some online capabilities such as saved-message folders and status flags that display message states; but using POP-3 in online mode usually requires the additional presence of some type of remote file system protocol.

IMAP-4 (IMAP stands for Internet Message Access Protocol) is the more advanced of the standards-based message access protocols. It follows the online model of message access, although it does support offline and disconnected modes. IMAP-4 offers an array of up-to-date features, including support for the creation and management of remote folders and folder hierarchies, message status flags, new mail notification, retrieval of individual MIME body parts, and server-based searches to minimize the amount of data that must be transferred over the connection.

21.1.5 Messaging Application Programming Interface

Messaging application programming interface (MAPI) is a standardized set of C functions placed into a code library known as a dynamic link library (DLL). The MAPI library is also available to Visual Basic application writers through a Basic-to-C translation layer. The functions were originally designed by Microsoft, but they have received the support of many third-party vendors. The standard library of messaging functions allows Windows applications developers to take advantage of the Windows messaging subsystem, supported by default with Microsoft Mail or Microsoft Exchange. By writing to the generic MAPI, any Windows application can become "mail-enabled," in which an e-mail can be sent from within a Windows application with the document you are working on as an attachment. Since MAPI standardizes the way messages are handled by mail-enabled applications, each such application does not have to include vendor-specific code for each target messaging system.

21.1.6 X.400

X.400 is a set of standards relating to the exchange of electronic messages (such as e-mail, fax, voice mail, or telex). It was designed to let you exchange e-mail and files with the confidence that no one besides the sender and the recipient will ever see the message, that delivery is assured, and that proof of delivery is available if desired.

One goal of X.400 is to enable the creation of a global electronic messaging network. Just as you can make a telephone call from almost anywhere in the world to almost anywhere else in the world, X.400 hopes to make that possibility a reality for electronic messaging. X.400 only defines application-level protocols and relies on other standards for the physical transportation of data such as X.25, frame relay, or ATM.

The address scheme that X.400 uses is called the originator/recipient address (O/R address). It is similar to a postal address in that it uses a hierarchical format. While a postal address hierarchy is country, zip code, state, city, street, and recipient's name, the O/R address hierarchy consists of countries, communication providers (administrative domain name), companies (private domain name), organization, surname, and given name. For example, while you might use Pat.Regan@acme.com for an Internet e-mail address, X.400 would use

```
G=Pat; S=Regan; OU=acme; PRMD=acme.com; ADMD=mci; C=us
```

21.2 MICROSOFT EXCHANGE

One popular e-mail server is **Microsoft Exchange,** which runs on a Windows NT or a Windows 2000 server. The largest unit, and the one at the top of the Exchange hierarchy, is the organization. All other Exchange structures are contained under this unit. The organization is parent to the sites. This relationship permits an administrator to configure something at the organization level and have it apply or be available at all sites. Because an organization encompasses an entire Exchange system, each company or business should create only one organization.

Sites are logical groupings of one or more Exchange servers. Even though resources reside on different servers in the site, the site groups all those resources without reference to their locations. This grouping makes using resources in the site very easy. For example, say that a certain mailbox physically resides on a site server, called the mailbox's home server. Senders do not need to know the physical location of the mailbox in order to send it a message. The principle applies to public folders in a site. The particular server on which a public folder is stored is of no concern to the users wanting to access it. From the user's perspective, a site creates a transparent messaging environment.

Exchange servers comprise the final main structure in the Exchange hierarchy. These computers run the Windows NT or Windows 2000 Server operating systems and the Exchange Server software. The Exchange servers are the physical location for mailboxes, folders, and other data and information for the site. Individual servers, while inheriting certain configuration parameters from the site, can also be individual configured. All Exchange objects and related processes are created and managed by the software components that comprise the Exchange products.

The Exchange components are executable programs that perform the Exchange functions. Some are in the form of EXE files, while others are in the form of DLLs. They are referred to as core components because they are necessary for Exchange to be operational. They are also referred to as services, because they run as services on the Microsoft Windows NT or Windows 2000 Server operating systems. The core components include the following:

- Directory Service (DS)
- Information Store (IS)
- Message Transfer Agent (MTA)
- System Attendant (SA)

21.2.1 Directory Service

The **Directory Service (DS)** component creates and manages the storage of all information about Exchange objects in an organization. Objects are resources such as mailboxes, distribution lists, public folders, and servers. The database that stores the objects is called the Directory. Objects in the Directory are organized in a hierarchical structure. The main purpose of the Directory is to allow users and administrators the ability to easily locate resources to use or configure. The other Exchange components also reference the Directory for information. The Exchange Directory is patterned after the X.500 standard.

21.2.2 Information Store

The **Information Store (IS)** creates and manages the message database on an Exchange server. All types of data can be stored in this database, including e-mail messages, electronic forms, word-processing documents, spreadsheets, and graphics. Users access the IS content through their mailbox and folders in their client applications. The Information Store is the back-end component that makes many of the Exchange features possible.

The **Public Information Store** contains all the public folders of a particular server. Public folders are containers for shared information. In a sense, they are like containers for shared information, or a public mailbox. Different types of data can be stored in a public folder, such as e-mail messages, word-processing documents, and graphic files.

The **Private Information Store** contains all user mailboxes. Security makes mailboxes accessible only to their owners and to others who have been given access permission. By default, a user's messages are stored in their mailbox in this database. Although mailboxes are always located in the Private Information Store on an Exchange server, messages do not have to be stored there. A user can specify a personal location for message storage. This type of folder is called a personal folder and has a PST file extension. A personal folder can be located on a server or on the client computer. When they are sent a message, the message still first goes to their mailbox in the Private Information Store, and then is rerouted to a personal folder. The advantage of storing messages in a personal folder, especially if it is on a local computer, is that the user can access it without logging on to the server. The advantages of keeping message storage in the Private Information Store, which is the default, are centralized backup and fault tolerance.

By default, Exchange servers contain both the Public and Private databases; but they can be configured to store only one of the two. If only the Private Information Store were configured, that server would act as a dedicated mailbox server. A dedicated folder server would only have the Public Information Store configured. Dedicated servers are created to improve the performance of information access.

21.2.3 Message Transfer Agent and System Attendant Service

The **Message Transfer Agent (MTA)** manages the routing of messages within a site, between sites, and to foreign message systems. The MTA uses other components, called connectors, to manage the transfer of data, while the MTA handles the routing functions. The MTA could be compared with a post office that receives mail and routes it to another post office. The MTA component is based on the X.400 standard.

When an MTA component receives a message to be forwarded, it first stores it, in order to determine a route. It then looks at the message recipients distinguished name (DN). If that address cannot resolve the next route, the MTA looks at the O/R address. In either case, the MTA compares the address with the various locations and paths in the routing table. If the MTA finds a path on which to send the message, it sends it on its way. Because there could be more than one path, the X.400 standard allows an administrator to assign costs to different paths. Costs allow priorities to be given to different routes. When faced with multiple paths, the MTA chooses the lowest-cost path. If the MTA cannot forward a message, a non-delivery report (NDR) is sent to the message originator.

The **System Attendant (SA) service** monitors and logs information about the other Exchange services. It also builds and maintains the routing tables for the site.

21.2.4 Optional Exchange Components

Whereas core components are required for normal operations, optional components provide additional servers that might be needed in your Exchange environment. Many of the optional components are connector components.

The **Internet Mail Service (IMS) connector** enables an Exchange system to interoperate with an SMTP mail system. The IMS permits Exchange users and SMTP mail users to send mail to and receive mail from each other. The IMS connector also permits an organization with geographically disperse locations to use the Internet as a messaging backbone. Messages can be sent from an Exchange user, through the Internet, to the Exchange recipient.

An Exchange system can interoperate with a Microsoft Mail system through the Microsoft Mail Connector. Microsoft Mail is a mail product for PC or AppleTalk networks. The **Microsoft Mail connector** can translate Exchange messages to and from the Microsoft Mail format. This permits Exchange users and Mail users to exchange mail.

A **Lotus cc:Mail connector** permits Exchange to interoperate with the Lotus cc:Mail system. Users of each system can exchange mail with the other system. As with the other connectors, this connector permits organizations to implement Exchange with another messaging system and still have interoperability.

Microsoft Schedule+ performs personal and group calendaring, scheduling, and task management. The group scheduling features allow a user who wants to schedule a meeting to view the schedules of other users. The user who is scheduling the meeting can see if a particular time period is unscheduled for the other users (a free period), or if it is already scheduled (busy period). The **Schedule+ Free/Bus connector** allows Microsoft Exchange with Schedule+ 7.5 and Microsoft Mail with Schedule+ 1.0 to share their incompatible free/busy information.

SUMMARY

1. Electronic mail (e-mail) is the transmission of messages over communications networks.
2. No matter which e-mail package or service you use to host the e-mail, your e-mail travels the same road as all Internet-based information including web page downloads.
3. The sender creates an e-mail message on an application. When the user sends the message, it is transmitted to the user's Internet mail server.
4. The client e-mail system is the user agent (UA).
5. Once the message reaches the Internet mail server, it enters the Internet's message transfer system (MTS).
6. The MTS relies on other Internet mail servers to act as message transfer agents (MTAs), which relay the message toward the receiving UA.
7. Once an MTA passes the message to the recipient's Internet mail server, the receiving UA can access the message.
8. E-mail messages have two parts, an envelope and its contents.
9. The envelope contains the e-mail address of the sender, the e-mail address of the receiver, and a delivery mode, which states that the message is to be sent to a recipient's mailbox.
10. The header contains detailed information about who sent the message, who received the message, and how the message got from the sending point to the receiving point.

In addition, the header displays the date of the message, the times at which the different MTAs received the message, and the unique ID of the message.

11. The body of the message contains the actual text that the sender typed.

12. The transmission of an e-mail message through the Internet relies on the simple mail transfer protocol (SMTP).

13. To figure out the address of a MTA, the mail system uses DNS.

14. Initially, the Internet e-mail system was limited to simple text messages because SMTP, the protocol used to transport mail across the Internet, could carry only 7-bit ASCII text.

15. To overcome the limitation of SMPT's 7-bit ASCII format, today's e-mail utilizes the multipurpose Internet mail extensions (MIME), which makes many data types appear as standard e-mail messages.

16. When an e-mail reaches the designated recipient's mail server, the message is placed in a message store, which is also called a post office. In its most basic form, the message store, located on a server, is usually some type of file system that holds delivered mail for access by users.

17. Once an e-mail message is delivered to the message store, it is ready for retrieval by the recipient. There are three basic models for message access: offline, online, and disconnected.

18. One of the first e-mail access protocols developed was the post office protocol (POP), the latest version of which is POP-3. In essence, POP-3 follows the offline model of message access and revolves around send-and-receive types of operation.

19. IMAP-4 is the more advanced of the standards-based message access protocols. It follows the online model of message access, although it does support offline and disconnected modes.

20. Messaging application programming interface (MAPI) is a standardized set of C functions placed into a code library known as a dynamic link library (DLL), which allows Windows applications developers to take advantage of the Windows messaging subsystem, supported by default with Microsoft Mail or Microsoft Exchange. By writing to the generic MAPI, any Windows application can become "mail-enabled," in which an e-mail can be sent from within a Windows application with the document you are working on as an attachment.

21. X.400 is a set of standards relating to the exchange of electronic messages (such as e-mail, fax, voice mail, or telex). It was designed to let you exchange e-mail and files with the confidence that no one besides the sender and the recipient will ever see the message, that delivery is assured, and that proof of delivery is available if desired.

22. One popular e-mail server is Microsoft Exchange, which runs on a Windows NT or a Windows 2000 server.

23. The largest unit, and the one at the top of the Exchange hierarchy, is the organization. Sites are logical groupings of one or more Exchange servers. Even though resources reside on different servers in the site, the site groups all those resources without reference to their locations.

24. The database that stores the objects is called the Directory. Objects in the Directory are organized in a hierarchical structure. The main purpose of the Directory is to allow users and administrators the ability to easily locate resources to use or configure.

25. The Information Store (IS) creates and manages the message database on an Exchange server.

26. The Public Information Store contains all the public folders of a particular server. Public folders are containers for shared information.

27. The Private Information Store contains all user mailboxes. Security makes mailboxes accessible only to their owners and to others who have been given access permission.

28. The Message Transfer Agent (MTA) manages the routing of messages within a site, between sites, and to foreign message systems. The MTA uses other components, called connectors, to manage the transfer of data, while the MTA handles the routing functions.

29. The System Attendant (SA) service monitors and logs information about the other Exchange services. It also builds and maintains the routing tables for the site.

1. E-mail can include which of the following? (Choose all that apply)
    a. text
    b. pictures
    c. video clips
    d. sound clips
    e. executable files

2. Which protocol is responsible for transmitting the e-mail message through the Internet?
    a. MTA
    b. SNMP
    c. SMTP
    d. DNS

3. When you send an e-mail message to PREGAN@ACME.COM, how does the MTA know which server to send the message to?
    a. It looks up the address using the WINS server.
    b. It looks up the address using the DNS server.
    c. It looks in its internal database to find the address.
    d. It sends it to the router and the router will figure out what to do with it.

4. Which protocol allows the embedding of pictures and video clips into an e-mail message?
    a. SNMP
    b. DNS
    c. IP
    d. MIME

5. Which protocol is responsible to receive an e-mail message and forward it to the appropriate mailbox?
    a. SNMP
    b. MIME
    c. DNS
    d. POP

6. What e-mail server is made to work with Windows NT and Windows 2000?
    a. Microsoft Exchange
    b. IIS
    c. Lotus Mail
    d. NetWare GroupWise

7. What are the two main parts of an e-mail message? (Choose two)
    a. envelope
    b. contents
    c. header
    d. sending address
    e. destination address

8. What component of Microsoft Exchange creates and manages the message database?
    a. Public Folder
    b. MTA
    c. X.400 Connector
    d. Information Store

Appendix

APPENDIX A   BINARY CONVERSION

The most commonly used numbering system is the **decimal number system.** In a decimal numbering system, each position contains 10 different possible digits. Since there are 10 different possible digits, the decimal number system involves numbers with a base of 10. These digits are 0, 1, 2, 3, 4, 5, 6, 7, 8, and 9. To count values larger than 9, each position away from the decimal point in a decimal number increases in value by a multiple of 10. See Table A1.

EXAMPLE 1   decimal number: 234

2	3	4
2×10^2	3×10^1	4×10^0
200	30	4

Therefore, the value is $200 + 30 + 4 = 234$

The **binary number system** is another way to count. The binary system is less complicated than the decimal system because it has only two digits, a zero (0) and a one (1). A computer represents a binary value with an electronic switch known as a transistor. If the switch is on, it allows current to flow through a wire or metal trace to represent a binary

TABLE A1 Decimal number system

7th Place	6th Place	5th Place	4th Place	3rd Place	2nd Place	1st Place
10^6	10^5	10^4	10^3	10^2	10^1	10^0
1,000,000	100,000	10,000	1,000	100	10	1

665

TABLE A2   One-digit binary number

Binary Equivalent	Decimal Equivalent
0	0
1	1

TABLE A3   Two-digit binary number

Binary Equivalent	Decimal Equivalent
0 0	0
0 1	1
1 0	2
1 1	3

TABLE A4   Four-digit binary number

Binary Equivalent	Decimal Equivalent
0 0 0 0	0
0 0 0 1	1
0 0 1 0	2
0 0 1 1	3
0 1 0 0	4
0 1 0 1	5
0 1 1 0	6
0 1 1 1	7
1 0 0 0	8
1 0 0 1	9
1 0 1 0	10
1 0 1 1	11
1 1 0 0	12
1 1 0 1	13
1 1 1 0	14
1 1 1 1	15

value of one (1). If the switch is off, it does not allow current to flow through a wire, representing a binary value of zero (0). See Table A2. The on switch is also referred to as a high signal, whereas the off switch is referred to as a low signal.

If you use two wires to represent data, the first switch can be on or off and the second switch can be on or off, giving you a total of four combinations or four binary values. See Table A3. If you use four wires to represent data, then you can represent 16 different binary values. See Table A4. Since each switch represents two values, each switch that is used doubles the number of binary values. Therefore, the number of binary values can be expressed with the following equation:

$$\text{Number of Binary Numbers} = 2^{\text{Number of binary digits}}$$

Therefore, one wire allows $2^1 = 2$ binary numbers, 0 and 1. Two wires allow $2^2 = 4$ binary numbers, 0, 1, 2, and 3. Four wires allows $2^4 = 16$ binary numbers.

TABLE A5   Binary number system

8th Place	7th Place	6th Place	5th Place	4th Place	3rd Place	2nd Place	1st Place
2^7	2^6	2^5	2^4	2^3	2^2	2^1	2^0
128	64	32	16	8	4	2	1

Question: How many values do 8 bits represent?

Answer: Eight bits, which is a byte of information, represent $2^8 = 256$ characters.

Much like decimal numbers, the binary digits have placeholders that represent certain values. See Table A5.

EXAMPLE 2   Convert the binary number 11101010 to a decimal number.

1	1	1	0	1	0	1	0
1×2^7	1×2^6	1×2^5	0×2^4	1×2^3	0×2^2	1×2^1	0×2^0
128	64	32	0	8	0	2	0

Therefore, the binary number of 11101010 is equal to the decimal number of $128 + 64 + 32 + 8 + 2 = 234$.

EXAMPLE 3   Convert the decimal number 234 to a binary number.

Referring to Table A5, you can see that the largest power of 2 that will fit into 234 is 2^7 (128). This leaves the value of $234 - 128 = 106$. The next largest power of 2 that will fit into 106 is 2^6 (64). This leaves a value of $106 - 64 = 42$. The next largest power of 2 that will fit into 42 is 2^5 (32), which is $42 - 32 = 10$. The next largest power of 2 that will fit into 10 is 23 (8), which is $10 - 8 = 2$. The next largest power of 2 that will fit into 2 is 2^1 (2), which is $2 - 2 = 0$.

Therefore, the binary equivalent is 11101010.

```
    234
  - 128    2^7
    106
  -  64    2^6
     42
  -  32    2^5
     10
  -   8    2^3
      2
  -   2    2^1
      0
```

1	1	1	0	1	0	1	0
2^7	2^6	2^5	2^4	2^3	2^2	2^1	2^0

APPENDIX B CREATING DOS BOOT DISK

The Microsoft Windows NT Network Client Administrator tool allows you to create a network installation start-up disk. This start-up disk allows a computer to start from a bootable floppy disk, which allows for an "over-the-network" installation of operating systems on computers that do not currently have an operating system or networking ability. The unfortunate aspect is that Windows 2000 does not include the Microsoft Windows NT Network Client Administrator tool. Therefore, you will need a Windows NT Server installation CD.

Making a Network Installation Start-up Disk

1. On your Windows NT Server computer, click the Start button, select the Programs option, select the Administrative Tools (Common) option, and click Network Client Administrator. To make a network installation start-up disk, click to select Make Network Installation Startup Disk in the Network Client Administrator. See Figure B1.

2. The Share Network Client Installation Files dialog box specifies the location of the network client installation files. In the Path text box, indicate where the files are located. If you previously used the Network Client Administrator and want to use an existing path

**FIGURE B1   Network Client
Administrator**

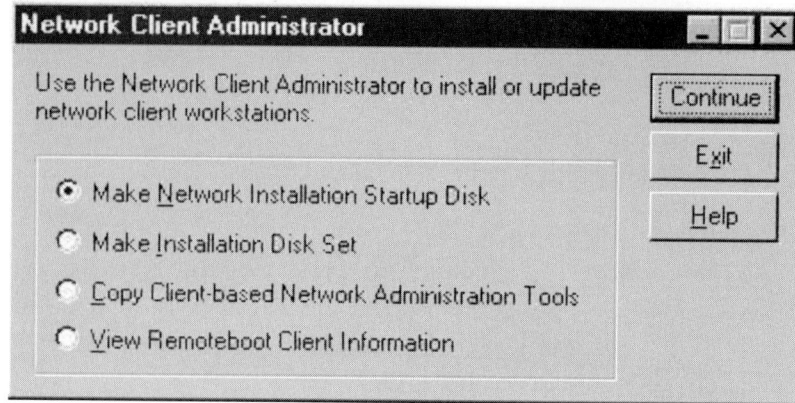

to the shared Network Client files, click to select Use Existing Path, and then click OK.
See Figure B2. If the files were not shared previously, use one of the following procedures for sharing the files:

a. To share the files contained on the CD, click to select Share Files.
b. In the Share Name text box, type the name of the new shared directory you want to
    create.

or

a. To copy the files to a new directory and then share the files, click to select Copy Files
    to a New Directory, and then click Share.
b. In the Path text box at the top of the window, type the destination to the Clients folder
    on the Windows NT Server CD.
c. In the Destination Path text box, indicate where you want to copy the files to. In the
    Share Name text box, type the name of the new shared directory you want to create.

To use files that have been previously copied and shared:

a. Click to select Use Existing Shared Directory.
b. In the Server Name text box, type the name of the network server in which the shared
    directory is located.
c. In the Share Name text box, type the name of the existing shared directory, and then
    click OK.

**FIGURE B2   Share Network
Client Installation Files dialog
box**

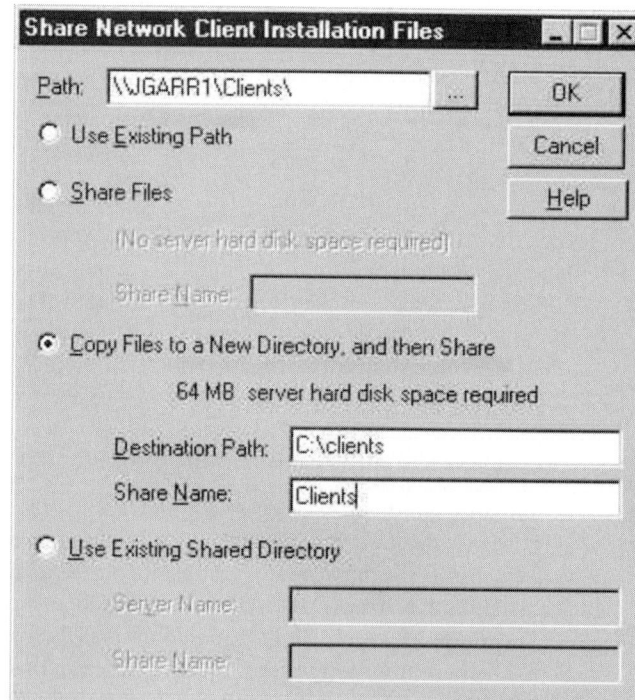

FIGURE B3 Target Workstation Configuration dialog box

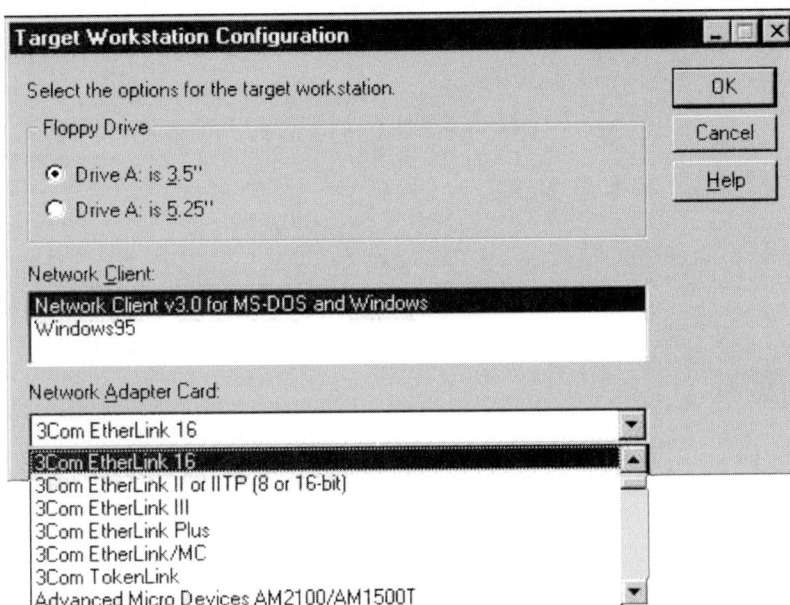

3. The Target Workstation Configuration dialog box appears. See Figure B3. It allows you to select the following:
   - Type of floppy disk
   - Operating system to be installed when the computer starts up
   - NIC installed on the target workstation (If another NIC is needed, see the "Configuring the Startup Disk for Network Interface Cards" section later in the appendix.)
4. After you make your selections and click on the OK button, you will see the Network Startup Disk Configuration dialog box. Set the following options:
   - **Computer Name:** The start-up disk must provide the workstation with a NetBIOS computer name that is unique on the network.
   - **User Name:** An optional setting that is used as the default logon name.
   - **Domain:** The name of the domain to log on to.
   - **Network Protocol:** Choose among NetBEUI, NWLink, or TCP/IP. When you click the Network Protocol dropdown list box, it may appear that you only have one protocol choice, but use the up and down arrows to scroll to the correct network protocol. The NWLink protocol defaults to 802.2 frame type.
   - **TCP/IP Settings** (if TCP/IP is selected as the network protocol): By default, the Enable Automatic DHCP Configuration checkbox is selected. To provide a static TCP/IP address, clear the Enable Automatic DHCP Configuration checkbox. Now manually provide an IP address and subnet mask. In a routed network, you must also provide a default gateway address.
   - **Destination Path:** Indicates the drive letter of your floppy disk drive. The floppy disk must already be formatted with the MS-DOS system files.

Configuring the Startup Disk for Network Interface Cards

When a network card is not listed, modify the start-up disk. This procedure requires installing the appropriate MS-DOS driver and editing two system files.

1. Install an NDIS2-compatible MS-DOS driver for the NIC. This is usually included with the floppy disk that is supplied by the manufacturer with its drivers. If no drivers are available, download the appropriate driver from the manufacturer's website. Appropriate drivers for the Microsoft Network Client for MS-DOS will always have a .dos extension. For example, the driver for Intel's EtherExpress Pro/10 EISA is Epro.dos. This driver should be placed in the Net directory on the computer (C:\Net, unless named differently) or on the MS-DOS start-up disk (A:\Net).

669

2. Modify the SYSTEM.INI file. The NIC driver must be referenced in the SYSTEM.INI file. This entry is found in the [network drivers] section, to follow:

```
[network drivers]
netcard=elnkii.dos
transport=ndishlp.sys,*netbeui
devdir=A:\NET
LoadRMDrivers=yes
```

For "netcard=," replace the current driver with the file name of the NDIS2-compatible driver placed in the Net directory (e.g., Epro.dos).

3. Modify the PROTOCOL.INI file. The NIC driver must be referenced in the PROTOCOL.INI file. This entry is found in the [ms$ *driver_name*] section (the driver name will reflect what was originally chosen in the network installation start-up disk process), as follows:

```
[ms$elnkii]
drivername=ELNKII$
;INTERRUPT=3
;IOADDRESS=0x300
;DMACHANNEL=1
;MAXTRANSMITS=12
```

For "drivername=," replace the driver listed with the file name of the NDIS2-compatible driver; use a dollar sign ($) to replace the .dos file extension (e.g., EPRO$). **Note:** Do not change the header (e.g., [ms$elnkii] in the previous example); the header is a pointer throughout the .ini file.

Making an Installation Disk Set

The installation disk set includes disks that are used to install the Network Client version 3.0 for MS-DOS and Windows machines. The Network Client version 3.0 for the MS-DOS and Windows programs will be used to install the necessary files to connect to the network and perform installations or downloads over the network. The Network Client version 3.0 for MS-DOS and Windows is also located on the Windows NT 4.0 installation CD in the MS-Client directory.

1. On your Windows NT Server computer, click Start, point to Programs, point to Administrative Tools (Common), and then click Network Client Administrator. To copy the Microsoft Network Client for MS-DOS Setup application to two floppy disks, click to select Make Installation Disk Set.

2. In the Path text box, indicate where the files are located and then click OK.

3. The Make Installation Disk Set dialog box appears. See Figure B4. Select Network Client v3.0 for MS-DOS and Windows, and verify that the destination drive contains a blank, formatted floppy disk (two floppy disks will be required). You may also select the Format Disks checkbox if you want to have the Network Client Administrator format the floppy disks before copying the setup files (remember that all data on the floppy disks prior to formatting will be lost). After the files are copied onto both disks, the installation disk set is ready for use.

FIGURE B4   The Make Installation Disk Set dialog box

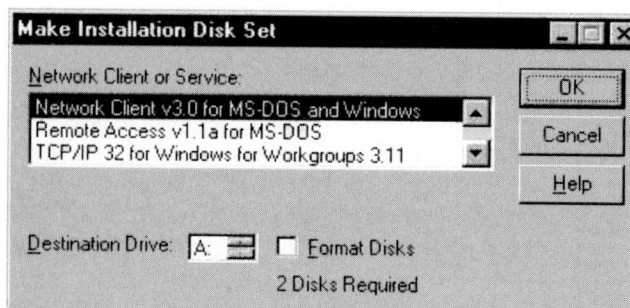

Running Setup on a Workstation

1. On the workstation that is already running MS-DOS, run Setup.exe from disk 1 of the installation disk set.

2. At the Setup screen, press ENTER to continue with setup. You will be prompted to provide a directory that will contain the networking files, and it is recommended that you use the default directory (C:\Net). Provide the directory, and then press Enter to continue.

3. You will be prompted to select a network interface card (adapter). Select your adapter from the list, and then press Enter. See Figure B5.

4. If you do not find your network interface card in the list, choose "Network adapter not shown on list below," and then press Enter. You will be prompted for the location of the network interface driver files. These files are provided by the manufacturer of the network interface card. The driver must be NDIS 2.0–compliant, and will be accompanied by an Oemsetup.inf file that is also provided by the manufacturer. Press Enter and continue.

5. You will be prompted for your user name. Type your user name, and then press Enter. This will take you to the Setup menu.

6. The Setup menu gives you three options:
   - **Change names:** Allows you to change the user name, computer name, workgroup name, and the domain name.
   - **Change Setup Options:** Allows you to change the redirector option, start-up options, and logon type and to set hot keys.
   - **Change Network Configuration:** Allows you to add and remove network adapters and protocols. It also allows you to change settings related to the protocol or network adapter.

   Use the up and down arrow keys to select an option, and then press Enter. You will be given a new menu associated with your previous choice. If all of the options have been set, select "The listed options are correct," press Enter, and the setup program will complete the installation process. Depending on your selections, you might not use disk 2 of the installation set.

   a. If you select "Change names in the Setup menu," you can then change the user name, computer name, workgroup name, and domain name. See Figure B6. Use the arrow keys to select an option, and then press Enter. Type the new information in the field that follows, and then press Enter. If all of the options have been set, select "The

FIGURE B5   The list that allows you to select the adapter. Note: The installation disk set allows you choose an unlisted network adapter.

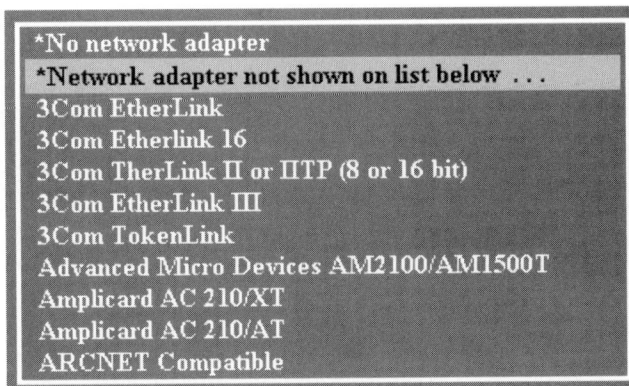

```
*No network adapter
*Network adapter not shown on list below ...
3Com EtherLink
3Com Etherlink 16
3Com TherLink II or IITP (8 or 16 bit)
3Com EtherLink III
3Com TokenLink
Advanced Micro Devices AM2100/AM1500T
Amplicard AC 210/XT
Amplicard AC 210/AT
ARCNET Compatible
```

FIGURE B6   Change Names List

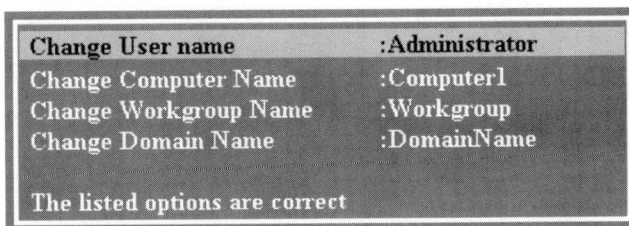

```
Change User name          :Administrator
Change Computer Name      :Computer1
Change Workgroup Name     :Workgroup
Change Domain Name        :DomainName

The listed options are correct
```

listed options are correct," and then press Enter to return to the setup menu for new options, or to complete the installation process.

b. If you select Change Setup Options in the Setup menu, you will see the following options:

- **Change Redir Options:** Choose either the full or basic redirector. The full redirector is used for logging on to a Microsoft Windows NT or LAN Manager domain, or for running programs that use advanced network functions such as named pipes. The basic redirector provides all standard workgroup functions such as connecting and disconnecting. It uses less memory and disk space than the full redirector. You must use the basic redirector if your computer has an 8088 processor.
- **Change Startup Options:** Allows you to configure the computer to automatically start the Microsoft Network Client for MS-DOS or both the Network Client and Pop-up Interface.
- **Change Logon Validation:** Allows you to choose whether to log on to a domain.
- **Change Net Pop Hot Key:** Allows you to select a particular key to use in conjunction with CTRL+ALT, to activate the pop-up interface.

Use the arrow keys to select an option, and then press Enter. Change the options as necessary, and then press Enter. If all options have been set, select "The listed options are correct," and then press Enter to return to the Setup menu for new options, or to complete the installation process. See Figure B7.

7. The Change Network Configuration menu allows you to change the adapter and protocol settings. Different from the previous options, the menu has a double window. The selections that are highlighted in the upper and lower windows are directly related. The active window is indicated by a double white outline. Use the Tab key to activate the upper or lower window. After the window is active, use the arrow keys to select an option. See Figure B8.

The upper window shows the installed adapters and protocols. The lower window has the following options:

- **Change Settings:** This option allows for settings to be made to the adapter or protocol that is currently highlighted in the upper window. If you are changing settings and a network adapter is selected, you may have the option to change some of the settings, such as driver name, IRQ, and I/O address. It is usually best to use the manufacturer's suggested settings if possible.

FIGURE B7 The Change Names List dialog box

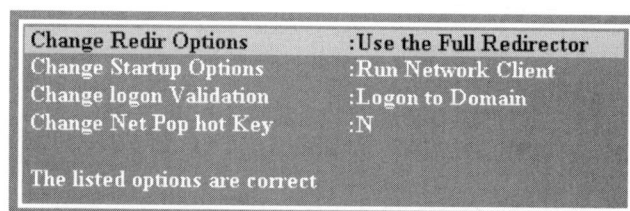

Change Redir Options	:Use the Full Redirector
Change Startup Options	:Run Network Client
Change logon Validation	:Logon to Domain
Change Net Pop hot Key	:N
The listed options are correct	

FIGURE B8 The Change Network Configuration dialog box

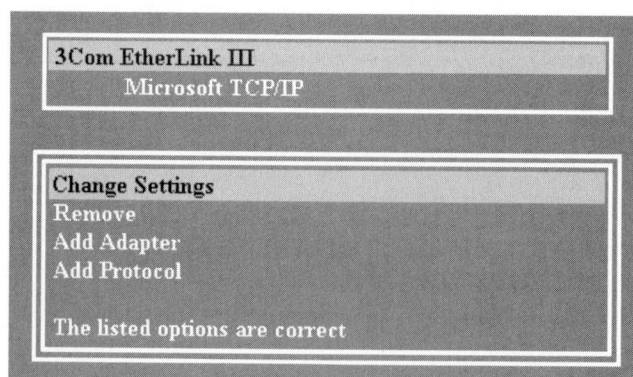

| 3Com EtherLink III |
| Microsoft TCP/IP |

| Change Settings |
| Remove |
| Add Adapter |
| Add Protocol |
| |
| The listed options are correct |

a. If you change settings to the NetBEUI protocol, change the Maximum Sessions or the NCBS setting.
b. If you change settings to the NWLink protocol, change the frame type from the default 802.2 (Ethernet_802.2) to any of the following: Ethernet_802.3, Ethernet_II, Ethernet_SNAP, or TOKENRING.
c. If you change settings to the TCP/IP protocol, enable or disable the domain host configuration protocol (DHCP); input the IP address, subnet mask, and up to two default gateways; and set the number of NetBIOS sessions. **Note:** The format of all IP addresses in this section is based on spaces which replace the dots that separate the octets (that is, 127 0 0 1 instead of 127.0.0.1).

- **Remove:** Remove whatever network adapter or protocol is currently highlighted in the upper window.
- **Add Adapter:** Allows you to add a network adapter, and then adds it to the upper window.
- **Add Protocol:** Allows you to add another protocol, and then adds it to the upper window. Use the arrow keys to select an option, and then press Enter. Change the options as necessary. If all options have been set, select "The listed options are correct," and then press Enter to return to the setup menu for new options, or to complete the installation process.

Using the NET.EXE Command

After the DOS client software is loaded, the most common network command is the NET.EXE command, which can be used to login to servers and access the shared resources located on those servers.

When using the NET.EXE command, you will find that NET.EXE has many options. If you perform NET HELP command at the prompt, you will see the following options.

```
For more information about a specific Microsoft NET
command, type the command name followed by /?
(for example, NET VIEW /?).


NET CONFIG      Displays your current workgroup settings.
NET DIAG        Runs the Microsoft Network Diagnostics program
                to display diagnostic information about your
                network.
NET HELP        Provides information about commands and error
                messages.
NET INIT        Loads protocol and network-adapter drivers
                without binding them to Protocol Manager.
NET LOGOFF      Breaks the connection between your computer and
                the shared resources to which it is connected.
NET LOGON       Identifies you as a member of a workgroup.
NET PASSWORD    Changes your logon password.
NET PRINT       Displays information about print queues and
                controls print jobs.
NET START       Starts services.
NET STOP        Stops services.
NET TIME        Displays the time on or synchronizes your
                computer's clock with the clock on a Microsoft
                Windows for Workgroups, Windows NT, Windows 95,
                or NetWare timeserver.
NET USE         Connects to or disconnects from a shared
                resource or displays information about
                connections.
```

NET VER	Displays the type and version number of the workgroupredirector you are using.
NET VIEW	Displays a list of computers that share Resources or a list of shared resources on a specific computer.

This book will only discuss the options that are the most useful in connecting to the network that is accessing the common servers.

During boot-up, it will usually ask for your user name and password so that you can log in to the network. The DOS client software can be configured to display a default user name. If you use the wrong user name and/or password, your access will be denied and you will have to log in again before accessing the network resources. If you perform the NET LOGON /? command at the prompt, the following screen will be displayed.

```
Identifies you as a member of a workgroup.


NET LOGON [user [password | ?]] [/DOMAIN:name] [/YES] [/SAVEPW:NO]


User          Specifies the name that identifies you in your
              workgroup. The name you specify can contain up to
              20 characters.
Password      The unique string of characters that authorizes
              you to gain access to your password-list file.
              The password can contain up to 14 characters.
?             Specifies that you want to be prompted for your
              password.
/DOMAIN       Specifies that you want to log on to a Microsoft
              Windows NT or LAN Manager domain.
name          Specifies the Windows NT or LAN Manager domain
              you want to log on to.
/YES          Carries out the NET LOGON command without first
              prompting you to provide information or confirm
              actions.
/SAVEPW:NO    Carries out the NET LOGON command without
              prompting you to create a password-list file.


If you would rather be prompted to type your user name and
password instead of specifying them in the NET LOGON command
line, type NET LOGON without options.
```

As you can see, to log on, you will use the following command:

NET LOGON

It will then ask you to input the user name and password. If you need to log on to another domain, you would then use the following command at the prompt:

NET LOGON /DOMAIN:domainname

Where domainname would be the actual name of the domain that you want to log in to.

The NET LOGON command allows you to log on as a user to the network using a user name and password. The user name can be up to 20 characters and the password can be 14 characters. The user name is not case sensitive whereas the password is; and both the user name and password can be up to 8 text characters long and are case insensitive.

You now know how to log on, so let us learn how to log off. The command to log off is

```
NET LOGOFF
```

The NET USE command is used to connect or disconnect to a shared resource or to display information about your connections. A shared resource is typically a drive, directory, or printer that has been shared or designated for access from clients.

To access a shared network drive or shared network directory, use the following syntax:

```
NET USE d: \\servername\sharename
```

Where d: is the local drive letter that you wish to assign which will point to the shared directory;servername (NetBIOS name) is the name of the server where the shared drive or directory resides; and sharename is the actual name given by the system administrator when creating the share. The \\servername\sharename is known as the universal naming convention (UNC). For example, to access the shared directory called SOFTWARE, which is located on the CLASSROOM server, type in

```
NET USE G: \\INSTRUCT\SOFTWARE
```

After this command is executed, as far as the client is concerned, he or she has a G drive. The G drive is actually a pointer that points to the SOFTWARE shared directory on the INSTRUCT server. Therefore, for the most part, the G drive acts just like any other local drive. Assuming that you have the proper rights or permissions, you can execute files, access data files, copy files from, or copy files to the G drive.

Another option within the NET USE command is the /HOME option, which points to the user's home directory. A home directory is designated by the network administrator and used for personal files. For example, to point the H to a person's home directory, type in the following command at the prompt:

```
NET USE H: /HOME
```

To disconnect from the shared resource, use the following syntax:

```
NET USE d: /D
```

Where d: is the local drive letter assigned to the shared drive or directory. Therefore, to disconnect from the G drive that was pointing to the SOFTWARE share, execute the following command at the prompt:

```
NET USE G: /D
```

The NET USE command can also be used to connect to a network printer. The proper syntax to connect to a network printer is

```
NET USE LPTx: \\servername\printername
```

Where LPTx is the local printer port that is assigned to the network printer; servername is the name of the print server; and printername is the name of the shared printer. For example, to connect to the IPRINTER network printer that is connected to the INSTRUCT print server using your local LPT2, use the following command at the prompt:

```
NET USE LPT2 \\INSTRUCT\PRINTER
```

Thus, every print job that you send to the LPT2 printer port will be redirected to the IPRINTER, even if you do not actually have an LPT2 port.

To disconnect from a network printer, still use the /D option. Therefore, to disconnect from the IPRINTER, use the following command at the prompt:

```
NET USE LPT2 /D
```

Lastly, the NET USE command can be used to display all connected shared directories, drives, and printers. The syntax to show the connected shared resources is

```
NET USE
```

The NET VIEW command is used to list the shared resources available on a specified computer. The syntax to do this is

```
NET VIEW \\servername
```

Therefore, to show all network resources on the INSTRUCT server, perform the following command at the prompt:

```
NET VIEW \\INSTRUCT
```

```
Displays a list of computers in a specified workgroup or the
shared resources available on a specified computer.

NET VIEW [\\computer] [/YES]
NET VIEW [/WORKGROUP:wgname] [/YES]


computer        Specifies the name of the computer whose shared
                resources you want to see listed.
/WORKGROUP      Specifies that you want to view the names of the
                computers in another workgroup that share
                resources.
wgname          Specifies the name of the workgroup whose computer
                names you want to view.
/YES            Carries out the NET VIEW command without first
                prompting you to provide information or confirm
                actions.


To display a list of computers in your workgroup that share
resources, type NET VIEW without options.
```

IPCONFIG and PING Commands

The client software includes IPCONFIG.EXE and PING.EXE commands. The IPCONFIG.EXE utility provides DHCP configuration information only. The version of IPCONFIG.EXE provided with the Microsoft Network Client for MS-DOS does not support command line switches for controlling DHCP address leases; you must use the DHCP Administration Utility instead.

APPENDIX C COMMON DOS COMMANDS

Regardless of the operating system (DOS/Windows 3.XX, Windows 95/98, Windows NT, or Windows 2000) you are using, there are certain DOS commands that you should know to set up Windows 2000 or to troubleshoot certain problems.

When performing DOS commands, there is always confusion as to when to use the back slash (\) and the forward slash (/). The back slash is always used to designate a location, whereas the forward slash is used to designate a switch (which has nothing to do with location). The backward slash is always in the same direction as the backslash in the prompt shown on the screen, because the prompt is showing the current drive and directory (location).

FORMAT A:/S	Makes a disk bootable (switch).
CD\DOS	Changes the current directory (location).
COPY C:\DOS\FORMAT.COM A:\DOS	Copies files from one location to another.
XCOPY C:\DATA*.* C:\DOS /S	Copies files from one location to another, but also includes subdirectories (switch).

DIR and CD Command

The DIR command, which displays the list of the files and subdirectories, is probably the most common used DOS command. When you use the DIR command without any parameters or switches, it displays the following:

1. Disk's volume label and serial number
2. Filename/directory name and their extensions
3. Date and time the file was created or last modified, and date and time of when a directory was made
4. File size in bytes
5. Total number of files listed, their cumulative size, and free space in bytes remaining on the disk

```
C:\.dir

Volume in drive D has no label
Volume Serial Number is 1CE8-1C7A
Directory of C:\

DOS             <DIR>            06-10-98   7:32p
WINDOWS         <DIR>            06-10-98   7:32p
COMMAND   COM          93,812   08-24-96  11:11a
CONFIG    SYS              91   06-06-98  11:03a
AUTOEXEC  BAT             183   06-06-98  10:44a
DATA            <DIR>            06-10-98   7:33p
3 file(s)            94,086 bytes
3 dir(s)        179,634,176 bytes free
The DIR command
```

The DIR command has many switches. Some of the more common ones used for a technician are shown in Table C1. Many of the DIR command switches can be combined.

TABLE C1   Helpful DIR command switches

Command Switch	Description
/P	Displays one screen of the DIR listing at a time. This comes in handy when there are too many files to be shown on the screen at once.
/W	Displays the listing in wide format (five columns). This comes in handy when there are too many files to be shown on the screen at once. It does not show the dates, times, and byte size of the files. Note: Directories are indicated with brackets [ ].
/ON	O stands for order; N stands for name. Therefore, when used with the DIR command, the files will be listed in alphabetical order by file name. This comes in handy when looking for a certain file.
/AH	A stands for attribute; H stands for hidden. Therefore, when used with the DIR command, it will show all hidden files. This comes in handy when verifying the boot files on a disk.
/AD	A stands for attribute; D stands for directories. It shows directories only.
/S	S stands for subdirectory. Therefore, when used, it will list every occurrence in the specified directory and all subdirectories. It comes in handy when looking for the location of a certain file or when trying to get the total byte size of a directory (including all subdirectories and files within the directory).

EXAMPLE 1   To show all files one screen at a time, perform the following command:

```
DIR /P
```

To show all files in alphabetical order, one screen at a time, perform the following command:

```
DIR /ON /P
```

To show the hidden files in the current directory, perform the following command:

```
DIR /AH
```

To show all files that have a GIF file name extension, perform the following command:

```
DIR *.GIF
```

To change the current drive, type in the drive letter followed by a colon (:) and press Enter. To change the current directory, use the CD command. To go up to the root directory, specify the backslash (\); and to go up to the parent directory (up one directory), specify the double dot (..).

EXAMPLE 2   To change to the TEST directory, which is located under the directory that you are currently in, perform the following command:

```
CD TEST
```

To change to the root directory, regardless of how many directories deep you are, perform the following command:

```
CD \
```

To change to the parent directory (one directory up), perform the following command:

```
CD . .
```

To change to the SYSTEM directory under the WINDOWS directory (located off the root directory) regardless of what directory you are in, use the following command:

```
CD \WINDOWS\SYSTEM
```

MD and RD Commands

The MD command is used to create directories; the RD command is used to delete empty subdirectories.

EXAMPLE 3   To create a TEST directory, type in the following command:

```
MD TEST
```

If the TEST directory is empty (no files or subdirectories), then use the following command to delete it:

```
RD TEST
```

DEL Command

The Del command is used to delete files. To delete the test.doc file, use

```
DEL TEST.DOC
```

The Copy Commands (COPY and DISKCOPY)

The two primary COPY commands used in DOS are COPY and DISKCOPY. Regardless of which command you use, they use the same three parts—command, source, and target. The source is what you are copying and the target is the location of where the copy is going.

The COPY command makes an exact duplicate of the file or files specified. The COPY command will not copy hidden or system files.

EXAMPLE 4   To make a copy of the RESUME.DOC file and call the copy JOB.DOC, perform the following command:

```
COPY RESUME.DOC JOB.DOC
```

To copy the AUTOEXEC.BAT file (located in the current directory) to the A drive, perform the following command at the prompt.

```
COPY AUTOEXEC.BAT A:
```

To copy the README.TXT file located in the C:\DOS directory, while you are in another directory, to the A:\DOS directory, perform the following command:

```
COPY C:\DOS\README.TXT A:\DOS
```

The command is COPY, the source is C:\DOS\README.TXT, and the target is A:\DOS.

To copy the README.TXT file located in the C:\DOS directory, while you are in another directory, to the A:\DOS directory and name the copy INFO.TXT, perform the following command:

```
COPY C:\DOS\README.TXT A:\DOS\INFO.TXT
```

Notice that all examples have three parts divided by two spaces.

EXAMPLE 5   To make an exact duplicate of a disk using the A drive, perform the following command at the prompt.

```
DISKCOPY A: A:
```

The REN and MOVE commands are similar to the COPY command, but not the same as the COPY command. The REN command allows you to rename a file and the MOVE command allows you to move a file from one location to another to rename a directory. Unlike the COPY command which leaves you with two files (the source and the target), the REN and MOVE commands leave you with only one.

EXAMPLE 6   To rename the README.TXT file to the INFO.TXT file, perform the following command:

```
REN README.TXT INFO.TXT
```

To move the README.TXT file (located in the C:\DOS directory) to the root directory of the C drive, perform the following command:

```
MOVE C:\DOS\README.TXT C:\
```

To rename the TEST directory to JOB directory, perform the following command:

```
MOVE TEST JOB
```

FORMAT and SYS Commands

The FORMAT command is used to prepare a disk for use or to erase everything on a disk. It creates the root directory (starting point on the disk) and the file allocation table (an index of files and directories on the disk and their location). In addition, if you use the /S parameter it also makes the disk bootable. To perform an unconditional format in which the disk cannot be unformatted or if you have received read and write errors, then use the /U switch.

EXAMPLE 7   To format the D drive, perform the following command:

```
FORMAT D:
```

To make a bootable floppy disk using drive A, perform the following command:

```
FORMAT A: /S
```

To make a disk bootable without reformatting the disk, use the SYS command. This will copy the IO.SYS, MSDOS.SYS, and COMMAND.COM and place them in the correct place, assuming that there is enough room for them to fit on the disk.

EXAMPLE 8 To make an important software diagnostic disk bootable using drive A, perform the following command (assuming you are not in the A drive):

```
SYS A:
```

The ATTRIB Command

The ATTRIB command displays, sets, or removes the read-only (R), archive (A), system (S), and hidden (H) attributes assigned to a file or directory. For more information on attributes, see Chapter 13. The archive attribute indicates whether a file has been backed up. To turn on an attribute, use a plus (+) sign and to turn off an attribute, use a minus (–) sign.

EXAMPLE 9 To show all files and their attributes, perform the following command:

```
ATTRIB
```

To make the AUTOEXEC.BAT read-only, perform the following command:

```
ATTRIB +R AUTOEXEC.BAT
```

or

```
ATTRIB AUTOEXEC.BAT +R (only on recent versions of DOS)
```

To turn off the read-only, system, and hidden attributes of the BOOT.INI file, perform the following command at the prompt. You must include spaces between each of the attributes.

```
ATTRIB -R -S -H BOOT.INI.
```

AUTOEXEC.BAT and CONFIG.SYS

The CONFIG.SYS is a special text file that executes special configuration commands that cannot be executed at the prompt. These commands configure your computer's hardware components so that MS-DOS and applications can use them and to activate and manage the various memory areas.

The HIMEM.SYS is the DOS's extended memory manager (RAM above 1 MB found on any newer machine), which makes the extended memory available to DOS programs. In addition, its purpose is to prevent two programs from using the same memory area in the extended memory. To load the HIMEM.SYS, use the DEVICE command in the CONFIG.SYS file. Any time that the DEVICE command is used to load a driver, always include the entire path (location) of the file specified and its file name extension. Since Windows 3.XX requires extended memory to run, Windows requires HIMEM.SYS to be loaded. In addition, if you add the /TESTMEM:OFF to the end of the HIMEM.SYS line, DOS will activate the extended memory but will skip the memory test.

The AUTOEXEC.BAT file is a special batch file that automatically executes during boot-up. It has a BAT file extension and it contains commands that you would normally perform at the prompt. In the AUTOEXEC.BAT files, you will usually find the PATH, PROMPT, and SET commands.

To enable hardware and to load useful utilities, you should know how to load device drivers and TSRs. A device driver controls how DOS and applications interact with specific items of hardware. They can be identified with a SYS file name extension, which is loaded in the CONFIG.SYS file.

The terminate and stay resident (TSR) loads instructions into the RAM to control some hardware device or to provide some useful function and give control back to the operating system. The TSR then performs its function quietly in the background while you can load other programs. A program that is not a TSR (e.g., EDIT or a word processor) is loaded in the RAM but typically does as not allow you to perform any other commands at the prompt until you exit the program. Since TSRs have a COM or EXE file name extension, they are loaded in the AUTOEXEC.BAT. Remember that the AUTOEXEC.BAT file is a batch file that can contain commands which you can execute at the prompt.

Most of the time when loading CD-ROM drivers, you insert the floppy disk into a floppy drive and you execute the SETUP.EXE, INSTALL.EXE, or similar file to automatically install and configure the drivers. For DOS, the EIDE CD drives require a device driver to be loaded in the CONFIG.SYS while SCSI drives usually require a device driver for the controller card and a device driver for the CD-ROM drive. The driver for the MITSUMI IDE CD-ROM drive is shown as

```
DEVICE=C:\CDROM\ ATAPI_CD.SYS /D:CD0001
```

Where /D: is the drive signature to identify the drive.

In addition to the driver, DOS requires the Microsoft Compact Disk Extension (MSCDEX.EXE) to be loaded in the AUTOEXEC.BAT so that DOS can read the CD file system.

```
C:\DOS\MSCDEX /D:CD0001 /M:12 /L:E
```

Where /D: indicates the drive signature, which must be the same as the one specified in the device driver loaded in the CONFIG.SYS. The /M: specifies the number of sector buffers and the /L: specifies the drive letter assigned to the CD-ROM drive.

SMARTDRV.EXE, a software disk cache that caches or buffers between the hard drive and the RAM, creates a buffer area called a cache area in the extended memory. The software controlling the cache area tries to anticipate what the microprocessor needs next. One method is to keep a copy of information that has been recently accessed. If it has already been accessed, there is a good chance that it will be accessed again. Another method is read ahead the additional sector after the one that has been accessed. In whatever method(s) used, when the disk needs to be accessed, it will look in the cache area first. If it cannot get what is needed, it will access the slower hard drive. The time to search the cache area (RAM) is almost negligible compared with the time it takes to access the hard drive. Therefore, if it finds the needed information in the RAM, the overall performance of the PC is also increased. Since SMARTDRV.EXE has an EXE file name extension, it is loaded in the AUTOEXEC.BAT file. SMARTDRV.EXE automatically loads into upper memory, if available.

EXAMPLE 10 To create a disk cache with the default values, use the following command in the AUTOEXEC.BAT:

```
SMARTDRV
```

To create a 4096-KB cache area (overriding the default values), use the following command in the AUTOEXEC.BAT:

```
SMARTDRV 4096
```

CONFIG.SYS
      DEVICE=C:\DOS\HIMEM.SYS
      DEVICEHIGH=C:\CDROM\ ATAPI_CD.SYS /D:CD0001

AUTOEXEC.BAT
      C:\DOS\MSCDEX /D:CD0001 /M:12 /L:E
      SMARTDRV 4096 128

APPENDIX D APPLE COMPUTERS ON THE NETWORK

A computer running Windows 2000 Server with AppleTalk network integration can store files so that users of both Intel-based and Macintosh computers can gain access to them. Users of Intel-based computers (including users of the MS-DOS, OS/2, Windows, Windows for Workgroups, Windows NT Workstation, Windows NT Server, Windows 2000 Professional, and Windows 2000 Server operating systems) look for shared files in a shared folder on the computer running Windows 2000 Server. Macintosh users look for shared files in the same folder; however, they see the folder as a volume, with familiar folders and files.

A Macintosh user shares a file with users of Intel-based computers by storing that file in a Macintosh-accessible volume on the computer running Windows 2000 Server. Likewise, a Macintosh user can mount a Macintosh-accessible volume on the desktop to use files stored in shared folders by users of Intel-based computers.

All Macintosh-accessible volumes must be created on an NTFS partition or on a compact disc file system (CDFS) volume. If you specify a CDFS volume, the Macintosh-accessible volume will provide read-only access. (In this case, CDFS volume refers to a hard disk volume.)

To install File Server for Macintosh, do the following:

1. Open the Windows Components wizard.
2. In Components, click Other Network File and Print Services (but do not select or clear its checkbox), and then click Details.
3. Select the File Services for Macintosh checkbox.
4. To install the print server for Macintosh, also select the Print Services for Macintosh checkbox.
5. Click OK, and then click Next.

You must have an NTFS partition installed before you install the file server for Macintosh. When File Server for Macintosh and TCP/IP are installed, the Apple filing protocol (AFP) over TCP/IP is enabled automatically. If you have not already installed the AppleTalk protocol, it is installed automatically when you install File Server for Macintosh.

To install the authentication files on the Macintosh client, do the following:

1. Open Chooser, click AppleShare, select the AppleTalk zone where the Windows 2000 Server is located, and then select that file server. Continue to connect and select the Microsoft UAM volume as the item you want to use.
2. On the Macintosh desktop, double-click the Microsoft UAM Volume to open it.
3. Double-click MS UAM Installer.

To create a Macintosh-accessible volume, do the following:

1. Open Computer Management.
2. In the Console tree, double-click the Shared Folders option, right-click Shares, and then click New File Share.
3. In Folder to Share, type the drive and path to the folder you want to make Macintosh accessible, or click Browse to find the folder.
4. In Share Name, type the name and, optionally, in Share description, type a description of the Windows 2000 Server share.
5. Select the Apple Macintosh checkbox, click Next, and then follow the instructions in the Create Shared Folder wizard.

A Macintosh-accessible volume is a folder on a computer that is running Windows 2000 Server and that is made available to Macintosh clients. Once a Macintosh-accessible volume has been created for Macintosh users and a share created for users of Intel-based computers, both types of users can exchange files. Macintosh clients need to run AppleTalk to access Macintosh volumes and AppleTalk printers. All Macintosh-accessible volumes must be created on an NTFS partition or on a CDFS volume. If you specify a CDFS volume, the Macintosh-accessible volume will provide read-only access. (In this case, CDFS volume refers to a hard disk volume.)

To install Print Server for Macintosh, do the following:

1. Open the Windows Components wizard.
2. In Components, click Other Network File and Print Services (but do not select or clear its checkbox), and then click Details.
3. Select the Print Services for Macintosh checkbox.
4. To install File Server for Macintosh, you can also select the File Services for Macintosh checkbox.
5. Click OK, and then click Next.

If you have not already installed the AppleTalk protocol, it is installed automatically when you install Print Server for Macintosh.

To create a printer, do the following:

1. Open Printers.
2. In the Printers dialog box, double-click Add Printer.
3. Follow the Add Printer wizard instructions and select Local Printer (select the "Automatically detect my printer" checkbox if the printer is directly attached to the Windows 2000 Server), and then click Next.
4. Click Create a New Port, click AppleTalk Printing Devices, and then click Next.
5. Expand the AppleTalk zone where the printer you want to connect to is located, and then select the printer.
6. Complete the Add Printer wizard.

If the printing device is on a network, click Add Port. In the Printer Ports dialog box, click AppleTalk Printing Devices, and then click OK. In the Available AppleTalk Printing Devices dialog box, select a zone and a printer, and then click OK.

APPENDIX E CONNECTING TO A NOVELL NETWORK

To connect to a Novell NetWare network on a Windows 2000 Professional computer, install the NWLink protocol and Client Services for NetWare. If you are running Windows 2000 Server, install the NWLink protocol and Gateway (and Client) Service for NetWare.

When you install Gateway (and Client) Service for NetWare or Client Service for NetWare, the NWLink protocol is installed automatically. NWLink supports Novell network routing and can support NetWare client-server applications. In networking environments that use IPX/SPX, the NWLink protocol allows computers running Windows 2000 to communicate with other computers or network devices that are using IPX/SPX. NWLink also provides client computers that are running IPX/SPX with access to applications that are designed for Windows 2000 Server, such as Microsoft SQL Server and SNA Server.

The process for installing and configuring Gateway (and Client) Service for NetWare and Client Service for NetWare is the same except for the gateway service. You install Gateway (and Client) Service for NetWare on Windows 2000 Server and Client Service for NetWare on Windows 2000 Professional.

NWLink and Client Services for NetWare are two key components of Microsoft networking and Novell NetWare networking interoperability. NWLink is a protocol that allows Windows 2000 clients to communicate with applications running in a Novell NetWare environment. Client Services for NetWare is the service that allows Windows 2000 clients to connect to network resources such as file and print resources on a NetWare server.

Client Services for NetWare also allows Windows 2000 clients to be authenticated by a NetWare server. Client Services for Netware is the implementation provided by Microsoft for the NetWare redirector. The redirector in both Windows 2000 and Novell NetWare environments is the service installed on the client that allows the client computer to communicate with the server service.

Gateway (and Client) Services for NetWare allows a Windows 2000 Server to act as a gateway for Windows 2000 clients who want to access NetWare servers via the Windows 2000 Server. Windows 2000 clients use this service for occasional or low usage to a NetWare server. Gateway (and Client) Services for NetWare creates a bridge through the Windows 2000 Server for the Windows 2000 Server and clients to access NetWare resources. This service allows a Windows 2000 Server to access a NetWare share and make it accessible to Windows 2000 clients.

The procedure for installing NWLink is the same as that used to install any network protocol in Windows 2000. To install NWLink, perform the following steps:

1. Right-click My Network Places, and then click Properties.
2. In the Network and Dial-up Connections window, right-click the icon that represents the local area connection that you want to configure, and then click Properties.
3. In the Local Area Connection Properties dialog box, click Install.
4. In the Select Network Component Type dialog box, click Protocol, and then click Add.

5. In the Select Network Protocol dialog box, in the Network Protocol list, click NWLink IPX/SPX/NetBIOS Compatible Transport Protocol, and then click OK.

6. In the Local Area Connection Properties dialog box, click Close. Install and configure Gateway (and Client) Service for NetWare (GSNW) to enable computers running Windows 2000 to make direct connections to file and printer resources on NetWare servers. GSNW installs NWLink by default when installed. However, it is recommended that you install, and if necessary configure, NWLink prior to installing GSNW.

To install Gateway (and Client) Service for NetWare, perform the following steps:

1. Double-click Network and Dial-up Connections in the Control Panel.
2. Right-click any local area connection, and then click Properties.
3. On the General tab, click Install.
4. In the Select Network Component Type dialog box, click Client, and then click Add.
5. In the Select Network Client dialog box, click Gateway (and Client) Services for NetWare, and then click OK.
6. In the Select NetWare Logon dialog box, choose either a preferred server from a list of NetWare servers or enter a default tree and context, and then click OK.
7. On the Local Network dialog box, click Yes to restart your computer.

When you install Gateway (and Client Services) for NetWare, it is installed for all connections. If you do not want Client Services for NetWare installed for a certain connection, view the properties of that connection and then clear the Client Service for NetWare checkbox. If you want to change the preferred server or the default tree and context, then you may do so with the GSNW icon in the Control Panel. Here you can also configure Print Options and Run Login Script. You can test that Gateway (and Client Services) for NetWare is initialized properly by running the NET VIEW /NETWORK:NW command. You should be able to see a list of available NetWare servers.

As with connecting to a Windows 2000 Server, connecting to a NetWare server can be done in several ways once CSNW/GSNW have been installed. You can access the server through Windows Explorer, map a network drive, or view the NetWare server through My Network Places.

APPENDIX F
SERVICES FOR UNIX

UNIX, including variations of Linux, use the Network File System (NFS). The NFS is an open operating system designed by Sun Microsystems that allows all network users to access shared files stored on computers of different types. NFS provides access to shared files through an interface called the Virtual File System (VFS) that runs on top of TCP/IP. Users can manipulate shared files as if they were stored locally on the user's own hard disk.

The Services for UNIX 2.0 include:

- Server for NFS allow NFS clients such as UNIX and Linux to access files on a Windows 2000 Professional or Server computer. To manage the Server for NFS, use the UNIX Admin Snap-in (sfumgmt.msc).
- Gateway for NFS (available in Windows 2000 Server only) allows non-NFS Windows clients to access NFS resources by connecting through an NFS-enabled Windows Server to NFS resources. It acts as a gateway/translator between the NFS protocol used by UNIX/Linux and the CIFS protocol used by Windows 2000.
- Server for PCNFS—Provides authentication services for NFS clients needed to access NFS files.
- Server for NIS—Allows server to act as the NIS master for a particular UNIX domain as it authenticates requests for NFS shares. The server for NIS must be installed on a Windows 2000 Server that is configured as a Domain Controller.

Microsoft Windows Services for UNIX is designed to facilitate interoperability between Windows-based and UNIX-based operating systems. To allow UNIX and Windows 2000 machines to communicate with each other, Windows 2000 uses the Common Internet File System (CIFS), which is an enhanced version of the Server Message Block (SMB) protocol. Client for Network File Systems integrates Windows Explorer so that you can browse and map drivers to NFS volumes and access NFS resources through My Network Places. Microsoft recommends this over installing Samba (SMB files services for Windows clients) on your UNIX server. The NFS shares can be accessed using standard NFS syntax or UNC syntax:

NFS share syntax   *servername:/pathname*
UNC syntax   *\\servername:/pathname*

If users' UNIX username and password differ from Windows username and password, click Connect Using a Different User Name option and provide new credentials.

Services for UNIX version 2.0 includes more than 60 of the most common and important UNIX command line utilities to provide a familiar environment for UNIX users and administrators. Some of the more popular UNIX utilities are installed along with the Client for NFS:

Utility	Description
grep	Searches files for patterns and displays results containing that pattern
ps	Lists processes and their status
sed	Copies files named to a standard output; edits according to a script of commands
sh	Invokes the Korn shell
tar	Used to create tape archives or add/extract files from archives
vi	Invokes IV text editor

The nfsadmin command-line utility is used for configuration and administration of the Client for NFS. Its options include:

Option	Description
fileaccess	UNIX file permission for reading, writing, and executing
mapsvr	Computer name of the mapping server
type	Mount type, HARD or SOFT
perf	Method for determining performance parameters (MANUAL or DEFAULT)
prefer TCP	Indicates whether to use TCP (YES or NO)
retry	Number of retries for a soft mount—default value is 5
rsize	Size of read buffer in KB
timeout	Timeout in seconds for an RPC call
wsize	Size or write butter in KB

APPENDIX G ADMINISTERING MICROSOFT OFFICE 2000

Microsoft Office 2000 provides you with the flexibility to customize and install Office in a number of different ways. The simplest methods of installing a customized version of Office include creating an administrative installation point on the network, or distributing a customized version of Microsoft Office 2000 Disc 1.

How to Install Office from a Network Server

The most common method of deploying a customized version of Microsoft Office 2000 to a large number of users is to create a central copy of Office on a network server. Then users can install Office on their computers over the network. This method provides several advantages over having users install Office individually from the CD:

- You can manage one set of Office files from a central location.
- You can create a standard set of Office features and options for all users.
- You can take advantage of flexible installation options, such as setting features to be installed on first use.
- You have more control when you upgrade Office in the future.
- When you install Office from a network server, you first create an administrative installation point and customize your version of Office Setup. Then you run Setup on users' computers.

The administrative installation point is a server share that contains all Office files. Users connect to the share and run Setup to install Office on their computers.

To create an administrative installation point for Office, do the following:

- Create a share on a network server for the administrative installation point. The network share must have at least 550 MB of available disk space.
- On a computer running Windows 95, Windows 98, Windows NT, or Windows 2000 that has write access to the share, connect to the server share.
- On the Start menu, select the Run option and click on the Browse button.
- Insert the Office Disc 1 in the CD-ROM drive and select the SETUP.EXE file and click the Open button.
- On the command line following SETUP.EXE, type /A DATA1.MSI and click OK, for example: D:\SETUP /A DATA1.MSI.
- When prompted by Setup, enter the organization name that you want to define for all users who install Office from this location.
- When prompted for the installation location, enter the server and share you created.
- Setup will copy all the files from Office 2000 Disc 1 to the network server, creating a hierarchy of folders in the root folder of the share. Setup also modifies the Windows installer package for Office (Data1.msi), identifying the network share as an administrative installation point. After creating the administrative installation point, make the share available to users by providing them with read access.

Customizing the Office Setup Files

Before users run Setup to install Office, modify the administrative installation point to customize the installation process and default settings. Following are three options for customizing Setup:

- Create a custom command line.
- Edit the Setup settings file.
- Create a Windows installer transform (MST file) by using the Microsoft Custom Installation wizard.

Setup command line options allow you to define properties that Setup uses to control the installation process. You also can specify whether Setup runs interactively or in quiet mode (without user interaction). Users install Office on their computers by running Setup with your command line options.

The Setup settings file (Setup.ini) is a text file in which you enter properties and values. You can edit the Setup settings file to specify the same properties as you do on the Setup command line—every command line option has a corresponding setting in the settings file. The advantage of modifying the Setup settings file is that your custom values are used whenever a user runs Setup from the administrative installation point with no command line options.

You can create more than one settings file with different values, and then you can specify which settings file you want to use with the /settings command line option. For example:

```
SETUP.EXE /SETTINGS NEWSETUP.INI
```

The Custom Installation wizard allows you to specify the same property settings that you can define on the command line or in the settings file, but the wizard also allows you to make many more customizations. For example, you can select the Office features that you want to install, modify shortcuts, customize options for Microsoft Internet Explorer 5 and Microsoft Outlook, and even add your own custom files to the installation. The wizard saves your selections in a transform.

When you run Setup, specify the transform that you want Setup to use. You can specify an MST file on the Setup command line, or you can set the TRANSFORMS property in the Setup settings file.

For example, this command line specifies the transform Custom.mst:

```
SETUP.EXE TRANSFORMS="CUSTOM.MST"
```

After you create and customize the administrative installation point, users can install Office by running Setup from the root folder of the server share. Users can run Setup themselves by using the command line options, settings file, or transform that you have chosen. For a more controlled installation, run Setup for them through a network logon script or systems management software, such as Microsoft Systems Management Server.

Glossary

10Base2 Also known as Thinnet, is a simplified version of the 10Base5 network. The name describes a 10-Mbps baseband network with a maximum cable segment length of approximately 200 meters (actually 185 meters).

10Base5 Also known as Thicknet, is the original type of Ethernet network. The name is derived because 10Base5 is a 10-Mbps baseband network that can have cable segments up to 500 meters long. It uses a 50-ohm RG-8 and RG-11 as a backbone cable.

10BaseT A form of Ethernet that uses UTP cabling, which is a physical star topology on a hub.

386 Protection Mechanism A mechanism used in Intel processors, as the 386 assigns data and instructions to a privilege ring or level. The four privilege levels are 0, 1, 2, and 3. The privilege level 0 is the highest and level 3 is the lowest. Programs in ring 0 can access programs in all rings, but the programs in ring 3 cannot access the programs in ring 0, 1, and 2. Today, some operating systems, such as Windows NT and Windows 2000, assign programs to level 0 and other programs to level 3. If a program becomes corrupt in ring 3, it will not affect the operating system.

Access Control List (ACL) A table that lists users and groups who have been given permission to an object and their permissions.

access method The set of rules that defines how a computer puts data onto the network cable and takes data from the cable.

access time The average amount of time it takes to move the hard drives read/write head to the requested sector. It is the sum of seek time and latency period.

Active Directory (AD) A directory service that uses the "tree" concept for managing resources on a Windows 2000 network. It stores all information about the network resources and services such as user data, printers, servers, databases, groups, computers, and security policies. In addition, it identifies all resources on a network and makes them accessible to users and applications.

active server pages (ASP) A specification for a dynamically created Web page with an .ASP extension that utilizes ActiveX scripting.

address bus Typically used in association with a data bus, is used to define where the data bus signals are going to or coming from.

address resolution protocol (ARP) A TCP/IP protocol used to obtain hardware addresses (MAC addresses) of hosts located on the same physical network.

administrative share A shared folder typically used for administrative purposes.

administrator An individual who is responsible for maintaining the network.

advanced configuration and power interface (ACPI) A relatively new power management specification developed by Intel, Microsoft, and Toshiba, in which the operating system can turn off peripheral devices, such as a controller card or network card, when not in use.

advanced power management (APM) An older power management specification that has system control power-saving features. See *advanced configuration and power interface.*

allocation unit Also known as a cluster, is the most basic storage unit for an operating system.

always on/dynamic ISDN (AO/DI) A networking application that uses the 16-Kbps ISDN D-channel X.25 packet service to maintain an always-on connection between an ISDN end user and an information service provider.

amplitude The peak voltage of the sine wave.

analog signal A signal that has infinite number values that are constantly changing. Analog signals are typically sinusoidal waveforms, which are characterized by its amplitude and frequency.

antivirus software A software package that will detect and remove viruses and help protect the computer against viruses.

AppleTalk The Apple network architecture that has been included in Apple Macintosh computers for several years.

application layer Part of the OSI reference model that represents the highest layer. It is responsible for interaction between operating systems and provides an interface to the system. It provides the user interface to a range of network-wide distributed services including file transfer, printer access, and electronic mail.

application server   A server that is similar to a file-and-print server, except it also does some of the processing.

ARC (advanced RISC computing) path   A string that is used to specify the location (partition) of an operating system.

ARCnet (Attached Resource Computer Network)   An early form of networking technology.

area border router (ABR)   A router that attaches an area to the backbone in an OSPF network.

asymmetrical DSL (ADSL)   A DSL standard that transmits an asymmetric data stream, with much more going downstream to the subscriber than coming back.

asynchronous signals   Intermittent signals, they can occur at any time and at irregular intervals. They do not use a clock or timing signal.

asynchronous transfer mode (ATM)   A LAN and a WAN technology, which is generally implemented as a backbone technology; a cell switching and multiplexing technology that combines the benefits of circuit switching and packet switching.

AT attachment packet interface (ATAPI)   An extension to EIDE (also called ATA-2) that enables the interface to support CD-ROM players and tape drives.

attenuation   When the strength of a signal falls off with distance over a transmission medium.

auditing   A feature of Windows NT and Windows 2000 that monitors various security-related events so that you can detect intruders and attempts to compromise data on the system. Some events that you can monitor include access to an object such as a folder or file, management of user and group accounts, and logging on and off a system.

autonomous area (AS)   A grouping of contiguous networks.

autonomous system boundary router (ASBR)   A router that exchanges routing information with routers belonging to other autonomous systems.

backbone   (1) Another term for bus, the main wire that connects nodes. The term is often used to describe the main network connections composing the Internet. (2) The central point that connects multiple networks together. (3) In OSPF, an address of 0.0.0.0 is a high-bandwidth logical area where all areas are connected.

backup   An extra copy of data and/or programs.

band   A contiguous group of frequencies, which are used for a single purpose.

bandwidth   The amount of data that can be carried on a given transmission media. Larger bandwidth means greater data transmission capabilities.

bandwidth allocation protocol (BAP)   Enhances multilink by dynamically adding or dropping links on demand.

baseband   A system that uses the transmission medium's entire capacity for a single channel.

basic authentication method   A widely used, industry standard method for collecting user name and password information for web servers.

basic disk   A physical disk that contains either primary partitions, extended partitions, or logical drives, which can be accessed directly using DOS, Windows 95, or Windows 98. Basic disks may contain volume sets, mirrored sets, striped sets, and stripe sets with parity if they were created using Windows NT 4.0 or earlier and upgraded to Windows 2000.

basic input/output system (BIOS)   Also known as firmware, ROM chips that control the boot process and hardware.

basic rate interface (BRI)   ISDN communications lines consisting of three independent channels, two bearer (or B) channels, each carrying 64 KB/s and one data (or D) channel at 16 Kbps. It uses T1 digital lines consisting of 23 B channels, each at 64 Kbps, and one 64-Kbps D channel for signaling.

basket systems   Baskets or trays used to hold and hide network cables without using cable ties.

bearer channel (B channel)   An ISDN channel that transfer data at a bandwidth of 64 Kbps.

bind   When a protocol is logically attached to a network card.

binding   The process when a protocol is linked to a protocol stack.

block cipher   Encrypts one block of data at a time.

boot partition   Contains the Windows 2000 operating system files.

BOOT.INI   A file that provides a Boot Loader Operating System Selection menu, which allows the selection between multiple operating systems.

bootstrap protocol (BOOTP)   A protocol, created for diskless workstations, enables a booting host to configure itself dynamically.

border gateway protocol (BGP)   A common exterior gateway protocol.

bridge   A data link OSI layer, is a device that connects two LANs and makes them appear as one or connects two segments of the same LAN. The two LANs being connected can be alike or dissimilar, such as an Ethernet LAN connected to a token ring LAN.

bridgehead server   A single server located in each site that is designated to perform site-to-site replication.

broadband   A system that uses the transmission medium's capacity to provide multiple channels by using frequency-division multiplexing (FDM).

Broadband-ISDN (B-ISDN)   A standard for transmitting voice, video, and data simultaneously over fiber-optic telephone lines.

brouter   Short for bridge router, is a device that functions as both a router and a bridge. A brouter understands how to route specific types of packets (routable protocols). For other specified packets (nonroutable protocols), it acts as a bridge, which simply forwards the packets to the other networks.

burst mode   A mode used in data transfers, in which sequential data are transferred faster by specifying one memory address followed by transferring multiple pieces of data.

bus mastering   Similar to direct memory address, which has an expansion card with its own processor, takes temporary control of the data and address bus to move information from one point to another.

bus topology   A topology that resembles a line, and data are sent along this single cable.

cable tester   A device specifically made to test a cable. An inexpensive cable tester will typically test for shorts and opens.

cable ties   Small devices used to bundle cables traveling together and to keep cables off the floor so that they will not get trampled or run over by office furniture.

cache   a special high-speed storage mechanism. It can be either a reserved section of main memory or an independent high-speed storage device. Two types of caching are commonly used in personal computers—memory caching and disk caching.

capacitor   A simple electrical device, similar to a battery, which is capable of storing a charge of electrons. When used as memory, the charge or lack of charge represents a single bit of data. If it is charged, it has a logic state of 1; if it is discharged, it has a logic state of 0.

carrier sense multiple access (CSMA)   The method that has each station listen for network traffic before it attempts to transmit. To avoid collisions, CSMA will use one of two specialized methods of collision management: collision detection (CD) and collision avoidance (CA).

catalog   A summary of the files and folders that have been saved in a backup set.

cell   Relatively small, fixed-size packets.

cellular topology   A topology used in wireless communication. A broadcast device is located at the center, broadcasting in all directions to form an invisible circle (cell). All network devices located within the cell communicate with the network through the central station or hub, which is interconnected with the rest of the network infrastructure. If the cells are overlapped, devices may roam from cell to cell while maintaining connection to the network as the devices.

central processing unit (CPU)   See *microprocessor.*

certificate authority (CA)   A server that issues an encrypted digital certificate that contains the applicant's public key and a variety of other identification information.

challenge handshake authentication protocol (CHAP)   The most common dial-up authentication protocol used, which uses an industry Message Digest 5 (MD5) hashing scheme to encrypt authentication.

channel   A part of the media's total bandwidth. In communications, the term refers to a communications path between two computers or devices.

channel service unit/data service unit (CSU/DSU)   The DSU is a device that performs protective and diagnostic functions for a telecommunications line. The CSU is a device that connects a terminal to a digital line. Typically, the two devices are packaged as a single unit.

chipset   A number of integrated circuits located on the motherboard and designed to perform one or more related functions.

circuit switching   A technique that connects the sender and the receiver by a single path for the duration of a conversation. Once a connection is established, a dedicated path exists between both ends (i.e., always consumes network capacity), even when there is no active transmission taking place (such as when a caller is put on hold). Once the connection has been made, the destination device acknowledges that it is ready to carry on a transfer. When the conversation is complete, the connection is terminated.

classical IP over ATM (CIP)   A mature standard that enables users to route IP packets over an ATM network or cloud using ATM as either a backbone technology or a workgroup technology.

classless interdomain routing (CIDR)   Used with supernetting, it collapses multiple network ID entries into a single entry.

clear text   Simple text without encryption.

client   A computer that requests services from a network server.

client software   used on workstations to attach and communicate to the network.

client-server network   Consists of servers and clients.

cluster   Also known as allocation unit, is the most basic storage unit for an operating system.

clustering   Connecting two or more computers, or nodes, in such a way that they behave like a single computer. It is used for parallel processing, load balancing, and fault tolerance.

coaxial cable (coax)   Consists of a center wire surrounded by insulation and then a grounded shield of braided wire. The shield minimizes electrical and radio frequency interference.

COM+   An extension of the component object model, is both an object-oriented programming architecture and a set of operating system services. It adds to COM a new set of system services for application components while they are running, such as notifying them of significant events or ensuring they are authorized to run.

command queuing   Used in SCSI devices to accept multiple commands and execute them in an order that is more efficient, rather than in the order received. Consequently, this increases the performance of computers running multitasking operating systems and makes it ideal for servers.

committed information rate (CIR)   The value specifies the maximum average data rate that the network undertakes to deliver under "normal conditions."

common gateway interface (CGI)   A standard for interfacing external applications with web servers.

communication server   A server that handles data flow and e-mail messages from one network to another network using modems or other dial-up technology.

compact disk (CD)   A 4.72-inch encoded platter, which is read by laser. To read a CD, you need a CD drive. They can hold large amounts of information (680 MB) and are inexpensive.

component object model (COM)   An object-based programming model designed to promote interoperability by allowing two or more applications or components to easily cooperate with one another, even if they were written by different vendors, at different times, in different programming languages, or if they are running on different computers that are running different operating systems.

concentric   Circles that share the same center much like rings in a tree.

contention   The method when two or more devices contend for network access. Any device can transmit whenever it needs to send information. To avoid data collisions (two devices sending data at the same time), specific contention protocols were developed requiring the device to listen to the cable before transmitting data.

Control Panel   A component of Windows that is used for configuration.

convergence   The process of agreement by all routers to which routes are the optimal routes.

cookie A message given to a web browser by a web server, which is typically stored in a text file on the PC's hard drive. The message is then sent back to the server each time the browser requests a page from the server.

count to infinity Typically occurs when a network has slow convergence, is a loop that forms when a link in a network goes down and routers on the network update their routing tables with incorrect hop counts.

crossover cable Connects one network card to another network card or a hub to a hub; reverses the transmit and receive wires.

current state method Periodically measures the digital signal for the specific state.

daisy chain A hardware configuration in which devices are connected one to another in a series.

data The raw facts, numbers, letters, or symbols that the computer processes into meaningful information.

data bus Can refer to a bus that carries data in to and out of a device such as the processor or expansion card.

data channels (D channels) An ISDN channel that uses a communications language called DSS-1 for administrative signaling, such as to instruct the carrier to set up or terminate a B-channel call, to ensure that a B channel is available to receive a call, or to provide signaling information for such features as caller identification.

data circuit-terminating equipment (DCE) Special communications devices that provide the interface between the DTE and the network. Examples include modems and adapters. Its purpose is to provide clocking and switching services in a network and actually transmit data through the WAN; therefore, it controls data flowing to or from a computer.

Data Encryption Standard (DES) A popular symmetric-key encryption method that uses block cipher.

data link control (DLC) A special, nonroutable protocol that enables computers running Windows 2000 to communicate with computers running DLC protocol stack such as IBM mainframes, IBM AS/400 computers, and Hewlett-Packard network printers connected directly to the network using older HP JetDirect network cards.

data link layer Part of the OSI reference model that is responsible for providing error-free data transmission and establishes local connections between two computers.

data terminal equipment (DTE) End systems devices that communicate across the WAN. They are usually terminals, PCs, and network hosts and are located on the premises of individual subscribers.

de facto standard The fact standard that has been accepted by the industry because it was the most common.

de jure standard The bylaw standard that has been dictated by an appointed committee such as the International Organization for Standardization (ISO).

decryption The process of converting data from encrypted format back to its original format.

cryptography The art of protecting information by transforming it (encrypting it) into cipher text.

demand paging A feature found in Intel processors which swaps data between the RAM and disk only on the demand of the microprocessor. Since the processor does not try to anticipate what data will be needed from the hard drive, the code and data stored on disk as virtual memory will remain on disk until the code or data are needed.

demand paging The type of virtual memory that is supported by the Intel processors. Its name comes from the fact that it swaps data between the RAM and disk only on the demand of the microprocessor. It does not try to anticipate the needs of the microprocessor.

demand priority An access method in which a device makes a request to the hub and the hub grants permission.

demarcation point (demarc) The point where the local loop ends at the customer's premises.

designator Keeps track of which drive designations are assigned to network resources.

device drivers Programs that control a device; serve as a translator between the device and programs that use the device.

Device Manager A component of Windows that allows users to view and configure the hardware devices on their computers.

DHCP relay agent A computer that relays DHCP and BOOTP messages between clients and servers on different subnets.

DHCP server Maintains a list of IP addresses called a pool. When a user needs an IP address, the server removes the address from the pool and issues it to the user for a limited amount of time. Issuing an address is called leasing.

dial-up networking When a remote access client makes a nonpermanent, dial-up connection to a physical port on a remote access server by using the service of a telecommunications provider such as an analog phone, ISDN, or X.25.

differential SCSI A SCSI specification that uses differential signaling, where each signal is actually carried by two different wires, each the mirror image of the other. In other words, when one of the two wires is a positive voltage, the other wire is at zero volts. As a result, the signal is much more resilient to interference and allows for a much longer cabling.

digest authentication method A method similar to the basic authentication method that offers the same features as basic authentication for web servers but involves a different way of transmitting the authentication credentials. The authentication credentials pass through a one-way process, often referred to as hashing.

digital certificate An attachment to an electronic message used for security purposes such as for authentication and to verify that a user sending a message is who he or she claims to be and to provide the receiver with a means to encode a reply.

digital envelope A type of security that encrypts the message using symmetric encryption and encrypts key to decode the message using public-key encryption. This technique overcomes one problem of public-key encryption—being slower than symmetric encryption because only the key is protected with public-key encryption, providing little overhead.

digital signals Computer language based on a binary signal system produced by pulses of light or electric voltages. The state of the pulse is either on/high or off/low to represent 1s and 0s. Binary digits (bits) can be combined to represent different values.

digital signature A digital code that can be attached to an electronically transmitted message that uniquely identifies the sender. Like a written signature, its purpose is to guarantee that the individual sending the message really is who he or she claims to be.

digital subscriber line (DSL)   A special communications line that uses sophisticated modulation technology to maximize the amount of data that can be sent over plain twisted-pair copper wiring, which is already carrying phone service to subscribers' homes.

DIP switches   Bank of tiny on–off switches found on many older computers to configure the motherboard or the expansion card.

direct memory access (DMA)   A data transfer mode used by IDE hard drives that uses bus mastering to transfer the data.

direct memory address (DMA) channels   Used by communications devices that must send and receive large amounts of information at high speed (sound cards, some network cards, and some SCSI cards), in which the DMA controller takes over the data bus and address lines to bring data from an I/O device to the RAM without any assistance or direction from the CPU.

directories   Part of a tree structure that can hold directories and files. Newer operating systems may refer to a directory as a folder.

directory number (DN)   The 10-digit phone number that the telephone company assigns to any analog line.

directory service   A network service that identifies all resources on a network and makes those resources accessible to users and applications.

directory services servers   Locates information about the network such as domains (logical divisions of the network) and other servers.

direct-sequence spread-spectrum (DSSS)   A system that generates a redundant bit pattern for each bit to be transmitted. This bit pattern is called a chip (or chipping code). The intended receiver knows which specific frequencies are valid and deciphers the signal by collecting valid signals and ignoring the spurious signals.

disk quotas   A system that tracks and controls disk space usage for a NTFS volume.

disk transfer rate   The speed at which data are transferred to and from the platters; usually measured in bits per second (bps) or bytes per second (B/s).

diskless workstations   Computers that have no floppy drives or disk drives. Since the computer does not have a hard drive, it boots by loading the instructions within a special ROM chip on the network card which connects to the network and loads all necessary files from the network.

distance-vector-based routing protocol   A dynamic routing protocol that periodically advertises or broadcasts the routes in the routing tables, but only sends it to the neighboring routers.

distinguished name (DN)   Uniquely identifies an object (name) in the Active Directory by using the actual name of the object plus the names of container objects and domains that contain the object; identifies the object and its location in a tree.

distributed component object model (DCOM)   A set of Microsoft concepts and program interfaces in which client program objects can request services from server program objects on other computers in a network.

distributed file system (DFS)   A significant improvement to the drive mapping process that will allow multiple network locations to be mapped to a single drive; shared folders that reside on the same or different computers joined in a single hierarchical tree structure.

distribution group   Used only for nonsecurity functions such as to distribute e-mail messages to many people.

DNS zone   A portion of the DNS name space whose database records exist and are managed in a particular DNS database file.

domain   A logical unit of computers and network resources that defines a security boundary. It is typically found on medium or large networks or networks that require a secure environment. A domain uses one database to share its common security and user account information for all computers within the domain. Therefore, it allows centralized network administration of all users, groups, and resources on the network.

domain controller   The computer that stores a replica (copy) of the account and security information of the domain and defines the domain.

domain local group   Primarily used to assign rights and permissions to network resources that are in the domain of the local group. Different from a global group, it can list user accounts, universal groups and global groups from any domain, and local groups from the same domain.

domain name system (DNS)   A hierarchical client-server-based distributed database management system that translates Internet domain names to an IP address. It is used because domain names are easier to remember than IP addresses. The DNS clients are called resolvers and the DNS servers are called name servers.

domain user account   Allows a user to log on to a domain to gain access to the network resources.

double data rate–synchronous DRAM (DDR SDRAM)   A type of SDRAM that supports data transfers on both the rise and fall of each clock cycle, effectively doubling the memory chip's data thoughput.

Dr. Watson   A program that starts automatically when a program error is detected; it records the information about the system and the program failure into a log file.

dual inline memory module (DIMM)   A RAM package type that uses a small circuit board consisting of several soldered DIP memory chips. Compared with SIMMs, DIMMs are larger, typically have a larger capacity, and have a larger data bus.

dumb terminal   A system consisting of a monitor to display the data and a keyboard to input the data used to access a mainframe. Different from a PC, it does not process the data.

duplex volume   A mirror disk that uses multiple controller cards.

dynamic data exchange (DDE)   A component introduced with Windows 3.0 to allow users to copy data between applications while maintaining a link. When data are changed at the source, they are also changed at the target, so DDE maintains this connection.

dynamic DNS   A method that performs automatic registration to a DNS name space.

dynamic HTML (DHTML)   Allows content to be displayed with more design flexibility and accuracy through the use of a cascading style sheet (CSS).

dynamic RAM (DRAM) chips   Memory chips (RAM) that use a storage cell consisting of a tiny solid-state capacitor and a MOS transistor.

dynamic routing algorithms   An algorithm that determines routes used in a network.

dynamic storage   A disk that is converted from basic to dynamic storage that contains simple volumes, spanned volumes, mirrored volumes, striped volumes, and RAID-5 volumes. Unlike basic disks, dynamic disks cannot contain partitions or logical drives or be accessed by DOS. The advantage of the dynamic storage is that they contain an unlimited number of volumes which can be extended to include noncontiguous space on available disks.

effective permissions   The actual permissions when logging in and accessing a file or folder. They consist of explicit permissions plus any inherited permissions.

electronic mail (e-mail)   A powerful, sophisticated tool that allows users to send text messages and file attachments (documents, pictures, sound, and movies) to anyone with an e-mail address.

emergency repair disk (ERD)   Contains information about your current Windows system settings. You can use this disk to repair your computer if it will not start or your system files are damaged or erased.

emulated LAN (ELAN)   A set of LAN emulation clients that reside as ATM endpoints.

encrypting file system (EFS)   A system that allows a user to encrypt and decrypt files that are stored on an NTFS volume.

encryption   The process of disguising a message or data in what appears to be meaningless data (cipher text) to hide and protect the sensitive data from unauthorized access.

enhanced interior gateway routing protocol (EIGRP) Combines the advantages of link-state protocols with those of distance-vector protocols.

enterprise   Any large organization that utilizes computers, usually consisting of multiple LANs.

enterprise WAN   A WAN that is owned by one company or organization.

error-correcting Code (ECC)   A type of error checking, typically done with memory. It uses a special circuitry for testing the accuracy of data as they pass in and out of memory and can be used to correct some errors in memory.

Ethernet   The most widely used LAN technology today that offers good balance between speed, price, reliability, and ease of installation.

Event Viewer   A utility that is used to view and manage logs of system, program, and security events on the computer. Event Viewer gathers information about hardware and software problems, and monitors Windows 2000 security events.

exception error   Occurs when an application tries to pass an invalid parameter to another program.

Exchange   An e-mail server made by Microsoft, which runs on a Windows NT or a Windows 2000 Server.

Executive Services   The component of Windows NT and Windows 2000 that consists of managers and device drivers.

expansion cards   Circuit boards that are inserted into expansion slots to expand the system.

expansion slot   Also known as the I/O bus, extends the reach of the microprocessor so that it can communicate with peripheral devices.

explicit permissions   Those granted directly to the folder or file.

extended ISA (EISA)   An older bus architecture developed to compete against IBM's MCA slots. Different from the MCA, the EISA was backward compatible with ISA cards; like the MCA, the industry never accepted the new architecture.

extensible authentication protocol (EAP)   A system that allows new authentication schemes to be plugged in as needed. EAP allows third-party vendors to develop custom authentication schemes such as retina scans, voice recognition, fingerprint identification, smart card, Kerberos, and digital certificates.

exterior gateway protocol (EGP)   A protocol for exchanging routing information between the neighbor gateway host (each with its own router) in a network of autonomous systems.

external IPX address   An eight-digit (4-byte) hexadecimal number used to identify a network on an IPX network.

fast Ethernet   An extension of the 10BaseT Ethernet standard that transports data at 100 Mbps yet still keeps using the CSMA/CD protocol used by 10-Mbps Ethernet.

fat client   Performs the bulk of the data processing operations.

FAT32   A file system that uses 32-bit FAT entries. Introduced in the second major release of Windows 95 (OSR2/Windows 95B), it is an enhancement of the FAT/VFAT file system. It uses space more efficiently, such as 4-KB clusters for drives up to 8 GB, which results in 15% more efficient use of disk space relative to large FAT drives.

fax server   Manages fax messages sent in to and out of the network through a fax modem.

FDISK   A DOS and Windows 95/98 utility used to partition hard drives.

fiber distributed data interface (FDDI)   A MAN protocol that provides data transport at 100 Mbps (much higher data rate than standard Ethernet or token ring) and can support up to 500 stations on a single network.

fiber-optic cable   Consists of a bundle of glass or plastic threads, each capable of carrying data signals in the form of modulated pulses of light.

file   A collection of related information that is referenced by name. Each file or directory on a volume can be uniquely identified by using the file name (including the file extension, if any) and the path (or location on the tree).

file allocation table (FAT)   (1) An index used to remember which file is located in which cluster. It lists each cluster, if the cluster is being used by a file, the name of the file using the cluster, and the next cluster if the file does not fit within the cluster. (2) The most common file system, FAT is a simple and reliable file system, which uses minimal memory. It supports file names of 11 characters, which include the 8 characters for the file name and 3 characters for the file extension.

file attributes   Several properties of the file. For example, they indicate whether the file is read-only, whether it needs to be backed up, and whether it is visible or hidden.

file server   Manages user access to files stored on a server.

file sharing   Allows users to access files, which are on another computer, without using a floppy disk or other forms of removable media.

file system   The overall structure in which files are named, stored, and organized.

file transfer protocol (FTP)   A TCP/IP protocol that allows a user to transfer files between local and remote host computers.

firewall   A system designed to prevent unauthorized access to or from a private network; can be implemented in both hardware and software, or a combination of both. Firewalls are frequently used to prevent unauthorized Internet users from accessing private networks connected to the Internet, especially intranets.

forest   A grouping of one or more trees that are connected by two-way, transitive trust relationships.

forward lookup zone   Part of the DNS system that allows users to perform name-to-address resolution (forward lookup queries).

frame relay   A packet switching protocol designed to use high-speed digital backbone links to support modern protocols that provide for error handling and flow control for connecting devices on a wide area network.

frame relay access device (FRAD)   Used to connect the DTEs to the frame relay network. It multiplexes and formats traffic for entering a frame relay network.

frequency hopping   A system used in wireless technology that quickly switches between predetermined frequencies, many times each second. Both the transmitter and receiver must follow the same pattern and maintain complex timing intervals to be able to receive and interpret the data being sent.

full initial transfer (AXFR)   When a DNS server obtains a full copy of the zone files.

full-duplex dialog   Used in communication that allows every device to both transmit and receive simultaneously.

fully qualified domain name (FQDN)   Sometimes referred to as simply domain name; used to identify computers on a TCP/IP network.

gateway   A hardware and/or software that links two different types of networks by repackaging and converting data from one network to another network or from one network operating system to another.

gigahertz (GHz)   one billion hertz.

global catalog   Holds replica information of every object in the Active Directory; but instead of storing the entire object, it stores those attributes most frequently used in search operations (such as a user's first and last names).

global groups   Primarily used to group people within a domain. They can list user accounts and global groups from their domain. The global group can be assigned access to resources in any domain.

global WAN   A WAN that is not owned by any one company and could cross national boundaries. The best known example of a global WAN is the Internet, which connects millions of computers.

globally unique identifier (GUID)   A 128-bit unique number used to identify objects in the Active Directory.

gopher   A system that predates the World Wide Web for organizing and displaying files on Internet servers. A gopher server presents its contents as a hierarchically structured list of files.

group   A collection of user accounts; tools that help simplify administration.

group scope   Defines how the permissions and rights are assigned to the group.

half-duplex dialog   Allows each device to both transmit and receive, but not at the same time; therefore, only one device can transmit at a time.

handshaking   The process by which two devices initiate communications. It begins when one device sends a message to another device indicating that it wants to establish a communications channel. The two devices then send several messages back and forth that enable them to agree on a communications protocol.

hard drives   Half electronic/half mechanical devices that use magnetic patterns to store information onto rotating platters. They are considered long-term storage device because they do not forget their information when power is disconnected.

Hardware Abstraction Layer (HAL)   A library of hardware manipulating routines that hide the hardware interface details. It contains the hardware-specific code that handles I/O interfaces, interrupt controllers, and multiprocessor operations so that it can act as the translator between specific hardware architectures and the rest of the Windows 2000 software. As a result, programs written for Windows 2000 can work on other architectures, making those programs portable.

Hardware Compatibility List (HCL)   A Microsoft listing that provides approved tested components that will work with Windows 2000.

hardware profile   A set of instructions that tell Windows 2000 which devices to start when you start your computer or what settings to use for each device.

hashing scheme   Scrambles information in such a way that it is unique and cannot be reversed to the original format.

hertz (Hz)   A unit of measurement for frequency (cycles per second) of a device or signal. Hertz is typically used to measure the clock speed of a electronic device such as the processor or the bandwidth of a signal. Today in computers, hertz is typically expressed in megahertz and gigahertz.

high-bit-rate digital subscriber line (HDSL)   A DSL standard that provides a high bandwidth in both directions over two copper loops. HSDL has proven to be a reliable and cost-effective means for providing repeaterless T1 and E1 services over two twisted-pair loops.

high-level formatting   The process of writing the file system structure on the disk so that it can be used to store programs and data; this includes creating a file allocation table (an index listing all directories and files and where they are located on the disk) and a root directory to begin.

hive A discrete body of keys, subkeys, and values; each has a corresponding registry file and .LOG file located in the WINNT\SYSTEM32\CONFIG folder.

home directory A folder used to hold or store a user's personal documents.

hop Data going through a router. Hop count is the number of a routers that a data packet travels between the source and destination points.

host (1) A computer system that is accessed by a user who is working at a remote location. Typically, the term is used when there are two computer systems connected by modems and telephone lines. The system that contains the data is called the host, and the computer at which the user sits is called the remote terminal. (2) A network connection to a TCP/IP network, including the Internet. Each host has a unique IP address. Besides computers, hosts may also be network printers and routers. (3) Provides the infrastructure for a computer service. For example, there are many companies that host web servers, which means they provide the hardware, software, and communications lines required by the server, but the content on the server may be controlled by someone else.

HOSTS file A text file that lists the IP address followed by the host name.

hot swapping The ability to add and remove devices to a computer while the computer is running and have the operating system automatically recognize the change.

hub A device that is a common connection point for devices in a network. Hubs, which have multiple hubs, are commonly used to connect segments of a LAN. When a packet arrives at one port, it is copied to the other ports so that all segments of the LAN can see all packets.

hypertext markup language (HTML) Used in web pages that utilize embedded code made of ordinary text to give the browser a general idea of how the information should be displayed.

hypertext transfer protocol (HTTP) A TCP/IP protocol that is the basis for exchange over the World Wide Web (WWW). WWW pages are written in the hypertext markup language (HTML)—an ACSII-based, platform-independent formatting language.

I/O address Used in the computer to identify a device.

incremental transfers (IXFR) When a secondary DNS server pulls only those zone changes it needs to synchronize its copy of the zone with its source.

Industry Standard Architecture (ISA) The bus architecture used in the IBM PC/XT and PC/AT. The AT version of the bus is called the AT bus and has became a de facto industry standard. The ISA bus is being replaced by the PCI bus.

infrared (IR) system A form of wireless technology based on infrared light (light that is just below the visible light in the electromagnetic spectrum).

inherited permissions Those that flow down from the folder into the subfolders and files, indirectly giving permissions to a user or group.

integrated drive electronics (IDE) interface A fast, low-cost hard drive interface. The IDE hard drive connects to a controller card (is an expansion card or is built into the motherboard) using a 40-pin cable.

Integrated Services Digital Network (ISDN) A digital network that was a planned replacement for POTS to provide voice and data communications worldwide using circuit switching while using the same wiring that is currently being used in homes and businesses.

Intellimirror A new management tool that replaces the NT 4.0 administration of user profiles and provides more availability and security to user's data applications and settings, and allows for roaming access.

interative query Gives the best answer it currently has back as a response to DNS servers. The best answer will be the address being sought or an address of a server that would have a better idea of its address.

interference When undesirable electromagnetic waves affect the desired signal.

interior gateway protocol (IGP) The routing protocol that manages routing information within an autonomous system.

interior gateway routing protocol (IGRP) An extension of the open RIP standard, which was developed in the mid-1980s by Cisco Systems, Inc. Cisco's principal goal in creating IGRP was to provide a robust protocol for routing within an autonomous system (AS) having arbitrarily complex topology and consisting of media with diverse bandwidth and delay characteristics.

internal IPX network number An eight-digit (4-byte) hexadecimal number used to identify a server on an IPX network.

Internet connection sharing (ICS) Allows a single dial-up connection to be shared across the network.

Internet control message protocol (ICMP) A TCP/IP protocol that sends messages and reports errors regarding the delivery of a packet.

Internet Explorer administration kit (IEAK) A software product that makes it easy to create a custom browser that can be deployed across a network or via Web download.

Internet group-management protocol (IGMP) A TCP/IP protocol that is used by IP hosts to report host group membership to local multicast routers.

Internet information services (IIS) Built-in web server found in the Windows 2000 Server family.

Internet protocol (IP) A TCP/IP connectionless protocol that is primarily responsible for addressing and routing packets between hosts.

Internet Protocol Security (IPSec) Creates a standard platform to develop secure networks and electronic tunnels between two machines. It also encapsulates each data packet in a new packet that contains the information necessary to set up, maintain, and tear down the tunnel when it is no longer needed.

internetwork A network consisting of several LANs, which are linked together.

Internetwork packet exchange (IPX) (1) A protocol suite that was heavily associated with Novell NetWare. (2) A connectionless IPX networking protocol used to interconnect networks.

Internetworking operating system (IOS) The operating system used on Cisco routers.

Interprocess Communication (IPC) A component of Windows NT and Windows 2000 that allows bidirectional communication between clients and servers using distributed applications. IPC is a mechanism used by programs and multiprocesses. IPCs allow concurrently running tasks to communicate between themselves on a local computer or between the local computer and a remote computer.

interrupt (IRQ)   A signal informing a program that an event has occurred. When a program receives an interrupt signal, it takes a specified action (which can be to ignore the signal) such as causing the program to suspend itself temporarily to service the interrupt.

intranet   A network based on the TCP/IP protocol, the same protocol that the Internet uses. Unlike the Internet, the intranet belongs to a single organization, accessible only by the organization's members. An intranet's websites look and act like any other websites, but they are isolated by a firewall to stop illegal access.

invalid page faults   An error that occurs when a program tries to access a memory area that belongs to another program.

IP next generation (IPng)   Formerly called IPv6, is the new TCP/IP standard to overcome some of the current TCP/IP limitations.

IPC$ An administrative shared folder used for remote administration of a computer and to view a computer's shared resources.

jumper   A small, plastic-covered metal clip that is used to connect two pins protruding from an expansion card. The jumper (same as an On switch) connects the pins, which closes the circuit and allows current to flow.

Kerberos   A security protocol that is used for distributed security within a domain tree or forest. This allows for transitive trusts and a single logon to provide access to all domain resources.

kernel   The central module of an operating system. It is the part of the operating system that loads first, and it remains in main memory. Because it stays in memory, it is important for the kernel to be as small as possible while still providing all the essential services required by other parts of the operating system and applications. Typically, the kernel is responsible for memory management, process and task management, and disk management.

kernel mode   A processing mode used in Windows NT and Windows 2000 that has programs or software components run in ring 0 of the Intel 386 microprocessor protection model. The kernel mode components are protected by the microprocessor. It has direct access to all hardware and all memory including the address space of all user-mode processes.

key   A string of bits used to map text into a code and a code back to text. To encrypt and decrypt a file, you must use a key.

kill process tree   The utility that has the ability to end all processes related to the one being terminated by the administrator.

knowledge consistency checker (KCC)   The Windows 2000 service that automatically creates and manages the replication path or site topology used to replicate Active Directory information.

LAN emulation (LANE)   An ATM standard that emulates an Ethernet or token ring LAN on top of an ATM network.

last mile   Also referred to as the local loop and subscriber loop, the cable carries analog signals on a twisted-pair cable.

latency period   The time it takes after the hard drive's read/write heads to move to the requested track, for the requested sector to spin underneath the read/write head. The latency period is usually half of the time for a single revolution of the disk platter.

legacy application   An application in which a company or organization has already invested considerable time and money. Typically, legacy applications are database management systems (DBMSs) running on mainframes or minicomputers. An important feature of new software products is the ability to work with a company's legacy applications, or at least be able to import data from them.

legacy cards   ISA cards, which are not plug-and-play cards.

lightweight directory access protocol (LDAP)   A set of protocols for accessing information directories. LDAP is based on the standards contained within the X.500 standard, but is significantly simpler; and unlike X.500, LDAP supports TCP/IP, which is necessary for any type of Internet access. Because it is a simpler version of X.500, LDAP is sometimes called X.500-lite.

line printer daemon (LPD)   A protocol that provides printing on a TCP/IP network.

link-state algorithms   Also known as shortest path first algorithms, link-state routers send updates directly (or by using multicast traffic) to all routers within the network. Each router, however, sends only the portion of the routing table that describes the state of its own links. In essence, link-state algorithms send small updates everywhere.

LINUX   A freely distributable implementation of UNIX that runs on several hardware platforms.

LMHOSTS   A text file that translates between NetBIOS names and IP addresses.

local area network (LAN)   A network with computers that are connected within a geographical close network, such as a room, a building, or a group of adjacent buildings.

local bus   A data bus that connects directly, or almost directly, to the microprocessor. Although local buses can support only a few devices, they provide very fast throughput.

local loop   See *subscriber loop*.

local procedure calls (LPCs)   A component of IPC, used to transfer information between applications on the same computer.

local user accounts   Allows users to log on at and gain resources on only the computer where you create the local user account.

logical link control (LLC)   A sublayer of the data link layer that manages the data link between two computers within the same subnet and defines service access points (SAPs) or ports.

logical topology   Describes how the data flow through the physical topology or the actual pathway of the data.

mail server   Manages electronic messages (e-mail) between users.

Mailslot   Connectionless messaging, which means that messaging is not guaranteed. Connectionless messaging is useful for identifying other computers or services on a network, such as the browser service offered in Windows 2000.

mainframes   Large, centralized computers used to store and organize data.

Manchester Signal Encoding   The type of encoding used on an Ethernet network.

master boot record (MBR)   Located in the first sector of the disk, contains a small program that reads the partition table (also located in the MBR), checks which partition is active (the partition to boot from), and reads the first sector of the bootable partition.

media access control (MAC) address   Also known as the physical device address, is a unique hardware address (unique on the LAN/subnet) burned on to a ROM chip assigned by the hardware vendors or selected with jumpers or DIP switches.

Media access control (MAC) sublayer   Part of the data link layer that specifies the communications directly with the network adapter card.

member server   A server that does not store copies of the directory database and therefore does not authenticate accounts or receive synchronized copies of the directory database. These servers are used to run applications dedicated to specific tasks, such as managing print servers, managing file servers, running database applications, or running as a web server.

memory address   (1) The set of all legal addresses in memory for a given application. (2) A memory area between 640 MB and 1 MB of RAM used as a working area for the expansion card or by the ROM BIOS chips on the card.

mesh topology   Links a computer to every other computer. Although not common in LANs, it is common in WANs where it connects remote sites over telecommunications links.

message transfer agent (MTA)   Part of an e-mail system that manages the routing of messages within a site, between sites, and to foreign message systems.

messaging application programming interface (MAPI)   A system built into Microsoft Windows that enables different e-mail applications to work together to distribute mail.

metadata   Data that describes how and when and by whom a particular set of data was collected, and how the data are formatted. Metadata is essential for understanding information stored in data warehouses.

metric   A standard of measurement, such as hop count, that is used by routing algorithms to determine the optimal path to a destination.

metropolitan area network (MAN)   Designed for a town or city, primarily uses high-speed connections such as fiber optics.

Micro Channel Architecture (MCA)   An older bus architecture developed by IBM for older PCs to replace the ISA slot. Although it had higher clock speeds, larger data bus, and other performance enhancements, the industry never accepted the new architecture.

Microkernel   The central part of Windows 2000, which coordinates all I/O functions and synchronizes the activities of the Executive Services. Much like the other kernels in Windows 3.XX and Windows 95/98, it determines what is to be performed and when it is to be performed, while handling interrupts and exceptions. Lastly, it is designed to keep the microprocessor(s) busy at all times.

microprocessor   Also known as the central processing unit (CPU), is an integrated circuit that acts as the brain of the computer.

Microsoft challenge handshake authentication protocol (MS-CHAP)   Microsoft's proprietary version of CHAP.

Microsoft Management Console (MMC)   A fully customizable administrative console used as a common interface for most administrative tasks.

miniport drivers   A component that controls the network interface adapter and connects the hardware device to the protocol stack.

mirror volume   Also known as RAID-1, is two identical copies of a simple volume on separate hard disks. When data are written to one disk, they are also written to the other disk. If one of the hard drives fail, the other hard drive will continue to function without loss of data or downtime.

modulation   To blend data into a carrier signal. At the receiving side, a device demodulates the signals by separating the constant carrier signals from the variable data signals.

motherboard   Also referred to as the main board or the system board, is the primary printed circuit board located within the PC.

multicast   To transmit a message to a select group of recipients.

multihomed server   Contains multiple network cards that act as the router.

multilink   A combined connection consisting of two or more physical communications links' bandwidth into a single, logical link used to increase remote access bandwidth and throughput.

multimode fiber   A type of fiber-optic cable that is capable of transmitting multiple modes (independent light paths) at various wavelengths or phases.

multipoint connection   Links three or more devices through a single communications medium. Because multipoint connections share a common channel, each device needs a way to identify itself and the device to which it wants to send information.

Multiprotocol over ATM (MPOA)   A mechanism to route protocols such as IP, IPX, and NetBIOS from traditional LANs over a switched ATM backbone.

multipurpose Internet mail extensions (MIME)   A specification for formatting non-ASCII messages so that they can be sent over the Internet. Many e-mail clients now support MIME, which enables them to send and receive graphics, audio, and video files via the Internet mail system. In addition, MIME supports messages in character sets other than ASCII.

multistation access unit (MAU)   A token ring network device that physically connects network computers in a star topology while retaining the logical ring structure.

My Computer icon   Represents the computer. It includes all disk drives and the Control Panel. From within the My Computer icon, you can access and manage all files on your drives.

named pipe   A connection-oriented messaging via pipes, which set up a virtual circuit between the two points to maintain reliable and sequential data transfer.

name space A set of unique names for resources or items used in a shared computing environment.

NetBIOS enhanced user interface (NetBEUI) Provides the transport and network layers for the NetBIOS protocol.

NetWare core protocol An IPX protocol that consists of the majority of network services offered by a Novell NetWare server.

NetWare Directory Services (NDS) A directory service, consisting of a global, distributed, replicated database that tracks users and resources and provides controlled access to network resources.

network Two or more computers connected to share resources such as files or a printer. For a network to function, it requires a network service to share or access a common media or pathway, to connect the computers and protocols, and to give the entire system common communications rules.

Network Basic Input/Output System (NetBIOS) A common program that runs on most Microsoft networks (Windows for Workgroups, Windows 95/98, and Windows NT).

network device interface specification (NDIS) A set of standards for protocols and network card drivers that allow multiple protocols to use a single network adapter card; the layer that provides a communication path between two network devices.

network file system (NFS) A TCP/IP protocol that provides transparent remote access to shared files across networks.

network interface card (NIC) Computers connect to the network by using a special expansion card.

network layer Part of the OSI reference model that specifies the addressing and routing that is necessary to move data from one network (or subnet) to another. This includes establishing, maintaining, and terminating connections between networks; making routing decisions; and relaying data from one network to another. For the networks to communicate with one another, the network addresses must be unique.

network operating system (NOS) Includes special functions for connecting computers and devices into a local area network (LAN), to manage the resources and services of the network and to provide network security for multiple users.

network service Network resources that are offered by a network to multiple users such as files or printers.

node (1) In networks, a processing location. A node can be a computer or some other device, such as a printer. Every node has a unique network address, sometimes called a data link control (DLC) address or media access control (MAC) address. (2) In tree structures, a point where two or more lines meet.

notification area A Windows component located on the right side of the taskbar is used for the clock and for any programs running in the background such as printers and modems. The rest of the taskbar is blank or holds the active program buttons, which can be used to switch between open programs.

Novell NetWare A LAN-based NOS, and a primary player in bringing networking to the PC arena.

NT File System (NTFS) A file system for Windows NT and Windows 2000, designed for both the server and workstation. It provides a combination of performance, reliability, and compatibility.

NTLDR The first Windows 2000 file read during boot-up, which switches the microprocessor from real mode to protected mode and starts the appropriate minifile system drivers (built into NTLDR) so that it can read the VFAT/FAT16, FAT32, or NTFS file systems.

Nyquist theorem Ensures accuracy by sampling a signal at least twice the rate of its frequency.

object A distinct, named set of attributes or characteristics that represent a network resource including computers, people, groups, and printers in the Active Directory.

object linking and embedding (OLE) A component of Windows that allows you to create objects with one application and then link or embed objects into another application. Embedded objects keep their original format and links to the application that created them.

open data link interface (ODI) A set of standards for protocols and network card drivers that allow multiple protocols to use a single network adapter card.

open shortest path first (OSPF) A link-state routing protocol used in medium and large networks that calculates routing table entries by constructing a shortest-path tree. Routers using OSPF periodically advertises itself to other routers.

open shortest path first (OSPF) A TCP/IP link-state route discovery protocol where each router periodically advertises itself to other routers.

Open Systems Interconnection (OSI) reference model The world's prominent networking architecture model.

organization unit (OU) Used to hold users, groups, computers, and other organization units in the Active Directory.

owner The person who controls how permissions are set on a folder or file and can grant permissions to others. When a folder or file is created, the folder or file automatically becomes the owner.

packet A piece of a message transmitted over a packet switching network. A key feature of a packet is it contains the destination address and the data. In IP networks, packets are often called datagrams.

packet switching A technique in which messages are broken into small parts called packets. Each packet is tagged with source, destination, and intermediary node addresses as appropriate. Packets can have a defined maximum length and be stored in RAM instead of on hard disk. Packets can take a variety of possible paths through the network in an attempt to keep the network connections filled at all times.

page A fixed number of bytes recognized by the operating system. It is typically used in memory such as virtual memory.

parity An error-checking scheme that looks at the number of bits and determines them to be odd and even. When data are checked for integrity, the number of bits would be counted a second time to determine if it is odd or even. Parity error checking is used with older memory and some forms of data communications.

partition A term used with hard drives, functions as if it were a separate hard disk (logical drive).

partitioning Defines and divides the physical drive into logical volumes called partitions.

password authentication protocol (PAP) The least secure authentication protocol because it uses clear text (plaintext) passwords.

patch panels A panel with numerous RJ-45 ports. The wall jacks are connected to the back of the patch panel to the individual RJ-45 ports.

peer-to-peer network Sometimes referred to as a workgroup, is a network that has no dedicated servers. Instead, all computers are equal, so they both provide and request services.

Pentium processor The minimum processor needed to load Windows 2000. It is based on an advanced superscalar architecture that allows the processor to execute two instructions at the same time.

peripheral component interconnect (PCI) A new bus that was developed by Intel to replace the ISA slot with a faster interface.

permanent virtual circuit (PVC) A permanently established virtual circuit that consists of one mode: data transfer. PVCs are used in situations in which data transfer between devices is constant.

permission Defines the type of access granted to an object or object attribute. The permissions available for an object depend on the type of object.

personal computer (PC) Used by one person and contains its own processing capabilities.

phase The measurement in degrees of two sinusoidals vary.

physical layer Part of the OSI reference model that is responsible for the actual transmission of the bits sent across a physical media.

physical topology Describes how the network actually appears.

pipe Connects two processes so that the output of one can be used as the input to the other.

plain old telephone service (POTS) The standard telephone service that most homes use. See *public switched telephone network*.

plenum the space above the ceiling and below the floors that is used to circulate air throughout the workplace.

plenum cable Special type of cable that gives off little or no toxic fumes when burned.

plug and play (PNP) The ability of a computer system to automatically configure expansion boards and other devices. You should be able to plug in a device and play with it, without worrying about setting DIP switches, jumpers, and other configuration elements.

plug-in A software module that adds a specific feature or service to the browser such as to display or play different types of audio or video messages.

point to point A topology that connects two nodes directly together.

point-to-point protocol (PPP) An advanced version on the SLIP protocol that is commonly used to connect to the Internet using a modem. Provides full-duplex, bidirectional operations between hosts and can encapsulate multiple network layer LAN protocols to connect to private networks.

point-to-point tunneling protocol (PPTP) Uses the Internet as the connection between remote users and a local network, as well as between local networks.

poison reverse A process that, used with split horizon, improves RIP convergence over simple split horizon by advertising all network IDs.

policies A tool used by administrators to define and control how programs, network resources, and the operating system behave for users and computers in the Active Directory structure.

polling An access method that uses a single device, such as a mainframe front-end processor, designated as the primary device, which polls or asks each of the secondary devices (or slaves) if they have information to be transmitted.

ports (1) An interface on a computer to which you can connect a device. (2) In TCP/IP and other protocols, an endpoint to a logical connection. The ports are used to identify the upper-layer protocols and acts as a switchboard to make sure the frames find their way to the right network layer process. (3) To move a program from one type of computer to another. To port an application, rewrite sections that are machine dependent, and then recompile the program on the new computer. Programs that can be ported easily are said to be portable.

POSIX A set of IEEE and ISO standards that define an interface between programs and UNIX.

post office protocol (POP) A TCP/IP protocol that defines a simple interface between a user's mail client software and e-mail server. It is used to download mail from the server to the client and allows the user to manage his or her mailboxes.

PPP multilink protocol (MP) An extension that is used to aggregate or combine multiple physical links into a single logical link.

preemptive multitasking Occurs when the operating system assigns time slices to threads, tasks, and applications; those with a higher priority get a larger time slice. At the end of the time slice, the kernel can regain control of the computer without permission from the thread or program.

presentation layer Part of the OSI reference model that ensures that information sent by an application layer protocol of one system will be readable by the application layer protocol on the remote system.

primary name server Stores and maintains the zone file locally. Changes to a zone, such as adding domains or hosts, are done by changing files at the primary name server.

print device A printer.

print server Manages user access to printer resources connected to the network, allowing one printer to be used by many people.

print sharing Allows several people to send documents to a centrally located printer in the office.

print spooler Software that accepts a document sent to a printer by the user and then stores it on disk or in memory until the printer is ready for it.

printer pool Two or more print devices that use the same driver which is connected to one print server and acts as a single printer.

private branch exchange (PBX) A private telephone network used within an enterprise. Users of the PBX share a certain number of outside lines for making telephone calls external to the PBX.

private key A key kept in a secure location and used only by you.

private-key encryption The most basic form of encryption, also known as the symmetric algorithm. This encryption system requires that each individual possess a copy of the key. Of course, for this to work as intended, you must have a secure way to transport the key to other people. If you must keep multiple single keys per person, this can be very cumbersome. Since private-key algorithms are generally fast and easily implemented in hardware, they are commonly used for bulk data encryption.

processor input/output (PIO) A data transfer mode used by IDE hard drives that is directly controlled by the processor.

program An executable outline that follows a sequence of steps. A program consists of code and data initialization, a private memory address space, system resources (such as files, communication ports, and Windows resources), and one or more threads.

protocol stack Sometimes referred to as a protocol suite, is a set of protocols that works together to provide a full range of network services.

protocols Rules or standards that allow the computer to connect to one another and enable computers and peripheral devices to exchange information.

proxy server Provides local intranet clients with access to the Internet while keeping the local intranet free from intruders.

public key A non-secret key that is available to anyone you choose, or made available to everyone by posting it in a public place. It is often made available through a digital certificate.

public-key encryption Also known as asymmetric algorithm, uses two distinct but mathematically related keys, public and private.

Public Switched Telephone Network (PSTN) The international telephone system based on copper wires (UTP cabling) carrying analog voice data. See *plain old telephone service*.

pull partner A WINS replication partner server that requests new database entries from its partner. The pull occurs at configured time intervals or in response to an update notification from a push partner.

punch-down block A device used to connect several cable runs to each other without going through a hub.

push partner A WINS replication partner server that sends update notification messages. The update notification occurs after a configurable number of changes to the WINS database.

quality of service (QofS) A networking term that specifies a guaranteed throughput level.

RAID-5 volumes Fault-tolerant striped volumes that also use an extra drive to store parity information.

rambus DRAM (RDRAM) RAM based around a rambus channel, a high-speed 16-bit bus running at a clock rate of 400 MHz.

rambus inline memory module (RIMM) A RAM package type that conforms to the standard DIMM form factor, but is not pin compatible. Since it runs at a higher clock speed than SIMMs and DIMMs, RIMMs typically have a heat spreader.

random access memory (RAM) Electronic memory chips, which store information inside the computer. It is so called because the information is accessed nonsequentially. RAM is volatile memory (loses its contents when the power is turned off).

read-only memory (ROM) Chips that do not lose their contents when the power is shut off.

Recovery Console A command line interface that will let you repair system problems using a limited set of command line commands, including enabling or disabling services, repairing a corrupted master boot record, or reading and writing data on a local drive (FAT, FAT32, or NTFS).

recursive query Asks the DNS server to respond with the requested data or with an error stating that the requested data does not exist or that the domain name specified does not exist.

Recycle Bin An icon in Windows that is used as a safe delete. When you delete a file using the GUI, Windows will store the file in the Recycle Bin.

redirector A small section of code in the network operating system that intercepts requests in the computer and determines if they should be left alone to continue in the local computer's bus or redirected out to the network to another server.

redundant array of inexpensive disks (RAID) A category of disk drives that employs two or more drives in combination for fault tolerance and performance. RAID disk drives are used frequently on servers but are not generally necessary for personal computers.

register A high-speed internal storage area that acts as the microprocessors short-term memory and work area.

Registry A central, secure database in which Windows 2000 stores all hardware configuration information, software configuration information, and system security policy. Components that use the registry include the Windows NT Kernel, device drivers, setup programs, NTDETECT.COM, hardware profiles, and user profiles.

Remote access policies A set of conditions and connection settings that give network administrators more flexibility in authorizing connection attempts for remote access.

remote access server A computer and associated software that is set up to handle users seeking access to the network remotely. A remote access server usually includes or is associated with a firewall server to ensure security and a router that can forward the remote access request to another part of the corporate network.

remote access service (RAS) Allows users to connect remotely using various protocols and connection types.

Remote Authentication Dial-In User Service (RADIUS) An industry standard client-server protocol and software that enables remote access servers to communicate with a central server to authenticate dial-in users and authorize their access to the requested system or service for authenticating remote users.

remote installation service (RIS) A network service that allows for remote automated installation of Windows 2000 Professional workstations.

remote procedure calls (RPCs) A component of IPC, are used to transfer information between applications that are on separate computers.

repeater   A network device used to regenerate or replicate a signal or to move packets from one physical media to another.

Request for Comments (RFC)   A series of documents that specify the standards for TCP/IP.

resource record (RR)   Entries of the DNS server database.

reverse lookup zone   Part of the DNS system that allows you to perform address-to-name resolution (also known as reverse lookup queries).

reverse query   Used with DNS servers, resolves an IP address and requests the host name.

right   Authorizes a user to perform certain actions on a computer, such as logging on to a system interactively or backing up files and directories.

ring topology   Connects all devices to one another in a closed loop. Each device is connected directly to two other devices.

root directory   Starting point of a volume, which is located at the top of the tree structure; holds files and/or directories.

root domain   Top of the DNS name space.

root hints   A file that normally contains the NS and A resource records for the Internet root servers.

root name server   Contains the resource records for all top-level name servers in the domain name space, such as the COM and EDU domain.

round-robin   A method used in DNS servers, rotates the order of resource records data returned in a query answer in which multiple resource records exist of the same resource record type for a queried DNS domain name.

router   A network OSI layer device that connects two or more LANs. In addition, it can break a large network into smaller, more manageable subnets.

routing information protocol (RIP)   (1) A TCP/IP protocol that is a distance-vector route discovery protocol in which the entire routing table is periodically sent to the other routers. (2) An IPX protocol that is a distance-vector route discovery protocol in which the entire routing table is periodically sent to the other routers.

RSA standard   Created by Ron Rivst, Adi Shamir, and Leonard Adleman, defines the mathematical properties used in public-key encryption and digital signatures. The key length for this algorithm can range from 512 to 2048, making it a secure encryption algorithm.

schema   The formal definition and set of rules for all objects and attributes of those objects in the Active Directory.

search engine   A program used on the Internet that searches documents for specified keywords and returns a list of the documents where the keywords were found.

secondary name server   Gets the data from its zone from another name server, either a primary name server or another secondary name server, for that zone across the network.

second-level domain names   Variable-length names registered to an individual or organization used within a DNS name space.

sectors   The smallest unit that can be accessed on a disk. It is typically 512 bytes.

secure HTTP (S-HTTP)   A protocol for transmitting data securely over the World Wide Web; designed to transmit individual messages securely.

secure sockets layer (SSL)   A protocol developed by Netscape for transmitting private documents via the Internet. SSL works by using a private key to encrypt data that are transferred over the SSL connection.

security group   Assigns permissions and gains access to resources.

security identifier (SID)   A hexadecimal string that uniquely identifies the computer, users, and printers.

seek time   The average time it takes the hard drives read/write heads to move to the requested track. It is usually the time it takes the read/write heads to move one-third of the way across the platter.

separator page   Sometimes known as a banner page, separates print jobs and typically states who sent the document to the printer and the date and time of printing.

sequenced packet exchange (SPX)   An IPX protocol that uses packet acknowledgments to guarantee delivery of packets.

serial line interface protocol (SLIP)   Sends packets down a serial link delimited with special END characters.

serial line protocols   Used for a computer to connect to a server, such as those used by an Internet service provider (ISP) via a serial line such as a modem to become an actual node on the Internet.

server   A computer that acts as a service provider to provide access to network resources.

server message block (SMB)   A protocol that defines a series of commands used to pass information between networked computers using NetBIOS.

server room   The work area of the IT department where the servers and most of the communication devices reside.

service   A program, routine, or process that performs a specific system function to support other programs.

service advertising packet (SAP)   An IPX protocol used to advertise the services of all known servers on the network, including file servers and print servers.

service pack   A collection of fixes bundled as a single upgrade.

service profile identifier (SPID)   A directory number and additional identifier used to identify the ISDN device to the telephone network.

session   A reliable dialog between two computers.

session layer   Part of the OSI reference model that allows remote users to establish, manage, and terminate a connection (sessions).

shared folder   A folder that is available to clients on the network.

shielded twisted pair   Similar to unshielded twisted pair, but is usually surrounded by a braided shield that serves to reduce both EMI sensitivity and radio emissions.

Shiva password authentication protocol (SPAP)   Shiva's proprietary version of PAP, offers a bit more security than PAP's plaintext password with its reversible encryption mechanism.

signaling   The method for using electrical or light energy to communicate; the two forms are digital signaling and analog signaling.

simple mail transfer protocol (SMTP)   Helps e-mail messages between servers. Most e-mail systems that send mail over the Internet use SMTP to send messages from one server to another; the messages can then be retrieved with an e-mail client using either POP or IMAP. In addition, SMTP is generally used to send messages from a mail client to a mail server. This is why you need to specify both the POP or IMAP server and the SMTP server when you configure your e-mail application.

Simple Network Management Protocol (SNMP) A TCP/IP protocol that defines procedures and management information databases for managing TCP/IP-based network devices.

simple volume Consists of disk space from a single physical disk (single region or multiple regions) linked together.

simplex dialog Used in communications, allows communications on the transmission channel to occur in only one direction. Essentially, one device is allowed to transmit and all other devices receive.

single inline memory module (SIMM) A RAM package type that uses a small circuit board consisting of several soldered DIP memory chips.

single-ended SCSI A SCSI specification that uses a conventional signaling in which a positive voltage indicates a 1 and ground or 0 volt indicates a 0. Each signal is carried on one wire.

single-mode fiber A type of fiber-optic cable that can transmit light in only one mode.

site One or more IP subnets connected by a high-speed link (128 Kbps or higher), typically defined by geographical locations. Sites are based on IP subnets of which any subnet can only belong to one site.

small computer systems interface (SCSI) Pronounced "skuzzy," is a high-speed bus capable of supporting multiple devices in and out of the computer.

small office/home office (SOHO) Small networks used primarily in home offices that might be part of a larger corporation but yet remain apart from it.

smart permanent virtual circuit (SPVC) A hybrid of a permanent and temporary virtual circuit.

software license The legal permission for an individual or group to use a piece of software.

solid cable Typically used for the cabling that exists throughout the building, including cables that lead from the wall jacks to the server room or wiring closet.

stranded cable Typically used as patch cables between patch panels and hubs and between the computers and wall jacks.

source routing bridge Used in Ethernet networks.

spanned volume A volume across multiple disks.

spanning tree bridge A bridge used in token ring networks.

split horizon A route-advertising algorithm that prevents the advertising of routes in the same direction in which they were learned.

spread-spectrum signals Wireless signals distributed over a wide range of frequencies and then collected onto their original frequency at the receiver.

star topology The most popular topology, it has each network device connect to a central point such as a hub, which acts as a multipoint connector.

Start button A component of Windows used to start programs. Using the Start button, you can open recently accessed documents and access the Control Panel and printer folder, find files, and get help for Windows 2000.

static routing algorithms Table mappings established by the network administrator prior to the beginning of routing.

steppings Internal version of the processor.

stop error An error that signifies that something unexpected has happened within the Windows environment.

Storage Area Network (SAN) High speed subnetwork of shared storage devices used to make all storage devices available to all servers on a LAN.

straight-through cable Used to connect a network card to a hub; it has the same sequence of colored wires at both ends of the cable.

stream cipher Encrypts each byte of the data stream individually.

strip set The combining areas of free space on 2 to 32 disks onto one logical volume. The difference is that a striped volume has the data allocated alternately and evenly in 64-KB stripes through the various drives. Therefore, when data are read or written, it writes to all drives simultaneously.

stub area In OSPF networks, an area with only one ABR. All routes to the destination outside the area must pass through the single router.

subdomain names Additional names that an organization can create that are derived from the registered second-level domain name of the DNS name space.

subnet Small networks that comprise a larger network.

subnet mask Numbers that resemble an IP address used to define which bits describe the network number and which bits describe the host address.

subnetworks Also known as subnets, smaller LANs that comprise an internetwork.

subscriber loop Also referred to as a local loop, is the telephone line that runs from your home or office to the telephone company's central office (CO) or neighborhood switching station.

supernet A group of networks identified by contiguous network addresses.

superscalar The ability of a processor to execute two instructions at the same time.

superscope A group of multiple scopes (child scopes) as a single administrative entity for a DHCP server. Using a superscope, the DHCP server computer can activate and provide leases from more than one scope to clients on a single physical network.

switched 56 lines Digitally switched or dial-up lines that provide a single digital channel for dependable data connectivity that is based on the DS-0 64-Kbps technology.

switched 64 lines Digitally switched or dial-up lines that provide a single digital channel for dependable data connectivity.

switched multimegabit data service (SMDS) A high-speed, cell-relay, wide area network (WAN) service designed for LAN interconnection through the public telephone network.

switching hub Sometimes referred to as a switch, is a fast multiported bridge, which actually reads the destination address of each packet and then forwards the packet to the correct port. A major advantage of using a switching hub is it allows one computer to open a connection to another computer (or LAN segment). While those two computers communicate, other connections between the other computers (or LAN segments) can be opened at the same time. Therefore, several computers can communicate at the same time.

symmetric multiprocessing (SMP) Involves a computer that uses two or more microprocessors that share the same memory. If software is written to use the multiple microprocessors, several programs can be executed at the same time, or multithreaded applications can be executed faster.

synchronous Devices use a timing or clock signal to coordinate communications.

Synchronous Digital Hierarchy (SDH) An international standard for synchronous data transmission over fiber-optic cables.

synchronous dynamic RAM (SDRAM) A popular form of RAM.

Synchronous Optical Network (SONET) A standard for connecting fiber-optic transmission systems, mostly used in North America.

system partition The active partition that contains the NTLDR and BOOT.INI file.

system ROM BIOS The central ROM chip found on the motherboard. It controls the boot-up procedure and power-on self-test, generates hardware errors during boot-up, and contains the CMOS setup program.

system state data A collection of system-specific data that can be backed up and restored. For Windows 2000, it includes the Registry, system boot files, Active Directory directory services database, and the SYSVOL directory.

System Volume (SYSVOL) A special folder that stores the server copy of the domain's public files, such as the Group Policy and Net Login (logon scripts), which are replicated among all domain controllers in the domain.

Tape drive A device, typically used to back up data onto long magnetic tape.

Task Manager A component of Windows NT and Windows 2000 that provides a list of the current programs that are running and the overall CPU (including number of threads and processes) and memory usage (including the amount of physical memory, size of the file cache, paged memory, and nonpaged memory). You can also use the Task Manager to switch between programs and to stop a program that is not responding.

Task Scheduler A program that comes with Windows 2000 that will schedule programs, batch files, and documents to run once at regular intervals or at specific times.

taskbar A component of Windows 2000 located at the bottom of the screen, which is used to switch between different programs.

T-carrier system The first successful system that converted the analog voice signal to a digital bit stream. Although the T-carrier system was originally designed to carry voice calls between telephone company central offices, today it is used to transfer voice, data, and video signals between different sites and to connect to the Internet.

Telecommunication Network (Telnet) A TCP/IP virtual terminal protocol (terminal emulation) allowing a user to log on to another TCP/IP host to access network resources.

switched virtual circuit (SVC) A virtual circuit that is dynamically established on demand and terminated when transmission is complete. Communication over an SVC consists of three phases: circuit establishment, data transfer, and circuit termination.

terminal emulation (Telnet) A program for TCP/IP networks such as the Internet. The Telnet program runs on your computer and connects your PC to a server on the network. You can then enter commands through the Telnet program and they will be executed as if you were entering them directly on the server console. This enables you to control the server and communicate with other servers on the network. To start a Telnet session, log in to a server by entering a valid user name and password. Telnet is a common way to remotely control web servers.

Terminal Services A fully integrated component of Windows 2000 Server, which enables a client to run a Windows application on a server.

terminating resistors A device that goes on the end of a cable or daisy chain to stop the signal from bouncing back.

termination A process in which two ends of a cable or daisy chain have resistors to stop the signal from bouncing back.

thin client Designed to be especially small so that the bulk of the data processing occurs on the server.

thread Part of a program that can execute independently of other parts. Operating systems that support multithreading enable programmers to design programs whose threaded parts can execute concurrently. This can enhance a program by allowing multitasking, improving throughput, enhancing responsiveness, and aiding background processing.

time-domain reflectometer (TDR) A device used to test cables by using a sonarlike pulse along a cable. Depending on the signal that is bounced back using a TDR, the reflectometer can detect a break or short and inform you how far the break or short is away.

token passing An access method that uses a special authorizing packet of information to inform devices that they can transmit data. These packets, called tokens, are passed around the network in an orderly fashion from one device to the next. Devices can transmit only if they have control of the token, which distributes the access control among all the devices.

token ring A type of network that uses token passing and forms a logical ring.

top-level domains The second level of the DNS name space to indicate a country, region, or type of organization. Three-letter codes indicate the type of organization; for example, COM indicates commercial (business) and EDU stands for educational institution. Two-letter codes indicate countries.

topology Describes the appearance or layout of the network.

tracks Concentric circles used on a disk, numbered starting with the outside track as track 0. The tracks are then further divided into sectors where each sector is typically 512 bytes of usable data.

transit area In OSPF networks, an area that contains more than one ABR.

transition state A method that represents data by how the signal transitions from high to low or low to high. A transition indicates a binary 1, whereas the absence of a transition represents a binary 0.

transmission control protocol (TCP) A TCP/IP protocol that provides connection-oriented, reliable communications for applications that typically transfer large amounts of data at one time or that require an acknowledgment for data received.

transport layer The middle layer of the OSI reference model that connects the lower to the upper layers. It is responsible for reliable transparent transfer of data between two endpoints.

trap An unsolicited message sent by an SNMP agent to an SNMP management system, when the agent detects that a certain type of event has occurred locally on the managed host.

trust relationship A link that combines two domains into one administrative unit that can authorize access to resources on both domains.

tunneling The method for transferring data packets over the Internet or other public network, providing the security and features formerly available only on private networks.

twisted-pair cable Consists of two insulated copper wires twisted around each other. Although each pair acts as a single communication link, twisted pairs are usually bundled together into a cable and wrapped in a protective sheath.

uniform resource locator (URL) A location of a web page or site. The first part of the address indicates what protocol to use, and second part specifies the IP address or the domain name where the resource is located. The last part indicates the folder and file name.

uninterruptible power supply (UPS) A device used to protect your system against power fluctuations including outages.

unique sequence number (USN) A 64-bit unique number that is used to track all changes to an Active Directory.

universal naming convention (UNC) A NetBIOS format for specifying the location of shared resources on a local area network (LAN). UNC uses the following format: \\server-name\shared-resource-pathname.

universal security group A type of group that can contain users, universal groups, and global groups from any domain and it can be assigned rights and permissions to any network resource in any domain in the domain tree or forest. It is only available in native mode.

universal serial bus (USB) This external port allows you to connect up to 127 external PC peripherals in series (daisy chain) and offers a data transfer rate up to 12 Mbps. The USB allows devices to be attached while the computer is running and requires no rebooting or reconfiguring when a peripheral is added or removed.

UNIX A popular multiuser, multitasking operation, is the grandfather of network operating systems, which was developed at Bell Labs in the early 1970s.

unshielded twisted-pair cable (UTP) Uses twisted-pair cable that does not include any additional shielding.

user account Enables a user to log on to a computer and domain with an identity that can be authenticated and authorized for access to domain resources.

user datagram protocol (UDP) A TCP/IP protocol that provides connectionless communications and does not guarantee that packets will be delivered. Applications that use UDP typically transfer small amounts of data at once. Reliable delivery is the responsibility of the application.

user mode A processing mode used in Windows NT and Windows 2000 that has Programs running in ring 3 of the Intel 386 microprocessor protection model. These programs are protected by the operating system. It is a less privileged processor mode that has no direct access to hardware and can only access its own address space.

user profile A collection of folders and data that stores the user's current desktop environment and application settings.

value-added network (VAN) A network with special services such as electronic data interchange (EDI) or financial services such as credit card authorization or ATM transactions.

VFAT An enhanced version of the FAT structure, which allows Windows 95, Windows 98, Windows NT, and Windows 2000 to support long file names (LFNs) up to 255 characters.

virtual (or logical) VLAN A local area network with a definition that maps workstations on some other basis than geographic location (for example, by department, type of user, or primary application). In other words, it is a network of computers that behaves as if the computers are connected to the same wire even though they may actually be physically located on different segments of a LAN.

virtual circuits Establish formal communication between two computers on an internetwork using a well-defined path. This enables two computers to act as though there is a dedicated circuit between two, even though there is not. The path the data take while being exchanged between the two computers may vary, but the computers do not know this nor do they need to. Virtual networks are sometimes depicted as a cloud, because the user does not worry about the path taken through the cloud, only about entering and exiting the cloud.

virtual link In OSPF networks, is a logical link between a backbone area border router, an unspecified number of routers, and a second border router that is not connected to the backbone.

virtual machine (VM) A self-contained operating environment that behaves as if it is a separate computer.

virtual memory Disk space pretending to be RAM, which allows the operating system to load more programs and data. For all programs and data to be accessed, parts are constantly swapped back and forth between RAM and disk. As far as the programs are concerned, the virtual memory looks and acts like regular RAM.

virtual private networking The creation of secured, point-to-point connections across a private network or a public network such as the Internet. A virtual private networking client uses special TCP/IP-based protocols, called tunneling protocols, to make a virtual call to a virtual port on a virtual private networking server.

virus A program that is designed to replicate and spread, generally without the knowledge or permission of the user.

volatile A term typically used with memory chips, meaning the device will lose its contents when the power is turned off.

volume A logical unit of disk space that functions as though it was a physical disk.

web browser The client program/software that you run on your local machine to gain access to a web server. It receives the HTML commands, interprets the HTML, and displays the results.

Web-distributed authoring and versioning (WebDAV) A Windows 2000 environment that extends the HTTP to allow clients to publish, lock, and manage resources on the Web.

Web server A computer equipped with the server software that uses Internet protocols such as hypertext transfer protocol (HTTP) and file transfer protocol (FTP) to respond to web client requests on a TCP/IP network via web browsers.

wide area network (WAN) Uses long-range telecommunications links to connect the network computers over long distances and often consists of two or more smaller LANs.

Win16 on Win32 (WOW) A component of Windows NT and Windows 2000 which translates or thunks 16-bit calls to 32-bit calls.

Windows 2000 Formerly known as Windows NT 5.0, an advanced, high-performance network operating system that is robust in features and services, security, performance, and upgradability. There are four versions of Windows 2000—Windows 2000 Professional, Windows 2000 Server, Windows 2000 Advanced Server, and Windows 2000 DataCenter Server.

Windows Explorer A component of Windows that is used to manage directories and files.

Windows NT A high-performance network operating system, which eventually led to Windows 2000.

Windows socket (WinSock) A port that provides the interface between the network program or service and the Windows environment.

WINS proxy agent A WINS-enabled computer configured to act on behalf of other host computers that cannot directly use WINS.

WINS replication partner A WINS server that synchronizes with another WINS server.

WINS server A name server that resolves NetBIOS names to IP addresses.

wireless LAN (WLAN) A data communication system that uses electromagnetic waves to transmit and receive data over the air, minimizing the need for wired connections.

workgroup Computers and devices on a peer-to-peer network are usually organized into logical subgroups called workgroups. In a workgroup, each computer has a local security database so that it tracks its own user and group account information. The user information is not shared with other workgroup computers.

X.25 A packet switched network that allows remote devices to communicate with each other across digital links without the expense of individual leased lines.

X.400 A set of standards relating to the exchange of electronic messages (e.g., e-mail, fax, voice mail, telex).

X.500 A common directory service that uses a hierarchical approach, in which objects are organized similar to the files and folders on a hard drive.

X.509 certificates The most widely used digital certificates.

Index